Macmillan Encyclopedia of Physics

MACMILLAN ENCYCLOPEDIA OF PHYSICS

John S. Rigden

Editor in Chief

Volume 2

MACMILLAN REFERENCE USA
Simon & Schuster Macmillan
NEW YORK

Simon & Schuster and Prentice Hall International
LONDON   MEXICO CITY   NEW DELHI   SINGAPORE   SYDNEY   TORONTO

Copyright © 1996 by Simon & Schuster Macmillan

Simon & Schuster Macmillan
1633 Broadway
New York, NY 10019

Library of Congress Catalog Card Number: 96-30977

PRINTED IN THE UNITED STATES OF AMERICA

Printing Number

1 2 3 4 5 6 7 8 9 10

LIBRARY OF CONGRESS CATALOGING-IN-PUBLICATION DATA

Macmillan Encyclopedia of Physics / John S. Rigden, editor in chief.
    p.   cm.
    Includes bibliographical references and index.
    ISBN 0-02-897359-3 (set).
    1. Physics—Encyclopedias.   I. Rigden, John S.
  QC5.M15   1996
    530′.03—dc20   96-30977
                                                          CIP

This paper meets the requirements of ANSI-NISO Z39.48-1992
(Permanence of Paper).

COMMON ABBREVIATIONS AND
MATHEMATICAL SYMBOLS

$=$	equals; double bond	$\|\,\|$	absolute value of
$\neq$	not equal to	$+$	plus
$\equiv$	identically equal to; equivalent to; triple bond	$-$	minus
		$/$	divided by
$\sim$	asymptotically equal to; of the order of magnitude of; approximately	$\times$	multiplied by
		$\oplus$	direct sum
		$\otimes$	direct product
$\approx, \simeq$	approximately equal to	$\pm$	plus or minus
$\cong$	congruent to; approximately equal to	$\mp$	minus or plus
		$\sqrt{}$	radical
$\propto$	proportional to	$\int$	integral
$<$	less than	$\oint$	contour integral
$>$	greater than	Σ	summation
$\nless$	not less than	Π	product
$\ngtr$	not greater than	∂	partial derivative
$\ll$	much less than	$\circ$	degree
$\gg$	much greater than	$^\circ$B	degrees Baumé
$\leq$	less than or equal to	$^\circ$C	degrees Celsius (centigrade)
$\geq$	greater than or equal to	$^\circ$F	degrees Fahrenheit
$\nleq$	not less than or equal to	$!$	factorial
$\ngeq$	not greater than or equal to	$'$	minute
$\cup$	union of	$''$	second
$\cap$	intersection of	∇	curl
$\subset$	subset of; included in		
$\supset$	contains as a subset	ϵ_0	electric constant
$\in$	an element of	μ	micro-
$\ni$	contains as an element	μ_0	magnetic constant
$\rightarrow$	approaches, tends to; yeilds; is replaced by	μA	microampere
		μA h	microampere hour
$\Rightarrow$	implies; is replaced by	μC	microcoulomb
$\Leftarrow$	is implied by	μF	microfarad
$\downarrow$	mutually implies	μg	microgram
$\Leftrightarrow$	if and only if	μK	microkelvin
$\perp$	perpendicular to	μm	micrometer
$\parallel$	parallel to	μm	micron

μm Hg	microns of mercury
μmol	micromole
μs, μsec	microsecond
μu	microunit
$\mu\Omega$	microhm
σ	Stefan–Boltzmann constant
Ω	ohm
Ω cm	ohm centimeter
Ω cm/(cm/cm^3)	ohm centimeter per centimeter per cubic centimeter
A	ampere
Å	angstrom
a	atto-
A$_s$	atmosphere, standard
abbr.	abbreviate; abbreviation
abr.	abridged; abridgment
Ac	Actinium
ac	alternating-current
aF	attofarad
af	audio-frequency
Ag	silver
A h	ampere hour
AIP	American Institute of Physics
Al	aluminum
alt	altitude
Am	americium
AM	amplitude-modulation
A.M.	ante meridiem
amend.	amended; amendment
annot.	annotated; annotation
antilog	antilogarithm
app.	appendix
approx	approximate (in subscript)
Ar	argon
arccos	arccosine
arccot	arccotangent
arccsc	arccosecant
arc min	arc minute
arcsec	arcsecant
arcsin	arcsine
arg	argument
As	arsenic
At	astatine
At/m	ampere turns per meter
atm	atmosphere
at. ppm	atomic parts per million
at. %	atomic percent
atu	atomic time unit
AU	astronomical unit
a.u.	atomic unit
Au	gold

av	average (in subscript)
b	barn
b.	born
B	boron
Ba	barium
bcc	body-centered-cubic
B.C.E.	before the common era
Be	beryllium
Bi	biot
Bi	bismuth
Bk	berkelium
bp	boiling point
Bq	becquerel
Br	bromine
Btu, BTU	British thermal unit
C	carbon
c	centi-
c.	circa, about, approximately
C	coulomb
c	speed of light
Ca	calcium
cal	calorie
calc	calculated (in subscript)
c.c.	complex conjugate
CCD	charge-coupled device
Cd	cadmium
cd	candela
CD	compact disc
Ce	cerium
C.E.	common era
CERN	European Center for Nuclear Research
Cf	californium
cf.	confer, compare
cgs, CGS	centimeter-gram-second (system)
Ci	curie
Cl	chlorine
C.L.	confidence limits
c.m.	center of mass
cm	centimeter
Cm	curium
cm^3	cubic centimeter
Co	cobalt
Co.	Company
coeff	coefficient (in subscript)
colog	cologarithm
const	constant
Corp.	Corporation
cos	cosine
cosh	hyperbolic cosine
cot	cotangent
coth	hyperbolic cotangent

cp	candlepower	e.u.	electron unit
cP	centipoise	eu	entropy unit
cp	chemically pure	Eu	europium
cpd	contact potential difference	eV	electron volt
cpm	counts per minute	expt	experimental (in subscript)
cps	cycles per second	F	farad
Cr	chromium	F	Faraday constant
cS	centistoke	f	femto-
Cs	cesium	F	fermi
csc	cosecant	F	fluorine
csch	hyperbolic cosecant	fc	foot-candle
Cu	copper	fcc	face-centered-cubic
cu	cubic	Fe	iron
cw	continuous-wave	fF	femtofarad
D	Debye	Fig. (pl., Figs.)	figure
d	deci-	fL	foot-lambert
$d.$	died	fm	femtometer
da	deka-	Fm	fermium
dB, dBm	decibel	FM	frequency-modulation
dc	direct-current	f. (pl., ff.)	following
deg	degree	fpm	fissions per minute
det	determinant	Fr	francium
dev	deviation	Fr	franklin
diam	diameter	fs	femtosecond
dis/min	disintegrations per minute	ft	foot
dis/s	disintegrations per second	ft lb	foot-pound
div	divergence	ft lbf	foot-pound-force
DNA	deoxyribose nucleic acid	f.u.	formula units
Dy	dysprosium	g	acceleration of free fall
dyn	dyne	G	gauss
E	east	G	giga-
e	electronic charge	g	gram
E	exa-	G	gravitational constant
e, exp	exponential	Ga	gallium
e/at.	electrons per atom	Gal	gal (unit of gravitational force)
e b	electron barn	gal	gallon
e/cm3	electrons per cubic centimeter	g-at.	gram-atom
ed. (pl., eds.)	editor	g.at. wt	gram-atomic-weight
e.g.	exempli gratia, for example	Gc/s	gigacycles per second
el	elastic (in subscript)	Gd	gadolinium
emf, EMF	electromotive force	Ge	germanium
emu	electromagnetic unit	GeV	giga-electron-volt
Eng.	England	GHz	gigahertz
Eq. (pl., Eqs.)	equation	Gi	gilbert
Er	erbium	grad	gradient
erf	error function	GV	gigavolt
erfc	error function (complement of)	Gy	gray
Es	einsteinium	h	hecto-
e.s.d.	estimated standard deviation	H	henry
esu	electrostatic unit	h	hour
et al.	et alii, and others	H	hydrogen
etc.	et cetera, and so forth	h	Planck constant

H.c.	Hermitian conjugate	ks, ksec	kilosecond
hcp	hexagonal-close-packed	kt	kiloton
He	helium	kV	kilovolt
Hf	hafnium	kV A	kilovolt ampere
hf	high-frequency	kW	kilowatt
hfs	hyperfine structure	kW h	kilowatt hour
hg	hectogram	$k\Omega$	kilohm
Hg	mercury	L	lambert
Ho	holmium	L	langmuir
hp	horsepower	l, L	liter
Hz	hertz	La	lanthanum
I	iodine	LA	longitudinal-acoustic
ICT	International Critical Tables	lab	laboratory (in subscript)
i.d.	inside diameter	lat	latitude
i.e.	id est, that is	lb	pound
IEEE	Institute of Electrical and	lbf	pound-force
	Electronics Engineers	lbm	pound-mass
if	intermediate frequency	LED	light emitting diode
Im	imaginary part	Li	lithium
in.	inch	lim	limit
In	indium	lm	lumen
Inc.	Incorporated	lm/W	lumens per watt
inel	inelastic (in subscript)	ln	natural logarithm (base e)
ir, IR	infrared	LO	longitudinal-optic
Ir	iridium	log	logarithm
J	joule	Lr	lawrencium
Jy	jansky	LU	Lorentz unit
k, k_B	Boltzmann's constant	Lu	lutetium
K	degrees Kelvin	lx	lux
K	kayser	ly, lyr	light-year
k	kilo-	M	Mach
K	potassium	M	mega-
kA	kiloamperes	m	meter
kbar	kilobar	m	milli-
kbyte	kilobyte	m	molal (concentration)
kcal	kilocalorie	M	molar (concentration)
kc/s	kilocycles per second	m_e	electronic rest mass
kdyn	kilodyne	m_n	neutron rest mass
keV	kilo-electron-volt	m_p	proton rest mass
kG	kilogauss	$M_\odot$	solar mass (2×10^{33} g)
kg	kilogram	MA	megaamperes
kgf	kilogram force	mA	milliampere
kg m	kilogram meter	ma	maximum
kHz	kilohertz	mb	millibarn
kJ	kilojoule	mCi	millicurie
kK	kilodegrees Kelvin	Mc/s	megacycles per second
km	kilometer	Md	mendlelvium
kMc/s	kilomegacycles per second	MeV	mega-electron-volt; million
kn	knot		electron volt
kOe	kilo-oersted	Mg	magnesium
kpc	kiloparsec	mg	milligram
Kr	krypton	mH	millihenry

mho	reciprocal ohm	No.	number
MHz	megahertz	Np	neper
min	minimum	Np	neptunium
min	minute	ns, nsec	nanosecond
mK	millidegrees Kelvin; millikelvin	n/s	neutrons per second
mks, MKS	meter-kilogram-second (system)	n/s cm^2	neutrons per second per square centimeter
mksa	meter-kilogram-second ampere	ns/m	nanoseconds per meter
mksc	meter-kilogram-second coulomb	O	oxygen
ml	milliliter	$o()$	of order less than
mm	millimeter	$O()$	of the order of
mmf	magnetomotive force	obs	observed (in subscript)
mm Hg	millimeters of mercury	o.d.	outside diameter
Mn	manganese	Oe	oersted
MO	molecular orbital	ohm^{-1}	mho
Mo	molybdenum	Os	osmium
MOE	magneto-optic effect	oz	ounce
mol	mole	P	peta-
mol %, mole %	mole percent	P	phosphorus
mp	melting point	p	pico-
Mpc	megaparsec	P	poise
mph	miles per hour	Pa	pascal
MPM	mole percent metal	Pa	protactinium
Mrad	megarad	Pb	lead
ms, msec	millisecond	pc	parsec
mu	milliunit	Pd	palladium
MV	megavolt; million volt	PD	potential difference
mV	millivolt	pe	probable error
MW	megawatt	pF	picofarad
mwe, m (w.e.)	meter of water equivalent	pl.	plural
Mx	maxwell	P.M.	post meridiem
mμm	millimicron	Pm	promethium
MΩ	megaohm	Po	polonium
n	nano-	ppb	parts per billion
N	newton	p. (pl., pp.)	page
N	nitrogen	ppm	parts per million
N	normal (concentration)	Pr	praseodymium
N	north	psi	pounds per square inch
N, N_A	Avogadro constant	psi (absolute)	pounds per square inch absolute
Na	sodium	psi (gauge)	pounds per square inch gauge
NASA	National Aeronautics and Space Administration	Pt	platinum
nb	nanobarn	Pu	plutonium
Nb	niobium	R (ital)	gas constant
Nd	neodymium	R	roentgen
N.D.	not determined	Ra	radium
NDT	nondestructive testing	rad	radian
Ne	neon	Rb	rubidium
n/f	neutrons per fission	Re	real part
Ni	nickel	Re	rhenium
N_L	Loschmidt's constant	rev.	revised
nm	nanometer	rf	radio frequency
No	nobelium	Rh	rhodium

r.l.	radiation length	tanh	hyperbolic tangent
rms	root-mean-square	Tb	terbium
Rn	radon	Tc	technetium
RNA	ribonucleic acid	Td	townsend
RPA	random-phase approximation	Te	tellurium
rpm	revolutions per minute	TE	transverse-electric
rps, rev/s	revolutions per second	TEM	transverse-electromagnetic
Ru	ruthenium	TeV	tera-electron-volt
Ry	rydberg	Th	thorium
s, sec	second	theor	theory, theoretical (in subscript)
S	siemens	THz	tetrahertz
S	south	Ti	titanium
S	stoke	Tl	thallium
S	sulfur	Tm	thulium
Sb	antimony	TM	transverse-magnetic
Sc	scandium	TO	transverse-optic
sccm	standard cubic centimeter per minute	tot	total (in subscript)
		TP	temperature-pressure
Se	selenium	tr, Tr	trace
sec	secant	trans.	translator, translators; translated by; translation
sech	hyperbolic secant		
sgn	signum function	u	atomic mass unit
Si	silicon	U	uranium
SI	*Système International* (International System of Measurement)	uhf	ultrahigh-frequency
		uv, UV	ultraviolet
sin	sine	V	vanadium
sinh	hyperbolic sine	V	volt
SLAC	Stanford Linear Accelerator Center	VB	valence band
		vol. (pl., vols.)	volume
Sm	samarium	vol %	volume percent
Sn	tin	vs.	versus
sq	square	W	tungsten
sr	steradian	W	watt
Sr	strontium	W	West
STP	standard temperature and pressure	Wb	weber
		Wb/m^2	webers per square meter
Suppl.	Supplement	wt %	weight percent
Sv	sievert	W.u.	Weisskopf unit
T	tera-	Xe	xenon
T	tesla	Y	yttrium
t	tonne	Yb	ytterbium
Ta	tantalum	yr	year
TA	transverse-acoustic	Zn	zinc
tan	tangent	Zr	zirconium

JOURNAL ABBREVIATIONS

Acc. Chem. Res.
Accounts of Chemical Research
Acta Chem. Scand.
Acta Chemica Scandinavica
Acta Crystallogr.
Acta Crystallographica
Acta Crystallogr. Sec. A
Acta Crystallographica, Section A: Crystal
Physics, Diffraction, Theoretical, and General Crystallography
Acta Crystallogr. Sec. B
Acta Crystallographica, Section B: Structural
Crystallography and Crystal Chemistry
Acta Math. Acad. Sci. Hung.
Acta Mathematica Academiae Scientiarum
Hungaricae
Acta Metall.
Acta Metallurgica
Acta Oto-Laryngol.
Acta Oto-Laryngologica
Acta Phys.
Acta Physica
Acta Phys. Austriaca
Acta Physica Austriaca
Acta Phys. Pol.
Acta Physica Polonica
Adv. Appl. Mech.
Advances in Applied Mechanics
Adv. At. Mol. Opt. Phys.
Advances in Atomic, Molecular, and Optical
Physics
Adv. Chem. Phys.
Advances in Chemical Physics
Adv. Magn. Reson.
Advances in Magnetic Resonance
Adv. Phys.
Advances in Physics

Adv. Quantum Chem.
Advances in Quantum Chemistry
AIAA J.
AIAA Journal
AIChE J.
AIChE Journal
AIP Conf. Pro.
AIP Conference Proceedings
Am. J. Phys.
American Journal of Physics
Am. J. Sci.
American Journal of Science
Am. Sci.
American Scientist
Anal. Chem.
Analytical Chemistry
Ann. Chim. Phys.
Annales de Chimie et de Physique
Ann. Fluid Dyn.
Annals of Fluid Dynamics
Ann. Geophys.
Annales de Geophysique
Ann. Inst. Henri Poincaré
Annales de l'Institut Henri Poincaré
Ann. Inst. Henri Poincaré, A
Annales de l'Institut Henri Poincaré,
Section A: Physique Theorique
Ann. Inst. Henri Poincaré, B
Annales de l'Institut Henri Poincaré,
Section B: Calcul des Probabilites et
Statistique
Ann. Math.
Annals of Mathematics
Ann. Otol. Rhinol. Laryngol.
Annals of Otology, Rhinology, & Laryngology
Ann. Phys. (Leipzig)
Annalen der Physik (Leipzig)

Ann. Phys. (N.Y.)
Annals of Physics (New York)
Ann. Phys. (Paris)
Annales de Physique (Paris)
Ann. Rev. Mat. Sci.
Annual Reviews of Materials Science
Ann. Rev. Nucl. Part. Sci.
Annual Review of Nuclear and Particle
Science
Ann. Sci.
Annals of Science
Annu. Rev. Astron. Astrophys.
Annual Reviews of Astronomy and Astrophysics
Annu. Rev. Nucl. Part. Sci.
Annual Reviews of Nuclear and Particle
Science
Annu. Rev. Nucl. Sci.
Annual Review of Nuclear Science
Appl. Opt.
Applied Optics
Appl. Phys. Lett.
Applied Physics Letters
Appl. Spectrosc.
Applied Spectroscopy
Ark. Fys.
Arkiv foer Fysik
Astron. Astrophys.
Astronomy and Astrophysics
Astron. J.
Astronomical Journal
Astron. Nachr.
Astronomische Nachrichten
Astrophys. J.
Astrophysical Journal
Astrophys. J. Lett.
Astrophysical Journal, Letters to the Editor
Astrophys. J. Suppl. Ser.
Astrophysical Journal, Supplement Series
Astrophys. Lett.
Astrophysical Letters
Aust. J. Phys.
Australian Journal of Physics
Bell Syst. Tech. J.
Bell System Technical Journal
Ber. Bunsenges. Phys. Chem.
Berichte der Bunsengesellschaft für
Physikalische Chemie
Br. J. Appl. Phys.
British Journal of Applied Physics
Bull. Acad. Sci. USSR, Phys. Ser.
Bulletin of the Academy of Sciences of the
USSR, Physical Series

Bull. Am. Astron. Soc.
Bulletin of the American Astronomical Society
Bull. Am. Phys. Soc.
Bulletin of the American Physical Society
Bull. Astron. Instit. Neth.
Bulletin of the Astronomical Institutes of the
Netherlands
Bull. Chem. Soc. Jpn.
Bulletin of the Chemical Society of Japan
Bull. Seismol. Soc. Am.
Bulletin of the Seismological Society of
America
C. R. Acad. Sci.
Comptes Rendus Hebdomadaires des Seances
de l'Academie des Sciences
C. R. Acad. Ser. A
Comptes Rendus Hebdomadaires des Seances
de l'Academie des Sciences, Serie A:
Sciences Mathematiques
C. R. Acad. Ser. B
Comptes Rendus Hebdomadaires des Seances
de l'Academie des Sciences, Serie B: Sciences
Physiques
Can. J. Chem.
Canadian Journal of Chemistry
Can. J. Phys.
Canadian Journal of Physics
Can. J. Res.
Canadian Journal of Research
Chem. Phys.
Chemical Physics
Chem. Phys. Lett.
Chemical Physics Letters
Chem. Rev.
Chemical Reviews
Chin. J. Phys.
Chinese Journal of Physics
Class. Quantum Grav.
Classical and Quantum Gravity
Comments Nucl. Part. Phys.
Comments on Nuclear and Particle Physics
Commun. Math. Phys.
Communications in Mathematical Physics
Commun. Pure Appl. Math.
Communications on Pure and Applied
Mathematics
Comput. Phys.
Computers in Physics
Czech. J. Phys.
Czechoslovak Journal of Physics
Discuss. Faraday Soc.
Discussions of the Faraday Society

Earth Planet. Sci. Lett.
  Earth and Planetary Science Letters
Electron. Lett.
  Electronics Letters
Fields Quanta
  Fields and Quanta
Fortschr. Phys.
  Fortschritte der Physik
Found. Phys.
  Foundations of Physics
Gen. Relativ. Gravit.
  General Relativity and Gravitation
Geochim. Cosmochim. Acta
  Geochimica et Cosmochimica Acta
Geophys. Res. Lett.
  Geophysical Research Letters
Handb. Phys.
  Handbuch der Physik
Helv. Chim. Acta
  Helvetica Chimica Acta
Helv. Phys. Acta
  Helvetica Physica Acta
High Temp. (USSR)
  High Temperature (USSR)
IBM J. Res. Dev.
  IBM Journal of Research and Development
Icarus.
  Icarus. International Journal of the Solar System
IEEE J. Quantum Electron.
  IEEE Journal of Quantum Electronics
IEEE Trans. Antennas Propag.
  IEEE Transactions on Antennas and
    Propagation
IEEE Trans. Electron Devices
  IEEE Transactions on Electron Devices
IEEE Trans. Inf. Meas.
  IEEE Transactions on Instrumentation and
    Measurement
IEEE Trans. Inf. Theory
  IEEE Transactions on Information Theory
IEEE Trans. Magn.
  IEEE Transactions on Magnetics
IEEE Trans. Microwave Theory Tech.
  IEEE Transactions on Microwave Theory and
    Techniques
IEEE Trans. Nucl. Sci.
  IEEE Transactions on Nuclear Science
IEEE Trans. Sonics Ultrason. Ind. Eng. Chem.
  IEEE Transactions on Sonics Ultrasonics
  Industrial and Engineering Chemistry
Infrared Phys.
  Infrared Physics

Inorg. Chem.
  Inorganic Chemistry
Inorg. Mater. (USSR)
  Inorganic Materials (USSR)
Instrum. Exp. Tech. (USSR)
  Instruments and Experimental Techniques
    (USSR)
Int. J. Magn.
  International Journal of Magnetism
Int. J. Mod. Phys. A
  International Journal of Modern Physics A
Int. J. Quantum Chem.
  International Journal of Quantum Chemistry
Int. J. Quantum Chem. 1
  International Journal of Quantum Chemistry,
    Part 1
Int. J. Quantum Chem. 2
  International Journal of Quantum Chemistry,
    Part 2
Int. J. Theor. Phys.
  International Journal of Theoretical Physics
Izv. Acad. Sci. USSR, Atmos. Oceanic Phys.
  Izvestiya, Academy of Sciences, USSR,
    Atmospheric and Oceanic Physics
Izv. Acad. Sci. USSR, Phys. Solid Earth
  Izvestiya, Academy of Sciences, USSR, Physics
    of the Solid Earth
J. Acoust. Soc. Am.
  Journal of the Acoustical Society of America
J. Am. Ceram. Soc.
  Journal of the American Ceramic Society
J. Am. Chem. Soc.
  Journal of the American Chemical Society
J. Am. Inst. Electr. Eng.
  Journal of the American Institute of Electrical
    Engineers
J. Appl. Crystallogr.
  Journal of Applied Crystallography
J. Appl. Phys.
  Journal of Applied Physics
J. Appl. Spectrosc. (USSR)
  Journal of Applied Spectroscopy (USSR)
J. Atmos. Sci.
  Journal of Atmospheric Sciences
J. Atmos. Terr. Phys.
  Journal of Atmospheric and Terrestrial Physics
J. Audio Engin. Soc.
  Journal of the Audio Engineering Society
J. Chem. Phys.
  Journal of Chemical Physics
J. Chem. Soc.
  Journal of the Chemical Society

J. Chim. Phys.
Journal de Chemie Physique

J. Comput. Phys.
Journal of Computational Physics

J. Cryst. Growth
Journal of Crystal Growth

J. Electrochem. Soc.
Journal of Electrochemical Society

J. Fluid Mech.
Journal of Fluid Mechanics

J. Gen. Rel. Grav.
Journal of General Relativity and Gravitation

J. Geophys. Res.
Journal of Geophysical Research

J. Inorg. Nucl. Chem.
Journal of Inorganic and Nuclear Chemistry

J. Lightwave Technol.
Journal of Lightwave Technology

J. Low Temp. Phys.
Journal of Low-Temperature Physics

J. Lumin.
Journal of Luminescence

J. Macromol. Sci. Phys.
Journal of Macromolecular Science, [Part B] Physics

J. Mater. Res.
Journal of Materials Research

J. Math. Phys. (Cambridge, Mass.)
Journal of Mathematics and Physics (Cambridge, Mass.)

J. Math. Phys. (N.Y.)
Journal of Mathematical Physics (New York)

J. Mech. Phys. Solids
Journal of the Mechanics and Physics of Solids

J. Mol. Spectrosc.
Journal of Molecular Spectroscopy

J. Non-Cryst. Solids
Journal of Non-Crystalline Solids

J. Nucl. Energy
Journal of Nuclear Energy

J. Nucl. Energy, Part C.
Journal of Nuclear Energy, Part C: Plasma Physics, Accelerators, Themonuclear Research

J. Nucl. Mater.
Journal of Nuclear Materials

J. Opt. Soc. Am.
Journal of the Optical Society of America

J. Opt. Soc. Am. A
Journal of the Optical Society of America A

J. Opt. Soc. Am. B
Journal of the Optical Society of America B

J. Phys. (Moscow)
Journal of Physics (Moscow)

J. Phys. (Paris)
Journal de Physique (Paris)

J. Phys. A
Journal of Physics A: Mathematical and General

J. Phys. B
Journal of Physics B: Atomic, Molecular, and Optical Physics

J. Phys. C
Journal of Physics C: Solid State Physics

J. Phys. D
Journal of Physics D: Applied Physics

J. Phys. E
Journal of Physics E: Scientific Instruments

J. Phys. F
Journal of Physics F: Metal Physics

J. Phys. G
Journal of Physics G: Nuclear and Particle Physics

J. Phys. Chem.
Journal of Physical Chemistry

J. Phys. Chem. Ref. Data
Journal of Physical and Chemical Reference Data

J. Phys. Chem. Solids
Journal of Physics and Chemistry of Solids

J. Phys. Radium
Journal de Physique et le Radium

J. Phys. Soc. Jpn.
Journal of the Physical Society of Japan

J. Plasma Phys.
Journal of Plasma Physics

J. Polym. Sci.
Journal of Polymer Science

J. Polym. Sci., Polym. Lett. Ed.
Journal of Polymer Science, Polymer Letters Edition

J. Polym. Sci., Polym. Phys. Ed.
Journal of Polymer Science, Polymer Physics Edition

J. Quant. Spectros. Radiat. Transfer
Journal of Quantitative Spectroscopy & Radiative Transfer

J. Res. Natl. Bur. Stand.
Journal of Research of the National Bureau of Standards

J. Res. Natl. Bur. Stand. Sec. A
Journal of Research of the National Bureau of Standards, Section A: Physics and Chemistry

J. Res. Natl. Bur. Stand. Sec. B
> Journal of Research of the National Bureau of Standards, Section B: Mathematical Sciences

J. Res. Natl. Bur. Stand. Sec. C
> Journal of Research of the National Bureau of Standards, Section C: Engineering and Instrumentation

J. Rheol.
> Journal of Rheology

J. Sound Vib.
> Journal of Sound and Vibration

J. Speech Hear. Disord.
> Journal of Speech and Hearing Disorders

J. Speech Hear. Res.
> Journal of Speech and Hearing Research

J. Stat. Phys.
> Journal of Statistical Physics

J. Vac. Sci. Technol.
> Journal of Vacuum Science and Technology

J. Vac. Sci. Technol. A
> Journal of Vacuum Science and Technology A

J. Vac. Sci. Technol. B
> Journal of Vacuum Science and Technology B

JETP Lett.
> JETP Letters

Jpn. J. Appl. Phys.
> Japanese Journal of Applied Physics

Jpn. J. Phys.
> Japanese Journal of Physics

K. Dan. Vidensk. Selsk. Mat. Fys. Medd.
> Kongelig Danske Videnskabernes Selskab, Matematsik-Fysiske Meddelelser

Kolloid Z. Z. Polym.
> Kolloid Zeitschrift & Zeitschrift für Polymere

Lett. Nuovo Cimento
> Lettere al Nuovo Cimento

Lick Obs. Bull.
> Lick Observatory Bulletin

Mater. Res. Bull.
> Materials Research Bulletin

Med. Phys.
> Medical Physics

Mem. R. Astron. Soc.
> Memoirs of the Royal Astronomical Society

Mol. Cryst. Liq. Cryst.
> Molecular Crystals and Liquid Crystals

Mol. Phys.
> Molecular Physics

Mon. Not. R. Astron. Soc.
> Monthly Notices of the Royal Astronomical Society

Natl. Bur. Stand. (U.S.), Circ.
> National Bureau of Standards (U.S.), Circular

Natl. Bur. Stand. (U.S.), Misc. Publ.
> National Bureau of Standards (U.S.), Miscellaneous Publications

Natl. Bur. Stand. (U.S.), Spec. Publ.
> National Bureau of Standards (U.S.), Special Publications

Nucl. Data, Sect. A
> Nuclear Data, Section A

Nucl. Fusion
> Nuclear Fusion

Nucl. Instrum.
> Nuclear Instruments

Nucl. Instrum. Methods
> Nuclear Instruments & Methods

Nucl. Phys.
> Nuclear Physics

Nucl. Phys. A
> Nuclear Physics A

Nucl. Phys. B
> Nuclear Physics B

Nucl. Sci. Eng.
> Nuclear Science and Engineering

Opt. Acta
> Optica Acta

Opt. Commun.
> Optics Communications

Opt. Lett.
> Optics Letters

Opt. News
> Optics News

Opt. Photon. News
> Optics and Photonics News

Opt. Spectrosc. (USSR)
> Optics and Spectroscopy (USSR)

Percept. Psychophys.
> Perception and Psychophysics

Philips Res. Rep.
> Philips Research Reports

Philos. Mag.
> Philosophical Magazine

Philos. Trans. R. Soc. London
> Philosophical Transactions of the Royal Society of London

Philos. Trans. R. Soc. London, Ser. A
> Philosophical Transactions of the Royal Society of London, Series A: Mathematical and Physical Sciences

Phys. (N.Y.)
> Physics (New York)

Phys. Fluids
Physics of Fluids
Phys. Fluids A
Physics of Fluids A
Phys. Fluids B
Physics of Fluids B
Phys. Konden. Mater.
Physik der Kondensierten Materie
Phys. Lett.
Physics Letters
Phys. Lett. A
Physics Letters A
Phys. Lett. B
Physics Letters B
Phys. Med. Bio.
Physics in Medicine and Biology
Phys. Met. Metallogr. (USSR)
Physics of Metals and Metallography
(USSR)
Phys. Rev.
Physical Review
Phys. Rev. A
Physical Review A
Phys. Rev. B
Physical Review B: Condensed Matter
Phys. Rev. C
Physical Review C: Nuclear Physics
Phys. Rev. D
Physical Review D: Particles and Fields
Phys. Rev. Lett.
Physical Review Letters
Phys. Status Solidi
Physica Status Solidi
Phys. Status Solidi A
Physica Status Solidi A: Applied Research
Phys. Status Solidi B
Physica Status Solidi B: Basic Research
Phys. Teach.
Physics Teacher
Phys. Today
Physics Today
Phys. Z.
Physikalische Zeitschrift
Phys. Z. Sowjetunion
Physikalische Zeitschrift der Sowjetunion
Planet. Space Sci.
Planetary and Space Science
Plasma Phys.
Plasma Physics
Proc. Cambridge Philos. Soc.
Proceedings of the Cambridge Philosophical
Society

Proc. IEEE
Proceedings of the IEEE
Proc. IRE
Proceedings of the IRE
Proc. Natl. Acad. Sci. U.S.A.
Proceedings of the National Academy of
Sciences of the United States of America
Proc. Phys. Soc. London
Proceedings of the Physical Society, London
Proc. Phys. Soc. London, Sect. A
Proceedings of the Physical Society, London,
Section A
Proc. Phys. Soc. London, Sect. B
Proceedings of the Physical Society, London,
Section B
Proc. R. Soc. London
Proceedings of the Royal Society of London
Proc. R. Soc. London, Ser. A
Proceedings of the Royal Society of London,
Series A: Mathematical and Physical Sciences
Prog. Theor. Phys.
Progress of Theoretical Physics
Publ. Astron. Soc. Pac.
Publications of the Astronomical Society of the
Pacific
Radiat. Eff.
Radiation Effects
Radio Eng. Electron. (USSR)
Radio Engineering and Electronics (USSR)
Radio Eng. Electron. Phys. (USSR)
Radio Engineering and Electronic Physics
(USSR)
Radio Sci.
Radio Science
RCA Rev.
RCA Review
Rep. Prog. Phys.
Reports on Progress in Physics
Rev. Geophys.
Reviews of Geophysics
Rev. Mod. Phys.
Reviews of Modern Physics
Rev. Opt. Theor. Instrum.
Revue d'Optique, Theorique et Instrumentale
Rev. Sci. Instrum.
Review of the Scientific Instruments
Russ. J. Phys. Chem.
Russian Journal of Physical Chemistry
Sci. Am.
Scientific American
Sol. Phys.
Solar Physics

Solid State Commun.
Solid State Communications
Solid State Electron.
Solid State Electronics
Solid State Phys.
Solid State Physics
Sov. Astron.
Soviet Astronomy
Sov. Astron. Lett.
Soviet Astronomy Letters
Sov. J. At. Energy
Soviet Journal of Atomic Energy
Sov. J. Low-Temp. Phys.
Soviet Journal of Low-Temperature
Physics
Sov. J. Nucl. Phys.
Soviet Journal of Nuclear Physics
Sov. J. Opt. Technol.
Soviet Journal of Optical Technology
Sov. J. Part. Nucl.
Soviet Journal of Particles and Nuclei
Sov. J. Plasma Phys.
Soviet Journal of Plasma Physics
Sov. J. Quantum Electron.
Soviet Journal of Quantum Electronics
Sov. Phys. Acoust.
Soviet Physics: Acoustics
Sov. Phys. Crystallogr.
Soviet Physics: Crystallography
Sov. Phys. Dokl.
Soviet Physics: Doklady
Sov. Phys. J.
Soviet Physics Journal
Sov. Phys. JETP
Soviet Physics: JETP
Sov. Phys. Semicond.
Soviet Physics: Semiconductors
Sov. Phys. Solid State
Soviet Physics: Solid State
Sov. Phys. Tech. Phys.
Soviet Physics: Technical Physics
Sov. Phys. Usp.
Soviet Physics: Uspekhi
Sov. Radiophys.
Soviet Radiophysics
Sov. Tech. Phys. Lett.
Soviet Technical Physics Letters
Spectrochim. Acta
Spectrochimica Acta
Spectrochim. Acta, Part A
Spectrochimica Acta, Part A: Molecular
Spectroscopy

Spectrochim. Acta, Part B
Spectrochimica Acta, Part B: Atomic
Spectroscopy
Supercon. Sci. Technol.
Superconductor Science and Technology
Surf. Sci.
Surface Science
Theor. Chim. Acta
Theoretica Chimica Acta
Trans. Am. Cryst. Soc.
Transactions of the American Crystallographic
Society
Trans. Am. Geophys. Union
Transactions of the American Geophysical
Union
Trans. Am. Inst. Min. Metall. Pet. Eng.
Transactions of the Amercian Institute of
Mining, Metallurgical and Petroleum
Engineers
Trans. Am. Nucl. Soc.
Transactions of the American Nuclear Society
Trans. Am. Soc. Mech. Eng.
Transactions of the American Society of
Mechanical Engineers
Trans. Am. Soc. Met.
Transactions of the American Society for
Metals
Trans. Br. Ceramic Society
Transactions of the British Ceramic Society
Trans. Faraday Society
Transactions of the Faraday Society
Trans. Metall. Soc. AIME
Transactions of the Metallurgical Society of
AIME
Trans. Soc. Rheol.
Transactions of the Society of Rheology
Ukr. Phys. J.
Ukrainian Physics Journal
Z. Anal. Chem.
Zeitschrift für Analytische Chemie
Z. Angew. Phys.
Zeitschrift für Angewandte Physik
Z. Anorg. Allg. Chem.
Zeitschrift für Anorganische und Allgemeine
Chemie
Z. Astrophys.
Zeitschrift für Astrophysik
Z. Elektrochem.
Zeitschrift für Elektrochemie
Z. Kristallogr. Kristallgeom. Krystallphys. Kristallchem.
Zeitschrift für Kristallographis, Kristallgeome-
trie, Krystallphysik, Kristallchemie

Z. Metallk.
Zeitschrift für Metallkunde

Z. Naturforsch.
Zeitschrift für Naturforschung

Z. Naturforsch. Teil A
Zeitschrift für Naturforschung, Teil A Physik, Physikalische Chemie, Kosmophysik

Z. Phys.
Zeitschrift für Physik

Z. Phys. Chem. (Frankfurt am Main)
Zeitschrift für Physikalische Chemie (Frankfurt am Main)

Z. Phys. Chem. (Leipzig)
Zeitschrift für Physikalische Chemie (Leipzig)

E

EARTHQUAKE

A dictionary definition of earthquake is the shaking caused by the passage of seismic sound waves generated by sudden and rapid movements within the earth. However, the scientific use of the term "earthquake" has come to be applied to the generating source itself, and there are several types of generating source. Earthquakes may accompany volcanic eruptions as a result of rapid movements of magma or the collapse of volcanic cavities, and they also may be caused by large landslides, but by far the most important, both in terms of numbers and size, are tectonic earthquakes, which are caused by the rapid slip of geologic faults.

The earth's outermost layers are, owing to their coolness, brittle and respond to tectonic forces by fracturing. Faults are shear fractures in which slip occurs in a direction parallel to the fault surface. This slip is resisted by friction, since the walls of faults are held together by compressive stresses that exist everywhere within the earth at depths greater than 1 or 2 km. The brittle layer is 10 to 50 km thick, and its deformation, in response to tectonic forces, primarily occurs by the slip of faults; this slip occurring almost entirely by the infrequent but very rapid movements in earthquakes. The underlying cause of this behavior is that the frictional properties of most rock types under these conditions of pressure and temperature result in a type of unstable behavior

known as stick-slip. Stick-slip friction is characterized by long periods in which the frictional surfaces are held stationary by friction, but when sliding commences, a dynamic instability occurs, accompanied by a large and rapid slip. This instability arises because the dynamic friction, which resists motion once sliding has begun, is smaller than the static friction that must be overcome to initiate sliding.

Once this instability is nucleated, it dynamically propagates over the fault surface at a rupture velocity close to the shear wave speed of the medium, typically about 3 km/s, and it will stop only where it cannot dynamically overcome the static friction. Within the earthquake, the velocity of slip of the two sides of the fault with respect to each other is typically of the order of 1 m/s. The average net slip within the earthquake scales linearly with the final dimensions of the rupture, with a constant of proportionality typically in the range 10^{-4} to 10^{-5}. The size of the earthquake is, therefore, determined by the size of the fault area over which it has propagated. The linear measure of an earthquake is the seismic moment $M_0 = \mu u A$, where μ is the shear modulus of the rock, u is the average slip, and A is the area of the fault ruptured by the earthquake. A commonly reported measure of earthquake size is magnitude (M), which is a logarithmic scale. Magnitude is related to moment by $M = 1.5 \log M_0 - 9.1$, in which the unit of M_0 is N·m. Therefore an increase of magnitude by one unit signifies an increase in moment of a factor of 30, and since the energy of the radiated seismic waves is proportional to the moment, this scales similarly with magnitude. To give some approximate examples, a magnitude 6 (M6) earthquake would be one that ruptures about 10 km of fault with a slip of about 40 cm, an M7 has dimensions of about 60 km and a slip of 2 m, and an M8 ruptures about 400 km of fault and has a slip of about 9 m.

Because earthquakes are the principal agent of tectonism, the map of global seismicity in Fig. 1 allows one to readily identify the earth's most active tectonic features and regions. The most prominent are the plate boundaries—the boundaries between the great tectonic plates along which most of the earth's deformation takes place. These are classified into three types. Divergent boundaries, along which the plates are moving apart, include the great midocean ridge system, shown by the narrow band of earthquake epicenters following the Mid-Atlantic Ridge and other such ridges near the center of many of the other ocean basins. There are also continental rift systems, most notably the one in East Africa. Earthquakes occur frequently along the midocean ridges, but do not attain large magnitudes, since there the brittle layer is thin and hot and most of the plate motions are taken up by the creation of

new sea-floor by submarine volcanic activity. The second class of plate boundary is the transcurrent boundaries, typified by the San Andreas Fault in California, in which the two adjoining plates move past one another in a direction parallel to their boundary. The third type is the convergent plate boundaries, of which there are two kinds. When an oceanic plate converges on either a continent or another oceanic plate, it sinks back into the mantle at what are called subduction zones, sliding on a frictional interface marked by the great oceanic trenches. Behind these trenches are the island arcs, formed by volcanism as a result of partial melting of the oceanic crust as it is subducted into the hot mantle. The most prominent of these subduction zones encircle the Pacific Ocean Basin, forming the the so-called Ring of Fire. The shallowly inclined frictional interface of these subduction zones can be very wide, as great as 200 km, and for this reason it is along these interfaces where the world's greatest earthquakes occur. The largest earthquake in the twentieth century occurred along the southern Chile Trench in 1960; it ruptured a section of that plate boundary 800 km long and 200 km wide and slipped 20 m. These undersea earthquakes produce vertical motions of the sea floor and therefore can

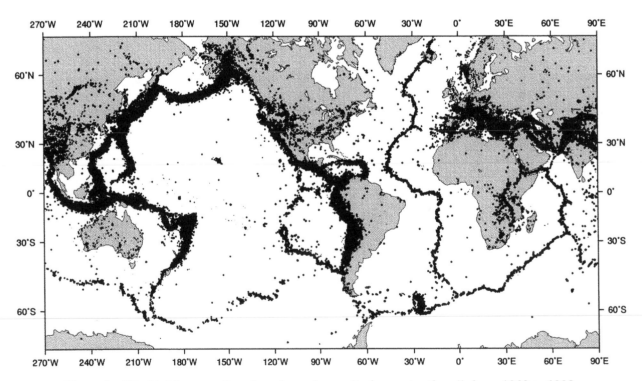

Figure 1 Worldwide map of earthquakes, of magnitude greater than 5, from 1962 to 1992.

produce tsunami, or seismic sea waves, which can be very damaging to coastal regions, even at great distances from the earthquake that caused them. However, at convergent plate boundaries involving two continental plates, the continents are too buoyant to sink into the mantle, and a broad zone of continental collision occurs, accompanied by the formation of high mountain ranges. The prime example of continental collision is the Alpine—Himalayan belt, where Africa and India are colliding with Europe and Asia. The zones of continental collision are very broad, shown by wide, diffuse bands of seismicity.

The major seismic belts in the United States occur along the Pacific Coast. In California, the San Andreas Fault marks the boundary between the Pacific and North American plates. There, the Pacific Plate moves at a rate of about 48 mm/yr to the northwest relative to the North American plates. The San Andreas Fault moves at a long-term rate of about 34 mm/yr, so it only takes up part of this motion, the remainder occurring by slip on many smaller, slower moving faults that are widely distributed from the California Coast as far inland as Salt Lake City. Earthquakes both on and off the main fault, the San Andreas, can be very damaging. The great M7.9 earthquake of 1906 produced about 3 m of slip on a 450-km-long section of the San Andreas Fault in northern California, and together with its resulting fires, destroyed San Francisco. The Northridge earthquake of January 1994, though much smaller at M6.7, caused considerable damage because of its location just beneath the heavily built-up San Fernando Valley near Los Angeles.

Along the coast of Oregon, Washington, and British Columbia there is a slow moving subduction zone that has produced the Cascade Range of volcanoes. There has not been a major earthquake along this subduction interface in historic times, but geologic data indicate that major earthquakes may occur there at intervals as short as 400 years. The other major seismic zone of North America occurs in Alaska, where the subduction zone of the Aleutian Arc extends into the Gulf of Alaska. A slip along this subduction interface produced the great Alaskan earthquake of 1964.

Earth deformation is not entirely restricted to the plate boundaries. Deformation also occurs within the plates, although at much slower rates. This produces the occasional intraplate earthquake, which can be seen as isolated epicenters on the map. Although rare, these intraplate earthquakes can be large. For example, the largest earthquakes that have occurred in historic times in North America were a series of $M > 8$ events centered around New Madrid, Missouri, in 1811 and 1812. These earthquakes occurred along a very slow moving rift zone that is followed by the central Mississippi embayment.

Earthquakes on one particular fault or section of fault follow a well-defined seismic cycle. Following a main shock, which ruptures the fault segment, a sequence of aftershocks occurs, with a frequency of occurrence that decays with the inverse of time since the main shock. These aftershocks, which are usually at least one magnitude unit less than the main shock, serve to relax the stresses redistributed by the main shock itself. On average the main shock has reduced the stress on the fault, so following the aftershocks the fault becomes quiet for a long period until the tectonic motions restore the stresses to high enough values to once again overcome friction, at which point the next main shock occurs. The length of this cycle depends on the rate of tectonic motion. For example, the 1906 San Francisco earthquake slipped on average about 3 m. In the San Francisco area, part of the San Andreas fault proper moves at a long-term geological rate of about 20 mm/yr. We should therefore expect 1906-type earthquakes to reoccur every 150 years, on average. Unfortunately, because of the complexity of the earth, this cycle is not perfectly periodic, so that it is not possible to accurately predict earthquakes on this basis. However, one can make useful statements regarding seismic risk based on the cycle. For example, when comparing the New Madrid Zone with California, we can say that because the rift in the Mississippi embayment is moving very slowly, earthquakes such as those in 1811–1812 probably recur at very long intervals, thousands or tens of thousands of years, as opposed to hundreds of years for more rapidly moving faults like the San Andreas.

Earthquakes also obey well-defined statistical laws that are useful in calculating earthquake hazards. In any given region and time period there are many more small earthquakes than large ones, following a very specific law. As an example of this law, suppose we take an area within which one M8 earthquake occurs every 100 years. Then in the same time period there will, on average, occur 10 M7 earthquakes, 100 M6, 1,000 M5, and so on. This is known as the Gutenberg–Richter law.

See also: GEOPHYSICS; SEISMIC WAVE; SEISMOLOGY; VOLCANO

Bibliography

Bolt, B. A. *Earthquakes: A Primer* (W. H. Freeman, San Francisco, 1978).

Scholz, C. H. *The Mechanics of Earthquakes and Faulting* (Cambridge University Press, Cambridge, Eng., 1990).

CHRISTOPHER H. SCHOLZ

ECLIPSE

An eclipse occurs when one celestial object passes through the shadow cast by another. The type of eclipse that occurs, solar or lunar, depends on the relative positions of the celestial objects involved.

A total lunar eclipse results when the Moon, in its relative orbit about Earth, passes through Earth's shadow. This shadow consists of the umbra, inside of which the view of the Sun is completely blocked from the Moon, and the penumbra, from which a portion of the Sun's surface is still visible from the Moon. Figure 1 illustrates the four possible alternatives for the orbital passage of the Moon through Earth's shadow. If the whole Moon moves through the umbra, (a) viewers on Earth see a total lunar eclipse. Partial passage through the umbra (b) results in a partial lunar eclipse. Passage through the penumbra (c) produces a penumbral eclipse in which the Moon will be dimmed, but not otherwise obviously shadowed. If the tilt of the Moon's orbit about Earth with respect to the Sun–Earth orbit car-

ries the Moon outside of Earth's shadow (d), no eclipse will occur. When a lunar eclipse does occur, however, Earth's umbra is about three times the diameter of the Moon. Thus, the Moon may be fully shadowed for a time as long as 1 hour and 40 minutes. Sunlight refracted through Earth's atmosphere often allows some illumination of the Moon, giving it a dark coppery appearance. If the orbit of the Moon about Earth (the lunar orbit plane) was identical to that of the orbit of Earth about the Sun (the ecliptic plane), a total eclipse would occur every 29.5 days at the time of the full moon. In reality, the lunar orbit is tilted by 5°9' to the ecliptic plane, therefore, there are times when the Moon passes above or below the midpoint of Earth's shadow. At such times a partial or a penumbral eclipse may occur. Anyone on the side of Earth facing the Moon is able to view the eclipse.

When the Moon passes in front of the Sun, its umbra reaches Earth's surface only if the Moon is near to perigee—the point in its orbit when the Moon is closest to Earth (Fig. 2a). At that time the umbra may be as wide as 270 km on Earth's surface. This shadow races across the surface at nearly 2,000 k/h; thus, a total solar eclipse, seen by people within the umbra, can last no more than about 7 minutes. The lunar penumbra is much wider and, hence, a partial solar eclipse may be viewed from a larger portion of Earth's surface located in the penumbra. When the Moon is not close to perigee, its apparent size is smaller than that of the Sun, so its umbra thus does not reach Earth's surface, and at best, an observer will view an annular eclipse (Fig. 2b).

For an eclipse to occur, the Sun, the Moon, and Earth must be along a line that nearly coincides with

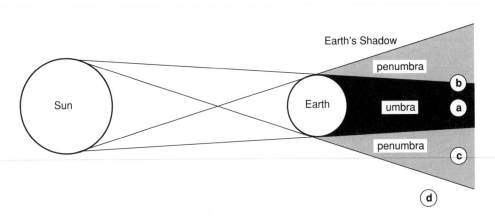

Figure 1 Lunar eclipses.

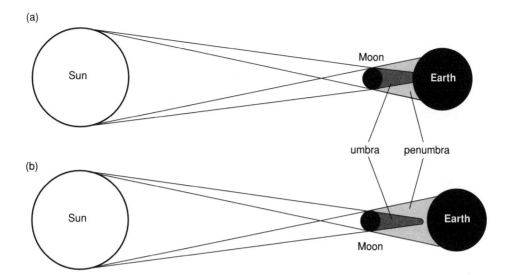

Figure 2 Solar eclipses.

the line of intersection of the lunar orbital plane and the ecliptic plane. This intersection, or line of nodes, completes one rotation in a westwardly direction every 18.6 years. Combined with the annual 365.25-day motion of the Sun from west to east around the sky and the 27.3-day orbital period of the Moon in the same direction, the potential for an eclipse exists every 173 days (the eclipse season). The actual number of eclipses visible each year varies from as few as two (both solar, as happened in 1969) to as many as seven (five solar and two lunar, which last occurred in 1935 and will occur again in the year 2160). If a lunar eclipse occurs in early January, a third lunar eclipse may occur in late December.

Various eclipse cycles have been recognized, the most famous of which is related to the saros, a period of 18 years 11 days and 8 hours that was first recognized 3,000 years ago by the Chaldeans. One saros period after the occurrence of a solar eclipse, the geometrical orientation of the Sun and Moon and the line of nodes will be similar, once again permitting the Moon to shadow the Sun, producing an eclipse; this eclipse is visible from a position 120° longitude west of the first eclipse. After three saros periods, a solar eclipse produced by a similar geometry will occur very nearly at the longitude of the original eclipse.

See also: KEPLER'S LAWS; MOON, PHASES OF; NEWTON'S LAWS; PLANETARY SYSTEMS; SUN; TIDES

Bibliography

ALLEN, D., and ALLEN, C. *Eclipse* (Allen and Unwin, Sydney, Australia, 1987).

LIU, B.-L., and FIALA, A. D. *Canon of Lunar Eclipses* (Willmann-Bell, Richmond, VA, 1994).

MEEUS, J.; GROSJEUN, C. C.; and VANDERLEEN, W. *Canon of Solar Eclipses* (Pergamon Press, New York, 1966).

U.S. NAVAL OBSERVATORY. *The Astronomical Almanac* (U.S. Government Printing Office, Washington, DC, reprinted annually).

CHARLES J. PETERSON

EDDY CURRENT

See CURRENT, EDDY

EDUCATION

Because physics has a reputation for being a difficult subject, many educators are searching for ways to make it easier to learn. People working in this area not only need to thoroughly understand the subject, but they also have to be familiar with educational methodology and psychology. Because of its multidisciplinary nature, this subfield of physics offers a large assortment of employment opportunities. These can range from teaching in a middle school science classroom to conducting research in a university, from designing museum displays to establishing policies for federal government agencies.

As one might expect, the kind of background required for these different types of jobs varies greatly. However, the jobs can be highly satisfying and are worth investigation by people interested in promoting the understanding of physics. The most obvious choice for many people completing graduate study in physics education is to continue working in an academic setting. A teaching certificate requires a bachelor's degree, depending on the state, and opens up the middle school or high school teaching profession.

Nearly a quarter of the nation's high school seniors are enrolled in or have completed a physics course. This fraction has been steadily rising since the 1980s. This fact, coupled with an expected 20 percent rise in the number of high school age students by the year 2000, indicates that there will be a great demand for teachers. (Since a solid background in physics is a good preparation for understanding all the sciences, high school physics teachers are often also qualified to teach courses in chemistry or mathematics. For most of these teachers though, physics is their favorite subject. One survey by the American Institute of Physics found that 92 percent of teachers with multiple courses would prefer to teach physics as much as possible.) A master's degree is often the ticket to a job teaching in a community college or junior college. Two- and four-year colleges almost always have some kind of physics offering. Larger universities, which normally require faculty to hold a doctorate in their field, often have dozens of physics courses. Obviously, the more courses a school offers, the greater the variation in the topics presented and the depth of coverage.

Physics education professors often find a university combination of teaching and research duties especially stimulating. One's research into learning provides input for improving teaching performance. Conversely, being in the classroom as a teacher presents many opportunities to conduct education research. Universities also offer opportunities for outreach service to the community. Faculty members might find themselves visiting elementary school classrooms to talk about rainbows, answering telephone questions from the general public, or perhaps appearing on television to discuss the physics of a current event such as the appearance of a comet or the dangers of selling nuclear technology to unstable foreign governments.

Besides academia, there are many other areas where physics education specialists can put their skills to use. A substantial number of state and federal government agencies are involved in education-related policies. People who not only have a strong grasp of technical topics but also have insight into how people gain understanding of scientific ideas are in high demand. Additional job prospects can be found in public educational facilities like museums and in industry. A person who knows how to relate complicated technical ideas in a clear fashion might make a good marketing manager or even become a public relations expert for a high-tech company. Independent consultants can specialize in evaluating instructional programs, writing textbooks, or developing new curricular materials. As more and more computers are finding their way into homes and classrooms, the instructional software market is growing rapidly.

An interesting facet of this particular field is the strong connection it promotes between education research techniques and the concepts of physics. At the very least, good teachers apply the findings of research into how people come to grasp physical concepts. But many teachers, and especially university faculty, are getting involved in finding answers to questions about how students learn a complex subject like physics. The laboratory for this type of work can be any classroom where physics is taught at any level. Research tasks include conducting interviews of students while they think about physical situations, developing new ways of assessing knowledge and attitudes, and designing and testing innovative teaching techniques. As you might guess, one studies how people learn physics by asking a lot of questions.

Of course, these questions stem from an awareness of what other researchers have done in the past. There are thousands of articles that have been published in scientific journals about many different aspects of physics education. Searching this enormous body of literature takes time, but the process has been simplified somewhat by recent advances in database technology. It is now possible for a researcher to scan a library's entire collection from the comfort of his or her office. Once this has been done and the papers related to the subject have been reviewed, an experimental design is developed. Often this will involve presenting students (either individually or in classes) with some sort of physical situation that focuses their attention on the concept of interest. During a series of interviews the researcher asks the students to explain what is going on in the situation in order to probe their understanding of the underlying concepts. Alternately, multiple choice testing of larger numbers of participants allows more generalizability of findings to students in other classrooms and schools. The cost of this extendability is limited resolution. (Interviews provide a much clearer picture of what an individual is thinking, but it is harder to assert that many students share those thoughts.) Many researchers are finding ways to combine these two research techniques. Qualitative interviews provide hints as to students' conceptualizations, while quantitative testing provides insight into how prevalent those ideas are in general. What we are finding is that students build their knowledge on a foundation of what they already understand about how the world works.

For the most part, these intuitive ideas are accurate; they have served their holders well for their entire lives. In a few critical cases, however, they make it difficult to learn new physics topics. A major outcome of these studies has been the realization that students can successfully complete a physics course and still have very basic misunderstandings of important topics. For example, after living their whole lives in a world where the effects of air resistance are so common that they unconsciously account for it in situations ranging from tossing out a piece of paper to playing a game of golf, students are faced with instruction that begins with "We will neglect the effects of friction." What often happens is that the connection between the real world and the artificial one presented in the classroom is never made. Although students can memorize formulas well enough to pass standard tests, they still have fundamental misconceptions. A major part of the physics education researcher's work is to find ways to assess these difficulties and then develop ways to help students come to grips with the problem. This requires a shift in thinking, a conceptual change, which is not an easy thing to do. It has been found that eliciting a student's existing concepts and then exposing them to situations conflicting with those ideas can help. Traditional instruction, where students passively sit back and are basically told the an-

swers, generally does not have a long-lasting effect on pre-existing intuitive thought patterns.

Besides investigating ways to address student misunderstanding of the basic concepts of physics, some researchers are looking at attitudes held toward physics and science in general. It is often felt that science involves nothing more than the memorization of a large number of well-established (perhaps even fossilized) facts. Real scientists, although they certainly have to know the central body of knowledge for their field of study, spend much of their time searching for answers to questions where the facts are not known. Many students are not aware of the fundamentally open-ended and non-authoritarian nature of science and that the measurements scientists make are always approximations. In other words, it is often felt that science provides absolutes that come from authority figures. Nothing could be further from the truth. Cutting-edge science is full of disagreements and scholarly debates. That is part of what makes it so fascinating.

Some physics education researchers are developing instructional activities that help students recognize how science really works, even including review of their work by their peers—the very foundation of the modern scientific enterprise. One of the advantages of being a physics education researcher is that the work can be conducted in many different settings. Because little equipment is needed, even elementary schools are potential experiment sites. A great deal of high-quality research and curriculum development has been done by high school teachers. Nonetheless, universities are where most of the investigations into how people learn physics are based.

This relatively new subfield of physics research is enjoying a period of substantial growth. During the 1980s, only two or three physics departments had graduate programs in this area. Today there are more than a dozen universities training researchers. This community of scholars meets regularly at conferences around the country. Many are invited to travel internationally, giving presentations and consulting at schools and businesses around the world. Even when these scientists are not actually in the same location, they continue to discuss ideas and share findings through heavy use of electronic mail and the World Wide Web. Collegiality is one of the greatest benefits of this type of professional endeavor.

See also: BASIC, APPLIED, and INDUSTRIAL PHYSICS; EXPERIMENTAL PHYSICS; INDUSTRY, PHYSICISTS IN

Bibliography

BEICHNER, R. "Testing Student Interpretation of Kinematics Graphs." *Am. J. Phys.* **62,** 750–762 (1994).

LAWS, P. "Calculus-Based Physics Without Lectures." *Phys. Today* **44** (8), 24–31 (1991).

McDERMOTT, L. C., and SHAFFER, P. S. "Research as a Guide for Curriculum Development: An Example from Introductory Electricity, Part I: Investigation of Student Understanding." *Am. J. Phys.* **60,** 994–1003 (1992).

REDISH, E. F. "Implications of Cognitive Studies for Teaching Physics." *Am. J. Phys.* **62,** 796–803 (1994).

REIF, F. "Millikan Lecture 1994: Understanding and Teaching Important Scientific Thought Processes." *Am. J. Phys.* **63,** 17–32 (1995).

ROBERT J. BEICHNER

EIGENFUNCTION AND EIGENVALUE

In many problems in physics and engineering, it is required to find a function which is not only a solution to a specified differential equation, but also has a specified behavior on the boundaries of some region of space. In mathematics, such problems are called boundary-value problems. An important class of boundary-value problems are the Sturm–Liouville problems. A Sturm–Liouville problem consists of a certain kind of second-order differential equation defined on an interval $a \leq x \leq b$ with the solutions also satisfying certain conditions to be specified below at a and at b. A Sturm–Liouville differential equation in the function $y = y(x)$ is one that can be cast into the form

$$(ry')' + qy + \lambda py = 0.$$

The prime sign indicates differentiation with respect to x. The coefficients p, q, and r are functions of x, and λ is a parameter. The only restrictions on the functions p, q, and r are that they be continuous over the interval $[a, b]$ and that r' is also continuous over that region. The Sturm–Liouville equation is very general and equations of that form occur often in physics and engineering.

There are three kinds of Sturm–Liouville problems: regular, periodic, and singular. The difference is in the form of the boundary conditions satisfied by the solutions. Regular boundary conditions are of the form

$$A_1 y(a) + A_2 y'(a) = 0$$

and

$$B_1 y(b) + B_2 y'(b) = 0,$$

where A_1 and A_2 are constants, not both zero, and B_1 and B_2 are also constants, not both zero. Periodic boundary conditions are of the form $y(a) = y(b)$ and $y'(a) = y'(b)$.

In general, solutions $y = y(x)$ may be found for a Sturm–Liouville equation for any value of the parameter λ. However, solutions that also satisfy the boundary conditions exist only for certain special values of λ. These special values are called characteristic values or eigenvalues, and are denoted λ_n. The subscript is called the index or quantum number. The solutions which correspond to particular eigenvalues are called eigenfunctions, and are labeled with the same index as the eigenvalue. That is, y_n is the function that satisfies the differential equation when $\lambda = \lambda_n$ and also satisfies the boundary conditions. While each eigenfunction corresponds to only one eigenvalue, it is possible that more than one eigenfunction may correspond to the same eigenvalue. If there are m different eigenfunctions that all correspond to the eigenvalue λ_n, that eigenvalue is said to be m-fold degenerate, and those m eigenfunctions are said to be degenerate with each other.

There are several theorems about the eigensolutions to Sturm–Liouville problems which demonstrate their importance. We list two of them here.

Theorem 1. If two eigenvalues λ_n and λ_m are distinct, then their corresponding eigenfunctions y_n and y_m are said to be orthogonal on the interval $[a, b]$ with respect to the weight function $p(x)$. This means that

$$\int_a^b p(x)\, y_m^*(x)\, y_n(x)\, dx = 0.$$

Here, the weight function $p(x)$ is the function $p(x)$ in the Sturm–Liouville differential equation and the asterisk indicates complex conjugation. This orthogonality theorem allows any function that satisfies the boundary conditions of the Sturm–Liouville problem to be expanded as a sum of eigenfunctions. The

Fourier expansion is the most well-known example of this result.

Theorem 2. The eigenvalues of a Sturm–Liouville problem are real. In quantum mechanics, eigenvalues correspond to quantities that are measureable, and hence physically meaningful. This theorem guarantees that measureable quantities are real valued and therefore can indeed be measured.

While Sturm–Liouville problems occur in many fields of physics and engineering, their role in quantum mechanics is central. The Schrödinger equation in one dimension may be written in Sturm–Liouville form as

$$\frac{\hbar^2}{2m}\frac{d^2\psi(x)}{dx^2} - V(x)\,\psi(x) + E\,\psi(x) = 0,$$

where $\hbar$ is the Planck constant, m is the mass of the particle being considered, $V(x)$ is the potential energy function under which the particle moves, and E is the particle's total energy. This is a Sturm–Liouville equation with $p = 1$, $q = -V(x)$, $r = \hbar^2/2m$, and the eigenvalue $\lambda = E$. The boundary conditions are determined by the potential energy function $V(x)$. In problems in which the potential energy binds the particle to some region of space, the boundary conditions are invariably either regular, periodic or singular, so that in bound state problems, energy quantization arises because the Schrödinger equation has solutions that satisfy the boundary conditions only for special values of the parameter E.

Another form of the eigenfunction equation is $\Omega y = \lambda y$, where y and λ are defined as before, but Ω is a linear second-order differential operator. The effect of applying the operator to the function is to return the same function back, multiplied by a constant. Three-dimensional problems in quantum mechanics are often solved by the technique of separation of variables. Each separation yields an equation in this form and a separation constant, which becomes an eigenvalue upon applying boundary conditions. The eigenvalue is described by a quantum number.

As an example, the three-dimensional, time-dependent Schrödinger equation for a particle (without spin) in a spherically symmetric potential $V(r)$ is

$$\left[\frac{-\hbar^2}{2m}\boldsymbol{\nabla}^2 + V(r)\right]\Psi(r,\theta,\phi,t) = i\hbar\,\frac{\partial}{\partial t}\,\Psi(r,\theta,\phi,t).$$

Since this equation involves the four variables, r, θ, ϕ, and t, the separation process must be applied three times, yielding four separated equations in one dimension each. The four one-dimensional equations involve three separation constants. Since each of these equations is an eigenfunction equation with appropriate boundary conditions, these three separation constants become eigenvalues, each described by its own quantum number. Hence, the properties of the motion of such a particle is described by a set of three quantum numbers. For example, for the states of an electron in an atom, neglecting spin, the solutions have three labels (n, l, m) that define their quantum properties. In classical mechanics, the motion of such a particle is also described by three constants. This demonstrates the fact that eigenvalues play a role in quantum mechanics similar to the role constants of the motion play in classical mechanics.

See also: ENERGY LEVELS; QUANTUM MECHANICAL BEHAVIOR OF MATTER; QUANTUM MECHANICS; QUANTUM NUMBER; SCHRÖDINGER EQUATION

Bibliography

ANDERSON, E. E. *Modern Physics and Quantum Mechanics* (Saunders, Philadelphia, 1971).

ARFKEN, G. *Mathematical Methods for Physicists*, 3rd ed. (Academic, Orlando, FL, 1985).

O'NEIL, P. V. *Advanced Engineering Mathematics*, 4th ed. (PWS-Kent, Boston, 1995).

JAMES R. HUDDLE

EINSTEIN, ALBERT

b. Ulm, Germany, March 14, 1879; *d.* Princeton, New Jersey, April 18, 1955; *relativity, quantum theory, Brownian motion, quantum statistics.*

Einstein's ancestors had long lived in small south-German towns. The well-to-do family of his mother Pauline was in the wholesale grain trade, and his father Hermann was a small businessman. Like many German Jews of their generation, his parents never denied their origins but were nonobservant and culturally quite assimilated. Always independent minded, and rather a "loner," young Albert was close to his sister Maja. A period of childhood religiosity ended at twelve when popular scientific literature made him a free thinker, but a feeling of wonder at the harmony of the universe never left him.

In 1880 the family moved to Munich, the largest south-German city, where Hermann and his brother Jakob, a trained engineer, started one of the city's first electrotechnical firms. Raised in a technological milieu, Albert was originally destined to take over the family business, which at first flourished. In 1894 competition from larger German firms led the brothers to relocate to northern Italy, where further business reverses soon led to the breakup of the partnership. Helped occasionally by Albert, Hermann's small, debt-ridden business ended with his death.

Einstein attended primary and secondary school in Munich. He found much of the curriculum and above all the instructional methods distasteful, later comparing most of his teachers to drill sargeants. His slow but thorough and methodical approach to the subjects in which he was interested earned him good but not outstanding grades. Encouraged by Uncle Jakob, he developed a bent for mathematics, especially geometry and calculus, mainly through self-study; a family friend stimulated a precocious interest in the natural sciences and philosophy. When the family left Munich in 1894, Albert stayed to finish school, but soon left for Italy after clashing with some of his teachers. He continued studying on his own, hoping to be admitted to the Poly (Swiss Federal Polytechnical School, a technical university) in Zurich, which recognized his talent but recommended he finish secondary school at the nearby Aarau Cantonal School. Its more liberal style of teaching and excellent scientific facilities soon changed his attitude towards schooling and made apparent his talent. In 1896 he entered the Poly as a physics student, earning generally good grades; but his independent attitude did not ingratiate him with his teachers. Happily working in the newly equipped physics laboratories, he remedied the dearth of advanced physics courses by self-study of theoretical physics, often joined by fellow physics student Mileva Maric, whom he married in 1903.

Unable to find a university or secondary school position in physics after graduation in 1900, he worked at a variety of temporary jobs until hired as an Examiner at the Swiss Patent Office. While there (1902–1909) he completed studies of the special theory of relativity, Brownian motion, and molecular dimensions (the topic for which he received his doctorate), all in 1905; published his first papers on quantum theory and started work on the general theory of relativity.

Growing recognition of his accomplishments by the physics community soon followed. He was appointed associate professor of physics at the University of Zurich in 1909 and two years later became a full professor in Prague (then part of the Austro-Hungarian Empire), but returned to Zurich—now at the Poly—in 1912. He was appointed to the Prussian Academy of Sciences in 1914, a full-time research position in Berlin, where he completed work on the general theory of relativity (1915). Separated from Maric in 1914, he married his cousin Elsa Einstein in 1919.

Einstein remained in Berlin throughout World War I, when his pacifist views set him against the mainstream of academic jingoism, and during the Weimar Republic, when his democratic views made him a hero to defenders of the republic and a target of the growing fascist movement. With Hitler's advent to power in 1933, Einstein left Germany for good and settled in Princeton, New Jersey, at the newly established Institute for Advanced Studies, where he continued to live and work for the rest of his life.

Einstein regarded the special and general theories of relativity and the search for a unified field theory as the central thread of his life's work. The special theory grew out of the problem of reconciling Newtonian mechanics, which implies the equality of all inertial (i.e., nonaccelerated) frames-of-reference (principle of relativity), with Maxwell's theory of electromagnetism, which was taken to imply the existence of only one frame (the "ether frame"), in which the speed of light is constant. For the classical law of addition of velocities implies that, with respect to any frame moving through the ether, the velocity of light should depend on that frame's velocity. But all attempts to detect such a variation in the speed of light with respect to Earth as it moves around the Sun had failed. A number of prominent scientists worked on this problem, but Einstein was the first to see clearly that the way out was to give up the classical law of addition of velocities by replacing the Newtonian concept of absolute time, that is, the same for all inertial frames, by that of a time that is relative to each inertial frame. The special theory preserves the equality of all inertial frames, but postulates that the speed of light is absolute, that is, the same in all inertial frames. The failure to detect a variation in the speed of light thus becomes evidence supporting the new relativistic kinematics. Such counterintuitive but well-established effects as time dilatation and the twin paradox provide further supporting evidence. Physical theories, such as Newtonian mechanics, had to be reviewed and modified to ensure compatibility with relativistic kinematics. Many surprising consequences followed from this review, notably the blending of the previously separate laws of mass and energy conservation into a single law of conservation of mass-energy (often known as the equivalence of mass and energy). The entire modern theory of elementary particles is built on the foundations of the special theory, as is the operation of high-energy particle accelerators.

In 1907 Einstein came to the conclusion that a modification of the theory was needed to incorporate gravitation because of a unique feature, already stressed by Galileo: Regardless of their mass, all bodies fall with the same acceleration in a gravitational field. Einstein realized this feature casts doubt on the privileged role of inertial frames because it implies that no mechanical experiment can distinguish between an inertial frame of reference with a uniform downward-acting gravitational field, and an accelerated frame-of-reference with no gravitational field, whose upward acceleration is numerically equal to that produced by the gravitational field in the first case. Einstein assumed that there is a complete equivalence between inertial frames with a gravitational field and accelerated frames without a gravitational field, and he made this principle of equivalence the foundation of an eight-year-long search for a relativistic theory of gravitation. In the resulting general theory of relativity, completed in 1915, all frames of reference are equally acceptable. Gravitation is an effect of matter on the structure of space and time, not a force pulling objects off their straight line (inertial) paths in a flat spacetime but a warping of spacetime in which objects attempt to follow the straightest possible paths. The same mathematical object that describes the structure of spacetime the metric tensor, also characterizes the gravitational field; the very structure of spacetime is now a dynamical field. With the new theory, Einstein was able to explain the hitherto anomalous portion of the precession of Mercury's perihelion and suggested a number of astronomical tests of his theory, such as the gravitational red shift of stellar spectra, the apparent deflection of light rays passing near a star, and the focusing effect that a concentrated massive object would have on light (gravitational lensing). All of these predictions have been confirmed with increasing accuracy by recent optical and radio wave observations. General relativistic corrections have proved important in the theory of such super-massive objects as neutron stars, and the theory also predicts novel phenomena such as black holes and gravitational waves, which have become the object of recent intense theoretical and observational study.

Starting in the 1920s, Einstein became increasingly absorbed by the search for a unified theory of the electromagnetic and gravitational fields. Always fascinated by the discovery of conceptual unity behind apparently different phenomena (Maxwell's unification of electricity, magnetism, and optics provided the outstanding example) and having developed a field theory of gravitation, Einstein felt that some generalization of the metric tensor should encompass the electromagnetic field. In the case of gravitation, the principle of equivalence provided a physical clue that led rather directly to the metric tensor, but Einstein never found a similar physical clue for its generalization, so he continued to explore a large number of mathematical possibilities on the basis of their formal simplicity until the end of his life, without ever really convincing himself or others that he was on the right track.

A major motivation for his decades-long search was the hope that such a unified field theory might have a discrete set of nonsingular solutions, thereby explaining the all-pervasive quantum effects he had been exploring since the turn of the century. Max Planck introduced the quantum of action, but Einstein first took the idea seriously enough to suggest in 1905 that electromagnetic radiation might consist of discrete quanta of energy. He was able thereby to offer simple, quantitative explanations of a number of puzzling phenomena involving the exchange of energy between matter and radiation, notably the photoelectric effect, mentioned in his 1921 Nobel Prize citation. Although Einstein continued to develop his concept of the quantum of radiation into that of a full-fledged particle, later named the photon, carrying momentum as well as energy, the idea was not taken seriously by most physicists—including Planck and Niels Bohr—until 1923, when the Compton effect turned the tide. In 1907 Einstein took Planck's idea of quantized material oscillators and developed it into the first quantum theory of the solid state, thereby providing an explanation for the anomalous low temperature specific heats of solids. This work, successfully tested almost immediately, was instrumental in making the study of quantum effects a central concern of the physics community. In 1924 Einstein made a major contribution to the development of quantum statistics by showing that the recent derivation of the blackbody radiation spectrum by Satyendranath Bose, who treated the radiation as a gas of light quanta (photons), was tacitly based on a method of counting particle configurations that differed from the classical one used by Boltzmann. The resulting Bose–Einstein statistics, as it came to be called, was later shown to hold for all particles with integral spins. By applying Bose's method to a gas of material particles, Einstein showed that it would undergo condensation at a certain temperature, thus providing the first theoretical model of a phase transition.

However, when the new quantum mechanics began to explain a number of quantum phenomena from 1925 on, Einstein found himself out of sympathy with the basic approach of the theory. At first he tried to find flaws in it but soon acknowledged that, within its theoretical framework and when given a statistical interpretation, quantum mechanics is the best explanation that can be given for these phenomena. What he continued to challenge was the theory's alleged completeness—the claim that the theory gave the most complete possible characterization of the state of an individual system—and the assertion that no other theoretical framework could be devised that would avoid what he regarded as objectionable features of quantum mechanics: The introduction of probability as an irreducible feature of reality and the continued entanglement of two quantum systems once they have interacted—no matter how far apart they may subsequently move. He continued to hope that a suitable classical unified field theory, which by its nature would avoid these features, could explain quantum phenomena, a hope shared by few physicists today.

The discovery of the weak and strong nuclear forces made obsolete Einstein's original program of unification confined to gravitation and electromagnetism. On the other hand, it has made the idea of a unification of these four fundamental interactions more attractive. Major successes have been achieved in the unification of the electromagnetic and weak interactions, and then the electroweak and strong forces, although these unfications differ from Einstein's attempts in that they are based on quantum

mechanics. But the general theory of relativity has so far resisted all attempts at conventional quantization, let alone its unification with the other fields. It is possible that Einstein was right to the extent that the unique features of gravitation—its character as a spacetime structure rather than a force—may require modifications of the quantum-mechanical formalism as well as of general relativity before any unification is possible.

While we have concentrated on those aspects of Einstein's work that go beyond classical physics, he was also a master of the latter, and developed many new applications of its methods. His explanation of Brownian motion and his method of estimating the size of molecules in a solution, both published in 1905, as well as his many studies of fluctuation phenomena over the years, provide outstanding examples.

After the successful testing of Einstein's prediction of the apparent deflection of light rays by two English solar eclipse expeditions in 1919, Einstein's name became well known to the nonscientific public. Indeed, he became the first scientific "superstar," often mobbed during his public appearances; as a consequence these became rarer and rarer over the years, especially after his move to the United States. While regarding his notoriety as a personal burden, it offered him a means of disseminating his views on a number of important political and social questions: whatever the great Einstein said was news. Increasingly identifying with the Jewish people as they became ever more frequent targets of anti-Semitic propaganda and physical attacks, first in Weimar Germany and especially after Hitler took power in 1933, he supported the Zionist ideal of a Jewish homeland in Palestine as a way of building up Jewish pride and self-confidence in the face of these attacks, and then as a place of refuge for Jews forced to flee Europe. After the Holocaust, he supported the establishment of Israel to ensure a haven for the remnants of European Jewry.

A convinced antimilitarist, he was first impelled to political action by his opposition to World War I. After the war, he supported the pacifist movement, advocating refusal of military service. When Hitler came to power, Einstein felt that pacifist tactics were powerless again fascism's ruthless threat to peace and democracy, and he advocated rearmament to deter and ultimately defeat aggression in World War II. When the development of nuclear weapons threatened the destruction of humanity, he advocated a world government as the only way to overcome national enmities and ensure disarmament. Ironically, he is often credited with—or blamed for—playing a major role in the development of the American atomic bomb, although his role was confined to alerting the American government in 1939 to the danger of Germany's doing so.

The economic chaos in Germany in the 1920s, followed by the worldwide economic crisis and collapse in the 1930s, convinced Einstein that the capitalist economic system needed drastic change, and he began to advocate a socialist reorganization of the economy. Well aware of the dictatorial features of the Soviet model, he stressed the need to preserve democratic political rights under socialism.

As may be imagined, Einstein's political and social views were not universally shared, and he became the object of intense personal attacks, often anti-Semitic in nature. He was denounced as unpatriotic for his stand against nationalism and war, and as a "red" for his social and economic views. This was true not only in Germany but also in the United States, especially during the "cold war" period, when he blamed the United States goverment for a large share of the rising tensions with Russia and urged resistance to all attempts at governmental inquisitions into individual beliefs. His defense of civil liberties in his adopted homeland was an inspiration to many during the McCarthy years.

See also: CONDENSATION, BOSE–EINSTEIN; EINSTEIN–PODOLSKY–ROSEN EXPERIMENT; MOTION, BROWNIAN; QUANTUM MECHANICS, CREATION OF; QUANTUM STATISTICS; QUANTUM THEORY, ORIGINS OF; RELATIVITY, GENERAL THEORY OF, ORIGINS OF; RELATIVITY, SPECIAL THEORY OF, ORIGINS OF; SPECIFIC HEAT, EINSTEIN THEORY OF

Bibliography

EINSTEIN, A. *Relativity: The Special and the General Theory*, 15th ed. (Crown, New York, 1961).

EINSTEIN, A. *Ideas and Opinions* (Modern Library, New York, [1954] 1993).

EINSTEIN, A., and INFELD, L. *The Evolution of Physics: From Early Concepts to Relativity and Quanta* (Simon & Schuster, New York, 1938).

PAIS, A. *"Subtle is the Lord . . .": The Science and the Life of Albert Einstein* (Oxford University Press, Oxford, Eng., 1982).

STACHEL, J., ed. *The Collected Papers of Albert Einstein* (Princeton University Press, Princeton, NJ, 1987).

JOHN STACHEL

EINSTEIN OBSERVATORY

In November 1978 NASA launched the second of its High Energy Astronomical Observatory (HEAO) satellites, HEAO-2. To commemorate the centennial of Albert Einstein's birth, which was being celebrated that year, the scientists and engineers at the Harvard-Smithsonian Center for Astrophysics, who were largely responsible for two of the four detectors as well as the x-ray telescope itself, renamed the mission the Einstein observatory. The Einstein observatory represented an enormous increase in sensitivity, by a factor of about a thousand, over any x-ray satellite mission launched before. Thus, in the sixteen years since the discovery of the first cosmic x-ray source in 1962 and the eight years since Uhuru (the first such satellite) was launched in 1970 to conduct the first sky survey for cosmic x-ray sources, x-ray astronomy had increased in sensitivity by nearly the same factor as the increase from Galileo's first telescope to the 200-in. telescope on Mt. Palomar. The increase was due to the power of direct imaging, which meant that the x-ray detector could be small (and thus enjoy low internal background), while the collecting area of the telescope, which focused the x-rays onto the detector, could be large. The collecting area of the Einstein telescope was nevertheless relatively small (only a few hundred square centimeters of effective area), so further large increases in total sensitivity are now being realized with much larger telescopes that also have much higher angular resolution. The Advanced X-Ray Astrophysics Facility (AXAF) will be more than 100 times as sensitive as the Einstein observatory.

The Einstein observatory, which had a 2.5-year operational lifetime, provided the first x-ray observations of virtually all classes of astronomical objects, from planets (Jupiter) to the most distant quasars. The advent of cosmic x-ray imaging meant that x-ray astronomy had finally joined the mainstream of astronomy, and the astrophysics of high-energy phenomena in the universe could be explored in detail. The high angular resolution achieved for the first time with the Einstein observatory meant that qualitatively new problems could be addressed. An example was the measurement of the precise positions of bright x-ray sources in globular clusters (dense star clusters in our galaxy in which the occurrence of x-ray sources is enhanced by a factor of 100 to 1,000 over their frequency in the Galaxy at large), which proved that they were indeed x-ray binaries containing neutron stars and not massive black holes that would have been at the precise centers of the clusters. The Einstein observatory made possible particularly significant discoveries for several classes of objects: supernova remnants, active galactic nuclei, and galaxy clusters, as well as normal stars and normal galaxies. Einstein deep survey observations provided new constraints on the origin of the diffuse cosmic x-ray background.

Normal Stars

Normal stars, such as the Sun, were finally detected in large numbers with the great sensitivity increase achieved with the Einstein observatory. Although the original discovery of solar x rays had prompted early expectations that stars would be the typical cosmic x-ray sources, their intrinsically low x-ray luminosities meant that the sensitivity of the Einstein imaging telescope was needed for their detection. Soft x-ray emission from essentially all classes of stars, from the very hottest (O and B stars) to the very coolest (M dwarf stars) was detected. The one exception was the A stars (like Vega) that are particularly faint, presumably because they are the transition between the radiative atmospheres of the hot stars (which have strong winds and outflows) and the convective atmospheres of the cool stars (which have considerable magnetic field reconnection and flaring events).

Normal Galaxies

Normal galaxies, such as the Large and Small Magellanic Clouds (satellite galaxies to our own galaxy), the nearby Andromeda (M31) Galaxy, and others in the Local Group, were studied for the first time in detail with Einstein images. The survey of the LMC system revealed a considerable population of supernova remnants, which allowed comparisons with those detected in our galaxy. The M31 observations detected more than 100 sources, including several binaries and pulsars, and a significant number (approximately twenty) of globular cluster sources, which then allowed the first measures of the total population and x-ray source luminosity function with a galaxy.

Diffuse X-Ray Background

The diffuse x-ray background radiation (along with the first cosmic point source, the x-ray binary Sco X-1) was discovered in the pioneering 1962 rocket flight of the small x-ray detector built by Riccardo Giacconi, Herbert Gursky, Frank Paolini, and Bruno Rossi. The nature and origin of this cosmic diffuse radiation remained a mystery until the deep survey observations of "blank" fields was carried out with the Einstein observatory. The discovery that enough quasars were detected to account for a substantial fraction (more than 30 percent) of the diffuse background radiation by their integrated emission made the alternative origin of the background, from diffuse hot gas filling intergalactic space, extremely unlikely. Indeed, this was finally ruled out with later precise measurements of the spectral shape of the cosmic diffuse microwave background radiation, which is the remnant afterglow of the big bang; the Cosmic Background Explorer (COBE) satellite found such perfect agreement with the blackbody radiation spectral shape that it could rule out significant distortion of this shape by an intervening hot gas, such as required for the cosmic x-ray background. Thus, the origin of the cosmic x-ray background radiation is primarily from the summed emission of active galactic nuclei (quasars), although other classes of objects undoubtedly contribute, and the typical luminosities and distances of the contributing sources are still not determined with certainty.

See also: ACTIVE GALACTIC NUCLEUS; ADVANCED X-RAY ASTROPHYSICS FACILITY; ASTROPHYSICS, X-RAY; BLACK HOLE; COSMIC BACKGROUND EXPLORER SATELLITE; COSMIC MICROWAVE BACKGROUND RADIATION; GALAXIES AND GALACTIC STRUCTURE; PULSAR; QUASAR; RADIATION, BLACKBODY; UHURU SATELLITE; X-RAY BINARY

Bibliography

FABBIANO, G. "X Rays From Normal Galaxies." *Annu. Rev. Astron. Astrophys.* **27**, 87–138 (1989).

FABIAN, A., and BARCONS, X. "The Origin of the X-Ray Background." *Annu. Rev. Astron. Astrophys.* **30**, 429–456 (1992).

GIACCONI, R. "The Einstein X-Ray Observatory." *Sci. Am.* **242** (2), 80–101 (1980).

GIACCONI, R.; GURSKY, H.; PAOLINI, F.; and ROSSI, B. "Evidence for X Rays from Sources Outside the Solar System." *Phys. Rev. Lett.* **9**, 439–443 (1962).

GRINDLAY, J.; HERTZ, P.; STEINER, J.; MURRAY, S.; and LIGHTMAN, A. "Determination of the Mass of Globular Cluster X-Ray Sources." *Astrophys. J. Lett.* **282**, L13–L16, (1984).

JONATHAN E. GRINDLAY

EINSTEIN–PODOLSKY–ROSEN EXPERIMENT

The Einstein–Podolsky–Rosen (EPR) experiment is a *gedankenexperiment* (thought experiment) that was put forward in 1935 by Albert Einstein, Boris Podolsky, and Nathan Rosen to demonstrate a supposed shortcoming at the very foundations of quantum mechanics. The EPR experiment concerns itself with the question, "Is quantum mechanics a complete theory?" By investigating through their model what is meant by a complete theory and physical reality, EPR concluded that quantum mechanics is not complete—although it may be correct—and additional theoretical elements might be required for it to fully describe physical reality. The same year, Niels Bohr answered this challenge to the then new quantum theory. He invoked his complementarity principle to argue that the supposed defect lay not in quantum theory but rather in the particular terminology introduced by EPR, and any apparent paradox was related to the issue of measurement.

While there is by no means unanimous assent among physicists that the EPR argument presents a problem at all, study of the experiment and the associated epistemological issues has yielded directly important investigations into the nature of quantum reality, including Bell's inequality, and the experiment serves to underscore the difference between classical and quantum measurement.

EPR set forth in the article requirements for a theory to be complete. They argued that a necessary condition is that the theory contain a description of every element of reality. A sufficient requirement for some property to be an element of reality would be that the property could be predicted without in any way disturbing the system. These definitions appear to be consistent with both classical and quantum ideas. However, Bohr would later base his counterargument to the EPR paper on the inadequacy of this definition of an element of physical reality as applied to a quantum mechanical situation.

As originally presented, the EPR experiment was formulated in terms of position and velocity measurements on two systems that at one time had been in contact with each other. The experiment has been formulated in other essentially equivalent ways. What is required are, say, two noncommuting observables—in quantum mechanics this means that we cannot know both exactly simultaneously. Conceptually, a model using spin components given by David

Bohm in his textbook is perhaps easier to understand. Suppose that in free space there is a molecule composed of two atoms and that the total spin is zero. At some point the molecule suddenly splits into its two constituent atoms (without changing the total spin), denoted A and B, and each atom moves off in opposite directions towards some sort of detector that is able to measure spin (such as a Stern–Gerlach apparatus). There is one detector for each atom. Let each atom have the spin 1/2, so that each can have the value either "up" or "down" when measured parallel to some detector angle. The atoms are in what is quantum mechanically known as a singlet state—that is, a certain superposition of the individual atomic spin up and spin down states that has total spin 0. Suppose, now, that we complete a measurement of the spin of A at the associated detector and find the value up. We can immediately conclude without using the detector for B that the spin at B (parallel to the direction of the detector for A) must be spin down, and as we have not disturbed atom B or its detector, we can say according to EPR that the value of the spin for B has an element of physical reality. Furthermore, it is apparent that if we suppose that no sort of signal is propagated between A and B, then this element of physical reality must have existed somehow already before the measurement was completed at A. Now, it is a property of spin that we can rotate the detector and measure the spin in another direction and we will then also find either spin up or spin down. So, let us carry out the experiment again with a rotated detector and suppose we now measure spin up in the new direction at A. Again, we may then be sure that the value of the spin is down for B along this new direction, and thus a corresponding element of physical reality again exists. Indeed, we can carry out the experiment at any such angle, and, supposing we find spin up for A at that angle, argue accordingly that we have spin down for B along the same angle. Again, since B was not disturbed in any way, EPR would argue that somehow these elements of physical reality must have pre-existed in B before we completed the measurement at A, as there is no communication between the systems. According to quantum mechanics, however, it is not possible to know simultaneously the value of the spin at more than one detector angle, but since according to the above argument all possible values must somehow be present in B—that is, they are elements of physical reality—quantum mechanics cannot be complete. EPR left open the possibility that quantum theory might be made complete by addition of theoretical elements, which later came to be called hidden variables. (We may of course carry out the above thought experiments on atom *B* and make analogous conclusions concerning the presence of elements of physical reality in *A*.)

Bohr argued that the resolution of the apparent incompleteness of quantum mechanics lay in the definition of an element of physical reality. The prediction of the spin value of *B* cannot be separated from the actual measurement. Thus, while we could predict two possible spin down values at two different directions for *B,* these conclusions are mutually exclusive as each measurement at *A* is indeed a separate experiment.

The puzzling aspect of the EPR experiment is perhaps illustrated another way by considering an example due to Bell. Suppose, in the dark, we take a coin and split it somehow lengthwise into a head piece and a tail piece and give each piece to a person. Each person then walks off without looking at their piece. After a while, one person looks at the side they were given and finds heads. Thus, it can be concluded without looking that the other person has tails. How does this experiment differ from the EPR experiment? This example is classical. In the quantum-mechanical example, we have observables that we are forbidden from knowing exact values for simultaneously (such as the horizontal and vertical spin components of *A*), and this is something the example of heads and tails does not possess.

One alternative to quantum mechanics that is consistent with the EPR conclusion is local realism, from locality (allowing the possibility that the result of the measurement at *A* might be somehow transmitted to *B,* but not faster than the speed of light) and from realism (the supposition that the outcome of the measurement must be a predetermined or objective property). John Bell showed that local realism was experimentally testable using the so-called Bell's inequality, which is obeyed by local hidden variable theories. Empirical measurements on various systems, in particular the atomic physics experiments of Alain Aspect and coworkers, have shown that nature violates the inequality for some forms of local hidden variable theories but is in agreement with the predictions of quantum mechanics.

See also: BOHR, NIELS HENRIK DAVID; EINSTEIN, ALBERT; QUANTUM MECHANICAL BEHAVIOR OF MATTER; QUANTUM MECHANICS; QUANTUM MECHANICS, CREATION OF; SPIN; STERN–GERLACH EXPERIMENT

Bibliography

BELL, J. S. *Speakable and Unspeakable in Quantum Mechanics* (Cambridge University Press, Cambridge, Eng., 1987).

BOHM, D. *Quantum Theory* (Prentice Hall, Englewood Cliffs, NJ, 1951).

BOHR, N. "Can Quantum Mechanical Description of Physical Reality Be Considered Complete?" *Phys. Rev.* **48,** 696–702 (1935).

D'ESPAGNAT, B. "Quantum Mechanics and Reality." *Sci. Am.* **241** (Nov.), 158–180 (1979).

EINSTEIN, A.; PODOLSKY, B.; and ROSEN, N. "Can Quantum Mechanical Description of Physical Reality Be Considered Complete?" *Phys. Rev.* **47,** 777–780 (1935).

JAMES F. BABB

EINSTEIN THEORY OF SPECIFIC HEAT

See SPECIFIC HEAT, EINSTEIN THEORY OF

ELASTICITY

The term "elasticity" has been used since the seventeenth century to describe the familiar behavior of objects deformed by an external force to regain their original shape once the deforming force is removed. Any solid spring, rod, block, or any other shape can be pulled, pushed, twisted, bent, or compressed some amount and will spring back to its original shape seemingly instantaneously upon removal of the external influence. Liquids and gases also exhibit elastic behavior, but seemingly only under compression.

The words have also been applied to less tangible concepts: for example, to economics (elasticity of prices and of supply and demand); philosophy and psychology (elasticity of mind, soul, good and evil, spirit, feelings, personality); law (implying that a law can be stretched and hence made lax)—indeed, to anything which can literally or figuratively be stretched and can rebound without permanent alteration.

It is also commonly understood that there is always a limit to elastic behavior: If the deformation is too great, the object will never rebound completely. The elastic limit has been exceeded.

Hooke's Law

The scientific and technical usages of the term are always applied to the limited set of phenomena related to the concept first stated explicitly in 1676 by the English physicist-philosopher Robert Hooke, commonly called Hooke's law. Hooke was a contemporary and rival of Isaac Newton; he was physically and emotionally an unattractive man, penurious and solitary, and irritable in temper. He actually made several major discoveries that are often credited to others, including a (primitive) wave theory of light that anticipated the phenomena of interference and diffraction (which he observed and explained); an approach to universal gravitation based on the motion of planets (quite independent of Newton); the invention of a wheel barometer for use in meteorological forecasting; optical telegraphy; the use of pendulums to measure gravity; and the invention of an escapement mechanism for clocks.

Hooke was always concerned that others would steal his ideas. For this reason, he initially stated his famous law as an anagram that had to be solved to elicit his discovery: "Ut tensio sic vis" (in modern language, "Stress is proportional to strain"). The term "stress" denotes a precisely defined measure of the deforming influence, while "strain" describes the deformation itself.

Hooke's law is hardly a real "law" of physics, such as Newton's laws of motion or the laws of conservation of energy and momentum. It is only an approximation, valid over a quite small range of strain. Yet it has made possible major fields of mechanics and structural analysis, because it describes a mathematically linear relationship that can be solved precisely for a wide variety of realistic problems.

Thus, Hooke, for all his faults and forgotten efforts, is immortalized by the law which bears his name. Indeed, he is even referred to adjectivally; we speak of "hookean" behavior to describe elastic phenomena that closely follow Hooke's law.

Limits to Elastic Behavior

The actual range of validity of Hooke's law is generally quite small, valid usually only for strains of far less than 1 percent. In most pure metals, for example, the elastic limit is reached for strains as small as 0.001 percent. For larger strains, the object never reverts exactly to its undeformed shape on removal of the stress but remains changed to some extent; it is said to be plastically deformed.

The simplicity of Hooke's law lies in the statement that the stress and strain are linearly proportional: Double the stress and the strain doubles; triple it, and the strain is three times as great. In fact, the stress may also depend to a smaller degree on other influences: on the square and higher powers of the strain; on the rate of strain; on time or temperature. Sometimes the strain does not occur all at once after a stress is applied, and an additional strain occurs slowly afterwards, often for a long time. This phenomenon is usually called creep, or, more technically anelasticity, or elastic after-effect.

Elastic Energy

When an object is deformed elastically by an external force, energy is stored as potential energy inside the object. When the stress is removed, the object will eventually regain its original shape and the stored elastic energy can be recovered. This is the means by which wound springs can be made lto run clocks or toy cars. But what happens if an object is deformed and then the external constraint is suddenly removed? Clearly, under the influence of the elastic restoring force (Hooke's law), the object tries to regain its initial shape, but to conserve energy, it must overshoot and then reverse itself over and over again. Think of a spring that you pull and let go—it just keeps bouncing up and down. Eventually this oscillatory behavior must end as the energy dissipates by heating up air molecules or just by internal friction due to inevitable nonlinear anelastic effects. The smaller the anelastic effects, the longer the object will oscillate; this is why a bell made of hardened bronze or steel will ring much longer than one made of soft lead or copper.

The same effects give rise to the propagation of elastic waves through media. If an object is deformed rapidly at one end, the deformation must move as a wave away from its starting point. The speed at which the wave moves will be greater for a medium of high stiffness (high elastic modulus) and low inertia (low density): The wave speed varies as the square root of the elastic modulus divided by the mass density. The most familiar forms of such waves are sound waves—compressional elastic waves transmitted through the air and waves in the strings, rods, diaphragms, and air columns of musical instruments.

Interactions

In a purely elastic deformation, all the stored energy is recoverable and the deformed object afterwards is exactly the same as it was before the deformation. If any energy gets converted internally to any irreversible form, it will not all be recovered and the object will be changed. We often generalize this phenomenon to many other interactions, for example, between elementary particles. If all the initial kinetic energy after an interaction is precisely the same as the kinetic energy before, we call the interaction elastic, and we know that no permanent changes have occurred. Conversely, if the kinetic energy afterwards is different from that before the interaction, we call the interaction inelastic and know that permanent changes have occurred. Thus, two electrons that approach one another and then fly apart due to their repulsion interact elastically; two atoms that approach one another and then combine to form a molecule interact inelastically.

See also: ELASTIC MODULI AND CONSTANTS; ENERGY, KINETIC; ENERGY, POTENTIAL; HOOKE'S LAW; OSCILLATION; STRAIN; STRESS

Bibliography

LOVE, A. E. H. *A Treatise on the Mathematical Theory of Elasticity*, 4th ed. (Dover, New York, 1944).

DAVID LAZARUS

ELASTIC MODULI AND CONSTANTS

Hooke's law defines elastic behavior by the statement that "stress is proportional to strain," a simple linear relationship relating deforming forces to the resulting deformation.

Stresses, Strains, and Moduli

To be precise, one must define the terms "stress" and "strain" to accord to known effects: It is easier to stretch a long spring by some distance than a short spring made of the same wire; it is easier to stretch a spring made of fine wire than one of the same length made of thicker wire; some materials make "softer" springs than others—a copper spring is easier to stretch than one made of steel.

We combine all these facts by proper definitions of stress and strain. We define stress not as just the deforming force, but as the force per unit area applied to the body; we define strain not as the total deformation, but as the relative deformation—for a spring, the change in length divided by the original length.

Denoting the stress by an uppercase S and the strain by a lowercase s, we can state Hooke's law as

$$S = Ms,$$

where the constant of proportionality, M, is the elastic modulus. Notice that since s is a ratio, it is dimensionless, so the units of M are the same as the units of S, force per unit area. The value of M depends on the type of strain and the type of material; it is larger for hard materials like glass or steel and smaller for soft materials like copper or lead.

For common deformations there are three special moduli defined for three deformations:

1. Young modulus, Y: for a simple extension, like pulling on a wire;
2. Torsion modulus, T: for simple twisting deformations, like twisting a wire or pulling a spring (which is really a twist);
3. Bulk modulus, B: for simple uniform compression, such as by applying a uniform external pressure by a gas or liquid.

These definitions are valid only for materials that have no large-scale internal structure, for example, fine-grained metals, glasses, or fluids. Such materials have no preferred directions for elastic behavior and are said to be isotropic. We sometimes define two additional concepts: Poisson's ratio and shear modulus.

When a material is stretched it gets longer, and its volume is increased. However, part of the volume increase is canceled by the fact that the material also gets smaller in cross section at right angles to the stretch direction: A stretched wire is reduced in diameter. Poisson's ratio is the relative reduction in cross-sectional area divided by the elongational strain.

When a rod or wire is twisted, the outside is deformed more than the middle, and the torsion modulus measures some average. Similarly, when a rod or bar is bent into a curved shape, the outside of the curve is deformed more than the inside. On a finer scale, the local deformation is what you get by applying a pair of opposed forces parallel to opposite faces of a cube, deforming the basic square cross section into a parallelogram, with no change in volume. This deformation is called a shear and the more fundamental modulus which describes it is the shear modulus.

Complications

Solids are made of atoms, which are generally arrayed in very regular structures called crystalline lattices, or just crystals. The crystals themselves are anything but isotropic and have very different elastic moduli in different directions. In most common materials, however, the crystals are very small and cannot be seen except with high-power microscopes and are arranged every which way. This is why solids made of anisotropic (not isotropic) crystals can behave isotropically as far as their elastic behavior is concerned. Anisotropic behavior is easily measured when you have large single crystals like quartz instead of fine-grained solids.

The anisotropic properties do not make much of a difference if all we want to do is stretch springs or bend bars, or perform similar practical large-scale deformations. They are very important, however, from the viewpoint of basic science and trying to understand the atomic basis for why solids behave as they do. Then, the simple form of Hooke's law gets *much* more complicated.

For three-dimensional anisotropic solids, we have to consider three separate directions of both stresses and strains, and the components of each both parallel and perpendicular to each direction; in other words, six separate stress components and six strains. To connect the six stresses with six strains we need a matrix of thirty-six terms called elastic constants. The mathematics can get very complicated, but if we choose the directions to correspond to the actual symmetry axes of the crystals, the mathematical complexity can be reduced enormously. For example, for a crystal with basic cubic symmetry, like copper, iron, or salt, the total number of elastic constants can be reduced to only three, instead of thirty-six. One of these constants is like Young's modulus, but without the complications of Poisson's ratio; another is a simple shear; and the third elastic constant is related to a second type of simple shear. These constants can be measured by pulling or shearing single crystals along special axes or by measuring the speed of elastic waves in special directions through the crystals (which is much easier to do). When these constants are known, we can calculate the values of the large-scale isotropic moduli, Y, T, and B, as well as Poisson's ratio and different shear moduli. However, we also know something very important about the forces between individual atoms in the solid and what makes them behave the way they do.

Most objects expand when they are heated and contract when they are cooled. Most objects also get hotter when they are compressed and cooler when they are expanded. This effect can get complicated because the values of the elastic moduli also depend on temperature, usually getting smaller for higher temperatures. Any deformation that changes the volume of a body also changes its temperature, unless the deformation is so slow that heat can flow in or out to keep the temperature constant. Only a pure shear can deform an object without changing its volume. For any other deformation, the volume must change.

If the deformation occurs very slowly so that the temperature stays constant, the strain is determined by the isothermal (constant temperature) elastic modulus. If the deformation is very rapid, or the body is totally isolated from the outside so no heat can flow in or out, it is determined by the adiabatic modulus, which is usually larger than the isothermal modulus. For elastic waves, the strain usually reverses so rapidly that heat does not flow in or out, so the sound speed is determined by the adiabatic modulus.

See also: ADIABATIC PROCESS; ELASTICITY; ISOTHERMAL PROCESS; STRAIN; STRESS

Bibliography

LOVE, A. E. H. *A Treatise on the Mathematical Theory of Elasticity*, 4th ed. (Dover, New York, 1944).

DAVID LAZARUS

ELECTRET

The term "electret," which was coined by the British electrical engineer Oliver Heaviside, refers to an insulating material, or dielectric, capable of sustaining an electric charge or internal polarization for a long time. This relaxation time may be of the order of seconds, years, or even centuries, depending on the internal losses, temperature, and nature of the sample. An electret was observed for the first time by Gentaro Eguchi in a resin compound.

Typical electret materials are inorganic substances like sulfur, ice, ceramics, and ionic solids, or organics substances such as hydrocarbons, waxes, polymers, and other insulators. Electrets are, in general, produced by the application of an external electric field to the sample for a certain time and at a certain temperature, usually above room temperature. After this polarization time, the temperature of the sample is lowered to room temperature or lower, "freezing" the induced charge or polarization. This polarization, with no field applied, is characteristic of the electret state. However, this is not stable as in the case of magnets, and slowly decays or is neutralized by ions in the surrounding atmosphere. Typical polarization charges are of the order of 10 to 100 $\mu C/cm^2$. The polarization may be due to trapped electrons, holes, dipoles, or ions.

There are other ways to produce electrets: by a spray of ions from a corona discharge (corona electret); the injection of charges from electron beams; trapping of charges produced by ultraviolet light (photoelectret); or ionizing radiation (x rays or gamma rays) in the material (radioelectret). Electret polarization was also found for biological materials such as proteins, DNA, cellulose, bones, hair, and others, and in this case, the name bioelectret is used.

Applications of electrets range from microphones, electrostatic filters, and transducers that use the external electric field of the electret to radiation dosimetry, where the electret field is used as a bias field and detector in a special ionization chamber. Electrets have also been used as implants for acceleration of bone consolidation in fractures and for avoiding coagulation in blood canules. The possibility exists that natural bioelectrets may have importance for biological processes such as those occurring in cell membranes. Enzyme action and macromolecular conformation are also discussed in the literature. The role of water bound to these biomolecules is important in that the large electric dipole moment of water contributes to the stored polarization. The interplay of water dipoles in the bioelectret state is another possible way for the bioelectret field to affect biological form and function.

When electrets are warmed to higher temperatures and the depolarization is monitored (voltage or current) typical peaks are observed. These thermal stimulated depolarization peaks (TSD) or spectra are the quantitative way to investigate electret and bioelectret phenomena. The corresponding theory for the spectra has been given in quantitative form by the so-called Gross equations. These spectra have been used for the analysis of transport and rheological phenomena in many materials from polymers to ceramics and insulators like silicon-oxides that are of great importance for materials in science and electronics.

See also: BIOPHYSICS; IONIZATION CHAMBER; POLARIZATION; TRANSDUCER

Bibliography

GROSS, B., ed. *Charge Storage in Solid Dielectrics* (Elsevier, Amsterdam, 1984).

LEWINER, J.; MORISSEAU, D.; and ALQUIE, C. *Proceedings of the Eighth International Symposium on Electrets* (IEEE, Piscataway, NJ, 1994).

SESSLER, G. M., ed. *Electrets,* 2nd ed. (Springer-Verlag, New York, 1987).

SERGIO MASCARENHAS

ELECTRICAL CONDUCTIVITY

Electrical conductivity is a measure of how easily an electric current is established in a material as a response to an applied electric field. It is defined through the equation

$$\mathbf{J} = \sigma\mathbf{E},$$

where $\mathbf{J}$ is the charge current density, or charge that traverses a unit of cross-sectional area in a unit of time; $\mathbf{E}$ is the electric field; and σ is the electrical conductivity, whose units in the International System are $\Omega^{-1}\text{m}^{-1} = \text{S/m}$. If the material is anisotropic, σ is a tensor ($\mathbf{J}$ and $\mathbf{E}$ need not be parallel).

In a gas, a strong electric field can generate ions, causing an appreciable conductivity (a spark is an extreme example). It is also ions, positive and negative, which are responsible for the conductivity in electrolytic solutions (water is a good conductor because of the minerals dissolved in it). Solids exhibit a more complex behavior, and quantum mechanics is needed to model the physical processes that determine whether solids behave like insulators, semiconductors, conductors, or superconductors.

In metals, which are excellent conductors, one or more electrons reside in the outermost subshells of the atoms and are consequently weakly bound to them. When atoms are brought together in a lattice, those electrons become detached and are almost free to wander through the crystal, as if they were a gas, with their motion interrupted only by collisions with other electrons or, more often, the lattice vibrations (phonons) and irregularities such as impurities. This very simplified picture, the free-electron Fermi gas (electrons obey the Pauli exclusion principle and therefore follow the Fermi–Dirac distribution), is extremely successful in explaining metallic properties, but cannot be extended to nonmetals.

A more comprehensive explanation, the band theory, develops when one considers that in a cluster of atoms forming a solid, the individual permissible energy levels become bands. If the uppermost populated band is partially filled, as in metals, then electrons can "move" within it, that is, gain energy and thus net momentum to carry their charge through the sample (these materials are good conductors). However, if the uppermost band is completely full, electrons cannot gain energy unless the change exceeds the energy gap separating the full band from the one above. These materials with wide energy gap are insulators, characterized by very low conductivity values. There is an intermediate case, that of semiconductors, in which the energy gap is small enough (of the order of 1 eV as in silicon and germanium) that thermal activation is able to raise electrons through it. However, the conductivity provided by these "intrinsic" carriers is quite low; by adding impurities which contribute energy levels within the energy gap, "extrinsic" carriers are provided which make a much more significant contribution to the conductivity. Semiconductors can be of n or p type, depending on whether the impurity contributes an electron to the conduction band, or it takes an electron from the valence band, leaving a positive hole to act as a carrier. (On the other hand, pure semiconductors have positive and negative carriers in equal numbers, since a hole is always created when an electron jumps to the conduction band, and both become carriers.) The addition of just one impurity atom per million can change the conductivity by a factor of a million, so such "doping" is an efficient and simple mechanism for controlling the conductivity. Transistors and diodes are examples of devices that were developed taking advantage of the property of semiconductors.

At room temperature the conductivity of the best conductors is 10^{25} times larger than that of the best insulators. However, in some metals, alloys, and compounds this factor becomes infinite at very low temperatures, due to the onset of superconductivity. In this case, electrical conductivity abruptly becomes infinite at a temperature called the critical temperature. A quantum mechanical behavior with no analog in classical physics produces an attractive interaction between pairs of electrons (Cooper pairs) which results in the superconducting state.

See also: CONDUCTION; CONDUCTOR; COOPER PAIR; FIELD, ELECTRIC; INSULATOR; PAULI'S EXCLUSION PRINCIPLE; SEMICONDUCTOR; SUPERCONDUCTIVITY

Bibliography

CRUMMETT, W. P., and WESTERN, A. B. *University Physics: Models and Applications* (Wm. C. Brown, Dubuque, IA, 1994).

GRAY, D. E., ed. *American Institute of Physics Handbook,* 3rd ed. (McGraw-Hill, New York, 1972).

KITTEL, C. *Introduction to Solid-State Physics,* 6th ed. (Wiley, New York, 1986).

SEARS, F. W.; ZEMANSKY, M. W.; and YOUNG, H. D. *University Physics,* 6th ed. (Addison-Wesley, Reading, MA, 1982).

MARIA CRISTINA DI STEFANO

ELECTRICAL RESISTANCE

The electrical resistance R of a circuit element is equal to the voltage V (in volts) that appears across it divided by the current I (in amps) flowing through it. When R is constant, for example, current is proportional to the applied voltage, the above relationship, $R = V/I$, is called Ohm's law after its discoverer German physicist Georg Ohm, who also lends his name to the unit of resistance, the ohm (abbreviated Ω). By Ohm's law, it is apparent that if V is constant, increasing R reduces I, hence resistance impedes the flow of current. The resistance of a component depends on its geometry and on its resistivity ρ (in units of Ωm) by the equation $R = \rho l/A$, where l is the length (in m) and A the cross-sectional area (in m^2) of the component. The resistivity, and hence the resistance, decreases with an increase in the number of electrons available to take part in the current flow. An insulator has most of its electrons tightly held in the chemical bonds between the atoms and hence has an extremely high resistance, whereas a metal has many electrons which are free to flow and therefore has a low resistance.

Resistivity also depends on how easily electrons can move through a material. Applying a voltage across a metal will accelerate the electrons that are free but they will tend to collide with the vibrating atoms (or phonons) in the material. The electrons will give up their energy in these collisions and will slow down. The average velocity of the electrons, and hence the current flow, will be smaller if the collisions are more frequent. Therefore, more collisions mean an increased resistance. If we increase the temperature of the metal, its atoms will gain kinetic energy and will vibrate more vigorously. This increases the likelihood of collisions between electrons and atoms. The resistance of a metal therefore increases with temperature. In insulators and semiconductors the opposite is true, as an increase in temperature gives more electrons sufficient energy to break free from the bonds and the resistance decreases due to the larger number of electrons available for conduction. We may also reduce the resistance of a semiconductor by adding small amounts of materials called dopants. These elements either provide more electrons for conduction or create "holes" that allow the more tightly bound electrons to move with greater freedom.

Transfer of energy from accelerated electrons to atoms appears as heat in a material. The electrical power P (in watts) transferred into heat is given by $P = I^2R$. To reduce such energy losses in electricity distribution systems, low resistance materials, such as copper and aluminum, are used. Many electrical circuits use resistive components to control current flow and to produce a particular voltage for a given current. These can be made from carbon, mixtures of metals and insulators, or doped semiconductors, depending on the application. The resistance of a component may be measured by applying a known voltage and dividing it by the value of the measured current through the device. Alternatively, resistance may be determined by balancing the current flow through the unknown component with that through a component of known resistance in a "bridge" arrangement.

See also: COLLISION; CONDUCTION; CONDUCTOR; CURRENT, ALTERNATING; CURRENT, DIRECT; ELECTRICITY; ELECTRON; ENERGY; ENERGY, KINETIC; HEAT; INSULATOR; METAL; OHMMETER; OHM'S LAW; PHONON; RESISTOR; SEMICONDUCTOR

Bibliography

DYOS, G. T., and FARRELL, T., eds. *Electrical Resistivity Handbook* (Peregrinus, London, 1992).

SCHWARTZ, S. E., and OLDHAM, W. G. *Electrical Engineering: An Introduction,* 2nd ed. (Saunders, Orlando, FL, 1993).

MICHAEL N. KOZICKI

ELECTRICAL RESISTIVITY

The electrical resistivity of a material characterizes the extent to which it impedes the flow of electrical charges. It is defined as the reciprocal of the electrical conductivity:

$$\rho = \frac{1}{\sigma}.$$

The units of ρ in the International System are Ωm. At room temperature, metals are the best conductors and have very low resistivities, of the order of 10^{-8} Ωm (see Table 1). Resistivities for good insulators can be as high as 10^{17} Ωm. Such an enormous range of variation has no match in any other physical quantity, and gives rise to a wide versatility in applications, like powerful superconducting magnets and insulating layers in microelectronic components.

The behavior of electrical resistivity under conditions of varying temperature differs among different materials, suggesting a diversity of mechanisms responsible for charge transport. In metals ($\rho < 10^{-6}$

Ωm), electrons are hindered in their motion under an applied field only when the host lattice deviates from a regular array. Departures from a perfectly ordered structure occur due to vibrations of the lattice ions, and to the presence of defects in the form of impurities or imperfections. The motion of an electron is visualized as proceeding randomly at a velocity determined by its thermal energy and as having a superimposed acceleration due to the external field; the motion starts anew when the electron collides with a phonon (a particle-like representation of lattice vibrations), an impurity, or an imperfection. At high temperatures, collisions with phonons are predominant, and since the number of phonons increases with temperature, the resistivity does too. At low temperatures, vibrations are reduced, and the most significant contribution to the resistivity is given by the defects, whose presence is obviously temperature independent. As a consequence, the resistivity is usually written as

$$\rho = \rho_0 + \rho(T),$$

where the residual resistivity ρ_0 is the resistivity at zero (kelvin) temperature, dependent on the purity and metallurgical history of the sample, and the temperature-dependent term $\rho(T)$ is characteristic of the metal (Fig. 1). The residual resistivity ratio RRR, defined as the ratio between the resistivities at room temperature and at liquid helium temperature (4.2 K), $RRR = \rho_{RT}/\rho_{4.2\,K}$, gives a measure of the

Table 1 Typical Values of Resistivity at Room Temperature

Substance	$\rho(\Omega m)$
Conductors	
Copper	1.7×10^{-8}
Gold	2.2×10^{-8}
Aluminum	2.6×10^{-8}
Tungsten	5.5×10^{-8}
Semiconductors	
Germanium (pure)	~0.5
Silicon (pure)	~10^2
Insulators	
Wood	$10^8 - 10^{11}$
Glass	$10^{10} - 10^{14}$
Quartz (fused)	7.5×10^{17}

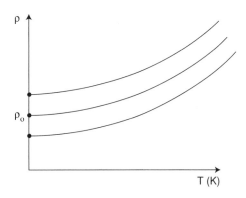

Figure 1 Resistivity as a function of temperature for metallic conductors. Samples of the same metal but with different degrees of purity would present different values for ρ_0 as shown.

purity (free of impurities and defects) of the metal. Values of RRR as high as 10^5 have been obtained for copper and aluminum.

For semiconductors ($\rho \sim 10^{-5} - 10^7$ Ωm), the temperature dependence of resistivity is quite different than for metals. Since the intrinsic carriers are created by thermal excitation across the energy gap between the valence and the conduction bands, increasing the temperature increases the number of carriers and therefore decreases the resistivity. (The same behavior is observed in insulators, for which $\rho > 10^7$ Ωm.) This effect can be quite pronounced, particularly at low temperatures, where semiconductors can be used as very sensitive thermometers. In fact, resistivity is a major thermometric property; there is such a diversity of temperature dependencies that it is always possible to find one adequate for a particular temperature range and application. Platinum resistance thermometers, for example, are used as standards to provide the interpolation between fixed points in the definition of the International Temperature Scale.

For a conductor of uniform cross section A and length l, resistance R is related to resistivity by

$$R = \frac{\rho l}{A}.$$

This relationship is often used in quantitative determinations of resistivity, since the resistance can be obtained from direct measurement of the potential difference and the current in the material ($R = \Delta V/I$).

See also: CONDUCTION; CONDUCTOR; ELECTRICAL RESISTANCE; INSULATOR; PHONON; SUPERCONDUCTIVITY

Bibliography

ASHCROFT, N. W., and MERMIN, N. D. *Solid-State Physics* (Holt, Rinehart and Winston, New York, 1976).

CRUMMETT, W. P., and WESTERN, A. B. *University Physics: Models and Applications* (Wm. C. Brown, Dubuque, IA, 1994).

KITTEL, C. *Introduction to Solid-State Physics,* 6th ed. (Wiley, New York, 1986).

MARIA CRISTINA DI STEFANO

ELECTRIC FIELD

See FIELD, ELECTRIC

ELECTRIC FLUX

The electric flux, that is, the flux of the electric field lines, is given by

$$d\Phi_E = \mathbf{E} \cdot d\mathbf{S},$$

where $d\mathbf{S}$ is an infinitesimal surface element. For a closed surface the flux is given by

$$\Phi_E = \oint \mathbf{E} \cdot d\mathbf{S},$$

where the integral sign $\oint$ signifies an integration over a closed surface.

Unlike the magnetic flux, the electric flux Φ_E for a closed surface does not always vanish. Positive and negative charges are sources and sinks of field lines and for closed surfaces enclosing a net charge $\Phi_E \neq 0$. The concept and terminology of electric flux are commonly employed in electrostatics as a means to describe the spatial density of electric field lines. Electromagnetic phenomena are organized in terms of the field concept into Maxwell's four fundamental equations. One of the equations represents Gauss's law, the integral form of which is given by

$$\varepsilon_0 \oint \mathbf{E} \cdot d\mathbf{S} = \oint dq.$$

The left-hand side represents ε_0 times the net electric flux out of a closed surface and the right-hand side represents a net charge enclosed within this closed surface. (For a magnetic flux, the surface integral of the left-hand side is zero, indicating the absence of magnetic monopole.)

The term "electric flux" should not be confused with electric flux density, which is used occasionally to represent electric displacement **D.**

See also: ELECTROSTATIC ATTRACTION AND REPULSION; GAUSS'S LAW; MAGNETIC FLUX; MAXWELL'S EQUATIONS

Bibliography

BLUM, R., and ROLLER, D. E. *Physics,* Vol. 2: *Electricity, Magnetism, and Light* (Holden-Day, San Francisco, 1982).

WANGSNESS, R. K. *Electromagnetic Fields* (Wiley, New York, 1979).

CARL T. TOMIZUKA

ELECTRIC GENERATOR

An electric generator (sometimes called a dynamo) is a device designed to transform mechanical energy into electrical energy. The mechanical motion of a conductor (usually a metal) in a magnetic field produces a changing magnetic flux through a current loop. The resulting induced electromotive force (emf) of the generator is thus seen to be a consequence of Faraday's law of electromagnetic induction, $E = -\Delta\phi/\Delta t$, where ϕ is the magnetic flux through the circuit. The magnetic flux is defined as the product of the magnetic field and the circuit area perpendicular to the field. The emf of a generator is measured in volts.

The simplest electric generator is a rod of length L moving with a constant velocity v in a constant magnetic field B directed perpendicular to the rod as shown in Fig. 1. In this system the moving rod

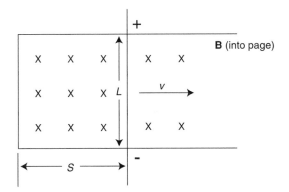

Figure 1 A simple generator. When the rod of length L is moved to the right with speed v in a uniform magnetic field B (directed into the paper), the emf produced across the rod is BLv. The top end of the rod is positive.

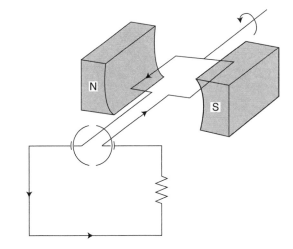

Figure 2 A single-plane coil with commutator contacts serves as a dc generator.

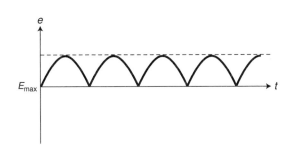

Figure 3 Plot of the induced emf for a single-loop dc generator.

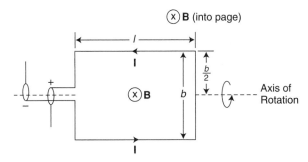

Figure 4 Side view of a coil rotating in a constant magnetic field with the direction of the induced current shown.

produces a changing magnetic flux through the circuit by changing the area intercepting the constant magnetic field. The rod will develop an electromotive force across its ends equal to BLv (this will be in volts if B is in telsa, L is in meters, and v is in meters per second). The mechanical energy supplied by the agent moving the rod is transformed to the electrical energy available at the ends of the rod.

The standard direct-current (dc) generator is designed as shown in Fig. 2. Here a rotating coil in the magnetic field intercepts a magnetic flux. This flux changes as the area of the coil perpendicular to the magnetic field varies with the rotation of the coil. The commutator produces a constant polarity pulsating dc voltage as shown in Fig. 3. Once a current begins to flow in the armature of a generator, it behaves like a motor producing a torque that opposes the external torque rotating the armature. The sym-

metry is completed when we note that a simple dc motor becomes a dc generator when the armature is rotated by an external agent and the output voltage is directly proportional to the rotation speed of the armature. For example, a wind speed indicator can be constructed by mounting a propeller or wind cup system on the armature shaft of a dc motor. The wind will supply the mechanical energy to rotate the armature and the output voltage will be directly proportional to the wind speed.

The standard alternating-current (ac) generator is a coil rotating in a constant magnetic field as shown in Figs. 4 and 5. The slip rings make contact with the brush system to produce the ac output at the terminal of the generator. Like the dc generator, the alternating emf is a result of the changing magnetic flux due to the varying area perpendicular to the field. The frequency of the ac voltage is

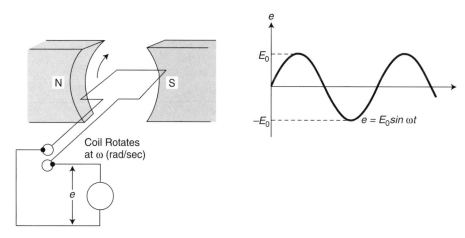

Figure 5 A rotating coil in a magnetic field becomes an ac generator.

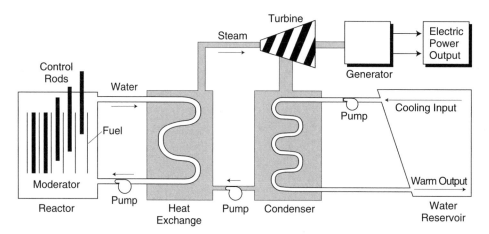

Figure 6 Schematic diagram of a nuclear power plant.

determined by the frequency of rotation of the coil. The output emf is given by $E = NBA \sin 2\pi ft$, where t is time in seconds, N is the number of loops in the coil, B is the magnetic field strength, A is the area of each loop, and f is the frequency of rotation of the coil. In the United States and Canada, this frequency is 60 Hz, and $2\pi f\text{-}\omega$ is 377 rad/s, whereas in much of the rest of the world this frequency is 50 Hz, and $2\pi f\text{-}\omega$ is 314 rad/s.

The ac generator is at the heart of ac power systems everywhere. In a hydroelectric power plant, the mechanical energy that rotates the armature is supplied by flowing water through a turbine connected to the coil of the generator; in a hydrocarbon-fueled power plant, steam generated by fuel combustion drives a turbine that turns the coil of the generator; in a nuclear power plant, the nuclear energy released in the fission process produces high tempera-

tures that produces the steam that drives the turbines connected to the generator. Figure 6 shows a schematic diagram of a nuclear power plant.

Thermoelectric generators are designed to convert heat into electrical energy through the application of the Seebeck effect. The thermocouple is a common thermoelectric device. In a thermocouple, one junction of two different conductors is held at a constant temperature and the other junction is held at a different temperature. The voltage generated between these junctions is proportional to the temperature difference. Usually this device is used to measure temperatures, but with the development of semiconductors and the discovery of their significant Seebeck effect, it has become feasible to make portable thermoelectric generators. These generators have no moving parts, and they are especially useful in remote locations where traditional generators and their required energy source present problems in size and transportation. They are popular generators in Siberia and the Arctic regions of the earth.

Cogenerator systems are new projects that have been introduced to address some of the energy and environmental concerns facing Earth's citizens. In the face of finite hydrocarbon fuel supplies, it is desirable to use these fuels that are more efficiently and environmentally as clean and pollution free as possible. These generators are designed to generate electricity and provide thermal energy for consumers instead of exhausting heat as thermal pollution into the environment. Cogeneration systems have been developed that are based on internal combustion engines that are suited for single households. Larger congeneration systems have been designed that incorporate a power plant that provides heat to its customers as well as electricity. This is an area of significant research interest for environmental scientists and engineers.

See also: CURRENT, ALTERNATING; CURRENT, DIRECT; ELECTROMAGNETIC INDUCTION, FARADAY'S LAW OF; ELECTROMOTIVE FORCE; ENERGY, MECHANICAL; ENERGY, NUCLEAR; FUSION POWER; MAGNETIC FLUX

Bibliography

HINRICHS, R. A. *Energy* (Saunders, Philadelphia, 1992).

RICHARD M. FULLER

ELECTRICITY

The basic properties of matter that most affect our daily life in diverse manners are mass and electric charge. Mass is related to gravitational phenomena. Electric charge is related to a broad class of phenomena under the general denomination of electricity. The phenomena associated with electric charges at rest, or at most in very slow motion, relative to the observer, constitute what is called electrostatics. The phenomena associated with charges in motion relative to the observer, which include magnetic effects, constitute what is called electrodynamics. The combined study of electric and magnetic phenomena is called electromagnetism.

Electric phenomena were first recognized in Greece, more than twenty-six centuries ago, when it was observed that an amber rod rubbed with a cloth or a fur attracted light objects such as feathers and bits of straw. The same phenomenon is observed when a glass rod is rubbed. Since the Greek name for amber is elektron ($\eta\lambda\epsilon\kappa\tau\rho o\nu$) William Gilbert, physician to Queen Elizabeth I, called the force exerted by the amber *vis electrica* (or amber force) to distinguish it from the much weaker force of gravitation. Today we recognize that when we rub an amber or a glass rod there is a transfer of electric charge between the rods and the cloth or the fur, a process called triboelectrification, (from the Greek *tribein*, to rub), which is rather common. For example, after combing our hair on a very dry day we find that the comb swiftly attracts tiny pieces of paper. Anyone who touches a piece of metal, such as a door knob, after walking over a rug on a dry day experiences an electric spark. (The phenomenon shows better on dry days because dry air is a poor conductor of electricity, and therefore, objects hold their electric charges better.) Triboelectrification has found an important application in modern copiers and printers, using a technique invented by Chester F. Carlson in 1938.

While the gravitational interaction between two masses is always attractive, the electric interaction between electric charges may be attractive or repulsive, a fact first recognized by Nicolo Cabeo in the seventeenth century. This indicates that there are two classes of electrification, or rather of electric charge, as can be seen by a simple experiment. When an electrified amber or glass rod is placed near a small pith or cork ball hung from a string, the ball is attracted toward the rod, but if both rods are placed simultaneously near the ball the attraction is small or there is no attraction. This indicates that the electrification of the amber and glass rods produce opposite effects on the ball. The first to recognize that there are two classes of electric charge was Charles F. de Cisternay du Fay, who proposed the name resinous for the amberlike charge and vitreous for the glasslike charge. Later, Benjamin Franklin proposed designating the vitreous charge as positive ($+$) and the resinous charge as negative ($-$). Franklin could have also made the opposite choice, which perhaps might have been better since the electrons came to have negative charge with his choice.

A second important electric property is that two bodies with the same kind of electrification (either positive or negative) repel each other, but if the two bodies have opposite kinds of electrification (one positive and the other negative) they attract each other. A simple experiment consists in touching two pith or cork balls with an electrified glass or amber rod. We observe that the two balls repel each other. If we touch one ball with an electrified glass rod and the other with an amber rod the balls attract each other. It also has been observed that the force of electric attraction or repulsion between two charges decreases in inverse proportion to the square of their separation $(1/r^2)$, a result known as Coulomb's law.

The amount of electric charge in an electrified body is measured in coulombs (C), named in honor of Charles A. Coulomb, who, in the eighteenth century, was the first to formulate correctly the law of electric attraction and repulsion. The coulomb is the amount of electric charge transported in 1s by a current of 1 A and is equivalent to the charge of about 6,250 quatrillions (6.25×10^{18}) of negatively charged electrons or positively charged protons.

To explain the electric interaction, we say that an electrically charged body produces an electric field in the surrounding space. The convention is that the field is directed away from a positive electric charge and toward a negative electric charge. Any other electric charge placed in the field experiences a force in the same direction as the field if the charge is positive and in the opposite direction if the charge is negative. For example, negatively charged electrons placed in an electric field move in a direction opposite to that of the field, while positively charged protons move in the direction of the electric field. The strength of an electric field is defined as the force on a unit charge and is expressed in newtons per coulomb (N/C).

It took several centuries to recognize that electric charge, like mass, is a fundamental property of matter. The Greek philosophers, such as Thales of Miletus in the sixth century B.C.E., had thought that electric and magnetic substances possessed some sort of nonmaterial (or spiritual) component. Du Fay proposed that electricity consisted of two fluids, one positive and the other negative, that could penetrate some bodies, thereby charging them. Franklin simplified this hypothesis by proposing that there was only one electric fluid (Franklin actually used the term electric fire), a terminology that remained in use until early in the twentieth century. A body would be charged positively or negatively depending on whether it had an excess or a defect of electric fluid. In 1874 the English physicist George John-

stone Stoney suggested that, instead of being a fluid, electricity consisted of negatively charged particles that he called electrons. He also assumed that electrons had negligible mass because no appreciable change in mass was detected when a body was charged. Stoney estimated the charge of the electrons, using experimental results from electrolysis, but his estimate was about one-tenth of the correct value because Avogadro's number was not well known in his time. Triboelectrification can be explained in terms of a transfer of electrons from one body to another. The body that receives electrons becomes negatively charged and the body that loses electrons becomes positively charged. Now we know that besides electrons there are other particles with negative charge and that some particles, such as protons, have positive charge.

The first direct experimental evidence of the existence of electrons came about in 1897 when Joseph J. Thomson proved that cathode rays consisted of negatively charged particles that were deviated by electric and magnetic fields in a form that was consistent with Newton's second law of motion. Thomson determined that the ratio of the charge to the mass of the electrons was $q/m = 1.7588 \times 10^{11}$ C/kg. Later on, in a series of delicate experiments initiated in 1909, Robert A. Millikan determined the charge of the electron as 1.6022×10^{-19} C. Consequently, the mass of the electron is 9.101×10^{-31} kg, which is about 1,850 times smaller than the mass of the atoms of the lightest element, hydrogen, which is 1.673×10^{-27} kg, confirming Stoney's hypothesis that electrons were very light.

Based on experimental evidence, such as chemical reactions and electrolysis, it has been concluded that hydrogen atoms consist of one electron orbiting about a positively charged nucleus called a proton. Using a mass spectrometer, it has been found that the ratio of the charge to the mass of the proton is $q/m = 9.5792 \times 10^7$ C/kg, which, combined with the mass of a hydrogen atom, gives the proton a positive charge equal to the negative charge of the electron. That amount of electric charge is called the fundamental charge, designated by $\pm e$. Protons are also found in the nuclei of all atoms. Another constituent of atomic nuclei are the neutrons, discovered in 1932 by James Chadwick. The neutron has a mass only sightly larger than that of the proton but it has no electric charge (see Table 1). All atoms consist of a positively charged nucleus, composed of protons and neutrons, and of negatively charged electrons orbiting around the positive nucleus under its force

Table 1 Charge of Fundamental Particles

Particle	Mass, m (kg)	Charge, q	q/m $(C \cdot kg^{-1})$
Electron	$m_e = 9.1091 \times 10^{-31}$	$-e$	1.7588×10^{11}
Proton	$m_p = 1.6726 \times 10^{-27}$	$+e$	9.5792×10^{7}
Neutron	$m_n = 1.6748 \times 10^{-27}$	0	0

$e = 1.6022 \times 10^{-19}$ C

of attraction. The number of electrons and protons in an atom is the same (it is called atomic number), so that atoms are electrically neutral.

Under certain conditions an atom or a molecule may gain or lose one or more electrons, becoming charged negatively or positively, constituting an ion. Ionization is rather common and gives rise to several electric phenomena. The electric fields of high energy charged particles (electrons, protons, α-particles, etc.) and electromagnetic radiation (ultraviolet, x, and gamma) ionize the atoms of the materials through which they pass by knocking out electrons. Cosmic rays ionize the gases in the atmosphere. The ions in the air may attract water molecules forming small droplets, which in turn become electrically charged clouds. Lightning consists of electric discharges between two clouds or a cloud and the ground.

From the above experiments and others, we may conclude that (1) atoms, molecules, and in general all matter, are composed of equal amounts of positive and negative charged particles, as well as some neutral particles, so that bodies are electrically neutral; (2) electric charges appear only as multiples of the fundamental charge e, that is, electric charge is quantized; (3) electric charge is conserved in all processes involving charged particles, so that an electrical balance is maintained in the universe; and (4) the properties of matter are determined by the electric forces between atoms and molecules.

Not all substances can be electrified in the same way, as was recognized by Gilbert in his experiments on triboelectrification. Gilbert used the term "electric" for the substances that could be electrified by rubbing and "non-electric" for the others. Later on Stephen Gray found that the substances called non-electric by Gilbert conduct electricity through their volume but could hold a charge (i.e., be electrified) if insulated, while the substances Gilbert called electric could be electrified rather easily (i.e., hold a charge) because they conduct electricity very poorly or not at all. Accordingly du Fay proposed to classify all substances as good and bad conductors. The current terms are "conductors" and "insulators" or "dielectrics," a name proposed by Michael Faraday. Whether a substance is a conductor or an insulator depends on its atomic structure.

Conductors are substances in which charged particles can move more or less freely through its volume. Metals (copper, silver, gold, etc.) are solid conductors composed of a lattice of positively charged ions with negatively charged electrons moving freely, somewhat like a gas, through the spaces between the ions. When a metallic conductor is placed in an electric field, the free electrons move through the conductor in the direction opposite to the field, constituting an electric current. This is what happens when the two poles of a battery are joined by a piece of copper wire. As the temperature of a metallic conductor is increased, its conductivity decreases because it becomes more difficult for an electric field to move the electrons.

When an insulated metallic conductor is charged by adding or removing electrons, all its charge appears on its outer surface, as demonstrated by Francis Hauksbee (in the seventeenth century), by Franklin in 1780, and by Faraday in 1837. Faraday's experiment consisted of touching an insulated charged metallic cylinder with a test conductor C. If C touches the outer surface it gets charged, but if C touches the inner surface it does not get any charge. The explanation is that the charges of a conductor spread over the outer surface until the electric field inside is zero. Even if the conductor has a cavity the electric field inside the cavity is zero. This property is used to insulate delicate instruments from stray electric fields in what are called Faraday cages.

If an insulated metallic conductor is placed in an electric field charges of opposite signs appear on opposite ends of the conductor, that is, the conductor becomes polarized, as shown by the following experiment. In Fig. 1a, charged body A is placed near an insulated conductor B. On the surface of B closest to A appear charges opposite to those in A, while the

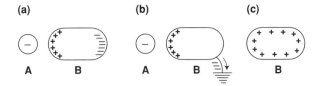

Figure 1 Behavior of an insulated conductor near an electric field produced by a charged body.

same kind of charges appear at the farther end. In the case illustrated, A is charged negatively and therefore repels the free electrons in B, pushing them as far as possible until the electric field inside the conductor is zero. When A is removed the polarization of B disappears. If B is connected to the ground, as in Fig. 1b, the electrons are pushed even farther away. B remains charged when the ground connection is removed while A is still present, as in Fig. 1c. This offers a way of charging a conductor. To charge a conductor it is necessary to spend some energy to overcome the repulsion of the charge already placed in the conductor. The energy of the conductor per unit charge is measured in volts, a name given in honor of Alessandro Volta.

The polarization of insulated metals is used in an instrument called an electroscope, which serves to determine if a body is charged and estimate its charge. A simple electroscope, invented by Abraham Bennet in 1787, consists of a conducting rod with a metal knob B at one end and two gold foils L and L' attached at the other end, as shown in Fig. 2. The rod is fastened through an insulating stopper to a

metal case M with a glass window. The gold foils repel each other and separate when the electroscope is charged by touching the knob with a charged conductor. When a charged body C is brought close to B, the polarization effect described in Fig. 1 charges the gold foils, which repel each other. The Faraday cage effect referred to above can be seen by placing an electroscope inside a hollow metallic conductor C, as in Fig. 3. When a charged body A is placed near C charging it, the gold foils of the electroscope do not move, which shows that no electric fields have been produced inside C. In the presence of ionizing radiation a charged electroscope is discharged by the ions in the air. For that reason electroscopes can be used to detect ionizing radiations. More elaborate instruments, such as electrometers, ionization chambers, Geiger counters, and dosimeters, serve to measure the amount and the rate of ionization. They are used wherever people are exposed to ionizing radiations (laboratories, hospitals, etc.).

Some conducting substances consist of a mixture of positive and negative ions that can move more or less freely through the substance. This is the case of molten salts, solutions of acids, bases, and salts (called electrolytes), and of ionized gases. If an electric field is set on these substances the ions move in opposite directions, according to their charges. When the field is produced by oppositely charged plates or electrodes the ions are attracted to the electrode with opposite charge and repelled by the electrode with the same charge. Ion separation by an electric field occurs in many important processes

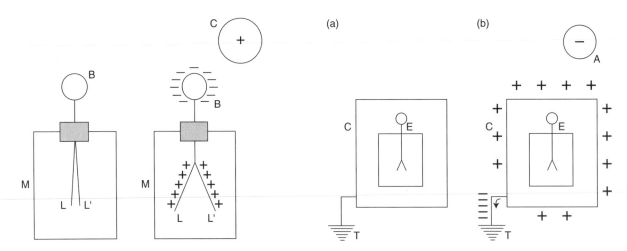

Figure 2 A simple electroscope.

Figure 3 Illustration of the Faraday cage effect.

(electroplating, batteries, ion transport across cell membranes, etc.).

An insulator is a substance in which the electrons are attached to its atoms or molecules and are not free to move from one place to another under the action of an electric field. Insulators are very poor electric conductors. Insulators can be solids, liquids, or gases. Widely used insulators are oil, rubber, glass, many ceramic materials, and some plastics. When an insulator is placed in an electric field, such as that produced by a charged body, it becomes polarized due to the deformation of the motion of the atomic electrons. To understand this effect consider Fig. 4. In the absence of an electric field the center of the motion of the electrons in an atom coincides with the nucleus, but in the presence of an electric field the motion of the negative electrons is deformed relative to the nucleus in a direction opposite to that of the field. Thus the center of motion of the negative electrons becomes slightly separated from that of the positive nucleus; it is said that the atom becomes an electric dipole. Therefore, in an insulator placed in an electric field all atoms or molecules become dipoles oriented in the same direction as the field so that opposite charges appear on opposite ends of the insulator; however, those charges are not free, as in the case of conductors. A polarized insulator tends to move in the direction in which the field is stronger. If the applied electric field is very strong some electrons may be pulled from the atoms or molecules in the insulator resulting in an electric breakdown or electric discharge. The polarization effect is used in capacitors that consist of an insulator or dielectric placed between two oppositely charged conductors. The electric charges on the surface of the dielectric compensate in part for the charges on the conductors. This is one reason why capacitors are widely used to store electric charges, sustain high voltages, or regulate the current in ac circuits and other electronic devices.

A third class of materials, intermediate between conductors and insulators, are called semiconductors. At low temperatures semiconductors are insulators or very poor conductors because they do not have many free electrons. However, contrary to what happens with metallic conductors, the conductivity of semiconductors increases as their temperature increases due to the thermal excitation of some electrons that gain enough kinetic energy to overcome the attraction of the positive ions and become free to move through the substance. Examples of semiconductors are silicon and gallium. When some im-

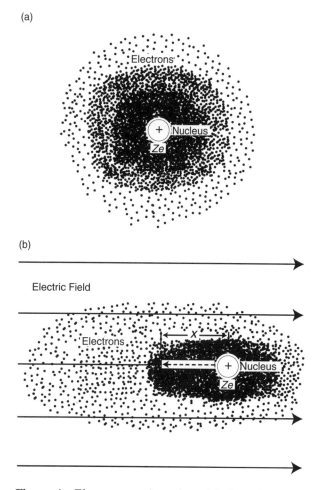

Figure 4 Electron motion when (a) there is no external field and (b) there is an external field.

purities, whose atoms have more or less electrons than those of the semiconductor, are added (a method called doping), the conductivity of the semiconductor changes profoundly. Semiconductors have found many applications in devices called transistors, widely used in solid state devices, such as chips, used in modern electronics and computers.

Our understanding of the electric structure of matter and our ability to manipulate and use the attraction and repulsion between charged particles, particularly electrons, has found many technological applications, such as solid state components (transistors, charge coupling devices, liquid crystal detectors, light emiting diodes, etc.) used in watches, radios, telephones, video cameras, television monitors, computers and communication systems, and photoelectric cells, as well as in particle accelerators for research and medical use.

See also: CATHODE RAY; CONDUCTOR; COULOMB, CHARLES AUGUSTIN; COULOMB'S LAW; ELECTRO-MAGNETISM; ELECTRON; ELECTROSTATIC ATTRACTION AND REPULSION; FARADAY, MICHAEL; FIELD, ELECTRIC; FIELD, MAGNETIC; FRANKLIN, BENJAMIN; INSULATOR; IONIZATION; LIGHTNING; SEMICONDUCTOR; THOMSON, JOSEPH JOHN

Bibliography

ALONSO, M., and FINN, E. J. *Physics* (Addison-Wesley, Reading, MA, 1993).

ASIMOV, I. *The History of Physics* (Walker, New York, 1984).

JONES, E. R., and CHILDERS, R. L. *Contemporary College Physics* (Addison-Wesley, Reading, MA, 1990).

MARCELO ALONSO

ELECTRIC MOMENT

In principle, at any point, there is an electric potential due to a group of electric charges or cloud of continuous-charge distribution by brute-force summation or integration. If the point P is in the neighborhood of the charge distribution, the potential is sensitive to the internal structure of the distribution. As point P moves away from region D populated by electric charges, the details of the charge distribution become less important, until finally, at a sufficiently large distance, the charge distribution approximates a point charge. However, the potential at P reflects the characteristic internal structure of the charge distribution. The electric potential in this limiting case can be described by electric multipoles; that is, by a linear combination of electric moments of different order, such as dipole or quadrupole.

This paradigm assumes that the charges are confined to the region D. If D is placed near the origin of the coordinate system and these point charges are designated $q_1, q_2, q_3, \ldots, q_N$, with position vectors $\mathbf{r}_1, \mathbf{r}_2, \mathbf{r}_3, \ldots, \mathbf{r}_N$, respectively, the electric potential $V(\mathbf{R})$ at P can be expressed as

$$V(\mathbf{R}) = (1/4\pi\varepsilon_0) \sum q_i/R_i, \qquad (1)$$

where $\mathbf{R}$ represents the position vector of P, and $R_i = |\mathbf{R} - \mathbf{r_i}|$ —that is, the distance between the i-th point charge q_i and P. Thus, the location of each charge is relative to its distance from P. The mathematical manipulation and approximation (that the distance R is very large compared to the size of the domain D) involved in the next few steps is beyond the scope of this article. The potential derived from these steps has the form:

$$V(\mathbf{R}) = (1/4\pi\varepsilon_0 R)\sum q_i + (1/4\pi\varepsilon_0 R^2)\sum q_i r_i \cos\theta_i$$
$$+ (1/4\pi\varepsilon_0 R^3)\sum(q_i r_i^2/2)(3\cos^2\theta_i - 1)$$
$$+ \ldots \text{(higher-order terms)}. \qquad (2)$$

This expression is called the multipole expansion of the potential. The angle θ_i represents the angle between vectors $\mathbf{R}$ and $\mathbf{r_i}$. Since both $\mathbf{R}$ and $\mathbf{r_i}$ vectors begin at the origin, these vectors and the line connecting P and q_i (of length R_i) form three sides of a triangle, and thus the side R_i is opposite angle θ_i. The terms in Eq. (2) are called, in the ascending power of $(1/R)$, the monopole term, the dipole term, and the quadrupole term. Since the dipole term varies with R as $1/R^2$ and the quadrupole terms as $1/R^3$, as P moves further away from D, the contributions from these higher-order terms become less significant. The functions of $\cos\theta$ in Eq. (2) are called Legendre polynomials.

Equation (2) can be conveniently written as

$$V(\mathbf{R}) = V_1(\mathbf{R}) + V_2(\mathbf{R}) + V_3(\mathbf{R}) + \ldots \qquad (3)$$

The Monopole Term $V_1(\mathbf{R})$

The first term in Eq. (2) is the monopole. Since $\sum q_i$ represents the net charge Q found in D, this term is reduced to $V_1(\mathbf{R}) = Q/4\pi\varepsilon_0 R$.

This is exactly the electrostatic potential due to a point charge Q at a distance R away from the charge. It shows that at a large distance, the most significant contribution to the potential is the net charge of the distribution. If the net charge is zero, there is no contribution from the monopole term, and the first nonvanishing contribution comes from the dipole term.

The Dipole Term $V_2(\mathbf{R})$

Geometrical considerations make it possible to rewrite the sum $\sum q_i r_i \cos\theta_i$ as $\breve{\mathbf{R}} \cdot (\sum q_i \mathbf{r_i})$, where $\breve{\mathbf{R}}$ is the unit vector in the direction of the vector $\mathbf{R}$ (i.e., $\check{\mathbf{R}} = \mathbf{R}/R$), and · signifies scalar or dot product between two vectors, $\check{\mathbf{R}}$ and $\sum q_i \mathbf{r_i}$. The vector quantity $\sum q_i \mathbf{r_i}$ is characteristic of the charge distribution and does not contain quantities related to the point of observation. It is defined as the dipole moment and commonly designated with a vector symbol $\mathbf{p}$; that is, $\mathbf{p} = (\sum q_i \mathbf{r_i})$.

In this context, the net charge Q, a scalar, in the monopole expression can be called the monopole moment. The dipole term $V_2(\mathbf{R})$ in Eq. (3) can then be written in terms of the dipole moment $\mathbf{p}$ as $V_2(\mathbf{R}) = \mathbf{p} \cdot \check{\mathbf{R}}/(4\pi\varepsilon_0 R^2)$, or, using the relation $\check{\mathbf{R}} = (\mathbf{R}/R)$, then $V_2(\mathbf{R}) = \mathbf{p} \cdot \mathbf{R}/(4\pi\varepsilon_0 R^3)$.

When the net charge (i.e., the monopole moment) is zero, the dipole term is the leading term in the multipole expansion of the potential. In other words, to an observer far away, this charge distribution looks very much like a dipole.

The Quadrupole Term $V_3(\mathbf{R})$

The summation $\sum (q_i r_i^2/2)\ (3\cos^2\theta_i - 1)$ in the quadrupole term results in the nine components of the quadrupole moment tensor. These symmetric tensor components Q_{jk}, where j and k represent two of the three coordinates x, y, and z, can be derived with some geometrical considerations from the above summation. The quadrupole term has the form

$$V_3(\mathbf{R}) = (1/8\pi\varepsilon_0 R^3)$$

$$\times \text{ (sum of the linear combinations}$$

$$\text{of terms first order in } Q_{jk}). \quad (4)$$

The Choice of Origin

In the definition of the dipole moment $\mathbf{p} = (\sum q_i \mathbf{r_i})$, $\mathbf{p}$ is not independent of the coordinate system because all $\mathbf{r_i}$'s are measured with respect to the origin of the coordinate system. A simple calculation shows that the dipole moment is independent of the choice of origin if the monopole moment Q vanishes.

The Multipole Field

Identifying the multipole potential by partial differentiation of the potential $V(\mathbf{R})$ with respect to x,

y, and z, or the polar coordinate r and θ, allows for calculation of the electric field.

In physics, the units of electric multipole moments are [charge] × [length]^l for the monopole and dipole moments ($l = 0,1$). For higher-order moments, the units are expressed as [length]^l by dividing the multipole moment by the electron charge in electrostatic units ($e = 4.8 \times 10^{-10}$ esu). Another unit associated with multipole moments that is not commonly used in general physics is the debye, which is defined as 10^{-18} esu cm. This is found from the order of magnitude of an estimate of the molecular dipole moment. Such a moment depends on the charge and distance. Using the electron charge ($e = 4.8 \times 10^{-10}$ esu) for charge and the atomic diameter (10^{-8} cm) for distance, the estimate is 4.8×10^{-18} esu cm.

Multipole moments can link macroscopic and microscopic theories of matter. For instance, to characterize a macroscopic dielectric medium, the macroscopic system can be broken up into idealized microscopic charge distributions at points through the space occupied by the physical medium. For each of these charge distributions, the electric potential can be represented as a sum of multipoles at that point. The potential of the macroscopic medium is then a linear superposition of these potentials.

Multipole moments are also important for the information they provide about the internal properties of the building blocks of matter. For instance, neutral elementary particles exhibit no measurable dipole moment, despite the highly sensitive measurements from such experiments at Grenoble, which place an upper limit on the neutron dipole moment of 10^{-26} esu cm.

Atomic nuclei, which are always charged objects, often have nonzero quadrupole moments. These moments imply that the force between the constituent protons and neutrons is noncentral, providing new information into the nature of the nuclear force.

See also: CHARGE; DIPOLE MOMENT; TENSOR

Bibliography

JACKSON, J. D. *Classical Electrodynamics,* 2nd ed. (Wiley, New York, 1975).

GAY B. STEWART

ELECTRIC POTENTIAL

An electric field can be described in two ways: by an electric-field vector **E;** and by a scalar quantity $V(\mathbf{P})$, an electric potential (at point **P**).

Electric potential difference, $V_\mathbf{B} - V_\mathbf{A}$, between two points **A** and **B** in an electric field is defined as the mechanical work $U_{\mathbf{AB}}$ required to move a positive test charge from **A** to **B** divided by the amount of electric charge q_0 (> 0) of the test charge. Thus,

$$V_\mathbf{B} - V_\mathbf{A} = \frac{U_{\mathbf{AB}}}{q_0} = -\frac{1}{q_0}\int_\mathbf{A}^\mathbf{B} \mathbf{F} \cdot d\mathbf{s} = -\int_\mathbf{A}^\mathbf{B} \mathbf{E} \cdot d\mathbf{s}, \quad (1)$$

where **F** is the force applied by an external agent to move the charge and $d\mathbf{s}$ the infinitesimal vector line segment along the line of path integration from **A** to **B**.

A standard reference point is chosen for **A.** This point is usually taken to be at infinity or at ground where the electric potential is set to zero. Thus, the electric potential $V(\mathbf{P})$ at $\mathbf{P}(x, y, z)$ can be defined as

$$V(\mathbf{P}) = U(\mathbf{P})/q_0, \quad (2)$$

where $U(\mathbf{P})$ is the work required to transport a test charge q_0 from the reference point to $\mathbf{P}(x, y, z)$. (This is the negative of the work done by the electric field on the test charge.)

Just as an object gains potential energy when it is raised against the gravitational force, a charged object gains electric potential energy when moved against the force of an electric field. Since, unlike gravity, the electric force can be attractive or repulsive, the direction of increasing potential depends on the sign of the charge of the particle, as well as the direction of the field.

The electric-field originates on positively charged objects and terminates on negatively charged objects (or infinity). It takes work to move a negatively charged particle in the direction of the electric field, since it is attracted to the source of the field, just as an object, when released from rest, is attracted to the earth by gravity. It also takes work to move a positively charged object in the direction opposite that of the field (toward the positively charged source). The positively charged object would move of its own accord in the direction of the field, thus converting potential energy into kinetic

energy. (The electric field would do positive work on the object.) Since field quantities are defined in terms of their effect on a positively charged particle, electric potential decreases in the direction of the electric field.

The mks unit for electric potential is the volt (V). Two points differ in potential by 1V if 1J of work is required to move 1C of charge from one point to the other.

Work equals force times displacement, where the force and displacement are in the same direction, so work per unit charge divided by distance moved is equal to the force per unit charge. Electric potential is the change in energy per unit charge, which is the negative of work per unit charge. Since force per unit charge is the definition of the electric field strength, electric potential divided by distance moved equals $-\Delta V/\Delta d$, which is equal to E.

The decreasing electric potential in the direction of the electric field necessitates use of the negative sign. This exact relationship only holds for constant electric fields. For electric fields that depend on distance, the limit must be taken for very small changes in distance, and thus partial differentiation is required to calculate the electric field from the potential as

$$E_x = -\partial V(\mathbf{P})/\partial x$$
$$E_y = -\partial V(\mathbf{P})/\partial y \quad (3)$$
$$E_z = -\partial V(\mathbf{P})/\partial z.$$

In general, it is preferable to find the electric potential first and derive the electric field from the potential, as the potential is a scalar quantity and therefore easier to calculate. Since the electric potential is not a vector, it carries less information than the electric-field vector **E.** The partial derivative operation in Eq. (3) uses information on the behavior of the function $V(\mathbf{P})$ in the neighborhood of the point $\mathbf{P}(x, y, z)$ in addition to its value at that location.

An object moved around on a level tabletop neither loses nor gains gravitational potential energy. There are similar surfaces for electric potential energy, called equipotential surfaces. In an equipotential surface, the electric potential energy does not change, so it takes no work to move a charged object along an equipotential. An equipotential is not a physical surface but a mathematical description. Since the electric potential decreases in the direction of the electric field, the lines, or surfaces, of equipotential must be everywhere perpendicular to the electric field. Since the electric field is everywhere perpendicular to the surface of a conductor in electrostatic equilibrium, the surface of such a conductor is always an equipotential surface. If this were not the case, the charges would move around on the surface until there was no force on them, which again results in an equipotential surface. When an object is grounded, it becomes an equipotential surface with electric potential equal to zero.

A common object, the battery, uses chemical energy to supply a potential difference between its terminals, raising the potential of positively charged particles by moving them from the negative to the positive terminal.

Examples

Potential due to a point charge q at distance r from the charge is given by $V = (1/4\pi\varepsilon_0)\ q/r$.

With r_i equal to the distance from charge i to point P, potential at point P due to a group of point charges is $V = U/q = (1/4\pi\varepsilon_0)\Sigma q_i/r_i$. The work required to assemble an arbitrary system of point charges in a space with no initial electric field from infinite separation is equal to the electrostatic potential energy of the system. The work required to bring the first charge in from infinity is zero, since, before any of the charges are brought in, there is no electric field. Each additional charge experiences the field because of the charges are already in place. This work can be expressed, with r_{ij} equal to the distance from charge i to charge j, as $W = qV = (1/4\pi\varepsilon_0)\Sigma_i([\Sigma_{j<i}(q_iq_j/r_{ij})])$.

Potential due to an electric dipole of moment $\mathbf{p}$: $V = (1/4\pi\varepsilon_0)\ \mathbf{p}\cos\theta/r^2$, where θ is the angle between the dipole and the line connecting the dipole and point P, and r is the distance between the dipole and P. (θ is the angle between the vectors $\mathbf{p}$ and $\mathbf{r}$.)

Potential due to an electric quadrupole of moment Q is $V = (1/4\pi\varepsilon_0)\ Q/r^3$.

See also: CHARGE; DIPOLE MOMENT; ENERGY, POTENTIAL; FIELD, ELECTRIC

Bibliography

SEARS, F.; ZEMANSKY, M.; and YOUNG, H. *College Physics,* 5th ed. (Addison-Wesley, Reading, MA, 1980).

GAY B. STEWART

ELECTRIC SUSCEPTIBILITY

When a dielectric material is placed in an electric field, the atoms in the material develop electric dipole moments. The average dipole moment per unit volume is called the polarization of the medium, which determines the internal-reaction electric field of the medium in response to the applied electric field. The total electric field in a dielectric medium is given by a linear superposition of the applied and reaction fields:

$$\mathbf{E} = \mathbf{E}_{\text{applied}} + \mathbf{E}_{\text{reaction}}, \qquad (1)$$

where $\mathbf{E}_{\text{reaction}}$ depends on polarization $\mathbf{P}$.

The polarization of the medium is in turn related to the total electric field in the medium by way of the electric sucseptibility, $\chi(\mathbf{r}, \mathbf{E})$,

$$\mathbf{P} = \epsilon_0\chi(\mathbf{r}, \mathbf{E})\mathbf{E}. \qquad (2)$$

The ϵ_0 in Eq. (2) is a fundamental constant called the permittivity of free space ($\epsilon_0 = 8.854 \times 10^{-12}$ $C^2/N \cdot m^2$). In Eq. (2), the electric susceptibility, $\chi(\mathbf{r}, \mathbf{E})$, is dimensionless. If all of this seems circular, it is. The relationship in Eq. (2) is programmed into nature, which means that when a field is applied to a dielectric, the dielectric responds in the only way it can until the system of the field and the material reaches equilibrium. This happens almost instantaneously, and once equilibrium is reached, the polarization and the electric field have precisely the correct values based on the susceptibility to maintain equilibrium.

The electric susceptibility carries all of the information about the response of a given dielectric material to an electric field. Each dielectric material is characterized by a $\chi(\mathbf{r}, \mathbf{E})$ relation.

The electric susceptibility depends on the detailed microscopic structure and dynamics of the medium. To calculate the electric susceptibility theoretically, a model must be adopted for the material. Theoretical predictions based on a particular model can be tested by experimental measurements of the electric susceptibility of the medium. In this way, electric susceptibility provides an important bridge between things that can be measured using macroscopic equipment and things, such as the dynamics of electrons and atoms, that form dielectric materials.

The relationship between the polarization and the electric field in the medium can be complicated. For instance, susceptibility can depend on the electric field in the medium in a complicated way. If susceptibility depends on the field at all, then polarization becomes a nonlinear function of the electric field. This is the case for nonlinear dielectrics. Another difficulty arises when susceptibility depends on position within the medium, in which case, the material is called an inhomogeneous medium. A further complication arises when the dielectric medium is a crystal, because a component of the polarization can depend not only on the parallel component of the electric field but also on other components of the electric field. In this case, the vector Eq. (2) is replaced by a set of three equations, one for each component of the polarization:

$$P_x = \epsilon_0(\chi_{xx}E_x + \chi_{xy}E_y + \chi_{xz}E_z) \tag{3}$$

$$P_y = \epsilon_0(\chi_{yx}E_x + \chi_{yy}E_y + \chi_{yz}E_z) \tag{4}$$

$$P_z = \epsilon_0(\chi_{zx}E_x + \chi_{zy}E_y + \chi_{zz}E_z). \tag{5}$$

In Eqs. (3)–(5), the nine quantities $\chi_{xx}, \chi_{xy}, \ldots, \chi_{zz}$ depend on the electric field and the position within the medium. These quantities behave together as a mathematical entity known as a tensor of rank two, or an electric-susceptibility tensor. This tensor is needed to describe a material because the material responds differently depending on the orientation of its electric field. Materials of this nature are called anisotropic dielectrics.

A special class of materials that is common in engineering and classroom applications is free of the complications described above or such complications can be ignored as long as the electric field in the medium is not too strong. For these linear, isotropic, homogeneous materials, the electric susceptibility is a constant. The polarization and the field in the material are simply related by

$$\mathbf{P} = \epsilon_0\chi_e\mathbf{E}. \tag{6}$$

In Eq. (6), χ_e is the constant electric susceptibility of the material. These types of materials are useful for making lenses, fiber-optic cables, prisms, and layers to insert into capacitors to increase the amount of charge they can store. Calculation of the electric field and the polarization in a medium is simplest in the case of linear, isotropic, homogeneous materials.

Equations (2) and (4) also hold in the case of time-varying electric fields, such as an electromagnetic wave, where polarization of the medium varies with time. In an electromagnetic wave, the polarization, electric field, and electric susceptibility are considered complex numbers for the sake of calculation. The imaginary part of the susceptibility is related to the absorption of radiation by a dielectric medium. When a quantum mechanical model is used for certain dielectric media, the absorption can be turned into an energy gain rather than an energy loss. When these conditions are met in practice, the result is called a laser.

Electric susceptibility, however complicated, contains everything about the response of a dielectric to an electric field. The actual form of the susceptibility is determined by microscopic dynamics in the atoms that form the dielectric itself.

See also: DIPOLE MOMENT; FIELD, ELECTRIC; POLARIZATION; TENSOR

Bibliography

GRIFFITHS, D. J. *Introduction to Electrodynamics* (Prentice Hall, Englewood Cliffs, NJ, 1989).

REITZ, J. R.; MILFORD, F. J.; and CHRISTY, R. W. *Foundations of Electromagnetic Theory* (Addison-Wesley, Reading, MA, 1979).

GAY B. STEWART

EDWIN E. HACH III

ELECTROCHEMISTRY

Electrochemistry is the science of the electronic processes that occur when an ionically conducting fluid contacts an electronic conductor. The subject has a long history. The physicist Alessandro Volta built the first battery in 1796, inspired by Luigi Galvani's observation that contact with dissimilar metals induced twitching in a frog's leg. Volta's pile consisted of plates of silver and zinc separated by brine-soaked cloth. Four years later, William

Nicholson and Anthony Carlisle were the first to use Volta's invention to electrolyze water, and in doing so they noticed that hydrogen appeared at one pole and oxygen at the other. In 1807 Humphrey Davy used electrolysis to prepare the alkali metals, sodium and potassium. Michael Faraday, who was a protégé of Davy, was the first to enunciate the relationship between the quantity of electricity used in electrolysis and the number of molecules or atoms produced as a product of the current and time (the ratio is the Faraday constant, F). Some other examples of such faradaic electrochemical processes are corrosion (undesired electrolytic oxidation of metals) and the transmission of information along neurons (carried by electrical impulses caused by transient calcium ion gradients across insulating membranes that enclose the neuron). These applications have placed the subject firmly into the realm of chemistry. However, more recent progress in understanding the structure of the metal–water interface and the invention of microscopes that can probe atomic structure under water have brought physicists back into the field after an absence of more than one hundred years.

When an electronegative atom (or other chemical species) accepts an electron from a less electronegative atom, the former is said to have been reduced, and the latter is said to have been oxidized. When equal numbers of reduced (R) and oxidized (O) atoms or molecules coexist in thermodynamic equilibrium, the work needed to take an electron from the system to a point at rest far from it (the vacuum) is an absolute measure of the potential of the system, lying, on average halfway between the ionization potential of O and R. If n electrons are transferred at the surface of a metal electrode, for example

$$O + ne^- \Leftrightarrow R, \qquad (1)$$

then the absolute potential of the metal surface is fixed at the redox potential for the reaction given in Eq. (1), E^0. If the concentrations of O and R are not equal, then this potential must be corrected for a consequent entropic contribution to the free energy. It is conventional to use the ideal gas formula for entropy (i.e., it is proportional to the logarithm of concentration) compensating for deviations by using corrected concentrations termed activities. Thus, the potential of the electrode is

$$E = E^0 + \frac{RT}{nF} ln \frac{\{O\}}{\{R\}}, \qquad (2)$$

where R is the gas constant, F the Faraday constant, and T the absolute temperature. The activities (corrected concentrations) of O and R appear in curly braces. This relationship was first derived by Walter Nernst, and it is called the Nernst equation in his honor. If two electrodes at different electrode potentials are placed into an electrolyte, the consequent potential difference is the familiar cell potential. Cells are discharged as electrolyte is used up [right-hand term of Eq. (2)] to equalize the potential of the two electrodes. Figure 1 shows an example where zinc is oxidized at one electrode (the anode), increasing the number of zinc ions in solution, while copper is reduced at the other electrode (the cathode) as the current flow reduces the number of copper ions in solution (and plates copper onto the cathode). The potential of the anode in this example is about 1.1 V closer to the vacuum than that of the cathode (before current is drawn). Thus, electrons flow from the anode to the cathode.

Although the absolute potential of an electrode is difficult to measure, the potential of an electrode can be made to be quite reproducible, and such electrodes are called reference electrodes. One example is the normal hydrogen electrode (NHE), which consists of platinum in equilibrium with equal concentrations of protons and hydrogen molecules. Redox potentials for other species are often quoted in terms of the potential difference with respect to an NHE. This is called the NHE scale. The reaction

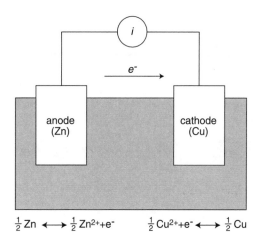

Figure 1 Oxidation of zinc.

$$\tfrac{1}{2}Zn^{2+} + e^- \Leftrightarrow \tfrac{1}{2}Zn$$

that forms one-half of the cell in Fig. 1 occurs at -0.763 V on this scale. The other reaction,

$$\tfrac{1}{2}Cu^{2+} + e^- \Leftrightarrow \tfrac{1}{2}Cu$$

occurs at 0.337 V. Together, they produce the ideal cell potential difference of 1.1 V. These potentials can be referenced to the vacuum using the fact that zero volt on the NHE scale corresponds to about 4.5–4.8 eV below the vacuum.

If an electrode does not react with atoms or molecules in solution (over some range of potentials) then its potential can be controlled over this range by means of a potentiostat and a three-electrode cell which incorporates a reference electrode. This apparatus is shown schematically in Fig. 2. The inert electrode is called the working electrode (W) and its potential is controlled by another inert electrode, the counter electrode (C) which is driven by a feedback control circuit so that a fixed potential difference (V) is set up between the reference (R) and working (W) electrodes. The study of faradaic processes as a function of the potential of the working electrode is called voltammetry. The region of potential over which species from the electrolyte are adsorbed onto the working electrode, but no faradaic processes occur, is called the double layer region. This is because charged species accumulate in a very narrow layer at the surface of electrode. The bulk of the potential difference between the electrolyte and the metal falls within this layer, which thereby acts as a capacitor, storing charge.

Such controlled polarized electrodes open up a new field of surface science. Because contamination molecules move slowly in a liquid, it is quite possible to operate electrodes that are as clean (i.e., free from unwanted contamination) as a surface in a vacuum of 10^{-12} Torr. The three-electrode cell may be used to alter the interfacial energy of such surfaces by an electron volt or more. To achieve the same thing in an ultrahigh vacuum environment (by changing temperature) would require a temperature change of 12,000 K. Many important atomic processes have been studied by using carefully prepared single crystals as working electrodes and measuring the current that flows (or the capacitance associated with the double layer) as a function of the electrode potential. Scientists are now using scanning tunneling and atomic force microscopes to image atomic processes as a function of electrode potential directly.

The physical basis of potential control is illustrated in a beautiful series of experiments described by E. R. Kötz, H. Neff, and K. Müller. These workers have shown that the double-layer set up on an electrode held under potential control can remain largely intact if the electrode is carefully pulled from the solution while potential control is maintained and transferred to vacuum. This *ex-situ* electrode is termed an emersed electrode. Thus, it has proved possible to determine the work function of emersed electrodes directly, by measuring the energy of the longest wavelength of light that will expel an electron from the metal by the photoelectric effect. The data show that the work function for all metals studied lies on a universal line. It is directly proportional to the electrode potential and is 4.8 eV at approximately zero volt on the NHE scale.

Understanding of electronic processes at the liquid-solid interface at the atomic scale is being pursued vigorously for the rewards it could bring in the form of better batteries, cleaner energy sources, new materials, better catalysts, and deeper insights into electron transfer processes in chemistry and biology.

See also: BATTERY; ELECTROLYTIC CELL; ELECTRON; FARADAY, MICHAEL; IONIZATION POTENTIAL

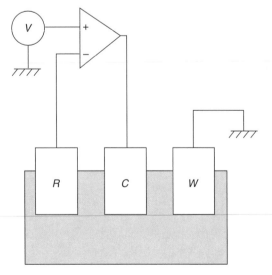

Figure 2 A potentiostat with a three-electrode cell.

Bibliography

BARD, A. J., and FAULKNER, L. R. *Electrochemical Methods, Fundamentals, and Applications* (Wiley, New York, 1980).

BOCKRIS, J. O'M., and REDDY, A. K. N. *Modern Electrochemistry* (Plenum, New York, 1973).

HAMELIN, A.; VITANOV, T.; SEVASTYANOV, E.; and POPOV, A. "The Electrochemical Double Layer on sp Metal Single Crystals: The Current Status of Data." *J. Electroanalyt. Chem.* **145,** 225–264 (1983).

KÖTZ, E. R.; NEFF, H.; and MÜLLER, K. "A UPS, XPS, and Work Function Study of Emersed Platinum and Gold Electrodes." *J. Electroanalyt. Chem.* **215,** 331–344 (1986).

SONNENFELD, R., and HANSMA, P. K. "Atomic Resolution Microscopy in Water." *Science* **232,** 211–213 (1986).

STOCK, J. T., and ORNA, M. V., eds. *Electrochemistry, Past and Present* (American Chemical Society, Washington, DC, 1989).

S. M. LINDSAY

ELECTRODYNAMICS

See QUANTUM ELECTRODYNAMICS

ELECTROLUMINESCENCE

Electroluminescence is the nonthermal emission of light from a nonmetallic solid in an applied electric field. This effect was discovered by Georges Destriau in 1936. He found that light was given off when large alternating voltage was placed across powdered ZnS dispersed in an insulator. The mechanism for electroluminescence in these materials is thought to be ionization at grain boundaries due to the impact of electrons accelerating in the nonuniform electric field at the boundary. Since Destriau's initial discovery, several mechanisms for electroluminescence have been discovered in forward- and reversed-biased *p-n* junctions; reverse-biased Schottky barrier diodes; and semiconductors fashioned into quantum wells, wires, and dots.

Modern electroluminescent devices are usually multilayered or nanostructured materials produced by some form of vapor deposition (e.g., Molecular-Beam Epitaxy, sputtering, or evaporation) onto a glass substrate. The first layer is typically an electrode made from a transparent conductor like indium-tin oxide (ITO). Semiconductor heterostructures are then deposited on top of the ITO layer. A second electrode is deposited on top of the semiconductor followed by a protective cap layer. For laser applications, the materials are usually engineered to form a quantum well to trap the electrons.

Devices based on electroluminescence first appeared in the late 1960s. The commercial market for electroluminescent devices has grown considerably since then. Such devices can be found in position sensors, computer displays, data storage devices, and telecommunication equipment.

Since the early 1990s, there has also been a flurry of excitement over blue-green light-emitting diodes (LEDs) and diode lasers made from wide bandgap semiconductors. For example, ZnSe-based semiconductors have been made into continuous-wave quantum well lasers at room temperature, and commercial high brightness blue and green LEDs have been made from GaN. These devices are important for applications like optical recording (where the information density is wavelength dependent) and high brightness, full-color computer displays.

Light-Emitting Diode (LED)

The most common electroluminescent device is the LED, which is based on injection electroluminescence wherein light is given off in a forward-biased *p-n* junction when electrons injected into the *p*-region and holes injected into the *n*-region recombine with holes and electrons, respectively.

This radiative recombination is most probable in direct-gap semiconductors like GaAs or InP. However, the probability of radiative recombination in indirect-gap semiconductors can be increased by introducing impurities that ensure momentum conservation during indirect transitions. For instance, nitrogen, sulphur, or zinc can be added to GaP to increase its electroluminescence.

Colored LEDs are made by adjusting the composition of the semiconductor to make the bandgap match the energy of light of the desired color. Red, yellow, and green LEDs are typically made from GaP with various impurities, $GaAs_{1-x}P_x$ of Al_xGa_{1-x}.

The intensity of light given off by an LED is roughly proportional to the electrical current through it. However, the best available LEDs are only about 5 to 10 percent efficient in turning electrical energy into light. The losses are due to nonradiative recombination processes and internal reflection of light at the surfaces of the device.

Diode Laser

Another device based on electroluminescence is the diode laser. The population inversion needed to induce laser oscillations in these materials are produced by filling most of the states near the bottom of the conduction band with electrons and most of the states near the top of the valence band with holes. Relatively inefficient laser diodes can be made from heavily doped GaAs LEDs.

More efficient lasers are made from heterojunction diodes. Electrons in these diodes are confined in quantum wells produced by depositing multiple layers of semiconductors on top of each other. The simplest structure is composed of a layer of one semiconductor sandwiched between two layers of another. The highest efficiency is found when the two materials have nearly the same lattice parameters (e.g., AlAs and GaAs). Otherwise, the point defects at the interfaces act as nonradiative recombination sites.

Further confinement is possible by producing nanostructures of lower dimensionality. This is usually accomplished by using various lithography techniques to make quantum well wires and dots. These devices tend to work only at temperatures below 100 K.

See also: DIODE; ELECTRON; INSULATOR; LASER; LIGHT; MOMENTUM, CONSERVATION OF; SEMICONDUCTOR; WAVELENGTH

Bibliography

ELLIOTT, R. J., and GIBSON, A. F. *An Introduction to Solid-State Physics and Its Applications* (Barnes and Noble, New York, 1974).

GOOCH, C. H. *Injection Electroluminescent Devices* (Wiley, New York, 1973).

KEVIN AYLESWORTH

ELECTROLYTIC CELL

An electrolytic cell is a device for transforming chemical energy into electrical energy (and also electrical energy into chemical energy). Such a cell is a source of potential difference to an external circuit. It often contains two different pieces of metal and a liquid; the metals are known as electrodes, and the liquid is called an electrolyte. The function of the electrolyte is to allow transfer of ions from one electrode to the other.

The first electrolytic cells were used by Luigi Galvani in 1780 for his famous experiment with the jumping frog leg. The severed legs jumped when two metals touched the frog's leg muscles; the electrolyte was the frog's muscle tissue. Giuseppe Volta, who began working with cells in the 1790s, designed the first electric "pile," or battery, in 1800.

If neighboring cells are brought into contact, the result is an increase in the potential difference that can be supplied to an external circuit. Such a combination of cells is known as a battery. The word has come to be used both for piles of cells and for individual cells (as in the "D-cell battery" commonly sold in stores, which is a single cell).

Metals are vessels full of electrons shared among all the metal atoms. Electrons are fermions and therefore must occupy different states. While at most two electrons in an individual atom will occupy one energy level (for example, in helium the two electrons occupy the lowest energy level, one having spin projection $\frac{1}{2}\hbar$, the other having spin projection $-\frac{1}{2}\hbar$), when several atoms are at about the same location the electrons must evince slight differences in order to distinguish themselves from the other electrons. The electronic quantum numbers cannot distinguish the electrons, so they must differ in energy. When large numbers of atoms of the same sort form a solid or liquid, every electron state (occupied or unoccupied) is slightly different in energy, and the energy levels spread to form bands of closely spaced energies. The highest filled energy state in a collection of many atoms is known as the Fermi level.

As with all bound systems, electrons bound in atoms have negative total energy. The Fermi level, the greatest energy an electron can have in the metal, is some energy below zero energy. The energy needed to raise the highest electron to zero energy (to make it not be bound to the metal) is known as the work function. The work function for a metal is analogous to the ionization energy for an atom. Different metals have different work functions, so if different materials are brought into contact, electrons that are higher in energy can decrease in energy by moving from one material to the other. The differences in work functions are commonly 1 to 2 eV, and so the cells may produce potential differences of 1 to 2 V. It is this fact that allows electrolytic cells to be constructed.

Cells are classified as primary or secondary. A primary cell can produce its charge only once. Most batteries sold for use in flashlights, radios, and cameras are primary cells. They are so-called dry primary cells, and, while not dry, use a paste electrolyte instead of a liquid electrolyte. A secondary cell may be recharged and discharged many times. Some cells that use nickel and cadmium metals as the electrodes and potassium hydroxide as the electrolyte (Ni-Cd batteries) allow chemical reversal of the discharge reaction and may be recharged many times. Most Ni-Cd cells are used in cordless electric appliances. The most commonly encountered secondary electrolytic cell is that in the automobile battery. The electrodes are lead and lead dioxide, and the electrolyte is sulfuric acid. During the discharge phase, lead is converted to lead sulfate and sulfuric acid to water. After the surfaces have been changed, the reaction ends (the battery has been discharged). If the current is run backwards through the battery, water and lead sulfate become lead and sulfuric acid again, respectively.

In addition, fuel cells are electrolytic cells in which the chemicals to the electrolytes are supplied separately. The simplest cells use hydrogen and oxygen as chemicals at the cathode (negative electrode) and anode (positive electrode), respectively, and water as the electrolyte; the hydrogen is decomposed into a hydrogen ion and an electron, and the ions migrate through the water to the anode where they accept electrons and bind with oxygen to make water. Many of the most efficient portable fuel cells built (for use in vehicles) use a molten electrolyte and must be insulated thermally from the surroundings.

See also: BATTERY

Bibliography

MANTELL, C. L. *Batteries and Energy Systems* (McGraw-Hill, New York, 1983).

VINCENT, C. A. *Modern Batteries: An Introduction to Electrochemical Power Sources* (Edward Arnold, London, 1984).

GORDON J. AUBRECHT II

ELECTROMAGNET

An electromagnet is a device used to provide high magnetic fields in a laboratory setting. It consists of two multiturn coils, each of which are wound around a pole piece made of a magnetically soft iron core. These in turn are mounted within a massive rigid iron frame known as a yoke. When the current is switched on, the iron cores become magnetized to near saturation and magnify the field of the coils by a factor of approximately 10^3. With a suitably designed power supply, electromagnets can provide a steady magnetic field in the air gap between the pole pieces. They are now superseded by nonconventional electromagnets and by superconducting solenoids beyond the field of about 20 kG or 20 T.

The magnetic field energy density is proportional to the square of the current used to excite the coils, and the magnetic field generated is proportional to the current. The Joule heat production is proportional to the resistance of the windings and to the square of the magnetizing current. There is an upper limit to the efficiency of Joule heat removal in the conventional electromagnet. Thus, if a higher field is desired, one is forced to reduce the volume between the pole pieces (accomplished by moving the opposing pole pieces closer together) to maintain the total energy constant. The available volume for a higher field, therefore, decreases as the field is increased.

A higher magnetic field can be achieved by water-cooling the conductors (Bitter magnets). Alternatively, a type-II superconductor, such as Nb_3Sn, can be used in a solenoid form immersed in liquid helium to provide high fields. To produce an even higher field, pulsed field method and explosives are employed.

The electromagnet was invented by William Sturgeon in England in 1824. In the United States, Joseph Henry constructed an electromagnet in

1832. Henry's electromagnet, which can be considered a prototype of a modern electromagnet, is a permanent exhibit at the Smithsonian Museum in Washington, D.C.

Half of a laboratory electromagnet, that is, a single pole piece with a coil, is used in industry for lifting heavy iron and steel pieces. Scrap metal weighing 50,000 lbs or more can be lifted by an industrial electromagnet.

See also: ELECTROMAGNETISM; JOULE HEATING; MAGNET; MAGNETIC BEHAVIOR; MAGNETIC POLE

Bibliography

HALLIDAY, D.; RESNICK, R.; and KRANE, K. S. *Physics,* 4th ed. (Wiley, New York, 1992).

HOLTON, G.; RUTHERFORD, F. J.; and WATSON, F. G. *Project Physics* (Holt, New York, 1975).

CARL T. TOMIZUKA

ELECTROMAGNETIC FORCE

An electric field exerts a force on a static or moving electric charge. A magnetic field exerts a force on a moving electric charge. The combined effect can be expressed as

$$\mathbf{F} = q\mathbf{E} + q\mathbf{v} \times \mathbf{B},$$

where $\mathbf{F}$ is the electromagnetic force, $\mathbf{E}$ and $\mathbf{B}$ are the electric and magnetic field vectors, respectively, q is the electric charge of the object, and $\mathbf{v}$ is its velocity vector. The multiplication notation " $\times$ " signifies a vector multiplication. It should be noted that the sign convention of the vector product follows the right-hand rule. This force $\mathbf{F}$ is referred to as the Lorentz force. When the electromagnetic field is nonstationary, one must include in $\mathbf{E}$ and $\mathbf{B}$ the fields generated by the moving particle itself.

See also: CHARGE; FIELD, ELECTRIC; FIELD, MAGNETIC; FORCE; LORENTZ FORCE

CARL T. TOMIZUKA

ELECTROMAGNETIC INDUCTION

In the spring of 1800, the Italian natural philosopher Alessandro Volta sent a letter to the president of the Royal Society of London in which he announced his invention of what came to be known as the voltaic pile. This combination of two dissimilar metals separated by blotting paper soaked in a slightly acid solution created the first electrical current. Almost immediately it was found that the current could decompose water and chemical solutions. All over Europe and America, chemists and other scientists constructed their own voltaic piles and theorized on the possible causes of the current. The study of electricity now became the center of scientific investigation.

Static electricity had been known since antiquity. Benjamin Franklin had proved in the middle of the eighteenth century that lightning was electricity on a grand scale, and it was early noticed that lightning strokes demagnetized compass needles on ships and magnetized bars of iron. The obvious conclusion was that electricity and magnetism were connected in some way. In the 1770s, this connection was strongly challenged and denied by the extremely accurate measurements of Charles Coulomb in France, who showed that the laws of attraction and repulsion of the electrical and magnetic forces, although similar in that they followed Isaac Newton's inverse square law, were completely different in terms of the constants required to give an accurate result. The search for a connection was almost entirely abandoned after Coulomb published his results.

Only a very small number of people still thought the search was worthwhile, and they based their faith on philosophy, not science. It was the German philosopher Immanuel Kant who provided the basis for this faith. He argued that all we could sense were forces, not the sub-sensible hypothetical material atoms that were generally assumed to be the seat of these forces. More important, he insisted that all the forces of nature were merely different manifestations of the fundamental forces of attraction and repulsion. It then followed that the various forces of nature—electricity, magnetism, heat, light, and gravity—were convertible one into another. Few of Kant's contemporaries were persuaded, but among those that were was a young Danish chemist, Hans Christian Oersted. Oersted devoted his early scientific career to a search for these conversions. As early as 1813, he insisted that electricity and magnetism *must* be convertible into one another. He was, however, unable to discover the experimental conditions to produce the conversion. Finally, in the late winter of 1819–1820, he succeeded. During the course of a public lecture, he was seized by an inspiration. Hitherto, those seeking the effect had placed a compass needle at an angle to the wire, confident that the effect would then move the needle parallel to the wire. Oersted had tried this many times and knew that no effect was evident. In his lecture demonstration, he positioned the wire so that it was directly over the needle and aligned with it. When he closed the circuit, the needle moved transversely to the wire. Electricity could be converted into magnetism just as Kant had insisted. In July 1820, Oersted announced his discovery to the scientific world.

The reception of Oersted's news was greeted with great skepticism in France for German-speaking nature philosophers (*naturphilosophen*) had recently published "astonishing" discoveries concerning relationships that no one else could find. When, however, Oersted's findings were shown experimentally to the members of the Academy of Sciences in Paris in early September, they also joined in the attempts to explain this phenomenon. The first and only respectable theory came from a middle-aged mathematician, André-Marie Ampère. He saw immediately that if the compass needle were shielded from the earth's magnetic field it would align itself perpendicularly to the current-carrying wire, which implied that the magnetic force was circular around the wire. From this it followed that magnetism was nothing else but electricity moving in circles. His first "theory" was that permanent magnets contained electrical currents rotating around their axes. If this were true, then certain other effects, such as heating, should accompany these currents, and they did not. Ampère then suggested that these currents circulated around the molecules of the magnet where the gross effects of electricity would not be evident. To decide between his two hypotheses, Ampère devised a simple experiment. He wound a coil of wire with many turns and connected the ends to a battery. Inside this wire, he placed, in his terms, *une lame* of copper turned into a circle. He argued that if the *lame* were affected by a permanent magnet brought up to the *lame* while current flowed in the coil, this would prove that the currents of electricity flowed around the molecules and not around the axis in permanent magnets. In January 1822, he did the experiment with a weak bar magnet and failed to detect an effect. In the summer of that year, he

repeated the experiment using a more powerful horseshoe magnet and found what he wanted. In his report to the Academy of Sciences he casually remarked that this showed that an electrical current could be induced by a magnet and dismissed the remark by saying, "But this is of no theoretical interest." This may have been of no interest to Ampère, but it certainly was to the rest of the scientific world that was searching for precisely this effect. Those who read Ampère's report tried to repeat the experiment but failed. Why?

Ampère's description of his experimental apparatus should serve as an object lesson to all experimentalists. Terms must be made crystal clear. *Lame* in French is generally translated as a sheet of solid metal, not as a ribbon, which is what Ampère turned into a circle. Great experimenters of the time, such as Michael Faraday, tried to reproduce the experiment by using a *disc* of solid copper and got no result whatsoever. Ampère's "discovery" was therefore summarily dismissed. Only after Faraday reported his discovery of electromagnetic induction did Ampère claim priority, but it was not to be given. He had not convinced anyone at the time and it is Faraday who is, quite rightly, credited with the first induction of an electric current by magnetic force.

Faraday had long believed in the unity of force, and from 1820 on he had sporadically attempted to induce an electrical current with magnets. Like Ampère, he had mapped the position of the compass needle around a current-carrying wire and found it to be circular. This struck him as both unique and important, for forces, up until then, had always been assumed to act in straight lines between the bodies involved. For Faraday, a circular "line of force"—the term is his—necessarily implied that the "force" emanating from the current in the wire was transmitted by the intervening medium and did not act at a distance. Could this supposition be proven by experiment? Faraday tried for years, with no success. It was in the spring of 1831 that he began to see how it might be done. His friend Charles Wheatstone, the inventor of the Wheatstone bridge, was primarily interested in the science of sound. Faraday asked him if he would give a lecture to the members of the Royal Institution of Great Britain, of which Faraday was the director, on the interesting patterns created by bowing glass or iron plates upon which light powder had been sprinkled. Wheatstone was working on these figures, and Faraday became interested in them. What he discovered after many months was that bowing one plate could cause patterns on that plate *and also on a neighboring one.* This was, if you will, acoustical induction. Immediately after his researches on sound, on August 29, 1831, Faraday wound a thick iron ring with two coils of wire. The first was connected to a voltaic battery, and the second was connected to a galvanometer. The iron ring served as the magnetic medium by which, he hoped, the electric current in the primary circuit would create a current in the circuit connected to the galvanometer. When he threw the switch on the primary circuit, the galvanometer jumped. Faraday had created an induced current from magnetism.

See also: AMPÈRE, ANDRÉ-MARIE; AMPÈRE'S LAW; ELECTROMAGNETIC INDUCTION, FARADAY'S LAW OF; ELECTROSTATIC ATTRACTION AND REPULSION; FARADAY, MICHAEL; FARADAY EFFECT; FRANKLIN, BENJAMIN; GALVANOMETER; LIGHTNING

Bibliography

GOODING, D., and JAMES, F. A. J. L. *Faraday Rediscovered* (Macmillan, London, 1985).

MARTIN, T. *Faraday's Discovery of Electro-Magnetic Induction* (Edward Arnold, London, 1949).

WILLIAMS, L. P. *Michael Faraday: A Biography* (Basic Books, New York, 1965).

L. PEARCE WILLIAMS

ELECTROMAGNETIC INDUCTION, FARADAY'S LAW OF

One of the fundamental laws of electromagnetism is the principle that an electromotive force (emf) is produced in an electric circuit in the presence of a time-varying magnetic field. That is, the galvanometer in circuit A will indicate a current when the current in circuit B changes, either by varying the resistance R or by replacing the dc battery $\mathcal{E}_0$ with an ac generator $\mathcal{E}(t)$, thereby creating a time-varying magnetic field in the vicinity of both circuits (see Fig. 1). Alternatively, the magnetic field in the vicinity of circuit A can be varied by moving circuit B, or by moving the permanent magnet C. Equivalently, circuit A can be moved with respect to a stationary circuit B or magnet C. The latter case is often called

a motional emf. The emf, or driving voltage, is said to be induced in circuit A. The term magnetic flux refers to the product of the magnetic field times an area, such as the area of circuit A. Quantitatively, Faraday's law states that the emf induced in a circuit is equal to the time-rate-of-change of the magnetic flux that passes through, or links, the circuit (with a coefficient of proportionality that depends on the system of units chosen).

This phenomenon, known as electromagnetic induction, was discovered in 1831 by Michael Faraday in London and, independently, by Joseph Henry in Albany, New York. Faraday also introduced the concept of lines of force as a symbolic way of picturing the magnetic field produced by an electric current (or of picturing the electric field produced by electric charge and by Faraday induction). Lines of force are drawn to follow the local direction of the vector field. For instance, Fig. 2 schematically illustrates some lines of magnetic field **B** produced by a current loop *I*. Furthermore, lines of force can be drawn (or imagined) in such a way that their density in a local region of space represents the local strength of the field—lines are close together where the field is strong, and far apart where it is weak. In this construction the magnetic flux that links circuit A in Fig. 1 can be measured by counting the number of lines, produced by circuit B or magnet C, that thread through the loop of circuit A. Now, typically, the magnetic field strength decreases with distance from a current loop or magnet. Increasing the current in circuit B creates new lines, which move out from the wire to become denser. All the cases described above, in which the flux changes through

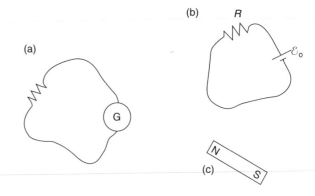

Figure 1 (a) Secondary circuit with galvanometer to detect induced current. (b) Primary circuit with current driven by emf $\mathcal{E}_0$. (c) Permanent magnet.

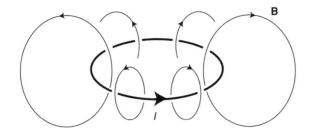

Figure 2 Representative lines of force of magnetic field **B** in the vicinity of a circuit loop carrying current *I.*

circuit A, have either lines of force moving with respect to the circuit, or the circuit moving with respect to the lines. Using Faraday's model, therefore, we can say that electromagnetic induction occurs when magnetic lines of force cut across a conductor, or when the conductor cuts across magnetic lines. (When a circuit moves in a region of uniform magnetic field, there is no change of flux and no induction. As many lines of force are "cutting in" as are "cutting out.")

Faraday induction forms the basis for the electric power industry, including the electric generators that convert mechanical energy (from fossil or nuclear fuels, water power, and so on) into electrical energy, and the transformers that change the ratio of voltage to current in the distribution system. It is also of importance in ac circuits: the magnetic field due to the changing current in one circuit induces a voltage in another, and vice versa. The two circuits are said to be inductively coupled, and the strength of the coupling is measured by the mutual inductance between the two circuits. A single circuit induces into itself, measured by its self-inductance. A coil of many turns of fine wire approximates a lumped inductor; that is, a device with strong internal magnetic field but relatively weak external field, which greatly increases the total self-inductance in its own circuit while minimizing the mutual inductance to neighboring circuits. A transformer is a pair of coils, usually wound on a ferromagnetic core, with a very large mutual inductance. When a time-varying magnetic field passes through an extended conductor, such as a sheet of metal, eddy currents are induced in the sheet.

The polarity of an induced emf is given by Lenz's law, which says that the effect opposes the cause. For instance, in the case of self-inductance, the partial current driven by the induced emf opposes the change in the original current that causes the induction; hence the effect is often called a "back-emf." An inductor is often used in filter circuits to suppress undesired time-variation of current.

The formal mathematical expression of Faraday's law illustrates the fundamental concepts of integral and differential calculus, together with an interesting exercise in vector geometry. Figure 3 illustrates a circuit loop made up of differential elements $d\ell$ [each element is a vector, having a direction and (infinitesimal) magnitude]. Imagine a surface that is bounded by the circuit loop (there are many such surfaces—pick any one). This surface consists of oriented elements of area **n** dA, where **n** is the unit vector normal to the element dA in the sense given by a right-hand rule (fingers curl around the loop in the direction of $d\ell$, and thumb points in the sense of **n**). The region in the vicinity of the circuit is filled with the magnetic field **B,** which is a vector function having magnitude and direction at each point. Now, the magnetic flux Φ_m that links the circuit can be written as the surface integral:

$$\Phi_m = \int_{surface} \mathbf{B} \cdot \mathbf{n} \, dA. \qquad (1)$$

That is, we sum up all the constituent products of **B** $\times$ dA, where the scalar ("dot") product counts only the component of the magnetic field that is parallel to **n** (i.e., normal to the patch of area dA). The formal statement of Faraday's law then is that the emf $\mathcal{E}$ is the derivative of the flux with respect to time:

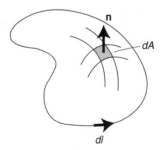

Figure 3 Geometry of magnetic flux: a surface consisting of elements dA is bounded by an electrical circuit consisting of elements $d\ell.$

$$\mathcal{E} = -\frac{d\Phi_{\mathrm{m}}}{dt}. \tag{2}$$

The negative sign is an expression of Lenz's law. If desired, the emf can be expressed as the line integral of the (Faraday) electric field **E** around the circuit:

$$\mathcal{E} = \int_{\mathrm{perimeter}} \mathbf{E} \cdot d\boldsymbol{\ell}. \tag{3}$$

That is, we sum up all the constituent products of **E** times $d\boldsymbol{\ell}$, where **E** is a vector function of position, and again the scalar ("dot") product counts only the component of the electric field that is parallel to $d\boldsymbol{\ell}$.

The integral form of Faraday's law [Eq. (2)] can be transformed to a partial differential equation, applying at each point in space, by using the curl operation of vector calculus:

$$\nabla \times \mathbf{E} = -\frac{\partial \mathbf{B}}{\partial t}. \tag{4}$$

This form of Faraday's law is one of the set of four equations constituting Maxwell's equations, which together are the fundamental statement of classical electromagnetism.

See also: ELECTRIC GENERATOR; ELECTROMAGNETIC INDUCTION; ELECTROMOTIVE FORCE; LENZ'S LAW; MAGNETIC FLUX; MAXWELL'S EQUATIONS; TRANSFORMER; VECTOR

Bibliography

PURCELL, E. M. *Electricity and Magnetism*, 2nd ed. (McGraw-Hill, New York, 1985).
TRICKER, R. A. R. *The Contributions of Faraday and Maxwell to Electrical Science* (Pergamon, New York, 1966).

MARK A. HEALD

ELECTROMAGNETIC INTERACTION

See INTERACTION, ELECTROMAGNETIC

ELECTROMAGNETIC LEVITATION

See LEVITATION, ELECTROMAGNETIC

ELECTROMAGNETIC RADIATION

Charged particles produce electric and magnetic fields that permeate the space around them. If the particles undergo acceleration, a part of the field detaches itself from the source charge and propagates out as an electromagnetic wave, carrying energy, momentum, and angular momentum. These detached fields are called electromagnetic radiation. The power radiated by a single point charge q with acceleration a is given by the Larmor formula: $P = \mu_0 q^2 a^2 / 6\pi c$, where μ_0 is the permeability of free space and c is the speed of light in a vacuum.

The nature of the acceleration serves as one way to characterize the radiation. A particle that slows down as it passes through matter emits bremsstrahlung (braking radiation); a particle traveling in a circle emits synchrotron radiation; a charge undergoing sinusoidal oscillation (as on a spring) emits dipole radiation (responsible for the blue sky); charges in random thermal motion give off heat radiation (blackbody radiation). When a charged particle travels through a transparent medium at a speed greater than that of light (in the medium), Cerenkov radiation is produced (in this case the particle itself need not be accelerating).

Electromagnetic radiation occurs over an enormous range of frequencies (ν) and wavelengths ($\lambda = c/\nu$), as indicated in Table 1. At low frequencies the source is typically an oscillating current; for example, household wires, which carry alternating current, radiate (very weakly) at 60 Hz, and an AM radio station drives currents up and down the antenna at around 1 MHz. Microwaves are produced by magnetrons and klystrons. Any warm object emits heat radiation in the infrared, and, if hot enough, the object can give off visible light (as in the filament of an ordinary light bulb). Electrons that are accelerated through a potential difference of several thousand volts and then hit a piece of metal will generate x rays by bremsstrahlung, and circular electron accelerators give off synchrotron radiation in the x-ray region.

In principle, all electromagnetic radiation is quantized, in the form of photons, though in practice the classical electrodynamics of James Clerk Maxwell ordinarily provides an adequate approximate account. But when the radiation is produced by discrete quantum transitions, its atomic-scale transitions, its quantum nature cannot be ignored. The photon emitted carries the energy (E) released in the transition, and the frequency of the radiation is given by Planck's law: $E = h\nu$ (where $h = 6.63 \times 10^{-34}$ J·s, which is Planck's constant). Hyperfine transitions typically generate radiation in the microwave region (such as the famous 21-cm line in hydrogen), as do molecular rotations; molecular vibrations typically radiate in the infrared; transitions between energy levels of the outer electrons in atoms generally radiate in (or near) the visible region, whereas inner electron transitions are in the ultraviolet or near x-ray range; nuclear transitions produce x rays, and elementary particle processes yield gamma rays. Particularly intense radiation in the microwave and visible region is produced by masers and lasers, respectively.

The means for detecting electromagnetic radiation also vary according to the frequency. At extremely low frequencies an oscilloscope connected to an antenna and a ground will do the job. To enhance the effect, and to select a particular frequency, the signal can be passed first to a resistor-inductor-capacitor (RLC) circuit. (An AM radio is, in essence, a tunable RLC circuit with a mechanism—diode and earphones—for converting the electrical signal to audio form.) Microwaves can be detected by the current they generate in an absorbing crystal. For visible light we use photographic emulsions, photomultiplier tubes, charge-coupled devices (ccd's), or the eye itself. Ultraviolet is most easily detected using fluorescent materials (which absorb ultraviolet and reradiate in the visible range). Photographic techniques can be used in the x-ray region. Gamma rays are most commonly detected by the secondary particles (such as electron-positron pairs) that they produce when passing through matter.

Electromagnetic radiation has many applications. Radio waves and microwaves are widely used in communication, from television to satellite links and cellular phones. Microwaves are also used in radar, and—tuned to a rotational transition in the water molecule at around 2.5 GHz—in microwave ovens.

Table 1 Electromagnetic Radiation

Type	Frequency (Hz)	Wavelength (m)	Photon Energy (eV)
	10^{22}	10^{-13}	10^{7}
Gamma rays	10^{21}	10^{-12}	10^{6}
	10^{20}	10^{-11}	10^{5}
X rays	10^{19}	10^{-10}	10^{4}
	10^{18}	10^{-9}	10^{3}
	10^{17}	10^{-8}	10^{2}
Ultraviolet	10^{16}	10^{-7}	10
	10^{15}	10^{-6}	1
Visible	10^{14}	10^{-5}	10^{-1}
	10^{13}	10^{-4}	10^{-2}
Infrared	10^{12}	10^{-3}	10^{-3}
	10^{11}	10^{-2}	10^{-4}
Microwave	10^{10}	10^{-1}	10^{-5}
	10^{9}	1	10^{-6}
TV, FM	10^{8}	10	10^{-7}
	10^{7}	10^{2}	10^{-8}
AM	10^{6}	10^{3}	10^{-9}
	10^{5}	10^{4}	10^{-10}
Radiofrequency	10^{4}	10^{5}	10^{-11}
	10^{3}	10^{6}	10^{-12}
ELF	10^{2}	10^{7}	10^{-13}
	10	10^{8}	10^{-14}

Infrared radiation emitted by heat lamps keeps food (and people) warm and allows for night vision goggles. Light itself makes possible not only sight but also photosynthesis and photovoltaic (solar) cells. Ultraviolet finds application in fluorescent lighting and the sterilization of hospital equipment, while x rays have familiar medical and dental applications. In addition, all forms of radiation are used in scientific research, from spectroscopy to astrophysics.

At low intensities, electromagnetic radiation below ultraviolet frequencies is relatively harmless (although very intense radio waves, especially those with wavelengths comparable to the size of the body, can produce deleterious internal heating; exposure to intense microwaves can damage the eye, strong infrared can cause burning, and extremely bright visible light—as from a laser or the Sun—can injure the retina). But ionizing radiation, from ultraviolet on up, is hazardous even at quite low intensity; it causes burns, mutations, and cancer. The Sun emits strongly over the entire the ultraviolet region, but only the low frequencies reach Earth's surface because wavelengths up to about 2.9×10^{-7} m are absorbed by a layer of ozone (O_3) about fifteen miles up in the atmosphere. Recent indications that the ozone layer is being inadvertently depleted by human activities are therefore reason for serious concern.

See also: BREMSSTRAHLUNG; ELECTROMAGNETIC WAVE; HYPERFINE STRUCTURE; LASER; RADIATION, BLACKBODY; RADIATION, SYNCHROTRON; RADIATION, THERMAL; RADIO WAVE; X RAY

Bibliography

HEALD, M. A., and MARION, J. B. *Classical Electromagnetic Radiation*, 3rd ed. (Saunders, Fort Worth, TX, 1995).

TSIEN, R. Y. "Pictures of Dynamic Electric Fields." *Am. J. Phys.* **40**, 46 (1972).

DAVID GRIFFITHS

ELECTROMAGNETIC SPECTRUM

James Clerk Maxwell's equations are accepted universally as the relativistically covariant formulation of the (observed) laws of electromagnetism. Maxwell's equations for the electromagnetic field in vacuum (no matter) and free space (no boundaries) allow transverse waves of all frequencies; thus the electromagnetic frequency spectrum, ν, extends from zero to infinity. The index to Maxwell's treatise does not contain the entry "spectrum" or "spectral," and while it lists "Fourier," the references are not to Fourier analysis. With a constant velocity of light c, the wavelength λ, associated with each frequency is given by $\lambda = c/\nu$. This entry is a historical overview of the generation, transmission, and detection of the electromagnetic radiation spectrum, and the acceptance that electromagnetic radiation is described by Maxwell's equations. Maxwell's equations are linear vector partial differential equations (PDEs), difficult to solve in general because they are vector PDEs; however, because they are linear, sometimes amenable to treatment by transform techniques (e.g., Fourier and Laplace transforms and/or Green's functions), this treatment often suffices for transmission problems. For the generation of electromagnetic radiation Maxwell's equations require sources. For fixed sources, those that are not affected by the radiation field that they generate, the fields are obtained by convoluting the source term with a Green's function. When the source is not fixed, Maxwell's equations become a set of nonlinear vector PDEs, and the linear techniques are inappropriate. The nonlinear case is omnipresent in laser physics.

For the detection of low-frequency electromagnetic radiation, one generally uses antennae feeding into devices with filters and amplifiers. The antenna must respond on a time scale less than $2\pi/\nu$ s, and so it is generally metallic, and for a wire antenna should have length $L \geq \lambda/2\pi$.

The first analysis of a portion of the electromagnetic spectrum was made by Issac Newton, who used the refractive properties of a prism to spatially spread out or disperse the optical spectrum emitted by the Sun. This occurred because after the solar radiation passed through the perpendicular surface of the prism, it struck the diagonal rear surface obliquely, and because the index of refraction varied as a function of wavelength, the light passing into air was bent to a greater or less extent depending on wavelength. This demonstration was part of a chain of reasoning that allowed Newton to overthrow a theory of the constitution of light that had come down from antiquity. After Newton, optics, the study of the electromagnetic spectrum and phenomena between 0.400 microns and 0.760 microns, or 4,000–7,600 Å, became a well established, separate branch of physics. Since the eye responds to the optical spectrum, it seems natural it would be the first portion of the electromagnetic spectrum to be studied. But how is the eye like an antenna?

In the eighteenth and early nineteenth centuries, experiments in electrostatics and magnetostatics, in effect, established three of Maxwell's equations, Coulomb's law, $\vec{\nabla} \cdot \vec{D} = 4\pi\rho$, Ampere's law, $\vec{\nabla} \times \vec{H} = (4\pi\vec{J})/c$, and $\vec{\nabla} \cdot \vec{B} = 0$, which do not depend explicitly on time and do not show a connection between the separate fields of electricity and magnetism. Such a connection is provided by a constitutive equation such as Ohm's law, $\vec{J} = \sigma \cdot \vec{E}$; however, the connection disappears at $\sigma = 0$. Clearly Ohm's law cannot be a law in the sense of the aforementioned three. The fourth equation, Faraday's law of electromagnetic induction, $\vec{\nabla} \times \vec{E} = -(\partial/\partial t)\,\vec{B}/c$, both relates electric and magnetic fields and introduces a time dependence and is clearly different from, and more complicated than, Ohm's law. Michael Faraday discovered his law of induction by wrapping two independent coils around an annular iron core, with one coil attached to a battery and switch, and the other to a galvanometer. On opening and closing the switch, Faraday observed a galvanometer response. While this effect is time dependent, it is a transient (decreases as an exponential function with a negative argument), for which a spectral analysis is inappropriate. The above four laws became Maxwell's equations when he modified Ampere's law to $\vec{\nabla} \times \vec{H} = [4\pi\vec{J} + (\partial/\partial t)\vec{D}]/c$, incorporating in it the continuity equation $\vec{\nabla} \cdot \vec{D} = \partial\rho/\partial t$.

The electrical engineering industry that developed in the third quarter of the nineteenth century used the physics in Maxwell's equations in generators, motors, and transformers, and in telegraphy and telephony, before Maxwell's treatise was published. While generators based on an armature rotating in a magnetic field produced an electrical signal at the frequency of rotation, and while there was an interest in spectral resolution for multiplexing telegraphy (sending many telegraph messages simultaneously, at different frequencies), it was the age of direct current (dc) in electrical technology: Thomas A. Edison's initial power plants for operating incandescent lights generated 110 V dc power. At the same time, Maxwell's theory, based on the same electromagnetic technology, is an enormous extrapolation from the relatively low frequency (very long

wavelength) phenomena, measured at his time, to the whole spectrum. In his treatise Maxwell began the integration of optics into electromagnetism by showing that the experimental propagation velocities for light and electromagnetism were the same and by attempting to show that the dielectric constant of a transparent medium was the square of its index of refraction. But, as mentioned above, Newton's spectral analysis relied on a wavelength-dependent difference in the index of refraction (it was not constant). Maxwell estimated an infinite wavelength index of refraction for paraffin by extrapolating from optical measurements and showed reasonable agreement between the index of refraction of paraffin and the square root of its dielectric constant. Maxwell argued that this agreement showed the identity of the electromagnetic medium and the luminiferous ether. Maxwell attempted to show a direct connection between optical and electromagnetic fields via the effect of Faraday rotation due to a magnetic field on the passage of polarized optical radiation through a birefringent crystal. This attempt was misguided, however, as there are acousto-optic effects in which sound waves affect the propagation of light, without implying that light is a longitudinal wave. It is difficult to see how a convincing case could be made for the applicability of Maxwell's theory to the whole electromagnetic spectrum if there was a luminiferous ether affecting measurements in an unknown way, and it is easy to see how Albert Einstein's theory of relativity, discarding the luminiferous ether, made twentieth-century physics possible.

For the electrical engineering industries that evolved from those of the 1870s, the full Maxwell's equations were largely irrelevant. On the other hand Heinrich Hertz in Germany in the 1880s attempted to test Maxwell's equations experimentally (Hertz accepted Maxwell's equations but rejected Maxwell's theory); for example, he tried to test the equivalence of $\vec{J}$ and $(\partial/\partial t)\vec{D}$ in $\vec{\nabla} \times \vec{H} = [4\pi\vec{J} + (\partial/\partial t)\vec{D}]/c$. To do this he developed resonant circuits with induction coils, with inductance L, and spark gap capacitors, with capacitance C, so that the current induced in a secondary coil would be a damped oscillation at the resonance frequency $\omega = 2\pi\nu = 1/(LC)^{1/2}$, providing $R^2 C/L \geq 4$, where R is the circuit resistance. Hertz was able to calculate the resonant frequency, and by using a sufficiently long straight wire in his secondary circuit, Hertz observed nulls and antinulls. Thus he measured the wavelength of the radiation, and showed that the product of the wavelength and frequency was close to the velocity of light, determined optically. Hertz showed that a single resonant circuit produced the same pattern of nulls and anti-nulls in free space, discovered that the power emitted per unit solid angle did not decrease significantly with distance of the detector from the source, and observed the diffraction of his waves around obstacles. Hertz verified Maxwell's equations, showed the existence of radiation propagating through free space, and provided the scientific fundamentals for modern radio and television. But he also verified that Maxwell's equations applied to the electromagnetic spectrum up to frequencies as high as 1.25×10^9 c/sec (1,250 Mc/sec), that is, wavelengths as short as 24 cm. It is probably impossible today to convey a sense of the impact of Hertz's work on European scientific thought at the close of the nineteenth century; but, for example, in the ten-volume collected works of the great mathematician Henri Poincare, half of Volume 10 is devoted to Poincare's papers on Hertzian waves.

While radio physics was emerging from the laboratory, other major discoveries (puzzles) in physics were made at the close of the nineteenth century: the electron, x rays (in 1896), and radioactivity. Arnold Sommerfeld points out that only in 1905 was the transverse nature of x rays proven. Clearly, while Hertz had established the probability that Maxwell's equations applies to the whole electromagnetic spectrum, sceptical physicists still demanded experimental proof.

One puzzle that was explained during this period was that of the spectrum of blackbody radiation. This was in the well-established subdiscipline of physics called "light and heat." A blackbody was a controlled source of radiation with perfectly absorbing (black) exterior walls. By Gustav Kirchhoff's law in thermal physics, the ratio of the power emitted to the power absorbed by any body is independent of the material, independent of wavelength, and dependent only on the temperature of the body. A blackbody is the ideal since it completely absorbs all incident radiation, and by Kirchhoff's law must be the maximum emitter at all wavelengths. By 1900 there had been two centuries of development of optical measurements, and almost a century of thermal measurements. With a small hole in the blackbody, a prism to disperse the optical radiation, and a small, calibrated, calorimeter or thermocouple to measure the spatially (spectrally) resolved intensities, one could spectrally resolve the absolute intensity of the blackbody radiation. This is a different radiation detector from the antenna discussed

earlier but similar to the eye. It was found that at low frequencies the power emitted by the blackbody was proportional to the first power of the temperature and the square of the frequency. One obtains this result from a model (the Lord Rayleigh–James Jeans formula) by counting the number of modes (standing wave patterns) in a cubical box with perfectly reflecting walls in a spectral region $d\nu$ about ν, proportional to ($\nu^2 d\nu$), and assigns an average energy kT, to each mode. The temperature is T, k is Ludwig Boltzmann's constant, and the average energy is that found from a Maxwell–Boltzmann distribution for an ideal gas. The reasoning here requires a "suspension of disbelief," as it requires that the modes be replaced by charged "oscillators," which may be replaced by an uncharged ideal gas. Since a blackbody is independent of material, the interior may be a vacuum. But then the ideal gas has no relation to the gas in the blackbody. The blackbody radiation must be generated at the walls. But the walls are not an ideal gas. At high frequency the blackbody power emitted was proportional to $\nu^3 e^{-C\nu/kT}$, where C is a constant. An explanation was proposed by Wilhelm Wien, who assumed an ideal gas oscillator model wherein the output power per unit frequency was proportional to the product of frequency and number of oscillators at that frequency, given by the Maxwell–Boltzmann distribution for an ideal gas. Max Planck proposed a new distribution connecting the low- and high-frequency regions, replacing the factor $e^{-C\nu/kT}$ in the Wien formula with the factor $1/(e^{h\nu/kT}-1)$, where h is Planck's constant. This leads to the equation for the spectral energy density in a blackbody:

$$\frac{d\varepsilon_\nu}{d\nu} = \frac{8h\pi}{c^3} \cdot \frac{\nu^3}{e^{h\nu/kT}-1}, \qquad (1)$$

which reduces to the Rayleigh–Jeans formula at low frequency and the Wien formula at high energy. The blackbody radiation story is vastly richer than this sketch; for example, Planck's new distribution antedates the Rayleigh–Jeans formula, and Planck's distribution requires a quantum hypothesis. The spectral energy density in Eq. (1), though developed from measurements in the infrared and optical regions, like Maxwell's equations has no restriction on frequency. The Stefan–Boltzmann equation for power radiated by a blackbody is $P = \sigma T^4$, where the experimental σ is the Stefan–Boltzmann constant. Integrating the spectral energy density times the velocity of light (which is the radiated power per unit area) over all frequencies gives the Stefan–Boltzmann expression with a prediction from theory of the Stefan–Boltzmann constant. The prediction agrees reasonably well with the experimental value. This, in effect, is an experimental proof that Maxwell's equations, which apply to the low-frequency blackbody spectrum, apply also to the optical region, and by implication to the whole electromagnetic spectrum. Note that there are two temperatures in the above discussion, the "color" temperature associated with the shape of the spectral distribution, and the "brightness" temperature in the Stefan–Boltzmann equation; these may differ as the blackbody may not be black at high frequencies.

While Planck's distribution described the blackbody emission spectrum, its physical foundations were continually under attack. In 1916 Einstein proposed a new approach to the underlying physics using the old quantum theory, that is, replacing the "oscillators" by identical atoms with a spectrum of energy levels. The radiation field, $\rho(\nu, T_R)$, and the atoms were assumed to be in thermal equilibrium, with the probability of occupation of the energy level ε_j by an atom being $g_j \cdot e^{-\varepsilon_j/kT}$, where g_j is a statistical weight factor. Today T, in the exponential, would be called the ion temperature, and T_R the radiation temperature, but if the atoms and radiation field were in equilibrium the temperatures would be equal. To maintain equilibrium the number of transitions from the atomic level m to another level n, of lower energy level, must be zero. The transition rate from the lower level to the higher (photoabsorption) is $B_{nm} N_n \rho$, where ρ is the radiation field density at the energy difference $\varepsilon_m - \varepsilon_n$, B_{nm} is an absorption coefficient from level n to m, and N_n is the number of atoms with energy ε_n. The transition rate from the higher level to the lower is $A_{mn} N_m + B_{mn} N_m \rho$, where A_{mn} is the spontaneous emission rate (Einstein introduces it as an analog to radioactive decay) and B_{mn} is the stimulated emission rate. Equating the rates leads to

$$A_{mn} \cdot N_m + B_{mn} \cdot N_m \cdot \rho = B_{nm} \cdot N_n \cdot \rho, \qquad (2)$$

which with the statistical relation

$$N_n/N_m = (g_n/g_m) \cdot e^{(\varepsilon_m - \varepsilon_n)/kT}, \qquad (3)$$

leads to

$$A_{mn} \cdot g_m = \rho \cdot (B_{nm} \cdot g_n \cdot e^{(\varepsilon_m - \varepsilon_n)/kT} - B_{mn} \cdot g_m). \,(4)$$

Einstein argues that as T goes to infinity ρ will go to infinity, and to satisfy the equation one must have the term in parenthesis go to zero, whence $B_{nm} \cdot g_n = B_{mn} \cdot g_m$, relating the absorption and stimulated emission coefficients, so that

$$\rho = \frac{A_{mn}/B_{mn}}{e^{(\varepsilon_m - \varepsilon_n)/kT} - 1}. \qquad (5)$$

In this clear and simple derivation, which has had an enormous impact on physics, Einstein obtained the Planck distribution with $h\nu = \varepsilon_m - \varepsilon_n$, introduced the idea of spontaneous emission, which is indispensible to the derivation, introduced the idea of stimulated emission, which is the scientific basis for masers and lasers, and reemphasized the importance of real atoms in the blackbody emission process. Einstein's derivation, too, is not without its critics. Since the atom number densities drop out of Eq. (4), does the derivation require only one atom in the blackbody? If more than one, how many? Today the blackbody radiation field is treated as a photon gas obeying Bose–Einstein statistics (the Planck distribution with a chemical potential to allow the photon number density to be variable) with the chemical potential set equal to zero because the photons do not interact. However, the treatment calls for "some matter, however little," that is, as vague as Einstein's original paper. Second, the infinite temperature–infinite photon density argument Einstein uses in going from Eq. (4) to Eq. (5) is dubious. As one increases the photon densities multiphoton processes become important, that is, one must replace $B_{nm}N_n\rho$ with $B_{nm} \cdot N_n \cdot \rho(h\nu) + N_n \cdot \int ds \cdot B_{nm}^2 \cdot \rho[h(\nu - s)] \cdot \rho(hs)$ for two photon absorption, which invalidates the use of rate equations linear in the photon density. The nonlinear terms have the effect of coupling photons of different frequency, invalidating the assumption of zero chemical potential. Third, to fit the measured blackbody distribution, one requires $A_{mn} = (8h\pi/c^2) \cdot \nu^3 \cdot B_{mn}$, which suggests that spontaneous emission is a form of stimulated emission. A sceptical reader of the literature comes away with the sense that blackbody theory amounts to fitting Eq. (5) to a measured blackbody distribution obtained for a narrow range of temperature, frequency, and photon density. With today's high-power lasers and nanotechnology, this old field of blackbody radiation could undergo a renaissance.

Marconi applied Hertz's results to the long-range propagation of radio waves. This is possible because of plasma layers in the ionisphere, which will reflect radiation with frequencies below $\nu_c = 28,500 \sqrt{n_e} \text{ cm}^{-3}$, where n_e is the electron density. The ionisphere and the conducting ocean act as the walls of a waveguide enabling intercontinental communications. Given the curvature of Earth, diurnal variations, and the solar cycle, there was significant effort to map n_e for the upper atmosphere. For a peak electron density of $10^6/\text{cm}^3$ the cutoff frequency is about 30 Mc/sec (10 m), well above the AM band, well below the FM band (87–108 Mc/sec or ≈ 300 cm), and in the middle of the short wave band. The VHF television band is 54–88 Mc/sec for channels 2–6, and 174–216 Mc/sec for channels 7–13; the UHF television band is 470–890 Mc/sec. Because of the high information transmission rate in television, each channel is alloted 6 Mc/sec bandwidth. In addition to reflection from the ionisphere there was static (noise) at frequencies above and below ν_c. Karl Jansky, with a detector at 20.5 Mc/sec, identified the noise as radiation from the Milky Way. This discovery led to radio astronomy. Thirty years later, with a horn antenna at 4,080 Mc/sec, Arno Penzias and Robert Wilson discovered the cosmic microwave background, a crucial element in contemporary cosmology.

While short wavelength radiation would not propagate by reflection from natural electron layers, it would be reflected from high electron density, human-made objects, that is, aircraft. This is the idea behind radar. The associated microwave ($\lambda \approx 0.03 - 30.0$ cm) technology had an enormous development during World War II. This technology greatly extended the communications bandwidth (with waveguide networks) and provided powerful new tools for physics experiments. The measurement of the Lamb shift with microwave techniques at 1,062 Mc/sec triggered the development of quantum electrodynamics. Today the use of artificial satellites as intermediate receivers and transmitters allows long range propagation up to optical frequencies. The maser and laser, based on Einstein's idea of stimulated emission, were developed in the 1950s and since then the laser has been extended in power to 10^{14} W and in wavelength to 35.6 Å.

Finally, research on the physics of the eye extends back over 150 years. On the retina there are rods (sensitive to black and white, that is, integrating over spectrum) and cones (sensitive to colors), of radius $\approx 10,000$ Å so that $L \gg \lambda/2\pi$, each cone with one of several different pigments. Each pigment maximally absorbs light at a different wavelength, feeding a signal to nerve fibre, and thence to the optic nerve. Details of this can be found in textbooks on the physiology of the eye, and the story of the evolution of the eye in mammals can be found in textbooks on developmental biology. Except for the eye, the body does not respond well to the electromagnetic spectrum. The dangers of excessive x-ray ($\lambda \approx 0.1$–10.0 Å) and ultraviolet radiation ($\lambda \approx 100$–4,000 Å) doses are well known, and high standards of eye safety are set for those working with lasers. Recently, there has been significant public contraversy over safety standards for microwave radiation and the possibility of cancer engendered by proximity to 50 and 60 c/sec near-in fields of transmission lines and cables. Arguments based on physical reasoning tend to dismiss the possibility of a significant interaction between the body and such low frequency fields.

See also: AMPÈRE'S LAW; COULOMB'S LAW; CURRENT, DISPLACEMENT; EINSTEIN, ALBERT; ELECTROMAGNETISM; FARADAY, MICHAEL; LIGHT; MAXWELL'S EQUATIONS; NEWTON, ISAAC; RADIATION, BLACKBODY; RADIATION, THERMAL; RADIATION PHYSICS; RADIO WAVE; VISION

Bibliography

BRODEUR, P. *Currents of Death* (Simon & Schuster, New York, 1989).

CLAYTON, D. C. *Principles of Stellar Evolution and Nucleosynthesis* (University of Chicago Press, Chicago, 1983).

KUHN, T. S. *Black-Body Theory and the Quantum Discontinuity, 1894–1912* (Oxford University Press, New York, 1978).

MAXWELL, J. C. *A Treatise on Electricity and Magnetism* (Dover, New York, 1954).

PLANCK, M. *The Theory of Heat Radiation* (Dover, New York, 1959).

SOMMERFELD, A. *Electrodynamics* (Academic Press, New York, 1952).

WESTFALL, R. *Never at Rest* (Cambridge University Press, Cambridge, Eng., 1980).

WESTFALL, R. *The Life of Isaac Newton* (Cambridge University Press, Cambridge, Eng., 1994).

ZEMANSKY, M. *Heat and Thermodynamics* (McGraw-Hill, New York, 1957).

EUGENE J. MCGUIRE

ELECTROMAGNETIC THEORY

See LIGHT, ELECTROMAGNETIC THEORY OF

ELECTROMAGNETIC WAVE

The physical process by which electromagnetic energy is transported through space or material substance is the electromagnetic wave. "Electromagnetic radiation," the classical term by which such energy is known, is produced whenever an electric charge undergoes accelerated motion. It also can be produced when aggregates of excited atoms or nuclei lose some of their internal energy. In such instances, however, the quantum theory of radiation must be used to describe the phenomenon fully. The spectrum of electromagnetic radiation extends from extremely long, very-low-frequency (VLF) radio waves, which often serve for maritime communication, to the extremely short, very-high-frequency cosmic rays that permeate interstellar space. It encompasses such seemingly diverse phenomena as gamma rays and x rays, ultraviolet and visible light, infrared radiation, microwaves and radar, and television and radio communication. These phenomena, which are ubiquitous in modern life, are all governed classically by the electromagnetic wave equation. They differ only in the wavelength λ or frequency ν that characterizes them.

In 1865, guided by the observations of a fellow Londoner, Michael Faraday, British physicist James Clerk Maxwell formulated a set of four equations that completely and concisely described the behavior of electric and magnetic fields. He observed that these equations could be combined into a single wave equation that predicted the propagation of electromagnetic radiation at the speed of light. Maxwell's formulation, as described in *Treatise on Electricity and Magnetism* (1873), is completely general, encompassing all forms of disturbances, including spherical, cylindrical, and plane waves. In all cases, the electromagnetic wave is characterized by time-varying electric and magnetic fields that are perpendicular to each other and to the direction of propagation of the wave. Furthermore, the amplitude of the time-varying electric field is greater than the amplitude of the magnetic field by a factor of

$1/\sqrt{\mu\varepsilon}$, where μ is the magnetic permeability of the medium in which the wave is traveling and ε is the dielectric permittivity. The factor $1/\sqrt{\mu\varepsilon}$ is the speed of propagation of the wave (wave velocity).

A plane-wave approximation, appropriate when the source of the disturbance is far from the observation point, serves to illustrate the principles of Maxwell's theoretical conclusions. If such a wave propagates in a vacuum along the $+z$ direction, the wave equation for the electric field takes the form

$$\frac{\partial^2 E_x}{\partial z^2} = \mu_0 \varepsilon_0 \frac{\partial^2 E_x}{\partial t^2},$$

where the source of the disturbance has been ignored and, for simplicity, the electric field has been taken as linearly polarized along the x axis. The electric vector then displays a harmonic behavior given by

$$\mathbf{E} = \mathbf{i} E_0 \sin(\omega t - kz),$$

where $\omega = 2\pi\nu$ is the angular frequency, $k = 2\pi/\lambda$ is the wave number or propagation constant, E_0 is the amplitude of the oscillation, and $\mathbf{i}$ is the unit vector along the x direction. The magnetic vector is similarly given by

$$\mathbf{B} = \mathbf{j} B_0 \sin(\omega t - kz),$$

where B_0 is the amplitude of its oscillation and $\mathbf{j}$ is the unit vector along the y direction. Figure 1 illustrates the appearance of the wave.

According to the form of the wave equation, the disturbance in a vacuum must propagate at a speed c given by $1/\sqrt{\mu_0 \varepsilon_0}$, where μ_0 and ε_0 are, respectively, the permeability and permittivity of free space. The amplitudes of the electric and magnetic oscillations are related by $E_0 = cB_0$. Evaluation of c yields a result of $\sim 3 \times 10^8$ m/s, which, as Maxwell noted, is the speed of light in a vacuum.

The instantaneous power or energy per unit time transported by an electromagnetic wave through a unit area perpendicular to the direction of propagation is given by the Poynting vector $\mathbf{S} = \mathbf{E} \times \mathbf{B}/\mu$. For a plane wave traveling in a vacuum, S reduces to $S = EB/\mu_0$. Its time-averaged value I, sometimes called the intensity, is $\varepsilon_0 c E_0^2/2$.

An electromagnetic wave also carries momentum, which can be observed in the form of radiation pres-

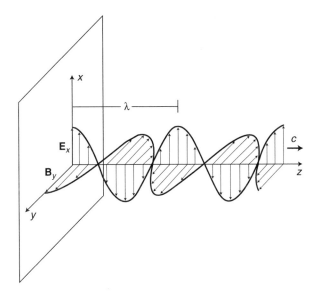

Figure 1 A linearly polarized plane wave propagating in the $+z$ direction, showing the sinusoidal behavior of the fields and the relative orientation of the electric and magnetic vectors.

sure. For the example of a plane wave in a vacuum, the momentum density dp/dV is given by S/c^2. The average pressure exerted by the radiation on a surface is then I/c for the case of total absorption and $2I/c$ for total reflection.

Although there are exceptions, synchrotrons and radio transmitters among them, most sources of electromagnetic radiation produce waves that are not polarized. For this general case, the electric field is randomly oriented in the plane perpendicular to the direction of propagation. Most sources also produce radiation that is incoherent, the phase of the electric field varying randomly in time across the radiation pattern. The laser is perhaps the most prominent exception. Its ability to generate coherent radiation allows it to deliver beams of light with extremely high brightness and small angular divergence. While unpolarized radiation can be polarized—by devices that selectively transmit or reflect electric fields in a given orientation—incoherent radiation cannot be made coherent in a similar fashion.

Maxwell is widely credited with establishing a sound mathematical and physical base for the properties of electromagnetic waves. Even though the interference and diffraction studies of Thomas Young, Augustin Fresnel, and Joseph von Fraunhofer, among others, had provided significant, if not overwhelming, evidence that light possesses wave-like properties, it

was Maxwell's seminal analysis that finally led to the universal acceptance of this principle. It is interesting to note that there is nothing in the wave equation he derived that requires a medium for the propagation of the radiation. Nonetheless, the scientific community's familiarity with mechanical waves, such as vibrating strings, water, and sound, had prejudiced it to demand the existence of a "luminiferous ether" as part of the price for accepting the wave theory of light. It took the null results of the Michelson–Morley experiments (1881–1887) to dispel this notion.

Electromagnetic waves pervade modern society: x rays are widely used in medicine and materials testing; lasers are found in supermarket scanners, compact disc players, fiber-optic telecommunications links, surgical tools, and military guidance systems; infrared radiation provides heating for homes, signatures of the earth's ecosystem, and targeting for heat-seeking missiles; microwaves are used for cooking food and in long-distance telephone communication; radar tracks air and sea craft, storm systems, and speeding automobiles; radio-frequency waves are used for telecommunication and astronomical observation; visible light needs no elaboration. The sources and detectors of electromagnetic waves are as numerous as their applications.

See also: DECAY, NUCLEAR; ELECTROMAGNETIC RADIATION; ELECTROMAGNETIC SPECTRUM; MAXWELL'S EQUATIONS; WAVE MOTION

Bibliography

HALLIDAY, D.; RESNICK, R.; and KRANE, K. S. *Physics*, 4th ed. (Wiley, New York, 1992).

HECHT, E., and ZAJAC, A. *Optics*, 2nd ed. (Addison-Wesley, Reading, MA, 1987).

JACKSON, J. D. *Classical Electrodynamics*, 2nd ed. (Wiley, New York, 1975).

YOUNG, H. D. *University Physics* (Addison-Wesley, Reading, MA, 1992).

MICHAEL S. LUBELL

ELECTROMAGNETISM

It had long been suspected that the phenomena of electricity and magnetism were connected to each other. But these connections began to be explored systematically only about 200 years ago. In a relatively short time thereafter, the combined sciences of electricity and magnetism were put on firm and sophisticated mathematical foundations. By 1864, when James Clerk Maxwell published his famous equations, he recognized that optics, which till then had been regarded as yet another branch of physics, could also be brought within the compass of his theoretical scheme. Considered as a consequence of Maxwell's equations, ordinary light turns out to be but a very narrow part of the full range of wavelengths of the electromagnetic spectrum. Radiant heat, ultraviolet and x rays, microwaves and radio waves, all constitute different parts of this spectrum. Every contemporary technological innovation owes the central aspects of its development to the synthesis that Maxwell codified. Moreover, in combination with quantum mechanics, the science of electromagnetism explains, and often gives a quantitative account of, nearly all everyday phenomena of the physical world; the major exception we ordinarily encounter is gravity. Electromagnetism, because of its wide reach and fundamental significance, is a major branch of theoretical physics. It is essentially the study of the interactions among electric charges. These interactions are understood to be mediated by electric and magnetic fields.

"Electric charge" is a primitive concept in electromagnetism and is therefore not derivable from previously familiar concepts. There are, however, some familiar phenomena whose explanation motivates the introduction of the idea of charge: If you walk on a carpeted floor on a dry winter day and touch a door knob, sometimes you experience a shock. Here we say that the objects involved are electrically charged. Empirical investigation of the behavior of charged objects reveals that they exert forces on each other. Moreover, the detailed study of these forces allows us to quantify the concept of charge.

Charge is an additive property of matter, and the amount of charge (on an object, or carried by an object) can be positive, negative, or zero. The current view of ordinary matter is that it is made up of microparticles (electrons, quarks), most of which are charged. Like-sign charges repel each other, while unlike-sign charges attract each other. The force of attraction or repulsion decreases inversely as the square of the distance separating the charges and is directed along the line connecting them. All of the above properties can be summarized mathematically in what has come to be called Coulomb's law:

$$\mathbf{F}_{1,2} = \frac{k_e Q_1 Q_2}{(r_{1,2})^2} \hat{\mathbf{r}}_{1,2}$$

where $\mathbf{F}_{1,2}$ is the force due to Q_1 on Q_2, $\mathbf{r}_{1,2}$ is the vector distance from Q_1 to Q_2, $\hat{\mathbf{r}}_{1,2}$ is the unit vector pointing from Q_1 to Q_2, and k_e is a constant.

Charge is conserved. If the net charge in any region of space is changing in time, it can only be through the flow of charge into or out of that region; net charge cannot be created or destroyed. Creation and destruction of equal amounts of positive and negative charges at one place is permitted.

Charge is invariant. The amount of charge on an object is the same whether the observer is at rest with respect to the object, or in any state of motion with respect to it. This rather more esoteric property is to be distinguished from conservation, which refers to the behavior of charge in time relative to a single observer. Conservation is fairly commonly used in elementary applications of the theory, whereas invariance is used in more advanced work discussing the form of electromagnetism in special relativity.

Charge is quantized. For reasons that are not deeply understood to date, in nature charges seem to occur only in integer multiples of a basic unit. This unit turns out to be one-third of the charge carried by the proton. This observation again does not find much use in most elementary discussions of the theory.

Returning now to Coulomb's law, upon reflection you might find it a bit mysterious, since it posits that a charge in one place acts directly on another charge some distance away. Today we may not be struck by the strangeness of objects exerting forces on each other without physical contact. But historically, Coulomb's law was viewed with grave suspicion on this score by physicists. In the case of electromagnetism, an alternative view, initiated by Michael Faraday in the early nineteenth century, was fruitfully developed to avoid distant actions. In this view, the charge Q_1 creates an electric field in all of space surrounding it, and the electric field, in turn, acts upon any other charge Q_2 at the location of the latter. The electric field created by a charge Q at a point a distance r away from it is given by $\mathbf{E} = (k_e Q / r^2)\, \mathbf{r}$, and the force on a charge q at a point where the electric field is $\mathbf{E}$ is given by $\mathbf{F} = q\mathbf{E}$. This combination clearly reproduces Coulomb's law. In the old days, the electric field (as well as the magnetic field) was imagined as an actual displacement of a subtle medium called ether. Thus the nineteenth century physicists could avoid talking about action at a distance by imagining that charge causes displacements in a medium that pervades even apparently empty spaces, and other charges elsewhere are affected by this displacement. This idea was carried further to associate energy and momentum with these hypothesized displacements in ether. It was found that not including the energy and momentum of the field would run afoul of the crucial conservation laws associated with these quantities.

Since Einstein's special theory of relativity (1905) dispenses with the ether as conceived in the nineteenth century, it might at first seem to take away the substantial basis for the concept of fields and thus spell serious trouble for the whole scheme. But the following considerations make acceptance of fields all the more compelling. If we wish to avoid the mysterious action at a distance, we have to take the fields seriously even in the absence of a mechanical support for them. Besides, in the spirit of special relativity, a charge at one location cannot instantaneously exert a force on a distant charge as given by Coulomb's law, since that would imply an infinite speed of propagation of an influence. One of the fundamental consequences of relativity is that forces (which are the means by which physical systems exchange energy and momentum) cannot propagate at speeds greater than the speed of light in a vacuum. The suggested modifications of Coulomb's law when moving charges are involved were already built into the nineteenth-century theory. Indeed, Einstein was motivated in large part by such considerations about Maxwell's theory when he discovered the theory of relativity.

This brings us to the question of moving charges. A moving charge constitutes an electric current. Nevertheless, we could also have a current in a neutral wire, if at every place there are equal and opposite charges, but say, only the negatively charged electrons are moving. There is a certain amount of charge crossing every section of the wire per unit time, but no part of the wire has a net charge. Hans Christian Oersted's accidental discovery in the early nineteenth century that a current-carrying wire affects the orientation of a magnetic compass needle led to the view that currents produce magnetic fields in their neighborhood, and these fields in turn exert forces on magnets and currents. Today we think of magnets themselves as constituted of persistent microscopic currents. In the microscopic realm, too, a neutral particle such as the neutron exhibits magnetic properties, because its constituent quarks are in motion. Since the magnetic field depends on the motion of charges, even motion at constant velocity, different observers in relative motion will find different magnetic fields. Thus what to one observer might seem to be a purely electric field, will be a combination of electric and magnetic fields to another. In the spirit of relativity, then, one speaks of the electromagnetic field rather than electric and magnetic fields separately. Though it requires a bit more mathematics than is assumed in this essay, an invariant formulation of Maxwell's theory is most economical and elegant.

Unfortunately, without a bit of vector calculus, the qualitative description of the magnetic fields due to currents is rather more complicated than the description of the electric field due to a point charge. A long, straight wire carrying a current produces a magnetic field that decreases as the first power of the distance away from the wire, and points around the wire. A tightly wound, long cylindrical coil of wire carrying a current produces a nearly uniform magnetic field in the interior of the coil and nearly zero field well outside the cylinder. Such a cylindrical coil behaves, as far as its magnetic field properties are concerned, just like a cylindrical bar magnet with a north pole at one end and a south pole at the other end. When a bar magnet is cut in two, one does not end up with a north pole piece separate from a south pole piece; instead one has two shorter magnets each with both poles. Similarly, if the long cylindrical coil were so reduced that one had only a single circular turn of wire carrying a current, one face of it would act like a north pole and the other face like a south pole. Indeed, a picture of permanent magnets as containing persistent atomic currents fits well with the observed absence of isolated poles. However, it should be noted that some of the best theories of microscopic matter require the existence of monopoles, but no reliable observational evidence for them is at hand.

The magnetic field, like the electric field, exerts a force on a charged particle. However, the magnetic force is proportional to the velocity of the charge, the quantity of charge, and the strength of the field. Besides, the magnetic force is in a direction at right angles to both the velocity of the charge and the direction of the magnetic field. Thus a charge at rest in a magnetic field, or one moving in the direction of the magnetic field, experiences no magnetic force.

There is yet another feature of electromagnetism that is logically independent of the foregoing properties. If a magnet is moved near a closed loop of wire, there is an electric current induced in the loop, just as if the loop were cut open and a battery connected across its ends. More generally, if the magnetic field in a certain region of space is varying in time, an electric field is induced in that region of space. And reciprocally, if an electric field is changing in time, a magnetic field is induced thereby.

Consider the following thought experiment that brings together many of the central features of electromagnetism. Imagine a neutral lump of matter in space. Let the negative charges briefly move apart from the positive charges and come back to their original configurations. Thus, except for a brief period of time, in this set-up there are no charges or currents. But during that brief period of time, the separating charges constitute a current that produces a magnetic field in the vicinity. Moreover, since the positive and negative centers of charge are not quite on top of each other, during that period there is also an electric field in that region. Further, since both the distribution of charge and current are changing in time, the fields they produce also vary in time. The time-dependent fields in turn create additional fields. The equations that Maxwell wrote down predict that this dance of electric and magnetic fields (typically) spreads out away from the original lump of matter at a speed determined by the theory. This speed turns out to be just the speed of light in a vacuum! After some time has elapsed, there is in space only a spreading pattern of electric and magnetic fields. They sustain themselves. Their sources (charges and currents) have ceased to exist. If another lump of matter is placed at the appropriate place, the electric and magnetic fields will cause the negative and positive charges in this second lump to separate. The arrival of the fields can be detected. In qualitative outline, this is how radio waves (and light, radiant heat, microwaves, and x rays) propagate. The original lump of matter is a transmission antenna, and the second is the reception antenna. The field picture has allowed us to give a causally continuous account of what goes on. If we avoid talking about the fields because they are not directly observed, we would have a rather more spooky picture of delayed action-at-a-distance. However, in the quantum theory, there are serious doubts cast on the ability of physical theory to give continuous causal accounts of phenomena. Even though the debate on this question is by no means settled, in the contemporary world of theoretical physics (which, with no apologies, is philosophically opportunistic), the idea of fields, and especially the theory of the electromagnetic field, continues to be very fertile.

See also: COULOMB'S LAW; ELECTROMAGNETISM, DISCOVERY OF; FIELD, ELECTRIC; FIELD, MAGNETIC; MAGNETIC POLE; MAXWELL, JAMES CLERK; MAXWELL'S EQUATIONS; RELATIVITY, SPECIAL THEORY OF

Bibliography

DAVIES, P. C. W. *The Forces of Nature,* 2nd ed. (Cambridge University Press, New York, 1986).

PURCELL, E. M. *Berkeley Physics Course,* Vol. 2: *Electricity and Magnetism* (McGraw-Hill, New York, 1973).

TIPLER, P. A. *Physics for Scientists and Engineers,* 3rd ed. (Worth, New York, 1991).

KANNAN JAGANNATHAN

ELECTROMAGNETISM, DISCOVERY OF

Two fundamental questions were raised in the Scientific Revolution of the seventeenth century: What was the nature of matter, and what caused material bodies to move? The first set of answers was offered by René Descartes. For Descartes the only essential quality of matter was extension in space. Therefore the cosmos was completely filled with matter and this aspect allowed Descartes to explain how bodies moved. They moved because they were pushed by neighboring pieces of matter; hence, all motion was the result of impacts. The influence of Descartes was great, for it permitted natural philosophers of the day to create a clear, mechanical picture of the universe. Unfortunately, when attempts were made to reduce the view to mathematics, it did not work. It was Isaac Newton who proved mathematically that the theory did not fit the observations of the motions of bodies. In his great work *Philosophiae Naturalis Principia Mathematica,* Newton laid down the three laws of motion that formed the foundation of modern physics. But what was matter and how did bodies interact? In *Principia* (1687) and *Opticks* (1704 in Latin, 1706 in English) he offered his answers. Matter consisted of hard, impenetrable, elastic atoms. Bodies were composed of these atoms and two bodies acted upon one another by gravitation, whose force varied inversely as the square of the distance between their centers. Precisely what this force was, Newton did not know, but he was able to express it in his famous law of universal attraction. This, of course, did not rule out impact; it merely broadened the horizons of physics.

The Newtonian system rapidly conquered the scientific world of the day. In the eighteenth century other areas besides ponderable matter were brought under its sway. Newton, himself, had led the way. His theory of light was interpreted as based on the idea of corpuscles of light whose differences in size accounted for the optical spectrum. Heat, in the second half of the century, was handled as a material, corpuscular substance that, unlike ponderable matter, was associated with a repulsive force, thus accounting for expansion with rise in temperature. In the 1770s Charles Coulomb in France showed that both electricity and magnetism obeyed Newton's inverse square law, and most natural philosophers simply assumed that they were corpuscular with two kinds of electricity (positive and negative) and two kinds of magnetism (north and south). In all these cases, force seemed to act at a distance across empty space and in straight lines between the centers of the bodies involved.

The first major challenge to this viewpoint came not from the scientists of the day but from a philosopher, Immanuel Kant. His view was not based on physics but upon epistemology, the study of how we know. He argued that all that we can sense is force in the form of attraction or repulsion. Thus, I know I am attracted to the earth because I can weigh myself, and I know repulsion from the fact that I cannot push my finger through my desk. From there, Kant went on to insist that matter did not consist of solid bodies that we could never know but of force, and force filled the cosmos. Where before there had been empty space, there was now a plenum of forces which could, under different circumstances, become manifest as electricity, magnetism, heat, light, and even gravity. Find the right circumstances, and forces could be changed into one another. These ideas are essentially the foundations of field theory. At the time, only a few natural philosophers took them seriously.

Among those (mostly in the Germanic world) who did was Hans Christian Oersted, who was obsessed by the idea of the convertibility of forces. In 1800 Alessandro Volta invented the voltaic pile that made current electricity available. For many years,

natural philosophers had been trying to see if electricity and magnetism were the same but could get no definitive results with static electricity. Oersted, as early as 1813, felt that current electricity *must* have some magnetic effect. It was not until the spring of 1820 that he was able to show that, indeed, it did. In the summer, he announced to an astonished world the fact that a current deflected a compass needle. What was truly astonishing was that, instead of pointing along a current-carrying wire, the needle pointed at a significant angle from it. Everywhere in every laboratory, men repeated Oersted's experiment and confirmed it. It was in France that the effect was clarified by André-Marie Ampère who, by eliminating the effect of the earth's magnetism, demonstrated that the needle lined up *perpendicular* to the wire. In England, Michael Faraday, an analytical chemist, at the request of a friend who edited a scientific journal, repeated all the experiments that had flourished in the wake of Oersted's announcement. While carefully mapping the positions taken by the needle around the wire, Faraday discovered that it was a circle and immediately leaped to the hypothesis that a single magnetic pole should, then, rotate around the wire. With his extraordinary experimental and manipulative skill, he was able to construct a little apparatus to illustrate this.

How could this extraordinary action be explained? To Faraday, it could not be explained by Newtonian central forces for here was clearly a curved "line of force," to use Faraday's expression. This phenomenon began Faraday's life-long pursuit of the nature of lines of force. With Kant and Oersted, Faraday believed in the unity and convertibility of force, and indeed, he was to illustrate it by his discoveries. But increasingly central to all his discoveries was his belief in the reality of the line of force.

What could a line of force possibly be? It could not, certainly, be the path of central forces that, according to Newton and the Newtonians, always acted in straight lines and at a distance. Since lines of force could be detected everywhere in the presence of electric charges, electric currents, and magnets, action-at-a-distance had to go. Faraday was also convinced that lines of force represented strains, because cutting them, as in electromagnetic induction, transformed the magnetic force into an electrical current. But what could be strained in empty space?

It should be remembered that Faraday was a chemist, and chemists needed singularities such as chemical elements and could not deal, as had most of the eighteenth-century Newtonians, in such abstractions as matter and space. Suppose atoms themselves were merely mathematical points surrounded by shells of alternating attractive and repulsive forces whose combinations led to unique and complex patterns of force such as the chemical elements must possess to account for their unique chemical properties. With such a supposition, many chemical and crystallographic phenomena could be understood and so, thought Faraday, could the line of force. In electrical science, the lines seemed to have ends, and this indicated that the line of force was a strain in the chain of contiguous particles, for all particles were linked by their forces. In magnets, however, the lines were closed curves and could not be explained in the same manner. Particles had to go, and Faraday made the daring and, at the time completely outrageous, suggestion that the strains existed in space itself. This was the first scientific field theory and it was to be the foundation upon which later scientists would build.

By the time Faraday had completed his field theory, important changes had taken place in the very nature of physics. Mathematics, which had been a tool in the eighteenth century for the expression of physical laws obtained by experiment, now, largely as the result of two generations of brilliant French mathematical physicists, became an instrument of discovery. The enunciation and acceptance of the principle of the conservation of energy in the mid-nineteenth century provided a quantitative measure by which physical (and all natural) processes could be analyzed. And, particularly in Britain, the growth of engineering focused attention on mechanical models as possible sources for the understanding of the subsensible world. All these influences came together in a young Scot, William Thomson (later Lord Kelvin). Thomson had corresponded with Faraday and was interested in Faraday's experimental results but could not accept his theories. Faraday's physics did not conform to the Newtonian laws of mechanics. The first problem for Thomson was to find some solution in Newtonian mechanics for Faraday's effects. He was confident that there had to be a material substance that made up the field, and he began to explore mathematically the properties that such a substance must have in order to propagate electrical and magnetic forces. There was immediately on hand the luminiferous ether that had been seen to be necessary in order to carry the light waves examined in mathematical detail by Augustin Fresnel in the second decade of the nineteenth century.

Because these waves were shown to be transverse, the ether had to have a very high coefficient of elasticity like that of tempered steel. That raised the first difficulty: How could the planets move, seemingly without resistance, through such a medium? Thomson set out to find a way. His method was typical of British science at the time. He made a mechanical model of the ether and then investigated its properties with sophisticated mathematics. His equations, in turn, pointed out possibly profitable new directions to take. Kelvin was influential but unsuccessful in his search. The same was not the case for his younger Scottish colleague, James Clerk Maxwell. It was Maxwell who created classical field theory.

Maxwell followed very much the same track as had Thomson. Also a skilled mathematician and acquainted with the mechanics of machines so prominent in the Industrial Revolution, and keenly aware of the brilliance of Faraday's work, he set out to reduce Faraday's experimental results to a coherent mathematical theory in physics. The problem was that the medium required for light propagation had to have a number of different properties. It must have vortex or rotational powers to account for Faraday's discovery of the rotation of the plane of polarized light in a magnetic field. It must be capable of suffering longitudinal and transverse strains. It must be capable of producing electrical currents and magnetic forces.

Maxwell's method, like Kelvin's, was to try to find a mechanical model whose properties could be assumed that would roughly represent the phenomenon to be explained, and then use known laws drawn from mechanics to analyze the model. His first essays were crude. Faraday's lines of magnetic force, for example, could be represented by fine tubes of variable sections carrying an incompressible fluid. The direction and the magnitude of the magnetic force at any given point in the field would be given by the direction and magnitude of the velocity of this fluid. Since the science of hydrodynamics was well developed by this time, the model would yield precise mathematical solutions. A much more detailed and sophisticated analysis was presented to the Royal Society of London on December 8, 1864, entitled, "A Dynamical Theory of the Electromagnetic Field." It was in this paper that he first published, in crude form, what have come to be known as Maxwell's equations. There were twenty of them. (What are today called Maxwell's equations are actually equations written in the language of vector analysis, which had not been invented in Maxwell's time, by an unschooled, eccentric telegrapher of mathematical genius, Oliver Heaviside).

One of the most surprising and important results of Maxwell's work was the discovery that electromagnetic waves traveled at the same speed as did light. If, in fact, light and electromagnetic radiation were one and the same, then a most important consequence followed: Newtonian action-at-a-distance must be discarded and one of the fundamental bases of Newtonian mechanics denied.

In 1873 Maxwell published his classic *A Treatise on Electricity and Magnetism.* Suddenly scientists on the continent woke up to what was going on in Britain. In particular, the great Hermann von Helmholtz supported Maxwell, since the only alternative was a theory of electrodynamics that violated the principle of the conservation of energy that Helmholtz had first enunciated in 1847. Helmholtz realized that if electromagnetic waves could be detected it would provide a powerful support for the new theory. He set his student, Heinrich Hertz, on this problem. After many trials, Hertz was successful and showed in 1887 that radio waves acted exactly like light. Maxwell's field theory appeared triumphant!

Science is never quite that simple. Although Maxwell's mathematics provided an excellent picture of electromagnetic phenomena, there were still some difficulties. The major one turned out to be the ether. Either the ether moved with Earth or it did not. In either case, there were certain observable phenomena that ought to settle the question, but they did not. Only a direct approach seemed capable of deciding. This involved actually trying to observe directly whether the ether was static or not. It was this experiment, developed by the Americans Albert Michelson and Edward Williams Morley in the 1880s that appeared to be delicate enough to decide the question. Earth's motion through the ether could not be detected. Hence, the ether must move with Earth. But if this were the case, then certain well-known phenomena, such as stellar aberration, could not be accounted for. The result of all this was a crisis in physics. Between 1880 and 1905, all kinds of hypotheses were suggested in order to save the ether and Maxwell's theory. None were successful until a young patent clerk in Switzerland, Albert Einstein, published his epoch-making paper, *"Zur Elektrodynamik bewegter Korper"* ("On the Electrodynamics of Moving Bodies"). Here Einstein boldly threw out the whole basis of Newtonian mechanics and introduced his own special theory of relativity. Matter, as a separate entity, disappeared to be replaced by energy; the energy was to be found in the field, and that field was to be one in which the new principles of special and general relativity reigned.

See also: ACTION-AT-A-DISTANCE; ELECTROMAGNETIC FORCE; ELECTROMAGNETIC INDUCTION, FARADAY'S LAW OF; ETHER HYPOTHESIS; FARADAY, MICHAEL; FIELD LINES; INVERSE SQUARE LAW; LIGHT, ELECTROMAGNETIC THEORY OF; MAXWELL, JAMES CLERK; MAXWELL'S EQUATIONS; MICHELSON–MORLEY EXPERIMENT; NEWTON, ISAAC; NEWTONIAN MECHANICS; NEWTON'S LAWS; RELATIVITY, SPECIAL THEORY OF, ORIGINS OF

Bibliography

BERKSON, W. *Fields of Force: The Development of a World View from Faraday to Einstein* (Routledge and Kegan Paul, London, 1974).

HARMAN, P. M. *Energy, Force, and Matter: The Conceptual Development of Nineteenth-Century Physics* (Cambridge University Press, Cambridge, Eng., 1982).

HESSE, M. B. *Forces and Fields: The Concept of Action at a Distance in the History of Physics* (Nelson, London, 1961).

WILLIAMS, L. P. *The Origins of Field Theory* (Random House, New York, 1966).

L. Pearce Williams

ELECTROMOTIVE FORCE

Devices such as batteries, generators (ac and dc), and thermocouples in a temperature gradient that maintain a potential difference between two points are referred to as seats of electromotive force (emf). The term "electromotive force" does not stand for a mechanical force. It is the potential difference between two terminals of an open circuit in the absence of a load.

In a circuit with stationary currents, emf is the algebraic sum of products of current I_i and resistance R_i of every branch of the circuit. In nonstationary cases, the sum $\sum I_i R_i$ is a good approximation of emf, provided the displacement current is small compared to the conduction current.

Emf can be provided by a number of different modes: electromagnetic (induction), thermoelectric, photovoltaic, and chemical as in batteries. Emf can vary with time in an alternating current circuit.

An ac generator provides the ac emf that changes its sign (direction) periodically, and its time average is zero.

See also: BATTERY; CURRENT, ALTERNATING; CURRENT, DIRECT; ELECTROMAGNETISM; THERMOCOUPLE; THERMOELECTRIC EFFECT

CARL T. TOMIZUKA

ELECTRON

In the late nineteenth century most physicists came to believe that electricity came in two forms: electrons, which were negatively charged and had a mass of 9.109534×10^{-31} kg, and a charge of $-1.602177 \times 10^{-19}$ C, and protons, which had a mass of 1.672623×10^{-27} kg and a charge of $+1.602177 \times 10^{-19}$ C. Atoms (and therefore molecules) were thought to be formed of combinations of electrons and protons. In the early 1930s, it was found that atomic nuclei (other than that of hydrogen) were composed of positive protons and neutral neutrons, which had mass of 1.675×10^{-27} kg and no charge. It was also found that positive electrons (positrons) existed (at least momentarily) with mass equal to that of the electron and charge equal to that of the electron, although positive.

A neutral atom, then, consists of a positively charged nucleus surrounded by orbital (negative) electrons. Nuclei are of the order of about 1/10,000 of the size of a typical atom, the rest of its volume being occupied by its orbital electrons. Conductors of electricity (usually metals) provide paths for the rapid transfer of electrons. Ions (positively and negatively electrically charged atoms or molecules in solutions) can also conduct electricity. Electricity can travel through air or other gases as a spark at atmospheric pressure, forced by a high-voltage source (a few thousand volts per centimeter of gap), or as a discharge at low pressure, as in a neon sign.

The number of electrons orbiting in an atom determine its chemical and optical characteristics.

Electrons exhibit spin. That is, they carry a specific quantum amount of angular momentum.

Electrons are copiously emitted by hot metals, and they can be accelerated in a high vacuum as cathode rays. These rays can be focused by electric and/or magnetic fields on a fluorescent screen in cathode-ray tubes. Related tubes are used in electron microscopes, computer monitors, and, of course, in television sets.

See also: ATOM; CHARGE, ELECTRONIC; ELECTRON, AUGER; ELECTRON, CONDUCTION; ELECTRON, DISCOVERY OF; ELECTRON, DRIFT SPEED OF; ELECTRON MICROSCOPE; INTERACTION, ELECTROMAGNETIC; NUCLEAR SIZE; POSITRON; PROTON

Bibliography

ANDERSON, D. L. *The Discovery of the Electron* (Van Nostrand, New York, 1964).

DAVID L. ANDERSON

ELECTRON, AUGER

An Auger electron is one type of free electron that has been emitted from an atom or an ion. The Auger electron is emitted by the rearrangement of the bound electrons in the initial atom or ion. The rearrangement occurs via the electron-electron interaction that produces a repulsive force that can overcome the positive attractive force due to the electron-nucleus interaction. The rearrangement, however, can occur only when there is at least one electron vacancy in a given energy level of the initial atom or ion and when there are at least two electrons in an energy level with a smaller binding energy than that of the vacancy. One of the loosely bound electrons falls into the more tightly bound vacant level while the other loosely bound electron is emitted as a free electron. By conservation of energy, the emitted Auger electron will have a well-defined kinetic energy, which is equal to the total binding energy of the initial atom or ion minus the total binding energy of the final ion.

A good example of an Auger electron is the electron emitted from doubly excited helium, He**. The energy of any state of He** is higher than that of He$^+$ (singly ionized He) plus a free electron. Therefore all states of He** can Auger decay. He** with both of its electrons in the $n = 2$ level, decays to the ground state of He$^+$, which has one electron in the $n = 1$ state, plus the emission of the other electron. The Bohr model predicts that the energy of the He** state is $E_i = 2 \times (-13.6\ Z^2/n_i^2)$ eV, neglecting screening of the nucleus by each electron, and the energy of the He$^+$ state is $E_f = -13.6\ Z^2/n_f^2$ eV. This yields the Auger energy $E_A = E_i - E_f = -27.2$ eV $- (-54.4$ eV$) = 27.2$ eV, using $Z = 2$, $n_i = 2$ and $n_f = 1$. This model neglects the effects of electron spin and angular momentum in addition to screening. When these effects are taken into account, He** with both electrons in the $n = 2$ level consists of many individual states, whereas He$^+$ (1s) has only one state. The Auger electron emission spectrum will therefore consist of many Auger electron energies. In this example, the Auger electrons are designated as *KLL* Auger lines, where the *K* refers to the *K*-shell ($n = 1$ level) of the initial vacancy and the *L*s refer to the *L*-shell ($n = 2$ level) of the initial states of the two electrons before Auger decay.

The energies of the Auger electrons are characteristic of the atom or ion and are therefore very useful in identifying atomic and ionic species. Auger electron emission is always in competition with x-ray emission. For example, a *K*-vacancy in an atom can be filled by the *KLL* Auger electron process, or by one electron filling the *K*-vacancy with the emission of a *K* x ray, which also conserves the total energy of the system. The Auger yield is defined as the probability for Auger decay and the fluorescence yield is defined as the probability for x-ray decay.

See also: AUGER EFFECT; ELECTRON; EMISSION; ENERGY LEVELS; NUCLEAR SHELL MODEL

PATRICK RICHARD

ELECTRON, CONDUCTION

The atoms in a metal are bound together by covalent bonds that are not highly directional and are spread out over many atoms. The most weakly bound (or valence) electrons of the atoms are therefore able to move throughout the metal. These mobile electrons, called conduction electrons, contribute to the electronic and thermal transport properties of metals.

Free Fermi Gas Model

For simple metals (e.g., Li, Na, Ka, Cs, Rb, Be, Mg, Ca, Sr, Ba, Al, Ga, In, Tl, and Pb), the conduction electron contribution to the thermal and electrical conductivity can be calculated by considering the conduction electrons as a gas of noninteracting fermions, and ignoring the potential energy due to the ionic cores. In this model the allowed energies of the conduction electrons are continuous, and there is a spherical Fermi surface at the Fermi energy ε_F.

Electronic Properties

When an external electric field is applied to a metal the conduction electrons begin to accelerate, but are retarded by collisions with impurities, phonons, and lattice imperfections. These processes produce a steady state where the average or drift velocity v of a conduction electron is

$$v = -eE\,\tau/m,$$

where e is the charge of an electron, E is the electric field, τ is the mean time between collisions (or relaxation time), and m is the mass of the electron. The expression for current density j then yields a form of Ohm's law:

$$j = nev = ne^2\,\tau E/m = \sigma E,$$

where n is the number of electrons per unit volume and σ is the electrical conductivity.

Since the collisions involve only electrons near the Fermi surface, the mean free path for conduction electrons l is

$$l = v_F\,\tau,$$

where v_F is the Fermi velocity (i.e., the electron velocity at the Fermi surface). Experimentally, very long (greater than 1 cm) mean free paths are observed in metals at low temperatures. This phenomenon is not explained by the free electron model, but is explained by a model that considers the perturbation of the valence electron wave functions by the periodic potential of the ionic cores. This model also predicts the formation of gaps between allowed energy bands that can cause the conduction electrons to behave as if they have effective masses larger or smaller than m, and charges of $-e$ or $+e$. This "band theory" also explains, qualitatively, the difference between metals, semiconductors, and insulators.

The density of states $D(\epsilon)$ for the free Fermi gas is

$$D(\epsilon) \equiv \frac{dN}{d\epsilon} = \frac{3N}{2\epsilon},$$

where ϵ is the energy of the electron and N is the total number of electrons with energy less than or equal to ϵ. The heat capacity $C_{\text{electronic}}$, at temperature T is given by

$$C_{\text{electronic}} = \frac{1}{3} \pi^2 D(\epsilon_F) k_B^2 T$$

$$= \frac{1}{2} \pi^2 N k_B \left(\frac{k_B}{\epsilon_F}\right) T = \frac{1}{2} \pi^2 N k_B \frac{T}{T_F},$$

where N is the number of conduction electrons, k_B is Boltzmann's constant, and T_F ($= \epsilon_F/k_B$) is called the Fermi temperature. The electronic contribution to the thermal conductivity $K_{\text{electronic}}$, in this model is given by

$$K_{\text{electronic}} = \frac{1}{3} C_{\text{electronic}} \, vl = \frac{\pi^2 n \, k_B^2 \, \tau T}{3m} \, .$$

In pure metals, the thermal conductivity is dominated by conduction electron transport. The phonon contribution can dominate in impure metals and disordered alloys.

The electronic contribution to the thermal properties calculated this way are of the right order of magnitude, but still differ from the observed values by a factor ranging from about 1/2 to about 2. This discrepancy is explained by the interactions of conduction electrons with the periodic potential of the lattice (band theory) and phonons, and interactions between conduction electrons, all three of which cause the effective mass of the conduction electrons to differ from m.

See also: COVALENT BOND; ELECTRON; ELECTRON, DRIFT SPEED OF; FERMI SURFACE; HOLES IN SOLIDS; INSULATOR; MEAN FREE PATH; OHM'S LAW; PAULI'S EXCLUSION PRINCIPLE; SEMICONDUCTOR

Bibliography

KEER, H. V. *Principles of the Solid-State* (Wiley Eastern Limited, New Delhi, India, 1993).

KITTEL, C. *Introduction to Solid-State Physics*, 6th ed. (Wiley, New York, 1986).

KEVIN AYLESWORTH

ELECTRON, DISCOVERY OF

The existence of "static electricity" was known in ancient times (amber or glass rubbed with fur or silk would attract bits of feather or paper), but electrostatic repulsion was not discovered until the 1600s, and the flow of electric charges (conductivity) was not discovered until 1729. The existence of two kinds of electricity (positive and negative) was deduced in 1733, and Charles Augustin Coulomb discovered the inverse square law governing electrical forces in 1785. In 1794 Luigi Galvani found that a flow of current resulted from touching dissimilar metals to dissected frogs' legs. Alessandro Volta substituted paper or cloth soaked with salt water or weak acids for animal tissue in 1801. By using stacks of metal disks ("batteries") he was able to give shocks. In 1820 Hans Christian Oersted discovered that a magnetic field surrounds any current-carrying wire. In the 1820s Georg Simon Ohm discovered his laws relating voltage, current, and resistance, and in the 1830s Michael Faraday discovered the induction of currents by changing magnetic fields and developed his laws of electrolysis.

Thus the principal laws governing electricity were known by 1850, but the conduction of electricity through low-pressure gases could not be studied satisfactorily until the invention of very good vacuum pumps by Heinrich Geissler in 1855. Before that time, it was known that electrical discharges through gases at low pressures produced a glow, the color of which depended on the gas as in a modern "neon sign." But with better vacuums, the dark space around the negative electrode (the cathode) extended further through the tube, even through the entire tube, if the vacuum were high enough. It was found that the glass at the opposite end of the tube would fluoresce because of "rays" that were emitted perpendicularly from the cathode—"cathode rays."

Eventually these rays were found to be (a) bendable by magnetic fields perpendicular to them, (b) independent of the kind of cathode metal, (c) capable of producing chemical reactions, (d) capable of heating thin foils (i.e., they convey energy), and (e) capable of exerting small forces.

In 1879 the distinguished British scientist William Crookes extended the earlier suggestion of Cromwell Varley that the rays were a stream of negative particles because of the direction in which they were bent by a magnetic field. He suggested that atoms hitting the cathode would pick up a negative charge and thus be repelled perpendicularly from the cathode. Crookes assumed that surrounding gas molecules would be "blown away" by the high speed torrent.

Crookes's theory accounted for all of the observable properties of cathode rays, except possibly for their ability to travel through even fairly high vacuums. (Air at 1/1,000 of atmospheric pressure would involve a collision about every hundredth of a centimeter if cathode rays were of molecular size.)

German physicists, however, pointed out that almost all of the properties of cathode rays were exhibited by electromagnetic waves such as light waves. Light waves (a) are emitted by hot (and some cold) surfaces, (b) travel in straight lines, (c) can cause glass to fluoresce, (d) exhibit properties (at least for hot emitters) that are independent of the nature of their source, and (e) can convey energy and exert forces. It is true that light is usually emitted at all angles from a source and is not bent by magnetic fields. But Maxwell's equations (about electromagnetic waves) were relatively new in the 1880s, and magnetic deflection was thought to be perhaps possible in some cases, for some kinds of light.

In 1883 Heinrich Hertz showed that most of the current in a cathode-ray tube did not necessarily follow the path of the cathode ray, and he carried out other experiments (plagued with difficulties) that he thought disproved the British theory that cathode rays were particles.

The English physicist Arthur Schuster, in 1884, wrote that he thought the rays were produced when molecules dissociated—the positive part being absorbed by the negative cathode, leaving the negative part to be repelled, forming the rays. He also knew that cathode rays of charge q and mass m, when moving in a perpendicular magnetic field of strength B, would move with velocity v in a circular path of radius R. Since the magnetic force on the particle would equal the centripetal force, he could write $Bqv = mv^2/R$, which can be rearranged to give

$$\frac{q}{m} = \frac{v}{BR}. \tag{1}$$

B and R were measurable, but v was not. Schuster was able to estimate upper and lower bounds for v, and thus suggested that q/m was less than 1×10^{10} C/kg, and more than 5×10^6 C/kg. He pointed out that hydrogen atoms in electrolysis were well within these limits.

In 1894 the young British physicist Joseph J. Thomson attempted to measure the velocities of the rays. [In the process he observed that the rays produced phosphorescence some feet from the cathode-ray tubes—no doubt due, we know now, to x rays. But x rays were not discovered (by Wilhelm Röntgen) until the end of the following year.]

In 1895 the French physicist Jean Perrin showed conclusively that cathode rays really do carry negative charges. Then in 1897 Thomson used Philipp Lenard's data (that cathode rays could penetrate air for short distances at atmospheric pressure) to argue that the rays must be *much* smaller than air molecules.

In 1897 Thomson performed experiments in which he sent cathode rays into a small receiver of known heat capacity for a given time. From the temperature rise, which was measurable, though small, he could determine the heat, H, transferred by n cathode rays and find an expression for v^2, from $H = n(1/2)mv^2$. From this, and the accumulated charge $Q = nq$, he could then find

$$\frac{q}{m} = \frac{2Qv}{H}. \tag{2}$$

Equations 1 and 2 can be combined to give numerical values for both v and q/m. Thomson found that v was around 30,000,000 m/s (about 1/10 the speed of light), and that q/m ranged from about 1.0 to 1.4 $\times 10^{11}$ C/kg.

Thomson also used a different method to determine the charge-to-mass ratio; namely, he sent the cathode-ray beam alternately through a perpendicular magnetic field and then (in the same space) through combined (crossed) magnetic and electric fields. By making the polarity and strength of the electric field the right size, the magnetic and electric deflections would cancel, and the ratio q/m could be calculated. His results were about 0.77×10^{11} C/kg. Eventually the discrepancies were found to be due to experimental errors.

After Thomson's work it was generally agreed that cathode rays were indeed constituents of all atoms, and have a mass close to 1/1,840 that of the hydrogen atom. They are, of course, used in cathode-ray oscilloscopes, and a vast number of electronic devices.

See also: CATHODE RAY; ELECTRON; FARADAY, MICHAEL; MAXWELL'S EQUATIONS; MILLIKAN, ROBERT ANDREWS; OHM'S LAW; OSCILLOSCOPE; RÖNTGEN, WILHELM CONRAD; STATICS; THOMSON, JOSEPH JOHN

Bibliography

ANDERSON, D. L. *The Discovery of the Electron* (Van Nostrand, New York, 1964).

MILLIKAN, R. A. *The Electron, Positive and Negative* (University of Chicago Press, Chicago, 1935).

THOMSON, J. J. *The Conduction of Electricity Through Gases* (Cambridge University Press, Cambridge, Eng., 1903).

DAVID L. ANDERSON

ELECTRON, DRIFT SPEED OF

Pure and doped semiconductors have "free" carriers (conduction electrons and holes) which are mobile and can move through the material fairly easily in response to an applied electric field. The numbers of conduction electrons and holes depends on the particular semiconductor, the amount and kind of doping, and the temperature, but typically there will be 10^{20} to 10^{26} electrons and/or holes per cubic meter in a useful semiconductor. In the absence of an electric field, the carriers move through the material in random directions so that there is no net electrical current.

If an electric field is applied, the carriers experience an electric force and are accelerated in the direction of the force, which gives rise to an electric current. However, the carriers interact with the atoms making up the substance by colliding with the atoms, and with defects such as impurities and dislocations. Such collisions tend to randomize the velocity of the electrons. Immediately after such a collision, the carrier is accelerated by the electric force and its velocity in the direction of the force continuously increases until the carrier undergoes another collision. Thus the electrons and holes have an average velocity in the direction of the electric force. This average velocity or drift speed of the carriers is thus determined by a balance between the electric force and the mean time τ between collisions. A simple equation can be derived for the drift speed V_D of the carriers:

$$V_D = \frac{e\tau E}{m^*}, \tag{1}$$

where E is the applied electric field in volts per meter, e is the electron charge (-1.6×10^{-19} C), and m^* the effective mass of the carrier. (Sometimes quite similar to the free electron mass $M_e = 9.1 \times 10^{-31}$ kg, but often much smaller. Electrons and holes in the same material may also have different effective masses.) The mean time between collisions is typically about 10^{-13} s at room temperature, and it becomes greater as the temperature is lowered. Note also that the heavier the electron (the larger the effective mass), the slower the drift speed for a given electric field, as a heavier object is harder to accelerate by the electric force. Equation (1) states that the drift speed is proportional to the electric field, so that for small electric fields the drift speed is also small. This proportionality does not hold for large electric fields (above about 10^5 V/m), and the drift speed saturates at about 40,000 m/s for germanium and silicon at room temperature.

Drift speeds of the carriers can be measured in a number of ways. A direct means of doing this is to use a point contact to inject free carriers at one point in the material, and then measure the time it takes the injected carriers to travel in the direction of the electric force to a voltage contact further down the sample. This kind of measurement was first carried out by James R. Haynes and William Shockley. A second method uses the Hall effect to determine drift speeds and is particularly useful if there is primarily only one kind of carrier (either electrons or holes). In this case, one can measure the transverse (Hall) voltage V_H across the width of the sample when there is a magnetic field H present. The Hall voltage is then given by

$$V_H = \frac{IB}{neh}, \qquad (2)$$

where I is the electric current in the material (in Amperes), H is the magnetic field applied in Tesla, h is the thickness of the material in meters, e is the charge on the carrier ($\pm 1.6 \times 10^{-19}$ C), and n is the carrier density (number of electrons or holes per cubic meter). Thus if the Hall voltage is measured, n can be determined. Then one can use the formula for the current density J in a conducting material

$$J = \frac{I}{A} = neV_D, \qquad (3)$$

where A is the cross-sectional area of the sample, to determine the drift velocity. This analysis is only correct when there is only one kind of carrier.

See also: CHARGE, ELECTRONIC; COLLISION; CONDUCTION; ELECTRICAL CONDUCTIVITY; ELECTRON; ELECTRON, CONDUCTION; FIELD, ELECTRIC; FORCE; HALL EFFECT; VELOCITY

Bibliography

HAYNES, J. R., and SHOCKLEY, W. "The Mobility and Life of Injected Holes and Electrons in Germanium." *Phys. Rev.* **81,** 835 (1951).

KITTEL, C. *Introduction to Solid-State Physics,* 6th ed. (Wiley, New York, 1986).

McKELVEY, J. P. *Solid-State and Semiconductor Physics* (Harper & Row, New York, 1966).

ROGER D. KIRBY

ELECTRON DIFFRACTION

See DIFFRACTION, ELECTRON

ELECTRONIC CHARGE

See CHARGE, ELECTRONIC

ELECTRONICS

Electronics is the field of science and engineering that focuses on the movement and control of electrons in devices. These devices form systems that are changing how we live, work, play, and communicate.

History

The term "electronics" comes from the German word *Elektronik,* first used in 1897 to describe the branch of physics that studied the electron.

In the 1920s, the adjective "electronic" and the noun "electronics" were first used to describe electron conducting devices. The first electronic device was the vacuum tube—an 8- to 10-cm tall quartz tube filled with a number of wire filaments and plates. One popular application using the vacuum tube was the radio, which amplified received electronic signals. Radios were composed of these bulky and unreliable vacuum tubes and required large batteries for portable operation.

In 1948, John Bardeen, Walter Brattain, and William Schockley invented the transistor. Because the transistor was smaller, used less energy, and was highly reliable, it started replacing the vacuum tubes in electronic systems. New products were invented. The new transistor radio had six transistors (as well as other electronic components such as diodes and capacitors), ran off small batteries, and was small enough to fit in a pocket.

In the early 1960s, another major advancement occurred when several transistors were connected together on a small single piece (die) of silicon. This discovery, plus improved manufacturing, allowed multi-transistor electronic circuits to be fabricated on a single die. At first, less than ten transistors were connected to make elementary logic functions like AND and OR. Then, up to one hundred transistors were interconnected to form complex operations like clocks and counters. Later, hundreds of transistors were fabricated together into more complex circuits, producing new products such as electronic calculators and special chips to store information (memories). At this stage, another major logic device was discovered—the microprocessor. It was similar to a calculator but could be programmed, making it more versatile.

In 1975, Gordon Moore, a pioneer in electronics, predicted that the number of transistors that could be fabricated on a die would double every two years. He was right, and this trend has continued for over twenty years, allowing previously unimagined numbers of transistors on a die.

In 1995, microprocessors contained more than nine million transistors; memories contained more than sixteen million transistors. Thanks to mass production and fabrication improvements, the cost of products has stayed relatively the same.

Circuits

Electronic systems are composed of circuits that are collections of transistors, resistors, and capacitors. Together they perform defined output responses. One output response they perform is amplifying signals. Many times the input signal to an electronic system is very small. For example, photon detectors (used to collect and convert light into electrical signals) can detect light from a distant star, but will give a signal of only one millionth of a volt. So, an amplifier is used to boost this signal. Figure 1 shows a circuit of a simple, single transistor amplifier that can boost an input signal several hundred times.

The circuit used to perform logic is a transistor switch, shown in Fig. 2. When the input is at ground potential, the transistor is nonconducting and is off. Since current cannot flow, the voltage at the output must rise up to the supply voltage. When the input voltage is equal to the supply voltage, the transistor is turned on, allowing a large amount of current to flow through the resistor and the transistor to ground. Because the resistance of a turned on transistor is small, the output voltage is small. If a large voltage (like the supply voltage) is defined as logical 1 or TRUE, and a small voltage as logical 0 or

FALSE, this circuit performs the NOT function. If the circuit is made with multiple inputs, this simple circuit can perform any logic function. Therefore, groups of these logic circuits can perform all logic operations.

The most popular use of the transistor is in making logic circuits, which are used to make decisions. For instance, if and only if A and B and C are TRUE, the output D is TRUE, then the AND function is performed. For example, A could be TRUE if the output of a sensor detects a person sitting on a car seat. B could be TRUE if a sensor indicates a seatbelt is not latched, and C could indicate if the engine is on. If the output D is TRUE, an alarm buzzer will sound. Electronic logic circuits can easily perform this operation.

Figure 2 shows an early version of the NOT gate that is no longer used because it took too much silicon real-estate (die area) to make the resistor. It's more efficient to use all transistors to make the circuit. By limiting the area that implements the logic gate, more logic gates will fit on a silicon die. As stated earlier, nine million transistors can be put on a 1-cm-square die (approximately the size of a thumb nail).

To design the circuit efficiently requires two types of transistors, called NMOS and PMOS. These stand for N channel and P channel metal-oxide semiconductor transistors, respectively. The N channel and P channel devices are complementary. The voltage required to turn on an NMOS device will cause a PMOS device to be turned off, and visa versa. Together, these devices are called complementary metal-oxide semiconductor (CMOS).

Figure 3 shows a circuit diagram of a two-input CMOS OR gate. An OR gate has the output logical 1 if any of the inputs are logical 1. When both in-

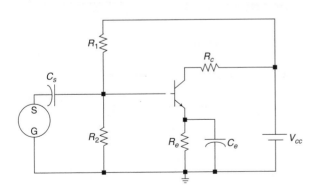

Figure 1 Transitor amplifier.

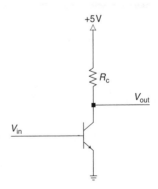

Figure 2 Digital switch.

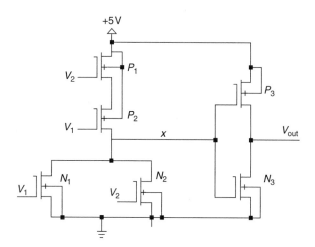

Figure 3 Two-input complementary metal-oxide semiconductor OR gate.

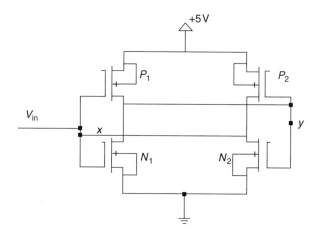

Figure 4 Complementary metal-oxide semiconductor static random access memory circuit.

puts are logical 0, the NMOS devices N1 and N2 are off and the PMOS devices P1 and P1 are on. P1 and P2 will increase the output to the supply voltage and the voltage at point x will be logical 1. The second part of the circuit is a CMOS inverter. When the voltage at point x is logical 1, N3 will be on, P3 will be off, and the output voltage will be logical 0. For any input voltage equal to logical 1, one of the combinations of N1 and N2 will be on and one of the combinations of P1 and P2 will be off, allowing the voltage at point x to be logical 0. The inverter will invert the signal at point x, making the output high for all these cases, implementing the OR gate. One application of an OR gate in computer programming is the command "push any key to continue."

The last example of a modern electronic circuit is the semiconductor memory element. This circuit helps store information in digital form. Figure 4 shows a CMOS static random access memory (SRAM) circuit. The circuit uses a teeter-totter (also called flip-flop) concept to store a single bit of information. The voltages at points x and y are complementary. If the voltage at point x is logical 1, then the voltage at point y will be logical 0. Therefore, N1 and P2 will be on and N2 and P1 will be off. Notice the self-reinforcement of this circuit. When the input voltage is removed, this circuit state will be retained. Therefore, the circuit is storing a logical 1 and will continue to hold this information until either a new bit is written or the power is turned off. Memories are a fundamental part of most electronic

systems and are used to store both instructions and data.

Applications

The availability of advanced, inexpensive electronic components has caused an explosive growth in electronic systems. Historically, most uses of a new electronic circuit are discovered after it enters the market place, causing an enormous proliferation of applications. This trend, plus the clever combination of electronic devices with software stored in memories, provides electronic systems for daily use. Wireless communicating is now easier with cellular phones. Home computers help track spending and control lights and home security systems. Automobiles have gone from a basic electrical system to a dozen computers controlling the operation of the engine. For entertainment, interactive multimedia games can be played on high-definition televisions.

Businesses use computers to handle finance, manufacturing, sales, planning, and communications. Wireless communications permit remote operations, from ordering food at a ball park to returning a rental car. Ships and planes use satellites to navigate. The military uses electronics to guide missiles and track aircraft. Satellite systems monitor for compliance with strategic arms limitation treaties.

In the medical field electronics have improved diagnosing illness and training medical personnel. Major advances have been made in collecting, accessing, and displaying large amounts of data.

Positron emission tomography (PET), computer to-mography (CT), and magnetic resonance imaging (MRI) systems use elaborate numerical algorithms running on fast electronic systems to display various functions of the body. Electronic regulating devices such as heart pacemakers can be implanted in the human body.

The Future

With the number of electronic devices on a silicon die doubling every two years, many people wonder when this will slow down. Theoretical and experimental research provide clues. At the start of the integration process in 1960, device lengths were 25 μm. Steadily, device lengths have been reduced to 10, 5, 3, 2, 1, 0.8, and, in 1995, to 0.5 μm. Some scientists forecast that this reduction will continue until the length is 0.1 μm (one thousand times thinner than a human hair). Using 0.1 μm length devices, we could place over ten billion on a die, providing enough memory to store 600,000 pages of text or 1,500 books. Complete electronic systems with computers, memory, and simple biological functions, such as vision and speech, would fit on one chip.

The 0.1 μm barrier is considered the quantum limit, meaning the devices will no longer behave as they did in the example circuits shown; they will have new behaviors caused by quantum effects. The barrier comes from not knowing how to design circuits using these new characteristics.

As with many perceived barriers in the growth of electronics, a clever person or persons will find a way to use the quantum effects to make smaller devices, potentially making systems with trillions of devices. These systems will be the electronics of tomorrow.

See also: BARDEEN, JOHN; CAPACITOR; CIRCUIT, INTEGRATED; DIODE; TRANSISTOR; TRANSISTOR, DISCOVERY OF

Bibliography

DeMasse, T., and Ciccone, J. *Digital Integrated Circuits* (Wiley, New York, 1969).

Ferry, D. K.; Akers, L. A.; and Greeneich, E. W. *Ultra-Large Scale Integrated Microelectronics* (Prentice Hall, Englewood Cliffs, NJ, 1988).

Keyes, R. *The Physics of VLSI Systems* (Addison-Wesley, Reading, MA, 1987).

Savant, C. J.; Roden, M.; and Carpenter, G. *Electronic Circuit Design* (Benjamin-Cummings, Reading, MA, 1987).

L. A. AKERS

ELECTRON MICROSCOPE

Electron microscopes use electrons instead of light for the imaging of objects. The physical principle behind this imaging is the wave nature of electrons, postulated by Louis de Broglie in 1924. When Hans Busch discovered that electrons could be focused like light by using magnetic fields, the basic principles of electron microscopy were born. It was only a few years until the prototype of modern electron microscopes was built by Max Knoll and Ernst Ruska. Today, electron microscopes find wide use in biology, metallurgy, materials science, physics, chemistry, and many technological applications.

In an optical microscope, light from a bright source is sent through an optical condenser lens onto the specimen, and then magnified using a second, or objective, lens. The contrast in the image is caused by local variations of the light absorption properties of the specimen, and for biological tissue is enhanced through appropriate staining techniques. The source of electromagnetic waves in a transmission electron microscope consists of a heated filament, usually made of tungsten, that emits electrons. The electrons are accelerated with a voltage of 50 to 100 kV. Again a condenser lens is used to illuminate the specimen; all the lenses in electron microscopes consist of specially shaped electromagnets producing a strong magnetic field. Behind the specimen, an objective lens followed by a very small aperture (~50 μm) is located in the focal plane of the objective lens. The objective lens forms a 100-times magnified image of the specimen, which is further projected and magnified by the intermediate lens and the projector lens. The electrons finally hit a fluorescent phosphorus screen, where they produce an image magnified up to 1,000,000 times over that of the specimen. The smallest distance that can be resolved with a modern transmission electron microscopes is approximately 2–5 Å (1 Å = 10^{-10} m); this is the typical separation between two atoms in a solid.

The mode of image formation is quite different from that of the light microscope. Electrons are strongly scattered due to interaction with the atoms in the specimen. The small objective lens aperture removes electrons that are scattered with a large scattering angle, and differential scattering across the specimen gives rise to the image contrast. For crystalline samples, the diffraction of the electron beam needs to be considered; as an alternative to a direct image, the electron diffraction pattern can be projected onto the phosphorus screen, allowing conclusions about the crystal structure of the specimen. Samples for transmission electron microscopes need to be very thin, typically less than 0.1 μm. This poses great challenges for appropriate sample preparation techniques.

The first part of a scanning electron microscope is very similar to the transmission electron microscope. The objective lens and objective lens aperture are used to form a finely focused beam that is scanned across the specimen with electromagnetic scanning coils. The electrons excite radiation that is detected, amplified, and used to modulate the intensity on a television screen (a cathode-ray tube). The horizontal and vertical deflection of the cathode-ray tube is coupled with the electronics of the scanning coils, and when the electron beam is scanned across the specimen at a fast rate, an image can be observed directly. The radiation used for image contrast can consist of backscattered electrons, secondary electrons, or x rays excited by the electron beam of the microscope. By filtering the x rays in a spectrometer and using x-ray lines characteristic for elements in the sample, the chemical composition rather than the morphology of the specimen can be used for imaging. The resolution of the scanning electron microscope is limited by the width of the exciting electron beam and by the interaction volume of electrons in a solid; a typical resolution of several tens of nanometers can be achieved. Specimen preparation techniques are far less sophisticated than for the transmission electron microscope. The only requirement is that the samples have to be conductive; for insulating materials, this is usually achieved by coating with a thin metallic film.

Some recent developments employing electrons for imaging techniques are the low-energy electron microscope for imaging surfaces and the scanning tunneling microscope.

See also: SCANNING TUNNELING MICROSCOPE; WAVE–PARTICLE DUALITY

Bibliography

BROGLIE, L. DE. "A Tentative Theory of Light Quanta." *Philos. Mag.* **47,** 446–458 (1924).

BUSCH, H. "Berchnung der Bahn von Kathoden-Strahlen im Exialsymmetrischen Elektromagnetischen Felde." *Ann. Phys.* (Leipzig) ser. 4, **81,** 974–983 (1926).

KNOLL, M., and RUSKA, E. Z. "Das Elektronenmikroskop." *Phys.* **78,** 318–336 (1932).

SLATER, E. *Light and Electron Microscopy* (Cambridge University Press, Cambridge, Eng., 1992).

ULRIKE DIEBOLD

ELECTRON SPIN

See SPIN, ELECTRON

ELECTROSTATIC ATTRACTION AND REPULSION

The forces that affect us most obviously and directly are gravitational and electric forces. Both forces are of long range but manifest themselves in different ways. Gravitation is the weakest of the two forces, it is associated with the property of matter called mass, it is always attractive, and it varies in inverse proportion to the square of the distance $(1/r^2)$ between the interacting masses. Because gravitation is always attractive it gives rise to important cummulative effects, such as the weight of a body, which are the aggregate result of Earth's gravitational attraction on all the atoms of the body. However, most of the phenomena we usually observe result from the much stronger electric forces, related to a property of matter called electric charge, that can be either positive or negative. The atoms of all substances are composed of charged particles, called electrons and protons, that carry equal negative and positive charge, and of neutral particles called neutrons. We do not feel the electric forces because most bodies are composed of the same amount of positive and negative charges and therefore appear electrically neutral. For electric forces to become appreciable, it is necessary to break that balance by transferring charges to or from a body, as is done by triboelectrification and by ionization; batteries and electric cells operate on the principle of separation of positive and negative charges. Electric forces in matter are also manifested when a partial separation of charges is produced by modifying the distribution of electric charges in atoms and molecules, an effect called polarization.

Electric charge is measured in coulombs, which is the amount of electricity transported in 1 s by a current of 1 A. The positive charge of a proton and the negative charge of an electron are designated by $+e$, where e is called the fundamental charge; its value is $e = 1.6022 \times 10^{-19}$ C.

The electric force between charges of the same sign (both positive or both negative) is repulsive, while the electric force between charges of opposite sign (one positive and the other negative) is attractive. Negatively charged electrons in an atom repel each other but are kept within the atom by the attractive electric force of the positively charged nucleus. The positively charged protons in the nucleus repel each other, but they are held together by a much stronger nuclear force. Although atoms are composed of the same number of positive and negative electric charges, the electric force between atoms is not zero because, due to the motion of the electrons around the nucleus, there is not a total cancellation of electric effects and residual electric forces exist, which are responsible for the many forms in which matter appears and for many physical processes. Atoms in molecules as well as atoms and molecules in solids and liquids are held together by attractive electric forces, which also play a fundamental role in other phenomena such as changes of phase (condensation and solidification) and chemical reactions. Intermolecular forces in gases are also electric. Electric forces play an important role in many processes in living organisms, such as in heartbeats and in the conduction of a signal by the nervous system. It is the balance between attractive and repulsive electric forces and the nuclear force that makes possible the existence of nuclei, atoms, molecules, and their aggregates, that is liquids, solids, gases, and even living beings.

The electric force between two charged particles is proportional to their electric charges and varies in inverse proportion to the square of the distance between the charges ($1/r^2$). This statement, which is correct only if the charges are at rest, and to a very good approximation if the charges move very slowly relative to the observer, is called Coulomb's law because it was first formulated in precise terms in 1785 by Charles A. Coulomb, although it had been inferred in 1766 by Joseph Priestley, who was also the discoverer of oxygen. Expressed in mathematical terms, the attractive or repulsive electric force between two charges q and q' separated the distance r is

$$F = k\frac{q\,q'}{r^2}, \qquad (1)$$

where k is a proportionality constant. The value of k depends on the units chosen to measure the other quantities and whether the charges are in a vacuum or immersed in a substance. In SI units, the charges are measured in coulombs, the distance in meters, the force in newtons, and, if the charges are in a vacuum, $k = 10^{-7}c^2 = 8.9874 \times 10^9$ N·m^2/C^2, where $c = 2.998 \times 10^8$ m/s, the velocity of light in vacuum. Since k is approximately 9×10^9 N·m^2/C^2, we can write $F = (9 \times 10^9)(q\,q'/r^2)$. The electric force between two equal charges of 1 C, separated 1 m, is 9×10^9 N, which is about the same as the weight of 900,000 tons. (In practice, it is very difficult to charge two bodies with 1 C and keep them 1 m apart.) For reasons related to further developments in electromagnetism, it is preferable to write Eq. (1) as $F = q\,q'/4\pi\varepsilon_0 r^2$, where $\varepsilon_0 = 10^7/4\pi c^2 = 8.854 \times 10^{-12}$, the vacuum permittivity. If the charges are immersed in a substance the vacuum permittivity is replaced by the permittivity ε of the substance. The ratio $\varepsilon_r = \varepsilon/\varepsilon_0$ is called the dielectric constant of the substance and $\chi = \varepsilon_r - 1$ is its electric susceptibility (see Table 1).

Table 1 Electric Susceptibilities

Substance	χ^a
Solids	
Mica	5
Porcelain	6
Glass	8
Bakelite	4.7
Liquids	
Oil	1.1
Turpentine	1.2
Benzene	1.84
Alcohol (ethyl)	24
Water	78
Gases (at 1 atm. and 20°C)	
Hydrogen	5.0×10^{-4}
Helium	0.6×10^{-4}
Nitrogen	5.5×10^{-4}
Oxygen	5.0×10^{-4}
Argon	5.2×10^{-4}
Carbon dioxide	9.2×10^{-4}
Water vapor	7.0×10^{-3}
Air	5.4×10^{-4}
Air (100 atm)	5.5×10^{-2}

$^a\chi = \varepsilon_r - 1$

To compare, consider two particles of masses m and m' with charges q and q' separated the distance r. The gravitational force is $F = G\,m\,m'/r^2$, where $G = 6.673 \times 10^{-11}$ N·m^2/kg^2, and the electric force is given by Eq. (1). The ratio of the electric force to the gravitational force is

$$\frac{kqq'}{Gmm'} = 1.35 \times 10^{20}\,\frac{qq'}{mm'}. \qquad (2)$$

For two protons $q = q' = e = 1.602 \times 10^{-19}$ C and $m = m' = 1.673 \times 10^{-27}$ kg, so that the ratio of the electric force to the gravitational force is 10^{36}, showing that the electric force is much stronger than the gravitational force. Therefore, the gravitational force can be ignored when discussing the structure of matter and related electric phenomena.

The exponent 2 in Coulomb's law has been measured with a precision of up to one part in one billion ($1/10^9$). Coulomb's own experiment was very simple, using a method originally designed by Henry Cavendish for verifying the inverse square law for gravitation. A glass rod with a metal ball attached to one end is suspended by its center of mass from the fiber of a torsion balance. The metal ball is charged and the electric force between the ball and a charged body (held at various distances from the ball) is obtained in each case in terms of the angle by which the fiber of the torsion balance is twisted. Priestley's prior inference of the inverse square law was based on the experimental result obtained by Cavendish that the electric charge of a conductor resides only on its outer surface. The Cavendish experiment consisted in charging a metallic sphere and covering it with two hollow metallic hemispheres. When the hemispheres are removed it is found that all the electric charge of the sphere has been transferred to the hemispheres. A more precise experiment was carried out years later by James Clerk Maxwell. A metallic sphere E' is held inside a larger sphere E by insulating supports. A charge is placed on E' through an aperture in E that has a cap to open and close it. The cap has a metalic wire that touches E' when the aperture is closed, so that both spheres are in contact. When the aperture is closed and opened again, it is found that the charge on E' has been transferred to the outer sphere. A mathematical analysis of the results of these experiments shows that they can be explained only if the electric force varies as $1/r^2$.

There are many other facts that support the inverse square law. For example, the structure of atoms and molecules can be explained assuming an attractive electric force between the electrons and the nuclei that follows Coulomb's law. The scattering of positive charged particles (protons, alpha particles, etc.) by positively charged nuclei can be explained in terms of a repulsive electric force that obeys Eq. (1). In fact, it was scattering experiments with alpha particles that allowed Ernest Rutherford to conclude that the positive charge in atoms resides in a small central nucleus. Also, Gauss's law is a direct consequence of the inverse square law.

For describing electric attraction and repulsion, it is convenient to introduce the concept of electric field, defined as the electric force on one unit of charge. Making $q' = 1$ C in Eq. (1), we have that the electric field produced by a charge q at the distance r is $E = kq/r^2$. The electric field is expressed in newtons per coulombs (N/C). The electric field produced by an electric charge is directed away from the charge if it is positive and toward the charge if it is negative. We may say that an electric charge acts on the charges in its vicinity through its electric field. The force on a positive charge is in the direction of the electric field and the force on a negative charge is in the opposite direction. The advantage of the notion of electric field is that if we know the field we do not need to take into account the position of the charges producing it to find the forces on other charges.

When an uncharged metallic conductor is placed in an electric field, its free negatively charged electrons tend to move in the direction opposite to the field so that one end of the conductor becomes charged negatively and the other end becomes charged positively, so that the conductor becomes polarized. The surface charges produce an electric field inside the conductor that is opposite to the external field. The charge separation continues until the resultant electric field inside the conductor is zero. The forces on the oppositely charged ends of the conductor tend to move each end in opposite directions relative to the electric field. If the forces are different, as happens when the field is not uniform, the conductor will experience a net electric force that will move it in the direction in which the field increases, even if the net charge of the conductor is zero. Similarly, when a metallic conductor is attached to the poles of a battery, the electric field between the poles keeps the electrons moving away from the negative pole and toward the positive pole, maintaining an electric current along the conductor.

In an insulator or dielectric, which has no free electric charges, an external electric field deforms the electronic motion relative to the nuclei in the atoms and molecules so that the atoms or the molecules are polarized, becoming electric dipoles oriented in the direction of the field. In many substances, if the molecules are already polarized, as is the case with water and carbon monoxide molecules, their electric dipoles become oriented in the direction of the applied field. The net effect in both cases is that the insulator becomes polarized, and positive and negative charges appear in opposite ends of the insulator, although the charges are not free to move through the insulator. The net electric field inside the insulator is reduced but does not become necessarily zero. The external electric field acts in opposite directions on both ends of the polarized insulator, that, although its net charge is zero, will move in the direction in which the field is stronger. The electric properties of matter serve to explain many phenomena observed in nature.

See also: CHARGE, ELECTRONIC; COULOMB'S LAW; DIELECTRIC PROPERTIES; ELECTRICITY; FIELD, ELECTRIC; GRAVITATIONAL ATTRACTION; INVERSE SQUARE LAW; POLARIZATION

Bibliography

ALONSO, M., and FINN, E. J. *Physics* (Addison-Wesley, Reading, MA, 1992).

FREEMAN, I. M. *Physics* (McGraw-Hill, New York, 1968).

JONES, E. R., and CHILDERS, R. L. *Physics* (Addison-Wesley, Reading, MA, 1990).

MARCELO ALONSO

ELECTROWEAK INTERACTION

See INTERACTION, ELECTROWEAK

ELEMENTARY PARTICLES

Elementary particles, sometimes called fundamental particles, are the simplest subunits of matter. By the standards now applied to test whether a particle is elementary, they should have no size at all, somewhat like geometric points. The current theory of elementary particles is known as the standard model, and while it has had considerable success the theory is not yet complete.

The idea of an eternal and indivisible atom of matter first appeared in the fifth century B.C.E. in the writings of Greek philosophers, especially Democritus of Abdera. The word "atom" is Greek for "uncuttable." The early atomists could only speculate about what atoms might be like, but they hoped that they would prove to have simple geometric shapes and that there would only be a few kinds of atoms, so that nature on the atomic level would be far simpler than the world we can see. Unfortunately, modern developments of atomism have done little to encourage that hope.

Atoms entered experimental science in the nineteenth century, when chemists such as John Dalton and Stanislao Canizzaro explained many of the known laws of chemistry by the assumption that each chemical element corresponded to a different type of atom. Although chemistry was able to establish the relative weights of the atoms of different elements, the absolute size or weight of an atom remained unknown.

In the first decade of the twentieth century the sizes and weights of atoms were finally measured to some accuracy. The diameters are typically somewhat less than a billionth of a meter (10^{-9} m, or 1 nm). By then the discovery of radioactivity had shown that atoms were not eternal and unchanging, and the discovery of the electron had made it clear that they had smaller parts. Physicists began to look for ways to uncover their internal structure.

In 1911 Ernest Rutherford discovered that most of an atom's mass is concentrated in a small central nucleus more than 1,000 times smaller than the atom. The size of the atom is determined by the motions of the electrons outside the nucleus, which were fully understood by 1926.

By 1932 it was clear that the nucleus was composed of two kinds of particles nearly equal in mass, positively charged protons and electrically neutral neutrons. Certain forms of radioactivity required the existence of a fourth particle, a neutral counterpart to the electron called the neutrino. Thus at that time it appeared that four elementary particles would be sufficient to account for all known forms of matter. Protons and neutrons are nearly 2,000 times heavier than electrons, while neutrinos have little or no mass.

Particle physicists usually state the masses of subatomic particles in terms of the equivalent energy, from Albert Einstein's formula $E = mc^2$. The customary energy unit is the electron volt (eV), equal to 1.6×10^{-19} J. Most particle masses are in the range of millions or billions of electron volts (MeV or GeV). On this scale the masses of the electron, proton, and neutron are 0.511, 938.21, and 939.51 MeV, respectively.

Starting in the late 1930s, as physicists sought to understand the forces that form the atomic nuclei from protons and neutrons, additional particles were discovered, some of which were heavier than protons and neutrons while others had masses between that of an electron and a proton. All were unstable, with none lasting more than a few millionths of a second before breaking up into lighter particles. In the 1950s, studies of the proton and neutron, carried out by bombarding them with electrons, revealed that they had quite complex internal structure, on a scale of roughly a quadrillionth of a meter. The neutron, for example, is not actually free of electric charge, but simply contains equal amounts of positive and negative electricity. As the list of known particles grew, many scientists working in this area began to doubt that all the objects they were studying were truly elementary.

By 1963 there were nearly 100 known subatomic particles. Murray Gell-Mann and George Zweig then independently proposed a scheme in which protons and neutrons were simply the stablest members of a vast family of particles composed of subunits that Gell-Mann called quarks. Electrons, neutrinos, and some related particles retained their elementary status, as members of a second family called leptons. Taken together, quarks and leptons are referred to as fermions. What distinguishes quarks from leptons is that quarks are sensitive to the strong interaction (a powerful force that binds them together to form the protons, neutrons, and other known particles), while leptons are not. Strong interactions between the quarks within the protons and neutrons are also responsible for the forces binding nuclei together.

In the 1960s, only three species of quarks and four of leptons were known. Since then, new discoveries have shown that each of these families has six members, and give strong indications that there are no more yet to be discovered.

The six quark species (the technical term employed by physicists is flavors) are designated by the letters d, u, s, c, b, and t, which stand for down, up, strange, charmed, bottom, and top. The d, s, and b carry an electric charge of $-\frac{1}{3}$ fundamental units,

Table 1 Quarks and Leptons

	Charge	Mass (MeV)
Quarks		
d	$-\frac{1}{3}$	~ 7
u	$+\frac{2}{3}$	~ 4
s	$-\frac{1}{3}$	~ 200
c	$+\frac{2}{3}$	~ 1,500
b	$-\frac{1}{3}$	~ 4,700
t	$+\frac{2}{3}$	~ 170,000
Leptons		
e	−1	0.51
ν_e	0	< 0.00002
μ	−1	106
ν_μ	0	< 0.25
τ	−1	1,784
ν_τ	0	< 70

while the charges of the u, c, and t are $+\frac{2}{3}$ A proton consists of two u and one d and the neutron of two d and one u. Combinations of quarks are known as hadrons.

The lepton family has three members with charge −1, denoted e, μ, and τ, for electron, muon, and tau. Each has a companion neutrino, ν_e, ν_μ, and ν_τ. The neutrinos are all electrically neutral and as yet have not shown any mass large enough to measure, though there are hints that they are not entirely devoid of mass. Table 1 lists the members of the quark and lepton families, with their masses in MeV. The neutrino masses are upper limits, and the quark masses are estimates rather than direct measurements for reasons that will be explained below.

A fundamental rule of physics asserts that every particle must have a corresponding antiparticle, which is equal in mass but opposite in electric charge and certain other properties. Fermions can only be created in conjunction with their antiparticles, a process called pair production. Although a heavy lepton (or quark) can be converted to a lighter one, it can only disappear if it meets an antilepton (or antiquark), a process called annihilation. Only the least massive flavors of leptons and quarks are stable, and the u, d, e, and ν_e are responsible for most of ordinary matter.

There is an additional family of particles, however, that can be freely created or destroyed as long as there is energy available. These are known as fundamental bosons. Their role in nature is to serve as "messengers" that transmit the forces that hold matter together. One particle emits a boson, which is absorbed by another, transferring a small bundle of energy and momentum. The familiar forces of our everyday world, such as gravity and electromagnetism, are the result of a huge number of tiny transfers of this sort. The theory that describes these processes is quantum field theory.

Electromagnetism is transmitted by photons, and gravity by gravitons. The strong interaction is transmitted by bosons called gluons. One additional force, called the weak interaction, is transmitted by three particles designated W^+, W^-, and Z^0. The primary role of this force is to transform one flavor of quark or lepton into another, so that the heavier particles can break up into lighter ones. Photons, gravitons, and gluons are massless, while the weak bosons are about 90 times heavier than protons. The electroweak theory, developed in the late 1960s by Steven Weinberg and Abdus Salam, shows that these bosons are simply massive cousins of the photon, and thus the weak interaction is related to electromagnetism. The fundamental bosons are listed in Table 2.

The more massive members of all families are terribly unstable, breaking up in a time too short to allow them to be observed directly. Their existence must be inferred from a careful study of the products of their disintegration. The most massive elementary particle in current theories is the t quark, discovered in 1995.

Because they are so unstable, all the more massive elementary particles are rarely found in nature. They must be created by the conversion of kinetic energy to matter in violent collisions of stable particles that have been raised to very high energy in particle accelerators.

Table 2 Fundamental Bosons

Field	Particle	Mass (MeV)
Electromagnetic	photon	0
Strong nuclear	gluon	0
Weak nuclear	$W^\pm$	80,220
	Z^0	91,190
Gravity	graviton	0

Accelerators not only create particles, they also permit tests of whether these particles are truly elementary. When two particles are close to one another, momentum can be transferred between them. The closer the encounter, the more the momentum transfer. The size of particles limits how close they can come to one another: If they have no size, there is no limit to how much momentum can be transferred. The energy imparted to particles by an accelerator determines how much momentum is available for transfer, and thus sets a limit on the minimum size that can be detected. In some of these experiments, the particles collide with one another, while in others they are produced together in a collision of other particles.

Studies of momentum transfer in the most powerful particle accelerators show that the lighter leptons and quarks, as well as the fundamental bosons, are still structureless on a scale of 10^{-19} m, about 10,000 times smaller than a proton. Whether they will show structure when more powerful accelerators reveal finer details remains to be seen.

Thus, while we do not know for certain whether quarks and leptons are the long-sought elementary particles, there is as yet no experimental evidence that suggests they are not. Furthermore, theories that attempt to explain the three copies of each charge of quark and lepton by introducing a substructure have not been successful.

One peculiar feature of quarks is that they are never found alone, but only in combination with one another. This proved an obstacle to widespread acceptance of the quark theory until 1975, when it received a perfectly natural explanation through a theory of the gluon field called quantum chromodynamics (QCD).

In addition to quarks, a hadron such as a proton also contains a large number of gluons. If energy is provided, these gluons can readily turn into quark-antiquark pairs. It is fairly easy to knock a quark free from a proton, for example, by bombarding it with very energetic electrons. However, the ever-present gluons quickly create new partners for the freed quark, so it never is observed alone.

QCD permits three kinds of quark combinations. Three quarks make a baryon, three antiquarks are an antibaryon, and one quark and one antiquark is called a meson. The proton is the lightest and stablest baryon. Because mesons contain a mixture of matter and antimatter, none are stable. As soon as the weak interaction changes the flavor of the quark or antiquark to match that of its companion, the meson can annihilate itself. It is the longer-lived mesons and baryons, rather than quarks themselves, that are directly observed in the laboratory. It is for this reason that the quark masses given in Table 1 are only approximations; the quark masses are not easily separated from the gluon field energy that is also present in a hadron.

This picture has one outstanding loose end, an explanation for the origin of mass, which represents the energy required to create a particle. One proposed theory suggests that particle masses are due to the action of a field transmitted by a particle called the Higgs boson, after the theorist Peter Higgs who first proposed it. Current estimates suggest that the mass of the Higgs boson will be from 300 to 500 GeV, which makes it impossible to detect with any existing particle accelerator. However, the Higgs theory is incapable of predicting the masses of the fundamental particles; it simply allows them to take any value.

The ideas presented constitute the standard model, and the picture was complete by the end of the 1970s. There have been several attempts to go beyond this model. One is the construction of unified field theories, which draw on the similarities between the photon, gluon, and weak fields, to construct a single theory embracing all three forces. Superstring theories are the most ambitious of all, attempting to incorporate gravity into the scheme, even though the effects of gravity are imperceptible on the subatomic scale. However, none of these theories to date has any experimental justification.

In any event, neither the standard model or any of its suggested replacements yet offers the sublime simplicity so earnestly sought by the early Greek atomists when they began the quest for the ultimate particles of nature 2,500 years ago.

See also: ACCELERATOR; ANTIMATTER; BOSON, HIGGS; CHARGE; COLLISION; ELECTRON; FERMIONS and BOSONS; FLAVOR; GRAVITON; HADRON; INTERACTION; INTERACTION, STRONG; INTERACTION, WEAK; LEPTON; LEPTON, TAU; MASS; MUON; NEUTRINO; NEUTRON; PARTICLE; PARTICLE MASS; PROTON; QUANTUM CHROMODYNAMICS; QUANTUM FIELD THEORY; QUARK; RUTHERFORD, ERNEST; SUPERSTRING

Bibliography

HALZEN, F., and MARTIN, A. D. *Quarks and Leptons: An Introductory Course in Modern Particle Physics* (Wiley, New York, 1984).

PARKER, B. R. *Search for a Supertheory: From Atoms to Superstrings* (Plenum, New York, 1987).

ROBERT H. MARCH

ELEMENTS

An element is a substance that cannot be further decomposed by chemical methods. The distinction between elements and chemical compounds formed from them was first stated by Robert Boyle.

The ancient Greeks thought that all substances could be represented as combinations of the four fundamental atoms or qualities: air (wet and hot), earth (dry and cold), fire (dry and hot), and water (wet and cold). As more understanding emerged, it was realized that mass was more important in differentiating substances rather than these qualities.

At the beginning of the nineteenth century, the concept that matter consisted of elemental units, or atoms, received experimental confirmation through the work of John Dalton and others. The study of the ways various elements combined to form chemical compounds led to the concept that different chemical elements have definite atomic weights. Hydrogen was assumed to be the lightest atom. The relative atomic weights of different elements were determined by noting the proportions in which they combined with others to form chemical compounds. The known elements were classified in a periodic table by the Russian chemist Dmitri Mendeleev.

At first the atoms of all elements were postulated to consist of integral numbers of hydrogen atoms. However, when accurate measurements of atomic weights became available, it was found that the atomic weights for many elements did not consist of integer multiples of the atomic weight of hydrogen. The explanation of this fact had to await the discovery that most elements are actually mixtures of several different kinds of atoms having the same chemical properties but different atomic weights. These different atomic species are known as isotopes. For example, the element chlorine was found to consist of about 75 percent atoms of mass 35 and 25 percent of atoms of mass 37.

The discovery of the electron by the English physicist Joseph John Thomson in 1897 indicated that atoms are not the most elementary, indivisible particles, but instead are complicated structures containing both positive and negative electric charges as well as mass. Since the negatively charged electron was shown to have a mass of approximately 1/1,800 of that of the mass of the simplest atom, the hydrogen atom (atomic number 1), it was concluded that the preponderance of an atom's mass must be associated with its positive charge. Thomson proposed a model of the atom in which the positive charges were distributed uniformly in a sphere, and the electrons were found embedded inside it, somewhat like raisins in a bun.

The definitive description of atoms had to await experiments by Ernest Rutherford, a physicist from New Zealand working in England. He worked with radioactive materials which served as sources of alpha rays. In previous studies he had already determined that alpha rays were doubly charged ions of helium, atomic number 2. He and his collaborators studied the scattering of these particles by thin metallic foils and found that some of them were deflected by angles greater than $90°$. This observation disagreed totally with the predictions of the Thomson model of the atom, and led Rutherford to postulate in 1911 that the positive charge of an atom is contained within an extremely small central object. His model was able to explain the large scattering angles when the central nucleus of the atom was ascribed a diameter of less than 10^{-12} cm. The electrons then had to be distributed about the positively charged nucleus in order to ensure that the atom itself was electrically neutral.

Subsequent experiments showed that the number of elementary positive charges in an atom is approximately equal to half its atomic weight. Quantitative comparisons with the predictions of Rutherford's model enabled the magnitudes of the positive charges on nuclei of different elements to be determined, in units of the charge on the electron. We now know that the electron charge has a magnitude of 1.6×10^{-19} C and is denoted by the symbol e. These experimentally determined charges coincided with the atomic numbers from the periodic table. Thus the atomic number Z gives the number of positive charges in the nucleus, and each element is distinguished by having a unique atomic number. In order to be electrically neutral, an atom consists of a nucleus with charge $+Ze$, surrounded by a cloud of Z electrons.

The explanation of isotopes became clear following the discovery of the neutron by the British physicist James Chadwick in 1932. The neutron has approximately the same mass as the proton (hydrogen nucleus), but it has no electrical charge. Atomic nuclei were then recognized as containing protons and neutrons. We now know that all atoms of the same element have nuclear charge Z and the same number of orbiting electrons. The different atomic weights of the isotopes are due to their possessing differing numbers of neutrons in their nuclei. For example, the isotope of chlorine of atomic weight 35 (denoted ^{35}Cl) has a nucleus containing seventeen protons and eighteen neutrons, with seventeen electrons orbiting around it, while the chlorine isotope of atomic weight 37 has the same number of protons and orbiting electrons (seventeen) and twenty neutrons.

Periodic Table of the Elements

Mendeleev produced his periodic table of the elements by observing that the physical and chemical properties of the elements are periodic functions of their atomic weights. This means that the properties of the elements repeat themselves at intervals as one goes to heavier and heavier atomic weights. We now know that the relevant parameter is atomic number rather than atomic weight. In the modern periodic table, the elements are classified into sixteen groups or families, and form seven periods through which the various properties cycle. All the elements in a group have similar chemical and physical properties, such as the rare gases (Group VIIIa), the alkali metals (Group Ia), the alkaline earths (Group IIa), and the halogens (Group VIIa). Although the original determinations for the element positions in the table were based on chemical and physical properties, it is now clear that the position of an element in the table depends on the configuration of its electrons, that is, its atomic structure. All the elements in a group turn out to have similar electron configurations.

The periodic table of elements was for many years thought to consist of ninety-two elements, with uranium being the heaviest. Bismuth, with atomic number 83, is the heaviest stable element. No stable isotopes exist for element number 43 (technetium) or element number 61 (promethium), although a number of their isotopes (all radioactive) have been produced and studied in the laboratory. Many of the elements with atomic numbers between 84 and 92 possess the long-lived isotopes that are found in nature.

Periodic Table of the Elements

Group IA																	VIIIA
1 **H**	IIA											IIIA	IVA	VA	VIA	VIIA	2 **He**
3 **Li**	4 **Be**											5 **B**	6 **C**	7 **N**	8 **O**	9 **F**	10 **Ne**
11 **Na**	12 **Mg**	IIIB	IVB	VB	VIB	VIIB	⎯ VII ⎯		IB	IIB		13 **Al**	14 **Si**	15 **P**	16 **S**	17 **Cl**	18 **Ar**
19 **K**	20 **Ca**	21 **Sc**	22 **Ti**	23 **V**	24 **Cr**	25 **Mn**	26 **Fe**	27 **Co**	28 **Ni**	29 **Cu**	30 **Zn**	31 **Ga**	32 **Ge**	33 **As**	34 **Se**	35 **Br**	36 **Kr**
37 **Rb**	38 **Sr**	39 **Y**	40 **Zr**	41 **Nb**	42 **Mo**	43 **Tc**	44 **Ru**	45 **Rh**	46 **Pd**	47 **Ag**	48 **Cd**	49 **In**	50 **Sn**	51 **Sb**	52 **Te**	53 **I**	54 **Xe**
55 **Cs**	56 **Ba**	57 **La**	72 **Hf**	73 **Ta**	74 **W**	75 **Re**	76 **Os**	77 **Ir**	78 **Pt**	79 **Au**	80 **Hg**	81 **Tl**	82 **Pb**	83 **Bi**	84 **Po**	85 **At**	86 **Rn**
87 **Fr**	88 **Ra**	89 **Ac**	104	105	106	107	108	109	110	111	112						

Lanthanides	58 **Ce**	59 **Pr**	60 **Nd**	61 **Pm**	62 **Sm**	63 **Eu**	64 **Gd**	65 **Tb**	66 **Dy**	67 **Ho**	68 **Er**	69 **Tm**	70 **Yb**	71 **Lu**
Actinides	90 **Th**	91 **Pa**	92 **U**	93 **Np**	94 **Pu**	95 **Am**	96 **Cm**	97 **Bk**	98 **Cf**	99 **Es**	100 **Fm**	101 **Md**	102 **No**	103 **Lr**

Man-Made Elements

Beginning in the 1940s, physicists and chemists began trying to produce elements with atomic number greater than 92. The first such transuranium element to be synthesized was neptunium, atomic number 93, found by bombarding uranium with energetic neutrons in a reactor. Subsequently, other elements have been produced by various means, including bombardment of other heavy elements by light charged particles by means of accelerators. Often the targets used in these experiments have themselves been man-made, long-lived transuranium elements.

Elements number 99 (einsteinium) and 100 (fermium) were first observed in the residue following thermonuclear bomb tests. These devices are also known as fusion or hydrogen bombs. An atomic, or fission, bomb is used to trigger the explosion of a thermonuclear bomb. Atomic bombs contain uranium or plutonium, not all of which is completely consumed by the fission explosion. In the subsequent explosion of the fusion device, this remnant material is subjected to bombardment by a huge number of neutrons in a very short time. The nuclei produced when uranium and plutonium nuclei capture upwards of a dozen neutrons are unstable. They eventually beta decay to long-lived isotopes of the new elements.

To create new elements with atomic number higher than 101, it has been necessary to bombard heavy targets with beams of ions such as boron (element 5), carbon (element 6), oxygen (element 8), neon (element 10), or even heavier ions, using an accelerator. Very sensitive equipment has been developed to make precision measurements on minute quantities of these elements. As of early 1996, the heaviest element to be observed is element number 112, of which two atoms were observed during an experiment lasting a total of three weeks. An isotope of this element, as yet unnamed, having mass 277 (112 protons plus 165 neutrons) decays by emitting an energetic alpha particle, and has a half-life of about 0.00024 s. It was produced by bombarding targets of ^{208}Pb with an intense beam of ^{70}Zn ions from the Unilac accelerator in Germany. In general, the half-lives of the elements decrease with increasing atomic number, although various theoretical models suggest that this trend may be reversed for elements around element number 114. It is clear that extremely sensitive equipment is needed in order to extend our knowledge of the heaviest elements beyond those already known. Future searches will likely involve the bombardment of ^{208}Pb and ^{209}Bi targets with beams of Zn, Ge, and Se isotopes to attempt the synthesis of element numbers 113–117.

Naming new elements is an interesting and often controversial process, often taking years. Official recognition of a new element is sanctioned by an international committee, which also decides on the name for the element. There are only a few laboratories with the capability of producing new elements, and often scientists from many countries are involved in the experiments.

Origin of the Chemical Elements

Much understanding was gained in the twentieth century regarding the origin and abundance distribution of the chemical elements found on Earth and in our solar system. The abundance of elements in the earth's crust is not representative of the general distribution of elements in the solar system, since the terrestrial material has been subject to chemical effects, weathering, and other processes. The best indications of the solar system abundances come from spectroscopic observations on light from the Sun and the study of material falling to Earth from space in the form of meteorites.

Hydrogen is the most abundant element in nature, making up approximately 75 percent of the mass of the universe. Helium is the next most abundant, weighing in at about 24 percent. The remaining elements comprise just 1 percent of the mass. It is believed that all the elements heavier than helium except for a small amount of lithium have been produced by nuclear reactions in stars. This has been accomplished by converting a small fraction of the primordial materials hydrogen and helium, the raw materials out of which stars are formed, into heavier elements. When this happens, energy is released by converting matter into energy via Einstein's famous equation $E = mc^2$. For example, in a series of reactions known as the pp chain, four protons eventually combine to produce one helium nucleus, with the conversion of 0.7 percent of the mass into energy. In fact, this process is the source of energy in stars like our Sun. Because small amounts of radioactive heavy elements with half-lives in the vicinity of 10^9 years are found in the earth's crust, it is believed that the material out of which our solar system condensed underwent its most recent nuclear processing some 4–5×10^9 years ago, the estimated age of the earth.

In Earth's atmosphere, the abundances of hydrogen and helium are negligible. This is due to the relatively weak gravitational force of Earth, which is only strong enough to keep heavier gases such as nitrogen and oxygen from escaping. However, the planet Jupiter does have a sufficiently strong gravitational field to bind hydrogen, and it is thought that this planet has a core of solid hydrogen.

See also: ATOM; ATOMIC NUMBER; ATOMIC THEORY, ORIGINS OF; ATOMIC WEIGHT; ELECTRON, DISCOVERY OF; ELEMENTS, ABUNDANCE OF; ELEMENTS, RARE EARTH; ELEMENTS, SUPERHEAVY; ELEMENTS, TRANSITION; ELEMENTS, TRANSURANIUM; ISOTOPES; NUCLEOSYNTHESIS

Bibliography

PHILLIPS, C. S. G., and WILLIAMS, R. J. P. *Inorganic Chemistry* (Oxford University Press, Oxford, Eng., 1965).

CARY N. DAVIDS

ELEMENTS, ABUNDANCE OF

The abundances of elements in the solar system are a combination of elemental and isotopic abundances derived from spectroscopic analyses of the solar photospheric light and of chemical analyses of the carbonaceous chondrite class of meteorites. Both the solar surface and carbonaceous chondrites are composed of material that appears to represent well the chemical mix of the original gas cloud that condensed to form the solar system, especially the Sun itself in which most of the solar system mass resides. Chemical compositions of individual planets vary widely and thus unfortunately cannot be used for this work. Planetary compositions contain valuable data for defining the severe chemical fractionation that occurred during formation of the planets, but give misleading information on the "primordial" solar system distribution of the elements.

The primordial abundance distribution has arisen from the combined efforts of several previous galactic stellar generations: stars were born, synthesized heavier elements from lighter ones in nuclear fusion reactions, and expelled some of these new elements into the interstellar medium, to be gathered into new stars possessing higher contents of the heavy elements. Different mass stars contributed different elements into the mix that eventually formed the solar system. Figure 1 plots this *relative* abun-

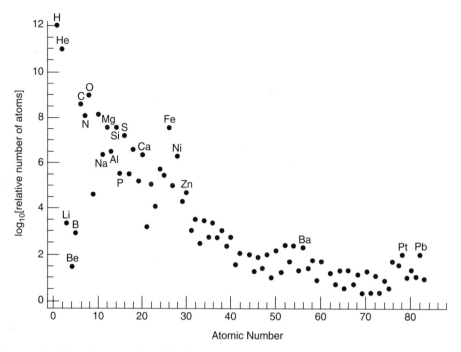

Figure 1 Abundances of stable elements in the solar system. The data are a combination of solar and carbonaceous chondrite meteoritic values.

dance distribution of stable elements, following astronomical convention in arbitrarily normalizing the distribution to elemental abundances per 10^{12} hydrogen atoms. Each point represents the total abundance of an element, including all isotopes of that element.

Solar photospheric and meteoritic abundances agree so well for almost all elements, that the solar system chemical composition given in the figure is a judicious merger of abundances from both types of analyses. In this distribution, the elements H and He clearly dominate: about 90 percent of the primordial solar system composition was H, and 9 percent He; the rest of the elements in the periodic table of elements comprise merely 1 percent of the number of atoms. Most of this 1 percent leftover is split between the relatively light elements C and O. The heavy metals such as Fe and Ni are mere trace elements in the solar system chemical composition.

Quite distinct trends exist in the abundance pattern of the heavy elements. Generally the abundances decline with increasing atomic number Z. Moreover, those elements with an even Z are more plentiful than neighboring elements with an odd Z—the effect is most striking among the lighter elements C ($Z = 6$) through Ca ($Z = 20$). Also, a sharp peak in the heavy element abundance distribution

occurs at Fe (and to a lesser extent Ni). The abundances of all elements with $Z > 30$ are quite low compared to the lighter elements. Finally, the three very light elements Li, Be, and B are extremely underabundant in the solar system distribution.

Remarkably, the abundance pattern of primordial solar system material is apparently a universal one: the vast majority of stars in our galaxy that have been analyzed (as well as some stars of nearby galaxies) have chemical compositions almost identical to the Sun's. The solar system abundance distribution is therefore often called the "cosmic" abundance pattern. But the deviations from this pattern, even if small, are repeated in enough stars to warrant close scrutiny. For example, in some stars the overall heavy element ($Z \geq 6$) content is lower than that of the Sun: These so-called metal-poor stars usually exhibit only mild (factors of 2 to 4) deficiencies of the heavy elements, but some stars have been found with less than a thousandth of the solar heavy element content. The element-to-element abundance pattern of metal-poor stars is slightly different than in the Sun. Also, highly evolved red giant stars have very different abundances of the light elements Li, C, N, and O. Finally, a tiny fraction of stars have distinct enhancements of some of the elements with $Z > 30$.

Very metal-poor stars are older stars in the sense that they were formed long before the birth of the Sun. Their overall heavy element abundances reflect the heavy element enrichment of the interstellar medium at the time of their birth. Thus the Galaxy's content of heavy elements has grown with time (the enrichment with time is complex and probably not monotonic), reflecting the enrichment caused by succeeding generations of stellar nucleosynthesis. This provides direct observational evidence for the synthesis of elements in stars.

The abundance pattern of the primordial solar system, the universality of that pattern, and the deviations from that pattern shown by some stars all provide clues to the origin and distribution of the elements. Decoding these clues begins with the understanding that stars synthesize nearly all elements via nuclear fusion reactions. Fusion of very light element nuclei generally is exothermic, and thus is the ultimate source of the energy emission of stars, both in photon and neutrino forms. The energy release from fusion provides the outward pressure support that prevents normal stars from gravitational collapse. However, fusion works efficiently only in environments in which the strong nuclear attractive forces, acting over very small distances, can overcome the electrostatic repulsions of the positively charged protons in nuclei. Thus nuclear fusion occurs almost exclusively in the extremely hot and dense conditions of stellar interiors.

The Sun currently is able to force the fusion of four hydrogen nuclei into one helium nucleus only in its deepest core regions, where $T \simeq 10^7$ K and $\rho \simeq 0.1$ kg·m^{-3}. The transmutation of hydrogen into helium cannot be done in a single high-energy collision of four hydrogen nuclei, but rather is a set of two-body fusion reactions (collectively known as the proton-proton cycle), that first create deuterium (H^2) out of ordinary hydrogen (H^1), then light helium (He^3) from deuterium, and finally ordinary helium (He^4) from light helium. By-products of this cycle include positively charged electrons, neutrinos, and gamma-ray photons. Gradually, the Sun's core is being turned from nearly pure hydrogen into helium. This process will occupy the Sun for nearly 90 percent of its life.

There are several variations to the proton-proton cycle, all yielding freshly minted He^4 from four H^1 nuclei, with the concomitant liberation of energy that powers the Sun's light output. One important variation, known as the carbon-nitrogen cycle, uses a C^{12} nucleus as a catalyst: successive captures of four H^1 nuclei by C^{12} creates C^{13}, N^{14}, N^{15}, and finally returns the C^{12} nucleus with a newly synthesized He^4 nucleus. Often this cycle does not go to completion, and so a gradual shift of C^{12} to C^{13} and N^{14} occurs in the solar core. Additionally, Li and Be nuclei are very susceptible to fusion reactions with H at modest interior temperatures, thus are expected to be completely depleted throughout the solar interior.

Fusion of more complex nuclei requires even hotter and denser conditions, which will be accomplished by contraction of the core regions under the influence of the Sun's gravity after the transmutation of H into He is complete. When the core attains $T \simeq 10^8$ K and $\rho \simeq 1$ kg·m^{-3}, fusion of three He^4 nuclei into a carbon nucleus (C^{12}) occurs in a near-simultaneous collision called the triple-alpha cycle. Capture of a single He^4 by a C^{12} nucleus then can create the major form of oxygen O^{16}. Concurrently, the outer envelope of the Sun will expand and become highly convective, allowing "dredge-up" of the fusion cycle products toward the solar surface.

While the details of these complex element synthesis and dredge-up processes are not perfectly understood, chemical composition analyses of stars confirm the broad outlines of this scenario. For example, many stars show severely depleted Li abundances; even the Sun's surface Li content is about 100 times less than the "primordial" value shown in Fig. 1. This accords well with the prediction that Li should be easily destroyed in stellar environments (the buildup of Li, Be, and B probably occurs not in stellar fusion cycles but rather in high-energy "spallation" breakdown of heavier elements in the interstellar medium). The observed nonsolar abundances and isotope ratios of the C, N, O element group (e.g., enhanced C^{13} and N^{14}) in red giant stars clearly indicate the dredge-up to the surface of elements that have participated in earlier interior H→He fusion cycles. Some highly evolved stars also manage to bring to their surfaces some C^{12} newly synthesized in the triple-alpha cycle.

The Sun and other relatively low-mass stars cannot manufacture elements heavier than C or O in the major nuclear fusion chains, because their central temperatures and densities and temperatures will never grow large enough to induce further merging of nuclei. Eventually, these stars will return a fraction of the new elements to the interstellar medium in wind outflows or as they eject their outer envelopes as so-called planetary nebulae. Many of the nuclear synthesis products, particularly C and O, will remain with the remnant white dwarf cores of the stars.

More massive stars, late in their evolutionary stages, achieve the requisite central densities and temperatures to pass beyond H→He and He→C,O fusion. Major fusion reaction networks build elements up to $Z = 30$ in so-called carbon burning, neon burning, and silicon burning cycles. Theoretical nucleosynthesis calculations can reproduce most of the qualitative and quantitative features of the solar system abundance distribution. Note again, for example, the large abundances of even-Z elements Mg, Si, S, etc., compared to neighboring odd-Z elements Na, Al, and P. The dominant isotopes of these even-Z elements are composed of collections of He^4 nuclei. Those nuclei are the original products of H fusion cycles. The advanced burning cycles then synthesize the even-Z elements more easily than they do the odd-Z elements.

The major nuclear fusion cycles in the cores of massive stars must end with the production of Fe-peak ($Z \leq 30$) elements. The Fe^{56} nucleus has the largest binding energy of all nuclei, and thus further attempts to fuse Fe nuclei into larger-Z nuclei are endothermic. Instead of releasing energy to support the star, energy is drawn in, and pressure support against gravity is lost in the core regions, resulting in a rapid collapse and rebound shock explosion of the core regions. Such cataclysms are observed as extremely energetic supernovae. In the supernovae outbursts, layers of new elements from previous fusion cycles are strewn into the interstellar medium, along with new elements synthesized in the few-second supernova explosion. Elements from O through Zn enter the heavy element mix of the galaxy in this fashion.

Elements with $Z > 30$ cannot be made in major fusion stages. They are built almost exclusively near the end of stellar lifetimes via captures of free neutrons onto Fe-peak seed nuclei. The neutrons can be liberated slowly by other fusion reactions that accompany the quiescent He→C fusion stage, or rapidly during a supernova detonation. Captures of these neutrons by the seed nuclei, followed by ejection of electrons (beta decay) by the heavier nuclei when they become too neutron-rich for stability, is sufficient to create all of the elements through Pt, Pb, and even Th and U. Nuclear physics theory and experiment can account well for the fractions of each element created via slow and rapid nuclear reactions.

See also: ACTIVE GALACTIC NUCLEUS; ASTROPHYSICS; ATOM; ATOMIC MASS UNIT; ATOMIC NUMBER; ATOMIC PHYSICS; DECAY, NUCLEAR; DENSITY; ELECTROSTATIC ATTRACTION and REPULSION; ELEMENTARY PARTICLES; ELEMENTS; ELEMENTS, SUPERHEAVY; ELEMENTS, TRANSITION; ENERGY, SOLAR; FUSION; FUSION POWER; NUCLEAR PHYSICS; NUCLEOSYNTHESIS; NUCLEUS, ISOMERIC; PROTON; RED GIANT; STARS and STELLAR STRUCTURE; SUN; UNIVERSE

Bibliography

ANDERS, E., and GREVESSE, N. "Abundances of the Elements: Meteoritic and Solar." *Geochim. Cosmochim. Acta* **53,** 197–214 (1989).

CLAYTON, D. D. *Principles of Stellar Evolution and Nucleosynthesis* (McGraw-Hill, New York, 1968).

CROSWELL, K. *The Alchemy of the Heavens* (Doubleday, New York, 1995).

SEEDS, M. A. *Horizons, Exploring the Universe,* 4th ed. (Wadsworth, Belmont, CA, 1993).

CHRIS SNEDEN

ELEMENTS, RARE EARTH

The rare earths comprise the following elements of the periodic table: scandium (^{21}Sc), yttrium (^{39}Y), lanthanum (^{57}La) and the fourteen so-called lanthanide elements of the periodic table, cerium (^{58}Ce), praseodymium (^{59}Pr), neodymium (^{60}Nd), promethium (^{61}Pm), samarium (^{62}Sm), europium (^{63}Eu), gadolinium (^{64}Gd), terbium (^{65}Tb), dysprosium (^{66}Dy), holmium (^{67}Ho), erbium (^{68}Er), thulium (^{69}Tm), ytterbium (^{70}Yb), and lutetium (^{71}Lu). Here the symbol and the atomic number of the element are specified. These elements were considered to have low natural abundance, hence the name. Except for Pm which has no stable isotope, these elements are found naturally in monazite, a material composed of mostly $CaPO_4$, but containing about 50 percent by weight rare earths, of which the most abundant are Ce, La, Nd, and Pr, and in bastnaesite, which has high abundance of Eu and other rare earths. The pure separated metals of these elements are bright and silvery, but the oxides of La, Ce, Pr, and Nd corrode quickly.

The elements Y and La are directly below Sc in the Periodic table. They each have one electron in an interior $(n\text{-}1)d$ subshell and a completely filled outer ns^2 subshell, where n is the principal quantum number and the superscript indicates the number of electrons in the subshell. The chemical behavior of the lanthanides is similar to La since they have outer subshells of $5s^25p^66s^2$ and a partially filled $4f$ subshell, with increasing number of electrons in the subshell for increasing atomic number. Specifically, the electronic structure of the lanthanides is given by Ce {[Xe] $4f^15d^16s^2$}, Pr {[Xe] $4f^35d^05f^06s^2$}, Nd {[Xe] $4f^45d^05f^06s^2$}, . . . , Yb {[Xe] $4f^{14}5s^25p^65d^05f^06s^2$}, Lu {[Xe] $4f^{14}5s^25p^65d^15f^06s^2$}, where [Xe] refers to a xenon closed shell, and . . . indicates the continuation of the series with increasing number of electrons in $4f$. Rare earths have many sharp optical transitions in the visible and near infrared. These transitions originate due to excitations of the $4f$ electrons.

One of the important recent applications of rare earths involve their use in solid-state lasers, where they are doped into crystals used as the gain medium of the laser. The transitions in doped rare earth ions remain sharp even though there is a strong local field affecting the ions, due to shielding of the f electrons by the outer $5s$, $5p$, and $6s$ (in the lanthanides) electrons. Upon doping, rare earths can have trivalent character (Nd^{3+}, Er^{3+}, Ho^{3+}, Tm^{3+}, Gd^{3+}, Eu^{3+}, Yb^{3+}) or divalent character (e.g., Sm^{2+}). The divalent ions give up their outermost $6s$ electrons, and the trivalent ones also give up their $5d$ electron if their is one, or one of its $4f$ electrons. One of the most prevalent solid-state lasers is Nd doped yttrium aluminum garnet (YAG), which lases at a wavelength of 1.06 μm.

The most extensive use of rare earths is for red phosphors in television sets. Eu and Y oxides are the compounds used for this purpose. Rare earths are also used as catalysts for cracking of crude petroleum and hydrogenation of ketones. A commonly used polishing agent for glass is Ce oxide. Also, La_2O_3 is used in doping glass to make high index of refraction and low dispersion (small $dn/d\omega$, where n is the index of refraction and ω is the light frequency) glasses.

See also: DOPING; ELEMENTS; ION; ISOTOPES; LASER; NUCLEAR SHELL MODEL

Bibliography

KOECHNER, W. *Solid-State Laser Engineering*, 3rd ed. (Springer, New York, 1992).

YEHUDA B. BAND

ELEMENTS, SUPERHEAVY

A group of relatively stable elements, with atomic numbers close to 114 and mass numbers close to 298, that are predicted to exist beyond our present periodic table are known as superheavy elements. This island of superheavy nuclei has not yet been discovered experimentally, but somewhat lighter nuclei on a rock of stability with atomic numbers close to 110 and mass numbers close to 272 have been synthesized in reactions between two fairly heavy nuclei. Superheavy elements serve as an ideal testing round for competing theories of nuclear structure and stability, providing new insight into how ordinary nuclei behave.

Stability

There occur in nature about 300 nuclei, representing isotopes of elements containing from one to at most 94 protons. Some 2,200 additional nuclei have been made artificially during the past seventy years. The heaviest nucleus produced thus far has 112 protons and a mass number of 277; this is also the heaviest element produced thus far. When the known nuclei are positioned in a chart according to the number of neutrons and the number of protons that they contain, the resulting chart of the nuclides forms a narrow peninsula running diagonally from the lower left-hand corner to the upper right-hand corner. This peninsula terminates because as additional protons are added to a nucleus, the disruptive electrostatic forces between the positively charged protons grow faster than the cohesive nuclear forces that hold the nucleons (protons and neutrons) together. The large electrostatic forces cause heavy nuclei to decay rapidly by the emission of alpha particles and by spontaneous fission.

The possible existence of nuclei with increased stability beyond the tip of the peninsula is associated with the closing of proton and neutron shells. Nucleons orbit around the center of a nucleus within imaginary shells of increasing radius, like the layers of an onion. Nucleons in the outer shells are more energetic than those in the inner shells. A given shell can hold only a certain number of protons or neutrons, and when a shell is filled, additional nucleons must go into the next shell of larger radius and higher energy.

A nucleus with a completely filled shell of either protons or neutrons is said to be "magic," because it is relatively more stable than nuclei with either a larger or a smaller number of nucleons. If both the proton shell and neutron shell are filled, the nucleus is said to be doubly magic. For the known nuclei, the proton magic numbers are 2, 8, 20, 28, 50, and 82; the neutron magic numbers are 2, 8, 20, 28, 50, 82, and 126. The higher shells contain fewer protons than neutrons because of the strong electrostatic repulsion between protons.

Magic nuclei are spherical in shape, whereas nuclei between the magic nuclei are generally deformed, with shapes that are either prolate like an egg or oblate like a pumpkin. Such deformed shapes arise because the intermediate nuclei can lower their energy somewhat, and hence increase their stability, by rearranging their protons and neutrons into deformed shells accommodating a different number of nucleons. The closing of these deformed shells leads to a different set of deformed magic numbers.

By use of theories that reproduce the magic numbers and other properties of known nuclei, theorists have predicted that the next spherical proton magic number beyond 82 is 114 and that the next spherical neutron magic number beyond 126 is 184. In addition, they have predicted a deformed proton magic number at 110 and a deformed neutron magic number at 162. These predictions are made by solving the Schrödinger equation of quantum mechanics with an appropriate single-particle potential to describe the motion of the protons and neutrons.

Because superheavy nuclei can decay by either spontaneous fission, the emission of alpha particles, or the emission or capture of electrons, all of these decay modes must be considered when calculating their half-lives. Such calculations are usually made by use of a two-part approach, with the smooth trends of the potential energy of a nucleus taken from an improved version of the liquid-drop model and the local fluctuations taken from a single-particle model. When all decay modes are considered, the nucleus with 110 protons and 180 neutrons (corresponding to a mass number of 290) is predicted to have a half-life of about 2,000 years, according to recent calculations. The half-life of the nearby nucleus with 110 protons and 178 neutrons is predicted to be about four years. In addition to such spherical superheavy nuclei, deformed nuclei in the vicinity of 110 protons and 162 neutrons are predicted to be sufficiently stable to constitute a rock of stability.

Production

There are two possible methods for producing superheavy nuclei: (1) the multiple capture of neutrons and (2) a reaction between two fairly heavy nuclei. In the first method, which is like trying to sail over to the island through the surrounding sea of instability, a nucleus successively captures one or more neutrons and then converts a neutron into a proton through the emission of an electron. This method, which has produced many of the naturally occurring nuclei, would lead to nuclei on the neutron-rich, or southeastern, shore of the island. In the second method, which is like trying to fly over to the island, two fairly heavy nuclei collide with sufficient energy to produce a nearly spherical or slightly deformed compound nucleus, which can de-excite by the emission of neutrons and gamma rays. This method would lead to nuclei on the proton-rich, or northwestern, side of the island.

Since 1968 many scientists have searched for superheavy elements, both in nature and in the products of reactions between fairly heavy nuclei at accelerators. Much of this research has taken place at the Lawrence Berkeley Laboratory in Berkeley, California, at the Gesellschaft für Schwerionenforschung in Darmstadt, Germany, and at the Joint Institute for Nuclear Research in Dubna, Russia. No conclusive evidence has been found for true superheavy elements, but nuclei on the rock of stability have been produced through such reactions as $^{208}Pb + {}^{70}Zn \rightarrow {}^{277}112 + {}^{1}n$. Some of the produced nuclei in this region have half-lives of several seconds. Experiments that could produce still heavier nuclei are planned for the future.

See also: ELECTROSTATIC ATTRACTION and REPULSION; ELEMENTS; NUCLEAR SHELL MODEL; NUCLEAR SIZE; NUCLEAR STRUCTURE; NUCLEON

Bibliography

ARMBRUSTER, P., and MÜNZENBERG, G. "Creating Superheavy Elements." *Sci. Am.* **260** (5), 66–72 (1989).

BEMIS, C. E., JR., and NIX, J. R. "Superheavy Elements: The Quest in Perspective." *Comments on Nuclear and Particle Physics* **7**, 65–78 (1977).

HERRMANN, G. "Superheavy-Element Research." *Nature* **280,** 543–549 (1979).

HOFMAN, S., et al. "The New Element 112." *Z. Phys. A* **354**, 229–230 (1996).

MÖLLER, P., and NIX, J. R. "Stability and Decay of Nuclei at the End of the Periodic System." *Nucl. Phys. A* **549,** 84–102 (1992).

MÖLLER, P., and NIX, J. R. "Stability of Heavy and Superheavy Elements." *J. Phys. G* **20**, 1681–1747 (1994).

PATYK, Z., and SOBICZEWSKI, A. "Main Deformed Shells of Heavy Nuclei Studied in a Multidimensional Deformation Space." *Phys. Lett. B* **256**, 307–310 (1991).

WALKER, F. W.; PARRINGTON, J. R.; and FEINER, F. *Nuclides and Isotopes,* 14th ed. (General Electric, San Jose, CA, 1989).

J. RAYFORD NIX

ELEMENTS, TRANSITION

The transition elements occupy the middle of the periodic table. They include the elements scandium through zinc (atomic numbers 21 through 30), yttrium through cadmium (atomic numbers 39 through 48), lanthanum (atomic number 57), and hafnium through mercury (atomic numbers 72 through 80). All transition elements are metals. They include many important metals such as iron, nickel, copper, zinc, silver, gold, platinum, mercury, chromium, and titanium. Several transition elements are found in nature as pure metals, while others are obtained from ores that contain the transition element combined with oxygen (oxides), sulfur (sulfides), or phosphorus (phosphates).

Several of the transition metals, such as copper, platinum, silver, gold, and mercury, are very useful as electrical conductors and connections, because of their high conductivity and their resistance to corrosion. Transition elements are also employed in catalysts for industrial processes. Platinum and rhodium are used in the catalytic converters in automobiles.

The transition elements are important in biochemistry of living organisms. Iron is essential to the biological activity of hemoglobin, the compound that binds oxygen and carbon dioxide in red blood cells. Vitamin B_{12} contains cobalt. Iron is also an important component of several of the enzymes that enable plants to convert sunlight into chemical energy.

Most of the transition elements can assume several different charge states (oxidation states) in compounds and in crystals. These states differ in the number of electrons associated with the transition element. A very large variety of compounds can be formed that have several chemical groups (ligands) attached to a central transition element atom. The ability of transition elements to coordinate in this way to various molecules and chemical groups, together with their ability to accept or give up electrons, plays an important role in the biological and catalytic activities of these elements.

Many transition element compounds absorb visible light and are brightly colored. Transition metal ions give many gems their color, and transition metal compounds are employed as pigments in many paints.

See also: CONDUCTOR; ELEMENTS; METAL

Bibliography

COTTON, F. A., and WILKINSON, G., eds. *Advanced Inorganic Chemistry,* 5th ed. (Wiley, New York, 1988).

MICHAEL F. HERMAN

ELEMENTS, TRANSURANIUM

Transuranium elements have an atomic number, Z, greater than the value for uranium (U, $Z = 92$). Although they might have been present at some early time in the universe, atoms of elements with an atomic number greater than uranium have no stable isotopes and their radioactive half-lives are so short that they are not found naturally on Earth today. There are only negligible traces of two transuranium elements, plutonium and neptunium, that today occur naturally on Earth. In order to be studied, transuranium elements have to be synthesized in nuclear reactions. Today, transuranium elements are usually produced artificially by bombarding heavy nuclei with lighter ones.

During an investigation of fission in uranium, Edward M. McMillan and Philip H. Abelson discovered the first transuranium element, neptunium (Np, $Z = 93$) in 1940. It was produced by neutron bombardment of uranium. Glenn T. Seaborg, McMillan, Joseph W. Kennedy, and Arthur C. Wahl discovered the second transuranium element, plutonium (Pu, $Z = 94$), using deuteron bombardment of uranium in 1941. Both discoveries used the University of California's 60-inch cyclotron at Berkeley.

Neptunium is the first transuranium element in the actinide series of the periodic table. The actinides have chemical similarities to the element actinium (Ac, $Z = 89$), just as members of the lanthanide, or rare-earth, series have chemical properties similar to lanthanum ($Z = 52$). The actinide elements include thorium (Th, $Z = 90$), protactinium (Pa, $Z = 91$), uranium, and the first eleven transuranium elements.

Further investigations of transuranium elements and related subjects were conducted in secrecy during World War II. The third member of the transuranium elements, americium (Am, $Z = 95$), was first produced by intense neutron bombardment of plutonium and identified by Seaborg, Ralph A. James, Leon O. Morgan, and Albert Ghiorso in late 1944 and early 1945 at the University of Chicago's Metallurgical Laboratory. Curium (Cm, $Z = 96$) was produced by helium-ion bombardment of ^{239}Pu (the superscript denotes the isotope's atomic mass number) using the Berkeley 60-inch cyclotron and identified by Seaborg, James, and Ghiorso in 1944, also at Metallurgical Laboratory. It was named after Pierre and Marie Curie.

The privilege of naming a chemical element has been traditionally given to its discoverer. The first transuranium element was named neptunium because it is just beyond uranium in the periodic table just as the planet Neptune is just beyond Uranus. Similarly, plutonium was named for the second planet beyond Uranus, Pluto. Berkelium (Bk, $Z = 97$) was produced by helium-ion bombardment of ^{241}Am and identified at Berkeley by Stanley G. Thompson, Ghiorso, and Seaborg in late 1949. It was named in honor of that city of Berkeley.

There was rapid progress in the discovery of transuranium elements during the 1950s, but as the atomic number of the new species increased, their lifetimes became shorter and they became progressively harder to produce. Californium (Cf, $Z = 98$) was first prepared by helium-ion bombardment of microgram quantities of ^{242}Cm by Thompson, Kenneth Street Jr., Ghiorso, and Seaborg in 1950 at Berkeley and named after the state of California.

Very heavy uranium isotopes can be formed by the action of the intense neutron flux on the uranium. These heavy uranium isotopes can then decay into elements 99 and 100 and other transuranium elements of lower atomic numbers. Einsteinium (Es, $Z = 99$) was unexpectedly discovered in early 1953 by Ghiorso and coworkers in the debris of the "Mike" thermonuclear explosion which took place on the Eniwetok Atoll in the Pacific in 1952. This was the first large-scale test of a thermonuclear device. Einsteinium was named after Albert Einstein.

Fermium (Fm, $Z = 100$) was isolated from the heavy elements formed in the same explosion in 1953 by Ghiorso and coworkers. Fermium was named for Enrico Fermi.

Transuranium elements beyond fermium are so unstable that they must be made, one atom at a time, using an accelerator to bombard a heavy element target with ions. Mendelevium (Md, $Z = 101$) was produced by bombardment of minute quantities of ^{253}Es with helium ions at the 60-inch Berkeley cyclotron in 1955. The discovery and identification was made by Ghiorso, Bernard G. Harvey, Gregory R. Choppin, Thompson, and Seaborg. It was named after Dmitri Mendeleev.

In 1957 an international group of scientists, working at the Nobel Institute for Physics in Stockholm, reported the discovery of nobelium (No, $Z = 102$) and named it after Alfred Nobel. However, their discovery has never been confirmed and the original claim was probably erroneous. Subsequently, Georgiy N. Flerov and coworkers, at the Kurchatov Institute of Atomic Energy in Moscow, reported a radioactivity they thought might be attributed to element 102. The more definitive work in the discovery of element 102 was performed in 1958 by Ghiorso, Torbjorn Sikkeland, John R. Walton, and Seaborg when they bombarded curium isotopes with ^{12}C ions in the Berkeley Heavy-Ion Linear Accelerator (HILAC). The Berkeley scientists probably have the best claim on the discovery of element 102, and they suggested retention of nobelium as its name.

Lawrencium (Lr, $Z = 103$) was discovered in 1961 by Ghiorso, Sikkeland, Almon E. Larsh, and Robert M. Latimer using the HILAC at Berkeley to accelerate boron ions for bombardment of a few micrograms of californium, which was itself produced in a nuclear reactor. It was named after Ernest O. Lawrence. Lawrencium is the last of the actinide series of the periodic table.

In 1964 Flerov and coworkers claimed the production of element 104, the first element beyond the actinide series, as a result of bombardment of ^{242}Pu with ^{22}Ne ions in their heavy-ion cyclotron at the Joint Institute for Nuclear Research (JINR) in Dubna, Russia. In 1969 Berkeley scientists Ghiorso, Matti Nurmia, James Harris, Kari Eskola, and Pirrko Eskola were, however, probably the first to identify element 104. The Berkeley researchers suggested that it be named rutherfordium after Ernest Rutherford. In 1994 a commission of the International Union of Pure and Applied Chemistry (IUPAC) recommended that the element be called dubnium (Db, $Z = 104$), named after the JINR at Dubna, Russia. Element 104 is the first member of the transactinide series of elements.

In 1968 Flerov and coworkers in Dubna reported production of element 105 in Dubna, but the small number of observed events and a discrepancy between the reported element 105 alpha-particle energies and those known to be correct place this claim in doubt. In 1970 Ghiorso, Nurmia, K. Eskola, Harris, and P. Eskola synthesized element 105 in the Berkeley HILAC and unambiguously identified it. They suggested the element be named hahnium (Ha, $Z = 105$) in honor of Otto Hahn. The IUPAC has recommended that the element be called joliotium (Jl, $Z = 105$) after Frederic Joliot-Curie.

Credit for the discovery of element 106 is shared between Ghiorso and coworkers at Berkeley and Flerov, Yuri T. Oganessian, and coworkers at Dubna. The Ghiorso group used the Super HILAC (which is the rebuilt HILAC) to bombard ^{249}Cf with ^{18}O in 1974. At nearly the same time, another isotope of element 106 was reported to be made at JINR in bombardment of a somewhat lighter target, ^{208}Pb with a much heavier projectile ^{54}Cr. The Lawrence Berkeley Laboratory group suggested that element 106 be named seaborgium (Sg), in honor of Glenn T. Seaborg.

Element 107 was unambiguously synthesized and identified in 1981 by a team at the Gesellschaft für Schwerionenforschung (GSI) in Darmstadt, Germany, under the leadership of Peter Armbruster and Gottfried Münzenberg. They used the nuclear fusion that occurs when ^{209}Bi target atoms are bombarded with ^{54}Cr ions. The same team synthesized and identified element 108 in 1984 and one atom of element 109 was produced in 1982. In 1987 scientists at Dubna claimed to have discovered element 110, but their finding has never been confirmed. Atoms of element 110 were detected in 1994 by the GSI group, lead by Armbruster and Sigurd Hofmann, as a result of bombarding atoms of lead with nickel ions. Later in 1994, the researchers at the GSI heavy-ion cyclotron announced the discovery of element 111 using nickel ions to strike a bismuth target. The detection and identification of individual atoms of elements 109, 110, and 111 were accomplished by examining the pattern of alpha particle emissions resulting from their radioactive decay.

The transuranium elements have been found to be less and less stable and increasingly difficult to produce as their atomic number increases. However, theoretical analysis suggests that there may be superheavy transuranium elements in the region of atomic number $Z = 114$, and neutron number, $N = 184$, that exhibit higher stability because of filling nuclear shells. Some theoretical calculations even suggest superheavy half-lives that would be comparable to the age of the universe.

Many of the transuranium elements are produced and isolated in large quantities in nuclear fission reactors by neutron capture. Plutonium is produced in ton quantities. Neptunium, americium, and curium are produced in kilogram quantities, berkelium is produced in 100 mg quantities, californium is produced in gram quantities, and einsteinium in milligram quantities.

Plutonium (^{239}Pu) undergoes fission following neutron capture and has uses as fuel for nuclear fission reactors and as an explosive ingredient in nuclear weapons. The heat produced by alpha-particle decay of the plutonium isotope, ^{238}Pu, is used in thermoelectric generators for space exploration such as in the Apollo missions to the Moon, the Viking Mars Lander, and the Pioneer and Voyager probes. The gamma radiation from americium is useful for a wide range of industrial gauging applications, the diagnosis of thyroid disorders, and in smoke detectors. When mixed with beryllium, americium can be used to make neutron sources that are used in daily oil well logging operations throughout the world.

See also: ACCELERATOR; ATOMIC NUMBER; CYCLOTRON; ELEMENTS; ELEMENTS, SUPERHEAVY; ION; ISOTOPES

Bibliography

SCHÄDEL, M. "Chemistry and the Single Atom." *Science Spectra* **1,** 26 (1995).

SEABORG, G. T. "The New Elements." *American Scientist* **68,** 279 (1980).

SEABORG, G. T., and LOVELAND, W. D. *The Elements Beyond Uranium* (Wiley, New York, 1990).

SEABORG, G. T.; LOVELAND, W. D.; and MORRISSEY, D. J. "Superheavy Elements: A Crossroads." *Science* **203** (4382), 711 (1979).

MICHAEL D. CRISP

EMF

See ELECTROMOTIVE FORCE

EMISSION

Emission of light refers to the process by which atoms produce photons. The electrons in an atom can exist only in discrete energy states, and they emit photons (light) when they go from a high energy state to a lower state. In 1913, Niels Bohr concluded that if an electron in an atom went from a state with energy E_n to another state with energy E_m, then the frequency ν of the light emitted would be $\nu = (E_n - E_m)/h$, where h is the Planck constant. Since the frequency of the photon depends on the energy levels of the atom, measuring the frequency of light emitted by an atom provides important information about atomic structure.

In order to emit a photon, an atom must first be "excited"; it must have an electron in a high energy (excited) state. The atom can be excited in several different ways: by absorbing one or more photons, by colliding with other atoms, or by colliding with charged particles (electrons, protons, etc.). For example, in a neon light, free electrons collide with the neon atoms and excite them, and when the neon atoms return to their unexcited (ground) state, a photon is emitted.

We can determine whether an atom has emitted a photon by detecting the photon. Photons emitted by atoms can be detected by various devices, but many depend on the photoelectric effect to convert light energy into an electrical signal. When a photon strikes such a detector, it will be absorbed by an atom in the detector and an electron will be ejected (a photoelectron). The photoelectron can then produce an electrical signal that will register on a meter, a computer, or another measuring device. Photomultipliers, which operate on this principle, are capable of detecting even a single photon.

See also: EXCITED STATE; GROUND STATE; PHOTOELECTRIC EFFECT; PHOTON

Bibliography

FISHBANE, P. M.; GASIOROWICZ, S.; and THORNTON, S. T. *Physics for Scientists and Engineers,* 2nd ed. (Prentice Hall, Englewood Cliffs, NJ, 1996).

MOORE, J. H.; DAVIS, C. C.; and COPLAN, M. A. *Building Scientific Apparatus,* 2nd ed. (Addison-Wesley, Reading, MA, 1989).

JUSTIN M. SANDERS

EMISSION, THERMIONIC

The term "thermionic emission" describes the emission of electrons from a solid when it is heated in a vacuum. It can be observed in an apparatus that maintains a vacuum around an electrode that is heated (the cathode) and an electrode that is not heated (the plate). When the cathode is biased negatively with respect to the plate, electrons that escape from the cathode are swept across the vacuum and collected by the plate. The resulting current is immeasurably small at room temperature, but increases rapidly with increasing cathode temperature, becoming large enough to be studied at temperatures of 500–1,500 K, depending on the type of material. This phenomenon is the basis of vacuum-tube electronics.

Thermionic emission occurs because, at any temperature except absolute zero, some electrons in a solid occupy states high enough in energy to escape from the solid. Under conditions of thermodynamic equilibrium, the probability that an electronic state is occupied is given by the Fermi distribution, as follows:

$$N(E)\big|_{\text{occupied}} = N(E)\ \frac{1}{1 + e^{(E - E_f)/kT}}, \qquad (1)$$

where $N(E)$ is the density of electronic states $(1/\text{eV}\cdot\text{M}^3)$, E is the energy of the electronic state (eV), E_f is the energy of the Fermi level (eV), k is Boltzmann's constant (eV/K), and T is the temperature (K). To escape from a solid, an electron must occupy a state with an energy of at least E_V, the energy of an electron at rest in the vacuum outside the solid, or equivalently

$$\frac{(E - E_f)}{kT} \geq \frac{(E_V - E_f)}{kT} = \frac{W}{kT}, \qquad (2)$$

where W is the work function of the solid's surface (eV). For different metal surfaces, W ranges from 1.9 to 5.4 eV. Thus (W/kT) is a large number and the exponential term in Eq. (1) dominates, giving thermionic emission its strong dependence on temperature.

To compute the thermionically emitted current, one must compute the current carried by all pertinent electronic states (those occupied, with energy above the vacuum level, and with electron velocity directed toward the surface) and one must compute what fraction of that current passes through the surface layer. During the early part of the twentieth century, theorists did this computation assuming two things: first, that electronic states are those of free electrons in a box; and second, that a fraction of the electrons can be reflected at the surface. The result is the Richardson equation

$$J = AT^2 e^{-W/kT},\qquad (3)$$

where J is the current density due to thermionic emission (A/m^2), and A is a constant, typically about $6 \times 10^5 (A/K^2 m^2)$. In this theory, A and W are constants. They are determined by fitting Eq. (3) to measurements of current versus temperature. In fact, W varies with temperature. On metals, the Fermi level is tied to a high density of partially filled states so W is approximately constant (it varies by less than 0.1 eV); but on semiconductors and insulators, the Fermi level often lies in a band gap so that W varies significantly. Thus, Eq. (3) was found to be useful mainly when working with experimental data obtained from metal cathodes. When used this way, Eq. (3) provided a framework for years of research and development aimed at improving the cathodes of vacuum tubes.

See also: CATHODE RAY; EMISSION; FERMI SURFACE; INSULATOR; SEMICONDUCTOR; STATES, DENSITY OF

Bibliography

ALBERT, A. L., *Electronics and Electron Devices* (Macmillan, New York, 1956).

DEKKER, A. J., *Solid-State Physics* (Prentice Hall, Englewood Cliffs, NJ, 1963).

MOTT, N. F., and SNEDDON, I. N. *Wave Mechanics and Its Applications* (Dover, New York, 1963).

SPANGENBERG, K. R. *Fundamentals of Electron Devices* (McGraw-Hill, New York, 1957).

T. M. DONOVAN

A. D. BAER

ENERGY

Energy is sometimes defined as the ability to do work. Whether or not a system or body is capable of performing a certain task depends on the amount of energy the system or object possesses. For example, if 100 J of work are required to move a nail 1 cm farther into a board, then a moving hammer must have at least 100 J of energy in order to drive the nail 1 cm, plus extra energy to account for the energy lost as heat when the hammer impacts the nail.

One type of energy is kinetic energy. This is the energy an object has because of its motion. The motion can be either translational or rotational motion. The hammer mentioned above has translational kinetic energy while a spinning fan blade has rotational kinetic energy. A pitched baseball has kinetic energy because of its movement toward the catcher and, if it is spinning, it has additional kinetic energy because of its rotational motion. A swinging bat has energy and has the ability to do work on the baseball.

Another type of energy is potential energy. This is stored energy; energy that can be retrieved. An illustration of work increasing potential energy and the subsequent conversion back to work is the operation of a pile driver where a heavy weight is repeatedly raised by a motor and allowed to drop onto a piling, driving the piling into the ground. A motor does work against gravity as it raises the heavy weight of the pile driver; the weight gains gravitational potential energy as it is pulled to a higher position. The higher it is, the greater is its ability to do work. When the weight is released, gravity does work on the weight and pulls it down. The weight loses potential energy as it falls. The lost potential energy is converted to kinetic energy. The total mechanical energy of the falling weight is the sum of its kinetic and potential energies. Upon impact the weight does work on the piling, driving it into the ground. At the end of the cycle when the weight and piling come to rest, all of the mechanical energy has been converted to heat.

A stretched or compressed spring has elastic potential energy that can be retrieved when the spring relaxes to its equilibrium length. When an acrobat who is jumping on a trampoline lands, he does work on the springs of the trampoline and his kinetic energy is transferred to elastic potential energy of the stretched trampoline. When the springs relax and do work on the acrobat, the potential energy is given back to the acrobat as kinetic energy as he is propelled upward again.

The conservation of energy means that the total energy of a system and its surroundings is constant. The pile driver at its highest point starts with potential energy that changes to kinetic energy as the weight falls. Upon impact with the piling much of the energy is converted to thermal energy from the inelastic collision between the driver and the piling; some energy is transferred to kinetic energy of the piling as it moves. The falling weight of the pile driver loses its mechanical energy, but the kinetic energy and the thermal energy of the piling increase along with the thermal energy of the weight (due to its collision with the piling) so that the total energy remains fixed. The movement of the piling is quickly converted to thermal energy because of the work done by friction on the piling and its surrounding earth as the motion is halted. The energy lost by the system is gained by its surroundings, where it is dissipated.

Energy can be released in chemical reactions. The energy needed for muscular contractions used for walking or the heart beating comes from chemical energy within the body being converted to mechanical energy. Chemical reactions within the muscle cells release the energy that is used for muscular contractions. The burning of gasoline, also a chemical reaction, is used to power automobiles. A spark ignites a mixture of gasoline vapor and air that has been compressed in a cylinder by a piston. The chemical reaction between the components of the gasoline vapor and air mixture releases energy as thermal energy. The sudden rise in temperature produces a sudden rise in pressure that pushes the piston down. Because of this sudden expansion the thermal energy is converted to the mechanical energy of the moving piston, which is transferred to the wheels of the automobile.

The first law of thermodynamics is a statement of the conservation of energy. The net heat that enters a system can be used by the system to do work, and what remains after work is done increases the internal energy of the system. The thermal energy released in burning the gas-air mixture in the internal combustion engine of the automobile pushes the piston down and does work. The energy that was not used in doing work remains as internal energy. The temperature of the exhaust is much higher than the temperature of the gas-air mixture prior to ignition, indicating that its internal energy has increased. Not all of the thermal energy released in the burn is transferred to work; heat is exhausted or dissipated and the opportunity to do work is lost.

Another form of energy is electrical energy. Devices such as batteries, generators, and alternators produce electric fields in conductors that do work on free electrons and cause electrons to move and, hence, produce currents. This electrical energy can be dissipated in resistive materials and produce thermal energy. This is the process in electrical appliances such as incandescent lamps, toasters, heaters, and irons. Or, if the currents exist in the presence of magnetic fields, then they experience torques that can turn an armature and produce mechanical energy as occurs in electric motors.

Batteries produce electrical potential energy by chemical means. Chemical reactions in the battery maintain an excess of positive charge on one terminal and negative charge on the other. While the circuit that is connected to the battery is expending energy, the battery supplies energy to the circuit from the chemical reactions in the battery. The work done by the battery is equal to the work done by the circuit to which it is connected. An automobile uses some of the mechanical energy of its motor to operate an alternator to produce electrical energy, which sends a reverse current through the battery to recharge or rebuild the chemicals of the battery for longer life.

The energy that is used to drive a generator to make electrical current for our homes and cities is quite often thermal energy from burning fuel or the mechanical energy from falling water at a hydroelectric dam. The thermal energy released in burning the fuel is converted to the mechanical energy of a spinning generator that produces electricity.

In chemical reactions energy is released when two atoms or molecules combine to form new molecules that are more tightly bound. For example, carbon atoms in molecules of a fossil fuel can combine with oxygen in air to form molecules of carbon monoxide. The combination of carbon with oxygen is a lower energy state than carbon in its molecular form in fuel. Energy is released in the reaction of carbon with oxygen as kinetic energy of the constituent particles of the reaction and as kinetic energy of nearby molecules. The increase of kinetic energy on the atomic scale manifests itself as an increase in temperature of the system.

Molecules are held together by attractive electrical forces. The amount of work that must be done in order to pull the molecule apart is called the binding energy or dissociation energy of the molecule. Work done against the attractive electrical forces in separating the molecule increases the potential energy of the system. When the atoms of the molecule are far apart, then it is said to have no potential energy. Therefore, the work done in separating the molecule increases its potential energy from a negative value to zero potential energy. In the bound molecular state the molecule has negative potential energy that is equal to its binding energy. When atoms combine to form bound molecules, the decrease of potential energy must be given to a close-by molecule as energy or be radiated away.

Another example of binding energy is the negative potential energy with which electrons are bound to atoms. Electrons bound on atoms have kinetic energy and negative potential energy. The total energy of the electron is the sum of its kinetic energy and potential energy. For a bound electron the magnitude of its potential energy is greater than its kinetic energy and, therefore, its total energy is negative. The energy that is required to raise the total energy of the weakest bound electron of an atom to zero is called the ionization potential of the atom. But, for the reverse process, if an electron is captured by an ion to form a neutral atom, then it is necessary for the electron to lose energy by some process such as radiation.

At the center of the atom is the nucleus, where protons and neutrons are bound together by nuclear forces. The binding energy of a nucleus is typically millions of electron volts (eV) compared with tens of electron volts for the binding energy of an electron on an atom. Because of the very large energies involved in nuclear reactions it is possible to observe changes of mass. The mass of a stable nucleus is less than the sum of the masses of the protons and neutrons that make up the nucleus. If Δm represents the difference of the sum of the masses of the protons and neutrons minus the mass of the nucleus, then the binding energy of the nucleus is $E = \Delta m c^2$ where c is the speed of light. The mass difference Δm is called the mass defect.

Certain large unstable nuclei undergo radioactive decay into another nucleus plus another particle. The energy released is $E = \Delta m c^2$ where Δm is the difference between the mass of the unstable nucleus and the sum of the masses of the two products. The released energy appears as kinetic energy of the products. Certain uranium nuclei can absorb neutrons and then undergo fission into smaller nuclei whose total mass is less than that of the splitting atom. This lost mass appears as kinetic energy of the fission fragments, and this kinetic energy is quickly converted to thermal energy of the surroundings by inelastic collisions. Nuclear reactors use fission reactions to produce heat to drive generators that produce electricity.

In a fusion reaction two small masses are combined into one nucleus that has a mass less than the total mass of the two initial particles. And, as in the case of fission, the mass defect appears as released energy.

The nuclear reactions illustrate that mass can be converted to energy and vice versa, so it becomes necessary to include mass in the expression for total energy. All particles at rest have a rest-mass energy given by $m_0 c^2$, where m_0 is the mass of the particle at rest. It is the rest-mass energy that can be converted to other forms of energy, and so it must be included in a statement of the conservation of energy. Total energy is the sum of the potential, kinetic, and rest-mass energy.

When the potential energy is zero, the total relativistic energy for a moving particle that has kinetic energy is given by $E^2 = p^2 c^2 + m_0^2 c^4$, where p is the momentum of the particle. This energy is the square of the sum of the relativistic kinetic energy and the rest-mass energy, or $E^2 = (K + m_0 c^2)^2$. When the momentum p is small, the kinetic energy is the nonrelativistic quantity $\frac{1}{2}mv^2$. In chemical reactions or in processes that are nonrelativistic, changes in the rest-mass energy are too small to be measured and it can be left out of the equation for total energy.

The source of energy for the earth is the Sun. It is the energy for growing plants that are the source of energy for animals or for plants that once lived and are now fossil fuels. It is the energy that drives the climate, produces wind, and evaporates water to produce rain. Solar energy can be collected and changed to thermal energy for heating homes and buildings. Or, it can be collected by solar cells to produce electrical energy. Solar energy is a source of energy that will be present for as long as life survives on Earth.

See also: ENERGY, ACTIVATION; ENERGY, CONSERVATION OF; ENERGY, FREE; ENERGY, INTERNAL; ENERGY, KINETIC; ENERGY, MECHANICAL; ENERGY, NUCLEAR; ENERGY, POTENTIAL; ENERGY, RADIANT; ENERGY, SOLAR; ENERGY AND WORK

Bibliography

ROMER, R. H. *Energy: An Introduction to Physics* (W. H. Freeman, San Francisco, 1976).

TIPLER, P. A. *Physics for Scientists and Engineers*, 3rd ed. (Worth, New York, 1991).

ALAN K. EDWARDS

ENERGY, ACTIVATION

The ability to observe products in a chemical reaction depends on whether significant amounts of the products are formed and on how rapidly they are formed. If a reaction proceeds extremely slowly, product levels may be difficult to detect. The rate of reaction indicates how quickly reactants disappear and is expressed as

$$\text{rate} = k[A]^m[B]^n,$$

where $[A]$ and $[B]$ are the molar concentrations of the reactants, k is the rate constant, and m and n are the order constants. The order constants describe how the reaction rate depends on changes in the concentration of any one reactant. The rate constant k is temperature dependent and is given by the Arrhenius equation:

$$k = Ae^{-(E/RT)},$$

where A is a proportionality constant, R is the gas constant ($8.314\ \text{J}\cdot\text{mol}^{-1}\cdot\text{K}^{-1}$), T is the temperature in Kelvins, and E is the activation energy. An increase in temperature increases the rate constant and therefore the reaction rate.

In a mixture of reactants an enormous number of collisions take place every second. At room temperature only a few of the collisions result in the formation of products. For collisions to be effective, the atoms of the reactants must be oriented correctly so that bond formation can occur and the reactants must collide with a certain miminum kinetic energy.

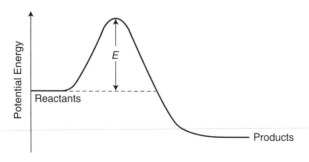

Figure 1 A potential energy diagram for an exothermic reaction.

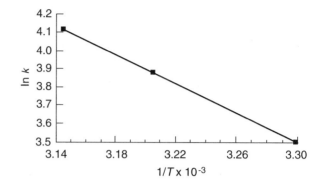

Figure 2 An Arrhenius plot.

This minimum kinetic energy is the activation energy E. The activation energy is the energy required to overcome the electrostatic repulsion of the reactants so that chemical bond breaking and formation can take place. The activation energy is shown in a potential energy diagram for an exothermic reaction (Fig. 1). The horizontal axis is called the reaction coordinate and indicates the degree to which the reactants have been converted into products. The reactants must have enough kinetic energy to overcome the potential energy barrier presented by the activation energy. In an exothermic reaction the potential energy of the products is less than that of the reactants. This decrease corresponds to an increase in the kinetic energy and consequently the temperature of the system.

The activation energy for a given reaction is determined from experimental data. Rate constants are determined for several different temperatures. If the Arrhenius equation is rewritten as

$$\ln k = \ln A - (E/R)(1/T),$$

a plot of $\ln k$ versus $1/T$ yields a straight line with slope $-(E/R)$. The activation energy is then determined directly from the slope and typically has units of Joules per mole or kilojoules per mole. An example of an Arrhenius plot is shown in Fig. 2.

See also: ELECTROSTATIC ATTRACTION and REPULSION; ENERGY, KINETIC; KINETIC THEORY

Bibliography

HILL, T. L. *An Introduction to Statistical Thermodynamics* (Addison-Wesley, Reading, MA, 1960).

HUMPHREYS, D. A.; COLIN BAIRD, N.; and ROBINSON, E. A. *Chemistry* (Allyn & Bacon, Boston, 1986).

CYNTHIA GALOVICH

ENERGY, CONSERVATION OF

In any physical process occurring in isolation from all outside influences, the total energy of the system in which the process occurs remains constant. This is the law of energy conservation. If the process occurs in a system that is *not* isolated, the energy of the system may change. It changes by an amount exactly equal to the energy that is added to or subtracted from the system during the change. In either case, observations support the conclusion that energy is neither created nor destroyed. This makes it one of the fundamental concepts of science.

Energy may, however, readily be changed from one form to another. These transformations occur at every level of science and technology: in elementary particle collisions, nuclear reactions, chemical reactions, human metabolism, transportation, electrical transmission, illumination, thunderstorms, and more. It is difficult to think of any process, physical or biological, that does not involve energy transformation.

The transformation of energy from one form to another may be accomplished directly, as when a radioactive nucleus emits a gamma ray, turning nuclear energy into electromagnetic energy. Or it may be accomplished indirectly, through the intermediary of work or heat.

Work is associated with a force moving something through a distance. It is measured simply as the product of the force and the distance: $W = Fd$. (In case the force is acting at an angle to the direction of motion, work is the magnitude of the force times the magnitude of the displacement times the cosine of the angle between the force and the displacement. This is the scalar product of the vector quantities force and displacement.) Work is often associated with forces and displacements in the large-scale, or macroscopic, world. One stage of the transformation of chemical energy in coal to electrical energy in a transmission line, for instance, is the work done by a steam turbine as it exerts a force on the rotating coils of an electric generator. A parent pushing a child in a swing is doing work that transforms chemical metabolic energy in the parent's body into mechanical energy of the swing and the swinging child.

Heat is the transfer of energy via the microscopic interactions of myriads of atoms or molecules. In the Sun, heat is an intermediary between the nuclear fusion source at the center of the Sun and the radiant energy emitted at the surface of the Sun. In the cooling of a cup of coffee, heat is the mechanism of energy transfer from the internal energy of the coffee to the internal energy of its surroundings. (Internal energy is sometimes called heat energy, a terminology dating from a time when the two concepts were not distinguished. Heat is now defined as a mode of energy transfer, not a measure of energy content.)

Energy associated with motion is called kinetic energy. The nonrelativistic formula for kinetic energy is $K = \frac{1}{2}mv^2$, one-half of the mass of an object times the square of its speed. (The relativistic definition differs appreciably from this definition only at speeds near the speed of light.) Because the speed is squared in this formula, the kinetic energy of an automobile moving at 60 mph is four times its kinetic energy at 30 mph. An object may gain kinetic energy by having work done on it. It may lose kinetic energy by doing work on something else.

Energy associated with the relative position of objects that exert forces on one another is called potential energy. A common form of potential energy is gravitational potential energy, related to the distance of an object from the center of the earth—or, for example, the distance of a planet from the Sun. Potential energy is also associated with mechanical and electrical forces. Stretching a spring adds potential energy to it. An electron gains potential energy as it moves farther from an atomic nucleus.

If a book is raised from the floor to a shelf, it gains potential energy because its distance from the center of the earth increases. If the book falls from the shelf back to the floor, it loses potential energy but gains kinetic energy during the fall. Just before it hits the floor, its kinetic energy is almost exactly equal to its loss of potential energy. This is an example of energy conservation. It is not quite exact, because as the book falls, there is a small amount of heat transfer to the air. When the book hits the floor, its kinetic energy is dissipated through atomic processes—that is, through heat—to the floor and the book itself, each of which gains some internal energy and becomes a little warmer. Some of the kinetic energy is also transformed into sound energy.

Kinetic and potential energy together are often called mechanical energy. For processes in which friction is small and other energy transformations are not occurring, the sum of kinetic and potential energies remains approximately constant. A swinging pendulum, for instance, regularly exchanges kinetic and potential energy as the sum of the two remains nearly constant. For a satellite coasting in a high orbit, where friction is negligible, the sum of its kinetic and potential energies remains accurately constant. If a rocket engine is fired, however, the satellite may gain or lose mechanical energy. In that case, chemical energy is transformed into energy of the satellite and energy of the rocket exhaust.

A collision in which kinetic energy is conserved is called an elastic collision. This can occur when one proton strikes another and the two rebound, the sum of their kinetic energies after the collision being the same as the sum before the collision. At very high energy, one proton striking another may create new particles. Such a collision is called inelastic. Total energy, of all forms, is conserved, but kinetic energy alone is not. The collision of two automobiles is inelastic. Their initial kinetic energies are transformed ultimately, via both work and heat, into internal energy of the automobiles, the pavement, and the air.

Not all of the forms of energy are truly independent of one another. Internal energy, for example, is really a measure of the kinetic and potential energies of all of the myriad submicroscopic constituents of a body. Chemical energy is, in effect, a form of potential energy—energy locked up within atoms and molecules that can be released in other forms through chemical reactions. Nuclear energy is likewise a form of potential energy. It is released through fission and fusion reactions and radioactivity. Radiant energy, or electromagnetic energy, is really kinetic energy of photons, the quantum particles that make up light. Electrical energy in an electric circuit is energy transmitted by the motion of electrons in wires. Despite the fact that some kinds of energy can be explained in terms of other kinds, there is good reason to keep the multiple designations. Energy is so manifold, showing itself in so many ways in so many places, with the ways of observing and measuring it so different from one realm to another, that it is appropriate to have specialized names reflecting specialized areas of application.

One important form of energy remains to be mentioned. In fact, it is the most fundamental form. It is mass. No doubt the most famous formula in all of science is $E = mc^2$, which states that mass multiplied by the square of the speed of light (c) gives the energy equivalent of that mass. The appearance of the speed of light in this formula is really to convert the unit of measurement of mass to the unit of measurement of energy. The physical essence of the formula is that energy equals mass. However, the large magnitude of c^2 by ordinary standards tells us that a relatively small amount of mass equates to a large amount of energy (in our standard units). In a nuclear reaction, the energy released is associated with a decrease in the mass of the reacting nuclei. The mass of a uranium nucleus, for example, is greater than the sum of the masses of the nuclei and neutrons into which it is transformed in a fission process. The mass that is lost makes its appearance as other forms of energy, possibly through heat transfer. In chemical reactions, too, the masses of the reacting atoms surely change, but the magnitudes of the change are too small to measure directly.

Mass is more than another form of energy. It is a complete measure of energy. If you have a box on a table and measure its mass hypothetically to great accuracy, you have measured its total energy content. If the box absorbs some radiant energy and gets warmer, it will also become more massive (in practice, by an amount too small to measure). If, within the box, a radioactive decay occurs and a gamma-ray photon shoots out of the box, the box loses some mass (again, an amount too small to measure). The Sun is such a "box." By radiating electromagnetic energy into space, it is losing four billion kilograms of mass every second.

The first suggestion that nature is governed by a general law of energy conservation came in the mid-nineteenth century. Julius Mayer in Germany and James Joule in England suggested that internal energy (then called heat) and mechanical energy are "equivalent"—meaning they can be transformed into one another, with the sum remaining constant. Mayer's and Joule's law of energy conservation united the fields of thermodynamics and mechanics.

Then, early in the twentieth century, Albert Einstein proposed that mass is equivalent to energy. The validity of his formula $E = mc^2$ was established first in the 1920s through measurements of nuclear reactions. Studies of elementary particles now confirm the formula every day in many laboratories. The idea that mass is a complete measure of the energy content of a system is a pillar of relativity theory.

Because energy is conserved, it can be a commodity. Most of the electrical energy transmitted from generating stations ends up in homes and factories (some is lost as heat in wires). The prices of oil, gas, and electricity can be compared based on their energy content. The transmission and utilization of energy is a central part of modern civilization. The energy used per person in industrialized societies is about 100 times the minimum energy needed to sustain life.

Energy can be assigned a "quality." All forms of potential energy—gravitational, electrical (including chemical) and nuclear—are of the highest quality. These forms are equivalent to internal energy at infinite temperature. The internal energy of a body has a quality dependent on its temperature—the higher the temperature, the higher the quality. (More exactly, the quality depends on the difference in temperature between the body and its environment.) As energy is used for practical purposes, it is normally degraded in quality. We think of it as being used up. Because it is conserved, energy is not used up. It is only transformed into less useful forms (i.e., lower quality).

The percentage of an energy source that is applied for a useful purpose is called the efficiency of use. Because energy has a cost and because many sources of energy are not renewable, it is desirable to achieve the highest efficiency of use. In common terminology, "energy conservation" means energy efficiency. A more efficient automobile or a more efficient refrigerator is said to conserve energy because it is more sparing in its use of an energy resource. Ultimately, most of the energy that we use is transformed via heat into the internal energy of our surroundings, where it cannot be tapped for further practical purposes.

Nuclear energy and the chemical energy in fossil fuels are non-renewable sources. Nuclear fusion energy, however, latent in the deuterium in the oceans, is sufficient to meet human needs for millions of years. Solar energy is a renewable source. The nuclear fusion that powers the Sun is expected to keep it shining for several billion years.

See also: COLLISION; ELECTROMAGNETIC RADIATION; ENERGY; ENERGY, INTERNAL; ENERGY, KINETIC; ENERGY, MECHANICAL; ENERGY, NUCLEAR; ENERGY, POTENTIAL; ENERGY, RADIANT; ENERGY, SOLAR; HEAT; HEAT, MECHANICAL EQUIVALENT OF; MASS-ENERGY

Bibliography

BONNER, F. T., and PHILLIPS, M. *Principles of Physical Science* (Addison-Wesley, Reading, MA, 1957).

FEYNMAN, R. P.; LEIGHTON, R. B.; and SANDS, M. *The Feynman Lectures on Physics*, Vol. 1 (Addison-Wesley, Reading, MA, 1963).

KENNETH W. FORD

ENERGY, FREE

In thermodynamics free energy is a function of state variables which reaches a minimum when the system is in equilibrium. It is the analog of the potential energy in mechanical systems.

The Helmholtz free energy F is such a function. For vapors, liquids, and solids the state variables are temperature T and volume V. When the two are fixed, internal energy U and entropy S obtain values that minimize $F \equiv U - TS$.

A special magnetic system clearly illustrates the opposing effects of U and S on F. The system consists of N moments of magnitude μ in a field B. The moments may align either along or opposite B. For this quantum system, internal energy is minimum if all the moments are along B, while entropy is maximum if half are along B, half opposite. Minimum $F(T,B)$ is achieved for intermediate alignment, which may be found by expressing U and S as functions of an order parameter, dependent on alignment. The magnetic moment of the system $M = \mu(N_+ - N_-)$, is such a parameter, where N_+ and N_- are the number of moments aligned along and opposite B. Then $U = -MB$, and

$$S = -\frac{Nk}{2}\left[\left(1 + \frac{M}{N\mu}\right)\ln\left(\frac{1}{2} + \frac{M}{2N\mu}\right)\right.$$
$$\left. + \left(1 - \frac{M}{N\mu}\right)\ln\left(\frac{1}{2} - \frac{M}{2N\mu}\right)\right],$$

where k is Boltzmann's constant. At equilibrium F is minimum, or $(\partial \Gamma / \partial M)_{B,T} = 0$, and $M = N\mu \tanh(\mu B/kT)$.

For systems at fixed temperature and pressure P, the Gibbs free energy $G(T,P)$ is minimized as U, S, and V obtain their equilibrium values. For example, the Gibbs energy of a system of N ideal gas molecules of mass m is

$$G = NkT \ln\left[\left(\frac{2\pi h}{m}\right)^{3/2} \frac{P}{(kT)^{5/2}}\right],$$

where h is Planck's constant. Equilibrium U, S, and V values are $(3/2)kT$, $-(\partial G/\partial T)_P$, and $(\partial G/\partial P)_T$, respectively.

For a real gas at a temperature below the critical value, the $G(T,P)$ surface will intercept that for the substance in its liquid or solid phase. Along the intercept the states are those of two-phase coexistence, and the projection of the coexistence curve is the vapor pressure $P(T)$. It is for systems with coexisting phases that the Gibbs energy is especially useful.

The parameter $\mu \equiv (\partial G/\partial N)_{P,T}$, where N is the number of molecules in a phase, provides a simple criterion for phase stability. To minimize G, that phase is stable for which the chemical potential μ is minimum. Phases coexist when their chemical potentials are equal.

Metastable phases, those having excess chemical potential, are common in nature, however, just as excess mechanical potential energy is common. Water vapor supercooled below the dew point is an example. A free energy barrier to condensation exists. The source is the work required to produce the surface of the embryo water droplet that serves as the seed for growth of the bulk liquid. Equivalent barriers exist wherever metastability occurs.

The Gibbs function facilitates the understanding of systems with more than one chemical species, such as solutions. Solvent species can exist in the vapor and liquid phases. Their chemical potentials are equal. Changes in the solute concentration affect solvent vapor pressure through the principle of minimum free energy as expressed by the Gibbs–Duhem equation. This principle is the basis of the Gibbs phase rule, which determines the number of degrees of freedom a system has. This is the number of variables whose values must be specified to know the state of the system. For the ideal gas it is two, P and T. For a saltwater system with precipitate it is one. When this one, the temperature, for example, is fixed, the division of water molecules between liquid and vapor phases and the division of salt ions between liquid and solid phases adjust until the Gibbs' energy is minimum.

Liquid or solid mixtures may be homogeneous at higher temperatures but separated at lower into two phases still mixed but differing in composition. For the latter, free energy varies nonlinearly with composition with the curve representing this variation having two inflection points. Unlike the water condensation with embryo droplet formation, phase separation may occur without nucleation as the system reduces its free energy. This spinodal decomposition is characterized initially by long-range composition fluctuations that may be revealed by light scattering similar to critical opalescence.

See also: ENTROPY; EQUILIBRIUM; GIBBS, JOSIAH WILLARD; MAGNETIC MOMENT; THERMODYNAMICS

Bibliography

KITTEL, C., and KROEMER, H. *Thermal Physics*, 2nd ed. (W. H. Freeman, San Francisco, 1980).

C. B. RICHARDSON

ENERGY, INTERNAL

Internal energy U, along with entropy S, are the two fundamental functions to describe a thermodynamic system in equilibrium. A macroscopic system, being an aggregate of atoms and molecules, is subject to the law of conservation of energy. Its macroscopic motion is subject to the conservation law of mechanical energy and, in the absence of electromagnetic fields, the remaining energy is conserved in an isolated system. This is the internal energy.

Historically, realization of the existence of a conservative (i.e., that which obeys the law of conservation) thermodynamic energy function had been developed prior to the emergence of the atomistic view of matter. Thus the concept had to be developed through strictly macroscopic experience. Early experiments and speculation as to the nature of heat by Benjamin Thompson (Count Rumford) led him to the conclusion that heat is a form of motion. Later experimental work by James Prescott Joule to establish the mechanical equivalent of heat, and the theoretical formulation by William Thomson (Lord Kelvin), conclusively showed that the energy of an isolated macroscopic system is conserved if heat is taken into consideration. This is the first law of thermodynamics, and the conserved function is the internal energy.

Conceptually, the difference in the internal energy between two states can be identified as the mechanical work done in taking the adiabatically enclosed (i.e., thermally insulated) system from an initial state (A) to the final state (B). If an adiabatic process $A \rightarrow B$ cannot be found (i.e., it is irreversible), a process $B \rightarrow A$ exists. Thus, mechanical method permits one to measure the internal energy difference between two states as long as there is not material leakage or addition. Naturally, internal energy difference is the only significant quantity, and the absolute value of the internal energy itself has no physical meaning.

An infinitesimal increase dU of internal energy of a system is the sum of work dW done to the system and the heat flow dQ into the system (i.e., the first law of thermodynamics) expressed as $dU = dW + dQ$, where a short bar on the differential symbol "d" as "d" signifies an inexact differential. This mathematical aberration is necessary to indicate that neither W nor Q are functions of the state. In other words, the quantities of heat and work "absorbed" by the system depend on the path the system traversed to reach its present (i.e., final) state from its initial state. In most cases, a state of a simple single-component system such as a gas can be represented by a rectangular coordinate system with axes representing pressure P and volume V, thus making it possible to describe a process with a path on the P-V diagram.

We can see the path-dependent nature of W and Q in an example such as shown in Fig. 1, which represents various states and processes of an ideal monatomic single-component gas on the P-V diagram. The curved path 1 from the initial state A to C

is an adiabatic compression during which the gas is compressed by an external agent (such as a hand pushing a piston) while the gas is insulated from its surroundings; thus, heat flow into the system is zero. The work done to the system ΔW_1 is represented by the area under the curve $C \rightarrow A$. For this path, $\Delta U = \Delta W_1$. Rectangular path (path 2) shown as $A \rightarrow B \rightarrow C$ represents two connected processes, $A \rightarrow B$ and $B \rightarrow C$. Of these, the first represents an increase in pressure while the volume is kept constant. This is obviously accomplished with an introduction of required quantity of heat ΔQ_1. Since the volume is kept constant, no work is done to the system. The second segment, $B \rightarrow C$, represents a process in which the pressure is kept constant during the compression. The compressive work done by an external agent here is given by the rectangular area under the line segment CB, which is $P_2(V_1 - V_2)$. In this case, the identical internal change ΔU is given by $\Delta Q_1 + P_2(V_1 - V_2)$.

Simple thermodynamic measurements of heat flow and changing temperature of molecular gas systems yield that portion of their internal energy that is due to rotation and vibration of the molecules, which are the degrees of freedom neglected for the ideal gas. From this, one can deduce molecular properties such as moments of inertia and discrete molecular vibration frequencies.

In liquids and solids, the internal energy is predominantly the potential energy of interaction among the atoms. Heat flow changes this by expansion and contraction due, in turn, to changing atom motion. This motion is collective and quantized in solids. Conducting solids add the energy of the electrons, kinetic and potential, to internal energy, and thermodynamics reveals quantum constraints on this energy, which are a consequence of the exclusion principle.

Internal energy may be changed by application of electric and magnetic fields to systems with permanent or induced dipole moments. Infinitesimal work is $E \cdot d\mathbf{\Pi}$ or $\mu_0 H \cdot d\mathbf{M}$, where E and H are the applied fields and $\mathbf{\Pi}$ and $\mathbf{M}$ are the electric and magnetic dipole moments of the system, respectively. These moments are the thermodynamic variables dependent on field values and temperature. An interesting example of internal energy change by magnetic work is cooling by adiabatic demagnetization. If the field is reduced to a small finite value, M and the entropy are constant. If M is constant, the magnetic work $\int H \cdot dM$ is zero. Since the process is adiabatic ($\Delta Q = 0$), $\Delta U = 0$. The temperature of the system

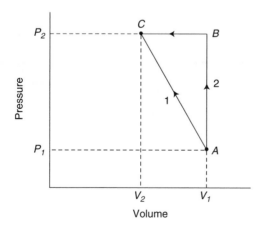

Figure 1 Pressure-volume diagram for a monatomic single-component gas.

must decrease as the negative magnetic energy increases to keep U constant.

A notable model system is electromagnetic radiation in a cavity of volume V. Each degree of freedom in the system has energy kT, according to thermodynamics. The number of degrees of freedom equals the number of modes in the cavity. In the wave frequency interval between ν and $\nu + d\nu$, the number is $(8\pi V\nu^2/c^3)\,d\nu$, where c is the speed of light. Mode frequencies range upward from near zero, giving, approximately,

$$U = (8\pi VkT/c^3)\int_0^\infty \nu^2 d\nu,$$

and the equation of state $PV = U/3$. Although neither result is physical, only an experiment in which internal energy is sampled by observing radiation escaping from a small hole in the cavity could reveal how the ultraviolet catastrophe contained in the expressions is avoided. This sampling shows that

$$U = (8\pi Vh/c^3)\int_0^\infty \frac{\nu^3 d\nu}{e^{h\nu/kT} - 1},$$

where h is Planck's constant. This is the first instance of internal energy measurements revealing quantum effects. More appropriately, these measurements showed, according to Albert Einstein, the existence of quantized radiation and photons and formed the basis of the quantum revolution.

See also: ADIABATIC PROCESS; DEGREE OF FREEDOM; ENTROPY; THERMODYNAMICS; THERMODYNAMICS, HISTORY OF

Bibliography

ZEMANSKY, M. W., and DITTMAN, R. H. *Heat and Thermodynamics: An Intermediate Textbook*, 6th ed. (McGraw-Hill, New York, 1981).

C. B. RICHARDSON

ENERGY, KINETIC

Kinetic energy is the energy that an object has because of its motion. If a body has mass m and velocity v then the magnitude of its kinetic energy is given by

$$K = \tfrac{1}{2}mv^2.$$

In the international system of units mass is given in kilograms, velocity in meters per second, and energy in joules. For a rotating object the kinetic energy is given by

$$K = \tfrac{1}{2}I\omega^2,$$

where I is the moment of inertia of the object about the axis of rotation and ω is its angular velocity. If the object has both translational motion and rotational motion, as does the wheel on a moving car, for example, then its kinetic energy is the sum of the two expressions.

If there is a net force acting on an object that produces a linear acceleration and/or a net torque that produces an angular acceleration, then its kinetic energy will change. The change in kinetic energy of the object is equal to the work done by the net force and/or torque.

Gottfried Wilhelm von Leibniz was the first to recognize the importance of kinetic energy in physics. In attempting to refute the arguments of René Descartes, who claimed that the product of mass times velocity, now known as momentum, was a conserved quantity in nature, Leibniz argued that the product of mass times the square of velocity (twice the kinetic energy) was the proper quantity to conserve. He gave the name *vis viva,* or living force, to this new concept and it became an integral part of future work on dynamical systems and in the field of heat and thermodynamics that was evolving from discussions in the late eighteenth century on perpetual motion machines. The Latin name that Leibniz had given to this quantity, double the kinetic energy, survived until the middle of the nineteenth century.

Thomas Young is credited with introducing the term "energy" to be used in place of Leibniz's *vis viva,* although this new expression did not gain common usage right away. In his lectures that were published in 1807 he suggested that the word energy be used rather than *vis viva* in order to better distinguish between the concepts of energy and force. In 1867 William Thomson (Lord Kelvin) and Peter

Gutherie Tait published volume one of *Treatise on Natural Philosophy*. In their book, which became the leading reference of its day, they introduced the modern-day phrase "kinetic energy" for the energy of a moving object and included the factor of one-half that Leibniz had omitted.

For objects traveling at velocities near the speed of light relativistic effects are important and kinetic energy is expressed as

$$K = \frac{m_0 c^2}{\sqrt{1 - (v^2/c^2)}} - m_0 c^2,$$

where m_0 is the rest mass of the object, v is its velocity, and c is the speed of light. In the limit where the object's velocity is small compared to the speed of light this expression reduces to the original $\frac{1}{2}mv^2$. This new expression for the kinetic energy is important in the design of high energy accelerators where particles reach relativistic speeds.

In 1924 Louis de Broglie postulated that particles had wave properties. According to de Broglie, the wavelength for a nonrelativistic particle is inversely proportional to the square root of its kinetic energy. Thus, if an electron is given sufficient kinetic energy, then it has a very short wavelength, shorter than visible light, and can be detected and used to form an image on a screen. Such properties are the basis of the electron microscope, which has a high resolving power because of the short wavelength of the energetic electrons.

The wave properties of particles lead to the quantization of kinetic energy on an atomic scale. If a particle is in a bound state, such as an electron bound to an atom, then its kinetic energy is no longer a continuous function of its mass and velocity but can have only finite values, which are determined by properties of the system.

Electron guns are used to produce beams of electrons in such devices as electron microscopes and television picture tubes. The cathode that is the source of electrons is set at a high negative electric potential in order to accelerate the electrons to a desired kinetic energy. These electrons can then be focused to a small spot and easily deflected. In a television picture tube the energetic electrons are swept across the colored phosphors of a screen in order to produce a picture. The kinetic energy of the electrons in electron guns or particles in accelerators are usually expressed in electron volts because their energy is established by accelerating them through known differences of electric potential that are expressed in volts.

Ion implantation is a commonly used technique to make semiconductor materials for many devices. Dopant materials are accelerated to energies ranging from tens-of-kilo-electron-volts to a few million electron volts and focused onto crystals such as silicon. The depth of the ion implantation can by controlled by controlling the kinetic energy of the dopant ions.

The kinetic energy of a large moving object of known mass, such as an automobile, can be determined by measuring its velocity. The kinetic energy of very small objects, such as electrons or other charged particles, can be measured by several different methods. Electrostatic analyzers use electric fields to measure kinetic energy by deflecting charged particles through a selected path. The path followed by the particle in an electric field depends on its kinetic energy. Magnetic analyzers use magnetic fields to deflect charged particles along a selected curved path that depends on the particles' momentum, which is related to the kinetic energy. Other techniques include solid state detectors that measure energy loss as a particle traverses certain materials and time-of-flight devices that measure velocity. Time-of-flight techniques can be used to measure the kinetic energy of neutral particles such as neutrons.

In the process of nuclear fission, energy is released in the form of kinetic energy of the fission fragments. As the fragments traverse the surrounding material, they give up their kinetic energy to the material and produce heat. This heat can be used to drive a generator to produce electricity.

See also: BROGLIE, LOUIS-VICTOR-PIERRE-RAYMOND DE; ELECTRON MICROSCOPE; ENERGY; ENERGY, CONSERVATION OF; ENERGY, POTENTIAL; FISSION; HEAT; KELVIN, LORD; MASS; MOMENTUM; VELOCITY; YOUNG, THOMAS

Bibliography

ELKANA, Y. *The Discovery of the Conservation of Energy* (Harvard University Press, Cambridge, MA, 1974).

LINDSAY, R. B., ed. *Energy: Historical Development of the Concept* (Dowden, Hutchinson, and Ross, Stroudsburg, PA, 1975).

TIPLER, P. A. *Physics for Scientists and Engineers,* 3rd ed. (Worth, New York, 1991).

ALAN K. EDWARDS

ENERGY, MECHANICAL

The concept of energy is surely one of humanity's greatest scientific creations. Any scientific study of the physical universe ultimately results in a study of matter and energy, which together form the stuff of which the universe is made. We acquire an intuition about matter, even some quantitative aspects, at a very early age. Ideas about energy, on the other hand, are more subtle and abstract. We generally cannot sense it directly; energy is not a "thing" that we can touch, hear, or see. Rather, we usually sense energy in an object (or wave) when it interacts with another object. We can sense the energy in a sound wave when the sound wave shatters a glass, for example. Or, we can sense the energy in a rolling bowling ball as the ball crashes into and scatters the bowling pins. Energy takes on a variety of forms, and it is usually observed when it changes from one form to another. Huge quantities of electrical energy, for example, are accumulated and stored in clouds as they are blown about during a thunder storm. But we do not actually sense this energy until a bolt of lightening strikes a nearby tree, and we sense light, sound, and, perhaps, even heat.

It is interesting to note that the concept of energy was not yet developed in the era of Isaac Newton, although the work–energy theorem is derived directly from Newton's second law.

It is virtually impossible to consider the concept of energy without considering a closely related concept: work. As described in detail elsewhere in this encyclopedia, work is the product of force $F_{||}$) exerted on an object and the displacement d of the object, where $F_{||}$ is the force component parallel to the displacement. Symbolically, $W = \mathbf{F} \cdot \mathbf{d}$ or $W = \int \mathbf{F} \cdot \mathbf{d}$. where W is work. This concept can best be appreciated by its application to a specific example such as the one below.

Mechanical energy—a form of energy that is distinguishable from other forms of energy such as electromagnetic, nuclear, atomic, or chemical energy—always involves the position or motion of matter or material objects. Mechanical energy can be divided into two forms: potential and kinetic. Let us treat mechanical potential energy first.

"Mechanical potential energy" is the term used to describe a condition in which an object or mass, by virtue of its location or position, has the potential of doing work. As an example of an object possessing potential energy by virtue of its position, consider Fig. 1.

Shown in Fig. 1 is a ball of mass m suspended directly above a nail that is partially driven into a block of wood. If the string that supports the ball is suddenly released, allowing the ball to fall on the nail, the ball will push the nail deeper into the wood. From a scientific point of view, we say that before the ball was released it had gravitational potential energy. Here the rationale for the term "potential energy" becomes clear; the energy possessed by the ball is not evident until after the ball is released. The ball has the potential of doing work (i.e., pushing the nail into the block), but it can do nothing until it is allowed to fall. Furthermore, the potential energy of the ball is said to be gravitational because as soon as the ball is released, the force of gravity mg accelerates the ball downward, eventually causing it to slam into the nail, pushing it deeper into the block. The quantity or amount of potential energy that the ball has, before it is released, is mgH_1. Notice that the original height of the ball, H_1, is measured from the center of the ball to the top of the block; this will be an important fact near the end of this discussion. The quantity mgH_1 is just the amount of *work* that the gravitational force would do *on the ball* if the ball were to fall from its position H_1 to the top of the block.

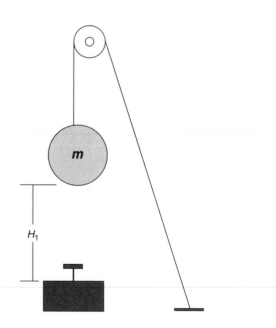

Figure 1 Illustration of an object possessing potential energy by virtue of its position.

As soon as the ball begins to fall it begins to acquire kinetic energy, or energy due to the fact that the ball is moving. Now the mechanical energy of the ball is suddenly very evident: surely anyone witnessing a massive ball falling quickly senses its energy. The amount of kinetic energy that the ball has is $\frac{1}{2}mv^2$, where v is the speed of the ball. Of course, as this ball falls we say it *loses* potential energy (since it loses height) and *gains* kinetic energy (since it is gaining speed). An important fact about these two forms of energy is that their sum is nearly constant; nearly constant because as the ball falls through air, a slight amount of energy is converted into heat as the ball collides with air molecules. (Objects that fall at very high speeds, such as meteorites, convert a great deal of their potential energy into heat.) Thus we can say that in the absence of air resistance, such as on the surface of the Moon or in a vacuum chamber, the sum of the potential and kinetic energy is constant. Mathematically $mgh + \frac{1}{2}mv^2 = mgH_1$, where h is the height of the ball above the top of the block when its speed is v. Finally we reconsider the connection between energy and work.

Figure 2 is a sketch showing the ball after it has come to rest on the nail head. The dotted lines show the initial position of the nail before being struck by the ball. That is, as a result of being struck by the ball, the nail was pushed into the block a distance d. During the time the nail was being pushed into the block, the ball exerted some average force against the nail; let's call this force F_{avg}. And, since this force F_{avg} was parallel to the displacement of the nail (downward), the work done on the nail by F_{avg} is written $F_{avg} \cdot d$. And, since the ball came to rest at a height H_2 above the block, we write the decrease in the potential energy of the ball as

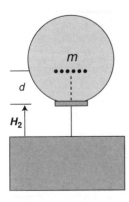

Figure 2 A ball at rest on the head of a nail.

$mg(H_1 - H_2)$. Some fraction of this decrease in the potential energy of the ball appeared as work done pushing the nail into the block; the remainder of the energy of the ball was converted to heat through friction between the nail and block, plus any deformation of the ball and nail.

See also: ENERGY; ENERGY, KINETIC; ENERGY, POTENTIAL; ENERGY AND WORK; FORCE

Bibliography

CRUMMET, W., and WESTERN, A. *University Physics: Models and Applications* (Wm. C. Brown, Dubuque, IA, 1994).

HALLIDAY, D.; RESNICK, R.; and WALKER, J. *Fundamentals of Physics*, 4th ed. (Wiley, New York, 1993).

FRED E. DOMANN

ENERGY, NUCLEAR

Since nuclei are made of neutrons and protons, they all have a positive electric charge. Thus, if a heavy nucleus like uranium were to split into two fragments, a process called fission, the electrical repulsion between the two would accelerate them to a total kinetic energy of 168 MeV. A normal uranium nucleus is restrained from this energy-releasing transition by a barrier; the initial separation requires about 6 MeV of extra energy to overcome the short-range attractive nuclear forces. It therefore undergoes such a spontaneous fission process only by penetrating this barrier, which gives it a half-life of 10^{15} years. However, when a neutron is captured by ^{235}U, its binding energy provides the required 6 MeV to overcome the barrier, and fission occurs in 10^{-14} seconds, which is usually before the alternative decay mode (gamma-ray emission) can take place; the reaction is (n,f). Neutron capture by ^{238}U falls short of providing the required 6 MeV, so gamma-ray emission occurs first and the reaction is (n,γ).

For the fission case, just after the split occurs the two fragments are highly deformed in shape. As the attractive nuclear forces pull them into a more spherical shape, the deformation energy is converted into nucleon excitation energy, enough to allow the most highly excited neutrons to escape. An average of 2.5 neutrons per fission are thereby emitted. When there is no longer enough energy for neutrons to escape, the remaining excitation energy is released as gamma rays.

^{235}U has 92 protons (p) and 143 neutrons (n), an n/p ratio of 1.55, which is far higher than for stable nuclei of half its mass, like ^{118}Sn with 50 protons and 68 neutrons, an n/p ratio of 1.36. The fission fragments must therefore undergo beta decays, changing neutrons into protons, to reach stability for their mass. These beta decays are frequently followed by gamma-ray emission or, in very exceptional cases, by neutron emission.

In addition to the 168 Mev in kinetic energy of the fission fragments, the (n,f) reaction releases 20 MeV as gamma rays and 8 MeV as beta rays, and the total kinetic energy of the 2.5 neutrons averages 5 MeV, a total of 200 MeV. As these particles interact in various ways with the surrounding material, their energy is converted ultimately into heat. This 200 MeV of energy release per 235 atomic mass units of fuel converts to 9 billion kilocalories per pound (or 1 MW-day per gram), 2,800,000 times the 3,200 kilocalories per pound from burning coal.

If an average of at least 1.0 of the 2.5 neutrons released in an (n,f) reaction, 40 percent, induces another (n,f) reaction, we satisfy the condition for a self-sustaining chain reaction. This equivalently requires that no more than 60 percent of the neutrons released induces reactions other than (n,f) or escapes out the sides of the assemblage. This last escape factor requires that a minimum quantity of uranium, called the critical mass, be present. If a critical mass is present, the chain reaction proceeds indefinitely, providing a constant generation of heat—this is then a nuclear reactor.

When a neutron encounters a ^{235}U nucleus, the probability for it to be absorbed (neutron cross section) is increased 200-fold if it has been slowed down (by a factor of 10^8) until it is in thermal equilibrium via collisions with atoms of the surrounding material. This is accomplished by introducing a moderator, material with nuclei of low mass to which the neutrons transfer their kinetic energy in elastic collisions, but these nuclei must have very small neutron cross sections. The most widely used moderators are carbon in the form of ultrahigh-purity graphite, heavy water (D_2O), and ordinary water (H_2O). The first two are such efficient moderators that natural uranium with 0.7 percent ^{235}U and 99.3 percent ^{238}U can be used as the fuel, but for H_2O the neutron cross section for H is uncomfort-

ably high, requiring that the fuel be enriched to about 3 percent ^{235}U.

The rate of energy release in a reactor is controlled by inserting control rods made of materials with high neutron cross sections. This control problem is greatly eased by the fact that 0.6 percent of the neutrons are not released in the fission reaction itself, but rather following beta decays of fission fragments with half-lives ranging from 0.2 to 56 seconds. Since the reactor normally has a critical mass only with the inclusion of these delayed neutrons, it responds to movement of control rods with a time constant of several seconds.

By withdrawal of control rods, a reactor's heat generation can be raised to any level, but adequate heat removal (cooling) must be provided. Where H_2O is used as the moderator, it is pumped through the reactor core to provide cooling. Water can also be used to cool graphite moderated reactors, but cooling by liquid metals (Na + K) or gases (He or CO_2) is more common.

Nearly all nuclear energy production is directed to generation of electricity via a steam cycle. In the boiling water reactor (BWR) the water used as moderator and coolant is converted inside the reactor vessel into steam at 550°F, 1,000 pounds per square inch (psi) pressure, although the fuel itself must always be surrounded by liquid water for adequate cooling. In the more common pressurized reactor (PWR), the moderator-coolant water is kept liquid at 600°F by high pressure (2,200 psi) and pumped to an external steam generator, where its heat is transferred to water to produce steam. Reactors cooled by liquid metals or gases also use steam generators, but they are not limited by the requirement of keeping water liquid and can therefore produce much higher temperature steam, 1,000°F or more.

In all systems, the steam is sent to a turbine that drives a generator to produce electric power, as in the generation of electricity from coal, oil, or gas. Effectively, the reactor plus steam generator replaces the boiler in those plants.

The great majority of uranium nuclei in the fuel are ^{238}U, and neutrons captured by them induce (n,γ) reactions to produce ^{239}U, which beta-decays rapidly to plutonium-239. This ^{239}Pu is also fissile like ^{235}U and can be used as a substitute reactor fuel. However, a more interesting application is in a fast reactor. If no moderator is present, the neutrons are not slowed down, so their (n,f) cross section is much reduced, but that problem can be overcome by using more highly enriched fuel, and the fast reactor has the advantage of yielding more neutrons per fission and fewer neutrons lost by (n,γ) reactions. It also can efficiently consume as fuel all the isotopes of uranium and plutonium. A chain reaction can be sustained, with steady generation of heat, with a wide variety of mixtures of plutonium and uranium fuel. Because of the current worldwide excess plutonium, the first mixtures to be used would be ones that result in a net consumption of plutonium. If, after present supplies of plutonium are exhausted, still more is wanted to fuel new power plants, the amount needed would be produced by adjusting the mixture so that more plutonium is produced by $^{238}U(n,\gamma)$ than is consumed by $^{239}Pu(n,f)$. The only material consumed will be ^{238}U, which is 140 times more abundant than ^{235}U. The raw fuel cost for breeder reactors is equivalent to gasoline priced at 40 gallons for a penny.

Uranium is leached out of rock and carried by rivers into the seas at a rate sufficient to provide 25 times the world's present total use of electricity. Using this, breeder reactors operating on uranium extracted from sea water can provide all the energy mankind will ever need without the cost of electricity increasing by even 1 percent due to raw fuel costs.

There has been much media publicity about dangers of reactor accidents and buried radioactive waste, but when these are treated with probabilistic risk analysis, they are found to cause less than 0.1 percent of the deaths now resulting from generating electricity by coal burning.

Worldwide, there are now 419 nuclear power plants in operation, capable of producing 332 MkW of power. In percentage of total electricity generation by nuclear power, France leads the world with 73 percent. The United States ranks eighteenth with 22 percent.

See also: CHAIN REACTION; FISSION; FUSION; NUCLEAR BINDING ENERGY; NUCLEAR REACTION; REACTOR, BREEDER; REACTOR, FAST; REACTOR, NUCLEAR

Bibliography

COHEN, B. L. *The Nuclear Energy Option* (Plenum, New York, 1990).

LAMARSH, J. R. *Introduction to Nuclear Engineering* (Addison-Wesley, Reading, MA, 1975).

MARSHALL, W. *Nuclear Power Technology* (Clarendon, Oxford, Eng., 1983).

BERNARD L. COHEN

ENERGY, POTENTIAL

Energy comes in several forms. One of them is potential, or stored, energy. How is this form like the others? How does it differ? How can we use it? Examples of potential energy are chemical energy, nuclear energy, gravitational energy, static electrical energy, and magnetic energy.

Potential energy can be important to us. When we turn on a television set to watch a space shuttle mission, we use electrical energy that started as potential energy (e.g., as gravitational potential energy of water stored behind a dam). That space shuttle mission "blasted off" from its launch pad by converting chemical potential energy in its rockets' fuel into kinetic energy. The batteries used in a camera's flash or in a portable radio, the gasoline used to run a car, and the food we eat all contain potential energy.

Given the importance of potential energy in science, in technology, and in our lives, we might suppose we have known and understood it for a long time. Not so. Our idea of force, which is closely linked to potential energy, originated with Isaac Newton in the seventeenth century. However, neither the idea of energy nor of its conservation arose until the nineteenth century. Much earlier, in the late seventeenth century, Christiaan Huygens hinted at potential energy in a discussion of motion. However, he did not use the term "potential energy" or grasp its importance. Early in the eighteenth century, Jacques Bernoulli described virtual work, a concept akin to potential energy, but he did not appreciate its importance. Nor did Joseph Lagrange, Pierre-Simon Laplace, Siméon-Denis Poisson, and George Green in the late eighteenth and early nineteenth centuries, who used electric potential (very closely related to electrical potential energy) in their mathematical formulations of electrical effects. Thomas Young mentioned conservation of energy at the start of the nineteenth century, although not in the sense currently used by physicists. An extensive debate about energy in physical science (including physics) occurred in the middle of the nineteenth century. Sadi Carnot, Julius Robert Mayer, Ludvig Colding, and particularly James Joule and Hermann von Helmholtz brought energy and its conservation to the forefront of physical science. They focused on mechanics and heat. Later discussions embraced all disciplines of physical science. Subsequently, an appreciation of the importance of potential energy evolved through the efforts of many engineers and scientists.

Potential energy is stored energy. Where is it stored? How is it stored? Potential energy is a *system effect*. It does *not* exist for a completely isolated object. An object has potential energy because of its location relative to other objects that exert a force upon it, or because of its location in a field that exerts a force upon it. *No one object* has the potential energy. Collectively, all the objects that mutually interact store the energy. A ball on the top of a table has gravitational potential energy that *both* the ball and Earth store. That energy arises because of the force exerted between Earth and the ball. Were the ball not there, or were Earth not there, then that gravitational potential energy could not exist. For a field, the potential energy is stored in the space where the field exists.

Changes in potential energy are important, not the actual value before or after the change. Though the location at which the potential energy is zero does not matter, sometimes there is a useful choice for this location, such as sea level on Earth's surface as a zero of Earth's gravitational potential energy, or the inner radius of a cylindrical capacitor as the zero for its stored electrical energy. Neither, however, is a necessary choice, because only the difference in potential energy between different locations matters. Its size never depends on how the change occurs, that is, the change is independent of the path. This is an essential feature of potential energy.

A change of potential energy may lead to kinetic, or electrical, or heat energy. Modern technology relies on this. The ease with which such a change occurs will depend on the stability of the stored energy. Consider three possible potential energy curves. Although not representative of every situation, these curves show how potential energy might vary with location. You could regard a small object, such as a marble, on an inverted bowl (unstable), in a bowl (stable), or on an inverted bowl with a rim (quasi-stable), as shown in Fig. 1. The outline of the bowl might represent a gravitational potential energy curve for the combination of Earth and a marble; or, if you regarded the object as a particle in a nucleus, then the "bowl" would be a nuclear potential energy curve. Change is unlikely in the stable case. *Extra* energy is needed to climb over the potential barrier (i.e., the rim) in the quasi-stable case—for instance, the extra energy might come from a spark that ignites the gasoline vapor in the cylin-

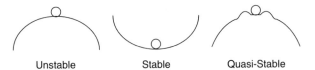

Figure 1 Potential energy as illustrated by a marble place on an inverted bowl, in a bowl, and on an inverted bowl with a rim.

ders of a car's engine. In some rare cases no extra energy is needed, as when a particle in the nucleus of an atom "burrows" through the barrier in a process called tunneling. Spontaneous change can occur in the unstable case. Modern technology favors quasi-stable equilibrium because the potential energy can remain "on hold" until we do something to activate it, such as turning on a transistor radio to convert a battery's chemical energy into electrical energy. Any change in potential energy leads to a force, for example, a gravitational force that would cause the marble to slip down the side of the bowl in the unstable case. We measure the size of the force by the slope of the curve: the steeper the slope, the larger the force.

Not all forces arise from changes in potential energy. Those that do are *conservative forces* like the gravitational force and the force in Coulomb's Law. For such conservative forces, $F = -dU/dx$ and $U = \int F dx$, where F is force, U is potential energy, and x is position. Those that do *not* arise from a potential energy change are *nonconservative forces*, such as friction and the electomotive force. For these forces no mathematical relationship to potential energy exists.

See also: CARNOT, NICOLAS-LÉONARD-SADI; ENERGY, CONSERVATION OF; ENERGY, KINETIC; ENERGY, NUCLEAR; FIELD; FORCE; HELMHOLTZ, HERMANN L. F. VON; HUYGENS, CHRISTIAAN; LAPLACE, PIERRE-SIMON; NEWTON, ISAAC

Bibliography

MASON, S. F. *A History of the Sciences* (Collier, New York, 1962).

TATON, R., ed. *A General History of the Sciences*, Vols. 2 and 3 (Thames and Hudson, London, 1964).

WOLSON, R., and PASACHOFF, J. M. *Physics for Scientists and Engineers*, 2nd ed. (HarperCollins, New York, 1995).

FREDERICK J. MORGAN

ENERGY, RADIANT

Radiant energy is emitted by a body in the form of electromagnetic radiation. This differs from energy transference by conduction and convection in that it requires no physical contact or mass motion of material. Radiant energy passes freely through a vacuum as in the example of the energy Earth receives from the Sun. Radiant energy is a broader term than thermal radiation since it includes the entirety of the electromagnetic spectrum and is not necessarily a continuous range of frequencies.

There are several mechanisms responsible for the stimulation of electromagnetic radiation:

1. Radio waves are emitted by the acceleration of electrons in a conductor carrying an alternating current. This occurs in devices such as broadcast antennae for radio and television.
2. Visible light and ultraviolet light are emitted when an electrical discharge is sent through a gas, as seen in fluorescent lighting tubes or neon signs.
3. Thermal radiation is emitted by all objects at a wavelength dependent on the object's temperature.
4. X rays are emitted by metals that are bombarded with electrons.
5. Gamma rays are emitted by some naturally radioactive substances and also by induced nuclear reactions.
6. Fluorescence is displayed by some substances after exposure to radiation from another source, such as sunlight.
7. The annihilation of particle-antiparticle pairs, such as electrons and positrons, results in the release of radiant energy in the form of gamma rays of energy equal to the sum of the kinetic energy and rest mass energy of the particle-antiparticle pair.
8. Radiation is emitted by a high-velocity charged object, such as an electron, when it is accelerated or decelerated. An example is when a proton, electron, or ion circulates in a cyclotron or synchrotron.

Radiant energy of a specific wavelength is usually the result of a transition between energy levels of an electron or a nucleon (proton or neutron). The radiant energy due to these atomic and nuclear transitions generally falls in the regions of the electromagnetic spectrum referred to as x rays and gamma rays, respectively. Energy levels on the atomic and nuclear scale are not continuous, but are separated by discrete (quantized) amounts of energy. The result of this energy level quantization is that transitions between energy levels require the input or release of energy exactly equal to the difference of the energy levels. The release of energy may be due to naturally occurring radioactive decay or may be caused by the bombardment of particles or electromagnetic radiation (light, x rays, gamma rays, etc.) For example, in a fluorescent lighting tube an electrical discharge excites the electrons of mercury atoms to higher energy levels. The electrons decay to lower energy levels by emitting electromagnetic radiation, mostly in the ultraviolet region of the spectrum. This ultraviolet light is absorbed by a coating on the tube walls. The electrons in the coating material move to higher energy levels dictated by the wavelengths of the absorbed ultraviolet light. The electrons drop back down to the lower energy levels by two successive, smaller transitions. This results in the release of electromagnet radiation of lesser energy and longer wavelength in each of the transitions. The longer wavelength radiation is in the visible spectrum. The main advantage of fluorescent lighting over incandescent lighting is that little energy is wasted in the production of invisible infrared light. Radiant energy of a specific wavelength may also be generated in the radio-wave region of the electromagnetic spectrum by using a specific frequency alternating current in a conductor. The resultant radio waves have the same frequency as the current.

Thermal radiation is a common form of radiant energy. All objects radiate a continuous spectrum of thermal radiation, with the temperature of the object determining which wavelengths predominate. Infrared light is predominant at lower temperatures, while at higher temperatures an object emits visible light. As an object is heated, one can feel the infrared radiation emitted and, as the temperature increases, the object eventually glows red (at about 500°C) and then white. An incandescent lightbulb is an example of a white-hot object. It emits a full spectrum of visible light and thus appears white. The bulb also emits in the infrared region, as is easily verified by putting one's hand near an operating incandescent light. The temperature of the universe may be inferred from observations of the isotropic background radiation. The average temperature of the universe computed from this data is 2.7 K.

The rate at which heat (Q) is radiated by an object (in joules per second, or watts) with surface area A, at temperature T (in kelvins), is described by the Stefan–Boltzmann law, named after its late nineteenth-century discoverers:

$$\frac{\Delta Q}{\Delta t} = \sigma e A T^4.$$

$\sigma = 5.6705 \times 10^{-8}$ W/m^{2}K^4 is a fundamental physical constant called the Stefan–Boltzmann constant. The constant e is the emissivity of the object and varies from 0 to 1. The emissivity of an object is determined by the roughness or smoothness of the surface and by the color of the object. The emissivity describes how well the object radiates energy. Dark, dull surfaces are the best emitters, while smooth, shiny surfaces with light colors are the worst.

William Herschel discovered that radiant energy existed outside of the visible spectrum in 1800. Herschel used a prism to disperse the spectrum of sunlight onto the surface of a table in a darkened room. By scanning the table with a sensitive thermometer, he was able to identify the invisible infrared light in the solar spectrum at a position beyond the red light.

Radiant energy is used in industrial applications to heat and dry materials. The infrared lamps used in these applications are often similar to incandescent lights but operate at a lower temperature and thus emit less visible light. Similar heat lamps are encountered in some hotel bathrooms. Radiant energy in the infrared portion of the spectrum is also used by the military and law enforcement agencies to "see" in total darkness. Infrared-sensitive video cameras are used to view the infrared light radiated by all objects and convert the image to visible light on a monitor. This type of device can also be used to determine where energy is being lost from a house or other structure by looking for bright regions on the structure in the infrared region. This type of thermal scanning is called thermography.

See also: CONDUCTION; CONVECTION; ELECTROMAGNETIC RADIATION; ELECTROMAGNETIC SPECTRUM; ELECTRON; FLUORESCENCE; HEAT; INFRARED; POSITRON; QUANTIZATION; STEFAN–BOLTZMANN LAW; THERMODYNAMICS

Bibliography

HEWITT, P. G. *Conceptual Physics,* 7th ed. (HarperCollins, New York, 1993).

JONES, E. R., and CHILDERS, R. L. *Contemporary College Physics,* 2nd ed. (Addison-Wesley, Reading, MA, 1993).

OHANIAN, H. C. *Principles of Physics* (W. W. Norton, New York, 1994).

YOUNG, H. D. *University Physics,* 8th ed. (Addison-Wesley, Reading, MA, 1992).

ZEMANSKY, M. W., and DITTMAN, R. H. *Heat and Thermodynamics,* 6th ed. (McGraw-Hill, New York, 1981).

JOHN C. RILEY

ENERGY, SOLAR

The amount of solar flux incident on only a very small fraction of Earth's surface is sufficient to supply all energy needs for humans in the foreseeable future. This fact makes solar energy an attractive alternative to fossil fuels, whose abundance is finite. To appreciate this, we will estimate the average (over a 24-hour day) solar flux on a square meter anywhere on Earth. This is called the insolation.

The only number that we need to know is the solor flux outside Earth's atmosphere as measured, say, by an orbiting spacecraft. This number, 1,353 W/m^2 is extremely stable; it varies by no more than a few percent over many years. In fact, since the Sun is what is known as a main sequence star and is stable in its current state of evolution, this number will be quite constant for many billions of years. Starting with this number, we first divide by 4 since the disc of Earth that intercepts the Sun's rays has an area of πr^2, while we are averaging over Earth's entire surface, which has an area of $4\pi r^2$. Next, we need to account for the atmosphere and cloud cover. Here, computer modeling can predict the average insolation at various locations using local weather data history. If we only want a rough estimate, then a factor of 2 is sufficient. This leads us to estimate approximately 170 W/m^2 for the average (24-hour day) insolation, approximately 4 kW h intercepted by 1 m^2 during a day. In fact, the average insolation varies by less than 50 percent from the northeast to the southwest of the United States. To put this in perspective, a large power plant is about 1,000 MW (1 GW). From our estimate, a comparable solar power plant operating at 10 percent efficiency would need to be about 60 km^2 (about 8 km, or 5 miles, on a side). However, solar energy is distributed and does not require central plants. Stating it another way, a modest house with a roof area of 100 m² intercepts an average of 15 to 20 kW, which at 10 percent conversion efficiency would provide about 2 kW, substantially more than the typical energy demand for one household.

Quality of Energy

To obtain useful work from solar energy, we need to collect it in the form of heat at elevated temperature or convert it into another form of energy, typically electricity. Direct conversion to electricity is done by a semiconductor diode called a solar cell. Efficiencies are typically 10 to 15 percent, which is good, considering the many uses of electricity. The chief barrier to the widespread use of solar cells is the cost, although use is increasing as costs are lowered. Today, the most economic use of solar energy is conversion to heat, which can be used for many purposes, including the generation of electricity. However, the heat has to be collected at elevated temperature, and generally speaking, the higher the

temperature, the higher the value. For this reason, the solar flux is often concentrated. Several types of solar thermal collecting systems are illustrated in Fig. 1. Concentrated solar flux is also useful in photovoltaic conversion by reducing the amount, and hence the cost, of solar cells required. However, there is a cost and complexity associated with the concentration of flux because there is a fundamental connection between concentration and angular acceptance of the concentrating device, or concentrator. To understand this, we need to consider the limits to concentration.

Maximum Concentration

The flux at the surface of the Sun (approximately 63 W/mm²) falls off inversely with the square of the distance to a value of approximately 1.37 MW/mm² above Earth's atmosphere and 0.8 to 1 MW/mm² on the ground. The second law of thermodynamics permits an optical device (in principle) to concentrate the dilute solar flux to attain temperatures up to but

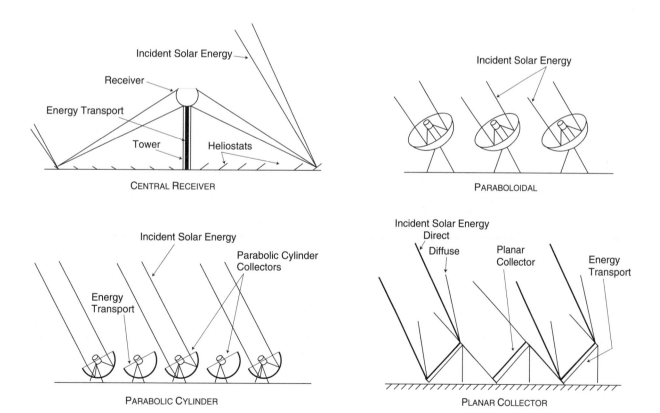

Figure 1 Typical solar thermal conversion system concepts.

not exceeding that of the Sun's surface. This places an upper limit on the solar flux density achievable on Earth and correspondingly on the concentration ratio of any optical device. This limiting concentration ratio is related to the Sun's angular size (2θ, where $\theta = 0.27°$ or 4.67 Mrad) by $C_{max} = 1/\sin^2\theta \approx 1/\theta^2$ (small angle approximation). This is called the sine law of concentration. When the target is immersed in a medium of refractive index n, this limit is increased by a factor of n^2, $C_{max} = n^2/\sin^2\theta$. This means that a concentration of about 100,000 will be the upper limit for ordinary ($n \approx 1.5$) refractive materials.

In experiments at the University of Chicago, a solar concentration of 84,000 was achieved by using a nonimaging design with a refractive medium (Sapphire). Experiments using conventional designs would not come close to this concentration, not for any fundamental reason, but because the imaging optical design is quite inefficient for delivering maximum concentration. Nonimaging optics began in the mid-1960s with the discovery that optical systems could be designed and built that approached the theoretical limit of light collection (the sine law of concentration). The essential point is that requiring an image is unnecessarily restrictive when only concentration is desired. Recognition of this restriction and relaxation of the associated constraints led to the development of nonimaging optics.

Nonimaging Concentration

A nonimaging concentrator is essentially a funnel for light. Nonimaging optics departs from the methods of traditional design to develop techniques for maximizing the collecting power of concentrating elements and systems. Nonimaging designs exceed the concentration attainable with focusing techniques by factors of 4 or more and approach the theoretical limit (ideal concentrators). The key is simply to dispense with image-forming requirements in applications where no image is required. One way to design nonimaging concentrators is to reflect the extreme input rays into the extreme output rays. This is called the edge-ray method. An intuitive realization of this method is to wrap a string about both the light source and the light receiver and then allow the string to unwrap from the source and wrap around the receiver. In introducing the string picture, we follow an insight of Hoyt Hottle, who discovered that the use of strings tremendously simplifies the calculation of radiative energy transfer

between surfaces in a furnace. The string is actually a "smart" string; it measures the optical path length (ordinary length times the index of refraction) and refracts in accordance with Snell's law at the interface between different refracting materials. The locus traced out turns out to be the correct reflecting surface.

Solar Geometry and Tracking

Because of Earth's rotation and the 23° angle its axis makes with the orbital plane, the apparent solar motion as observed from the ground is both azimuth (east-west) and elevation. A two-dimensional (trough-shaped) concentrator can accommodate the azimuth by aligning the axis in the east-west direction. However, the change in elevation angle will limit the concentration unless the collector is tracked. With seasonal adjustment, the sine law permits concentration factors of about 10 for nearly daily adjustments, 3 for biannual adjustment, and about 2 for no adjustment at all. Nowhere is the contrast between imaging and nonimaging concentration more dramatic than the case of the fixed collector, where imaging predicts deconcentration, while nonimaging actually allows factors close to 2, which is a very useful concentration allowing for moderate-temperature (200°C) collection.

Outlook and Prospects

Solar power plants for generating electricity have been built up to several hundred megawatt size. Perhaps the most successful of these is the Kramer Junction Power Plant in the Mojave Desert of Southern California. This plant uses tracking parabolic trough technology and is a marvel of engineering. It is nonpolluting and does not disturb the environment. Nevertheless, present-day economics is not favorable for this kind of plant because of the low cost of electricity produced by burning natural gas. Other options for solar power are less centralized, more modular units that provide heat close to the source. A major advantage of solar power is the obvious fact that solar flux is distributed and need not be converted in central facilities remote from the point of use. Figure 2 shows a solar thermal plant on the roof of a building. The heat is used to power heat-driven (absorption) air conditioners for cooling the building. The collectors employ nonimaging concentration and are nontracking. In fact, a promising

Figure 2   Rooftop solar thermal plant used to power heat-driven airconditioners for cooling the building.

application of solar heat is the cooling of buildings, because the good match between the demand for cooling and the availability of high solar flux mitigates (but does not eliminate) the need for storage. Combining solar heat with heat derived from fuel produces a hybrid system, such as a solar-gas hybrid, which is likely to be an attractive application of solar energy for the near future.

See also: ENERGY; ENERGY, RADIANT; HEAT TRANSFER; SNELL'S LAW; SOLAR CELL

Bibliography

DUFFIE, J. A., and BECKMAN, W. A. *Solar Engineering of Thermal Processes,* 2nd ed. (Wiley, New York, 1991).

WELFORD, W. T. *High Collection Nonimaging Optics (Academic Press,* New York, 1989).

WINSTON, R. "Nonimaging Optics." *Sci. Am.* **264** (3) 76–81 (1991).

ROLAND WINSTON

ENERGY AND WORK

Energy comes in many forms: mechanical, thermal, electrical, nuclear, and so on. It is measured by the amount of work that it can supply. Hence, a clear understanding of energy also involves a clear understanding of work.

Work, as used in physics, has a specific meaning. It is defined as (force) × (the distance moved in the direction parallel to the force). If one represents the applied force as the vector **F** and the displacement as the vector **S**, then work W is the dot product of the two vectors and is given by $W = \mathbf{F} \cdot \mathbf{S}$. Thus, work is a scalar and is equal to (the magnitude of the force) × (the magnitude of the displacement) × (the cosine of the angle between the force and the displacement). Hence, in order for work to be done, the object or system must be subjected to a net force, and there must be motion, of which at least a component must be along the direction of the force.

In the metric system, the unit of work is the joule. Symbolically, the joule is represented by the letter J, and it is equal to $(1 \text{ N}) \times (1 \text{ m})$. The British system has foot-pound, abbreviated as ft.lb, as its unit of work. A foot-pound is the work done by a force of 1 lb resulting in a motion of 1 ft along the force. In the cgs system the unit force is the dyne and the unit of distance is the centimeter; the unit of work is the erg, which is equal to $(1 \text{ dyn}) \times (1 \text{ cm})$.

A simple illustration of work can be seen in the case of weight lifting. Consider a weight of 200 N, a force that the earth exerts on the object, which is lifted straight up through a distance of 2 m. In lifting the weight, the work done is $(200 \text{ N}) \times (2 \text{ m}) = 400 \text{ J}$. It is interesting to note that, if the object is being held up in the lifted position, the work done is zero, as there is no motion!

With a clear notion of what work is, one can discuss the concept of energy. Energy is the ability to do work. Therefore, the units of work and energy are the same—joule, in the metric system, and foot-pound, in the British system. One of the practical units of energy is the kilowatt hour (kW h), energy consumption for 1 h at the rate of 1kW (1 W being 1 J/s); 1 kW h is equal to 3.6 million J. Another unit of energy that is very familiar to weight watchers is the calorie. A calorie is equal to 4,186 J. Finally, electron volt (eV) is another commonly used unit of energy and is approximately equal to 1.6×10^{-19} J.

One of the most common forms of energy is mechanical energy. Mechanical energy appears in two forms—kinetic and potential energy. Kinetic energy is due to motion, whereas potential energy is due to the position or state of an object relative to another object. For most ordinary cases, kinetic energy K may be expressed as $K = \frac{1}{2}mv^2$, where m is the mass and v the velocity. More exactly, it is the difference between the total energy, E, of an object and its rest-mass energy. Using the famous Einstein mass-energy relation, $E = mc^2$, where m is the mass of the object and c is the speed of light, $K = mc^2 - m_0c^2$, where m_0 is the mass of the object when at rest. All moving objects have kinetic energy—a moving car, a moving bullet, a moving electron, and so on.

Potential energy U on the other hand, is expressed in terms of a configuration. To put it differently, one must define a reference level or "zero." A good example of potential energy is the energy that an elevated object has. If the height from the reference level is h, then the potential energy, with respect to the reference level, is given by $U = mgh$,

where m is the mass of the object and g is the acceleration due to gravity.

Energy in one form may be converted to another form. A good example is an elevated object. The elevated object at rest has only potential energy, with respect to ground level. After the object is dropped and just as it is ready to hit the ground, it has no potential energy. However, because it is moving, it has kinetic energy. Thus, the potential energy the object had before being dropped was converted to kinetic energy, a clear case of energy conversion. Similarly, mechanical energy can be converted to heat energy, electrical energy to light energy, and so forth.

The fact that energy is measured by the work done is, in part, stressed by a theorem called the work-energy theorem. This theorem simply states that the net (or resultant) work done is equal to the change in energy. The example of an elevated object at rest can again be used. When dropped in the presence of gravity, gravity does the work, $W = (mg) \times (h)$. The object, in turn, picks up a kinetic energy (equal to $\frac{1}{2}mv^2$). Noticing that the initial kinetic energy of the object was zero, the change in kinetic energy is the final kinetic energy. Thus, the work done is equal to the change in kinetic energy—hence, the work-energy theorem.

One of the most important laws of nature is the law of conservation of energy. The law states, "Energy can neither be created nor destroyed." Included in this statement is the mass of objects, as is expected from the Einstein's mass-energy relation.

See also: CONSERVATION LAWS; ENERGY, KINETIC; ENERGY, MECHANICAL; ENERGY, POTENTIAL; MASS-ENERGY

Bibliography

SEARS, F. W.; ZEMANSKY, M. W.; and YOUNG, H. D. *College Physics*, 7th ed. (Addison-Wesley, Reading, MA, 1991).
SERWAY, R. R. *Physics for Scientists and Engineers*, 4th ed. (Saunders, Philadelphia, 1996).

SANKOORIKAL L. VARGHESE

ENERGY LEVELS

Classical systems of particles may have any value of energy. In quantum mechanics, bound systems of particles may have only certain discrete energies. The observation and modeling of these "energy levels" has been instrumental in the development of quantum mechanics, and continue to provide insight into the physics of atoms, molecules, and nuclei.

Spectroscopy

The most widely exploited experimental technique used to study energy levels in microscopic systems is spectroscopy. Spectroscopic observation of the absorption and emission of discrete wavelengths of light is such a powerful investigative technique because it has high resolution and sensitivity, and can be used in any wavelength region of the electromagnetic spectrum.

Bohr Model

In 1913, Niels Bohr published a description of his model of the structure of the hydrogen atom. Preceding this was the establishment of the quantization of light energy, by Max Planck (blackbody radiation theory) and Albert Einstein (photoelectric effect), and the discovery of the atomic nucleus by Ernest Rutherford. Bohr presented the first comprehensive physical model of the hydrogen atom in which he assumed that the electron has stationary (nonradiating) orbits only for quantized angular momenta $L_n = nh/2\pi$, where n is an integer greater than zero and h is Planck's constant. If the absorption of energy quanta of light were responsible for increasing the energy of the atom, then the atom's energy would not change unless the light energy $h\nu$ were equal to the difference between two allowed energy levels. In excellent agreement with experimental observations, Bohr's model predicted the energy levels of a one-electron system of nuclear charge Z to be

$$E_n = -\frac{m_e}{2}\left(\frac{kZe^2}{n\hbar}\right)^2,$$

where m_e is the mass of the electron, k is the electrostatic force constant, e is the electron charge, and $\hbar = h/2\pi$. The energy of a quantum of light emitted when the atom drops from a higher level n_2 to a lower level n_1 is just $h\nu = E_{n2} - E_{n1}$, in exact agreement with the empirical formula found by Johannes Rydberg in 1890.

Inner-Shell Transitions

A slightly modified Bohr model may be applied to specific inner-shell transitions between $n = 2$ and $n = 1$ (called K_α) for larger atoms. The combined force of the spherically symmetric cloud of outer electrons on the $n=_2$ electron is very small, and the nucleus is screened by one electron in the innermost $n = 1$ shell. Consequently, the Bohr formula with an effective nuclear charge $Z - 1$ gives correct K_α energies for elements of atomic number Z. This trend was established experimentally by Henry Mosely in 1913. Energies of K_α photons increase rapidly with atomic number Z, ranging from about 10 eV for hydrogen to over 100 keV for the heaviest elements.

Schrödinger's Model

Solutions to the equation derived by Erwin Schrödinger are interpreted to be probability amplitudes, the absolute squares of which are probability densities for the location of a particle. The requirements that these wave functions and their derivatives be everywhere finite and continuous led to the prediction of the same discrete energy levels as those of the Bohr model for one-electron systems. Schrödinger's theory included new quantum numbers l and m_1 that contain information about angular momentum. The quantum states of the system are identified by the triplet of quantum numbers (n, l, m_1). Because the energy levels depend only on n, there may exist a multiplicity of states having the same energy. Different states with the same energy are called degenerate states, and the existence of a plurality of states having the same energy arises from spatial or geometric symmetry in the interaction potential (in this case the radial dependence of the electrostatic force binding the electron to the nucleus).

Schrödinger's model facilitates the analysis of energy levels in more complex atomic systems, although solutions for the probability amplitudes must be estimated when the potential energy includes the mutual Coulomb repulsion of numerous electrons. Approximate solutions to the Schrödinger equation and the energy levels of these complex systems may be obtained numerically.

Schrödinger's model also predicts how degeneracies may be removed when symmetry is removed from the potential energy. For example, the very small magnetic interaction between the spinning electron and the magnetic field produced by the relative motion of the electron and the nucleus results in slightly different energies for states of differing quantum numbers l. This fine-structure splitting of the energy level is observed spectroscopically and is not predicted by the Bohr model.

Energy Levels in Molecules

Molecules possess quantized energy levels for the electrons surrounding the nuclei and exhibit absorption and emission of radiation-like atoms. Molecular spectra are much richer in structure, however, because they also have modes of vibration and rotation. Spectroscopic observations of molecular energy levels enable researchers to work backward to find the effective potential energies of bound molecules as they depend on internuclear separations. Spacings between adjacent vibrational energy levels are typically on the order of 0.1 eV. Rotational energy levels are generally much closer together yet, typically separated by a few milli-electron-volts.

Nuclear Energy Levels

Forces binding nuclei are much stronger than the electrical forces that bind atoms and molecules. Consequently, nuclear energy-level spacings and transition energies are relatively large. Nuclei absorb and emit gamma (γ) radiation, ranging from tens of kilo-electron-volts to tens of mega-electron-volts. While nuclear energy levels should be obtainable from Schrödinger's theory, the problem is complicated by the fact that the details of nuclear potentials are not as well understood as those for electromagnetic interactions. Therefore the study of energy levels in nuclei enables physicists to better understand the nature of nuclear forces.

Energy-Level Spacings and the Uncertainty Principle

The noted trends in the magnitudes and separations of adjacent allowed energy levels in quantum systems are governed by the uncertainty principle. The smallest separations occur for more massive particles, larger spatial scales, and weaker binding forces (molecular rotation). The largest level spacings are found in small, tightly bound systems (inner-shell electrons and nuclei).

See also: BOHR, NIELS HENRIK DAVID; BOHR'S ATOMIC THEORY; NUCLEAR SHELL MODEL; QUANTUM MECHANICS; RADIATION, BLACKBODY; RUTHERFORD, ERNEST; SCHRÖDINGER, ERWIN; SCHRÖDINGER EQUATION; UNCERTAINTY PRINCIPLE; WAVE MECHANICS

Bibliography

BOORSE, H. A., and MOTZ, L. *The World of the Atom* (Basic Books, New York, 1966).

HERZBERG, G. *Spectra of Diatomic Molecules* (Van Nostrand, Princeton, NJ, 1950).

SCHRÖDINGER, E. *Four Lectures on Wave Mechanics* (Blackie & Sons, London, 1928).

TAYLOR, J. R., and ZAFIRATOS, C. D. *Modern Physics for Scientists and Engineers* (Prentice Hall, Englewood Cliffs, NJ, 1991).

TIPLER, P. A. *Elementary Modern Physics* (Worth, New York, 1992).

MARK GEALY

ENGINE, EFFICIENCY OF

An engine is a device that operates in cycles converting heat into mechanical work. In each cycle, let the engine receive heat Q_h from a high-temperature heat reservoir, use some of this energy to perform mechanical work W, and eject heat Q_c to a low-temperature reservoir. With the sign convention that heat is positive going *into* the engine and that work done *by* the engine is positive, conservation of energy gives

$$Q_h = W + |Q_c|. \qquad (1)$$

The efficiency of an engine is the ratio of the work done per cycle ("what you want") to the heat input per cycle ("what it costs"),

$$e \equiv \frac{W}{Q_h}. \qquad (2)$$

Suppose a certain engine receives 600 units of energy per cycle from its heat source but converts only

200 units of it into work, producing an efficiency of $200/600 = 1/3$. The other 2/3 of the energy (400 units per cycle) is ejected into a heat sink in the surroundings.

According to the second law of thermodynamics, for any engine $|Q_c|$ must be nonzero. This means that e is strictly less than 1. How close to unity can the efficiency of an engine be if it is limited *only* by the second law? The answer is obtained by making an idealized mathematical model of the engine, whose working medium is always in thermal equilibrium with the surroundings, so that equations of state exist for the engine at all times. The real engine's efficiency e will be less than the ideal engine's efficiency e_0 because real engines have processes that are out of thermal equilibrium.

Consider, for example, the Carnot engine, an idealization of the steam engine, operating between the "hot" temperature T_h and the "cold" temperature T_c. One calculates its efficiency to be

$$e_0 = 1 - \frac{T_c}{T_h} \qquad (3)$$

where T_c and T_h are measured in Kelvins. Any real engine operating between the temperatures T_h and T_c has an efficiency e that is less than e_0.

The difference between e_0 and e is responsible for the increase in the entropy of the surroundings. Consider a real engine that operates between the temperatures $T_h = 1,200$ K and $T_c = 600$ K. In each cycle, let the engine receive $Q_h = 1,000$ J of heat and perform work $W = 200$ J, giving an efficiency $e = (200 \text{ J})/(1,000 \text{ J}) = 1/5$, thereby exhausting the heat $|Q_c| = 800$ J to the heat sink. By contrast, a Carnot engine operating between these temperatures corresponds to the ideal efficiency $e_0 = 1 - (600 \text{ K})/(1,200 \text{ K}) = 1/2$, performing work $e_0 Q_h = 500$ J, so that in principle only 500 J of heat need be ejected to the heat sink. Comparing the real engine to the Carnot engine, we see that the real engine ejects to the surroundings 300 J of heat *beyond* the requirements of the second law. This is energy that is unnecessarily dumped as heat when in principle it could have been converted into work. The surroundings at the temperature of $T_c = 600$ K receives this burden of excess heat. For the two-temperature engine, the entropy increase of the surroundings is $(e_0 - e)Q_h/T_c$ in each cycle, which in our example is 1/2 J/K.

See also: ENERGY, MECHANICAL; HEAT; HEAT, MECHANICAL EQUIVALENT OF; HEAT ENGINE; HEAT TRANSFER; THERMODYNAMICS

Bibliography

FEYNMAN, R. P.; LEIGHTON, R. B.; and SANDS, M. *The Feynman Lectures on Physics,* Vol. 1 (Addison-Wesley, Reading, MA, 1963).

HALLIDAY, D.; RESNICK, R.; and KRANE, K. S. *Physics,* 4th ed. (Wiley, New York, 1992).

REIF, F. *Fundamentals of Statistical and Thermal Physics* (McGraw-Hill, New York, 1965).

SERWAY, R. A. *Physics for Scientists and Engineers with Modern Physics,* 3rd ed. (Saunders, Fort Worth, TX, 1990).

DWIGHT E. NEUENSCHWANDER

ENSEMBLE

In classical mechanics, we are able to predict the precise behavior of any given mechanical system as it starts from a well-defined state. In statistical mechanics, however, we need to deal with systems consisting of a very large number of particles, such as a macroscopic piece of solid or a gas in a container. We do not know the precise initial conditions of individual particles, and thus, we cannot specify all the microscopic details of the entire system. We have some knowledge of the system, such as its volume and temperature, but certainly not enough to describe the complete behavior of the entire system and, moreover, no reasonable person would want or need such pieces of information. (The time required to print out and read positions and velocities of 10^{23} particles would be comparable to the age of the universe.) What we need and can use are the averages of some of these data, such as pressure, temperature, and volume, to characterize the system as a whole. To accomplish this goal, we begin with a hypothetical collection of a large number of systems having the same structure as our system and that are found in all possible states permissible under the conditions our real system is subjected to. Josiah W. Gibbs, an American theoretical physicist of the early twentieth century and one of the founders of statistical mechanics, called this an ensemble of systems.

The theory is based on the concept of phase space, which is defined by a set of coordinate axes, one for each dynamical variable of the system. For example, for all the particles of a monatomic gas system, three position coordinates and three momentum coordinates describe each atoms so that the phase space for a system of N atoms requires $6N$ axes to specify the microscopic state of the system by the coordinates of a single point. As the atoms move and collide with the container walls and each other, the point in phase space moves, following a path that is too cumbersome to compute even with complete knowledge of the equations of motion as applied to individual atoms. In time, the phase point motion reveals a pattern determined by the microscopic dynamics and the system constraints.

Because of the way an ensemble is constructed, each member of an ensemble is not like a gas molecule, which interacts with its neighbors. Being a representative of an entire system, each member of the ensemble instead follows a trajectory of its own in phase space totally independently by obeying the laws of mechanics. (There are some specific cases, such as heat transfer, that require some modification on this point.)

Each member of the ensemble will, in general, be in a different microscopic state at the instant those states are imagined to be sampled. With $\mathcal{N}$ members of the ensemble, there are $\mathcal{N}$ distinct points in phase space lying within the pattern revealed by the motion of the single point. These $\mathcal{N}$ points may be described by a density $\rho(q,p)$, where q and p symbolize the $6N$ or more variables of the microscopic description. A dynamic quantity $f(q,p)$ then has an average value of

$$\langle f \rangle = \frac{\int f(q,p)\,\rho(q,p)\,dq\,dp}{\int \rho(q,p)\,dq\,dp}$$

Measurement of this quantity for the real system yields an average over the duration of the measurement, the thermodynamic value. For large $\mathcal{N}$, this value and $\langle f \rangle$ are equal, according to ensemble theory.

We might question at this point whether or not the density varies with the passage of time. In other words, should the density be written as $\rho = \rho(q_1, q_2, ..., p_1, p_2, ..., t)$, where q's and p's all together (there are $3N$ q's and $3N$ p's) represent a point in $6N$-dimensional phase space. The q's come from the coordinates of all the N constituent particles and p's

come from their momenta. The Hamiltonian form of classical mechanics conveniently provides the proof that this density does not change with time when we consider the rate of change of density ρ in the neighborhood of any moving member of the ensemble (phase point) rather than sitting on a fixed point in phase space. This is known as Liouville's theorem and can be expressed in a number of different ways.

We can further restrict ourselves to the distribution of ρ, which is permanently independent of time at all points in the phase space. In this case, the average value of a physical property of a system represented by the ensemble is time-independent. Such an ensemble is said to be in statistical equilibrium. We can generally set up ensembles in statistical equilibrium by taking ρ to be a function only of a constant of motion such as the total energy of the system.

To describe a various system in statistical equilibrium, there are usually three standard types of ensemble, that is, choices of ρ. In a conservative system (i.e., a system in which energy is conserved), we can create an ensemble distribution that is a function only of energy E. We can define a function ρ such that it is constant within a narrow range, δE, of values of E, that is, between E and $E + \delta E$, and is equal to zero outside of this range. Such an ensemble is called a microcanonical ensemble. We can picture this ensemble as occupying a thin skin of a constant energy surface in phase space. This type of ensemble is of theoretical importance in areas such as the endeavor to provide the fundamental justification and validity of statistical mechanics (ergodic theory). Actual calculation based on this ensemble to describe the properties of a system is fraught with computational tedium.

A large number of thermodynamic systems of practical importance are systems in contact with a heat reservoir. It is expedient in such cases to employ the ensemble defined by

$$\rho = Ne^{(\psi - E)/\theta},$$

where N is the total number of systems in the ensemble, E is the energy of the system, and ψ and θ are parameters to fine-tune the description of the system of interest. This type of ensemble, known as a canonical ensemble, was first introduced into the theoretical framework of statistical mechanics by Gibbs. The physical and mathematical properties of this ensemble are such that most of the systems in the ensemble find themselves in the neighborhood of the average. In most cases of practical interest in statistical mechanics, adoption of canonical ensemble formalism renders the computation significantly simpler than those with microcanonical ensemble.

The grand canonical ensemble is used to describe systems with a varying number of particles. In these cases, the system is thought to be in contact with a much larger system through a wall that allows heat (i.e., energy) and material (i.e., particles) flow. The grand canonical ensemble consists of a weighted collection of canonical ensembles, each representing a possible value of the total number of particles of the system of interest.

See also: ERGODIC THEORY; GIBBS, JOSIAH WILLARD; HEAT TRANSFER; STATISTICAL MECHANICS; THERMODYNAMICS

Bibliography

GIBBS, J. W. *Elementary Principles in Statistical Mechanics* (Yale University Press, New Haven, CT, 1914).

REIF, F. *Fundamentals of Statistical and Thermal Physics* (McGraw-Hill, New York, 1965).

TOLMAN, R. C. *The Principles of Statistical Mechanics* (Oxford University Press, Oxford, Eng., 1938).

CARL T. TOMIZUKA

ENTHALPY

Enthalpy is not a familiar concept to most students of physics. Everyone has an intuitive understanding of pressure, temperature, and even perhaps the concept of internal energy. Enthalpy is a combination of these thermodynamic variables that is especially useful for analyzing processes occurring at constant pressure.

The first law of thermodynamics can be written as $dQ = dU + dW$, where d refers to *difference;* this statement contains within it the concepts of heat (Q), internal energy (U), work (W), and the conservation of energy. Internal energy is a result of the complex motions and interactions of the molecules that make up a system. The infinitesimal transfer of heat (dQ) into the system can be caused by changes in the internal energy (dU) and by work (dW). Work is done by the system as the volume changes by an amount dV, where $dW = P \cdot dV$, with P being the pressure in the system. By substitution, the first law becomes $dQ = dU + P \cdot dV$.

Now let the system undergo a change from internal energy U_1 to internal energy U_2 and from volume V_1 to volume V_2, while the pressure remains constant at P (i.e., an isobaric process). The first law then becomes

$$
\begin{aligned}
dQ &= dU + P \cdot dV \\
&= (U_2 - U_1) + P(V_2 - V_1) \\
&= U_2 - U_1 + P \cdot V_2 - P \cdot V_1) \\
&= U_2 + P \cdot V_2 - U_1 - P \cdot V_1) \\
&= (U_2 + P \cdot V_2) - (U_1 + P \cdot V_1) \\
&= d(U + P \cdot V) \\
&= dH,
\end{aligned}
$$

where $H = U + P \cdot V$ is the enthalpy of the system. Therefore, in a system with constant pressure, $dQ = dH$ means that the heat into the system equals the change in enthalpy. For this reason, enthalpy is sometimes called the heat content.

Consider a flow of various materials entering a chamber. Inside the chamber complex processes can occur; afterward the reaction products flow out. Let the reaction proceed at constant pressure for this example. It is possible to look up the tabulated enthalpies of all the inputs (H_{in}) and the products (H_{out}). Any change in enthalpy indicates a release of heat: $Q = H_{out} - H_{in}$.

Examples

The familiar steam radiator is a very simple example of a reaction chamber; only one material flows through the chamber. For this illustration, specify that the steam temperature drops from 300 to 250°F as it passes through the radiator while the pressure remains at 20 lb/in.^{2}, using engineering units. To determine the amount of heat transferred to the room simply look up the enthalpies under these conditions in so-called steam tables: roughly, $H_{in} = 660$ and $H_{out} = 650$ cal/g. $Q = H_{out} - H_{in} = 660 - 650$, and thus each gram of steam passing through this radiator releases about 10 cal of heat.

Another important application of enthalpy is the throttling process, or the Joule–Thomson effect. Briefly, throttling involves pumping gas at high (but constant) pressure through a constriction and maintaining low (but constant) pressure beyond the constriction. The enthalpy is constant in this variable pressure process because the pressures are kept constant both before and after the throttling.

Surprisingly, a gas can either cool or warm after throttling, depending on the starting conditions. If cooling is desired, the throttling process must begin with a gas precooled below a certain temperature T_{INV} for the gas; this temperature is known as the inversion temperature. It is possible to repeat the throttling again and again until the gas reaches the required temperature, perhaps even down to liquification.

Measurements

Finally, how are enthalpies determined? Historically, dQ has also been defined by a quantity called the specific heat: $dQ = mc_P dT$, where m is the mass of material, c_P is the specific heat (at constant pressure), and dT is the temperature difference. Since it has already been established that $dQ = dH$, it must be true that $dH = mc_P dT$. A process taking the system from an initial state i to a final state f results in an enthalpy change of

$$H_f - H_i = m \int_i^f c_P dT.$$

Careful calorimetry experiments are performed to determine c_P; results are readily available for many substances over a wide range of conditions. The above relation can be integrated, providing extensive tabulated values of H.

See also: CALORIMETRY; ENERGY, CONSERVATION OF; ENERGY, INTERNAL; HEAT; JOULE–THOMSON EFFECT; SPECIFIC HEAT; THERMODYNAMICS

Bibliography

CALLEN, H. B. *Thermodynamics and an Introduction to Thermostatistics*, 2nd ed. (Wiley, New York, 1985).

KARLEKAR, B. V. *Thermodynamics for Engineers* (Prentice Hall, Englewood Cliffs, NJ, 1983).

KEENAN, J. H., and KEYES, F. G. *Thermodynamic Properties of Steam* (Wiley, New York, 1965).

MARTIN, M. C. *Elements of Thermodynamics* (Prentice Hall, Englewood Cliffs, NJ, 1986).

ZEMANSKY, M. W., and VAN NESS, H. C. *Basic Engineering Thermodynamics* (McGraw-Hill, New York, 1966).

JOHN MOTTMANN

ENTROPY

In 1850 Rudolf Clausius and Lord Kelvin formulated the second law of thermodynamics (essentially the impossibility of the perpetual motion machine of the second kind). They also derived the Carnot theorem on the basis of the mechanical theory of heat instead of the caloric theory as Nicolas-Léonard-Sadi Carnot did in 1824, thus resolving the question of the nature of heat by debunking the caloric theory forever. Still, there are questions on the nature and origin of the second law.

Clausius emphasized the property of heat transport that heat always flows from the hot to the cold. In fact the Clausius version of the second law was formulated along this line. To have a firmer grasp of this elusive property of heat transport, Clausius introduced the concept of entropy (meaning transport in Greek) and formulated the principle of increase of entropy as an alternate form of the second law. The concept of entropy widened thermodynamics from engineering to physics and chemistry and made it a basic science. Again, one may ask the nature and origin of the esoteric entropy law.

Attempts to derive it from Newtonian mechanics failed. Finally Ludwig Boltzmann interpreted entropy as probability in the atomic structure of matter. In this view the entropy law is considered to be a manifestation of the mathematical probability law in natural phenomena, supplementing the deterministic Newtonian laws to form a complete theoretical system that is known as classical physics.

According to the Clausius–Kelvin statement of the Carnot theorem on the efficiency of a heat engine,

$$\frac{Q_1 - Q_2}{Q_1} \leqq \frac{T_1 - T_2}{T_1},$$

where Q_1 is the heat absorbed from the boiler, Q_2 is the heat rejected to the condenser, and T_1, T_2 are the absolute temperatures of the boiler and condenser. The equality/inequality sign for the efficiency refers to reversible/irreversible engines. It follows that

$$\frac{Q_2}{Q_1} \gtreqless \frac{T_2}{T_1}.$$

Then

$$\frac{Q_1}{T_1} - \frac{Q_2}{T_2} \leqq 0.$$

For a working substance undergoing a cyclic change, the cycle may be sliced into many narrow Carnot cycles. The above equation, added up over all such cycles, gives the general result

$$\oint \frac{dQ}{T} \leqq 0,$$

where $\oint$ denotes the sum over the entire closed cycle. This is the Clausius inequality, an equivalent of the second law. Three consequences follow from this inequality.

1. Consider reversible processes, for which the equality sign applies. Define dQ/T in the reversible process as the change of entropy dS of the substance because of the reversible absorption of heat dQ. The equation with the equality sign then implies that the entropy S is a function of the thermodynamic variables of the state and is independent of the integration path. Also, for a change dS of the substance in any reversible process there is an equal and opposite change of the entropy of the environment that supplies the heat. Thus the total change of entropy of the combined systems, which together form an isolated system, is zero. Thus we arrive at the first conclusion: Entropy does not change in reversible processes in isolated systems.

2. Consider irreversible processes in a closed cycle. Now dQ/T cannot be defined to be the entropy change of the working substance because the process is not reversible. But dQ/T is the negative change of entropy of the environment, and therefore the inequality says that the total change of entropy of the environment is positive. On the other hand, the total changed of entropy of the substance over any *closed* cycle is zero, as we have determined

above that entropy is a function only of the state of the system and is unchanged, in a closed cycle. Thus the total change of entropy of the combined (isolated) system of working substance plus environment is positive for irreversible processes over a closed cycle. From this we can prove that the total change of entropy of an isolated system in any (not closed) irreversible process is positive, which is the second conclusion from the Clausius inequality. The way to prove it is to close the open process with a reversible process, for which the change of entropy is zero. Now the closed cycle, which is irreversible, is known to have positive change of entropy. Subtracting zero from it, we still have a positive change of entropy for the open irreversible process.

3. Many natural processes (heat conduction, gas diffusion) occur spontaneously. All of them are irreversible (reversible process is an abstraction and does not happen in the real world). Thus we arrive at the third conclusion from the Clausius inequality that in all natural processes entropy increases for isolated systems.

The above three conclusions together form the contents of the principle of increase of entropy, an equivalent to the second law. A corollary is that the change of an isolated system will stop and the system will reach an equilibrium when entropy has reached maximum. Maximum entropy is thus the equilibrium condition.

The first two of the three statements are related to the heat engine, but the third is not. Instead, it concerns physical and chemical processes that are new additions to thermodynamics due to the introduction of entropy. However, in our practical use of the equilibrium condition we take it as a necessary condition and use it to derive the equilibrium characteristics, such as the law of mass action, whereas in the above derivation we have only proved it to be the *sufficient* condition. In other words, we proved only that if a process occurs entropy will increase. We have not proved that the process must occur and entropy must increase if entropy is not yet maximum. That is a completely different problem and the second law with a passive disposition (what cannot happen) has never asserted actively on what must happen.

The Leiden physicists contended that the entropy principle actually contains three independent parts in the three statements as listed above, and the necessary condition of equilibrium has never been proved. In practical application to physics and chemistry we smuggled in the necessary condition without proof, which is not satisfactory logically. There are also other aspects of thermodynamics that aroused suspicion and many attempts of reformulation have been made.

Most criticisms may be explained away, such as the use of heat engines to start a new science, once the thoughts of the great genius Clausius are fully grasped. Yet there is no excuse for the missing necessary condition, which is the *only* unavoidable shortcoming. Most reformulations pay no attention to it, which is not surprising because the necessary condition involves new physics not contained in the Clausius–Kelvin second law and cannot be obtained by merely a churning of the old material. The new physics is the law of spontaneous processes. It is a new macroscopic theory of equilibrium based on a methodology first developed in the microscopic theory. It has been worked out and published as a new formulation of thermodynamics.

The problem was over-shadowed by Boltzmann's momentous discovery of the probabilistic interpretation of entropy. Probability tends to increase to the maximum spontaneously. The necessary condition is thus tacitly implied. There are many spontaneous processes, such as the free expansion of a gas into the vacuum and interdiffusion of two gases, in which entropy increases but the molecular mechanisms of these processes are obviously controlled by probability. An association of the entropy S with the probability W, the number of microscopic states corresponding to a macroscopic state, is not only plausible but also unavoidable.

It was Boltzmann who made the bold suggestion that entropy is a function of probability $S = F(W)$. Since entropy is additive and probability is multiplicative, the function F must be logrithmic:

$$S = k \ln W$$

where k is a constant. By treating the free expansion of a gas into the vacuum as a process of entropy increase, we obtain the left-hand side of the above equation and by probabilistic treatment, the right-hand side. Thus, we determine the constant k, which turns out to be the universal gas constant divided by the Avogadro number, or the gas constant per molecule, which is now called the Boltzmann constant.

The establishment of a connection between entropy and probability not only explains the origin and nature of entropy (its tendency to increase) in terms of a well-known concept probability but also enables a microscopic development of an otherwise macroscopic theory. Thermodynamic functions can now be calculated from molecular physics. Thus, physical and chemical properties derivable from the laws of thermodynamics can be calculated from thermodynamic functions so obtained from molecular physics. This opens up a very promising field of statistical thermodynamics. The successes of statistical thermodynamics is evidenced by the voluminous calculations of atomic and molecular properties based on the Boltzmann equation.

The connection of entropy with probability just discussed can be illustrated by examples of very simple systems. Suppose we consider a group of six coin-shaped dice with values of 0 and 1 on the two sides. For a given throw of the dice, the total value T of the six dice may be equal to 0, 1, 2, 3, 4, 5, or 6, and there are 1, 6, 15, 20, 15, 6, and 1 possible combinations to produce the respective totals. The numbers of combinations represent the relative probabilities of the values of T to show up on a given throw. Thus, the entropies of each value of T are $k \ln 1$, $k \ln 6$, $k \ln 15$, $k \ln 20$, $k \ln 15$, $k \ln 6$, $k \ln 1$, respectively, the maximum of which falls on $T = 3$, which determines the final equilibrium state.

This may be compared with a gas of six molecules with the face values corresponding to possible energies. A macroscopic state recognizes only the total energy T. For each macroscopic state there are many microscopic states corresponding to the combinations that have the same total value T. When we start with a state with a T value equal to the maximum 6 and keep on shaking the dice, we will see the average of all the T values gradually tends to 3, the one with maximum probability. From the macroscopic point of view we see the entropy increases from $k \ln 1$ to the maximum $k \ln 20$. As the number of coins increases, the maximun peak becomes narrower and narrower and the final equilibrium state can be sharply defined. This is the statistical interpretation of the principle of increase of entropy.

Since all natural processes tend to increase probability, the corresponding increase of entropy points out an arrow of time by which the universe is progressing. The underlying principle is a change from the ordered to the disordered. The world is becoming more and more disorganized. Ultimately the universe will become completely disordered. According to the entropy principle, there will be no useful work and no life of any kind, a state appropriately called heat death.

Bibliography

FONG, P. *Foundations of Thermodynamics* (Oxford University Press, Oxford, Eng., 1963).

PETER FONG

EQUILIBRIUM

Equilibrium is a very broadly applied term, meaning that a system does not change in time because external influences that would cause it to change are absent or are in balance. As such, there are many different types of equilibriums. The precise meaning of the term depends on the particular type of equilibrium under consideration.

Thermal Equilibrium

When two solid bodies (or confined liquid or gaseous bodies) having different temperatures are placed in contact with each other, thermal energy will flow from the one at higher temperature to the one at lower temperature. As this energy transfer proceeds, the body at the higher temperature will have its temperature decrease, while the one at lower temperature will increase in temperature. Eventually, the temperatures of the two bodies become equal and the energy transfer ceases. In this case, it can be said that the bodies are in thermal equilibrium. More formally, two bodies are in thermal equilibrium if, upon being placed in contact with each other, no thermal energy flows from one to the other. The temperature of a body is a parameter that helps determine if the two bodies are in equilibrium without placing them in contact with each other: If the temperatures of the two bodies are equal, then they are in thermal equilibrium. In

order for the concept of temperature to be meaningful, the so-called zeroth law of thermodynamics must be invoked, which states that if each of two bodies, A and B, are in thermal equilibrium with a third body, C, then A and B are in thermal equilibrium with each other. As an illustration of this, let C be a thermometer, which is reading a temperature T. When C is brought into contact with body A, no energy flows, so A and C are in thermal equilibrium. Similarly, when B is brought into contact with C no energy flows, so B is in thermal equilibrium with C. By the zeroth law then, A and B are in equilibrium with each other, and it makes sense to assign the same temperature T to both A and B.

The discussion in the preceding paragraph has been limited to the simple case of confined bodies. More generally, when parts of a system are bought into contact, there exists the possibility of transfer of mass as well as thermal energy, and also chemical reactions, phase changes, and so on. The situation is thus much more complicated. Thermal equilibrium is defined by the condition that no mass transfer, thermal energy transfer, phase changes, or changes in chemical composition are taking place. The condition for this more general equilibrium (sometimes called thermodynamic equilibrium) is that an appropriate quantity called the free energy or thermodynamic potential is a minimum. The free energy is analogous to the potential energy in the case of mechanical equilibrium.

Mechanical Equilibrium

In mechanics, one is concerned with the motion of bodies. In this context, the state of equilibrium means that the system under consideration does not move or moves uniformly (i.e., with constant speed in a fixed direction). First, consider a single particle moving in one dimension, under the action of a force that can be derived from a potential energy function $V(x)$. At points where the derivative of $V(x)$ vanishes, the net force on the particle is zero. At such points, a particle at rest may remain at rest. At points where the derivative is nonzero, there will be a net force, and the particle must move, in accordance with Newton's second law, $F = ma$. Now equilibrium points, that is, points where the derivative of $V(x)$ is zero, may be further classified as points of stable or unstable equilibrium, depending on whether the second derivative of $V(x)$ is positive or negative. In the case that the second derivative is positive, $V(x)$ has a local minimum. If the particle is displaced slightly from this point, it will perform oscillatory motion when released, remaining always near the point of stable equilibrium. The force near the point of equilibrium always pushes the particle back toward that point (a restoring force). On the other hand, if the second derivative is negative, the function $V(x)$ has a relative maximum. If the particle is displaced slightly from its equilibrium position, it will feel a force pushing it further away from the equilibrium point, and it will move away from this point. This is unstable equilibrium. Note that if the second derivative is zero, higher derivatives must be examined. The motion may be stable (remaining always near the equilibrium point) or unstable (wandering far). The equilibrium point may also be a point of inflection, as for $x = 0$ when $V = x^3$. For positive displacement in this case, there is a restoring force so the motion appears stable, but for negative displacement, the force pushes the particle further from the equilibrium point; the net result is that it is unstable. It is also possible for the function $V(x)$ to have a horizontal straight line portion. In this case, all the points on this portion are equilibrium points, and a condition called neutral stability exists. The considerations of this paragraph may be extended to three-dimensional motion by using the techniques of vector calculus. The condition for equilibrium becomes the vanishing of the vector sum of the forces on the particle.

A related concept of mechanical equilibrium occurs in consideration of the motion of a rigid body. In this case, if the total external forces on the body vanish, the center of mass of the body will not move (or will continue in uniform motion). However, the body may not be in equilibrium, because it may still be free to rotate about its center of mass. The necessary condition to prevent this rotation is that the total torque applied by the external forces must vanish. To summarize, the necessary and sufficient condition that a rigid body be in equilibrium is that the total external force and total external torque must both vanish. As before, either stable or unstable (or neutral) equilibrium might exist. As an example, consider a ruler suspended in a vertical position and acted upon by gravity. If the point of suspension is at the top of the ruler, it will be in stable equilibrium. A small displacement will cause the ruler to oscillate like a pendulum about the vertical equilibrium position. If the point of suspension is at the bottom, the equilibrium will be unstable, since a small displacement will cause the ruler to fall. Finally, if the point of suspension is at the

center of the ruler, the equilibrium will be neutral; a small displacement will simply reorient the angle of the ruler, which may have any value.

See also: ENERGY, POTENTIAL; HEAT; NEWTON'S LAWS; TEMPERATURE

Bibliography

BECKER, R. A. *Introduction to Theoretical Mechanics* (McGraw-Hill, New York, 1954).

PIPPARD, A. B. *The Elements of Classical Thermodynamics* (Cambridge University Press, Cambridge, Eng., 1957).

ZEMANSKY, M. W., and DITTMAN, R. H. *Heat and Thermodynamics*, 6th ed. (McGraw-Hill, New York, 1981).

MICHAEL LIEBER

EQUIPARTITION THEOREM

The equipartition theorem is a rather general statement that relates the internal energy of a thermodynamic system to the system's temperature (a uniform temperature implies that the system has reached thermodynamic equilibrium). There are many ways of stating the equipartition theorem. One serviceable statement is as follows:

Equipartition theorem: In equilibrium there is a mean energy $\frac{1}{2}kT$ per molecule associated with each independent quadratic term in the molecule's energy.

Those independent quadratic terms may be quadratic in coordinate, velocity component, angular velocity component, or other quantities that when squared are proportional to energy.

It should be stressed that the equipartition theorem is *not* an exact result, but rather an approximation that is quite good in a wide number of applications in which classical statistics apply. It can only be applied to classical thermodynamic systems and should not be used when quantum statistics dominate.

Application to Ideal Monatomic Gases

Consider an ideal monatomic gas, at a temperature T sufficiently higher than its boiling point so that classical statistics are valid. Then in equilibrium the distribution of velocities will follow the Maxwell–Boltzmann formula,

$$f(v) = C e^{-\frac{1}{2}\beta m v^2},$$

where C is a constant, $\beta \equiv 1/kT$, m is the atomic mass, and k is Boltzmann's constant. If the factor v^2 is rewritten in terms of the (rectangular) velocity components as $v^2 = v_x^2 + v_y^2 + v_z^2$, one can then use elementary means to compute the mean square value of any one of the velocity components (they are clearly all the same). Such a calculation gives the result

$$\overline{v_x^2} = \frac{kT}{m}.$$

This result implies that there is a mean kinetic energy $\frac{1}{2}m v_x^2 = \frac{1}{2}kT$ associated with the x component of each atom's motion, consistent with the statement of the equipartition theorem. The results will be the same for the other two rectangular components y and z, and when these are combined the mean kinetic energy per atom should be $3\left(\frac{1}{2}kT\right) = \frac{3}{2}kT$.

Comparison with Molar Specific Heats

Predictions made using the equipartition theorem can be tested experimentally by measuring molar specific heats. The internal energy U of one mole (Avogadro's number, N_A of atoms) of a monatomic ideal gas is simply N_A times the mean energy per atom, or $U = \frac{3}{2}N_A kT$. The molar specific heat capacity $c_v \equiv \partial U/\partial T = \frac{3}{2}N_A k = \frac{3}{2}R$, where the result is expressed in terms of the molar gas constant $R \equiv N_A k$. Using the numerical value $R \cong 8.3145$ J·mol^{-1}·K^{-1}, the expected molar specific heat capacity becomes $c_v \cong 12.47$ J·mol^{-1}·K^{-1}. Experimental values for monatomic ideal gases such as the noble gases (e.g., argon and krypton) are extremely close to the predicted result, generally within 1 percent.

Application to Other Gases

The measured molar specific heats of diatomic gases tend to be significantly higher, and less uni-

form. For a number of diatomic gases, including N_2 and O_2, this value is very close to $\frac{5}{2}R$ at room temperature. The equipartition theorem can be successfully applied to diatomic molecules by considering the molecule to be a rigid rotator. The two atoms can be thought of as joined by a massless, rigid rod that is free to rotate about either of the two axes perpendicular to the rod. Rotations about an axis through the rod are insignificant; this subtle fact is due to quantum theory. In classical mechanics the kinetic energy of a rigid rotator is proportional to the square of the angular velocity, and therefore the equipartition theorem may be applied to the rotations. For each rotation axis a term $\frac{1}{2}kT$ is added to the molecule's mean kinetic energy, and therefore as in the preceding analysis the molar specific heat capacity increases by $2 \times \left(\frac{1}{2}R\right) = R$. The total is then $c_v = \frac{3}{2}R + R = \frac{5}{2}R$, consistent with the experimental results mentioned above.

However, other diatomic gases have significantly higher molar specific heats, and even those with experimental results close to the equipartition theorem's prediction show significant temperature dependence. A striking example is H_2 gas, with $c_v = \frac{3}{2}R$ just above its boiling point, $c_v = \frac{5}{2}R$ at room temperature, and c_v approaching $\frac{7}{2}R$ when the molecule dissociates at very high temperatures. The equipartition theorem provides a reasonable explanation of the $c_v = \frac{7}{2}R$ result, if one assumes that a vibrational energy mode is excited at elevated temperatures. In that case it is necessary to take into account the oscillator's kinetic energy, which is quadratic in velocity, and its potential energy, which is quadratic in the linear coordinate. These two additional terms add $2\left(\frac{1}{2}R\right)$ to the molar specific heat, bringing the total (translation plus rotation plus vibration) to $c_v = \frac{7}{2}R$.

Application to Solids

At high temperatures (generally taken to mean $kT \geq \hbar\omega$, where ω is the average oscillation frequency of atoms) simple atomic solids tend to have molar specific heats very close to $3R$. This is the well-known Dulong–Petit law, first presented in 1819. Notice that in a solid the translational and rotational modes are no longer available, and only vibration need be considered. In this case, however, each atom can oscillate in three dimensions. As described for the diatomic gas, a one-dimensional oscillator has two quadratic energy terms, one kinetic and one potential. Therefore in three dimensions the molar specific heat becomes $3 \times 2 \times 1/2R = 3R$, in reasonable agreement with experimental results.

As is always the case when applying the equipartition theorem, the range of applicability is limited, and other factors should be taken into account. At low temperatures quantum phenomena will dominate, and the correct solution for c_v in this case was given by Albert Einstein in 1907, and later modified by Peter Debye. Also, there is a small contribution to c_v due to the conduction electrons, typically around $0.02\,R$ at room temperature.

Summary

The equipartition theorem is widely applicable in the study of gases and solids as a way of estimating the mean atomic or molecular energy of a system in equilibrium. One must keep in mind, however, that it is applicable only in the classical regime. Even then, the accuracies of predictions made using the equipartition theorem, while often quite close to measured values, are only in general approximate.

See also: EQUILIBRIUM; MAXWELL–BOLTZMANN STATISTICS; SPECIFIC HEAT

Bibliography

BAIERLEIN, R. *Atoms and Information Theory* (W. H. Freeman, San Francisco, 1971).

KITTEL, C. *Elementary Statistical Physics* (Wiley, New York, 1958).

THORNTON, S. T., and REX, A. *Modern Physics for Scientists and Engineers* (Saunders, Philadelphia, 1993).

WANNIER, G. *Statistical Physics* (Wiley, New York, 1966).

ANDREW F. REX

EQUIVALENCE PRINCIPLE

Four hundred years ago, Italian physicist Galileo Galilei found that acceleration rates of bodies falling toward Earth were independent of the material composition of the bodies—different woods, rocks, metals, and so on all fell at a common rate (neglecting air resistance) of about 32 ft·s^{-2}. Decades later when Isaac Newton formulated his fundamental laws of mathematical physics, Galileo's discovery therefore obliged him to use the concept of inertial mass in his law of the gravitational force. Having assumed that every body has an inertial mass that determines the acceleration (rate of change of velocity) of the body when acted upon by any force,

Force = (Inertial) Mass × Acceleration.

Newton then added that the gravitational force acting between any two bodies is proportional to the product of the (inertial) masses of the bodies (and inversely proportional to the square of the distance between the bodies):

Gravitational Force$_{AB}$ ~

$$\frac{(\text{Inertial})\,\text{Mass}_A \times (\text{Inertial})\,\text{Mass}_B}{D_{AB}^2}.$$

In this way the gravitational acceleration of any body became independent of the body's mass, as Galileo had found. By the early twentieth century, experiments performed with ever-increasing accuracy had established the universality of free-fall rates of different materials to about a part in a billion precision.

Albert Einstein noted in 1906 that this universality of gravitational free-fall rates meant that gravity was locally equivalent to being in an accelerated coordinate system. Suppose, he reasoned, one was in a gravity free, but accelerated, room. All objects set free in the room would appear to "fall" to the floor at precisely the same rate, since, in fact, it would be the floor of the room that would be accelerating up to reach the various objects. Einstein generalized this simple observation into his equivalence principle—there should be no differences, whatsoever, between physical phenomena as observed in local gravitational situations and as observed in gravity-free, but accelerated frames of reference. By analyzing some simple hypothetical situations in accelerated frames of reference, Einstein was able to make two predictions concerning the effect of gravity on the propagation of light and on the behavior of clocks. His principle indicated that light does not propagate in straight lines but rather is deflected toward a source of gravity. And perhaps more revolutionary, it required that clocks run slower than other identical clocks simply by reason of being located closer to a source of gravity.

Today the apparent angular positions of stars and galaxies are precisely observed (to a thousandth part of Einstein's predicted deflection) to change when their lines of sight pass through the strong gravity near the Sun. The actual deflections are found to be twice that predicted by the equivalence principle, the extra deflection being caused by yet another startling aspect (the warping of Euclidean space) of Einstein's subsequently formulated (1916) complete theory of gravity—general relativity—which, as one might expect, also fully incorporated his equivalence principle.

In the 1970s an atomic clock was put on a high but sub-orbital rocket flight, during which it was tracked by clock-regulated radio signals to compare its rate to that of identical atomic clocks on Earth. The equivalence principle's predicted speedup of this clock (because it moved farther from Earth's gravity) was confirmed to about a part in ten thousand of the effect.

These phenomena are very minute in normal gravitational environments. A light ray passing close by the Sun deflects about an arc-second. A clock in very-high Earth orbit goes faster than a ground clock by about a part in a billion. But black hole objects have been hypothesized to exist in the galaxies, perhaps even to be commonplace in the universe, and are being actively sought experimentally. These objects have such strong gravity that they completely capture light that passes too near. A clock that approaches a black hole ticks at a rate brought arbitrarily close to zero as observed by clocks further away from these gravitational holes.

The equivalence principle played a key role in guiding Einstein toward the final formulation of his general relativity theory of gravitation. Since gravity is locally equivalent to accelerated coordinate frames and gravity affects the rate of all clocks, Einstein eventually perceived that gravity actually established the global structure of time and space itself, with the absolute time and absolute Euclidean space of Newton becoming, in Einstein's theory, an intertwined, non-Euclidean spacetime arena whose structure is determined by the matter-generated gravity of the universe. Einstein knew from his first groundbreaking theory of space and time—special relativity—that a body's inertial mass was equivalent to its total energy content:

$$E = Mc^2.$$

Therefore,

$$(\text{Inertial})\,\text{Mass} = \frac{\text{Energy}}{c^2},$$

where c is the speed of light. But needing to incorporate and explain the universality of free fall, Einstein was led to use also matter's energy content as the strength with which matter both produces and responds to gravity (gravitational mass).

The most precise modern test of the equivalence principle is provided by lunar laser ranging. By measuring the round trip travel time of laser light pulses, originating at and returning to Earth by bouncing off reflectors placed on the Moon's surface by Apollo astronauts, the shape of the Moon's orbit relative to Earth is determined to a precision of about a centimeter and is found to be in accord with the predictions of the general relativity theory. These measurements also indicate that Earth (in part iron core) and the Moon (primarily silicates) fall toward the Sun at the same rate to a precision of better than a part in a trillion. Earth is a sufficiently massive celestial body to contain almost a part in a billion of its total energy content (mass) in the form of gravitational binding energy. So to a part in a thousand precision (a billion divided by a trillion), it is experimentally established that Earth's gravitational energy content contributes equally to its inertial and gravitational masses. Gravity "pulls" on gravitational energy just as it "pulls" on all other forms of energy, confirming the specific nonlinear, self-coupling nature of Einstein's theory. This has been called by some the strong principle of equivalence.

Contemporary laboratory experiments also confirm the universality of free-fall rates for various small test bodies to comparable precision as the lunar orbit test. In addition, space satellite experiments have been proposed and designed to test the principle's foundation to yet five orders of magnitude higher precision.

See also: ACCELERATION; EINSTEIN, ALBERT; FRAME OF REFERENCE; GRAVITATIONAL ATTRACTION; GRAVITATIONAL FORCE LAW; INERTIAL MASS; NEWTON, ISAAC; NEWTONIAN MECHANICS; NEWTON'S LAWS; RELATIVITY, GENERAL THEORY OF; RELATIVITY, GENERAL THEORY OF, ORIGINS OF; RELATIVITY, SPECIAL THEORY OF; RELATIVITY, SPECIAL THEORY OF, ORIGINS OF

Bibliography

NORDTVEDT, K. "From Newton's Moon to Einstein's Moon." *Phys. Today* **49** (5) 26–31 (1996).

WILL, C. M. *Theory and Experiment in Gravitational Physics*, rev. ed. (Cambridge University Press, Cambridge, Eng., 1993).

KENNETH NORDTVEDT

ERGODIC THEORY

In the course of the historical development of statistical mechanics, some of the pioneers in the field, especially James Clerk Maxwell and Ludwig Boltzmann, struggled with the problem of validity of statistical mechanics itself. They hoped to justify the method of statistical mechanics by showing that the time average of a physical quantity of the system would agree with the ensemble average. The postulate that would enable one to reach this desired proof was called the "ergodic hypothesis" by Boltzmann. Since both the ensemble concept and ergodic theory can best be described in the framework of phase space, the concept of phase space will be introduced.

To begin with, the dynamical state of a single particle can be specified completely by three components of its position vector and three components of its momentum vector, altogether six parameters. A closed system of N particles representing an equilibrium state can thus be represented by a point in hypothetical space of $6N$ dimensions. This is called the phase space and, more specifically, the Γ space. A state of a N-particle system can be represented by a point in this phase space. If we consider a large number of identical isolated systems (in equilibrium) of given energy E with varying initial conditions, these representative points at time $t = 0$ will be distributed uniformly over the hypersurface S_E of this $6N$-dimensional space corresponding to constant energy E. (This collection of states forms a microcanonical ensemble.) These points traverse trajectories in space following the laws of mechanics.

Let us now look at one of these representative points and let us assume (without losing the generality of the argument) that the system consists of N molecules of gas. Since molecular motion is completely random at time $t = 0$, it should follow a trajectory sampling a large number of points on the surface S_E. At the time when the question was raised as to the validity of statistical mechanics, Boltzmann perceived that the representative point in phase space would traverse all the points on the surface S_E and return to the starting point. This is the ergodic hypothesis, which guarantees the equality of time and ensemble averages. Efforts to prove this hypothesis gave rise to various attempts to relax the strict ergodic constraint. (With the exception of one-dimensional case, the original ergodic hypothesis was found to be mathematically untenable by Arthur Rosenthal and Michel Plancherel in 1913.) French mathematician-physicist Henri Poincaré showed that such a trajectory can return to the infinitesimally close neighborhood of the original point from which the system started at $t = 0$ (Poincaré's recurrence theorem). It is only sufficient to assume that the trajectory passes arbitrarily close to any point on the energy surface S_E, This is called the quasi-ergodic hypothesis. Mathematician Hermann Weyl proved the quasi-ergodic hypothesis for a highly limited geometry of a point on a torus that is topologically equivalent to a moving ball on a billiard table. Study of "ergodicity" is not limited to completely disordered systems. For instance, a simple harmonic oscillator is ergodic and its elliptical trajectory in phase space closes on itself. A collection of harmonic oscillators (as in crystal lattice) is not ergodic. It had been speculated that a small nonlinear interaction among the oscillators would render the system ergodic. A rigorous mathematical proof by Andrej Kolmogorov, Vladimir Arnold, and Jurgen Moser (KAM theorem) showed that this is not the case. Mathematical analysis of the ergodic nature of various mechanical systems is still an active area of mathematical physics that has become more challenging with the addition of aspects of chaos.

See also: BOLTZMANN, LUDWIG; CHAOS; ENSEMBLE; MAXWELL, JAMES CLERK; PHASE SPACE

Bibliography

FORD, J. "How Random is a Coin Toss?" *Phys. Today* **36** (4), 40 (1983).

BAUMAN, R. P. *Modern Thermodynamics with Statistical Mechanics* (Macmillan, New York, 1992).

TOLMAN, R. C. *The Principles of Statistical Mechanics* (Oxford University Press, Oxford, Eng., 1938).

CARL T. TOMIZUKA

ERROR, EXPERIMENTAL

Experimental error is the difference between the truth and the conclusion a scientist infers from imperfect or indirect physical measurements. The word "error" here does not carry the negative connotations it often carries in everyday conversation. Experimental errors are not the result of mistakes, blunders, or oversights. They arise from effects beyond the control of experimenters and are an inevitable source of uncertainty in scientific inferences. Good experimenters devote much of their attention to the description and estimation of experimental errors in their inferences.

"Experimental error" is usually used in reference to *estimates* of the true or actual errors in an inference. If the actual error were known, one could simply correct for it, and draw perfectly accurate conclusions. For example, if a voltage measurement is 1.1 V, but one knows the voltmeter always provides measurements 0.1 V too large, one can correct for the actual error of 0.1 V to find the correct voltage of 1.0 V. But since the causes of experimental errors are typically beyond control, one cannot know the actual error made in a particular inference. Nevertheless, detailed knowledge of an experiment's design usually allows the experimenter to quantitatively estimate the error.

Since experimental error can usually only be estimated, inferences are uncertain, and one must provide a measure of the degree of confidence one can have in an inference. The mathematical discipline concerned with quantifying such uncertainty is statistics. Statistics applies probability theory to the analysis of experimental data. Two approaches to statistics exist, each differing from the other in its choice of how probabilities should be used to quantify uncertainty.

The earliest formal approach to statistics is the Bayesian approach, developed by astronomers and mathematicians of the late eighteenth and early nineteenth centuries in order to reconcile apparently discrepant data in astronomy and geodesy. In this approach, the viability of a hypothesis is quantified by calculating the probability for the hypothesis given all the information available that is relevant to assessing the hypothesis. One can be uncertain of the truth of a hypothesis only if there are alternative hypotheses that may be true in its place; also, the extent to which the evidence supports a particular hypothesis must depend on how well competing hypotheses account for the evidence. Thus, an important part of the information needed to calculate a probability is specification of all relevant alternative hypotheses; together they comprise the hypothesis space. The probability for a hypothesis is a number from zero to one indicating how strongly the available information prefers the hypothesis over the specified alternatives. A probability of zero indicates that the hypothesis is not compatible with the evidence, so that one of the alternatives must be true. A probability of one indicates that the hypothesis is compatible with the evidence and that all specified alternatives are inconsistent with the evidence; the hypothesis must then be true (unless one has neglected alternatives that are compatible with the evidence). More commonly, experimental errors prevent one from saying definitively whether the evidence is consistent or inconsistent with a hypothesis, and the probabilities for the various hypotheses take on values between zero and one.

If the hypothesis space consists of N hypotheses denoted by the symbols H_i ($i = 1$ to N), one writes $p(H_i | I)$ for the probability for hypothesis H_i conditioned on information I. This symbol is also commonly read as the probability for H_i *given I*, or the probability for H_i *supposing I* to be true. Bayesian inference supplies two sets of rules for calculating the numerical values of probabilities: (1) rules for directly assigning numerical values to simple probabilities based on symmetries in the problem and (2) the rules of probability theory that tell one how to calculate more complicated probabilities from simple probabilities. One of the most important rules of probability theory is Bayes's theorem, named after the Thomas Bayes, who first wrote down a special case of the theorem. Bayesian inference derives its name from this theorem. Bayes's theorem applies to situations where the available information can be separated into two parts: newly observed data D_{obs} and background information I specifying the hypothesis space and any other information available about the experiment and the hypotheses apart from D_{obs} (possibly including the results of other experiments). In the common situation where the hypotheses are exclusive and exhaustive (one and only one must be true), Bayes's theorem specifies how to calculate the probability for each hypothesis conditioned on all the information in terms of simpler probabilities:

$$p(H_i | D_{obs}, I) = \frac{p(H_i | I)\, p(D_{obs} | H_i, I)}{}.$$

The first factor in the numerator is the probability for H_i conditioned only on the background information; it is called a prior probability, since it describes the implications of the evidence available prior to consideration of the data. The second factor, the probability for the data presuming H_i is the true hypothesis, is called the likelihood for the hypothesis or the sampling probability for the data. It implies that hypotheses that make the data more probable tend to be more probable themselves. The denominator is simply a normalizing constant that assures that the probabilities for all hypothesis sum to one (since by assumption one of them must be true). Bayes's theorem describes learning: how to update the prior probability to give the posterior probability, $p(H_i | D_{obs}, I)$. In words, the posterior is proportional to the prior times the likelihood. Combined with rules for assigning prior probabilities and likelihoods appropriate to various experimental situations, Bayes's theorem allows one to quantify the uncertainty due to experimental errors by calculating the posterior probability for the hypotheses under consideration.

Bayesian inference fell out of favor in the late nineteenth and early twentieth century. Practitioners agreed on the validity of Bayes's theorem and on rules for assigning likelihood functions in many common problems, but many disagreed over the rules one should use to assign prior probabilities for hypotheses. The frequentist approach to statistics gradually evolved as an alternative to Bayesian inference; it is the approach most commonly used by physical scientists today.

The fundamental notion behind frequentist statistics is that of a random event. The term "random" here has a technical meaning: A random event is an outcome that has a chance of occurring during an experiment, but whose occurrence cannot be predicted with certainty. The frequentist probability for a random event is defined to be the relative frequency or fraction of the time that the event occurs in an infinite number of repetitions of an experiment. With this definition, one cannot usefully speak of the probability for a hypothesis. A hypothesis is either true or false for every repetition of an experiment, so its frequentist probability is either zero or one. The presence of experimental error prevents us from knowing with certainty whether this probability is zero or one, so probability must be used in some other way to assess hypotheses.

Many sources of experimental error lead to data that change unpredictably when an experiment is repeated, due to random changes in factors beyond the control or knowledge of the experimenter. Such data are random events, and frequentist probabilities can be meaningfully assigned to them. Frequentist statistics attempts to use probabilities for random data to quantify uncertainty about hypotheses. One must first single out a particular hypothesis as being of special interest; it is called the null hypothesis, and is denoted H_0. Given the null hypothesis, one can calculate the probability for each possible hypothetical set of data D from the sampling distribution $p(D | H_0, I)$. One then constructs a procedure that selects one hypothesis (the null or one of its alternatives) as being true, based on some function of the data $S(D)$ called a statistic. Using the sampling distribution, one calculates how well the procedure works in the long run, averaged over all possible data. The procedure is then applied to the observed data D_{obs}, and the long-run behavior of the procedure is taken as a measure of the confidence one can have in the decision based on the observed data.

A fundamental difficulty with this approach is that we must apply it in situations where we do not know which hypothesis is true. Thus we cannot know which hypothesis to choose as the null when calculating sampling probabilities. An important goal of frequentist statistics is therefore to find procedures with properties that are independent of which hypothesis is true. Some such procedures are known for a number of common inference tasks. Unfortunately, though, no general procedures are available, and often an experimenter is forced to condition calculations on an hypothesis that may be false but that is hopefully similar enough to the true hypothesis to give accurate results.

The sampling distribution that plays a key role in frequentist statistics is identical to the likelihood function used in Bayesian inference. However, frequentist statistics focuses on how $p(D | H_i, I)$ behaves as a function of D with H_i fixed at the null hypothesis, whereas Bayesian inference focuses on how it behaves as a function of H_i with D fixed to be the observed data D_{obs}. Some common sampling distributions (such as the bell-shaped Gaussian or Normal distribution) have a symmetry between the data and parameters indexing the hypothesis space, so that Bayesian and frequentist calculations using these distributions can give similar or identical results. But in general Bayesian and frequentist results will differ, and there is currently some controversy over which approach is superior. The Bayesian approach has a superior axiomatic foundation, but subjectivity in the assignment of prior probabilities still leads many experimenters to prefer the frequentist approach.

Another important difficulty with the frequentist approach is that it is difficult to justify its application to studies of unique events. When an experiment or observation cannot be repeated, it can be difficult to justify application of an approach based on the notion of randomness. Also, some sources of experimental error do not manifest themselves as randomness in the data. For example, if a measuring device is miscalibrated by an unknown amount, all data obtained with the device is affected in the same way, leading to systematic errors in one's inferences, rather than random errors. Frequentist methods cannot readily quantify systematic errors. Finally, frequentist methods cannot readily account for quantitative prior information, because they have no counterpart to the prior probabilities of Bayesian inference.

Thus, difficulties arise in the practical application of both frequentist and Bayesian methods for estimating experimental errors. Frequentist methods have dominated statistical practice in the physical sciences for most of the twentieth century. However, there has been a resurgence of interest in Bayesian methods in the past decade, particularly in astronomy, where one commonly observes unique events and the concept of randomness may not apply.

See also: ERROR, RANDOM; ERROR, SYSTEMATIC; ERROR and FRAUD; PROBABILITY

Bibliography

EADIE, W. T.; DRIJARD, D.; JAMES, F. E.; ROOS, M.; and SADOULET, B. *Statistical Methods in Experimental Physics* (North-Holland, Amsterdam, 1971).

JAYNES, E. T. "Bayesian Methods: General Background" in *Maximum Entropy and Bayesian Methods in Applied Statistics,* edited by J. H. Justice (Cambridge University Press, Cambridge, Eng., 1986).

JEFFREYS, H. *Theory of Probability* (Clarendon Press, Oxford, 1961).

LEE, P. M. *Bayesian Statistics: An Introduction* (Oxford University Press, New York, 1989).

PARRATT, L. G. *Probability and Experimental Errors in Science* (Wiley, New York, 1961).

STIGLER, S. M. *The History of Statistics: The Measurement of Uncertainty Before 1900* (Harvard University Press, Cambridge, MA, 1986).

TAYLOR, J. R. *An Introduction to Error Analysis: The Study of Uncertainties in Physical Measurements* (University Science Books, Mill Valley, CA, 1982).

THOMAS J. LOREDO

ERROR, RANDOM

Random errors are experimental errors that change in a random fashion when an experiment is repeated; we here use the word "random" in its technical sense. These errors result from random variations in factors that affect the experimental outcome but that are beyond the control or knowledge of the experimenter; such random variations are commonly called noise. By repeating an experiment and observing how the results vary, one can get a sense of the typical size of the noise, but the actual amount of noise present in each result will be unknown. As a result, inferences drawn from noisy data are uncertain, and experimenters must use statistical methods to quantify this uncertainty. Such methods can also be used to determine how best to combine the results of repeated experiments to reduce the uncertainty in one's inferences.

Both frequentist and Bayesian methods exist for quantifying uncertainty due to random errors. In both approaches, it is useful to distinguish between two kinds of inferences, each with its own set of methods for describing the confidence one can have in inferences drawn from noisy data.

The first type of inference is called hypothesis testing. It arises when one must decide about the truth of a particular hypothesis for a phenomenon based on the outcome of an experiment. Experimental errors can lead one to make an incorrect decision, so one must provide some measure of the strength of evidence in favor of or against the hypothesis being considered. In the Bayesian approach, one calculates the posterior probability for the hypothesis given the data and any other relevant information, including specification of alternative hypotheses. In the frequentist approach, one decides whether to accept or reject the hypothesis (called the null hypothesis H_0) based on the value of some test statistic. The confidence one can have in the decision is specified by two probabilities indicating the fraction of the time the test leads one to make each of two kinds of errors: rejection of the null when it is in fact true (a Type-I error) and acceptance of the null when it is in fact false (a Type-II error). The Type-I error probability depends on the statistic used and on the null hypothesis, but it does not depend on what alternatives there may be to the null. The Type-II error probability depends on the statistic used and on the specification of a single alternative to the null, but it does not depend on the null itself. The Type-I error probability is sometimes called the significance of the test, but some investigators use this term to refer to one minus the Type I error probability.

In practical applications, it may be difficult to specify an alternative hypothesis to the null hypothesis. Without at least one specific alternative, a Bayesian calculation is impossible. One can still calculate a frequentist Type-I error probability, and it is common practice to do so without also calculating a Type-II error probability. But the confidence one can have in the frequentist test is not adequately described by a Type-I error probability alone, and this practice is not recommended.

The second type of inference is called parameter estimation. It arises when one must infer the size or amplitude of one or more physical quantities, called parameters, from noisy measurements. Here one must provide not only an estimate of the parameters but also estimates of how large the random errors in the estimates may be. In the Bayesian approach, one calculates the posterior probability density function, or posterior pdf, for the parameters, using Bayes's theorem. If there is a single parameter θ, the posterior pdf based on the observed data D_{obs} and background information I is commonly written as $p(\theta \mid D_{obs}, I)$. It is a "density" of probability in the sense that $p(\theta \mid D_{obs}, I)\,d\theta$ is the probability that the true parameter value is in the interval $[\theta, \theta + d\theta]$. Thus, once the posterior pdf is known, the probability that the true value lies in any particular interval can be calculated simply by integrating the posterior pdf over the interval. When the posterior pdf has a simple, well-localized shape, it is common to summarize it by providing a parameter estimate (such as the value with the highest density, or the average value, weighted by the density) and an estimated error (such as the width of the smallest region containing, say, 95 percent of the probability—a 95 percent credible region). These concepts generalize in a straightforward way when there are two or more parameters.

In the frequentist approach, parameter estimation is broken into two separate problems: (1) point estimation, where the goal is to provide a single "best" value for the uncertain parameter, and (2) interval estimation, where the goal is to provide an interval in which the true parameter value is likely to lie and a measure of the confidence with which it lies in that interval. For point estimation, one constructs a function of the data, $\hat{\theta}(D)$, called an estimator, and applies it to all possible hypothetical data sets that might arise for a particular choice of θ. The quality of the estimator is determined by measuring how closely the estimates cluster around the true value. Ideally, the quality of the estimator should not depend on the true value of θ used in the calculation, since we do not know what the true value is in the experiment. For some problems, such ideal estimators can be found, but in general the quality of the estimator will depend on the true value of the parameter, and one must settle for results based on a chosen value that is hopefully close enough to the true value to give accurate results. Once one identifies an estimator with good average behavior, it is evaluated using the observed data to give the parameter estimate $\hat{\theta}(D_{obs})$.

For interval estimation, one constructs an interval-valued function of the data and applies it to all possible data sets that might arise for a particular choice of θ. The confidence level for the interval is the relative frequency that the intervals contain the true value; an interval-valued statistic that contains the true value, say, 95 percent of the time is called a 95 percent confidence region. Ideally, the confidence level should be independent of the true value of θ, but this is not always possible; one must often settle for an approximate confidence region.

Despite the obvious differences between Bayesian and frequentist approaches to parameter estimation, they produce the same final estimates for the three simplest and most common types of random error.

When estimating a parameter from real-valued data, such as a meter reading, it is often possible to consider the data to be the sum of the true parameter value and an unpredictable error or noise value that can be either positive or negative with equal probability. In such cases, repeated measurements in situations where the true parameter value is known often display a distribution of errors that closely follows the bell-shaped Gaussian or normal pdf, shown in Fig. 1. This pdf is symmetric with its peak at zero error. Its width is specified by an additional parameter, the standard deviation σ, which we here presume is known (it has been taken to

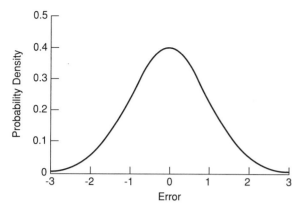

Figure 1 The Gaussian or normal error distribution for errors that can take on positive or negative real values. The curve shows the pdf, so that the probability for an error value between any two points on the horizontal axis is given by the area under the curve between those points; for example, the probability for an error between -1 and 1 is approximately 0.683. The pictured distribution is for $\sigma = 1$; for other values of σ, error values on the horizontal axis should be multiplied by σ and pdf values on the vertical axis should be divided by σ.

equal one in Fig. 1). It has the meaning that σ^2 is the average of the squared errors one would find in infinitely many repetitions of the measurement (so that σ is the root-mean-square or rms error). If N measurements are made, giving data values x_i ($i = 1$ to N), then both Bayesian and frequentist methods indicate that the best estimate for θ is the sample mean $\bar{x}$, found by averaging all the x_i values:

$$\bar{x} = \frac{1}{N} \sum_{i=1}^{N} x_i.$$

The region within a distance of $\sigma/\sqrt{N}$ of $\hat{\theta}$ is both the Bayesian 68.3 percent credible region and the frequentist 68.3 percent confidence region for this problem. These results are often summarized by writing $\theta = \bar{x} \pm \sigma/\sqrt{N}$. One should note that these results depend on the shape of the error distribution. If it is not symmetric, or if it does not fall off as rapidly as the Gaussian distribution, the sample mean may be a poor estimator, and the width of credible or confidence regions may not decrease in size with N by the factor $1/\sqrt{N}$.

Often, real-valued parameters such as event rates or fractions must be estimated using data consisting of integer-valued counts of discrete objects. Two such situations are common in physics. In the first, one wishes to infer the fraction f of objects that have some property by examining a sample of N objects from the entire population. For example, an astronomer may want to infer the fraction of all stars that have a temperature within 500 K of the Sun's temperature by measuring the temperatures of 1,000 stars. If the size of the sample is much smaller than the size of the entire population, and if each member is chosen without regard to the properties of other members, then the probability for seeing n objects with the desired property is given by the binomial distribution, illustrated in Fig. 2. For this sampling distribution, the best estimate for f is simply the fraction of samples that have the property $\hat{f} = n/N$. As long as n, $N - n$, and N are much bigger than one, the 68 percent credible and confidence regions are approximately bounded by $\hat{f} \pm \sqrt{n}/N$.

The second situation where counting data is common occurs when one must estimate a rate or density by examining how many events occur in a specified interval of time or space. For example, a particle physicist may need to estimate the rate of production of gamma rays by a radioactive source by counting how many are produced in a particular finite interval of time. When the event rate is known, it is often true that the occurrence of an event at a

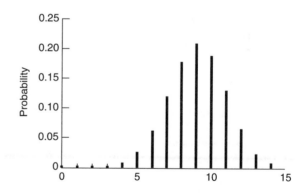

Figure 2 The binomial distribution, giving the probability for observing n objects with some specified property in a sample of N objects when the fraction of objects with the property is known to be f. The sample size N should be much smaller than the total size of the population to justify the use of this distribution. The pictured distribution is for $f = 0.6$ and $N = 15$. The probability is positive for all possible values of n from zero to fifteen, although it is very small for some extreme values.

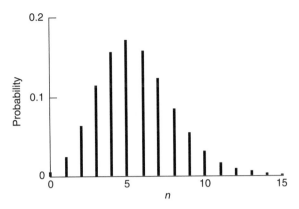

Figure 3 The Poisson distribution, giving the probability for seeing n events in some specified interval when the average number of events expected in that interval, μ, is known. The expected number is often written as the product of a rate or density and the size of the interval. The pictured distribution is for $\mu = 5.5$. The expected number can be any nonnegative real number; the probability is positive for all possible values of n from zero to infinity, although it is very small for values that are very different from μ.

Bibliography

EADIE, W. T.; DRIJARD, D.; JAMES, F. E.; ROOS, M.; and SADOULET, B. *Statistical Methods in Experimental Physics* (North-Holland, Amsterdam, 1971).

JAYNES, E. T. "Bayesian Methods: General Background" in *Maximum Entropy and Bayesian Methods in Applied Statistics,* edited by J. H. Justice (Cambridge University Press, Cambridge, Eng., 1986).

LEE, P. M. *Bayesian Statistics: An Introduction* (Oxford University Press, New York, 1989).

PARRATT, L. G. *Probability and Experimental Errors in Science* (Wiley, New York, 1961).

STIGLER, S. M. *The History of Statistics: The Measurement of Uncertainty before 1900* (Harvard University Press, Cambridge, MA, 1986).

TAYLOR, J. R. *An Introduction to Error Analysis: The Study of Uncertainties in Physical Measurements* (University Science Books, Mill Valley, CA, 1982).

THOMAS J. LOREDO

particular instant does not help us predict when the next event will happen; our probabilities for events in adjacent intervals are independent of one another. In such cases, the Poisson distribution, illustrated in Fig. 3, gives the probability for seeing various numbers of events in a specified interval when the rate is known. It depends only on the expected number of events, $\mu = rT$, given by the product of the rate r and the size of the interval, T. Using this sampling distribution, if n events are observed in a time T, both Bayesian and frequentist methods estimate the event rate as $\hat{r} = n/T$, with approximate 68 percent credible and confidence regions given by $\hat{r} \pm \sqrt{n}/T$.

Note that in both the binomial sampling problem and the Poisson rate problem, the uncertainty in the parameter was proportional to the square root of the number of events. Uncertainties that arise from counting discrete events and that have this root-n behavior are generically called counting uncertainties. Not all inferences arising from counting discrete events display this behavior, however.

See also: ERROR, EXPERIMENTAL; ERROR, SYSTEMATIC; ERROR and FRAUD; PROBABILITY

ERROR, SYSTEMATIC

Systematic error is an experimental error that exhibits some regularity when an experiment is repeated, rather than varying randomly. An experimenter who has not verified that systematic error is negligible and who has used analytical methods that assume only random error is present, can seriously underestimate the uncertainties present in inferences. Thus detection and reduction of systematic error can be a crucial part of an experiment.

The simplest example of systematic error is the constant experimental error present in measurements made with a poorly calibrated instrument. A scale that provides mass measurements that are always 10 mg too small introduces a 10 mg systematic error in all such measurements. By repeatedly measuring the mass of a sample whose true mass is unknown, random errors (due, say, to motion of air around the scale) will be revealed by variation in the results; the 10 mg offset, however, will not produce any variation, and will thus remain hidden. Averaging the repeated measurements will reduce the random errors but will have no effect on the 10 mg systematic error. If the random errors are substantially larger than 10 mg, the systematic error is inconsequential and can be ignored. But if the random errors are comparable to or smaller than the systematic error, estimates of the accuracy of the reported measurement will be misleading.

This simple example reveals several aspects of systematic error. First, the uncertainty due to systematic error may not be revealed by repeated measurement of an unknown quantity. Thus systematic errors usually can be quantified only by measuring known quantities (i.e., by calibrating the experiment). In the example just described, the 10 mg error could be identified by measuring the mass of calibration masses whose masses are reliably known to high accuracy. Second, methods used to reduce random error (such as averaging repeated measurements) may not reduce systematic error, and may even exacerbate it. Finally, the best way to deal with systematic error is simply to make sure it is smaller than the random error. If it is not, the instrument should be recalibrated in an effort to make the systematic error negligibly small compared with random error.

Systematic error can be more complicated than a simple constant offset. For example, an electrical instrument may heat up over an extended time after it is first turned on, and this heating may affect its calibration. As a result, measurements made at different times during the warm-up period may have different systematic errors. Averaging measurements taken after the warm-up period with measurements taken during that period can produce results with greater uncertainty than that of a single measurement taken after the warm-up period. Another common source of systematic error is contamination of a measurement by undesirable external sources. A common such source in the laboratory is the 60 Hz electromagnetic radiation produced by equipment using alternating current. A good example of how subtle such sources of contamination can be is provided by the Cosmic Background Explorer (COBE) satellite. The sensitivities of the instruments on the COBE satellite depend somewhat on the strength of the magnetic field at the satellite. During its orbit of Earth, this field strength changes as the orientation of the spacecraft with respect to Earth's magnetic poles changes, leading to systematic errors that change in time. Knowledge of the spacecraft's position and of the geometry of Earth's magnetic field allows experimenters to remove this subtle effect from the COBE data.

If systematic error cannot be reduced to negligibly small levels, it must be quantified (by analysis of calibration measurements) and reported. It should be reported separately from any random error rather than combined with it. For example, one might report a mass measurement as $m = 3.271 \pm 0.008$ (random) ± 0.010 (systematic) g if the random error had a magnitude of 8 mg and the systematic error a magnitude of 10 mg. One should also report any other pertinent information available about the systematic error component (such as whether it is the same in measurements of different objects).

Scientists commonly have to perform calculations based on the results of imperfect measurements, and must account for how the uncertainties in the measurements affect the outcome of the calculations. Such an accounting can be seriously complicated by the presence of significant systematic errors. In the frequentist approach to statistical inference, one can quantify the effects only of random errors, so no formally correct approach for dealing with systematic errors exists. Nevertheless, it is not uncommon to base calculations on an intuitive measure of the total (random plus systematic) error, typically found by combining the errors in quadrature (that is, by taking the square root of the sum of the squares of estimates of the sizes of the random and systematic errors). This rule is a practical device for providing a measure of uncertainty that is larger than the random uncertainty but smaller than the sum of the random and systematic uncertainties, since it is possible that these uncertainties will be in opposite directions and hence might cancel somewhat.

From the Bayesian point of view, it is possible to treat systematic errors rigorously by using probability theory, since Bayesian methods use probability to describe any kind of uncertainty, not merely random uncertainty. This can be accomplished by introducing a model for the source of the systematic error into one's calculations. In some cases, these calculations provide a rigorous justification for the practical quadrature formula just described. In the example discussed above, one could model the mass measurements as being the sum of the true mass value, a random error, and a fixed but unknown systematic error. The systematic error would enter the calculations as an offset parameter in addition to the parameter representing the true mass value we want to measure. The offset parameter could be estimated by considering calibration data in addition to the data provided by measurements of the unknown mass. The effect of uncertainty in the offset parameter (due, perhaps, to limited precision of the calibration masses or random error in their measured masses) on the inferred value of the unknown mass can then be found from the rules of probability theory. The resulting calculations are complicated by the presence of the new parameter. It thus remains true that the simplest practical method for dealing with systematic error is to ensure that it is negligibly small by carefully calibrating instruments. But when systematic error cannot be sufficiently reduced, the complication associated with additional parameters needed to model the error in a Bayesian calculation may be unavoidable.

See also: COSMIC BACKGROUND EXPLORER SATELLITE; ERROR, EXPERIMENTAL; ERROR, RANDOM; ERROR AND FRAUD

Bibliography

JEFFREYS, H. *Theory of Probability* (Oxford University Press, Oxford, Eng., 1961).

PARRATT, L. G. *Probability and Experimental Errors in Science* (Wiley, New York, 1961).

TAYLOR, J. R. *An Introduction to Error Analysis: The Study of Uncertainties in Physical Measurements* (University Science Books, Mill Valley, CA, 1982).

THOMAS J. LOREDO

ERROR AND FRAUD

Science is a human activity. Hence unintentional mistakes and some rare intentional mistakes will occur in science. In their education, scientists are taught methods which are designed to avoid errors and then to check for errors. More important, they are taught scientific ethics where errors are expected but should be searched for and corrected. Fraud is abhorred since science is built on trust, and fraud destroys confidence.

To appreciate the problems and evaluate solutions, some case histories of wrong results will be described. In 1953 Irving Langmuir described this as pathological science and gave the following list of six symptoms:

1. The effect produced by the source is always small, even when the source is varied.
2. The measured effect is always at the limit of detectability.
3. Claims of fantastic accuracy.
4. Fantastic theories contrary to experience.
5. Criticisms are met by ad hoc excuses thought up on the spur of the moment.
6. Ratio of supporters to critics rises to 50 percent and then falls gradually to oblivion.

It is important to note that a claim that fulfills one of the symptoms, does not necessarily mean the claim is wrong—for example, a major discovery often fulfills number 4. It is because several of the symptoms, not just one, are fulfilled that the claim may be doubted.

Since Langmuir, several other examples of wrong results have occurred and the number of symptoms has increased—particularly because of cold fusion. General considerations including legal ones, and the frequency of error and fraud, are discussed in a major report, "Responsible Science: Ensuring the Integrity of the Research Process," issued by the National Academy of Sciences, National Academy of Engineering, and the Institute of Medicine.

Case Histories

N rays. In 1902 Henri Blondlot of Nancy, France, announced the discovery of N rays. He produced them initially in a cathode ray tube but later from a gas burner in a steel box with a thin aluminum window. The N rays were said to produce a brightening on a calcium sulphide screen. They could be dispersed with a quartz prism and their refractive index measured with great accuracy.

Blondlot found that N rays could traverse aluminum, black paper, and wood. Many sources of N rays were found including the Sun, lamps, and secondary sources such as a quartz lens and metals. N rays were stopped by water, but not salt water. N rays were shown to be waves as diffraction patterns and Newton's rings were observed. Quickly, some 300 papers were published by more than 100 authors.

Dynamic effects were important—humans and dogs gave off N rays, particularly from brain, muscles, and nerves, but curare deadened the effects. Vegetables gave off N rays, but if chloroform was added, there were no more N rays—which is consistent with the idea of a living dynamic process. But if a metal gave off N rays, then was it reasonable to claim that adding chloroform to the metal killed the N rays? It is characteristic that if people believe one miracle, then they uncritically accept other miracles.

An American, R. W. Wood, was asked to investigate. He was well received by a confident Blondlot, who was passing N rays through two slits and dispersing them by a prism onto a calcium sulphide screen. He was measuring the N rays with great accuracy. All this was happening in darkness, so when Wood asked for the measurements to be repeated, he was able to remove the prism secretly. Despite this loss, Blondlot did the impossible and repeated the measurements and managed to get the same accurate values. This embarrassing event killed N Rays, for Wood wrote an account and sent it to two journals.

Despite this decisive demonstration that stopped most work on N rays, Blondlot himself wrote a book published two years later, which consists mainly of his published papers but does not contain any retraction or excuse.

Polywater. In 1962 the group of B. V. Deryagin in the USSR reported that water in small glass capillary tubes exhibited anomalous properties. It was closely associated with the capillaries, but once formed could exist on its own—there were suggestions that it could have structure. In 1969 interest in the subject rose steeply, partly through a U.S. government agency, and also through the enthusiastic support of E. R. Lippencott, a distinguished professor of chemistry. Its unusual properties and the thought that it could have biological implications in the thin capillaries in living bodies created great media interest. Lippencott obtained spectra different from that of water and suggested that the water was polymerized—hence the name "polywater." Only small quantities were ever produced. Many theoreticians assumed that the experimental reports in favor of polywater were true and then developed theories, often complicated ones, to justify and explain it— but they ignored the many papers that reported not finding polywater.

During 1970 experiments were reported where contamination by organic substances (e.g., perspiration) was observed to give the effects claimed for polywater. What was crucial was that Lippencott now said that he could reproduce his previous spectra when the liquid was contaminated, but on removing these contaminants, he could not now find the spectra. Nonetheless, most people did not change their opinion. Many military organizations were interested and research flourished.

In 1971 and 1972 opinion turned strongly against polywater although Deryagin rebuked doubters. Finally in 1973 Deryagin retracted and said that impurities must be the explanation of polywater effects.

Heavy 17-keV neutrino. In 1975 John Simpson claimed that a small distortion in the energy spectrum of electrons coming from the decay of tritium gave evidence for the existence of a heavy neutrino with a mass of 17 keV. Evidence against the heavy neutrino appeared to kill it, but Simpson claimed that this evidence was flawed and he produced confirming evidence with other decays. In early 1991 two major laboratories showed evidence of a heavy neutrino, in each case of 17-keV mass. Although this result was contrary to astrophysical arguments, it was taken seriously and many careful experiments were designed to verify the findings. When these second generation experiments found no effect, several of the proposers then returned and checked their older experiments and found unsuspected errors— they then withdrew their earlier claims. This chain of events is a good example of how the scientific method is meant to work.

The fifth force. In 1986 Ephraim Fischbach proposed that there was evidence of a fifth force that was much weaker than gravity. A first series of experiments gave mixed results, which were interpreted as evidence for a fifth force even though they were mutually contradictory. A second generation of careful, large experiments showed no evidence for such a force, and Fischbach withdrew. Again this is how one would like the scientific method to work by eventually correcting a wrong idea.

Cold fusion. Since the 1920s it has been understood that the Sun's energy results from the conversion of hydrogen to helium (i.e., fusion). The dream of duplicating this process on Earth at lower temperatures has, from that time, been attractive to the scientific community. In fact, some great scientists, including John Cockcroft, have even claimed success, but their results were later proved wrong and the findings were publicly retracted.

On March 23, 1989, Martin Fleischmann and Stanley Pons announced at the University of Utah that they had achieved cold fusion in a tiny electrochemical cell with a palladium cathode and using heavy water, D_2O. They claimed to have observed excess heat and nuclear products, 40,000 neutrons per second and comparable numbers of tritons per second, that showed the reaction was fusion. The announcement was made at a well-attended press conference and was followed by a series of escalating claims that maintained media interest. Confirming results were also reported in the media—the first was by Steven Jones of Brigham Young University.

The basic idea proposed in 1986 by Jones was that since palladium could absorb large amounts of hydrogen, by applying great pressure—for example, by electrolysis—the Coulomb barrier could be overcome and fusion produced. However, the protagonists did not seem to have checked that when deuterium is forced into palladium, on the contrary, the deuterium nuclei are further apart than previously, and the probability of fusion is vanishingly small.

Thousands of experiments were quickly started to repeat this simple table-top experiment. Most found no excess heat and no nuclear particle production, but a few found a positive effect and, although these effects were often mutually contradictory, each was counted as a confirmation. Careful second generation experiments were performed and found no evidence, so the majority of scientists concluded that there was no reasonable evidence for cold fusion—Jones is one of those who has retracted his claims. However, a small, determined group who believed in cold fusion have continued—they have meetings that are attended mainly by fellow believers, and they have an active media presence. As with N rays, some remarkable claims have been made, which are not criticized by other believers—examples are black hole formation, transmutations, and most surprising, several groups have claimed cold fusion using normal water and not heavy water, in contradiction to the claims of Fleischmann, Pons, and others that the process was fusion because the excess heat was found with heavy water and not with light water.

There were some new pathological features of cold fusion. One was the use of threats from lawyers against those who found unwelcome results. Also it was observed that positive results were found in some parts of the world and only null results in other parts—this regionalization of results is in contradiction to the belief that the laws of science are universally applicable. Furthermore, a characteristic of science is that experiments are reproducible, but almost all believers find that cold fusion results are not reproducible. These constitute new symptoms to be added to the six of Langmuir.

Science and the Law

In general, science and the law are incompatible. In a legal case each lawyer defends his own case and tries to discredit the other side's case. This approach is very different from the scientific method whereby all scientists try to look at *all* the evidence and make a judgement on the totality of facts. However, science has been increasingly involved in legal proceedings, particularly in cases in the United States involving misuse of funds by fraud, and in settling damage claims. These actions have resulted in the Carnegie Commission proposing that the previously used Frye rule be replaced by a three-step test in which the judges ask of a scientific claim:

1. Is it testable?
2. Has it been empirically tested?
3. Has the testing been carried out according to scientific methodology?

A negative answer on any one of these three points should disqualify the evidence. These three criteria provide a good, simplified basis and work well with the above examples. These criteria are applicable to theories as well as to experiments.

Responsible Science

The report on "responsible science" issued by the National Academy of Sciences, the National Academy of Engineering, and the Institute of Medicine notes that the world "fraud" legally involves the question of intention, which is difficult to prove. Hence they prefer the phrase "misconduct in science." They define this as being of three types: fabrication; falsification; or plagiarism in proposing, performing, or reporting research. "Fabrication is making up data or results, falsification is changing data or results, and plagiarism is using words or ideas of another person without giving appropriate credit." The report makes a clear distinction between this misconduct and questionable research practices. The latter include "(a) refusing to give peers reasonable access to unique research materials or data that support published data, (b) misrepresenting speculations as fact or releasing preliminary research results, especially to the public media, without providing sufficient data to allow peers to judge the validity of the results or to reproduce the experiment."

Frequency of Errors and Fraud

Errors are inevitable in the research process. Unfortunate statistical fluctuations must occur. Errors in judgment, lack of knowledge of important factors, and so on happen. As the report on responsible science stated, "Errors are an integral part of the process of attaining scientific knowledge."

It is difficult to estimate precisely how frequently errors or fraud occur since there is no general agreement on the terms. A first attempt made at a major laboratory gave an estimate of about a few errors per thousand in published results—this is reasonable considering that statistical fluctuations and unforeseen effects must occur. In general, factors responsible for these wrong results appear to be a lack of self-criticism, a lack of desire to try and prove oneself wrong, for example, by trying critical experiments, a willingness to accept uncritically and without control results from assistants and colleagues, an eagerness to accept confirming results and views and a rejection of null results and of criticism. As Richard Feynman once said, "The easiest person to deceive is oneself."

Fraud or "misconduct in science," seems to be very rare but does occur. Estimates in the United States range between 40 and 100 cases during the period from 1980 to 1990.

Peer review is one of the bases of the scientific process. In general it works well but there are many cases (though small in percentage) where friendship or other considerations affect a peer's judgment. Thus the fact that a paper is peer reviewed and published does not guarantee its lack of error—it still must be evaluated in combination with all other knowledge. Thus peer review is "the worst of all possible systems, except all the others."

Estimate of Errors

A scientific measurement means little unless an estimate of the errors is also given. There are two types of error usually called "statistical" and "systematic" but more descriptive names might be "combination of known experimental errors" and "combination of uncertainties in theory."

The statistical error contains errors such as those from limited statistics, the resolution of the instruments, and the system. Normally the distribution of each error is assumed to be Gaussian and the individual errors are added in quadrature.

The systematic errors are often difficult to estimate since they are rarely known precisely. If there are several errors, they are often assumed to be of a Gaussian shape and added in quadrature. However, they could be asymmetric or have long tails, hence the more cautious scientist often adds them linearly, making them bigger for safety. The most complete way is to quote both.

If the experiment is repeated exactly, thus increasing the numbers, the statistical error will decrease but the systematic error will not change. In recent years it has become the standard practice in many fields to publish with both statistical and systematic errors.

It occasionally happens that someone makes errors unnaturally small, so that the experiment looks better than others, or so that the difference between theory and experiment looks more significant. It is recommended that attention be paid as to how the errors are actually calculated.

Conclusions

Errors occur normally as part of the research process and are usually corrected by testing the hypothesis by repeating or extending the experiment. Questionable research practices happen and cause problems, but they eventually get corrected, in particular by the peer-review system that, while not perfect, is the best available. Fraud or "misconduct in science" is very rare. However, it must be recognized that such fraud does occur. When there are major contradictions, fraud is one possible solution, along with error and questionabled research practices, as well as a new scientific explanation. Continual awareness and teaching, especially by personal example, of scientific ethics and integrity may be the solution.

See also: CATHODE RAY; ERROR, EXPERIMENTAL; ERROR, RANDOM; ERROR, SYSTEMATIC; ETHICS; FIFTH FORCE; FUSION; NEUTRINO, HISTORY OF; N RAY; SOCIETY, PHYSICS AND

Bibliography

BLONDLOT, R. *The N-Rays* (Longmans, Green, and Co., London, 1905).

CARNEGIE COMMISSION ON SCIENCE, TECHNOLOGY, AND GOVERNMENT. *Science and Technology and the President* (Carnegie Commission, New York, 1991).

CLOSE, F. E. *Too Hot to Handle: The Race for Cold Fusion* (W. H. Allen, London, 1990).

FRANKLIN, A. *The Rise and Fall of the Fifth Force* (American Institute of Physics, New York, 1993).

FRANKS, F. *Polywater* (MIT Press, Boston, MA, 1981).

LANGMUIR, I. "Pathological Science." *Phys. Today* **42** (10), 36–48 (1989).

MORRISON, D. R. O. "The Rise and Fall of the 17-keV Neutrino." *Nature* **366**, 29–32 (1993).

NATIONAL ACADEMY OF SCIENCES, NATIONAL ACADEMY OF ENGINEERING, and INSTITUTE OF MEDICINE. *Responsible Sci-*

ence: Ensuring the Integrity of the Research Process (National Academy Press, Washington, DC, 1992).

TAUBES, G. Bad Science: The Short Life and Weird Times of Cold Fusion (Random House, New York, 1993).

DOUGLAS R. O. MORRISON

ESCAPE VELOCITY

The work energy theorem relates the work done by a force on an object to the change in its kinetic energy:

$$\text{Work} \equiv \int \mathbf{F} \cdot \mathbf{dr} = \tfrac{1}{2}mv_f^2 - \tfrac{1}{2}mv_i^2.$$

The gravitational force on a projectile that has been launched from the surface of a planet ($r = R$) with initial speed v_i is given by

$$\mathbf{F} = \frac{-GMm\mathbf{r}}{r^3}.$$

Thus the work done by the gravitational force as the projectile travels from the surface of the planet to an infinite distance, where the gravitational force is negligibly small, is

$$W = \int_R^\infty \frac{-GMm\,dr}{r^2}$$

$$= \frac{-GMm}{R}$$

$$= \tfrac{1}{2}m(v_f^2 - v_i^2).$$

Therefore, the speed of the object at infinity is given by

$$v_f = \sqrt{v_i^2 - 2GM/R}.$$

A real solution for v_f exists only for $v_i \geq \sqrt{2GM/R}$. The lower limit, $v_i = \sqrt{2GM/R}$, is the escape velocity, the minimum speed sufficient for the object to move beyond the gravitational field of the planet. Note that v_i is actually a speed (the scalar magnitude of the vector velocity) although it is called a velocity (which normally has both magnitude and direction); the same initial speed is needed whether the projectile is launched radially or at an angle relative to the vertical. For Earth, the value of the escape velocity is given by $v_i = 11.2$ km/sec, whereas for the moon only 2.4 km/sec is required. This result applies only when all other forces on the object can be ignored; the escape of a projectile from Earth should take air resistance into account, thus increasing the necessary speed and eliminating the angle independence.

The escape velocity from an astronomical body determines how rapidly gases in its atmosphere will be lost to space. Molecules escape from the exosphere, the outer region of an atmosphere where molecules only rarely collide, and therefore move on ballistic trajectories. On planets with weak gravity, some light molecules have sufficient thermal velocity at the base of the exosphere to escape. The speeds of these molecules are spread out according to the Maxwell–Boltzmann distribution, and it is necessary for only a small fraction of the molecules in the high-speed tail of the Maxwell–Boltzmann distribution to have $v \geq v_{esc}$ in order to remove a gas over geologic times. This mechanism, called Jeans escape, sets an upper limit for the lifetimes of hydrogen and helium in Earth's atmosphere of a few million years; that these species are found in the atmosphere at all is due to replenishment via photodissociation and outgassing, respectively.

More massive gas molecules whose thermal speeds are too low to escape may acquire the necessary kinetic energy in exothermic chemical reactions in the exosphere. Since in both cases speeds are inversely proportional to the square root of the molecular mass, escape of a particular element increases the proportion of heavy isotopes over time. On low-gravity Mars, for example, preferential loss of N^{14} atoms produced in chemical reactions has resulted in a ratio of N^{15} to N^{14} in the Martian atmosphere 1.64 times that of Earth, implying a possible loss of a large fraction of the initial nitrogen inventory.

See also: ENERGY, KINETIC; ENERGY AND WORK; GRAVITATIONAL ASSIST; GRAVITATIONAL FORCE LAW; MAXWELL–BOLTZMANN STATISTICS; WORK FUNCTION

Bibliography

BARTH, C. A.; STEWART, A. I. F.; BOUGHER, S. W.; HUNTEN, D. M.; BAUER, S. J.; and NAGY, A. F. "Aeronomy of the

Current Martian Atmosphere" in *Mars,* edited by H. H. Kieffer, B. M. Jakosky, C. W. Snyder, and M. S. Matthews (University of Arizona Press, Tucson, 1992).

CHAMERLAIN, J. W., and HUNTEN, D. M. *Theory of Planetary Atmospheres,* 2nd ed. (Academic Press, Orlando, FL, 1987).

OHANIAN, H. C. *Physics,* 2nd ed. (W. W. Norton, New York, 1989).

PHILIP B. JAMES

ETHER HYPOTHESIS

The laws of classical physics (that is, pre-1905 mechanistic physics) dictate that waves must propagate through a medium. For example, sound propagates through matter in the form of an oscillating pressure wave. The observed speed at which a wave travels through a medium depends on many factors, including the relative motion between the observer and the source of the wave. In addition, wave speed is a function of the physical properties of the medium through which it propagates. The speed of sound, for example, depends on the elasticity of the medium as well as other parameters. One can think of elasticity as a sort of stiffness; the stiffer the medium, the faster the sound will travel. Thus, the speed of sound is faster in water than it is in air. It follows that if the medium is removed, the wave cannot propagate. Sound, therefore, does not travel in the vacuum of space, despite what science fiction movies would have you believe.

Physicists have known since the middle of the nineteenth century that light is a wave-like phenomenon (in the classical sense, that is, before photons were discovered). Like other waves, light waves interfere, diffract, and refract. Consistent with the mechanistic interpretation of other wave phenomena, when physicists discovered that light was a wave, they hypothesized that for light waves to propagate, there must be some ubiquitous medium through which they travel. This hypothesized medium was known as ether, and was thought to be the background medium of the universe through which all light, from starlight to candlelight, must travel.

To conform this mechanistic view of light propagation with Newtonian and Galilean physics, the ether was further hypothesized to be the inertial reference frame in which light travels at its ultimate velocity, **c**. It naturally followed that for another reference frame moving at velocity **u,** with respect to the ether, the observed velocity of light with respect to the moving frame would be

$$\mathbf{v} = \mathbf{c} - \mathbf{u}. \tag{1}$$

Note that this result is simply a Galilean transformation of the velocity of light from that in its rest frame to its velocity in an inertial frame moving at some velocity relative to the rest frame.

The ether hypothesis immediately presented physicists with difficulties. The most pressing issue was that no one had ever observed any ether. Its properties were never measured, and none was ever detected. In addition, one particularly vexing observation was that light travels through an evacuated glass jar. Therefore, if the ether does exist, either it cannot be removed by a vacuum pump or its density is already so low that a vacuum pump cannot remove any more. Physicists also have long known that nothing exceeds the speed of light in a vacuum. The flash from a distant rifle shot was observed to arrive instantaneously (or so it seemed, until more sophisticated measurements were made), whereas the rifle's report would arrive seconds later. If the speed of a wave depends on the elasticity of the medium through which it travels, then light's high speed of propagation means that the ether must be very stiff. Yet, on Earth and throughout the observable universe, there is no evidence for this resilient, ubiquitous medium.

Even more troublesome than the elusiveness of the ether is the fact that Maxwell's equations are not invariant under a Galilean transformation. Maxwell's equations describe light propagation through any medium. In the mid-nineteenth century, however, it was assumed that they applied only in the reference frame at rest with respect to the ether (the ether frame). Yet, if Maxwell's equations were valid only in a unique reference frame, then by definition they were not invariant under a Galilean transformation to any other inertial reference frame.

Despite these formidable challenges to the ether hypothesis, physicists continued to accept the reality of the ether for lack of any better understanding of light propagation. In 1887, however, two physicists, Albert A. Michelson and Edward W. Morley, designed an experiment to test not the existence of the ether but its effect on how light travels through it.

Michelson and Morley hypothesized that they could measure the ether drift, or the speed of Earth as it moved through the ether, by measuring the difference in time it takes a beam of light to travel two perpendicular paths of the same length, d. Specifically, Michelson and Morley sought to find different times, t and t', for light to travel distance d, assuming that t was elapsed in a reference frame at rest with respect to the ether frame and t' was elapsed in a frame moving with respect to the ether (i.e., Earth). Using perpendicular beams of light enabled them to exploit the vector addition of relative light velocities given by Eq. (1); rotating their apparatus by 90° would guarantee that one beam would be at rest in the ether frame while the other was moving relative to the ether.

Michelson and Morley assumed that Earth was moving through the ether at velocity **u,** where **u** is essentially the velocity of Earth as it orbits the sun (about 1/10,000 the speed of light). Of the two beams, one would be moving parallel (or antiparallel) to Earth's path through the ether, and the other beam, being perpendicular to the first, would not have any change in relative velocity. By estimating **u,** Michelson and Morley could estimate the time of travel for each wave, using the formula that time is equal to distance divided by speed. For the parallel wave (that is, t'), the elapsed time was hypothesized to be

$$t' = \frac{d}{(c+u)} + \frac{d}{(c-u)} = \frac{2d\gamma^2}{c}, \qquad (2)$$

where

$$\gamma^2 = \left(1 - \frac{u^2}{c^2}\right)^{-1} \le 1. \qquad (3)$$

For the perpendicular wave (i.e., t),

$$t = \frac{2d}{c}. \qquad (4)$$

Thus, t' should be greater than t if there is a unique reference frame at rest with respect to the ether. Yet, because c is so much greater than u, the difference between t' and t should be very small and difficult to measure directly. However, Michelson and Morley had the insight to use the Michelson interferometer

to measure the time difference by measuring a shift in interference fringes that would result from a slight difference in phase between the two beams.

Michelson and Morley rotated their apparatus through many orientations and performed their experiment at several different times of year, yet they detected no measurable shift in the interference fringes of the two beams. Their negative result indicated that, to the precision of their instrument, they could not measure Earth's motion through the ether.

Many possible interpretations were offered for Michelson and Morley's negative result. One explanation, the ether drag, proposed that the moving Earth dragged the ether along with it locally, similar to a ball dragging air along with it as it flies, thus eliminating any relative motion between the ether frame and Earth. Ether drag was rejected for several reasons, one of which was that this differential motion of the ether would have altered stellar observations in ways not consistent with actual observations.

Another explanation was offered by George FitzGerald and Hendrik A. Lorentz, who proposed that the motion of the interferometer through the ether caused a length contraction of the apparatus in the direction parallel to the motion relative to the ether. This contraction of the distance between the two mirrors was hypothesized to be just enough to cancel any difference between t and t'. The FitzGerald–Lorentz explanation, while plausible, was considered unsatisfactory because of its ad hoc nature.

The Michelson–Morley result was not reconciled with accepted physical principles until 1905, when Albert Einstein reexamined the laws of light propagation in the context of Maxwell's equations, the Galilean transformation, and the persistent inability of physicists to detect a unique reference frame through which light travels at velocity c. Einstein based his special theory of relativity on two postulates: (1) all inertial reference frames are physically equivalent, so no experiment can be designed to distinguish between them, and (2) the velocity of light is independent of the motion of its source. Given the first postulate, Einstein was compelled to modify either Maxwell's equations or the Galilean transformation; based on several reasons, Einstein preserved Maxwell's equations and modified the Galilean transformation to a new form under which Maxwell's equations were invariant. The resulting transformation, known as the Lorentz transformation, is the cornerstone of relativistic physics in inertial reference frames.

The postulates of special relativity led to a consistent theory of the behavior of light (or electromagnetic radiation) that resolved Maxwell's equations with the other physical theories known at the time. Given Einstein's first and second postulates, the ether hypothesis was no longer necessary, for the mechanistic view of light propagation had yielded to Einstein's theory, which itself was a precursor to the development of field theory.

See also: EINSTEIN, ALBERT; FRAME OF REFERENCE; GALILEAN TRANSFORMATION; LORENTZ TRANSFORMATION; MAXWELL'S EQUATIONS; MICHELSON, ALBERT ABRAHAM; MICHELSON–MORLEY EXPERIMENT; RELATIVITY, SPECIAL THEORY OF, ORIGINS OF

Bibliography

EINSTEIN, A. *Relativity: The Special and the General Theory.* (Bonanza Books, New York, 1961).

EISBERG, R., and RESNICK, R. *Quantum Physics of Atoms, Molecules, Solids, Nuclei, and Particles,* 2nd ed. (Wiley, New York, 1985).

HALLIDAY, D., and RESNICK, R. *Physics,* 3rd ed. (Wiley, New York, 1978).

JENKINS, F. A.; and WHITE, H. E. *Fundamentals of Optics,* 4th ed. (McGraw-Hill, New York, 1976).

RICHTMEYER, F. K.; KENNARD, E. H.; and LAURITSEN, T. *Introduction to Modern Physics,* 5th ed. (McGraw-Hill, New York, 1955).

P. W. "BO" HAMMER

ETHICS

Although he won the 1923 Nobel Prize, in part for his studies on electronic charge, in 1913 Robert A. Millikan published a paper based on "all of the drops" on which he experimented. Half a century later, historians discovered that Millikan discarded, without comment, about a quarter of his drops, even though he claimed not to have done so. Although discovery of the Millikan misrepresentation had few effects, beginning in the 1960s and 1970s a series of biomedical scandals involving fabricated experiments and deception of research subjects helped to generate a body of literature on scientific ethics.

In 1974 the U.S. Congress required institutions receiving federal research grants to have research ethics committees—institutional review boards. By 1983, major universities throughout the world had guidelines for dealing with misconduct in research. Scientists seeking funding from U.S. government agencies now must explicitly address questions of research ethics, such as whether humans will be experimental subjects, or whether an institutional review board has evaluated the research procedures.

Processes and Products

Two broad categories of ethical problems arise in connection with science: those related to processes and those related to products. As long ago as 1831, Charles Babbage described some of the unethical processes of doing scientific research, including trimming, cooking, and forging. Trimming consists of smoothing irregularities in the data to make them look consistent. Cooking occurs when one discards data—as Millikan did—and retains only evidence that fits the theory. Forging is inventing some or all of the research data reported. Because not all data gathered in a study can be used, most scientists must simplify them in order to make research processes manageable. Hence, one scientist's cooking may be another researcher's simplification.

Scientific products likewise may be ethically questionable if they harm people or arise from biased methods. For example, radioactive fallout from U.S. above-ground nuclear-weapons testing has caused epidemics of leukemia and other cancers among downwinders in Nevada, Utah, Arizona, and California.

When the Bureau of Drugs of the U.S. Food and Drug Administration (FDA) carried out 496 inspections of clinical studies, it found numerous problems with research processes and practices. These problems included lack of patient consent, inadequate drug accountability, protocol nonadherence, record inaccuracy, and record nonavailability. Since 1962, FDA investigations undertaken in response to reports of suspected research misconduct have resulted in the disqualification of one in every four investigated researchers.

If the FDA's statistics are typical, then unethical research practices, processes, and consequences may cause serious harm. A 1986 General Accounting Office report revealed that 90 percent of the Department of Energy's 127 nuclear facilities—many of which conduct research—had contaminated groundwater that exceeded regulatory standards by a factor of up to 1,000. The report also revealed that cleanups, caused by environmental violations at U.S. Department of Energy and Department of Defense installations, will cost $300 billion.

General Rules of Scientific Ethics

As Babbage's discussion of trimming and cooking reveals, scientists have a clear ethical obligation not to do biased research. The weapons' tests in Nevada show that they also have obligations not to expose people or the environment to unjustified dangers and to obtain free informed consent from those who might be put at risk from scientific work.

Because science often generates financial benefits, researchers likewise have a duty not to convert public monies to private gain. Scientists have enjoyed rich consulting arrangements and taxpayer funding for projects such as developing laser technology and improved x-ray equipment. Although much technical work is funded by taxpayers through grants or university salaries, sometimes most of the profits on these projects go to scientists. One Nobel Laureate, for example, owns more than a million dollars in shares in a biotechnology company designed to commercialize inventions that he developed as a university faculty member.

In general, practicing scientific ethics requires physicists and other researchers to follow at least five basic ethical principles:

1. Scientists ought not do biased research.
2. Scientists ought not do research that causes unjustified risks to people.
3. Scientists ought not do research that violates norms of free informed consent.
4. Scientists ought not do research that unjustly converts public resources to private profits.
5. Scientists ought not do research that seriously jeopardizes environmental welfare.

As Jacques Monod affirmed, the first principle is the most basic. Truthfulness is the cement that holds science together.

Objectivity, Bias, and Fraud in Science

How should researchers pursue truthfulness? Obviously they ought to avoid frauds such as plagiarizing information, fabricating it, and stealing ideas. Lack of objectivity also occurs when scientists fail to acknowledge the areas of uncertainty in research methods or conclusions, when they fail to give credit to individuals for work done, when they fail to carry out research in a conscientious manner, or when they fail to publish a retraction after conclusive evidence exposes an error in earlier work. Such acts sometimes stem from an unethical mentality of "success at any cost." More controversial kinds of scientific fraud include loose authorship (removing or inserting names of persons who may or may not have helped in the work) and duplicate publication. Duplicate publication is controversial because sometimes scientists wish to bring their results to a wider audience, while in other situations, authors want to increase the perception of productivity.

Although all cases of research misconduct violate principles of objectivity, scientists often disagree over whether a situation involves fraud or mere error, sloppiness, or overstatement. Because of such disagreement, it may be easier to spot misconduct than to determine a researcher's ethical culpability. For this reason, practicing scientific ethics typically requires careful analysis of the processes and products of scientific work. Oliver Cromwell's famous plea to the assembly of the Church of Scotland provides a model for the ethical behavior of scientists: "I beseech you . . . think it possible you may be mistaken." Just as the best scientists are those who continually evaluate their methods so as to improve them, so the most ethical scientists may be those who continually analyze their work on the basis of the five basic ethical principles.

See also: ERROR, EXPERIMENTAL; ERROR, RANDOM; ERROR, SYSTEMATIC; ERROR AND FRAUD; MILLIKAN, ROBERT ANDREWS; SOCIAL RESPONSIBILITY; SOCIETY, PHYSICS AND

Bibliography

BAYLES, M. *Professional Ethics* (Wadsworth, Belmont, CA, 1981).

COURNAND, A., and ZUKERMAN, H. *The Code of Science: Analysis and Reflections on Its Future* (Columbia University Institute for the Study of Science in Human Affairs, New York, 1970).

FRADEN, R., and BEAUCHAMP, T. *A History and Theory of Informed Consent* (Oxford University Press, New York, 1986).

FRADKIN, P. L. *Fallout: An American Nuclear Tragedy* (University of Arizona Press, Tucson, 1989).

HOOK, S.; KURTZ, P.; and TODOROVICH, M., eds. *The Ethics of Teaching and Scientific Research* (Prometheus, Buffalo, NY, 1977).

MILLER, R. L. *Under the Cloud: The Decades of Nuclear Testing* (Free Press, New York, 1986).

MONOD, J. *Chance and Necessity* (Knopf, New York, 1971).

PAYNE, S. L., and CHARNOV, B. H., eds. *Ethical Dilemmas for Academic Professionals* (Thomas, Springfied, IL, 1987).

SHRADER-FRECHETTE, K. *Ethics of Scientific Research* (Rowman & Littlefield, Lanham, MD, 1994).

KRISTIN SHRADER-FRECHETTE

EVAPORATION

Evaporation is the escape of molecules from a surface, especially from a liquid. The concept is usually extended to include boiling, in which the liquid is converted into bubbles of gas within the liquid and the gas bubbles subsequently escape from the surface of the liquid. Evaporation from the surface of a solid is called sublimation.

Evaporation is a very effective and important means of cooling. Molecules of a liquid attract each other strongly, so that a molecule attempting to leave the liquid surface is pulled backward into it. It can escape only if it has a much greater speed, and hence kinetic energy, than the average. Therefore, the process of evaporation removes the molecules of highest kinetic energy from the surface, leaving behind molecules of lower kinetic energy and, hence, a liquid at a lower temperature. Human beings are cooled by evaporation of water from the skin. Evaporation of one liter of water per day would remove about one-third of the typical energy intake from food (2,000 cal = 8×10^6 J). More rapid cooling (e.g., to bring down a fever or prevent swelling) may be achieved by sponging or spraying a person with a liquid that evaporates more rapidly, such as alcohol or ether. Dogs and many other animals are cooled by evaporation of water from the mouth and tongue.

The rate of evaporation depends on the surface area exposed. Evaporation would be much more rapid than usually observed except that molecules return to the surface of the liquid from the vapor. For example, water evaporates until the humidity reaches the point at which the number of returning molecules exactly balances the number leaving. Air molecules do not significantly affect either process, except to slow both escape and return by deflecting water molecules backward. If there is air circulation, the evaporating molecules are swept away and evaporation (and cooling) is more rapid.

The dynamic character of evaporation and condensation may be illustrated by some simple calculations based on kinetic theory. The average speed $\bar{v}$ of a water molecule at room temperature is about 600 m/s (1,300 mph), calculated from $\sqrt{8kT/\pi m}$, where k is the Boltzmann constant. To escape from the surface, a molecule must have an energy of about 7.4×10^{-20} J (nearly 2,500 J/g), or a speed of about 2,200 m/s (nearly 5,000 mph). From the Maxwell–Boltzmann equation, the probability of a molecule's having such a speed is roughly 1/20,000 (5×10^{-5}) of the probability of its having its average speed. However, there are so many molecules reaching the surface so often, that approximately 10^{22} molecules escape from 1 cm^2 of surface area per second (a figure that is most easily calculated from the number of gas molecules reaching the surface at equilibrium, which is given by $\frac{1}{4}An\bar{v}$, where A is the area and n is the number per volume in the gas). If no molecules returned and the temperature remained constant, all of the molecules (3×10^{22}) in 1 cm^3 of·water would evaporate in about three seconds. In practice, most molecules would return; if they did not, the water would quickly freeze, slowing the rate of evaporation.

Application of the kinetic theory to liquids and solids is very limited, because of the strong interactions between molecules, but from the more powerful methods of statistical mechanics, it is known that the kinetic energies resemble those of gases at the same temperatures. Because of the strong energy dependence of evaporation, the rate of evaporation increases rapidly with temperature.

See also: KINETIC THEORY; LIQUID

Bibliography

Hecht, E. *Physics* (Brooks/Cole, Pacific Grove, CA, 1994).

TABOR, D. *Gases, Liquids, and Solids* (Penguin, Baltimore, MD, 1969).

ROBERT P. BAUMAN

EVENT

An event is anything that happens at a particular point in space and at a particular time. Figure 1 shows three different events, each involving a flashing lightbulb. Event A shows a lightbulb that flashes at position 1 and at time 0. Event B occurs at the same time but at a different point in space. Event C occurs at the same position as B but at a later time. The actual units used to measure position or time are not important, but it is useful, in this context, to choose measuring units with the speed of light as unity. For example, this would occur if time is measured in years and position is measured in light-years (LY). Each event shown in Fig. 1 is unique and is uniquely described by its location in spacetime. Event B and event C are not necessarily related, but they could be related. For example, the lightbulb at C might be the same as the one at B if it were stationary and programed to flash every four years. Multiple events like B and C that could be causally related (what happens at B could cause something to happen at C) are said to be timelike separated. The situation for A and B is much different. For example, they could not involve the same lightbulb since that would mean the same bulb would have to be in two different places at the same time. Such events are called spacelike separated. Spacelike events can never be causally connected. A pair of events are timelike separated if the magnitude of the change in time is greater than the magnitude of the change in position ($|\Delta T| > |\Delta P|$, or $|c\Delta T| > |\Delta P|$, in units where the speed of light c is not equal to one). When this condition is satisfied, it is possible for the earlier event to influence the latter one by sending a signal at the speed of light or less. Since signals at greater than the speed of light are impossible, pairs of events that do not satisfy this condition (i.e., spacelike separated) can never be causally connected.

Are events A and C timelike separated or spacelike separated? Since $\Delta T = 4$ and $\Delta P = 3$, they are timelike separated (i.e., the bulb at C could be the same as the one at A). For example, event A might be the beginning event of a bulb moving along a line from the position of 1 LY to 4 LY and flashing its beacon every half year. Figure 2 depicts this scenario as a series of the individual events. In space the Moon orbits Earth in very close to a circular path. From the point of view of spacetime events, the picture is much different. If time is pictured as being perpendicular to the plane of Earth's orbit, the Moon as it moves in a circle in space, moves upward in time, producing a spiral path in spacetime. A similar situation plays an important role in modern physics. A bubble chamber is an instrument used for the detection and study of elementary par-

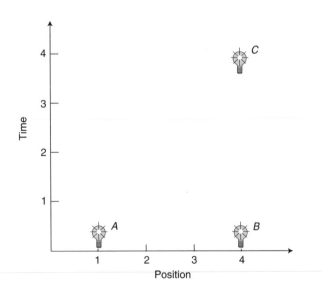

Figure 1 Three different events, each involving a flashing lightbulb.

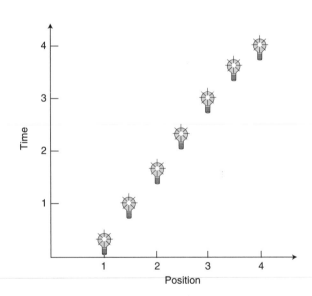

Figure 2 Series of individual events as a lightbulb moves in position and time.

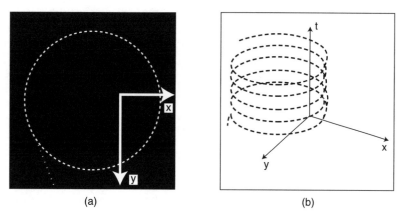

(a) (b)

Figure 3 A representative picture of particle detection by a bubble chamber.

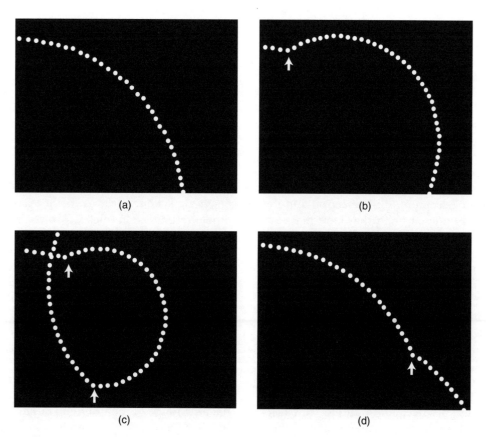

(a) (b)

(c) (d)

Figure 4 Four different outcomes when on four occasions a pion is injected into a bubble chamber at the same place and with the same energy.

ticles (electrons, mesons, and many others). In the bubble chamber, the paths of charged particles passing through liquid hydrogen are detected due to a series of bubbles that are formed along the particles' paths. A representative picture is shown in Fig. 3a. In this picture, the tail on the circle represents a pion (a type of meson) that at the rim of the circle has decayed into another subatomic particle called a muon. A superimposed magnetic field causes the muon to move in a circle in space. Note that the bubbles are more dense along the circular path than they are along the short path of the pion. This is easily understood by examining the spacetime diagram of the same process, as displayed in Fig. 3b, in which the pion decay is marked by the arrow. This spacetime figure is reminiscent of the one previously described for the Moon's motion; both the muon and the Moon travel in a circle in space, constantly overlaying their path, but in spacetime they produce an upward spiraling series of events.

Unlike classical physics, in modern subatomic physics there is not a deterministic connection between a present event and all subsequent future events. In classical physics, in principle, if enough information is known about the conditions affecting an initial event the outcome is entirely deterministic. Figure 4 shows four different outcomes from four identical initial events. In all four cases a pion is injected into the bubble chamber at the same place (entering at the left) and with the same energy. With the exception of the electron and the proton, all subatomic particles that have mass are unstable and will eventually decay into other particles. However, the specific lifetime, before it decays, of any particular unstable particle cannot be predicted. In Fig. 4a the pion lives long enough to escape the chamber. In Fig. 4b it lives a short time before decaying (arrow). And in 4c it decays in the chamber but after a longer time. In both Fig. 4b and Fig. 4c the resulting particle is a muon, a particle with a relatively longer life than the pion. In both of these cases the muon escapes the chamber. However in Fig. 4d a relatively rare event occurs; the muon undergoes decay into an electron (second arrow) before it can escape the chamber. (Note that in all cases other chargeless particles that cannot be directly detected by the chamber might be present.) Clearly the lifetime of both particles varies from trial to trial. There is no deterministic chain of events that connects the decay of a specific subatomic particle back to its inception.

However, a very reliable average lifetime can be determined for particles like the muon and the pion. The tools of modern physics, such as quantum mechanics, deal only in probabilities and averages. This lack of certainty in terms of the uncertain outcome of any one particular chain of events appears to underpin all of nature.

See also: Bubble Chamber; Muon; Particle Physics, Detectors for; Space and Time; Spacetime

Bibliography

Boas, M. L. "Event as the Key to a Graphic Understanding of Special Relativity." *Am. J. Phy.* **47,** 938–942 (1979).

Taylor, E. F., and Wheeler, J. A. *Spacetime Physics,* 2nd ed. (W. H. Freeman, San Francisco, 1966).

Carl G. Adler

EXCITATIONS, COLLECTIVE

Collective modes in many-body systems may be density waves (plasmons in solids, zero sound in the helium liquids), spin density waves (magnons in magnetic systems), or, as in the case of the giant dipole resonance in nuclei, involve the out-of-phase motion of one component (neutrons) against another (protons). Unlike individual particle excitations (electrons in solids, ^{3}He or ^{4}He atoms in the helium liquids), collective modes do not exist in the absence of interactions between particles in a many-body system. The frequency of the collective modes depends on the strength and range of the particle interaction; the restoring force comes from the action on a given particle of the averaged self-consistent potentials produced by the other particles in the system. Thus in systems of electrons, because of the long range of the Coulomb interaction, the density fluctuation collective mode at long wavelength is a plasma oscillation of constant frequency, while in the helium liquids, ^{3}He and ^{4}He, which possess a strong short-range interaction, it obeys a linear dispersion relation, $\omega(k) = c_0 k$. Despite its superficial resemblance to ordinary sound, zero sound is a different kind of mode. Zero sound and other collective modes are collisionless modes, while ordinary sound is a hydrodynamic mode, for which the frequent particle collisions that bring about local equilibrium provide the restoring force. These same collisions act to disrupt (damp) zero sound, which depends on the organized or coherent superposition of particle fields for its existence.

Perhaps the best known collective mode in solids, and the first to be identified as such, is the plasmon, the quantized plasma mode characteristic of electron systems. In the course of their work on electron interaction in metals in the early 1950s, David Bohm and David Pines showed that plasmons, unlike the collective modes found in systems with short-range interactions, would always possess a finite energy. $\hbar \omega_p = \hbar (4\pi n e^2 m)^{1/2}$, where n, e, and m are the electron density, charge, and mass, in the limit of long wavelengths. For a collective mode to be a well-defined elementary excitation, it must possess an energy that lies above the energy spectrum of the individual particles in the system. For valence electrons in a metal, the characteristic energy for a single-particle (intraband) excitation of wave vector $\mathbf{k}$, corresponding to the excitation of an electron of momentum $\mathbf{p}$ below the Fermi surface to states $\mathbf{k} + \mathbf{p}$ above it, is $\sim \hbar k v_F$, where v_F is the Fermi velocity. As a result, individual electrons moving within a given band cannot exchange energy and momentum with a long-wavelength plasmon, and hence cannot act to damp it. Plasmons thus exist as a well-defined, distinct excitation mode above the spectrum of intraband excitations until their wavelength becomes sufficiently short that for $k_c \cong \omega(k)/v_F \sim (\omega_p/v_F)$, the condition for simple-particle damping is met, and the plasmons become immersed in the continuum of single-particle excitations.

In the course of developing his general description of particle interactions in a Fermi liquid, which became known as Landau–Fermi liquid theory, Lev D. Landau showed that for systems of fermions with repulsive interactions, a collective density mode, zero sound, would, like the plasmon, lie above the single-particle spectrum at long wavelengths and hence exist as a well-defined elementary excitation. The subsequent discovery of this mode in liquid ^{3}He by John Wheatley and his collaborators was one of the key experimental tests of the Landau theory.

For systems of interacting fermions, Landau theory can be regarded as a generalization of the random phase approximation (RPA) developed by Bohm and Pines to identify collective modes in quantum liquids. In the RPA, the force on a given particle that gives rise to collective behavior of the density fluctuation ρ_k is given by $V_k \langle \rho_k \rangle$, where V_k is the Fourier transform of the particle interaction $V(r)$, and $\langle \rho_k \rangle$ is the average self-consistent density fluctuations produced by the other particles. In polarization potential theory, a post-Landau, post-RPA theory proposed by Pines, that restoring force is replaced by $f_k \langle \rho_k \rangle$, where f_k is the Fourier transform of a phenomenological potential that incorporates the influence of strong short-range particle correlations on V_q. The energy of a long-wavelength collective mode is then $\omega_k = (nk^2 f_k/m)^{1/2}$. For an electron system, the RPA is always valid in the long-wavelength limit, so that $f_k \equiv V_k \equiv 4\pi e^2/k^2$, and one recovers the plasmon spectrum; for a system with strong-repulsive short-range interactions, such as the helium liquids, the limit of f_k as k goes to zero is f_0, the average strength of that interaction, and one recovers the strong-coupling Landau result for the zero sound velocity, $c_0 = (n f_0/m)^{1/2}$. One can also understand in this way the existence of a zero sound mode in the normal state of the Bose liquid, ^{4}He, and, on taking into account the momentum dependence of f_k, the phonon-roton spectrum found for density fluctuations in superfluid ^{4}He.

There is a second class of collective modes associated with the presence of long-range order in many-body systems: phonons in solids, magnons in ferromagnets and antiferromagnets, and so on. Here it is the long-range order that not only provides the restoring force but acts to suppress the single-particle excitations that would otherwise act to damp the mode. In these systems it is the interaction between the collective modes (anharmonicity associated with phonon-phonon interactions, magnon-magnon coupling in magnets, roton-roton interactions in superfluid ^{4}He above 1K, etc.) that acts to damp the modes.

For systems in which the restoring force is strong, such as the helium superfluids, the distinction between these two kinds of collective modes become blurred. Zero sound in superfluid ^{3}He goes over smoothly into zero sound in normal ^{3}He, while the main change that occurs in the phonon mode of superfluid ^{4}He when the system becomes normal is that it is much more readily damped, since the single-particle excitations that are suppressed in the superfluid state act to damp the mode in the normal state. A similar crossover is found in both ferromagnets and antiferromagnets, when the spin wave restoring force is sufficiently strong that magnons persist as a well-defined excitation above the transition temperature to the ferromagnetic or antiferromagnetic phase.

See also: MAGNON; PLASMON

Bibliography

PINES, D., and NOZIÈRES, P. *The Theory of Quantum Liquids* (Addison-Wesley, Reading, MA, 1966).

DAVID PINES

EXCITED STATE

The quantum state is a concept that plays a key role in the modern theory of atomic structure. For any quantum system, such as an atom or molecule, a complete set of states provide a full description of the system. In bound systems there exists a particular quantum state, called the ground state, for which the total energy of the system is a minimum. Most of the atoms and molecules in the universe are in their ground state. However, when energy is transferred to the system either through a collision with other atomic particles, or with a photon, the system may momentarily exist in an excited quantum state.

An excited state is defined to be any quantum state, available to the system, in which the system energy is greater than its ground state energy. Once an atom is in an excited state it can decay to a state of lower energy and, in the process, release energy to the environment in the form of electromagnetic quanta (photons). The process can continue until the atom reaches its ground state, at which point it can no longer radiate photons.

The representative time that it takes for a quantum system to relax from an excited state, through photon emission, is called the radiative lifetime of the excited state. Many atomic lifetimes are extremely short, the 2p excited state of the hydrogen atom has a lifetime of only 1.6×10^{-9} s. Excited states with long lifetimes are called metastable states. For example, the 2s state of atomic hydrogen decays, primarily, by the emission of two photons, resulting in a significantly long lifetime of $\frac{1}{7}$ s. Metastable states of the constituent atoms or molecules of a gas are

necessary for population inversion to occur in a lasing medium. Perhaps the most beautiful manifestation of excited states in atoms is the optical aurora, or the northern lights. In aurora, energetic, charged particles collide with the oxygen and nitrogen atoms and molecules in the upper atmosphere. The collisions excite some of the atoms, which subsequently decay and emit radiation in the visible wavelength region. An example is the monochromatic lines at wavelengths of 630.0 and 636.6 nm seen in some aurora, and caused by emissions from the metastable^2 *D* state of excited atomic oxygen.

In nuclear physics the concept of an excited quantum state is also useful. The nucleus of an atom is a quantum system that, in many cases, allows a description that is similar to that of the atomic model. However, the forces between particles in the nucleus are much more complex than the electromagnetic forces that bind atoms, and several decay mechanisms exist. *Isomeric nuclei* are nuclei that have the same atomic and mass numbers but are distinct since one of the isomers is an excited metastable state and the other is the ground state of a given nuclide.

See also: ATOM; COLLISION; GROUND STATE; NUCLEAR FORCE; NUCLEUS, ISOMERIC; PHOTON; QUANTUM

Bibliography

BETHE, H. A., and SALPETER, E. E. *Quantum Mechanics of One- and Two-Electron Atoms* (Academic Press, New York, 1957).

BLATT, J. M., and WEISSKOPF, V. F. *Theoretical Nuclear Physics* (Wiley, New York, 1952).

CHAMBERLAIN, J. W. *Theory of Planetary Atmospheres: An Introduction to Their Physics and Chemistry* (Academic Press, New York, 1975).

FEYNMAN, R. P.; LEIGHTON, B. R.; and SANDS, M. *The Feynman Lectures on Physics,* Vol. 3 (Addison-Wesley, Reading, MA, 1965).

BERNARD ZYGELMAN

EXCLUSION PRINCIPLE

See PAULI'S EXCLUSION PRINCIPLE

EXPANSION

See THERMAL EXPANSION; UNIVERSE, EXPANSION OF

EXPERIMENTAL PHYSICS

Physics is an exact science. It is the study of the forces or interactions between matter and the consequences of these forces. Modern physics is subdivided into two broad groups, experimental and theoretical physics. Experimental physics is the empirical approach or method to determine the forces and resulting behavior of matter, whereas theoretical physics describes observations in mathematical terms to understand underlying forces and develop fundamental laws of nature. These theories representing observed behavior have predictive properties that can be tested by further experiments to evaluate how fundamental and complete these laws are. Experimental physics focuses on quantitative measurements, ultimately related to forces or interactions, in space and time in a laboratory. The physics laboratory can be a room with table tops, benches, and scientific equipment and facilities, such as electric power, gases, water, and vacuum pumps. It can be a vast terrain of buildings and devices such as a great high-energy laboratory, tunneling underground through miles of earth in great rings around which subatomic particles are accelerated; it can be the sky observed by the "eyes" of a physicist on Earth or in a satellite; it can be a mound of sand crumbling because of a small avalanche; or it can be a kitchen sink to observe hydrodynamic flow.

Experimental physics can be broadly divided into high-energy and low-energy physics. High-energy physics includes the study of the existence of elementary particles, that is, particles that cannot be further subdivided, and the forces between these elementary particles. A unified theory of forces includes the comparatively weak gravitational forces between masses. Experimental high-energy physicists study elementary particles such as electrons and quarks. Quarks are held together by forces to form composite particles such as the proton, consisting of three quarks. Such composite particles are bound together to form nuclei. Electrons are bound to nuclei to form atoms, which transcends to the arena of low-energy physics. Forces bind atoms to form molecules and these form large aggregates of matter: gases, liquids, and solids. Low-energy experimental physics is the study of atomic and condensed matter physics. Intermediate classes of study are nuclear physics, the study of forces within the nucleus and plasma physics, the study of neutral fluids of charged particles.

The experimental physicist develops observational methods that allow the underlying forces of nature to be revealed. The high-energy particle physicist accelerates particles to high energy and then scatters them off of other particles. The scattering pattern reveals the forces between these particles. Similar experiments can be done with atoms or molecules at lower energy. The interaction of light from a laser with an atom, molecule, or solid reveals the forces that hold these substances together. Simple experiments such as compressing a gas, liquid, or solid yield a response of a change in the volume. This response is controlled by the forces between the particles. X rays scatter off of crystalline solids with a pattern that reveals the arrangement of the atoms. The list of techniques can go on, but in general they are developed to reveal the forces of nature in a simple way, easy to interpret.

The laws of nature are independent of where they are observed, and all phenomena should be reproducible by different observers. An exact science is one in which phenomena can be measured quantitatively. However, exact does not mean with absolute precision. Measurements have limitations of accuracy. The most basic limitations are due to fundamental fluctuations found in nature. These fluctuations themselves are an important area of study. As an example, the light from the glowing filament of a light bulb heated by an electric current is emitted in quantized entities called photons. The brightness of the light bulb is measured by the number of photons emitted per unit of time. However, the number of photons emitted per unit of time fluctuates so that the brightness fluctuates about a mean value. These fluctuations are fundamental in that they cannot be suppressed. The consequence of these fluctuations is a fundamental precision in an observation of the brightness. A less fundamental source of error which limits precision would be a fluctuation of the current that heats the filament. This can be suppressed by regulating the electric power supply. Still more variations can accrue from the instrument that measures the brightness. These variations about a mean value are called the measurement error.

The fields of study considered as experimental physics are broad and multifaceted. Physicists have learned that the universe is a complex system. It may be possible to follow the trajectories of planets and heavenly bodies, but it is not possible with modern techniques to follow the trajectories of billions and billions of atoms that form a gas or liquid. Thus, physicists resort to measurements of statistical properties of these many-body systems. The measure of the vapor pressure of a gas is a statistical phenomena. It is the result of billions of collisions per second of atoms with the area of the container, giving an average force per unit of area. As systems become more complex, experimental physicists try to simplify the observations to understand the fundamental controlling forces.

With the development of modern high speed computers, a new area of computational physics has developed. The physicist can create a small group of particles (of order 100), representative of very large aggregates of particles, interacting under forces believed to represent those found in nature. These particles are allowed to undergo a large number of collisions or interactions with each other and the trajectories are followed in time by the computer. After sufficient time the properties of the system, such as the state of matter (gas, liquid, solid), the geometrical arrangement of particles, the pressure, conductivity, and other important physical properties, can be determined. This is then compared to the behavior of real systems to see if they can be reproduced, which implies an understanding of the forces. Such simulations are often thought of as experimental laboratories in a computer.

What subjects of inquiry fall within the subject of experimental physics? Centuries ago astronomy was part of physics; it is now a field of its own. Physics evolves with time. Most generally, physics is what a physicist does when practicing the trade.

See also: ATOMIC PHYSICS; CONDENSED MATTER PHYSICS; NUCLEAR PHYSICS; PARTICLE PHYSICS; PHYSICS, HISTORY OF; PHYSICS, PHILOSOPHY OF; PLASMA PHYSICS; THEORETICAL PHYSICS

ISAAC F. SILVERA

EXPONENTIAL GROWTH AND DECAY

Exponential growth produces nonintuitive results. A sheet of paper, folded in half only fifty times (only about seven times is actually feasible), would be thick enough to extend to the Sun and back . . . five thousand times! Starting with paper of thickness d, each fold multiplies the previous thickness by 2, giving successive thicknesses of $2d$, $2^2 d$, $2^3 d$, $2^4 d$, and so on. After fifty foldings the thickness equals $2^{49} d$, or about 10^{15} m for $d = 0.01$ in.; this is 10,000 times the distance from Earth to the Sun.

A famous legend illustrates how exponential doubling rapidly gets out of control. A king offered his daughter's hand to whoever would give him enough wheat to put one grain of wheat on the first square of a chess board, two on the second, four on the third, and so on for all sixty-four squares. Is this a reasonable dowry? The number of grains on the last square alone equals $2^{63} \approx 10^{19}$ grains, which equals the total on the first sixty-three squares, plus one. The total number of grains is about 2×10^{19}, a truly astronomical number, and equals several hundred times the world's annual wheat harvest.

Without limits to growth, populations increase exponentially. Recently, the world's population has been doubling about every forty years. Any quantity that doubles after a fixed length of time, called the doubling time, is increasing exponentially. Mathematically, exponential growth after n doublings is described by the function 2^n. The thickness of the paper after n folds is $2^n d$; the number of wheat grains on the nth square after the initial one is 2^n; and the population after n doublings (the population of the nth generation following the initial one of population N_0) is $2^n N_0$.

An uncontrolled nuclear chain reaction is another example of exponential behavior. When ^{235}U fissions, the nuclei release an average of 2.2 neutrons each. Each neutron can strike another ^{235}U nucleus, causing it to fission and releasing another 2.2 neutrons, and so on, in an exponentially increasing chain reaction. Actually, some of the 2.2 neutrons per nucleus are lost through the surface of the uranium material and some are absorbed by impurities. The exact number of neutrons that produce additional fissions depends on the purity, mass, and shape of the uranium sample. Assuming for simplicity that each fission leads to exactly two new fissions, the number of fissions in the nth "generation" is 2^n (where the original generation corresponds to $n = 0$). There are $(1{,}000 \text{ g}/235 \text{ g} \cdot \text{mol}^{-1})$ $(6.024 \times 10^{23} \text{ nuclei/mol}) = 2.5 \times 10^{24}$ nuclei in 1 kg of ^{235}U. Since $2^{81} = 2.5 \times 10^{24}$, it takes only about eighty generations to fission 1 kg of ^{235}U completely. Half of the energy is released in the final generation, but since each fission takes only 10^{-8} s (except for a tiny percent of "delayed neutrons"), the entire process takes less than a millionth of a second (0.8 μs); this short time scale produces the explosive nature of the energy release.

In nuclear reactors control rods are used to prevent exponential growth by absorbing some of the neutrons, thereby reducing the number of neutrons that produce fissions in the next generation. Exactly one neutron remains to produce a fission in the next generation: the nth generation produces the same number of fissions as the previous one. Here, the growth law becomes $1^n = 1$ for all n-values, representing no growth at all: the reaction rate is stable and constant, not exponential. If the rods are inserted farther, the number of fissions in the next generations will be multiplied by f^n for $f < 1$. The reaction rate will die out exponentially instead of growing exponentially. This is a form of exponential decay.

Half-life

Carbon is common to all life, both plants and animals. Most carbon atoms have carbon-12 nuclei (^{12}C), a stable nucleus of six protons and six neutrons. A small fraction of carbon nuclei—called carbon-14 or ^{14}C—have two additional neutrons. ^{14}C nuclei are radioactive: They will eventually emit an electron and change to ^{14}N (nitrogen-14), consisting of seven protons and seven neutrons. We say that the ^{14}C has decayed to ^{14}N. Many other kinds of nuclei are radioactive although they may decay by emitting a gamma ray or an alpha particle instead of an electron. ^{14}C is important because of its use in radiocarbon dating and as a biological tracer.

The half-life of a radioactive substance is the time required for half of it to decay away; the half-life of ^{14}C is 5,730 years. Starting with 1 g of ^{14}C, 0.5 g will remain after 5,730 years, 0.25 g will remain after the next 5,730 years, and so on. A plot of the remaining nuclei as a function of time (Fig. 1) shows a steady decrease as the curve tends to, but never actually reaches, zero. This kind of behavior is called exponential decay. Mathematically, it is

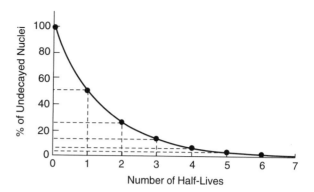

Figure 1 Exponential decay of radioactive nuclei.

similar to exponential growth, except that exponential decay is described by the half-life instead of the doubling time. The half-life is a kind of halving time. The fraction of the original material remaining after n generations is $(1/2)^n$, instead of 2^n for exponential growth.

The Exponential Law

^{14}C is used to date ancient artifacts. As long as a plant or animal is alive, it continually replenishes the decaying ^{14}C by taking in ^{14}C from its food (animals) or from the CO_2 in the air (plants). When an organism dies, its ^{14}C begins to decay away. By measuring how much remains today, one can tell how long ago the object died. For example, if exactly half of the original ^{14}C remains, the object is 5,730 years old (within the error of measurement). After n half-lives have elapsed, the fraction F that remains of the original ^{14}C is

$$F = (1/2)^n$$

The number of half-lives elapsed is

$$n = -\frac{\log F}{\log 2},$$

which is valid is even when n is not an integer. The sample is $5{,}730n$ years old.

Except for a difference in sign, exponential growth and decay follow the same law:

$$\pm \, dN = \lambda \, N \, dt,$$

where the "+" sign is used for growth and the "−" sign for decay. In population growth, for example, N represents the population at time t, and dN is the change in population during a short time dt; the constant λ is related to the doubling time. For radioactive decay, N represents the number of original nuclei and $(-dN)$ is the number of decays during dt; dN is always negative for decay since N is decreasing. The negative sign is used in the equation for exponential decay in order to make λ positive; the plus sign is used for exponential growth so that λ will be positive there also.

For radioactive decay the law says that the number of decays is proportional to both the time interval, dt, and to the number of undecayed nuclei N at the beginning of that interval. This means that, for example, if the number of nuclei is doubled, there will be twice as many decays during dt, a reasonable expectation, or that there will be twice as many decays for $(2\ dt)$ as for dt, also very reasonable (if dt is small compared to the half-life).

Rewriting the law as a derivative gives

$$\frac{dN}{dt} = \pm \lambda N.$$

The only function whose derivative equals itself multiplied by a constant is the exponential function,

$$N = N_0\, e^{\pm \lambda t}.$$

For one doubling time, t_2, N equals $2N_0$ so that $2N_0 = N_0\, e^{+\lambda t_2}$. Solving for λ in terms of the doubling time gives $\lambda = (\ln 2)/(t_2)$, an important relation between λ and t_2. Similarly, for half-life,

$$\lambda = \frac{\ln 2}{t_{1/2}} = \frac{0.693}{t_{1/2}}.$$

A convenient way to write this exponential equation is

$$N = N_0\, e^{\pm (\ln 2)\, t/\tau},$$

where τ is either t_2 ("+" sign) or $t_{1/2}$ ("−" sign). Since $e^{\ln 2} = 2$, this becomes $N = N_0\, 2^{\pm t/\tau}$, which is the same as the previous equations for growth (base 2) or decay (base $1/2 = 2^{-1}$) with $n = t/\tau$. Thus, the exponential law can be written using a base of either 2 or e, whichever is convenient. Only the exponential function has the property of doubling or halving, shown in Fig. 1 for decay, when the exponent increases by a fixed amount for any starting point on the curve.

Why use a base of e instead of 2 (or $1/2$) in the exponential law? From our discussions of doubling times and half-lives, it seems more natural to write the law as 2^n or $(1/2)^n$, as was done earlier. But there are two good reasons why it is frequently more natural to write it as $e^{\pm x}$.

First, the $e^{\pm x}$ form arises naturally in solving $dN/dt = \pm \lambda N$. This is why the logarithm to the base e is called the "natural logarithm."

Second, the $e^{\pm x}$ form arises automatically in other situations. An everyday example is the accumulation of principle P in a savings account with compound interest. If the interest is compounded annually at the rate I the principle after one year will be the original principle P_0, plus interest IP_0: $P = P_0 + IP_0 = (1 + I)P_0$. Since each year the principle increases by the factor $1 + I$ after t years, $P_t = (1 + I)^t P_0$. If the principle is compounded every six months at the semiannual rate of $I/2$, after one year the principle will be

$$P = \left(1 + \frac{I}{2}\right)P_0 + \frac{I}{2}\left(1 + \frac{I}{2}\right)P_0 = \left(1 + \frac{I}{2}\right)^2 P_0.$$

Again, after t years, $P_t = (1 + I/2)^{2t}\, P_0$. For compounding n times a year, the result is $P_t = (1 + I/n)^{nt}\, P_0$. For "continuous compound interest" the principle after a year is $P_t = P_0 \lim_{n\to\infty} (1 + I/n)^{nt} = P_0 \lim_{n'\to\infty} (1 + 1/n')^{n'It}$, where $n' = n/I$ replaced n in the last step. Comparing this expression with the definition of $e = \lim_{n\to\infty} (1 + 1/n)^n$, gives $P_t = P_0\, e^{It}$. Thus, with continually compounded interest the principle grows exponentially and e appears naturally as the base.

Doubling Time and Rate of Growth

The growth of population, energy use, and so on is frequently expressed in terms of the annual percent growth. For example, the world's population has been increasing by 1.6 percent per year recently. Given the growth rate, there is a simple way to relate it to the doubling time. Since the law for exponential growth is $dN/dt = \lambda N$, the fractional growth rate is $(1/N)(dN/dt) = \lambda = (\ln 2)/(t_2)$. Multiplying by 100 to convert to percent and solving for t_2 gives

$$\text{Doubling time} \approx \frac{69.3}{\text{Percent growth rate}}.$$

If the percent growth is expressed on an annual basis, the doubling time is in years. This expression is valid only if the doubling time is long compared to a year (or whatever the basis for the percent growth is). An annual growth of 1.6 percent corresponds to a doubling time of forty-three years. Since forty-three years is large compared to one year, the approximation is valid.

The expression also is valid for continually compounded interest. Thus, it takes $t_2 = 69.3/I(\%)$ years to double the principle, where I is the annual interest rate. For 10 percent interest, the doubling time is seven years. For a 20 percent annual interest rate the principle increases by 22 percent per year; the approximation is still pretty good, but becomes increasingly poor for higher percentages. For an exact result, use $t_2 = (\log 2)/[\log(1 + I/100)]$.

Consequences of Exponential Growth

Exponential growth can seem quite ordinary until its final, explosive stage. As the population of a bacterial culture grows in a petri dish, doubling each generation, there is plenty of room to grow until the dish is suddenly filled. Two generations before this moment, the dish is only one-fourth full, and it is still only half full after the next generation; to the bacteria there is no problem with continued growth. Then, in one final generation the dish is full and the dynamic of population growth changes drastically as the bacteria starve and die.

A similar thing happens with other plant and animal populations unless exponential growth changes to less rapid growth, or even to a constant population. A sense of the explosive nature of exponential growth can be seen by noting that the population of any particular generation equals the cumulative population of all preceding generations. There are as many people living now as have lived and died before our time. With exponential growth in energy use, the transition from seemingly abundant energy supplies to the exhaustion of those resources occurs within one doubling time. This property of exponential growth makes it very difficult for societies to anticipate and prevent crises in population, resource depletion, and so on.

See also: CHAIN REACTION; CHAOS; DECAY, ALPHA; DECAY, BETA; FISSION

Bibliography

BARTLETT, A. A. "Forgotton Fundamentals of the Energy Crisis." *Am. J. Phys.* **46** (Sept.), 876–888 (1978).
PAULOS, J. A. *Beyond Numeracy* (Knopf, New York, 1991).

LAWRENCE A. COLEMAN

F

FABRY–PÉROT INTERFEROMETER

See INTERFEROMETER, FABRY–PÉROT

FALLOUT

When a nuclear bomb is detonated near the earth's surface large amounts of earth or water enter the rising fireball and mix with the radioactive debris of the explosion at an early stage. As the violence of the explosion subsides, the contaminated particles and droplets gradually descend to earth. This phenomenon is referred to as "fallout." The same name also applies to the particles themselves as they reach the ground. This fallout is the main source of residual radioactivity after a nuclear explosion.

Formation in a Nuclear Explosion

Within less than a millionth of a second after the detonation of a nuclear explosion the residues of the device radiate large amounts of energy mainly as invisible x rays that are absorbed within a meter in the surrounding atmosphere. This leads to the formation of an extremely hot, highly luminous spheri-cal mass of air and gaseous residues known as the fireball. Immediately after its formation the fireball begins to grow in size, engulfing the surrounding air. At the same time, the fireball rises like a hot air balloon at speeds of hundreds of kilometers per hour. Within several tenths of a millisecond, the fireball from a 1-megaton explosion would be more than 120 m across; by ten seconds after the burst it would grow to a maximum size of more than 2 km. By one minute after the burst, it will have risen to about 8 km above the point of detonation and will have cooled sufficiently that it no longer emits visible radiation.

While the fireball is still luminous, the temperature in the interior is so high that all of the materials of the original nuclear device are in the form of vapor. This includes the radioactive fission products produced in the explosion, uranium (or plutonium) that may not have fissioned in the explosion, the device casing, and any other original materials. As the fireball increases in size and cools, the vapors condense to form a cloud containing solid particles of the bomb debris, as well as other material sucked into the rising fireball.

When the explosion occurs near the surface of the earth (less than about 150 m above the ground), the fireball in its rapid initial growth may actually touch the surface of the earth. The intense heat in the fireball will then vaporize and incorporate some of the rock, soil, and other material on the ground, leaving behind a crater. Additional material may be melted (but not vaporized), or partially melted on its surface. This material, along with dust, dirt, and other particles, can be sucked into the fireball by the strong after winds as the fireball rises. These particles in the fireball range in size from very small ones produced by condensation to the very large ones that were raised by the afterwinds. The exact composition of the rising cloud will depend upon the surface materials present and the extent of their contact with the fireball. If the explosion occurs near water, large amounts of water are vaporized and carried up into the rising radioactive cloud. At high altitudes, the vapor condenses to form water droplets similar to those in an ordinary cloud.

The Radioactive Cloud

Quite early in the ascent of the fireball, cooling of the outside by radiation and drag of the air through which it rises frequently bring about a change of shape. The sphere becomes a toroid (or doughnut) shape as depicted in Fig. 1. The eventual height reached by the cloud depends upon the heat generated by the explosion and the atmospheric conditions, but is most strongly influenced by the boundary between the lower atmosphere (called the tropopause) and the stratosphere above.

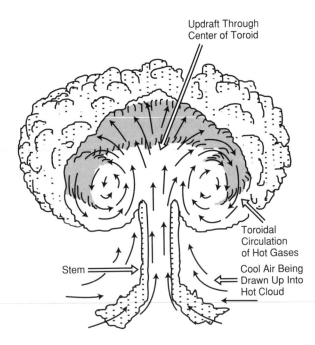

Figure 1 Radioactive cloud resulting from a nuclear explosion.

When the cloud reaches the tropopause, it tends to spread out laterally, that is, sideways. If sufficient energy remains, however, a portion of it will penetrate the tropopause and ascend into the stable air of the stratosphere. The cloud attains its maximum height in about ten minutes and is then said to be stabilized. It continues to spread out, however, into the characteristic mushroom shape. The cloud will continue to be visible for about an hour or more before being dispersed by winds into the surrounding atmosphere.

Once penetrating into the stratosphere, the radioactive cloud can be carried long distances. Even for particles in the lower atmosphere, the residence time of the smaller particles (less than 1 μm in diameter) can be years.

The extent and nature of the fallout can range between wide extremes. In an air burst, for example, occurring at an appreciable distance above the ground, the magnitude of the fallout is far less than that for a surface burst. Thus, for example, on the one hand, there were essentially no injuries from fallout from the bursts at Hiroshima and Nagasaki, since the explosion occurred about 0.5 km above ground zero. On the other hand, the 15-megaton thermonuclear device (BRAVO test) exploded only 2.1 m above the surface of a coral reef at Bikini Atoll on March 1, 1954, caused substantial contamination over an area of more than 19,000 km^2 (about the size of the state of New Jersey) extending as far as 570 km downwind from the explosion. Figure 2 illustrates the total accumulated dose at ninety-six hours after the explosion. The 50 percent lethal dose of 400 rad extend to greater than 300 km from ground zero, while substantial health effects would result from exposures out to more than 600 km downstream.

Such fallout patterns are mainly affected by directions and speeds of the winds over the fallout area from the ground up to the top of the cloud, which may be as high as 100,000 ft. It should be noted that fallout is a gradual process occurring over a period of time. For example, in the 1954 Bikini Atoll explosion, particles began to fall after about ten hours, at a time when the cloud could no longer be seen. Nevertheless, the area of contamination that represents the most serious radiation hazard involves the larger particles, ranging from that of a fine grain of sand (approximately 100 μm) to pieces up to about 1 cm in diameter. Particles in this size range fall to the ground within a day after the explosion and will not travel more than a few hundred km. This is referred to as the early or local fallout. Early fallout poses the

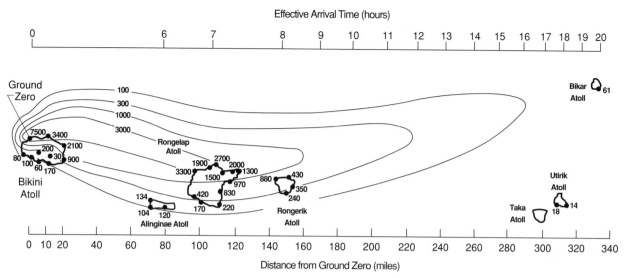

Figure 2 Estimated accumulated dose contours, measured in rads ninety-six hours after the BRAVO test explosion.

largest biological hazard due to its high radioactivity and concentration near the explosion site. The smaller particles descend very slowly and are deposited over large areas of the earth's surface. This is referred to as delayed (or worldwide) fallout.

The term "scavenging" describes the various processes by which the radioactive particles are removed from the cloud and deposited on the ground. One of these processes is the condensation of fission products and other radioactive vapors on the dirt and debris, which then rapidly fall to Earth. Another scavenging process is that due to rain falling through the explosion debris and carrying contaminated particles down with it. This is one mechanism for the production of "hot spots," that is, areas on the ground that have much higher radioactivity than the surroundings in both early and delayed fallout patterns. Another effect that rain may have during or after deposition on the ground is to wash radioactive debris over the surface of the ground, thus cleaning some areas and causing hot spots in other (lower) areas.

Radioactivity in Fallout

The radioactivity in fallout arises from a combination of fission products from the explosion, radioactive uranium, plutonium, and tritium (which were a part of the device), and neutron induced radioactivity in soil and other materials present in the vicinity of the explosion. The fission products constitute a complex mixture of more than 300 isotopes. Most of these isotopes are radioactive, decaying by the emission of gamma and beta radiation. About 2 oz of fission product are formed per kiloton of fission energy yield. The initial radioactivity is enormous, but falls off with time. The largest biological hazard is from the early fallout, which will continue to deposit hazardous radiation doses up to twenty-four hours after the explosion. The greatest danger from less hazardous delayed fallout is from the possible ingestion of long-lived radioactive isotopes of iodine, strontium, and iodine in food, especially milk. The neutron induced activities include radioactive isotopes of carbon, sodium, magnesium, silicon, and aluminum. For example, it is estimated that the nuclear testing in the 1960s increased the amount of natural carbon-14 in the atmosphere by 1.6 tons or nearly 2 percent.

See also: ATMOSPHERIC PHYSICS; ENERGY, NUCLEAR; FISSION; FISSION BOMB; FUSION; FUSION BOMB; ISOTOPES; NUCLEAR BOMB, BUILDING OF; NUCLEAR REACTION; RADIOACTIVITY

Bibliography

GLASSTONE, S., and DOLAN, P. J. *The Effects of Nuclear Weapons,* 3rd ed. (U.S. Government Printing Office, Washington, DC, 1977).

GRANT J. MATHEWS

FARADAY, MICHAEL

b. Newington near London, England, September 22, 1791; *d.* Hampton Court, Middlesex, England, August 25, 1867; *electricity, magnetism.*

Faraday was born the third of four children and the second son of James Faraday, a blacksmith in ill health, and the former Margaret Hastwell of Yorkshire, who had only recently left their ancestral homes to seek economic security in London.

The Faraday family was brought up within the congregation of a fundamentalist Christian sect known, as Faraday later put it, "if known at all, as the Sandemanians." Sandemanians lived by the letter of the Bible and the words of Jesus. The church stressed love, charity, and sacrifice for one's fellows, rather closely defined as other Sandemanians. It was here that Faraday's character was formed. He learned the rudiments of reading, writing, and ciphering in a church school; he learned, as well, that worldly riches were to be shunned and, many years later as one of the most famous scientists in Europe, he still gave most of his income to the needy of his congregation, attended them when they were sick, and succored them in their troubles. One month after he married Sarah Barnard, the daughter of a Sandemanian elder in June of 1821, he formally became a member of the church and it was to guide him through the rest of his life. These were the two loves of his life. The Faradays had no children, so his natural affection was poured out on his wife, his nieces who spent a good deal of time at the Royal Institution (RI) in London, and his fellow Sandemanians.

Although he said that his religion did not influence his science, it is perfectly clear that it did. It gave him confidence that the cosmos was rational and comprehensible and, more importantly, that God's works were to be seen throughout nature and were beneficent. It also made him suspicious of scientific explanations that depended on subsensible matter, such as John Dalton's material atoms, which acted solely by ordinary forces, for this implied atheism or, at the very least, the kind of rational deism that many besides Faraday in England felt lay behind the attack on Christianity in the French Revolution.

At the age of twelve, Faraday was apprenticed to a French émigré bookbinder who also loaned out newspapers. Faraday's thirst for learning was partially slaked by being able to read the books that came in for binding. There was, however, no method involved. He read everything from *The Arabian Nights* to the *Encyclopaedia Britannica,* but nothing was wasted; the first stimulated his imagination while the second introduced him to facts of the natural world. He was particularly excited by an article on electricity and promptly constructed an electrostatic machine out of old bottles and other scraps. His passion for science was heightened when, in 1812, one of his master's clients presented him with tickets to the lectures on chemistry given by Sir Humphry Davy at the Royal Institution of Great Britain. Again, he immediately built an electrochemical cell and decomposed water and other compounds. He now desired, above all, to enter the mansion of science but there were absolutely no prospects for him. In October of 1812 his apprenticeship was over and with a heavy heart he entered the world of trade. In that very month, however, fate intervened. Davy was temporarily blinded by an explosion and Faraday was hired as his secretary, a purely temporary position. In February 1813 one of the assistants in the laboratory was fired for brawling, and Davy sent for Faraday. He was hired in March at the RI, an institution in which he was to spend his entire professional life.

Faraday's scientific education was now provided for him by his association with Davy, the most eminent chemist in Great Britain. It was Davy who taught him those manipulative skills that made Faraday a superb analytical chemist and it was Davy who introduced him to the theories of matter then current. Davy also introduced Faraday to the larger world. In 1813 Davy was awarded a medal and prize that had been created by Napoleon in honor of Alessandro Volta, the inventor of the voltaic cell and the discoverer of current electricity. Davy immediately set out for France, with Faraday serving somewhat reluctantly as his valet. They traveled through France, Italy, and Switzerland, where Faraday met many of the eminent scientists of the day, learned French and a smattering of Italian, and helped Davy revise his 1812 publication entitled *The Elements of Chemical Philosophy* in which Davy probed deeply the philosophical and metaphysical foundations of the current matter theories. It was undoubtedly during the discussions with Davy on these subjects that Faraday's lifelong interest in and search for an adequate theoretical foundation for the rather unconventional hypotheses that guided his experiments began.

Two basic views dominated Faraday's later thought. One was that "the various forms under which the forces of matter are made manifest have one common origin; . . . or, in other words, as so directly related and mutually dependent, that they are convertible as it were, one into another, and possess equivalents of power in their action." This idea was at the very center of the German eighteenth-century philosophy of nature (*Naturphilosophie*). Davy was exposed to this early in his career when he worked in Bristol for a physician who was intensely interested in *Naturphilosophie*. It was also in Bristol that Davy met and became fast friends with Samuel Taylor Coleridge, the poet and essayist who had just returned from Germany filled with enthusiasm for this new way of looking at nature. The second had to do with the difficulties that followed from Isaac Newton's and John Dalton's views of atoms as infinitesimally small, absolutely hard, completely elastic bodies in empty space. The first objection was that elastic collisions between atoms was ruled out, for elastic rebound depends on deformation and recovery of the colliding bodies and, by definition, atoms were perfectly hard and therefore incapable of deformation. The second objection was more theological in nature, for if changes in atomic motions could not be by collisions, they had to be caused by occult forces that could, as with the case of Newton, lead dangerously close to pantheism, that is, the supposition that God filled the cosmos. In the *Philosophiae naturalis principia mathematica*, generally known simply as the *Principia*, Newton had suggested that space was the sensorium of God, thus universalizing the deity.

There seemed to be only two ways out of this dilemma: either forces acted in some mysterious ways at a distance or all changes of atomic and ordinary motions must be due to impacts. Newton had proven in the *Principia* that impact physics simply did not fit the mathematical laws that he enunciated. Action at a distance violated the old philosophical dictum that bodies could not act where they were not. A "solution" to this difficulty was suggested in the middle of the eighteenth century by a Jesuit, Roger Joseph Boscovich. It involved nothing less than the redefinition of what a body was. Boscovich suggested an atomic theory containing nothing known conventionally as matter, but consisting solely of force. This was to give what philosophers call "ontological status" to forces. Boscovich's atoms consisted of mathematical points around which attractive and repulsive forces varied with their distance from the center.

Resultant molecules also had a property of great importance for Faraday. They can be "stretched" or, to put it another way, they can undergo strains without decomposing.

It should always be kept in mind that Faraday was a chemist, one of the best chemists of his time. He thought in chemical terms, not in the mathematical formulas of the physicist. Indeed, he was a mathematical illiterate and could not follow the arguments of later physicists like William Thomson or James Clerk Maxwell. By 1819 he was well known as a first-rate analytical chemist, earning money for himself and the RI by analyzing waters from wells and municipal water systems. From 1818 until 1831 he was concerned almost exclusively with matters that could roughly be considered chemical. He worked with a cutler on improving the quality of steel and determined that the properties of steel owed as much to its crystalline structure as it did to its composition. By introducing the practice of examining the structure of polished steel etched by mild acid, Faraday founded the science of metallography. In 1825, at the request of the Royal Society to which he had been recently elected, he began tedious research attempting to produce a clear and dense glass for astronomical purposes. He did succeed but the glass proved useless at that time. It was, however, to play a central part in Faraday's discovery of the action of magnetism on light. From 1820 to 1826 he moved into organic chemistry, discovering a number of organic compounds such as benzene, iso-butene, tetrachloro-ethene, and hexachlorobenzene. He also observed that metals such as mercury were surrounded by their own vapor, thus casting doubt on the existence of a separation between the gaseous, liquid, and solid states. His liquefaction of chlorine in 1823 was the first liquefaction of a so-called permanent gas. It also could be nicely explained by the continuous Boscovichean curve.

Faraday retained his interest in electricity but had little time away from his professional duties to do any active research. From 1820 until 1831 he could only work on this subject sporadically, with one exception. In the summer of 1820, Hans Christian Oersted announced that he had been able to get magnetic effects from a current-carrying wire. This transformation of electricity into magnetism fit in perfectly with the speculations of the *Naturphilosophen* and it was no coincidence that Oersted was an ardent adherent of this philosophy. The announcement in Latin in all the leading scientific journals of the day set off a feverish spurt of activity with theories and weird observations turning up almost daily. A friend of Faraday's who edited one such journal asked him if he would check these reports and Faraday agreed. Oersted had noted that a magnetic needle held near a current-carrying wire tended to align itself at an angle to the wire, not parallel to it. In France, André-Marie Ampère had shown that the needle always stood at right angles to the current-carrying wire when the effects of the earth's magnetism were nullified. By careful mapping of needle positions, Faraday discovered, somewhat to his surprise, that the line that the needle followed was a circle with the wire at the center. This discovery was the origin of Faraday's concept of the line of force, which was the foundation of field theory. Except for the works of a very few of those who had published on the subject, Faraday fairly well demolished most of what had been written. His results were published anonymously in the *Annals of Philosophy* in the autumn of 1821 and winter of 1822. Beyond clarifying the exciting new field of electromagnetism, these papers thrust Faraday deeply into electrical and magnetic phenomena.

Using his discovery of the circular line of force, Faraday then invented a clever demonstration,

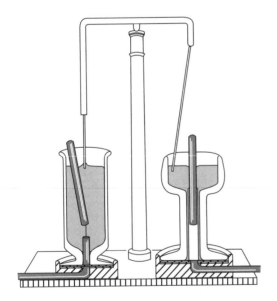

Figure 1 On the left, a current-carrying wire dips into a glass filled with mercury, at the bottom of which a permanent magnet is attached in such a way that the top can rotate around the wire. The right is the opposite; the wire can rotate and the magnet is fixed. This dramatically illustrated the circular nature of the magnetic force around a current.

which was, in fact, the first electric motor, shown in Fig. 1.

After 1822, except for occasional forays into electrical matters, Faraday did nothing of much significance in this field until 1831. Then, as he noted in his laboratory journal for August 29, he did an epoch-making experiment. He took a thick iron ring and wound one side with wire connected to a battery. The other he wound and connected to a galvanometer. When he threw the switch on the battery, he noted that the galvanometer needle swung—he had created one current by making the iron ring a magnet that then created another current. This induction of a current by magnetism completed the symmetry of Oersted's creation of a magnet by a current. Faraday's paper announcing this discovery was published in 1832 and began the magisterial *Experimental Researches in Electricity* that were to extend for almost thirty years and literally create a new science of electricity and magnetism and firmly establish field theory.

There was one thing that puzzled Faraday in this experiment. When he closed the circuit, the galvanometer needle jumped to the right and then fell back; when he broke the circuit, the needle jumped to the *left*. He hypothesized that closing the circuit threw the secondary wire into a state of strain, which was relieved only when the primary current ceased. It was like pulling on a rope and then suddenly moving the end of the rope up and down: A wave passes down the rope, but the rope is still in tension. When one lets go of the rope, another wave passes as the strain is relieved. Although he could not detect it then (and never could), Faraday named this state the electrotonic state. It was to be the unifying idea in his theory of electricity.

Faraday did not stop here. Knowing that magnetism could produce an electric current, he now began to try to find the conditions under which a permanent bar magnet would succeed in doing the same thing. After many trials, he discovered that when a magnet was in motion with respect to a closed circuit so that the wire of the circuit cut the lines of force surrounding the magnet that was made visible by iron filings sprinkled on a piece of paper over the magnet, a current resulted. From here, it was just a short step to producing a continuous current by rotating a copper disk between the poles of a horseshoe magnet and taking leads off the circumference and center of the disk so that the leads cut the lines at different rates. When the arrangement was changed so that the leads provided a current in the wires and the disk, the disk rotated between the magnetic poles. This was the first generator and first real electric motor.

Before continuing this electromagnetic research, Faraday had to meet a challenge that stated that there was more than one kind of electricity. His test was to try all the different kinds—animal from electric fish, static electricity, voltaic electricity, thermal electricity—and see if they caused electrochemical decomposition. Some were just too weak, but static electricity seemed quite capable of proving his case. In the course of these experiments, he was able to cause decomposition, but without any poles or centers of attraction as had been supposed in the voltaic cell. Instead, simply by passing electricity from the static electricity generator through a piece of blotting paper containing sodium iodide (NaI) and starch water as a detector and into the air, he produced free iodine detected by the starch. He was also able to devise what he called a volta-meter that pemitted him to compare the current that passed in electrochemical decomposition, either by static or current electricity. This led him to Faraday's first law of electrochemistry: The quantity of decomposition products is directly proportional to the quantity of electricity that passes through the solution. The second law states that the amounts of different substances liberated in an electrochemical process are proportional to their chemical equivalents. Faraday had now successfully refuted the older theories of electrochemistry that depended upon electricity as a material substance, poles and action of the electrical particles on the elements involved at a distance. Instead, he viewed it as the setting up and breaking down of the electrotonic state that occurred when the elements of the compounds involved were separated by the strain and passed along to neighbors who had suffered the same breakdown with electropositive and electronegative particles moving in opposite directions, as in a square dance, where partners move past one another but are always attached to someone.

To avoid the kinds of theoretical trap that the old language of Newtonian physics had created, Faraday consulted William Whewell and others who were experts in Greek and, with them, created the modern terms of electrochemistry (anode, cathode, anion, cation, electrolysis, etc.).

The specificity of electrochemical action in which the chemical substances all displayed specific "breaking points" that allowed them to move past one another suggested to Faraday that insulators might also have specific capacities for charge before they, too, broke down. Again, with careful experimentation, Faraday was able to show that, indeed, there was a specific inductive capacity for such substances.

After eight years of sustained mental effort and clever and accurate experimentation, Faraday possessed the elements for his own theory of electricity. The electrical force acted on matter by placing the force atoms of which materials were composed under various degrees of strain. Conductors were those substances that could not bear much strain, so that when an electrical force was imposed on them, a strain was momentarily created. However, it broke down rapidly, only to be followed immediately by the reimposition of a strain, and so on. As this wave of strains and relaxations passed down the conductor, it formed the electric current. In electrochemical decomposition, the strain was relieved by the passage of the ions by one another on their way to the electrodes. Insulators could bear much stronger strains and their breakdown was accompanied by a physical and quite perceptible mechanical effect, as when a hole is punched in a glass plate when high voltages are applied to it. In all these operations, the line of force marked the presence of the electrical power.

In 1839 Faraday's line of force ruptured and he suffered a severe mental breakdown forcing him to abandon serious research, seek rest and recreation in places like Switzerland, and generally pull back. From 1839 until his recovery in 1846, little original science emerged from his laboratory. There was, again, one exception. Faraday was called upon unexpectedly and suddenly to fill in for his friend Charles Wheatstone, who was unable to give his scheduled Friday Evening Discourse. Faraday had had no time to prepare, and his lecture only occupied a part of the allotted time. To fill it up, he offered some speculations, later entitled "Thoughts on Ray Vibrations," which were published soon after. In this remarkable little paper of only seven pages, he used Boscovich's concept of matter to suggest that every particle of matter is infinite in extent since the force of gravity extends to infinity. All particles, then, are contiguous and touch each other by their forces, so that the cosmos is an infinite and complex web of forces. Could not this web, he asked, suffice to transmit radiation? Might not the lines of magnetic force, as well as those of gravitation, be quite suitable for the purpose? James Clerk Maxwell, many years later, was to make this speculation scientifically respectable.

The third, and last, of the scientific innovations for Faraday began in 1845 and lasted until his mind failed him in the late 1850s. He was still haunted by the electrotonic state that he had never been able to detect, and for the arch-experimenter who insisted that hypotheses should be backed up by solid experimental evidence, this was intolerable. So he tried once more to reduce this elusive hypothetical strain to measurement. It had long been known that when plane polarized light was passed through transparent substances such as glass under mechanical strain, the light was affected. When he placed a piece of the heavy glass that he had made in the 1820s in as powerful an electric field as he could devise, he still could get no effect when light was passed through it, even though he knew that there *must* be a strain. It was the young and upcoming physicist, William Thomson, later Lord Kelvin, who suggested that he substitute the much more powerful field of the large horseshoe electromagnet at the RI for the electric field. On September 13, 1845, Faraday was successful; the plane of polarization of a light was rotated by passing through the glass placed across the poles of the electromagnet where the field was most intense. This phenomenon later came to be known as the Faraday effect. Further experiments revealed a peculiar situation: Objects, such as glass, aligned themselves perpendicular to the magnetic lines of force, rather than parallel to them as is the case with iron. A whole new world of magnetism opened up. Faraday named this transverse magnetism, diamagnetism, and the usual kind, paramagnetism magnetism, or for the more magnetically powerful substances such as iron, nickel, and cobalt, ferromagnetism.

As Faraday delved ever deeper into the nature of magnetism in general, he made another startling discovery. Unlike the electrical lines of force—electrostatic, electrochemical, and current—all of which had "ends" just like a stretched cord, the magnetic lines of force did not. Nowhere could he find the magnetic equivalent of positive and negative charges that were associated with the termini of the electrical lines of force. The analogy of the stretched cord was not applicable to magnetism for the particles of the cord are distorted by the strain and have "poles" along the line. Hence, he concluded that the magnetic line of force could not be a line of strained particles. This ruled out the luminiferous ether that had been invented to account for the wave nature of light. But if the particles were not strained, what was? There was no doubt that the magnetic force did exist in space. Faraday's suggestion was literally unthinkable at the time. Might not space itself be the medium by which the magnetic force was transmitted? With this idea, Faraday was now able to explain both paramagnetism and diamagnetism. Space conducted the lines of paramagnetic force better than did the surrounding medium of material substance and worse for diamagnetics. This was why paramagnetics move into and along the magnetic field and diamagnetics try to move out of it and across the lines. This explanation had few converts at the time, but the general theory of relativity propounded by Albert Einstein in 1916 gave it respectability.

After his work on magnetism, Faraday began to fade both mentally and physically. His last sustained research in the late 1850s was on colloidal gold, making him one of the founders of colloidal chemistry. He also attempted again to detect the influence of magnetism on light—a failure that later inspired Peter Zeeman to try it once again and led to the discovery of the Zeeman effect.

From 1859 on, Faraday rapidly lost his intellectual powers until, near the end, he could do little more than sit in his chair and be cared for like an infant. Having turned down a knighthood earlier in his career, he simply remained Mr. Faraday to his death.

See also: ELECTRICITY; ELECTROMAGNETIC INDUCTION; ELECTROMAGNETIC INDUCTION, FARADAY'S LAW OF; FARADAY EFFECT; KELVIN, LORD; POLARIZATION; ZEEMAN EFFECT

Bibliography

AGASSI, J. *Faraday as a Natural Philosopher* (University of Chicago Press, Chicago, 1971).

CANTOR, G. *Michael Faraday: Sandemanian and Scientist* (St. Martin's Press, New York, 1991).

FARADAY, M. *Chemical Manipulation* (John Murray, London, 1827).

FARADAY, M. *Experimental Research in Electricity* (Dover, New York, [1839–1855] 1965).

FARADAY, M. *Experimental Research in Chemistry and Physics* (Taylor and Francis, London, 1859).

FARADAY, M. *Faraday's Diary, Being the Various Philosophical Notes of Experimental Investigation Made by Michael Faraday*, edited by T. Martin (G. Bell, London, 1932–1936).

THOMAS, J. M. *Michael Faraday and the Royal Institution* (Adam Hilger, Bristol, Eng., 1991).

THOMPSON, S. P. *Michael Faraday, His Life and Work* (Cassell, London, 1898).

TYNDALL, J. *Faraday as a Discoverer* (Longmans, London, 1879).

WILLIAMS, L. P. *Michael Faraday, A Biography* (Chapman & Hall, London, 1965).

L. PEARCE WILLIAMS

FARADAY EFFECT

First discovered by Michael Faraday, the Faraday effect is the general designation for the induced change in the response of a medium to the different forms of circularly polarized radiation, right and left, upon the application of a magnetic field. The effect arises as a consequence of the fact that the energy levels for the different components of the angular momentum J of a state are split when the material is placed in the field (Zeeman effect), and the dispersion of circularly polarized radiation reflects the interaction of the radiation with the angular momentum components. The dependence of the dispersion on the form of circular polarization is expressed as a difference in the indices of refraction for right- and left-polarized radiation, which manifests itself in the rotation of the plane of polarization of linearly polarized radiation passing through the medium (Faraday rotation). The angle θ through which the rotation occurs depends on the material, the strength of the applied magnetic field B and the distance ℓ in the material traversed by the radiation. This dependence is expressed by the simple relationship

$$\theta = V \cdot B \cdot \ell,$$

where V is an empirically determined constant known as the Verdet constant. The direction of the rotation depends on the direction of the applied magnetic field and will always be in the same direction regardless of the direction of propagation of the radiation through the medium. This is in contrast to natural circular dichroism in which the rotation is in different directions for opposite directions of propagation through the material. For positive V the material is levorotatory when the light is propagated parallel to the field and dextrorotatory when the propagation is antiparallel to the field, and conversely for negative V.

The Faraday effect has application in molecular spectroscopy in the determination of the energy levels of magnetically active excited states. The magneto-optic rotation spectrum is recorded by placing the molecular species in a solenoid, which generates the magnetic field. Linearly polarized radiation is made to pass through the sample, and the transmitted radiation is viewed through a linear polarizer at right angles to the first. In the absence of the Faraday effect no radiation will be transmitted; the rotation induced by the applied field produces bright lines at those energies at which the magnetic field has shifted the energy levels. The Faraday effect can also be used to advantage to design an amplitude modulator for laser radiation, where the modulation occurs through the relative transmission of the rotated linearly polarized light and a fixed polarizer as determined by the law of Malus. The modulation follows the signal from an alternating current that is used to generate the magnetic field inside a solenoid in which the material is placed.

See also: DICHROISM, CIRCULAR; FARADAY, MICHAEL; POLARIZATION; POLARIZED LIGHT; POLARIZED LIGHT, CIRCULARLY; SPECTROSCOPY; ZEEMAN EFFECT

Bibliography

FOWLES, G. R. *Introduction to Modern Optics,* 2nd ed. (Holt, Rinehart and Winston, New York, 1975).

HECHT, E. *Optics,* 2nd ed. (Addison-Wesley, Reading, MA, 1987).

STEINFELD, J. I. *Molecules and Radiation: An Introduction to Modern Molecular Spectroscopy* (MIT Press, Cambridge, MA, 1978).

C. DENISE CALDWELL

FARADAY'S LAW

See ELECTROMAGNETIC INDUCTION, FARADAY'S LAW OF

FAST REACTOR

See REACTOR, FAST

FERMI, ENRICO

b. Rome, Italy, September 29, 1901; *d.* Chicago, Illinois, November 28, 1954; *statistical mechanics, nuclear physics, particle physics.*

Many people know Fermi as the architect of the atomic age—the scientist who developed the first nuclear chain reaction, the basis for the peaceful uses of atomic energy. However, his contributions to the fundamental theories of atomic physics are of even greater importance. Fermi is an example of that rare type of scientist who is both a superb experimentalist and a brilliant theorist.

Fermi's father, Alberto, was a chief inspector in the Railway Ministry; his mother, Ida de Gattis, was trained as a school teacher and taught in the elementary schools most of her life. Maria, Fermi's older sister, was born in 1898. His brother, Giulio, who was two years older, was his most important playmate and teacher. It was a severe blow to fourteen-year-old Fermi when Giulio died. The children were raised as agnostics. They attended a strictly secular school with other pupils from lower-middle-class families.

Fermi spent a great deal of time studying mathematics and physics from secondhand books and books he borrowed from his father's friend Adolfo Amidie, an engineer. It was Amidie who urged Fermi to compete (successfully) for a fellowship for a free education at the Scuola Normale Superiore in Pisa. While still a student at Pisa, he published a significant paper that established him as the leading expert in Italy on Albert Einstein's general theory of relativity. After being awarded his doctorate magna cum laude from Pisa on July 7, 1922, he received a fellowship to spend a year at Göttingen. He returned to Italy to teach first at Rome in 1923 and then at Florence.

In January 1926, Fermi wrote a paper on one of his most significant theoretical contributions: the Fermi statistics, a method for calculating the behavior and properties of a system of particles that obey the Pauli exclusion principle. It was the culmination of two years of effort trying to explain the experimental values of the specific heats of certain (monoatomic) gases. Several months later, using a very different approach, Paul Dirac derived the same statistical mechanics. Fermi's paper was much more than the announcement of just a theoretical principle; it gave energy and momentum distributions. Almost immediately, Fermi's theory was applied to conduction electrons in metals and provided an understanding of the specific heat of metals. The theoretical foundation for solid-state physics (modern condensed matter physics) is the Fermi–Dirac statistical mechanics. The particles that obey these statistical laws are called fermions. In particle physics, all of the basic building blocks of matter, the quarks and leptons, are fermions.

Another very important theory developed by Fermi was his explanation of the beta decay of nuclei—the radioactive process in which a neutron changes to a proton by creating an electron and a neutrino. Fermi's theory of beta decay introduced a fourth fundamental force, called the weak interaction, to be added to the previously known three: the gravitational force between masses, the electromagnetic force between charges, and the strong force between the particles in nuclei. A new fundamental constant, called the Fermi constant, determines the strength of the weak interaction. In the 1960s the electromagnetic and the weak interactions were combined in a unified theory, and the well-measured Fermi constant plays a major role.

A special professorship in theoretical physics was set up for Fermi at the University of Rome in 1927, and at age twenty-six he became the youngest professor in Italy since Galileo. The following year he married Laura Capon, whose family was Jewish. In 1938, when Mussolini introduced his anti-Semitic laws, Fermi realized he had to leave Italy. The trip to Sweden to receive the Nobel Prize in November 1938 provided the opportunity. On January 2, 1938, Enrico and Laura Fermi and their two children, Nella and Giulio, arrived in New York City.

Fermi followed his achievements in theoretical physics with equally significant discoveries in experimental physics. In 1934 he and a group of young associates in Rome began to produce new artificially radioactive isotopes by bombarding target elements with neutrons. In the course of these experiments, Fermi discovered that neutrons slowed down by collisions with hydrogen nuclei in paraffin or water were more likely than fast neutrons to produce a nuclear reaction. Fermi's discovery of the effect of slow neutrons on nuclear reactions was central to his later work in the development of a nuclear chain reaction. It was also the basis of the Nobel Prize in physics that he received in 1938.

In January 1939, shortly after Fermi had started working at Columbia University, Niels Bohr arrived with the sensational news of the discovery of uranium fission; when a uranium nucleus absorbs a neutron, it splits into lighter nuclei, with the release of enormous amounts of energy. Fermi, Leo Szilard, and others immediately realized the significance of this discovery; if the fission of uranium by neutrons released other neutrons, these in turn could produce additional fissions, leading to a chain reaction of tremendous potential power. Szilard had secretly patented the chain reaction process in 1936. Fermi and Herbert L. Anderson and, independently, Szilard and Walter H. Zinn quickly showed that several neutrons are emitted in the fission process. Anderson, Fermi, Szilard, and Zinn joined forces and worked together first at Columbia and then after April 1942 at the University of Chicago, where studying the feasibility of the chain reaction become a large wartime project. Fermi worked out the detailed theory for the chain reaction in large test assemblies (piles) of uranium and graphite. A total of thirty experimental piles were tested over a two-year period. During the construction of the final pile in November 1942, Fermi gave detailed daily instructions on where to place the different materials. From daily measurements during construction, Fermi extrapolated that, on the night of December 1, the pile would allow the control of a self-sustaining nuclear chain reaction at the fifty-seventh layer. On December 2, Fermi made measurements of the neutron intensity over six hours with the control rods at a series of positions to determine the final operating position. At about 3:30 P.M., he instructed that the control rod be set to the calculated position where there would be a self-sustaining chain reaction. After twenty-five minutes of measuring the rising neutron intensity, he stopped the reaction by putting the control rods into the pile. In a very quantitative experiment, Fermi established that man could control nuclear chain reactions.

From 1943 to 1947, using beams of neutrons from nuclear reactors, Fermi established the field of neutron optics as a powerful research tool that is very important in many fields of science. The most expensive National Experimental Research facility planned in the United States, The Advanced Neutron Source, is based on Fermi's pioneering research papers in neutron optics. In the last decade of his life, his experiments used a synchrocyclotron for studying the interaction of mesons. His theoretical work in astrophysics and cosmic rays included a theory on the origins of cosmic rays.

Fermi's nonphysics activities and fun were mostly in sports such as swimming, skiing, hiking, and ten-

nis. He was very competitive in sports, and he usually was the first one to the top of a hill when hiking. By contrast in physics, he was patiently supportive of his colleagues and a fabulous teacher. For Fermi, and with Fermi, physics was both a challenge and a source of joy.

See also: Chain Reaction; Decay, Beta; Fermions and Bosons; Fermi Surface; Nuclear Reaction; Reactor, Nuclear; Statistical Mechanics

Bibliography

Anderson, H. L. "Assisting Fermi" in *All in Our Time,* edited by J. Wilson (Educational Foundation for Nuclear Science, Chicago, 1975).

Fermi, L. *Atoms in the Family* (University of Chicago Press, Chicago, 1954).

Segrè, E. *Enrico Fermi Physicist* (University of Chicago Press, Chicago, 1940).

Wattenberg, A. "The Birth of the Nuclear Age." *Phys. Today* **46,** 44–51 (1993).

Wattenberg, A. "The Fermi School in the United States." *European Journal of Physics* **9,** 88–93 (1988).

Albert Wattenberg

FERMIONS AND BOSONS

Elementary particles are classed into two types, bosons and fermions, according to their quantum statistics. This classification turns out to coincide with what at first sight appears to be an unrelated one, into particles with integer spins or half-odd integer spins. It is observed that particles with integer spin are bosons, while particles with half-odd integer spin are fermions.

Before considering the concrete classification, a remark on the notion of "elementary particle" is in order. For many purposes, one would like to be able to regard (for example) different helium atoms as indistinguishable elementary particles. Yet we know that a helium atom can be further analyzed into protons, neutrons, and electrons—and the protons and neutrons in turn could be further analyzed into quarks and gluons. How, then, can it possibly be valid to consider helium atoms as elementary? And how can they be indistinguishable? Answers to both these questions is supplied by the quantum theory of matter. According to this theory, helium atoms can assume a discrete number of states, with the lowest or ground state separated by an energy barrier from higher ones. Thus in all processes involving insufficient energy to excite the ground state into any higher energy state, helium atoms are in a uniquely determined state and can be regarded as strictly indistinguishable. Considerations of this kind underlie any rigorous use of the concept of elementary particle in modern physics.

Of the particles which appear elementary, even at the highest energies so far probed, the photon, the gauge bosons $W^{\pm}$ and Z that mediate the weak interactions, and the color gauge bosons that mediate the strong interactions are bosons. Likewise the graviton, which mediates gravitational interactions, is predicted to be a boson (although the prospects of detecting individual gravitons are remote). Quarks and leptons, including electrons, are fermions. Bound states of an odd number of fermions, and any number of bosons, are fermions; while bound states of an even number of fermions and any number of bosons are bosons. This rule for bound states follows directly from the definition of quantum statistics in terms of particle interchange, since each interchange of an underlying fermion introduces a factor -1 multiplying the wave function. Equally, it follows from the rule that fermions have odd half-integer spins ($\frac{1}{2}, \frac{3}{2}, \frac{5}{2}, \ldots$) while bosons have integer spins.

Thus protons and neutrons, which can be regarded as bound states of three quarks, are fermions. An extremely interesting and important distinction arises between two types of helium atoms. Helium atoms whose nuclei represent the isotope ^{3}He (two protons, one neutron) are fermions, while helium atoms whose nuclei represent the more common isotope ^{4}He are bosons. Because of this distinction, even though both species have essentially the same electronic structure and chemistry (both are inert) the behaviors of ^{3}He and ^{4}He liquids at low temperature are dramatically different.

Properties of Bosons

The most distinctive properties of bosons arise from the fact that many identical bosons can, and in some sense would like to, occupy the same quantum state. Three major phenomena should be considered from this perspective.

1. Bosons can form classical fields, that is, states in which the fundamental quantum particulate nature of the underlying matter can be safely neglected. Thus, for example, many photons can give rise to the sort of electromagnetic field that is the subject matter of classical electromagnetic theory. This occurs if many photons occupy the same quantum state, because then the granular structure—that is, whether there is one photon more or less—becomes relatively insignificant, and one may replace properties of the probability distributions for slightly different, large numbers by average values. Similarly, many gravitons can give rise to the sort of gravitational field which is the subject matter of classical general relativity.

2. Bosons—specifically, photons—can exhibit laser action. The physical basis of laser action is the stimulated emission phenomenon, predicted by Albert Einstein in 1917. Suppose that one has an atom in an excited state, that in vacuum would decay into the ground state by emitting a photon. The essence of the stimulated emission phenomenon is that the rate for this decay is enhanced if the decay takes place not in vacuum, but rather in a state where there are already many photons of the same kind present. Since the probability is proportional to the square of the amplitude, one finds an enhancement factor of $N + 1$ over the rate one would have if the photons were distinguishable. The enhancement occurs because, once a state with N identical photons exists, an additional photon can be added to it in $N + 1$ ways, just as with a list of N objects, there are $N + 1$ locations in the list to put the next entry. Since all the orderings are indistinguishable, the probability amplitudes for each of these $N + 1$ possibilities must be added and the sum squared to give the probability of the process. The resulting probability is proportional to $(N + 1)^2$ rather than $N + 1$, which would be the result for distinguishable orderings. Thus the more photons there are in the state, the more likely it will be that a transition that would add an additional identical photon to the state will occur.

3. At low temperatures, bosons can exhibit the properties of Bose condensation and superfluidity. An important example is the behavior of ^{4}He at low temperatures. As liquid ^{4}He is cooled, the number of atoms in the lowest available energy state—ordinarily, the state of rest—increases. For reasons quite similar to those discussed for laser action, the existence of a large number of bosons already in a state makes it more favorable for additional atoms to be in that state. Thus ^{4}He below about 4 K, under its own vapor pressure, condenses into a new state of matter in

which a macroscopic fraction of atoms all occupy a single quantum state. If one perturbs the fluid, so that the most energetically favorable state becomes, for a single atom, something other than simply to be at rest, these many atoms respond coherently—all rapidly adapt to the most favorable state together, by laser action. This behavior underlies the remarkable phenomena of superfluidity. Because its atoms are fermions, ^{3}He fluid behaves very differently at such low temperatures, even though it is chemically identical and extremely similar in its physical properties to ^{4}He fluid at higher temperatures.

Properties of Fermions

The most distinctive properties of fermions arise from the fact that identical fermions cannot occupy the same state. This rule is known as the Pauli exclusion principle. Three major phenomena should be considered from this perspective.

1. Bound systems of fermions build up groups of states, for example, those known as orbitals or shells for the electrons in an atom. Their low-energy properties depend sharply upon the precise number of fermions involved. Atoms, with different numbers of electrons, and nuclei, with different numbers of protons and neutrons, form major examples. To analyze atomic structure, one first considers the behavior of single electrons under the influence of the electric field of the nucleus. One finds that there is a discrete spectrum of bound states for a single electron. If the electrons were bosons they would all tend to occupy the same, lowest-energy state, but this is forbidden for fermions. Instead they must occupy different states of higher energy, going up the spectrum as each level, or shell, is filled. This simple picture can be fleshed out to yield a qualitative understanding of the periodic table of elements. In more rigorous or quantitative work, however, it is necessary to take into account that the electrons exert electric forces on one another. To solve for the energy levels for each additional electron in the "self-consistent field," one must take into account the forces generated by electrons occupying the already filled energy levels. This consideration is vital in justifying a similar approach to the structure of nuclei, since in that case there is no central field source independent of the protons and neutrons themselves; instead, it is the attractive forces between the nucleons themselves that create the binding force.

2. The major properties of *metals* are consequences of the fermion character of electrons.

3. The hardness of matter is largely due to the resistance of ensembles of large numbers of electrons to compression. This resistance stems from their Fermi statistics, which requires that they occupy different quantum states. The quantitative measure of this resistance is called the Fermi pressure, which exists even at low temperatures (unlike classical pressure due to the kinetic motion of atoms). It is Fermi pressure of electrons that supports white dwarf stars against the force of their own gravity, and Fermi pressure of neutrons supports neutron stars. Under sufficient loads this pressure fails, so that sufficiently massive stars, as they cool, cannot avoid collapse to black holes

See also: CONDENSATION, BOSE–EINSTEIN; EINSTEIN, ALBERT; ELECTRON; ELEMENTARY PARTICLES; GRAVITON; INTERACTION; INTERACTION, STRONG; INTERACTION, WEAK; ISOTOPES; LEPTON; NEUTRON; PARTICLE; PARTICLE PHYSICS; PROTON; QUANTUM MECHANICAL BEHAVIOR OF MATTER; QUANTUM STATISTICS; SPIN; SPIN AND STATISTICS

Bibliography

LANDAU, L. D., and LIFSHITZ, E. M. *Statistical Physics,* 3rd ed. (Pergamon Press, Oxford, Eng., 1980).

PEEBLES, P. J. E. *Quantum Mechanics* (Princeton University Press, Princeton, NJ, 1992).

FRANK WILCZEK

FERMI SURFACE

The Fermi surface is a theoretical concept in solid state physics that describes the ground state of electrons in metals. Electrons are fermions and therefore Pauli's exclusion principle applies to them. This means that no two electrons in a metal can have the same quantum numbers. At a temperature of 0 K, a piece of metal is in its ground state and all of its electrons will have different quantum numbers corresponding to electron energies ranging from the lowest possible value up to some maximum value, the latter being defined as the Fermi energy, ε_F. According to quantum mechanics or wave mechanics, these electrons behave as waves and their allowed energy levels, $\varepsilon(k_x, k_y, k_z)$, depend on three wave numbers $k_x = 2\pi/\lambda_x$, $k_y = 2\pi/\lambda_y$, $k_z = 2\pi/=\lambda_z$, where the λ_x, λ_y, and λ_z are the wavelengths in the x, y, and z directions, respectively. These wave numbers are quantum numbers for the system. The *Fermi surface* is the constant-energy surface ($\varepsilon = \varepsilon_F$) in k *space* separating the occupied from the unoccupied energy levels for the ground state of the metal.

Metals typically solidify as crystals. They can be described as a crystal lattice of metal ions (each consisting of protons and core electrons) filled with the valence electrons, which are relatively free to move throughout the lattice. Electrons in isolated metal atoms have discreet energy levels described by atomic physics. When these atoms are brought together to form a crystal lattice, these discreet levels broaden forming electron energy bands (because of the interaction between the crystal lattice and electrons), which are often separated by gaps of forbidden energies.

The quantum mechanical problem of *one* electron in a metallic lattice has solutions that are special waves, called Bloch waves, for which the energy levels depend on the quantum numbers n, k_x, k_y, k_z, and s, where n is the energy band number and s is the spin. For each set of (k_x, k_y, k_z) values there are two spin values and an infinite number of energies, $\varepsilon_n(k_x, k_y, k_z)$—one for each band. Finally, a finite piece of metal will have a finite number of discreet k values equal to the number of lattice sites. All of these allowed k values are contained within any one Brillouin zone.

To construct the ground state of a metal with N electrons, these one-electron levels are filled, starting with the first band, until all N electrons are accounted for. The energy of the electron in the highest level is the Fermi energy, which will always be in the middle of an energy band for a metal. As each energy level, $\varepsilon_n(k_x, k_y, k_z)$, is filled a corresponding point is placed in a three-dimensional k space. The surface in k space separating the filled from the unfilled levels at ε_F is the Fermi surface.

At finite temperatures, $T > 0$ K, some of the electrons in the metal acquire thermal energy and are excited to levels outside of the Fermi surface (i.e., to levels higher than the Fermi energy ε_F). Since the amount of thermal energy per electron per degree of freedom is $k_BT/2$, where k_B is Boltzmann's constant, only electrons within about k_BT of the Fermi surface are excited. Typical values of e_F are 5–10 eV, which correspond to Fermi temperatures ($T_F =$

ε_F/k_B) of the order of 10^4–10^5 K. Thus, even at room temperature (300 K) only electrons very close to the Fermi surface can be excited and they are excited to levels that are also very close to the Fermi surface. Therefore, in calculating the electronic properties of metals, for example, the electronic contribution to the specific heat, over most temperatures of interest only those electrons near the Fermi surface need be considered. This usually simplifies the calculations considerably.

The Fermi surface also plays an important role in considering metals in the presence of externally applied electric and magnetic fields. Again the energy of electrons in the metal is increased as they are accelerated by the fields giving rise to many interesting phenomena such as high electrical and thermal conductivity, the Hall effect, the magnetoacoustic effect, cyclotron resonance, and the de Haas–van Alphen effect. The latter three provide means for measuring the Fermi surfaces of metals with the most important being the de Haas–van Alphen effect. For typical fields, only electrons near the Fermi surface are important which again simplifies calculations.

One of the earliest simple models for metals which included the Pauli exclusion principle is that due to Arnold Sommerfeld. He considered the valence electrons in a finite piece of metal to be free electrons. According to quantum mechanics, free electrons behave as simple waves, such as the functions $\sin(k_xx + k_yy + k_zz)$ and $\cos(k_xx + k_yy + k_zz)$, and have energies and momentum components given by

$$\varepsilon(k_x, k_y, k_z) = \frac{\hbar^2(k_x^2 + k_y^2 + k_z^2)}{2m},$$

$$p_x = \hbar k_x, \qquad p_y = \hbar k_y, \qquad p_z = \hbar k_z,$$

where m is the electron mass and $\hbar$ is Planck's constant divided by 2π. It is obvious from the free-electron $\varepsilon(k_x, k_y, k_z)$ relation that constant energy surfaces in k space are spheres. The Fermi surface for the free-electron case is thus a sphere with radius

$$k_F = \sqrt{k_x^2 + k_y^2 + k_z^2} = \frac{\sqrt{2m\varepsilon_F}}{\hbar}.$$

For metals with one atom per lattice site and one valence electron per atom the number of electrons,

N, is equal to the number of lattice sites. This describes the alkali metals such as Li, Na, K, Rb, and Cs, and to a fair degree the noble metals Cu, Ag, and Au. Since the number of electrons that can be contained in any Brillouin zone is $2N$, due to the twofold spin degeneracy of electrons, the Fermi surface for these metals is contained entirely within the first Brillouin zone. It has roughly the spherical shape of the free-electron case but with deviations in regions where this "sphere" is closest to the Brillouin zone edge.

Di-valent, tri-valent, and other multivalent metals have quite complicated overlapping energy bands and Fermi energies that cross several of these bands. As a result their Fermi surfaces involve several higher order Brillouin zones and are rich in structure reflecting the symmetry of the crystal.

See also: BRILLOUIN ZONE; CONDENSED MATTER PHYSICS; CYCLOTRON RESONANCE; DEGENERACY; ELECTRICAL CONDUCTIVITY; ENERGY LEVELS; GROUND STATE; HALL EFFECT; HAAS–VAN ALPHEN EFFECT, DE; PAULI'S EXCLUSION PRINCIPLE; QUANTUM NUMBER; SPECIFIC HEAT; THERMAL CONDUCTIVITY; WAVELENGTH

Bibliography

ASHCROFT, N. W., and MERMIN, N. D. *Solid-State Physics* (Saunders, Philadelphia, 1976).

BLAKEMORE, J. S. *Solid-State Physics*, 2nd ed. (Saunders, Philadelphia, 1974).

BURNS, G. *Solid-State Physics* (Academic Press, San Diego, 1990).

KITTEL, C. *Introduction to Solid-State Physics*, 6th ed. (Wiley, New York, 1986).

WILLIAM F. OLIVER III

FERRIMAGNETISM

Ferrimagnets are materials with a nonzero spontaneous magnetization below some critical temperature that, unlike ferromagnets, have atomic moments that are not all aligned. In the simplest case, ferrimagnets have two sublattices of antiparallel atomic moments of unequal magnitudes or numbers. Examples of some possible ferrimagnetic spin arrangements are shown in Fig. 1.

Ferromagnetism was first found in ferrites with a spinel structure and general composition Fe_2O_3MO, where M = Fe, Mn, Co, Ni, and so on. The spontaneous magnetization versus temperature curves for these materials, when extrapolated to absolute zero, indicate that the spontaneous magnetization is not equal to the sum of the sublattice magnetizations. Louis Néel was the first to explain the magnetic behavior of these materials as arising from sublattices of spins pointing in opposite directions.

Mean Field Theory

To explain the temperature dependence of the magnetic properties of the spinels, Néel employed the mean field theory. In the case of a ferrimagnet the molecular field acting on the *A* sublattice is given by

$$\mathbf{H}_A = N_{AA}\mathbf{M}_A - N_{AB}\mathbf{M}_B,$$

where $\mathbf{M}_A$ and $\mathbf{M}_B$ are the average sublattice magnetizations and N_{AA} and N_{AB} are proportional to the appropriate exchange interaction constants. Likewise, the molecular field on the *B* sublattice is

$$\mathbf{H}_B = N_{BA}\mathbf{M}_A - N_{BB}\mathbf{M}_B.$$

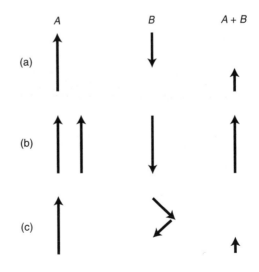

Figure 1 Examples of ferrimagnetic spin orientations for a two-sublattice system.

The A sublattice magnetization at thermal equilibrium is given by

$$\mathbf{M}_A = \sum_i n_i g_i \mu_B S_i B_{S_i}(\chi_A),$$

where

$$\chi_A = \frac{g_i \mu_B S_i}{k_B T} \mathbf{H}_A,$$

and

$$B_{S_i}(\chi_A) = \frac{2S_i + 1}{2S_i} \coth\left(\frac{2S_i + 1}{2S_i}\chi_A\right) - \frac{1}{2S_i} \coth\left(\frac{\chi_A}{2S_i}\right)$$

is the Brillouin function for spin S_i with spectroscopic splitting factor g_i, μ_B is the Bohr magneton, k_B is Boltzmann's constant, T is the temperature, and n_i is the number of spins of type i in the A sublattice. The spontaneous magnetization on the B sublattice is given by substituting the appropriate values into the above equation.

For T less than a critical temperature T_N, defined as the Néel temperature, there are nontrivial solutions to the above equations. Since, in general, the temperature dependencies of the sublattice magnetizations are not identical the magnetization versus temperature curves for ferrimagnets sometimes have unusual shapes. Two such curves are depicted in Fig. 2.

Magnetization versus temperature curves like Fig. 2b are found in some alloys, compounds, and multi-layered films containing rare earth elements and Fe, Ni, and Co. The compensation temperature, T_c of these materials can often be raised or lowered by varying the relative concentration of rare earths and transition metals. Ferrimagnets of this type with T_c's close to room temperature are used in magneto-optic computer disks.

For temperatures larger than T_N, the ferrimagnets become paramagnetic. According to the mean field theory, the magnetic susceptibility χ at low fields is given by

$$\frac{1}{\chi} = \frac{T}{C_A + C_B} + \frac{1}{\chi_0} - \frac{\sigma}{T - \theta},$$

where

$$C_A = \sum_i n_i g_i^2 \mu_B^2 S_i(S_i + 1)/3k_B$$

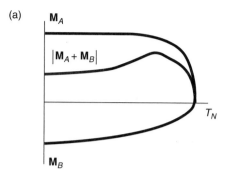

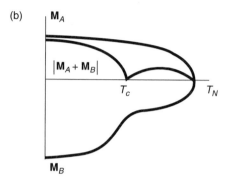

Figure 2 Examples of unusual spontaneous magnetization versus temperature curves seen in some ferrimagnets. T_N is the ordering or Néel temperature, and T_C in curve (b) is called the compensation temperature.

is the Curie constant for the A sublattice, C_B is calculated as above for the B sublattice, and χ_0, σ, and θ are constants that depend upon the Curie and exchange constants. At high temperatures the susceptibility is given by

$$\frac{1}{\chi} = \frac{T}{C_A + C_B} + \frac{1}{\chi_0}.$$

As T approaches T_N the $1/\chi$ curve rapidly deviates from linearity.

For many purposes the mean field theory provides an adequate description of the magnetic properties of ferrimagnets. However, more exact descriptions of their low temperature and microwave properties can be obtained by including a rigorous treatment of quantum mechanical effects.

See also: BOHR MAGNETON; BOLTZMANN CONSTANT; FERROMAGNETISM; MAGNETIC SUSCEPTIBILITY; PARAMAGNETISM

Bibliography

ASHCROFT, N. W., and MERMIN, N. D. *Solid-State Physics* (Holt, Rinehart and Winston, New York, 1976).

CHIKAZUMI, S., and CHARAP, S. H. *Physics of Magnetism* (Wiley, New York, 1964).

KITTEL, C. *Introduction to Solid-State Physics* (Wiley, New York, 1986).

MORRISH, A. H. *The Physical Principles of Magnetism* (Wiley, New York, 1965).

KEVIN AYLESWORTH

FERROELASTICITY

Ferroelasticity is a crystal property characterized by a lattice that is capable of reorientation under an applied mechanical stress; it was first recognized explicitly as such in 1969. Ferroelastic effects in the form of mechanically twinned crystals, however, were well known earlier. Ferroelasticity is a relatively common property closely related to the existence of pseudosymmetry in the underlying crystal structure. It may be present either as an independent property or coupled to one or more other crystal properties. In the latter case interchange of crystal axes may also reorient the coupled property with consequent useful device potential.

A crystal in a given phase is ferroelastic if it contains two or more equally stable orientational states, in the absence of mechanical stress, and if one state can be transformed reproducibly into another by the application of compressive, tensile, or shear stress about an appropriate direction. Transformation of states, commonly referred to as ferroelastic switching, is detectable by means of direction-sensitive properties within the switching plane, for example, by conoscopic observation (i.e., observation of interference under convergent polarized light) in biaxial crystals of a rotation of the optic axis figure. Ferroelasticity is always associated with a small distortion from a higher symmetry state that may be only hypothetically stable. The spontaneous strain e_s is a measure of this distortion, typically with a magnitude on the order of 10^{-3}. The minimum ferroelastic stress required for switching is the coercive stress E_{ij}, where i,j denote the stress and transformed axis directions, respectively. Experimentally, E_{ij} is a function of temperature, pressure, and crystal defect distribution, with magnitude in the range $10^4 < E_{ij} < 10^8$ N·m^{-2} (moderate finger pressure on a 1 mm^2 area produces a compessive stress on the order of 10^6 N·m^{-2}). Since e_s is a second rank tensor, its component magnitudes depend on the unit cell setting; a recent literature survey shows a range from about 0.003 for $NaH_3(SeO_3)_2$ to the unusually large 0.257 for $KClO_3$. In case the transition from a ferroelastic to a higher symmetry (generally paraelastic) phase is without integral change in lattice dimensions, the ferroelastic phase forms a subgroup of the paraelastic point group.

As-grown ferroelastic crystals generally contain numerous domains that have equal volumes of each permitted orientational state, for a macroscopically net zero spontaneous strain. Domains of different orientational state are separated by walls similar to the domain walls observed in ferroelectric crystals. The application of compressive (or other) stress, with magnitude greater than coercive, along a direction able to transform one state into another by moving domain walls out of the crystal, necessarily increases the net strain. As the applied stress continues to increase and the residual number of domain walls approaches zero, the spontaneous strain saturates at the maximum expected value of e_s. Incremental reduction of the applied stress through zero to values with opposite sense until the spontaneous strain again saturates (with a sign reversed from that of the previous spontaneous strain), followed by an increasing stress having the original sense, traces a ferroelastic hysteresis loop analogous to that characterizing ferroelectric crystals. Crystals with saturated spontaneous strain (i.e., without domain walls) are described as having been "detwinned."

Ferroelasticity is often accompanied simultaneously by formation of other tensor properties, including pyroelectricity, ferroelectricity, ferromagnetism, and a number of optical effects, at the transition from a higher symmetry phase. Coupling between properties results from a coupling between e_s and the thermodynamic order parameter. If the properties are coupled, then ferroelastic reorientation of e_s causes the other property to be reoriented, and vice versa. $Tb_2(MoO_4)_3$ is an example of a fully coupled ferroelastic-ferroelectric crystal in which reorientation of e_{12} simultaneously interchanges the a and b axes as the sense of P_s (the spontaneous polarization) is reversed along [001], an effect that has led to the production of such devices as optical switches, adjustable optical slits, and pattern generators.

Reorientation of e_s is the result of atomic distribution reorientation. Typical atomic displacements in the course of ferroelastic switching are on the order of 1 Å, although some displacements have been reported as much as an order of magnitude larger.

See also: CRYSTALLOGRAPHY; CRYSTAL STRUCTURE; FERROELECTRICITY; HYSTERESIS; PHASE TRANSITION; SYMMETRY

Bibliography

ABRAHAMS, S. C. "Structure Relationship to Dielectric, Elastic, and Chiral Properties." *Acta Crystallogr. Sec. A* **50**, 658–685 (1994).

KLASSEN-NEKLYUDOVA, M. V. *Mechanical Twinning of Crystals* (Consultants Bureau, New York, 1964).

SALJE, E. K. H., ed. *Phase Transitions in Ferroelastic and Coelastic Crystals* (Cambridge University Press, Cambridge, Eng., 1990).

S. C. ABRAHAMS

by applying an electric field, the local structure would remain the same. Also, the new structure could be obtained by moving the + charge from the left end to the right end of the line, without reversing any dipoles, so this one-dimensional NaCl structure is not ferroelectric.

If the crystal lacks an inversion center, as in the following one-dimensional structure:

$$+-0+-0+-0+-0+-0,$$

it is ferroelectric provided an electric field applied toward the right can reverse these dipoles to the structure

$$-+0-+0-+0-+0-+0.$$

Crystals that lack an inversion center may not be ferroelectric because dielectric breakdown can occur below the coercive field required to reverse the polarization. Such crystals are called pyroelectric,

FERROELECTRICITY

As implied by the name, ferroelectricity is the electrical analog of ferromagnetism. Ferroelectrics have electric dipoles whose direction can be reversed by an electric field E. This reversal of polarization P can be detected by the Sawyer–Tower circuit shown in Fig. 1. Ferroelectric behavior is characterized by an S-shaped curve that encloses area in the E-P plane, as shown in Fig. 2.

Ferroelectricity is associated with crystalline order. A centrosymmetric crystal with inversion symmetry such as NaCl cannot be ferroelectric. A one-dimensional line of ionic charges in NaCl is arrayed as

$$+-+-+-+-+-+-+-+-+-.$$

This line has inversion symmetry (looks the same in opposite directions) at each plus and minus charge. Even if one could reverse these dipoles to the configuration

$$-+-+-+-+-+-+-+-+-+$$

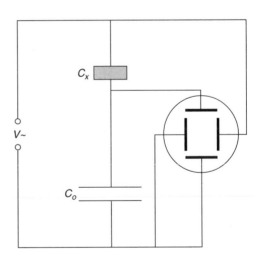

Figure 1 Modified Sawyer–Tower circuit for displaying ferroelectric hysteresis loops consisting of polarization P plotted against applied field E. The alternating voltage source V is connected to the crystal of average capacitance C_x in series with a much larger capacitance C_o. The voltage across C_o (proportional to P) is applied to the oscilloscope's vertical deflection plates and the voltage V (proportional to E for large C_o) is applied to the horizontal deflection plates.

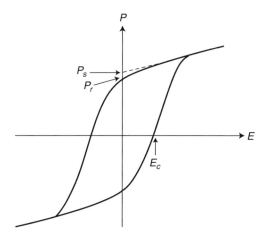

Figure 2 Ferroelectric hysteresis loop showing polarization P as a function of electric field E for a large low-frequency alternating field. The coercive field E_c is the field required to bring a previously negative polarization to zero. The spontaneous polarization P_s corresponds to the polarization within a single ferroelectric domain at zero field. The remanent polarization P_r corresponds to the actual average polarization in zero field, which may be lower than P_s because domains of opposite polarization may appear already at zero field.

because heating the crystal changes its dimensions and thus changes its dipole moment.

A related property is piezoelectricity. Any ferroelectric crystal will change its dipole moment as a result of any kind of pressure (hydrostatic, tension/compression, or shear), and thus is piezoelectric. In addition, the structural symmetry of some crystals allows them to be piezoelectric but not ferroelectric.

To summarize the relation of crystal structure to ferroelectricity, pyroelectricity, and piezoelectricity, the following hierarchical diagram is helpful:

<all crystals [piezoelectrics
{pyroelectrics (ferroelectrics)}]>

Some ferroelectric crystals can be grown easily from aqueous solution, such as Rochelle salt, triglycine sulfate, and KH_2PO_4. Rochelle salt, a mixed potassium/sodium tartrate was the first crystal found to be ferroelectric by Joseph Valasek in 1921. Next, potassium dihydrogen phosphate (KH_2PO_4) was found to be ferroelectric by Georg Busch and Paul Scherrer in 1935. In 1941 John Slater explained its

ferroelectricity in terms of ordering of the protons in their hydrogen bonds. Crystals with such ordering, usually of protons, are called order-disorder ferroelectrics.

Another ferroelectric family, the perovskites, have displacive transitions. The prototype of this family, $BaTiO_3$, was found to be ferroelectric by Bentsian Wul and I. M. Goldman in 1945. Its ferroelectric transition involves displacement of the positive Ti ion from the center of its cage of surrounding negative charges, thus creating a dipole moment.

In the 1950s and 1960s, ferroelectricity was found in many more crystals, by Bernd Matthias, Raymond Pepinsky, and others. In 1960 William Cochran developed a soft mode theory for ferroelectric transitions. Also, such transitions became a testing ground for Kenneth Wilson's renormalization group theory, which improved on the Landau theory by taking polarization fluctuations into account.

Ferroelectricity also occurs in the crystalline portions of semicrystalline polymers such as poly(vinylidene fluoride) (PVDF), formula $(CH_2CF_2)_n$, and in certain liquid crystals.

For a long time, most applications of ferroelectrics did not specifically use the ferroelectric property, but rather other properties based on the low crystal symmetry. Rochelle salt was used in the past as a piezoelectric pickup in phonographs. Now, KH_2PO_4 and related crystals are used in optical frequency doubling, for instance to turn red light into green. The pyroelectric property (sensitive heat detection) is used for intruder detection and infrared imaging. The piezoelectric property of PVDF is now employed in high-frequency "tweeters" to produce sound, and in sonar devices to produce and then detect sound to locate objects such as submarines and fish. Piezoelectric ceramics, often based on $BaTiO_3$-family crystals, have many applications as sensors, actuators, and as vibration isolators.

The ferroelectric property has been used specifically in computer memory devices based on ferroelectric thin films. The individual memory element is the polarization of a small pixel (a litle square) on a thin ferroelectric ceramic layer. Ferroelectric liquid crystals contain polar molecules having a twisted arrangement which can be uncoiled by an electric field, providing a basis for liquid crystal displays.

Some exciting future applications for $BaTiO_3$ and other ferroelectrics are based on the photorefractive effect. This effect can provide optical phase conjugation, which allows restoration of optical signals distorted in passing through an inhomogeneous medium. It may also provide the basis for very fast optical computers.

Interesting applications are foreseen for "smart materials" based on "relaxor ferroelectric" mixed crystals in the perovskite family. The electrostrictive effect, which is a displacement proportional to the square of the applied electric field, is large in these materials. The smartness consists in automatic adjustment of the size of the electrostrictive coefficient in response to changing parameters such as the mass to be vibration-isolated or changing frequency spectrum of the vibration source.

See also: CRYSTALLOGRAPHY; CRYSTAL STRUCTURE; DIPOLE MOMENT; ELECTRIC MOMENT; PIEZOELECTRIC EFFECT; PIXEL; PYROELECTRICITY; THIN FILM

Bibliography

JONA, F., and SHIRANE, G. *Ferroelectric Crystals* (Macmillan, New York, 1962).

LINES, M. E., and GLASS, A. M. *Principles and Applications of Ferroelectrics and Related Materials* (Clarendon Press, Oxford, Eng., 1977).

XU, Y. *Ferroelectric Materials and Their Applications* (North-Holland, New York, 1991).

V. HUGO SCHMIDT

FERROMAGNETISM

Ferromagnetism is the ability of certain materials to exhibit a magnetic induction field, called the spontaneous magnetization, in the absence of an external magnetic field. Among the elements, only iron, cobalt, nickel, and gadolinium are ferromagnetic at ordinary temperatures. Numerous alloys and compounds, both naturally occurring and synthesized, are ferromagnetic, and these materials have enormous technological importance. Every motor, actuator, transformer, most recording devices and loudspeakers are based on ferromagnets, and these materials are used extensively in household appliances, automobiles, novelties, and toys. Ferromagnetism only appears at temperatures below a value T_C, called the Curie temperature, that depends on the nature of the material. Table 1 lists the Curie temperatures of a variety of ferromagnetic substances on the Kelvin temperature scale (to convert to Celsius, subtract 273.15 K). Naturally occurring ferromagnetic oxides such as magnetite in the earth's crust become ferromagnetic as they cool following volcanic activity, thereby recording the direction of the earth's magnetic field at that epoch. The record of continental drift is written in the ferromagnetic materials of the earth's crust, just as music and data are recorded on a strip of magnetic tape. These naturally occurring ferromagnets have been known since antiquity and, through their use in the compass, played a major role in navigation and exploration.

Ferromagnetism arises from two related fundamental properties of the electron: its magnetic dipole moment and the requirement that it obey Pauli's exclusion principle. The electron's dipole moment is similar to that produced by a current flowing in a loop of wire, but it is not correct to think it a consequence of the rotation of the electron's charge. Rather, it is a fundamental quantum mechanical property of the electron. Its magnitude is very small, and is measured in terms of the energy difference between alignment parallel and antiparallel to an external magnetic field. The value is 9.23×10^{-24} J/T. Pauli's exclusion principle, which is responsible for the structure of atoms and the buildup of the periodic table, tends to favor the parallel alignment of the magnetic dipole moments of electrons in unfilled atomic shells to the extent it is possible, and this causes many atoms to exhibit magnetic moments, a combination of the intrinsic moment of the electrons and that which arises from their motion around the nucleus.

Frequently, but certainly not always, magnetic atoms retain their magnetic moments when they are incorporated into solid materials. If a solid containing such atoms is held at a temperature T and placed in a magnetic field H, the magnetic dipole moments of the atoms—if they are independent of each other—will tend to align with the magnetic field, giving rise to the magnetization, defined as the magnetic dipole moment per unit volume. The magnetization is given by $M(H,T) = CH/T$, where the constant C depends on the magnetic dipole moment of the atoms in the solid (squared) times the number per unit volume. This expression is known as Curie's law, named for Pierre Curie, and the constant is called the Curie constant. In 1907 Pierre Weiss realized that, for the magnetization to exist in the absence of an applied magnetic field, a bootstrap mechanism is required. To the external magnetic field H that acts on each magnetic atom must be added a "molecular field" that depends on the value of the magnetization already present, giving a total field $H + \lambda M$, where λ is called the molecular field constant. When this is substituted into the Curie law and the equation is solved for M, a new expression, the Curie–Weiss law, arises, written as

$$M = \frac{CH}{(T - \lambda C)}.$$

No matter how small H is, as the temperature approaches the value $T_C = \lambda C$, there will be a large magnetization. That temperature is known as the Curie temperature, and it marks the onset of ferromagnetism. The Curie–Weiss Law holds only above T_C; at lower temperatures, M appears spontaneously, increasing as $(T_C - T)^{1/2}$ just below T_C, and then becoming constant. A plot of the actual data of Pierre Weiss and R. Forrer for nickel is shown in Fig. 1. At T_C, the magnetization increases as the $1/3$ power of the applied field. Very close to T_C, the molecular field approach becomes invalid because the magnetization ceases to be uniform throughout the material. As a result of these deviations from uniformity, the Curie–Weiss law is changed to

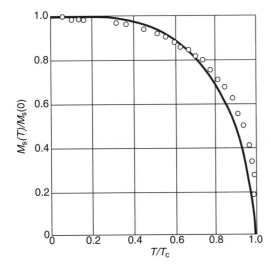

Figure 1 Saturation magnetization of nickel as a function of temperature, together with the theoretical curve for $S = 1/2$ on the molecular field theory. [Experimental values from P. Weiss and R. Forrer, *Ann. Phys.* **5,** 153 (1926).]

$$M = \frac{CH}{T_C(T/T_C - 1)^\gamma}; \qquad (1)$$

for many ferromagnets γ turns out to be approximately 1.4, rather than 1.0 as in the Curie–Weiss law. Other properties exhibit similar changes, all of which are known collectively as "critical behavior," with the Curie temperature a typical phase transition, or critical, point.

An explanation of the source of the effective magnetic field λH had to await the development of quantum mechanics. In 1928 Werner Heisenberg demonstrated that Pauli's exclusion principle can also work to favor either parallel or antiparallel alignment of the individual atomic magnetic moments in solids. He was able to express this effect as an interaction among the magnetic dipole moments of the neighboring atoms, with a multiplicative constant he called J, and which we now call the Heisenberg exchange constant. The molecular field constant can be directly related to Heisenberg's J. Thus, ferromagnetism is probably the most readily observable direct consequence of quantum mechanics at work on the human scale. Those materials in which Heisenberg's J favors antiparallel alignment are known as antiferromagnets. In some cases, neighboring atoms have different magnetic dipole moments, in which case antiparallel alignment of neighbors leaves a net magnetization. Such materials are known as ferrimagnets. They share many properties with ferromagnets, and are technologically important.

Although ferromagnetism is defined by the existence of a magnetization M in the absence of a magnetic field, a sample of a ferromagnetic substance may exhibit only a small external magnetic induction field. While they should exist for a uniformly magnetized object, such external fields are energetically unfavorable for most sample shapes. A lower energy state arises if the magnetic material divides itself into small regions called domains, each with the full value of the magnetization, but arranged in such a way as to minimize the magnetic field outside the material. The size of the domains and the ease of their formation vary widely among ferromagnetic materials. Domain formation represents a compromise between the energy lost by forming domain walls between regions magnetized in different directions and the energy gained by reducing the external field. Those materials in which domains can readily form are termed soft and, correspondingly, those in which domain formation is suppressed are called hard. Both hard and soft magnets have important applications.

Whether a magnet is hard or soft, the application of a magnetic field at first rotates the magnetization within each domain and then, as the field is increased, sweeps out all domain walls, and drives the magnetization to its full value, termed saturation. When the saturating magnetic field is removed, domains form in soft materials and the magnetization returns to a small value. In hard materials, however, domain formation is inhibited, and the magnetization remains high. That value is known as the remanent magnetization. Hard magnets with large remanence are often called permanent magnets. To destroy the remanent magnetization, it is necessary to apply a field in the opposite sense to that which caused saturation. The field required to drive the remanent magnetization to zero is called the coercive field, H_c. If the reversed field is increased further, the magnetization can be saturated in the opposite sense, and when the field is removed, a reversed remanent magnetization results. When the entire process is carried back to the original saturated state, a graph of the magnetization, or of the magnetic induction $B = \mu_0 H + M$, versus the applied field traces a closed curve called the hysteresis loop. That the magnetic state depends on its history of exposure to magnetic fields is a fundamental characteristic of ferromagnets. Figure 2 shows a schematic of a hysteresis loop for a hard magnet. Soft magnets also exhibit hysteresis, but with much smaller values of the remanent magnetization and coercive fields. For permanent magnet applications, both the remanent magnetization and coercive field should be as large as possible. The maximum product of B times H achieved on the hysteresis loop serves as a figure of merit for hard magnetic materials called the energy product. The record holder is the compound $Nd_2Fe_{14}B$, which has an energy product greater than 40 MGOe. The properties of a variety of ferromagnetic and ferrimagnetic substances are listed in Table 1.

From the laws of electromagnetism, it follows that the area enclosed by the hysteresis loop represents work done by the magnetic field on the substance, and it appears as heat. For many applications, transformers, and motors in particular, an open hysteresis loop represents the undesirable loss of energy to heat. The heating occurs because the domain walls move during the magnetization process and, in most materials, this motion is fric-

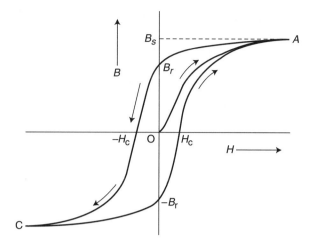

Figure 2 Diagrammatic hysteresis loop.

tional due to the existence of impurities and boundaries between crystalline grains. Of the crystalline materials, the softest magnetic materials are based on the alloy $Ni_{80}Fe_{20}$, known commercially as Permalloy. It has proven possible to make ferromagnetic alloys by a rapid cooling process that leaves them in a state very similar to window glass, in which there are no crystalline boundaries. Such materials, known as Metglas alloys, are finding applications in transformers and other devices that involve cycling the magnetic field between positive and negative saturation values. In certain materials, small stable domains called bubble domains can be formed and manipulated, and these form the basis of a computer memory device known as bubble memory. Magneto-optical data storage is also based on domain formation.

In addition to the strong interactions identified by Heisenberg that give rise to ferromagnetism, there are a number of weaker forces that affect the properties of such materials. Along with the intrinsic magnetic moment of the electron, the orbital motion of the electrons contribute to magnetism, and the two sources are coupled together within the atom by forces known as spin-orbit coupling. As a consequence, the magnetization is sensitive to the local arrangement of atoms in a material, causing some directions to be more favorable than others for the orientation of the magnetization. If a single direction is favored, the material is called uniaxial, and the favored direction is called the easy axis. A considerable part of the energy required to form a domain wall occurs because the atomic moments

Table 1 Properties of Ferromagnets[a]

Substance	Curie Temperature $(T_c\,K)$	Saturation Magnetization $(M_s\,T)$	Remanent Magnetic Induction $(B_r\,T)$	Coercive Field $(\mu_0 H_c\,T)$	Energy Product $(BH_{max}\,MGOe)$[b]
Iron (Fe)	1,043	2.19		1.0×10^{-4}	
Cobalt (Co)	1,388	1.82		1.0×10^{-3}	
Nickel (Ni)	627	0.64		7.0×10^{-5}	
Gadolinium (Gd)	293	2.49			
Magnequench ($Nd_2Fe_{14}B$)	600	1.62	1.30	1.15	40.0
Samarium-Cobalt ($SmCo_5$)	1,020	0.90	0.84	0.87	18.5
Permalloy ($Ni_{78}Fe_{22}$)	355	0.73		1.4×10^{-6}	
Metglas ($Fe_{72}Co_8Si_5B_{15}$)	650	1.60		0.7×10^{-6}	
Ferrite ($MnFe_2O_4$)	573	0.70			
Magnetite (Fe_3O_4)	858	0.64			

[a]Ferrite and magnetite are actually ferrimagnets.
[b]1 MGOe = 1 kJ/m^3.

within the wall are forced to point away from the easy axis. In some materials, one direction is strongly disfavored over others, leading to an easy plane of magnetization. These same considerations often favor a change in the local arrangement of atoms in response to the magnetization, an effect called magnetostriction. The forces involved in magnetostriction are enormous, and this effect has been used to produce devices for the sensitive measurement of strain and other sensing applications.

Although the magnetic moments of all the atoms in a ferromagnet become aligned along the easy direction of a perfect sample at temperatures far below the Curie temperature, it is still possible for the moments to undergo small oscillations around that direction. Because the moments are coupled through the exchange interaction, the oscillations travel through the ferromagnet as waves, known as spin waves. These are the magnetic analog of sound waves, and like sound waves, they behave quantum mechanically. The quantized waves are known as magnons, by analogy with quantized sound waves or phonons. The presence of such waves can be detected directly when neutrons are scattered from a ferromagnet, or by resonance experiments. The excitation of spin waves finds applications in a number of devices used in radar and cellular communications.

See also: ANTIFERROMAGNETISM; ELECTROMAGNETISM; HEISENBERG, WERNER KARL; PAULI'S EXCLUSION PRINCIPLE

Bibliography

CHAKRAVARTY, A. S. *Introduction to the Magnetic Properties of Solids* (Wiley, New York, 1980).

IBACH, H., and LÜTH, H. *Solid-State Physics* (Springer-Verlag, Berlin, 1991).

KITTEL, C. *Introduction to Solid-State Physics,* 7th ed. (Wiley, New York, 1996).

M. B. SALAMON

FEYNMAN, RICHARD PHILLIPS

b. Far Rockaway, New York, May 11, 1918; *d.* Pasadena, California, February 15, 1988; *quantum theory, quantum electrodynamics, relativity.*

Feynman, one of the greatest and most original physicists of the second half of the twentieth century, grew up on Long Island and attended both junior and senior high school in Far Rockaway, where he was fortunate to have some very competent and talented teachers for his chemistry and mathematics courses. He entered the Massachusetts Institute of Technology in the fall of 1935 and was recognized as an unusually gifted individual by all of his teachers. In 1939 he went to Princeton University as a graduate student in physics and was assigned to be John A.

Wheeler's assistant—a propitious event in retrospect. Wheeler, who had just come to Princeton as a twenty-six-year-old assistant professor in the fall of 1938, proved to be an ideal mentor for the young Feynman. Full of bold and original ideas, a man who had the courage to look at any problem, a fearless and intrepid explorer of ideas, Wheeler gave Feynman viewpoints and insights into physics that would prove decisive in later research. In the spring of 1942, Feynman obtained his Ph.D. and immediately thereafter started working on problems related to the development of an atomic bomb. In 1943 he was one of the first physicists to go to Los Alamos. He was quickly recognized by Hans Bethe, the head of the theoretical division, and by J. Robert Oppenheimer, the director of the laboratory, to be one of the most valuable members of the theoretical division. He was perhaps the most versatile, imaginative, ingenious, and energetic member of that community of outstanding scientists. In 1944 he was made a group leader in charge of the computations for the theoretical division. He introduced punch card computers to Los Alamos and continued to develop his lifelong interest in computing and computers.

While at Los Alamos, Feynman accepted a Cornell University appointment as an assistant professor and joined its department of physics in the fall of 1945. In 1951 he left Cornell to become a member of the faculty of the California Institute of Technology, remaining there until his death of stomach cancer in 1988.

One aspect of Feynman's genius was that he could make explicit what was unclear and obscure to most of his contemporaries. His doctoral dissertation and his 1948 *Reviews of Modern Physics* article, which presented his path integral formulation of nonrelativistic quantum mechanics, helped clarify in a striking manner the assumptions that underlay the usual quantum mechanical description of the dynamics of microscopic entities. In Feynman's approach, a (nonrelativistic) particle in going from the spatial point x_1 at time t_1 to the spatial point x_2 at time t_2 is assumed to be able to take *any* path that joins x_1 at time t_1 to x_2 at time t_2, and each path is assigned a "probability amplitude." His reformulation of quantum mechanics and his integral over paths may well turn out to be his most profound and enduring contributions. They have deepened understanding of quantum mechanics and have significantly extended the systems that can be quantized. His path integral has enriched mathematics and has provided new insights into spaces of infinite dimensions.

Feynman was awarded the Nobel Prize for physics in 1965 for his work on quantum electrodynamics (QED). In 1948, simultaneously with Julian Schwinger and Sin-Itiro Tomonaga, he showed that the infinite results that plague the usual formulation of QED could be removed by a redefinition of the parameters that describe the mass and charge of the electron in the theory, a process that is called renormalization. Schwinger and Tomonaga had done this by building on the existing formulation of the theory. Feynman, on the other hand, invented a completely new diagrammatic approach that allowed the visualization of space-time processes, simplified concepts and calculations enormously, and made possible the exploration of the properties of QED to all orders of perturbation theory. Using Feynman's methods it became possible to calculate quantum electrodynamic processes to amazing precision. Thus, the magnetic moment of the electron has been calculated to an accuracy of 1 part in 10^9 and found to be in agreement with an experimental value measured to a similar accuracy.

In 1953 Feynman developed a quantum mechanical explanation of liquid helium, which justified the earlier phenomenological theories of Lev Landau and Laslo Tisza. Because a ^{4}He atom has zero total spin angular momentum, it behaves as a Bose particle: The wave function describing a system of N helium atoms is unchanged under the exchange of any two helium atoms. The ground state of such a system is described quantum mechanically in terms of a unique function that is everywhere positive. When in this state the system—even when N is of the order 10^{23} and the system is macroscopic—behaves as one unit. This is why helium near 0 K is superfluid, acting as if it had no viscosity. Near 0 K, pressure waves are the only excitations possible in the liquid. At somewhat higher temperatures, around 0.5 K, it becomes possible to form small rings of atoms that can circulate without perturbing other atoms; these are the rotons of Landau's theory. With increasing temperature, the number of rotons increases and their interaction with one another gives rise to viscosity. An assembly of rotons behaves like a normal liquid, and this liquid moves independently of the superfluid. At a certain point, when the concentration of normal liquid becomes too large, a phase transition occurs and the whole liquid turns normal. This was Feynman's quantum mechanical explanation why at any given temperature helium could be regarded as a mixture of superfluid and normal liquid.

560

In 1956 T. D. Lee and Chen Ning Yang analyzed the extensive extant data on nuclear beta decay and concluded that mirror (parity) symmetry is not conserved in these interactions. This was soon confirmed experimentally by Chien Wu, Ernest Ambler and Evans Hayward, and others. Subsequent experiments further indicated that the violation of parity is the maximum possible. On the basis of these findings, Robert Marshak and George Sudarshan, and somewhat later and independently Feynman and Murray Gell-Mann, postulated that only the "left-handed" part of the wave functions of the particles involved in the reaction enter in the weak interactions. Furthermore, Feynman and Gell-Mann hypothesized that the weak interaction is universal, that is, that all the weak particle interactions have the same strength. This hypothesis was later corroborated by experiments.

In the late 1960s, experiments at the Stanford Linear Accelerator on the scattering of high energy electrons by protons indicated that the cross section for inelastic scattering was very large. Feynman found that he could explain the data if he assumed that the proton was made up of small, point-like entities that interacted elastically with electrons. He called these sub-nuclear entities "partons." The partons were soon identified with the quarks of Gell-Mann and George Zweig. The study of quarks and their interactions and, in particular, explaining their confinement inside nucleons and mesons were important components of Feynman's research during the 1980s.

Feynman disliked pomposity and made fun of pretentious and self-important people. He was always direct, forthright, and skeptical. His uncanny ability to get to the heart of a problem—whether in physics, applied physics, mathematics, or biology—was demonstrated repeatedly. Thus, while sitting on the presidential commission that investigated the Challenger disaster, he pinpointed the central problem by dropping a rubber O-ring into a glass of ice water to demonstrate its shriveling. In his physics, Feynman always stayed close to experiments and showed little interest in theories that could not be experimentally tested. He imparted these views to undergraduate students through his justly famous *Feynman Lectures on Physics* and to graduate students through his widely disseminated lecture notes for the graduate courses he taught. His writings on physics for the interested general public, *The Character of Physical Laws* and *QED: The Strange Theory of Light and Matter,* convey the same message.

See also: FEYNMAN DIAGRAM; INTERACTION, WEAK; LANDAU, LEV DAVIDOVICH; LIQUID HELIUM; OPPENHEIMER, J. ROBERT; QUANTUM ELECTRODYNAMICS; QUANTUM THEORY, ORIGINS OF; QUARKS, DISCOVERY OF; RENORMALIZATION

Bibliography

FEYNMAN, R. P. *The Character of Physical Law* (MIT Press, Cambridge, 1965).

FEYNMAN, R. P. *QED: The Strange Theory of Light and Matter* (Princeton University Press, New Haven, CT, 1985).

FEYNMAN, R. P.; LEIGHTON, R. B.; and SANDS, M. *The Feynman Lectures on Physics,* 3 vols. (Addison-Wesley, Reading, MA, 1963–1965).

GLEICK, J. *Genius: The Life and Science of Richard Feynman* (Pantheon Books, New York, 1992).

S. S. SCHWEBER

FEYNMAN DIAGRAM

Feynman Diagrams are pictures that physicists draw to represent particle processes. The diagrams represent the terms in a calculation of the rate for a physical process. They also provide an intuitive picture of what that calculation is about.

Feynman diagrams were introduced by the physicist Richard Feynman in 1948 as a tool for calculating in the theory known as quantum electrodynamics (QED), which is the quantum field theory that describes the interactions of charged particles with one-another via electric and magnetic fields. At a historical conference in 1980 Julian Schwinger, another of the physicists who helped develop the early understanding of quantum electrodynamics, remarked that "the Feynman diagram, like the hand calculator, brought computation into the realm of the masses." By this he meant that before the diagrams were introduced only he and one or two others could do the calculations, but the diagrams provided a way to teach other physicists the calculational method.

Physicists refer to the Feynman rules for a theory. The rules are derivable from the underlying field theory. If you know these rules, you can then construct the diagrams for any possible process in that theory. If you cannot draw any diagram for a process, then that process cannot happen. If there is a diagram, then the rules also tell how to calculate the probability for the process to occur by assigning a mathematical expression to each element of the diagram. Let us take the original example of electromagnetism and develop the Feynman rules for constructing the Feynman diagrams for some simple processes.

Each line in a Feynman diagram represents a particle or antiparticle moving with a definite momentum. We use here a time-ordered convention for drawing diagrams to distinguish between electron and positron lines. In this convention, all lines are to be interpreted as starting at the left side of the diagram and ending at the right. (The horizontal axis is time, and vertical position means nothing; the diagrams do not display momentum direction.) A vertex where three lines meet represents a possible sub-process in the theory. To describe the electro-

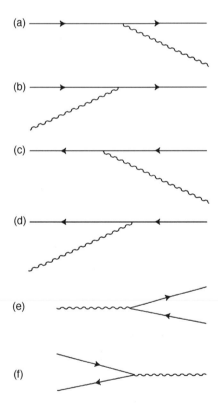

Figure 1 Feynman diagrams representing (a) an electron emitting a photon, (b) an electron absorbing a photon, (c) a positron emitting a photon, (d) a positron absorbing a photon, (e) a photon producing an electron and a positron (an electron positron pair), and (f) an electron and a positron meeting and annihilating (disappearing), producing a photon.

magnetic interactions of electrons and positrons we need three types of lines: (1) a straight line with an arrow on it that points to the right representing an electron; (2) a straight line with an arrow pointing to the left representing a positron, the antiparticle of the electron (do not confuse the arrow with a direction of motion; every diagram goes left to right with time and does not show position); and (3) a wiggly line representing a photon, the massless particle associated with a quantum excitation of the electromagnetic field.

In quantum electrodynamics the only allowed vertex has one wiggly photon line coming from a straight line with one arrow into the vertex and one arrow out. This one vertex can be drawn with six possible orientations (see Fig. 1).

Positrons have electric charge of the same magnitude as, but opposite sign to, that of the electron. Electric charge conservation requires that the number of electrons minus the number of positrons is the same at the beginning and end of any process. By looking at the sub-diagrams in Fig. 1, you can readily see that each of them has this property. Thus, no matter how complicated a diagram you build by connecting these pieces together this rule will still apply. All the conservation laws of particle types and charge found in particle physics are built into the diagrams in this way.

Conservation of energy and of momentum are not obvious from the diagrams. The calculation rules that go along with the diagrams ensure that any process that violates these conservation laws results in a zero contribution to the probability for the process. The rules for the flow of energy and momentum in the diagram say that at every vertex the momentum and energy entering (on the left) must equal the momentum and energy leaving (on the right).

Any processes involving only these three types of particles can be represented by connecting together two or more of these sub-diagrams. Figure 2 shows some examples of possible diagrams that represent processes where an incoming electron and positron interact and exchange some momentum (just like a collision of two billiard balls).

The strange truth is that not one of the six processes shown in Fig. 1 can occur for actual (physicists say "real") isolated electrons and positrons. They all violate the conservation of energy and momentum. A particle of momentum p and mass m has energy

$$E = \sqrt{p^2 c^2 + m^2 c^4} = \gamma m c^2$$

(for a massive particle at rest this is the familiar $E = mc^2$). However, none of the diagrams in Fig. 1 can satisfy this condition for all three particles at once.

Now let us consider the diagrams shown in Fig. 2. These represent some contributions to a process where an electron and a positron with momentum p_1 and p_2, respectively, exchange some momenta and emerge with momenta p_3 and p_4, respectively. This overall process certainly can occur within the rules of conservation of energy and momentum; it is just like a collision between two balls. But all the sub-processes that make up this process cannot occur. Clearly something a little odd is going on here; the culprit is quantum mechanics.

No matter which of the diagrams you investigate, if you apply energy and momentum conservation at the vertex you will find the photon in the intermediate stage of the process does not have the right relationship between its energy and its momentum for a particle of zero mass ($E = pc$). Any real, that is observable, photon must have this relationship right. The photon in this diagram is not real; it can never be observed. It is a fiction of the calculational method. Physicists call this a "virtual" photon. The intermediate stage of the process occurs over such a short time that the mismatch between the available energy and the mass and momentum of the particle is less than the minimum uncertainty in the energy in that short time period, as defined by the Heisenberg uncertainty principle, which can be stated in the form $\Delta E \Delta t \geq h/(4\pi)$. The uncertainty in the energy, ΔE, times the uncertainty in the time at which this energy is measured, Δt, is always greater than or equal to the quantity on the right of the equation, where h is Planck's constant.

The many possible diagrams must be combined to calculate the probability of the process that they represent. This calculation also has quantum mechanical features. If we label the diagrams $i = a, b, c \ldots$, the Feynman prescription determines a complex number A_i associated with each diagram. This is called the amplitude. The probability of the process occurring is given by the absolute value of the square of the sum of all contributing amplitudes

$$P = |\Sigma_i A_i|^2.$$

Now notice that

$$|\Sigma_i A_i|^2 \neq \Sigma_i |A_i|^2.$$

This shows that we cannot interpret the probability of the overall process as a sum of probabilities associated with individual sub-processes.

Since this is so, it is meaningless to ask which sub-process occurred when one observes the scattering. The intermediate stages of the diagrams are not observable stages in a set of events. The diagrams are nothing more than pictorial representations of a mathematical calculation. If one sets up an experiment to observe the intermediate stage, the very act of observation must be included in the calculation. Then diagrams must include whatever additional particles carry information to the observer. This gives quite different results.

Physicists invented the term "virtual" for the unobservable particles that occur in intermediate stages of a Feynman diagram. We say that electrons and positrons interact via the exchange of virtual photons. This is just a description of the Feynman diagrams; there is no classical picture that can explain, for example, why like charges attract (and unlike charges repel) via such exchanges, but it is indeed the result of the calculation based on the diagrams. Notice also that in Fig. 2c a quite different

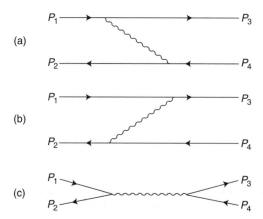

Figure 2 Three diagrams representing single-photon contributions to a process where an incoming electron and positron interact and exchange some momentum. In (a) the electron emits a virtual photon, which is later absorbed by the positron. In (b) the positron emits a virtual photon, which is later absorbed by the electron. In (c) the electron and positron meet and disappear (annihilate each other) to produce a virtual photon, which subsequently decays to produce an electron and a positron.

process occurs; at the intermediate stage there is only a photon.

The diagrams in Fig. 2 all contain only one intermediate photon. Clearly one could draw many diagrams with two exchanged photons, even more with three, and so on. In the calculations associated with these diagrams, a factor e occurs for every vertex where a photon line begins or ends. Hence, every photon line contributes with a factor of $\alpha = e^2/4\pi \approx 1/137$, which is a small number, so the contribution from diagrams with additional photons is small. This keeps the calculation managable. One can obtain a very accurate prediction for processes by calculating only the terms with low powers of α.

The most accurately calculated quantity with this method is a property of the electron known as its gyromagnetic ratio g. The calculations include up to α^4. There is agreement of experiment and theoretical calculation to better than ten significant figures. This is just one of many calculations that have used Feynman diagrams to predict physical processes and given results that compare well with experiment. When particle physicists say the standard model is a well-established theory, they mean that many such comparisons of calculation and experiment have been made.

Feynman diagrams can be drawn for all particle processes. The diagrams give amplitude contributions proportional to the number of vertex coupling constants, such as the factors of $\alpha = e^2/4\pi \approx 1/137$ for the photon lines in the QED example above. The size of the coupling constant represents the strength of the interactions; the rates of the processes grow with the size of the coupling constant.

The Feynman diagram technique is useful only when the terms with higher powers of the coupling constant, arising from diagrams with larger numbers of exchanged particles, get successively smaller. Otherwise, we would have to calculate an infinite number of diagrams involving arbitrarily large numbers of exchanged particles to determine a rate; that is clearly impossible. The coupling is small for electromagnetism, as discussed above, and also for weak interaction theory. However, for strong interaction processes it is not generally true. Not all strong interaction processes can be calculated using the Feynman diagram approach.

Fortunately, there is a regime where strong interactions actually have a small coupling strength, namely in any situation where the particles have very high relative momenta. This allows us at least some cases where we can compare the predictions of the

strong interaction theory with measurements. Without a direct comparison of our theory with experiment we would be unable to call it more than a speculation. Its success in predicting the outcome of many high-energy experiments makes it instead a part of the well-tested standard model of particle processes.

See also: ANTIMATTER; INTERACTION, ELECTROMAGNETIC; INTERACTION, STRONG; INTERACTION, WEAK; PHOTON; QUANTUM ELECTRODYNAMICS; QUANTUM MECHANICS

Bibliography

FEYNMAN, R. P. *QED: The Strange Theory of Light and Matter* (Princeton University Press, Princeton, NJ, 1985).

HELEN R. QUINN

FIBER OPTICS

Fiber optics is the study of the transmission of light through optical fibers. This includes the development of the components for a superior communications system and also the fast emerging area of fiber optic sensor technology. The use of fiber optics has revolutionized communications. Scientific and technical advances have provided enhanced performance and, along with mass commercial production, have lowered the costs both of optical fibers and the necessary optoelectronic components. Fiber optics is now routinely used in communications systems, allowing data transmission rates of greater than 100 Mbits/s or 1,000 voice channels per fiber. The primary advantage of light-based communications stems from the high frequency of the electromagnetic waves. The rate at which information can be transmitted increases with increasing frequency of the carrier. The frequency of visible light is about 10^{14} Hz while that for microwaves, for example, is around 10^{10} Hz, so a visible light system can offer a 10,000-fold improvement in information carrying capability over a microwave system. Other advantages over conventional coaxial-cable-based systems come from the very small diameters of the optical fibers, 10 to 50 μm, which are comparable to, or less than, that of a human hair. This size saves space, reduces weight, and enhances cable flexibility. Additionally, fiber optic systems do not leak electromagnetic radiation. Thus, they eliminate crosstalk and interference effects and offer signal security.

Light has always been recognized as a useful communications media. In 1880 Alexander Graham Bell is reported to have transmitted speech by reflecting sunlight off the diaphragm of one of his recently invented telephones. However, direct optical communication was hampered by the lack of a suitable light source and the problems of being restricted to line-of-sight and atmospheric transmission, which are severely affected by rain, snow, dust, and atmospheric turbulence, but its potential for achieving high transmission rates meant interest in light transmission persisted.

The invention of the laser in 1962 provided an excellent light source for optical communications. Proposals for optical communication based on light passing through optical fibers, to overcome the problems of direct transmission through air, soon followed. These were based on a principle, demonstrated in 1870 by the Irish scientist John Tyndall, that light could be guided inside the arc of a water spout. This principle was total internal reflection of the light from the inside surface of the water spout caused by the difference in index of refraction between water and air. In the same way light can be passed through thin threads, or fibers, of transparent material such as glass or some plastics.

A typical fiber optics communication line now consists of an optical fiber and optoelectronic components, such as a light transmitter, light receiver, and possibly some repeaters along the length of the fiber. The light transmitter is generally a semiconductor laser or light-emitting diode (LED). The intensity of the light output from a semiconductor LED or laser depends on the electrical current passing through it. So, fluctuations in an electrical signal can be easily transferred via the laser or LED to the intensity of the light beam in the fiber. The information can be carried either in analog form, where the light intensity varies over a continuous range, or digital form, where the light intensity has just two values, zero and one. The receiver is a semiconductor photodetector whose electrical output depends on the intensity of the incoming light. A repeater also has a photodetector whose output is amplified to drive a laser or LED and so accurately regenerate a strengthened version of the incoming light signal.

In communications systems the optical fiber is generally made from SiO_2 glass. A uniform surface for total internal reflection is provided by making the optical fiber from two concentric cylinders. The core is surrounded by a cladding that has a slightly (about 1%) lower index of refraction than the core. The index of refraction of the glass can be raised or lowered precisely by adding small, known concentrations of germanium or boron, respectively. When materials such as neodymium and erbium are added, the optical fiber can exhibit lasing. This offers the possibility of an all-optical communication system where signals are modulated and strengthened without first requiring conversion to electrical signals.

An optical fiber will always scatter and absorb some of the passing light. The magnitude of these inherent losses of light intensity depends on the material and quality of the optical fiber and the wavelength of the light. The ability to extrude thin fibers of very pure SiO_2 glass of uniform density and with no flaws has significantly reduced losses and has been crucial to the development of fiber optics. In SiO_2 the wavelength dependence of the loss and scattering effects combine so that they are at a minimum at 1,300 and 1,550 nm. Therefore, light sources operating at these wavelengths are most frequently used in optical communication systems. At 1,550 nm only 5 percent of the original light intensity will be lost after passage through one kilometer of the fiber.

Optical fibers are also used in medicine and engineering to view otherwise inaccessible regions of the human body or machinery. These "endoscopes" can also be used to deliver light, for surgery, to light-activate drugs or epoxies, or to collect light for spectroscopic chemical analysis.

A range of fiber optic sensors or transducers have been and are now being developed. Interferometric optical fiber sensors are used as gyroscopes and for acoustic and electric and magnetic field measurements. The effect to be measured produces a frequency, velocity, or path length change in light traveling in a fiber. The light waves from the fiber and a reference fiber, which does not experience the effect, are compared and the interference between the two allows the magnitude of the effect to be accurately determined. Other fiber optic sensors use changes in the intensity of light flowing in or reflected into a fiber to measure a range of parameters from temperature, pressure, position, and vibration to acceleration.

Fiber optics has provided the basis of today's high speed communication systems, making possible the voice, video, and data transmission rates required to fuel the information revolution. It also promises to revolutionize the way in which we monitor our bodies and our environment.

See also: COAXIAL CABLE; ELECTROMAGNETIC WAVE; INTERFERENCE; LASER; OPTICAL FIBER

Bibliography

HECHT, J. "Fibre Optics." *New Scientist* **128** (1738), 1–4 (1990).

HECHT, J. *Understanding Fiber Optics,* 2nd ed. (Sams, Indianapolis, IN, 1993).

SENIOR, J. M. *Optical Fiber Communications: Principles and Practice,* 2nd ed. (Prentice Hall, London, 1992).

UDD, E., ed. *Fiber Optic Sensors* (Wiley, New York, 1991).

UNGAR, S. *Fibre Optics: Theory and Applications* (Wiley, Chichester, Eng., 1990).

ZANGER, H., and ZANGER, C. *Fiber Optics: Communications and Applications* (Merrill, New York, 1991).

WILLIAM G. GRAHAM

FIELD

The term "field" is used in physics to denote a quantity given as a function of space and time associated with a particular physical property. The quantity may be a scalar, such as the pressure defined at every point. In aerodynamics and hydrodynamics the velocity vector $\mathbf{v}(x, t)$ of a fluid element describes the velocity field. The general deformation of a solid body can be described by a tensor, the strain field, at each point.

The most well known fields are force fields characterized by a vector at each point of space. The gravitational field gives the force that would act on a unit mass due to the attraction of all other masses. Similarly the electric (magnetic) field gives the force that would act on a unit charge (unit magnetic pole) due to other charges (magnets or currents).

Force fields at a given time can be pictured in terms of lines of force. Thus one may picture gravitational lines of force emerging radially in all directions from the surface of the sun. The decrease in the density of these lines can be associated with the decrease in the strength of the field in accordance with the inverse square law. In the case of a bar magnet the lines of force emerge from one pole and curve around and enter the opposite pole. In this case the lines of force in two dimensions can be visualized using iron filings. The force field at any point in space due to a set of point sources can be calculated by taking the vector sum of the force fields due to each source separately.

The concept of field, when introduced in electrostatics, appears at first rather useless. One can just as well describe the forces in terms of Coulomb's law without ever introducing the electric field. The true importance of the field concept becomes clear with the introduction of Maxwell's equations. These are partial differential equations from which the electric and magnetic fields as functions of space and time can be derived from specified currents and charge distributions. Of particular importance are the solutions that correspond to waves propagating with the velocity of light that occur as a result of an oscillating charge. These are the electromagnetic waves that have a spectrum that goes from radio waves to light waves to x rays as the frequency is increased.

While the wave character of light was established in the early part of the nineteenth century, the composition of the wave was unknown. For a long time it was considered as a wave in a medium called the ether that filled all space. The Michelson–Morley experiment and the special theory of relativity ended the possibility of an ether. It became clear that light and other electromagnetic waves were nothing else than time-varying electric and magnetic fields in the vacuum. The concept of the field is absolutely essential.

The general theory of relativity is a theory of gravitation in which the field concept plays the essential role. The gravitational field is described by a tensor that is associated with the curvature of spacetime. There exist time-dependent solutions of the field equations that correspond to gravitational waves propagating with the velocity of light.

The transition in physics from the emphasis on matter to the emphasis on fields is of great importance for modern physics. Thus some physicists have raised the possibility that all physics could be expressed in terms of fields:

Could we not reject the concept of matter and build a pure field physics? What impresses our senses as matter is really a great concentration of energy into a comparatively small space. We could regard matter as the regions in space where the field is extremely strong. . . .

There would be no place in our new physics for both field and matter, field being the only reality [Einstein and Infeld, 1938, pp. 242–243].

A somewhat different use of the field concept emerges in quantum electrodynamics (QED) or more generally in quantum field theory. In QED the term "field vector" is used for the four-vector $A_\mu(x, t)$, corresponding to the classical vector potential. When the electromagnetic field is quantized the field vector A_μ becomes an operator that creates and annihilates photons. There is also a four-component Dirac spinor field $\psi(x, t)$ describing electrons. This becomes an operator that annihilates electrons and creates positrons. (ψ^* annihilates positrons and creates electrons.) More generally, in quantum field theory there also exist scalar fields that describe spin-zero particles such as pions.

The use of the term "field" to describe other functions of space and time in physics is somewhat arbitrary. In ordinary quantum mechanics the function $\psi(x, t)$ is simply referred to as the wave function. On the other hand the related quantity in relativistic field theory is referred to as a field. It should also be noted that the term "field" is used in a completely different sense in formal mathematics.

See also: GRAVITATIONAL WAVE; MAXWELL'S EQUATIONS; QUANTUM ELECTRODYNAMICS; QUANTUM FIELD THEORY; RELATIVITY, GENERAL THEORY OF; TENSOR

Bibliography

ADAIR, R. K. *The Great Design: Particles, Fields, and Creation* (Oxford University Press, New York, 1987).

BERKSON, W. *Fields of Force* (Wiley, New York, 1974).

EINSTEIN, A., and INFELD, L. *The Evolution of Physics* (Simon & Schuster, New York, 1938).

LANDAU, L. D., and LIFSHITZ, E. M. *The Classical Theory of Fields*, 4th rev. English ed. (Pergamon, Oxford, 1975).

SOPER, D. E. *Classical Field Theory* (Wiley, New York, 1976).

LINCOLN WOLFENSTEIN

FIELD, ELECTRIC

When an object with a nonzero electric charge is placed near another charged object, we observe that the two objects exert a force on one another. If we remove the charge from either of the objects, this force vanishes. It is therefore logical to conclude that this force is of electrical origin and that it results from the charge on one object as sensed by the charge on the other. Here we are concerned with electrostatics, that is, electric forces between charged particles arising only from their relative locations. Forces due to the motion of charged objects fall within the realm of magnetic phenomena.

Originally, this force was regarded as a direct and instantaneous interaction between the two objects. Any change or movement in one charge was, according to this view, instantaneously felt by the other as a change in the force. This view is not consistent with the theory of relativity, according to which no signal can propagate at a greater speed than the speed of light c.

The modern view of this interaction is based on the field concept. According to this view, one charge establishes an electric field and the other charge interacts with that field. The field serves as the intermediary between the two charges—instead of two charges interacting directly with one another, each interacts with a field that is established by the other.

This point of view traces its origin to Michael Faraday, whose nonmathematical approach to electric forces led him to visualize the forces by means of flux lines, the tangent to which indicates the direction of the force at any particular point. The magnitude of the force is represented by the density of flux lines: the force is large where the lines are close together and small where they are far apart. We still use such figures today as convenient graphical representations of the electric field, although we have supplemented Faraday's visual techniques with a formal mathematical analysis that was given its most coherent and modern formalism by James Clerk Maxwell. However, Faraday's influence on the evolution of the field concept is still manifest in that the field vectors were called by Maxwell "flux densities," a name that persists today.

The problem of analyzing the electrostatic effect of any collection of charges breaks down into two steps: (1) calculating the electric field at any location due to the collection of charges and (2) calculating the force due to the field on a charge at that location.

The electric field strength or intensity at any location is represented by the vector $\mathbf{E}$ and can be measured from the force on a positive test charge at that location. The magnitude of $\mathbf{E}$ is determined by the magnitude of the force per unit test charge, and the direction of **E** is determined by the direction of the force on the test charge, or

$$\mathbf{E} = \frac{\mathbf{F}}{q_0}, \qquad (1)$$

where q_0 represents the test charge. Equation (1) can be regarded as the defining equation for the electric field. The test charge is usually specified to be small, so that it does not change the field in which it is placed; Eq. (1) is often written in the limit in which the test charge becomes vanishingly small.

Dimensionally, the electric field has units of force/charge, such as newton/coulomb (N/C); however, the field is more often expressed in the equivalent units of volts/meter (V/m). Representative values of the strengths of various electric fields range from 10^{21} V/m near the surface of an atomic nucleus, to 10^9 V/m near the outer electronic orbits in atoms, to 10^5 V/m in devices such as television picture tubes, to $10^2 - 10^3$ V/m near an object such as a plastic comb that has acquired a static charge.

Point Charges

Using Coulomb's law and Eq. (1), we find the electric field at a distance r from a point charge q to be given in magnitude by

$$E = \frac{1}{4\pi\varepsilon_0} \frac{q}{r^2}, \qquad (2)$$

where the combination of constants $1/4\pi\varepsilon_0$ has the numerical value of approximately 8.99×10^9 N·m²/C². The constant ε_0, called the permittivity constant, has the exact value $\varepsilon_0 = 8.85418781762 \times 10^{-12}$ C²/N·m². The direction of the field is radially outward if the point charge is positive, as shown in Fig. 1. If q is negative, the field has the same magnitude but the opposite direction.

The field due to a collection of point charges is obtained by the (vector) addition of the fields due to the individual charges, treating each as if the others were not present:

$$\mathbf{E} = \mathbf{E}_1 + \mathbf{E}_2 + \mathbf{E}_3 + \ldots. \qquad (3)$$

For example, we can find the electric field due to the combination of a positive and a negative charge

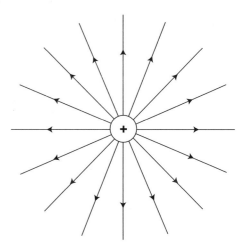

Figure 1 Electric field lines from a positive point charge.

of equal magnitude, as shown in Fig. 2. This combination, called an electric dipole, has important applications in nuclear, atomic, and molecular processes. At large distances from an electric dipole, the electric field varies as r^{-3}, in contrast to the r^{-2} variation that characterizes the single point charge. Figure 2 shows the field lines associated with an electric dipole. Note that the field lines emanate from the positive charge and terminate on the negative charge.

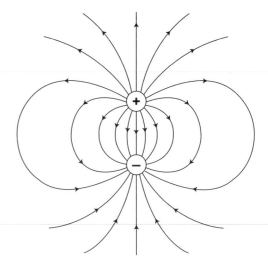

Figure 2 Electric field lines from a dipole (positive and negative charges of equal magnitude).

Continuous Distributions

When the charges are distributed continuously, Eq. (2) cannot be used to calculate the electric field. In this case the techniques of integral calculus must be used. The charge distribution is divided into infinitesimally small elements dq, and the magnitude dE of the contribution to the electric field at the observation point due to each charge element is determined as if it were a point charge:

$$dE = \frac{1}{4\pi\varepsilon_0} \frac{dq}{r^2}. \tag{4}$$

The direction of the field element $d\mathbf{E}$ is determined by the direction of the force due to dq on a positive test charge at the observation point. The net electric field is found by integrating (summing) over all charge elements dq, taking care that the directions of the individual field elements $d\mathbf{E}$ are accounted for:

$$\mathbf{E} = \int d\mathbf{E} = \frac{1}{4\pi\varepsilon_0} \int \frac{dq\,\hat{\mathbf{r}}}{r^2}, \tag{5}$$

where $\hat{\mathbf{r}}$ is a unit vector along the direction from dq to the observation point.

Using this method, it is possible to find the electric field for a variety of continuous charge distributions. For example, at locations close to a uniform linear charge distribution, in which a total charge q is distributed uniformly on a long, thin, straight string of length L, the electric field is directed radially outward (if q is positive; otherwise, inward) and has magnitude:

$$E = \frac{q}{2\pi\varepsilon_0 rL}, \tag{6}$$

where r is the distance from the string (we assume $r \ll L$).

An important case of practical interest is the large uniformly charged sheet. Suppose the sheet, of area A, carries a total charge q uniformly distributed over its surface, so that the surface charge density $\sigma = q/A$ is constant over the surface. Then, at points close to the sheet and far from any edge, the electric field is

$$E = \frac{\sigma}{2\varepsilon_0} = \frac{q}{2\varepsilon_0 A}. \tag{7}$$

Note that this result is independent of the distance from the sheet (as long as the distance is small compared with the dimensions of the sheet). The field lines point outward from the sheet if the charge is positive and inward if it is negative.

A practical application of this result occurs for the parallel-plate capacitor, which consists of parallel sheets of positive and negative charge, separated by a distance that is small compared with their dimensions (Fig. 3). The field in the interior is uniform in both magnitude and direction, which provides the experimenter with a useful way of creating a field with minimal spatial variation. If the capacitor is connected to a battery that maintains a potential difference V, then the magnitude of the electric field in the interior is

$$E = \frac{V}{d}, \tag{8}$$

where d is the separation between the plates. Near the edges of the capacitor, the field loses its uniformity, and the fringing field extends beyond the edges of the plates.

Induced Electric Fields

So far we have discussed only electrostatic fields, which are associated with spatial distributions of charge. It is also possible to produce an electric field in another way, through the action of a changing magnetic field. Such fields are called induced electric fields.

If we place a loop of wire in a constant magnetic field, there is no current in the loop. However, if the

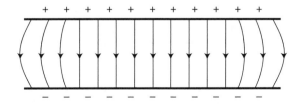

Figure 3   Electric field lines from a parallel-plate capacitor.

magnetic field changes with time, charge flows in the loop. For charges to be accelerated into motion, there must be an electric field present. This induced electric field, which results from the time-varying magnetic field, would exist in the region of the changing magnetic field even if the loop of wire were not present. Any test charge placed in that region would experience an electric force $q_0\mathbf{E}$ due to the induced electric field. Because the field does not result from charges, the lines of the field form closed loops, because there are no charges on which the field lines can terminate. The electric field that results from the variation in a magnetic field can be computed from Faraday's law of electromagnetic induction.

Effects on Point Charges

When a point charge q is placed in an electric field $\mathbf{E}$, it experiences a force that can be computed according to Eq. (1):

$$\mathbf{F} = q\mathbf{E}. \qquad (9)$$

The force is parallel to $\mathbf{E}$ if q is positive and antiparallel if q is negative.

This simple phenomenon gives rise to an enormous number of practical applications. In a television picture tube or an oscilloscope, a beam of electrons passes between the plates of a parallel-plate capacitor, in which the electric field is transverse to the motion of the electrons. The electric force deflects the electrons, which then leave an image when they strike a fluorescent screen. By changing the deflecting signal rapidly as the beam sweeps across the screen, a picture can be produced. An ink-jet printer works in a similar fashion—a beam of charged ink drops is deflected by a set of plates to form the letters on the page. An electrostatic precipitator removes particulate air pollutants from industrial exhausts by first charging the particles and then sweeping them out of the exhaust stream using an electric field.

Accelerators used for fundamental research in nuclear and particle physics also depend on electric fields to produce energetic particle beams. In an electrostatic (Van de Graaff) accelerator, a charged ion is accelerated once from a terminal at a potential of as much as twenty million volts. The acceleration takes place over tens of meters, so the electric field is of the order of 10^5–10^6 V/m. In a cyclic accelerator such as a cyclotron or synchrotron, particles circulate in a ring (under the influence of a magnetic field, which does not increase their energy) and are accelerated once or twice in each cycle by passing through a region in which there is an electric field along the direction of motion. An individual particle may make thousands of cycles before it reaches the maximum energy. An accelerator that uses induced electric fields is the betatron, which produces beams of electrons of energies up to 100 MeV that can be used to generate x rays for medical diagnosis and treatment.

If an electric dipole is placed in a uniform electric field, there is no net force on the dipole, since the forces on the (equal in magnitude) positive and negative charges are equal and opposite. However, as shown in Fig. 4, there can be a net torque on the dipole that tends to rotate it. We define the electric dipole moment $\mathbf{p}$ as a vector whose magnitude is equal to the product qd of the magnitude of either charge q and their separation distance d, and whose direction is from the negative charge to the positive one. The torque exerted by the field on the dipole is then

$$\tau = \mathbf{p} \times \mathbf{E}. \qquad (10)$$

The effect of this torque is to rotate the dipole so that $\mathbf{p}$ tends to align with $\mathbf{E}$.

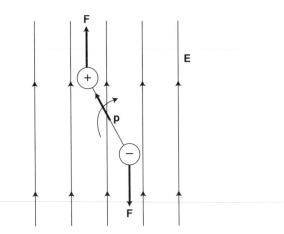

Figure 4 The forces on a dipole placed in a uniform external electric field.

Energy Storage in an Electric Field

If we connect a capacitor to a battery, the capacitor will become charged as current flows in the circuit. When the capacitor is fully charged, the current stops. Energy has been expended by the battery; some (but not all) of this energy has been dissipated in the wires of the circuit. The remainder of the energy must be stored somewhere, and it is reasonable to assume that the energy is stored in the electric field in the region between the capacitor plates. If the plates are released, they will accelerate toward one another as the stored energy is converted into kinetic energy. This confirms the hypothesis that the system has stored energy. We regard the energy as being stored in the electric field itself, throughout the region in which the field acts. At any point, the energy density (energy per unit volume) associated with a field **E** is

$$u = \tfrac{1}{2}\varepsilon_0 E^2. \tag{11}$$

Like the electric field, the energy density changes from point to point.

The total electrostatic energy of any charge distribution can be found by integrating the energy density over the entire region in which the field acts. This energy stored by the charges can equivalently be regarded as the energy that must be supplied by an external agent to construct the charge distribution from component charges located at infinite separations.

Electric Field in Matter

The behavior of matter when an electric field is applied determines whether the material is classified as a conductor, in which charges are free to move under the action of an electric field, or a nonconductor (or dielectric), in which charges are not free to move. We are interested in electrostatic situations, so we examine the material a long enough time after it has been placed in the field that the charges have settled into their static equilibrium configurations.

If we transfer a net charge to a conductor, we are in effect adding an excess of electrons to the material or taking electrons away from it. Electrons are free to move throughout the conductor. If there is an excess of electrons, the repulsive Coulomb forces between them result in their distribution with the largest possible separation; a net negative charge will then reside on the outer surface of the conductor. If there is a lack of electrons, the remaining electrons flow so that the equilibrium condition is a distribution of positive ions again on the surface. In either case, there is no net charge anywhere within the material of the conductor. Application of Gauss's law then shows that, under static conditions, the electric field must be zero everywhere inside a conductor.

If an uncharged conductor is placed in an external electric field, the same situation applies. The free electrons in the conductor move in the direction opposite to the field, giving a net excess of positive charge on one side of the conductor and a net excess of negative charge on the opposite side. These charges arrange themselves so that, in the interior of the conductor, the field due to the charges exactly cancels the original applied field, giving a net field of zero. Outside the conductor, the net field is the vector sum of the original applied field and the field due to the charges on the surface of the conductor (Fig. 5).

When a nonconductor is placed in an external field, such as between capacitor plates, no charges are free to move in the material. However, it often happens that the molecules of the material behave like electric dipoles. (They may have permanent electric dipole moments, or they may acquire temporary dipole moments induced by the applied field.) The applied electric field rotates the dipoles into alignment with the field. There are now two contributions to the electric field in the material: the original applied field $\mathbf{E}_{app}$ and the induced field $\mathbf{E}_{ind}$ due to the aligned dipoles, which can be regarded as due to the positive and negative surface

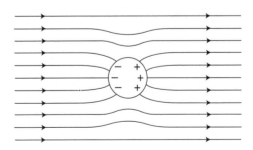

Figure 5 Electric field lines near an uncharged conductor placed in an originally uniform field.

charges on the two sides of the material. The net field **E** inside the material is

$$\mathbf{E} = \mathbf{E}_{app} + \mathbf{E}_{ind} \qquad (12)$$

or, in terms of magnitudes, $E = E_{app} - E_{ind}$.

The induced field is usually represented as the polarization field **P,** defined as the net electric dipole moment per unit volume and computed from the vector sum of the fields due to all the aligned dipoles in the material. Dimensionally (in SI units) the polarization field is related to the electric field by $\mathbf{P} = -\varepsilon_0\mathbf{E}_{ind}$; the minus sign occurs because the dipole moment vector is defined as pointing from the negative charge to the positive charge, while the electric field vector is defined in the opposite sense. The externally applied field is usually represented in terms of the electric displacement field **D,** which also differs in units and dimensions from the electric field, such that $\mathbf{D} = \varepsilon_0\mathbf{E}_{app}$. In this notation, the net field, including both the displacement (applied) field and the polarization (induced) field, is written

$$\mathbf{E} = \frac{1}{\varepsilon_0}(\mathbf{D} - \mathbf{P}). \qquad (13)$$

In Eq. (13), **D** is the field due to "free" charges (such as those on capacitor plates surrounding the nonconductor), **P** is the field due to "bound" charges (such as those on the surface of the nonconductor), and **E** is the field due to all charges.

Since the dipoles are aligned by the action of the applied field, the polarization field **P** is a result of the displacement field **D,** and so a functional relationship between **D** and **P** is expected. For some materials, **P** varies linearly with **D,** but in general there is not a simple relationship between **D** and **P.** In fact, the two vectors may not even be parallel.

See also: ACTION-AT-A-DISTANCE; COULOMB'S LAW; FARADAY, MICHAEL; ELECTROMAGNETIC INDUCTION, FARADAY'S LAW OF; FIELD, MAGNETIC; GAUSS'S LAW; MAXWELL, JAMES CLERK; VAN DE GRAAFF ACCELERATOR

Bibliography

GRIFFITHS, D. J. *Introduction to Electrodynamics,* 2nd ed. (Prentice Hall, Englewood Cliffs, NJ, 1989).

HALLIDAY, D.; RESNICK, R.; and KRANE, K. *Physics,* 4th ed. (Wiley, New York, 1992).

WILLIAMS, L. P. *Michael Faraday* (Basic Books, New York, 1965).

KENNETH S. KRANE

FIELD, GRAVITATIONAL

The concept of field was initially introduced in connection with electromagnetic phenomena. In the case of electromagnetism, one could, for example, ask, "How does a charged particle exert its influence on another charged particle when they are separated in space and not in contact?" A similar question arose earlier in connection with Isaac Newton's theory of gravitation. It could be that both electromagnetic and gravitational forces were examples of action-at-a-distance, the idea that one object can influence another at a distance without direct contact. Physicists, however, found the concept of action-at-a-distance quite unacceptable. Newton himself wrote: "That one body may act upon another at a distance through a vacuum without the mediation of anything else, by and through which their action and force may be conveyed from one to another, is to me so great an absurdity, that I believe no man, who has in philosophical matters a competent faculty of thinking, can ever fall into it" (Smoot and Davidson, 1993, p. 31).

It was the English experimental physicist Michael Faraday who found a way of replacing action-at-a-distance in electromagnetic phenomena with the concept of a force field. In the field view, one charged particle sets up its force field in space—that is, space itself becomes imbued with the presence of a force field—and another charged particle responds directly to the force field at its own position. The interaction between charged particles thus becomes a direct, or "local," interaction.

In the field description of gravitation, the gravitational force likewise becomes a direct interaction. One object sets up a gravitational field in space, and another object responds directly to the gravitational field at its own position. The gravitational field produced by an object of mass M assumes a value at a distance R given by

$$g = \frac{GM}{R^2},$$

where g denotes the gravitational field and G is the universal gravitational constant. Another object of mass m, located at R from the source of the field, then experiences a gravitational force F given by

$$F = mg = \frac{GMm}{R^2}.$$

This is, of course, Newton's law of universal gravitation. The concept of the gravitational field thus replaces action-at-a-distance with the notion of a local, or direct, interaction.

In order to further explore the meaning of the gravitational field, let us recall Newton's second law of motion,

$$F = ma,$$

which implies that if a force F is applied to an object of mass m, the object accelerates with an acceleration a, which is proportional to F and inversely proportional to m.

Because mass in this case is a measure of inertia, that is, the resistance to a change in velocity, it is also referred to as the inertial mass, denoted by m_I. When an object of mass m responds to a gravitational field g, on the other hand, this mass is referred to as the "gravitational" mass, denoted by m_G. With this distinction of inertial and gravitational masses, the gravitational force acting on m_G can now be rewritten as

$$F = m_G g,$$

where g is the gravitational field at the position of m_G. The acceleration of the object in response to such a force, on the other hand, can be expressed as

$$F = m_I a.$$

Equating these two expressions yields a very interesting result:

$$m_G g = m_I a.$$

We can now ask the next crucial question: How does the gravitational mass m_G of an object compare with its inertial mass m_I? This is essentially an experimental question. Numerous experiments have been performed, and their results indicate that the gravitational and inertial masses are, in fact, equal to within better than 1 part in 10^{12}. If we assume that they are indeed equal, then we have the following very important result:

$$g = a.$$

That is, when an object falls in a gravitational field g, its acceleration a simply equals the gravitational field at the position of the object.

This equality has a rather astounding implication: since the acceleration of an object falling under gravity is independent of its mass, all objects must fall at the same rate under gravity. This is just what Galileo is supposed to have demonstrated from the Leaning Tower of Pisa. This fact actually turns out to be a unique property of gravity; only the gravitational force accelerates all objects at the same rate. Under an electric force, for example, two equally charged objects of different masses accelerate at different rates.

The equality of g and a further implies that an object in a freely falling frame of reference would be weightless. For example, a space shuttle orbiting Earth in response to Earth's gravitational field g accelerates toward Earth's center at a. Suppose an astronaut drops an apple in such a spacecraft. The apple would accelerate toward Earth's center at the same rate as the astronaut and the spacecraft. Consequently, the astronaut would observe the apple to be at rest, and would thus conclude that the apple is free of external forces (and hence weightless). Indeed, the orbiting space shuttle provides just such a forcefree environment, in which the gravitational force mg is canceled by the inertial force ma.

Earth is another example of a freely falling reference frame. Earth orbits the Sun in response to the Sun's gravitational field g_S and accelerates toward the Sun at a_S. We are thus in free fall, and do not feel the Sun's gravitational pull. We would be weightless except for the fact that we experience Earth's gravitational pull, and to this alone we owe our weight.

The fact that an object in a freely falling reference frame is weightless further implies that gravity and acceleration produce similar effects; that is, gravity can be canceled by accelerated motion. Albert Einstein took this as the foundation of his theory of gravitation, general relativity. Called the principle of equivalence, it can be stated thus: The effects of gravity are indistinguishable from the effects of acceleration.

Let us now work out a simple implication of the principle of equivalence. Consider a spacecraft coasting in gravity-free space, far from all gravitating bodies. In such a reference frame, an astronaut would clearly be weightless. Moreover, if the astronaut shone a laser beam, it would follow a straight-line path. The situation, however, changes dramatically if the spacecraft accelerates. The astronaut would no longer be weightless, and the laser beam would no longer follow a straight-line path but instead a curved path. The principle of equivalence would then imply that in a gravitational field, not only would the astronaut have weight but light would follow a curved path.

Einstein went one step further in analyzing the bending of light by a gravitational field. Let us assume that light follows the shortest path between any two points. If it follows a curved path in a gravitational field, then we must surmise that the shortest path in a gravitational field is a curved path. Since light travels on the three-dimensional "surface" of space, space itself would have to be curved if the shortest path in space is to be a curved path. (If one considers only the two-dimensional surface of Earth, the shortest path on its curved surface is certainly a curved path.) Einstein thus deduced that space would be curved by a gravitational field.

The gravitational field of Earth is too weak to bend the path of light to any appreciable degree. The Sun's gravity, by contrast, is much stronger. Einstein showed in 1916 that starlight passing by the surface of the Sun would be bent by 1.75 seconds of arc. The bending of light by the Sun's gravitational field was first observed in 1919 by Arthur Eddington during a total solar eclipse.

A contracting body provides a particularly interesting example of the curving of space. The gravitational field of a body increases on its surface as its size decreases, and thus light passing by such an object suffers a large degree of bending. In fact, when a body of mass M contracts to a radius called the Schwarzschild radius R_S given by

$$R_S = \frac{2GM}{c^2},$$

its gravitational field curves space so as to close in on itself, and light cannot escape such a body. Such a body is called a black hole.

In relativity, space and time are inseparable and form a single entity called spacetime. Gravity must therefore be viewed as a curvature of not only space but also of time. Curved time is also called "warped" time, and means that gravity slows down time. An astronaut approaching a black hole, for example, would age much more slowly as observed from Earth. In the modern view of gravity, then, the gravitational field emerges as geometrical property of spacetime.

See also: ACTION-AT-A-DISTANCE; BLACK HOLE; FIELD, ELECTRIC; GRAVITATIONAL ATTRACTION; GRAVITATIONAL CONSTANT; GRAVITATIONAL FORCE LAW; NEWTON'S LAWS; RELATIVITY, GENERAL THEORY OF; SPACETIME; WEIGHT; WEIGHTLESSNESS

Bibliography

FEYNMAN, R. P. *The Character of Physical Law* (MIT Press, Cambridge, MA, 1965).

NARLIKAR, J. V. *The Lighter Side of Gravity* (W. H. Freeman, San Francisco, 1982).

SMOOT, G., and DAVIDSON, K. *Wrinkles in Time* (Morrow, New York, 1993).

SUNG KYU KIM

FIELD, HIGGS

Forces govern the interactions of matter. Most forces, like friction, are not elementary but result from the cooperative action between bodies as a response of more fundamental forces. At present we know of only four such fundamental forces. Two of these, gravity and electromagnetism, were recognized early on because they have macroscopic manifestations. The other two, the strong and the weak force, became apparent only after probing matter at the subatomic level. It is possible that other fundamental forces will reveal themselves as we probe matter to even shorter distances. Indeed, attempts to unify the known forces at short distances, in so-called grand unified theories (GUTs), always require the existence of some such additional forces.

Remarkably, all the known fundamental forces are associated with symmetry transformations. Gravitation, as first recognized by Albert Einstein, is deeply connected to general coordinate transformations of space and time. The symmetries underlying the other three forces are more abstract, reflecting invariance of the laws of nature under specific transformations among the elementary force carriers—the quarks and leptons. Each of the quarks carries an attribute (called, whimsically, color) which takes three different values, so that an up quark u really is a triplet of objects: $u = \{u_1, u_2, u_3\}$. The strong force is associated with the invariance of the laws of nature under, essentially arbitrary, reshufflings at any spacetime point of the quark triplets among each other. [Technically, the strong force is connected to unitary transformations of unit determinant belonging to the group $SU(3)$]. The electromagnetic and weak forces jointly are combined into an electroweak force and are associated with similar arbitrary reshufflings of pairs of quarks and leptons among each other. For instance, the up quark and the down quark form an electroweak doublet: (u, d). (Technically, the transformations connected with the electroweak force belong to the group $SU(2) \times U(1)$, where the extra $U(1)$ factor reflects an additional freedom of phase transformations.)

Theoretically, the freedom of performing arbitrary spacetime dependent transformations (local transformations) among the quarks and leptons necessitates the introduction of compensating fields, known as gauge fields, whose interactions are totally fixed by the symmetries. For each independent parameter of the group of transformations there is a separate gauge field. Thus there are eight gauge fields associated with the strong force and four gauge fields connected with the electroweak force. Pictorially, one can think of the gauge fields as the entities that mediate the fundamental forces. Indeed, this is a familiar interpretation for electromagnetism, where the force among charged particles results from the interaction of the electromagnetic field with the charged particle currents.

The theory of the strong force that binds quarks together to form protons, neutrons, and other strongly interacting particles (hadrons) is known as quantum chromodynamics (QCD). The electroweak interactions are described by a theory developed in the 1960s by Sheldon Glashow, Abdus Salam, and Steven Weinberg. Both QCD and the electroweak theory are gauge theories, with their respective gauge fields mediating the forces that act on the quarks and leptons. However, there is a significant difference between how these theories are realized in nature. The $SU(3)$ symmetry of QCD is exactly preserved, with the color interactions becoming so strong at large distances among color carrying objects—the quarks and the $SU(3)$ gauge fields—that these excitations are permanently confined. In contrast, the $SU(2) \times U(1)$ electroweak symmetry is spontaneously broken to an overall phase symmetry. This remaining $U(1)$ symmetry and its gauge field are associated with electromagnetism, while the three spontaneously broken symmetries and their gauge fields are connected to the weak force.

The gauge fields of local symmetry transformations have particle-like excitations associated with them. If the symmetry is manifestly realized, the symmetry forces these particles to have zero mass. However, if the symmetry is spontaneously broken, these gauge particles can acquire mass. This phenomena, known as the Higgs mechanism, was proposed theoretically in the early 1960s. The spontaneous breakdown of the $SU(2) \times U(1)$ electroweak symmetry to the $U(1)$ symmetry of electromagnetism is the most spectacular practical manifestation of the Higgs mechanism. The photon—the gauge particle associated with the unbroken symmetry—is massless and is responsible for the long-range Coulomb force among charged particles. In contrast, the $W^{\pm}$ and Z bosons—associated with the spontaneously broken electroweak symmetries—acquire a very large mass and are responsible for the short-range nature of the weak force.

For a symmetry of nature to suffer spontaneous breakdown it is necessary that the dynamics be such that some direction in symmetry space is preferred. An example is provided by superconductivity, where the interactions of the electrons with the lattice produce a superconducting ground state containing pairs of electrons with opposite spin. These Cooper pairs, being condensates of two electrons, break electron number spontaneously since they carry electron number equal to two, not zero. Analogous phenomena are necessary for the spontaneous breakdown of any other symmetry, with the formation of some condensates by the underlying dynamics triggering the symmetry breakdown.

Not all broken symmetries, however, necessitate the formation of condensates of fermions. In particular, the symmetry breakdown in the electroweak theory of Glashow Salam and Weinberg, necessary to give mass to the $W^{\pm}$ and Z bosons, is thought to arise as a result of a nonzero expectation value of a scalar field—called, appropriately, a Higgs field. It is easy to arrange the nonlinear interactions of this field so that dynamically an asymmetric ground state is favored. Hence, postulating the existence of such a self-interacting Higgs field provides the most economical mechanism for breaking down the electroweak symmetry. Whether this is truly the way nature works remains to be seen. However, there are ways in which future experiments can check the validity of these speculations.

To break down the symmetry of the electroweak theory, it turns out that a single Higgs field does not suffice. In fact, the minimal structure necessary involves a Higgs field that is a complex doublet, containing four independent fields. As a result of the symmetry breakdown, three of these four Higgs fields transmute themselves into the longitudinal polarization components of the spin-1 $W^{\pm}$ and Z bosons. This transmutation is the basis for the Higgs mechanism, since a massive spin-1 particle requires three polarization components and the gauge particles of the theory—like the photon—originally only had two. One of the Higgs fields, however, remains in the theory after the breakdown. The spin-0 particle associated with this leftover field is called a Higgs boson. Its experimental discovery would provide compelling evidence that the Higgs field is the underlying agent for the symmetry breakdown of the electroweak symmetry.

The electroweak theory determines the interactions of the Higgs boson to all other particles. However, the theory does not fix the Higgs boson mass since this mass depends on the unknown strength of the Higgs field self-interactions. The Higgs boson of the electroweak theory has been the subject of intense experimental search, but has not yet been discovered. High statistics data from the LEP e^+e^- collider at the CERN Laboratory in Geneva, Switzerland, give a bound for the Higgs rest mass $M_H c^2 > 60$ GeV. Even though this is an impressive limit—over sixty times the mass of the proton—it is still well below the value of around 800 GeV, which is the estimated theoretical upper bound for $M_H c^2$. Higgs bosons with masses as high as this should eventually be detectable in the Large Hadron Collider (LHC) under construction at CERN. This proton-proton collider should become operational around the year 2004.

See also: GAUGE THEORIES; GRAND UNIFIED THEORY; LEPTON; QUARK; SYMMETRY BREAKING, SPONTANEOUS

Bibliography

VELTMAN, M. J. G. "The Higgs Boson." *Sci Am.* **255** (Nov.), 76 (1986).

ROBERTO D. PECCEI

FIELD, MAGNETIC

When two charged objects are in motion, there can be an additional force between them that cannot be accounted for by Coulomb's law alone. This additional force disappears if the charge is removed from either object, and it also disappears if one of the objects is at rest. Furthermore, we find that the force varies in direct proportion to the magnitude of either object's velocity, that it always acts at right angles to the velocity, and that it varies with the relative orientation of the velocities, ranging from a maximum for one orientation to zero for another.

We analyze this situation in analogy with our analysis of the electric force between two static charges. The first moving charge sets up a field, and the second moving charge interacts with that field. We call such a field the magnetic field **B.** As with the electric field, the analysis of the magnetic forces due to charges in motion breaks down into two parts: (1) calculating the magnetic field at any location due to a collection of moving charges, and (2) calculating the force due to the field on a moving charge at that location.

The magnetic field strength or intensity at any location can be measured from the force on a moving test charge q_0. By varying the magnitude and direction of the velocity of q_0, we can measure the resulting force and thus map the magnetic field. Experimentally, we find that the magnetic force vanishes if the velocity of the charge is along one particular axis. This axis defines the direction of the magnetic field.

These experimental findings can be summarized in the force equation

$$\mathbf{F} = q_0 \mathbf{v} \times \mathbf{B}, \qquad (1)$$

where the symbol "$\times$" stands for vector product. The relative directions of **F, v,** and **B** are thus de-

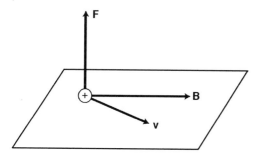

Figure 1 The relative directions of **F, v,** and **B** for a positive charge, as determined by the right-hand rule.

termined by the right-hand rule, as indicated in Fig. 1. The force **F** is perpendicular to the plane of **v** and **B.**

The field **B** is often called the magnetic flux density or the magnetic induction. It is becoming common practice to refer to **B** simply as the magnetic field, in analogy with the electric field **E.**

The modern concept of the magnetic field as the intermediary for magnetic interactions was originated by Michael Faraday, who developed a geometric means of visualizing the magnetic interactions through lines of force or flux lines. Faraday ascribed a reality to the lines of force that made them the agent of transmission of electric or magnetic forces. He regarded the lines as representing ropes in tension, such as rubber bands, in which the tension provides the means for transmitting the force through space. Although we do not give the lines of force the same reality that Faraday did, we still use them to represent a magnetic field, such that the tangent to the field lines gives the direction of the field and the density of field lines gives the intensity or strength of the field (which is the origin of the term "flux density" for the magnetic field vector). Faraday's ideas were given mathematical form by James Clerk Maxwell, who developed the field equations in the form in which they are presently used.

Dimensionally, the magnetic field has units of force/charge · velocity or (equivalently) force/current · distance, such as newton/ampere · meter. This unit is known as the tesla (T) in the SI system. A non-SI unit frequently used is the gauss, equal to 10^{-4} T. Commonly encountered magnetic fields range from a few T in the vicinity of a superconducting magnet, to 10^{-2} T near a small bar magnet, to 10^{-4} T at Earth's surface.

Point Charges

The development of the formalism for magnetic fields does not quite parallel that for electric fields because of one significant difference: isolated magnetic "charges," known as magnetic monopoles, do not appear to exist. If monopoles did exist, we could write a force law similar to Coulomb's law and define the magnetic field for a monopole in similar fashion to the electric field of a point charge.

Instead, we can find the magnetic field of an electric charge q moving at velocity **v:**

$$\mathbf{B} = \frac{\mu_0}{4\pi} \frac{q\mathbf{v} \times \hat{\mathbf{r}}}{r^2}, \tag{2}$$

where $\hat{\mathbf{r}}$ is a unit vector in the direction from the charge q to the observation point (Fig. 2). The direction of **B** is perpendicular to both **v** and $\hat{\mathbf{r}}$, determined by the right-hand rule. The permeability constant μ_0 has the exact value $4\pi \times 10^{-7}$ T·m/A.

Force on Point Charge

A point charge moving in a magnetic field experiences a force given by Eq. (1). This force always acts at right angles to the path of the moving particle. If the field is uniform, the force is of constant magnitude, and as a result the path of the particle is a circle of radius

$$r = \frac{mv}{q\mathbf{B}}. \tag{3}$$

The plane of the circle is perpendicular to the field direction. A velocity component along the field di-

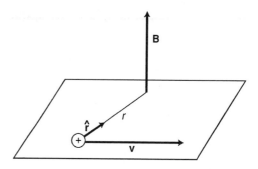

Figure 2 The magnetic field at the observation point P due to a positive charge moving with velocity v.

rection is unaffected by the field; if there is such a component, the motion of the particle is a combination of a constant drift along the field direction and a circular path at right angles to it, that is, a spiral.

The circular motion of a particle in a magnetic field is the basis of the operation of the cyclotron. Spiral motion is found in devices that use magnetic confinement to contain a beam of charged particles; for example, magnetic mirrors or magnetic bottles are used to confine plasmas that are used to generate fusion power. A similar spiral motion characterizes the electrons and protons that are trapped in the Van Allen radiation belts by the magnetic field of Earth; this effect also causes the trapping of charged particles in the intense magnetic fields of a neutron star, and the acceleration of these particles is believed to account for the radiation bursts we identify with pulsars.

Force on Current Distributions

Since an electric current can be regarded as a stream of point charges, the force on another current distribution can be obtained by analogy with Eq. (1). Consider a current distribution in which a current i flows through a length $d\ell$ of a medium such as a conducting wire. The direction of $d\ell$ is that of a positive current. If a magnetic field $\mathbf{B}$ is present at the location of $d\ell$, the current element experiences a force given by

$$d\mathbf{F} = i\, d\ell \times \mathbf{B}. \tag{4}$$

In exact analogy with Fig. 1, the force acts perpendicular to the plane containing both $d\ell$ and $\mathbf{B}$, and the force vanishes if $d\ell$ and $\mathbf{B}$ are parallel.

If the current path is a straight line of length ℓ and the field is uniform over that length, the total force on the current-carrying element is

$$\mathbf{F} = i\,\ell \times \mathbf{B}. \tag{5}$$

The directional information contained in Eq. (5) indicates that parallel wires carrying currents in the same direction attract one another, while if the currents are in opposite directions they repel. The force is small—for two wires each carrying a current of 1 A and separated by a distance of 0.1 m, the force per meter of wire length is only 2×10^{-6} N,

about the weight of a grain of sand. Nevertheless, the force can be measured with great precision, and in fact such measurements are used to define the SI base unit for current, the ampere.

Field of Current Distributions

The magnetic fields due to various current distributions can be obtained in analogy with Eq. (2). Specifically, we again consider a current i flowing through a length $d\ell$ of its carrier (a conducting wire, for instance). The contribution of this current element to the magnetic field is

$$d\mathbf{B} = \frac{\mu_0}{4\pi}\,\frac{i\,d\ell \times \hat{\mathbf{r}}}{r^2}, \tag{6}$$

where, in analogy with Eq. (2), $\hat{\mathbf{r}}$ is in the direction from $d\ell$ to the observation point P. Equation (6) is known as the Biot–Savart law. Note the similarity between Eqs. (2) and (6)—with $i = dq/dt$, Eq. (6) reduces to Eq. (2). The total magnetic field at P due to current i is determined by summing (integrating) all contributions from the current elements:

$$\mathbf{B} = \int d\mathbf{B} = \frac{\mu_0}{4\pi} \int \frac{i\,d\ell \times \hat{\mathbf{r}}}{r^2}, \tag{7}$$

taking into account that different current elements may contribute to the field in different directions. Using this method, we can find the field due to a long straight current-carrying wire

$$B = \frac{\mu_0 i}{2\pi d}, \tag{8}$$

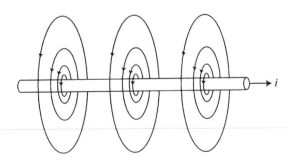

Figure 3 Magnetic field due to a long, straight, current-carrying wire.

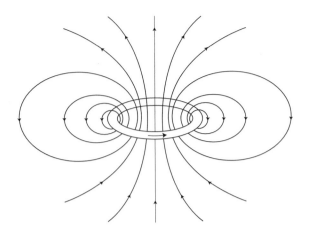

Figure 4 Magnetic field due to a circular loop of current.

where d is the distance from the wire to the observation point. We can also find the field due to a circular loop of current of radius R at a point on the axis of the loop and a distance d from its center:

$$B = \frac{\mu_0 i R^2}{2(R^2 + z^2)^{3/2}}. \qquad (9)$$

Figures 3 and 4 show the magnetic field lines for these two cases.

A close-packed helical winding of wire carrying a current i forms a solenoid. The magnetic field due to a solenoid can be found using the Biot–Savart law or using a simpler method based on Ampere's law, which, like Gauss's law for electrostatics, is useful for calculating the field in situations having a high degree of symmetry. The field of the solenoid is

$$B = \mu_0 i n, \qquad (10)$$

where n is the number of turns per unit length of the winding.

Figure 5 shows the field of the solenoid. Note that the field is approximately uniform inside the solenoid near its center. In this respect, the solenoid is the magnetic analog of the parallel-plate capacitor, which produced a uniform electric field at interior locations far from its edges. In the ideal solenoid, the field is uniform in the interior and negligibly small outside the windings.

If a solenoid is bent into a circle and the ends joined together, the resulting donut-shaped object is called a toroid. For an ideal toroid, the magnetic field vanishes except in the "cake" part of the donut ($B = 0$ at all points outside). In the interior, the field is

$$B = \frac{\mu_0 i N}{2\pi R}, \qquad (11)$$

where R is the distance from the center and N is the total number of turns of wire wound around the toroid. Note that, in contrast to the ideal solenoid, B is *not* constant in the interior of the toroid. Toroidal windings are used to confine hot plasmas in the fusion power research device known as the tokamak.

The Dipole Field

In the absence of a magnetic monopole, which would set up a magnetic field of the same form as the electric field of a point charge, the simplest magnetic field is the dipole field, established (in analogy with the electric dipole) by a pair of magnetic poles of opposite types, which we conventionally call north (N) and south (S) poles. Figure 6 shows the field of a simple bar magnet, a good example of an approximately dipole field, which varies with distance from the dipole as r^{-3}, in analogy with the field of the electric dipole. Just as the electric field lines emanate from positive charges and terminate on negative charges, the magnetic field lines (external to the magnet) emanate from north poles and terminate on south poles. Unlike electric field lines, which for a single point charge have no termination, all magnetic field lines that originate with a pole at

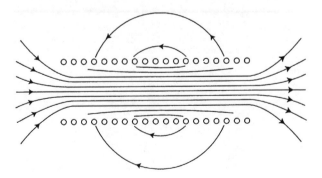

Figure 5 Magnetic field of a solenoid, representing a slice through the axis of the solenoid.

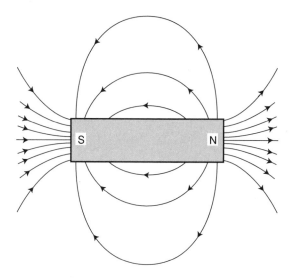

Figure 6 Magnetic field of a simple bar magnet.

one end must terminate with a pole at the other end. The magnetic field lines may be made visible with sprinkles of iron filings, as shown in Fig. 7.

The magnetic field of a circulating current loop (Fig. 4) has a similar shape and can also be regarded as a dipole field. As Eq. (9) shows, the field of a current loop also varies at large distances as the inverse cube of the distance from the loop. In fact, the field of a permanent magnet such as that of Fig. 6 can be regarded as originating from the alignment of the current loops of the electrons in the atoms of

Figure 7 Magnetic field lines made visible by sprinkles of iron filings.

the material. Circulating electrons in the atoms can be regarded as current loops, with an associated magnetic field. Note the similarity of the external field of the solenoid, Fig. 5, to that of the dipole magnet.

When two dipole magnets are placed close together, there is an attraction or a repulsion between the poles in analogy with the force between electric charges. The rule for magnetic poles is that like poles repel and unlike poles attract. That is, two N poles or two S poles repel one another, but a N pole and an S pole attract one another.

If a small dipole magnet, such as a magnetic compass, is placed in an external magnetic field, the effect of the field is to rotate the dipole into alignment with the external field. The N end of the magnet is attracted to the equivalent S end of the external field. We can define the magnetic dipole moment of the probe magnet as a vector that points from its S pole to its N pole. The effect of the external field is then to rotate the magnet such that its dipole moment vector becomes parallel with the externally applied field.

Induced Magnetic Fields

So far we have discussed only the magnetic fields established by moving charges. As proposed by Maxwell, it is also possible to produce a magnetic field through a changing electric field. For example, only an electric field exists under steady conditions in the interior of a charged parallel plate capacitor. If the capacitor is charging or discharging, the electric field in its interior changes with time; this changing electric field establishes a magnetic field in the interior of the capacitor. Because there are no magnetic poles there, the field lines cannot terminate but instead must form closed loops.

Again, note the symmetry between electric and magnetic fields: A changing magnetic field produces an electric field, and a changing electric field produces a magnetic field. Beyond a mere physical or mathematical symmetry, this coupling is responsible for the existence of electromagnetic waves.

Energy Storage in a Magnetic Field

If we hold two permanent magnets close to one another, the magnets exert forces on one another. If we release them, the attractive or repulsive force causes the magnets to move. Clearly, the system has acquired kinetic energy, and the source of that energy must be the magnetic field of the magnets. We regard this energy as being stored in all space in the magnetic field rather than at any particular site. The energy density (energy per unit volume) stored in the field **B** is

$$u = \frac{1}{2\mu_0} B^2. \tag{12}$$

The energy density may change from point to point as the magnetic field changes.

The total magnetic energy of any current distribution can be found by integrating the energy density over the entire region in which the field acts. Equivalently, the stored energy in any current distribution can be regarded as the energy that must be supplied by an external agent to assemble the current distribution from isolated current elements located at infinite separations. In effect, the work done by the external agent is stored as the energy of the field.

Magnetic Fields in Matter

The response of magnetic materials to an external field depends on the atomic structure of the material and whether the atoms can be regarded as permanent magnetic dipoles that can be aligned by the field. Even in "nonmagnetic" materials, the external magnetic field can induce a dipole behavior in atoms that are magnetically neutral. In some materials, the alignment of the dipoles is preserved even when the external field is removed.

The presence of the material modifies the field so that it is different from its value in the absence of the material. If we regard the atomic structure as being composed of magnetic dipoles, then the effect of the field is either to align the existing dipoles or to induce atomic dipoles that did not exist in the absence of the field. The sum of the fields of these aligned dipoles gives another contribution to the net magnetic field. In analogy with the behavior of electric fields, we can call this an induced magnetic field $\mathbf{B}_{ind}$, which adds to the applied field to give the net field:

$$\mathbf{B} = \mathbf{B}_{app} + \mathbf{B}_{ind}. \tag{13}$$

It is customary to represent the induced field as the magnetization field **M** of the material, computed as the net magnetic dipole moment per unit volume by taking the sum of the fields of all the atomic dipoles. In this notation and using SI units, $\mathbf{M} = \mathbf{B}_{ind}/\mu_0$. The externally applied field is usually represented as the magnetizing field **H**. (Unfortunately, **H** is sometimes also called the magnetic field intensity, which allows confusion as to the meaning of the various magnetic field vectors. This same nomenclature often refers to **B** as the magnetic induction, which overworks the term "induction" and obscures the nature of **B** as the fundamental magnetic field vector.) More unfortunately, it is also customary to assign to **M** and **H** dimensions and units different from those of **B,** so that the relation between these vectors is

$$\mathbf{B} = \mu_0(\mathbf{H} + \mathbf{M}). \tag{14}$$

In this interpretation, **H** is the field due to free currents (such as electrons flowing in conducting wires), **M** is the field due to induced currents (such as electrons circulating in aligned atomic dipoles), and **B** is the field due to all currents. These three magnetic vectors have their counterparts in the three electric field vectors.

It is usually not possible to give a simple relationship between the magnetizing field (applied field) **H** and the resulting magnetization **M**. Often **M** depends not only on the particular value of the applied field but also on the entire past magnetic history of the material. In some materials, **M** and **H** may not even be parallel.

Magnetic Fields and Relativity

If two charges are at rest in a certain reference frame, they exert a force of purely electric origin on one another. However, to an observer who moves at constant velocity in a direction perpendicular to the line joining the charges, there is a magnetic interaction between them. That is, an interaction that is purely electric in one frame of reference can appear partly magnetic in another frame of reference. Similarly, if an interaction is purely magnetic in one frame of reference, it can have both electric and magnetic components in another frame—the magnetic field measured at rest with respect to a current-carrying wire, for example, may appear to another observer in relative motion to be partly an electric field.

It is therefore tempting to regard a magnetic field as nothing more than an electric field that has been transformed to another frame of reference. The type of interaction that a charged particle might experience therefore depends on the reference frame of the observer. The mathematical formalism for transforming the fields from one reference frame to another is provided by the Lorentz transformation.

See also: AMPÈRE'S LAW; BIOT–SAVART LAW; CYCLOTRON; FIELD, ELECTRIC; FIELD LINES; FARADAY, MICHAEL; GAUSS'S LAW; INDUCTANCE; MAGNET; MAGNETIC FLUX; MAGNETIC MATERIAL; MAGNETIC MOMENT; MAGNETIC MONOPOLE; MAXWELL, JAMES CLERK; RIGHT-HAND RULE; SOLENOID; TOKAMAK

Bibliography

ELLIOTT, R. S. *Electromagnetics* (IEEE Press, New York, 1994).

HALLIDAY, D.; RESNICK, R.; and KRANE, K. S. *Physics,* 4th ed. (Wiley, New York, 1992).

WILLIAMS, L. P. *Michael Faraday* (Basic Books, New York, 1965).

KENNETH S. KRANE

FIELD LINES

Field lines are useful as an aid in visualizing electric and magnetic fields. While electric and magnetic fields exist, field lines are a convenient fiction. Electric and magnetic fields are fundamental, but field lines are not. In the past, electric field lines were sometimes referred to as lines of force.

Associated with the distribution of electric charges is an electric field. At any point in the space between charges, the electric field has both magnitude and direction; that is, the electric field is a vector field. With any combination of electric currents, there are magnetic fields that are also vector fields. Field lines provide a means of visualizing the direction and magnitude of either of these fields over an entire region; field lines are not intended for quantitative work.

In order to represent both aspects of the vector nature of the field, field lines are drawn taking into account two rules, one for direction and one for magnitude. The rule regarding direction must always be followed. The rule regarding magnitude is followed when an overall representation of the field is desired. However, it is possible to select a legitimate set of field lines that do not follow the second rule in all regions of the area displayed.

According to the direction rule, at any point along a field line, the tangent to the line gives the direction of the field. Typically, an arrowhead along the line shows the sense of the field. A single field line gives information about direction but not magnitude. Magnitude is represented by relative spacing of the field lines. Where the lines are closer together, the field is stronger; where the lines are farther apart, the field is weaker. Fields typically vary in three dimensions. Thus, according to the magnitude rule, the number of field lines per unit cross-sectional area perpendicular to the field gives a measure of the strength of the field. In any given field representation, there are only a few lines, otherwise the picture would be too cluttered. The field lines in any drawing represent a subset of the infinite set of lines that would be necessary to represent the field everywhere.

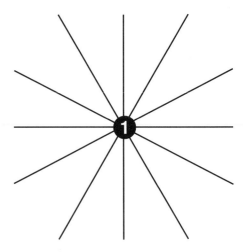

Figure 1 Electric field lines emerge from a positive point charge. The field radiates outward. The distance between field lines increases as the distance from the charge increases, which indicates that the magnitude of the field is decreasing with increasing distance from the source.

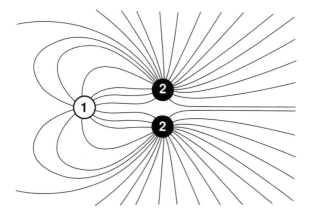

Figure 2 The electric field associated with three charges (+2, +2, −1). The field lines permit an immediate visualization of the field and the direction of the field at any point within the region. In addition, one gets a rough sense of the variation in magnitude as well, for example, how weak the field is in the region between the two positive charges.

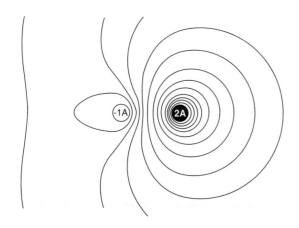

Figure 3 The magnetic field associated with two straight-line unequal-magnitude currents perpendicular to the page. Once current is into the page; the other is out of the page. Close to the line currents, the field lines are nearly circular, reflecting the minimal influence of the distant current. In the region between the line currents, the field is enhanced, while in the outer regions the field is reduced. The concentration of lines provides a qualitative sense of the magnitude of the field.

Electric field lines originate from positive charges and terminate on negative charges. Depending on the location and magnitude of the charges, the electric field can vary widely in direction and magnitude. But no matter how complex the arrangement of charges, at any given point in space, the electric field has only one magnitude and points in only one direction. Consequently, there can only be one field line through one point; field lines cannot intersect.

Magnetic field lines associated with electric currents do not begin and end as electric field lines do. Rather, they form closed loops. However, just as in electric fields, once all the contributing magnetic fields are superposed, the resultant field has a unique direction and magnitude, and the lines cannot intersect.

See also: FIELD, ELECTRIC; FIELD, MAGNETIC

Bibliography

FEYNMAN, R. P.; LEIGHTON, R. B.; and SANDS, M. *The Feynman Lectures on Physics,* Vol. 2 (Addison-Wesley, Reading, MA, 1964).

JACKSON, J. D. *Classical Electrodynamics* (Wiley, New York, 1975).

LORRAIN, P., and CORSON, D. R. *Electromagnetic Fields and Waves* (W. H. Freeman, San Francisco, 1970).

DENIS P. DONNELLY

FIELD THEORY

See ELECTROMAGNETISM; INTERACTION, ELECTROMAGNETIC; INTERACTION, ELECTROWEAK; QUANTUM CHROMODYNAMICS; QUANTUM ELECTRODYNAMICS; QUANTUM FIELD THEORY; QUANTUM MECHANICS AND QUANTUM FIELD THEORY

FIFTH FORCE

In recent years considerable theoretical and experimental effort have been devoted to formulating and testing theories whose objective is to unify the four known forces (strong, electromagnetic, weak, and gravitational). One outgrowth of these attempts has been the recognition that many of the proposed theories predict the existence of additional forces that are mediated by new ultra-light bosons, just as electromagnetic interactions arise from the exchange of photons. If the mass μ of any such boson is sufficiently small (typically $\mu \leq 10^{-4}$ eV/c^2), the force it produces will extend over macroscopic distances and can thus simulate gravity over these distances. In 1986 a specific example of such a force was advanced to explain the results of the Eötvös experiment and became known as the "fifth force." This name has since come to describe any of a variety of similar forces whose existence has been conjectured in the framework of unified theories.

Much of the current interest in new macroscopic forces such as the fifth force dates from the work of Yasunori Fujii in the early 1970s. He noted that the effect of such a force would be to modify the Newtonian expression for the potential energy $V(r)$ of two point masses m_1 and m_2 to read

$$V(r) = \frac{-Gm_1 m_2}{r}\left(1 + \alpha e^{-r/\lambda}\right), \qquad (1)$$

where $G = 6.67259(85) \times 10^{-11}$ m³kg⁻¹s⁻² is the Newtonian gravitational constant, r is the distance between the masses, and the constants $\lambda = \hbar/\mu c$ and α respectively describe the range of the new force and its strength relative to gravity. The more interesting physical quantity is the force $\mathbf{F}(r)$ and from Eq. (1) this is given by

$$\mathbf{F}(r) = -\boldsymbol{\nabla}V(r) = -G(r)\, m_1 m_2 \frac{\hat{r}}{r^2}, \qquad (2)$$

where $G(r) = G[1 + \alpha(1 + r/\lambda)e^{-r/\lambda}]$. It follows from Eq. (2) that in the presence of a new force ($\alpha \neq 0$), $G(r)$ is not a constant and hence the usual inverse square law of gravity no longer holds.

Fujii's work motivated a number of experimental searches for deviations from the inverse square law in the 1970s and 1980s. Most of these found no evidence for any such deviations, although Frank

Stacey and collaborators initially found indications of a variation in $G(r)$ in data obtained in mines. At present there is no evidence in such experiments for any new forces, although in principle a new force characterized by appropriate values of α and λ could still be compatible with existing data, as shown in Fig. 1.

In 1986 Ephraim Fischbach and collaborators noted that in some theories of macroscopic forces the parameter α in Eq. (1) was not a universal constant but depended on the chemical compositions of m_1 and m_2,

$$\alpha = -\xi(B_1/\mu_1)(B_2/\mu_2). \qquad (3)$$

Here B_1 and B_2 are the baryon numbers of the masses (the sum of the numbers of neutrons and protons), μ_1 and μ_2 are the masses (expressed in terms of the hydrogen mass), and ξ is a constant. The immediate implication of Eq. (3) is that two dissimilar objects falling in the Earth's gravitational field would accelerate at different rates and would thus appear to violate the weak equivalence principle (WEP). These authors went on to demonstrate that there was suggestive experimental evidence for just such a dependence on chemical composition in

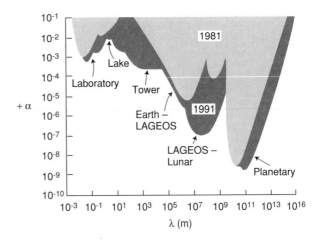

Figure 1 Results of tests of the inverse square law of gravity. Values of α and λ in the shaded region are excluded by experiment. The lighter shading illustrates the progress made in the decade between 1981 and 1991.

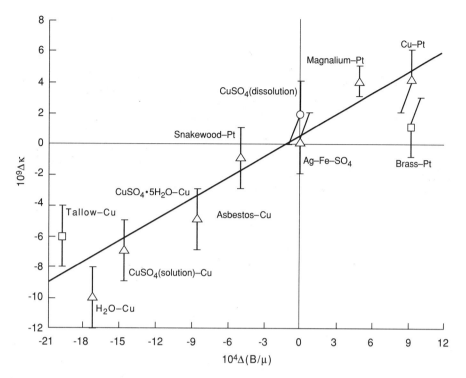

Figure 2 Results of the 1986 reanalysis of the Eötvös experiment by Fischbach and his colleagues. The linear pattern that emerges, when the fractional acceleration differences $\Delta\kappa = (a_1 - a_2)/a_1$ are plotted against the differences in baryon number-to-mass ratio $\Delta(B/\mu)$, lends support to the fifth force hypothesis.

data obtained by Baron Roland von Eötvös in 1922, who compared the accelerations of different pairs of materials (Fig. 2).

The Eötvös data, and the implication of a new fifth force, stimulated a large number of experimental searches for violations of the WEP, in which the accelerations of various pairs of test objects were compared. Many innovative experiments have been carried out, including a laboratory version of the famous (and perhaps apocryphal) experiment of Galileo comparing the accelerations of two balls dropped from the Leaning Tower of Pisa. The most extensive series of experiments to date has been carried out by Eric Adelberger and collaborators at the University of Washington. In their experiments, which are more accurate modern versions of the Eötvös experiment, the samples whose accelerations toward Earth (or another source) are being compared are suspended from fine torsion fibers. Any acceleration difference would appear as a torque on

the fiber. Although fractional acceleration differences of order 10^{-12} could be seen in this way, no evidence for any differences have yet been detected either by Adelberger or by other workers carrying out similar experiments. Thus, the suggestive result from the earlier Eötvös experiment has been ruled out by the more accurate results of these experiments. As in the case of tests of the inverse square law, the results of such experiments can be summarized in an exclusion plot (Fig. 3), in which the shaded region in the $\xi - \lambda$ plane is excluded by experiment.

There is at present no compelling experimental evidence for a breakdown of either the inverse square law or of the weak equivalence principle, and hence, there is no experimental support for the fifth force hypothesis. At the same time, theoretical arguments for such forces continue to be advanced, and they are serving to stimulate the continuing experimental searches for new macroscopic forces.

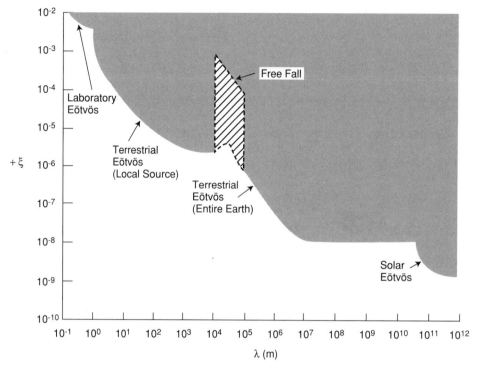

Figure 3 Results of searches for composition-dependent acceleration differences. Values of ξ and λ in the shaded region are excluded by experiment.

See also: GRAVITATIONAL ATTRACTION; GRAVITATIONAL FORCE LAW; INTERACTION, ELECTROMAGNETIC; INTERACTION, STRONG; INTERACTION, WEAK; INVERSE SQUARE LAW

Bibliography

ADELBERGER, E. G.; HECKEL, B. R.; STUBBS, C. W.; and ROGERS, W. F. "Searches for New Macrocopic Forces." *Ann. Rev. Nucl. Part. Sci.* **41**, 269–320 (1991).

FISCHBACH, E.; GILLIES, G. T.; KRAUSE, D. E.; SCHWAN, J. G., and TALMADGE, C. "Non-Newtonian Gravity and New Weak Forces: An Index of Measurements and Theory." *Metrologia* **29**, 213–260 (1992).

FISCHBACH, E.; SUDARSKY, D.; SZAFER, A.; TALMADGE, C.; and ARONSON, S. H. "Reanalysis of the Eötvös Experiment." *Phys. Rev. Lett.* **56**, 3–6 (1986).

FUJII, Y. "The Theoretical Background of the Fifth Force." *Int. J. Mod. Phys. A* **6** 3505–3557 (1991).

GIBBONS, G. W., and WHITING, B. F. "Newtonian Gravity Measurements Impose Constraints on Unification Theories." *Nature* **291**, 636–638 (1981).

HOSKINS, J. K.; NEWMAN, R. D.; SPERO, R.; and SCHULTZ, J. "Experimental Tests of the Gravitational Inverse Square Law for Mass Separations from 2 to 105 cm." *Phys. Rev. D* **32**, 3084–3095 (1985).

NIEBAUER, T. M.; McHUGH, M. P.; and FALLER, J. E. "Galilean Test for the Fifth Force." *Phys. Rev. Lett.* **59**, 609–612 (1987).

EPHRAIM FISCHBACH

FINE-STRUCTURE CONSTANT

The fine-structure constant is a fundamental constant that plays a very important role in atomic and particle physics. It is a dimensionless or pure number, approximately equal to 1/137, that is obtained by combining three other fundamental constants of nature: the elementary electric charge e; Max Planck's quantum of action divided by 2π $(h/2\pi = \hbar)$; and the speed of light c. The fine-structure constant is usually represented by the symbol α. In the Gaussian system, where e is measured in electrostatic units (esu),

$$\alpha = \frac{e^2}{\hbar c},$$

whereas in the SI system, where e is measured in coulombs ($e = 1.60 \times 10^{-19}\,\text{C}$),

$$\alpha = \frac{1}{4\pi\epsilon_0} \frac{e^2}{\hbar c},$$

where ϵ_0 is the permittivity of vacuum.

The fact that α is dimensionless implies that its numerical value is independent of the choice of units of physical quantities, such as length, time, mass, or electric charge. Arnold Sommerfeld introduced α in 1915 when he generalized Niels Bohr's model of the hydrogen atom to include the effects of Albert Einstein's theory of special relativity. The Sommerfeld calculation gives the following expression for the position of emission and absorption lines in the spectrum of hydrogen or of hydrogen-like ions of nuclear charge Z:

$$E_{n,k} = mc^2 \left[1 + \frac{(Z\alpha)^2}{(n - k + \sqrt{k^2 - (Z\alpha)^2})^2} \right]^{-1/2},$$

where m is the mass of the electron, n is a strictly positive integer quantum number, k is a nonzero integer quantum number, and the absolute value of k is smaller than or equal to n.

In particular this formula predicts the so-called fine structure of the spectrum, namely, the existence of several closely positioned spectroscopic lines, where the nonrelativistic Bohr model predicts only one line. The magnitude of the displacement of the lines with respect to their predicted location in the Bohr model can be seen when the Sommerfeld formula is expanded in a series in powers of α^2. The first term represents the rest energy of the electron, the second term is the Bohr energy, and the third term gives rise to the fine structure:

$$E_{n,k} = mc^2 \left[1 - \frac{(Z\alpha)^2}{2n^2} + \frac{(Z\alpha)^4}{n^4} \left(\frac{3}{8} - \frac{1}{2k} \right) + \dots \right].$$

The truncation of the series is made possible by the small numerical value of α^2. Successive terms decrease like α^2 or $(1/137)^2$, indicating the relative magnitude of the fine structure.

Experimental evidence of atomic fine structure dates back to Albert Michelson in 1887. Erwin Schrödinger's nonrelativistic wave equation provided an improvement of the Bohr model but could reproduce fine structure only by introducing three relativistic correction terms known as the kinematical correction, the spin-orbit interaction, and the Darwin term. Paul Dirac's relativistic wave equation for the electron (1928) recovered the Sommerfeld formula and in addition gave the correct interpretation of the angular quantum number k by including spin.

Further structure in the spectrum of hydrogen was established by the discovery of the Lamb shift and the polarization of the vacuum in 1947. These effects can be explained when the radiation field surrounding the atom and the field of the electron are properly quantized. In this field theory, known as quantum electrodynamics (QED), α appears prominently when one considers the strength of the interaction between charged matter and the electromagnetic radiation. The constant α is therefore also known as the electromagnetic coupling constant.

The small numerical value of α has led to the development of QED using perturbation theory, which calculates predictions as a power series expansion in α. Elaborate calculations, as illustrated by Feynman diagrams, have matched experimental results with remarkable precision, leading to QED's status as a paradigm for theories of elementary particles and to the introduction of weak and strong coupling constants, α_W, and α_S, in analogy with α.

An explanation for the actual numerical value of α presently eludes physicists although it was already considered one of the most important questions in physics by Wolfgang Pauli. Modern theories introduce the notion that the numerical value of α varies depending on the energy scale of the problem, and hence assign less importance to the actual numbers. Renormalization theory and grand unified theories in particle physics are concerned with the effective value of the different αs at different energy scales.

Quantum electrodynamical determinations of α can be obtained by comparing experimental values of physical quantities appearing in bound state problems with their theoretical values that are expressed in terms of α. Examples of such quantities include the measurement of the anomalous magnetic moment of single electrons or positrons bound to electromagnetic traps fixed to Earth (geonium), the Lamb shift in the levels of hydrogen and other atomic systems, and spin effects such as the hyperfine structure in hydrogen, and, in muonium, the bound state of an electron and an antimuon. Bound states of a particle and its antiparticle, such as positronium, also are used, since α appears in the expression for the rate at which positronium decays.

Determination of α from high-energy reactions such as the scattering of electrons and positrons, pair annihilation, or muon pair production in particle colliders leads to less precise values because the effect of nonelectrodynamical interactions becomes important at high energies. These interactions introduce large uncertainties because the weak and strong parameters are typically less well known than electrodynamical ones.

The value of α also can be determined by measurements that are based not on quantum electrodynamics but on condensed matter physics. In the quantum Hall effect, a measurement of the quantized values of the Hall resistance, defined as the ratio between the voltage across a layer and the current along the layer, is combined with the permeability of the vacuum and the speed of light to yield a precise value for α. In the ac Josephson Effect measurement, the voltage across the junction between two superconductors and the frequency at which a pair of electrons tunnels through the junction are combined with measurements of the gyromagnetic ratio of the proton, the magnetic moment of the proton, and the Rydberg constant to give α. A combination method using the two previous methods has the advantage that the dependence on the resistance standard cancels out and yields the value

$$\alpha^{-1} = 137.0359840(51).$$

The condensed matter measurement and the quantum electrodynamical measurements of α agree with each other to the level of one part per 10 million. To determine the value of α to a higher precision might necessitate studying quantum electrodynamical contributions in the condensed matter effects.

See also: GRAND UNIFIED THEORY; JOSEPHSON EFFECT; LAMB SHIFT; POSITRONIUM; QUANTUM ELECTRODYNAMICS; RENORMALIZATION

Bibliography

KINOSHITA, T., ed. *Quantum Electrodynamics* (World Scientific, Singapore, 1990).

PAIS, A. *Inward Bound* (Oxford University Press, New York, 1986).

JEAN-FRANÇOIS VAN HUELE

FISSION

Nuclear fission is a process in which the nucleus of a heavy atom splits into two parts, with the release of energy. With the heavy element uranium as raw material, the phenomenon is exploited to generate a major part of the world's electricity.

Fission can be spontaneous, or it can be induced in certain heavy-metal nuclei by adding a neutron. To understand the fission process, we need first to be familiar with some terms used in describing various aspects of the atomic nucleus.

Nucleons, Atomic Number, and Mass Number

An atomic nucleus is a cluster of nucleons. A nucleon can be either a proton or a neutron. Those two particles are nearly the same size and have a mass of approximately one atomic mass unit (1 amu $= 1.66 \times 10^{-24}$ g). A neutron is electrically neutral, while each proton carries a unit positive electric charge that balances the unit negative charge of an electron.

A chemical element is defined by the number of protons (its atomic number), but it can exist in several isotopic forms determined by the number of neutrons. The total number of nucleons is the mass number, which is approximately equal to the atomic weight in amu.

In modern notation, the symbol for an isotope (or nuclide) is written with the mass number as a superscript preceding the chemical symbol (e.g., ^{238}U, which is read as uranium-238). The atomic number can be indicated explicitly by writing it as a preceding subscript, $_{92}$U. To specify both, one can write $^{238}_{92}$U.

Nuclides

"Nuclide" is a commonly used term that means almost the same thing as isotope. Both refer to a nuclear species that is distinguished by its atomic number and its number of neutrons. "Nuclide" is used when nuclear properties are the focus, the chemical element being more or less irrelevant. For instance, one could say that ^{60}Co and ^{240}Pu are two radioactive nuclides, whereas ^{240}Pu and ^{241}Pu are two isotopes of plutonium or two nuclides (but *not* two nuclides of plutonium).

Transuranics

Atomic numbers go from 1 (hydrogen) to more than 100, including the human-made transuranic elements—the ones beyond 92 (uranium). Mass numbers range from 1 (hydrogen's lightest isotope, with only one proton and no neutrons) to more than 250. Although many transuranic nuclides were formed when the elements making up our solar system were created, their half lives are short enough that all of the original ones have long since decayed. The most important transuranic nuclide is probably $^{239}_{94}$Pu, which has a half-life of 24,390 years. ^{239}Pu is fissile like ^{235}U—making it, along with ^{235}U, an excellent fuel for a nuclear reactor, as well as efficient material for a nuclear weapon. Some other transuranic elements are $_{93}$Np (neptunium), $_{95}$Am (americium), and $_{96}$Cm (curium).

Actinides

An element whose atomic number is greater than 89 (actinium) but not greater than 103 (lawrencium) is called an actinide (sometimes actinium itself is considered an actinide). All the actinides are heavy metals, and all of them have a tendency to become unstable and split (fission) upon absorbing a neutron of appropriate energy.

The fission threshold of a nuclide is the energy an incoming neutron must have to induce the nucleus to fission. Although any nucleus with mass greater than 1 amu can be disrupted if you hit it with enough energy, the cases of practical interest are those in which the fission threshold is less than a few million electron volts (MeV). Such a nucleus is said to be fissionable. A nucleus that undergoes fission upon absorption of a slow, or thermal, neutron (one with very low kinetic energy—a fraction of an electron volt) is said to be fissile. (Fissionable includes fissile as a subset.) Fissions triggered by thermal neutrons and fast neutrons are called thermal fissions and fast fissions, respectively.

All the actinide nuclides are fissionable. As fuel in fast-reactor systems that efficiently recycle their spent fuel, they can be almost totally consumed—but not in today's nuclear power plants, virtually all of which use thermal reactors. Those reactors get their power from thermal fissions in fissile nuclides and produce more transuranic actinides than they can consume. Consequently, plutonium and other long-lived radioactive transuranics are accumulating around the world.

Actinide nuclei that have an odd number of neutrons tend to be fissile; after ^{235}U and ^{239}Pu, the most important ones (because they have relatively long half-lives and can be created in nuclear reactors) are ^{233}U and ^{241}Pu. The fissionable nuclides with an even number of neutrons, such as ^{238}U and ^{240}Pu, are called fertile—because, in a process sometimes called breeding, absorption of a neutron transmutes them into fissile nuclides: ^{240}Pu becomes ^{241}Pu, and ^{238}U becomes ^{239}Pu after two successive beta decays ($^{239}_{92}$U decays with a 23.5-min half-life to $^{239}_{93}$Np, which in turn decays with a 2.3-day half-life to $^{239}_{94}$Pu).

Binding Energy

In spite of the electrostatic force that causes protons to repel one another, the nucleons in a nucleus are held together by short-range attractive nuclear forces that overcome the electrostatic repulsion. When nucleons fuse (are driven together to form a nucleus, which happens in stars), those short-range forces cause binding energy to be released in the form of radiation of various kinds. The resulting nucleus has a mass that is measurably less than the sum of the masses of the free nucleons. The relationship between the binding energy and the mass deficit is given by Albert Einstein's famous equation $E = mc^2$, where c is the speed of light and E is the amount of energy that is equivalent to the mass m. One amu is equivalent to 931 MeV.

Figure 1 shows how the binding energy per nucleon changes with mass number. The binding-energy curve peaks near mass 56, which means that the mass per nucleon is at a minimum there. Thus, additional binding energy can be released by forcing together (fusing) light nuclei or by splitting (fissioning) heavy nuclei. The first three points at the left in Fig. 1 are ^{1}H, ^{2}H, and ^{4}He; it is mainly the fusing of hydrogen into helium that powers stars and hydrogen bombs.

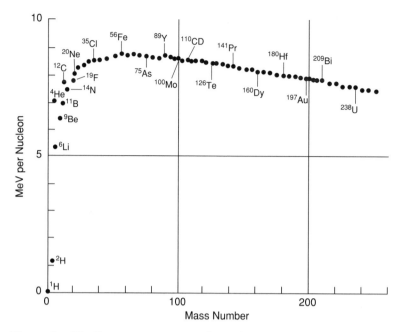

Figure 1 Binding energy per nucleon (Krane and Halliday, 1988, with permission of the publisher).

Fission power exploits the high-mass end of the binding-energy curve. If a passing neutron is absorbed by a ^{235}U nucleus, for instance, the resulting ^{236}U nucleus is likely to split into two fission fragments, one with around 96 nucleons and the other about 140. As Fig. 1 shows, the nuclei in those mass regions are more tightly bound (have greater binding energy) than uranium; the binding energy released when an actinide nucleus fissions amounts to about 200 MeV. In the fission process, two or three fission neutrons are promptly emitted with fairly high velocity (their energies are distributed around 1 MeV).

Chain Reaction and Critical Mass

A fission neutron is sooner or later absorbed by one of the nuclei it encounters, perhaps after being slowed down by scattering (colliding with various atoms). If the absorbing atom is a heavy metal, its nucleus might then split, releasing still more neutrons that can cause more fissions, and so on, in a chain reaction. To avoid having the chain reaction peter out, a large enough fraction (~60%) of the fission neutrons must avoid both escaping from the assembly and being captured in some way that does not lead to fission. When the balance is such that the neutron population remains constant (each fission leading, on average, to exactly one more fission), the configuration is said to be critical.

Nuclear Reactors

To produce power at a steady rate, as in a nuclear reactor, the chain reaction must be controlled. That means the assembly must be kept precisely in a configuration that is just barely critical. This can be done with neutron-absorbing control rods that move in and out to control the rate of the reaction, exploiting the fortunate occurrence of delayed neutrons.

Atomic Bombs

A fission explosive (atomic bomb) differs from a nuclear reactor in many ways (materials, design, control, etc.). While waiting to be used, the warhead is in a subcritical configuration. To set it off, the parts are rapidly "assembled" by driving them forcefully together by detonating a chemical high-explosive. The resulting configuration is highly supercritical, and the chain reaction is uncontrolled: It proceeds until the consumption and dispersal of the fissionable nuclei have reached the point where the fraction of neutrons lost to capture and leakage is large enough that the configuration is again subcritical. That takes only about a microsecond (a millionth of a second) or less, depending on the bomb's yield (explosive energy), but by the end of that brief interval its nuclear energy has been released and the explosion is under way.

The Fission Process

In outline, the process of fission begins when a neutron approaches a fissile nucleus, say ^{235}U, and is absorbed. The resulting ^{236}U nucleus is unstable and, about 80 percent of the time, promptly separates into two energetic fission fragments, with the immediate release of two or three excess neutrons. Why is the ^{236}U nucleus unstable? Remember that the incoming neutron loses mass when it joins the nucleus, adding an amount of energy equivalent to the lost mass (about 6 MeV). According to the liquid-drop model, the resulting commotion among the 236 nucleons causes the cluster to stretch out of shape. Once that starts, the protons' electrostatic repulsion tends to elongate the nucleus further. Soon the shape resembles a dumbbell, and when the ends get far enough apart that the short-range internucleon attraction can no longer override the repulsive force between the positively charged ends of the dumbbell, the two ends are sent rapidly on their way by electrostatic repulsion. That is how the fission fragments get their kinetic energy.

One might expect that the two fission fragments (the fission product nuclides) would tend to have about the same mass (~118 amu). However, in nuclei that are heavy enough to be fissionable (mass number greater than 230 or so), it happens that factors related to nuclear stability heavily favor unsymmetrical fission; the lighter fragments are grouped around mass 96 and the heavier ones around mass 140.

The ratio of neutrons to protons is considerably greater in actinides than in stable nuclides in the fission-product region. Thus the primary fission-product nuclei are neutron rich, which makes them radioactive: Two or three prompt neutrons are immediately emitted, and somewhat later there is the occasional (but important) delayed neutron, but most of the extra neutrons are converted one-by-one into protons in a sequence of beta-particle emissions.

Spontaneous Fission

The vast majority of fissions in a reactor or a bomb are induced by the absorption of a neutron. There are some heavy nuclides, however, that undergo spontaneous fission from time to time without having it triggered by an external event such as an incoming neutron. This is a tunnel-effect phenomenon. Two such nuclides are ^{240}Pu and ^{242}Pu, in which the probability of spontaneous fission is high enough that the neutrons they emit can seriously interfere with the performance of a nuclear weapon if those nuclides constitute more than a few percent of the plutonium in the warhead. (The neutrons cause pre-ignition by triggering the chain reaction before the parts are fully driven to the desired degree of supercriticality, and the result is an unpredictable, low-yield explosion called a fizzle. Spontaneous fission does not degrade the performance of a nuclear reactor.

Fission Energetics

When a nucleus fissions, the released energy is ultimately converted to heat (which can be used to make a destructive explosion, or to turn water into steam for driving electric generators). Most of the released energy appears initially as kinetic energy of the fission fragments. The fragments are slowed down in collisions with atoms that are in their way, causing local heating and ionization, and soon come to rest.

The two fission-fragment nuclei are initially in a highly excited state, and they decay with various half-lives to more stable nuclei by emitting energetic particles and photons—beta particles, gamma rays, neutrons, and neutrinos. Except for the neutrinos (essentially all of which escape without interacting), the emitted particles lose their kinetic energy by colliding with atoms, creating more heat. The photons (the gamma rays, along with the x rays and other photons emitted by orbital electrons as the excited and ionized atoms return to their ground states) transfer their kinetic energy mainly to electrons of atoms they encounter, thus contributing their share to the heating. Some of the neutrons escape from the reactor (or bomb) and deposit their energy in surrounding materials.

The portion of fission energy carried off by the various particles and photons varies considerably from one fission to the next. Roughly, however, the 200 MeV or so of energy released in a fission event is initially distributed as listed in Table 1.

Table 1 Approximate Initial Distribution of Fission Energy

Energy Carrier	MeV	Percent
Kinetic energy of the lighter fission fragment	93	47
Kinetic energy of the heavier fission fragment	66	33
Prompt gamma rays	8	4
Prompt neutrons	7	3
Beta particles from radioactive fragments	19	10
Gamma rays and delayed neutrons from radioactive fragments	7	3

A fission neutron that does not cause another fission is eventually absorbed by the nucleus of one of the atoms it encounters, a process called radiative capture; the result is a radioactive atom called an activation product.

Importance of Nuclear Fission

The phenomenon of nuclear fission permits electric power to be generated by nuclear reactors, tapping the vast amounts of nuclear energy locked up in the earth's endowment of the fissionable elements thorium (rarely used) and uranium. Many scientists are convinced that increasing amounts of electricity will be needed around the world to satisfy a rapidly growing energy demand; they predict that nuclear fission will eventually be generating most of the world's electricity because environmental concerns will force cutbacks in the use of fossil fuels (coal, oil, and natural gas).

By far the more important of the two naturally occurring fissionable elements is uranium, almost all of which consists of the fertile isotope ^{238}U; the fissile ^{235}U amounts to only seven parts per thousand. Because they rely heavily on fissile nuclides, today's reactors exploit only a small fraction of the energy stored in the uranium they use. However, scientists have learned how to build reactors that can extract virtually all of the energy from ^{238}U when the need arises. The energy in the earth's uranium is many times larger than the total available from fossil fuel.

Energy in large quantities is readily extracted from uranium because nuclear energy is far more concentrated than the chemical energy that comes from oxidizing (burning) fossil fuels. Consider the two main ways by which electric power is generated: burning coal and fissioning uranium. In one year a typical large power plant generates about 8.8×10^6 MW h (megawatt hours) of electricity. If it is a coal plant, it produces that energy by burning three or four million metric tonnes of coal; if it is a nuclear plant, it produces that energy by consuming only one metric tonne of uranium. That means, among other things, that the volume and weight of the waste from fossil power is enormously larger than from fission power—an important consideration in comparing environmental impacts.

Fission Energy Versus Fusion Energy

In ^{238}U there is far more energy available than people are likely ever to use, but there is also the possibility of another inexhaustible source of energy: nuclear fusion. Both fission and fusion power can be generated with no emission of greenhouse gases, acid-rain precursors, or ozone-layer destroyers.

Ongoing research is bringing fusion power closer to practicality every year, but the extent to which it will ever replace fission is not yet clear. The predictions are that fusion reactors, to be economical, will have to be much larger than fission reactors—maybe 3,000 MWe or more, whereas fission power reactors can be economical when as small as 300 MWe, or even less.

The radioactivity problems of fission and fusion are different but probably of comparable magnitude—assuming not the current generation of fission reactors but high-efficiency fast reactors that consume the long-lived actinides. Like fission plants, fusion plants will need effective neutron shielding, since operating fusion reactors will produce a large flux of high-energy (14 MeV) neutrons. Fission products are not produced in pure fusion plants, but neutron-activation products are; per unit of power produced, the amount of radioactivity from a fusion reactor will perhaps be less, but the volume of radioactive materials will be greater. A significant aspect of fusion plants is that large quantities of tritium (^{3}H, an isotope of hydrogen that beta-decays with a 12.3-year half-life) will have to be contained and handled safely.

There is also the possibility of hybrid power plants, which might offer some advantages. Such a plant would consist of a central fusion reactor surrounded by an actinide-fueled subcritical assembly—a "blanket" in which fissions are induced by otherwise wasted fast neutrons from the fusion reaction. The transuranic actinides bred in the blanket could be extracted and consumed as fuel in fast reactors.

In individual cases, the deciding factor will be economics: If there comes a time when a fusion or hybrid plant will generate cheaper steam than a cluster of five or ten nuclear reactors, the fusion option is likely to be chosen when a large installation is wanted. In any event, the consensus of experts in the field is that fission power will be a vital part of the world's energy mix for the foreseeable future.

See also: CHAIN REACTION; DECAY, NUCLEAR; ELEMENTS; ELEMENTS, TRANSURANIUM; ENERGY, NUCLEAR; FISSION BOMB; FUSION; FUSION BOMB; FUSION POWER; NUCLEAR BINDING ENERGY; NUCLEAR REACTION; NUCLEON; REACTOR, BREEDER; REACTOR, FAST; REACTOR, NUCLEAR

Bibliography

KRANE, K. S., and HALLIDAY, D. *Introductory Nuclear Physics* (Wiley, New York, 1988).

LEACHMAN, R. B. "Nuclear Fission." *Sci. Am.* **213** (2), 49–59 (1965).

GEORGE S. STANFORD

FISSION BOMB

The first requirement for a fission bomb is that the fissionable, or fissile, material be capable of sustaining a chain reaction. To satisfy this requirement, every incident neutron must produce more than one secondary neutron when the nucleus fissions. For example, suppose two neutrons are liberated. These neutrons are then each available to fission two other nuclei, which in turn each release at least four more neutrons; those four beget eight, and so

on. By the nth generation, at least 2^n neutrons will have been produced. In this way the process can spread rapidly through the mass of fissile nuclei. Isotopes of only a few elements in the periodic table are subject to fission by slow (thermal) neutrons. They include uranium 233 (^{233}U), uranium 235 (^{235}U), and plutonium 239 (^{239}Pu). Most fission bombs employ ^{235}U, ^{239}Pu, or a composite of the two, organized in concentric rings. The so-called composite core makes more efficient use of both materials than is otherwise possible.

A second requirement for a fission bomb is the assembling of a critical mass of fissile material. A critical mass must have a large enough volume of this material so that the chance of a fission-liberated neutron escaping before fissioning another atom of active material is low. The actual amount of material required for a critical mass depends on the material's shape and density. The moment criticality is achieved, the explosion process begins. Therefore, the critical mass cannot be assembled ahead of time but must be brought together only at the instant of detonation at a predesignated target.

It is essential that during the assembly process, no neutrons be present; otherwise, fissioning will begin prematurely, increasing the temperature and pressure inside the assembling volume. This will slow and then reverse the assembly process, allowing the escape of neutrons. The result is a feeble explosion.

At the instant of complete assembly, neutrons must be supplied to ensure detonation. The device that provides those neutrons is called the initiator. Typically, an initiator is composed of two chambers, one containing the element polonium (Po), an alpha particle emitter, and the second containing beryllium (Be). The wall between the compartments is composed of some material that melts at a relatively low temperature, for example, aluminum. At the instant a critical mass is achieved, by design the wall melts, allowing the mixing of the Po and Be. When Be absorbs alpha particles, the Be emits neutrons.

Efficiency

The efficiency of a fission bomb is a function of the fraction of fissionable material that participates in the explosion. Once the fissioning process starts, there is a very rapid accumulation of heat energy. The fission process is very rapid, on the order of 10^{-8} s. Within just a few tenths of a microsecond the active material has vaporized and begun to expand at the speed of sound. Once the density of the fissioning material is reduced sufficiently, the reaction stops.

The efficiency of the explosion can be increased by encasing the active material within a hollow shell of dense material called a tamper. The tamper serves two purposes. First, it reflects back into the active material any neutrons that escape the fissioning volume. Second, it retards the rate of expansion of the vaporized critical mass for a fraction of a microsecond, allowing several additional generations of neutrons to participate in the explosion than would otherwise be possible. Materials known to have been used for tampers include ^{238}U and tungsten carbide. The outside surface of the tamper can be coated with a thin shell of a strong neutron absorber, such as boron-10, as a way of preventing extraneous neutrons from pre-detonating the fissile material.

Assembly

An obvious technique for assembling a critical mass is the use of a cannon to fire one subcritical mass of fissionable material into another. This is called gun assembly. A variety of shapes for the target and projectile might work. The actual shapes used are closely guarded secrets, although it is known that in the bomb that was used at Hiroshima, Japan, in World War II, the target was a solid cylinder and the projectile a hollow cylinder. The cannon's required muzzle velocity for gun assembly is of the order of magnitude of 10^5 cm/s.

Gun assembly cannot be used with plutonium. The nature of the manufacturing process used to produce plutonium necessarily results in the production of ^{239}Pu with a small amount of ^{240}Pu mixed in. The latter fissions spontaneously, that is, in the absence of neutrons or any other triggering mechanism. The gun assembly technique would require an unrealistically high muzzle velocity to ensure that assembly took place in the absence of significant numbers of neutrons.

The technique that is used is called implosion assembly. A sphere of subcritical plutonium, containing at its very center the initiator, is surrounded by shaped charges of two different kinds of explosives, called explosive lenses. When simultaneously detonated, these lenses produce a powerful, uniformly compressive shock wave that squeezes the pluto-

nium into a fraction of its original volume, thereby making its mass supercritical. The resulting rise in temperature melts the initiator components at just the right time. Implosion assembly is several orders of magnitude faster than gun assembly. Most modern fission weapons use implosion assembly.

Later Innovations

Besides the use of composite cores, two other major innovations after World War II resulted in significant improvements in efficiency and yield of fission weapons. The first of these is levitation. Rather than placing the tamper shell in direct contact with the shell of fissile material, an air gap is established between the two. As a result, the tamper acquires considerable momentum as it implodes and therefore compresses the fissile material to a much greater extent than would otherwise be the case. This results in significantly higher efficiency.

A second innovation is the so-called boosted fission weapon. A segmented hollow sphere of plutonium surrounds a mixture of deuterium and tritium gas. The high pressure of implosion and the high temperatures produced by the fission of the plutonium cause a tritium-deuterium reaction (fusion), producing some energy and, more important, neutrons, which increase considerably the efficiency of the fission process, resulting in higher yields or the same yields using considerably less fissile material. As the critical mass begins to expand, both the fission and fusion processes shut down.

Yield

Since each fission generation takes but 10^{-8} s, the typical fission bomb's generation of energy takes less than a microsecond. The effects then spread out from the source at their typical speeds—the speed of light for x and gamma radiation, the speed of sound for the shock wave, and so on. By comparison, the typical stick of TNT requires about 10^{-4} s to explode. Comparison of the energy release in nuclear fission to the energy release in combustion (as in TNT) reveals that the fission process is far more energetic than the combustion process. A single kilogram of ^{235}U or ^{239}Pu gives an energy equivalent of more than 10^4 tons of TNT. The bombs that were dropped on Hiroshima and Nagasaki contained, respectively, 200 lb (90 kg) of ^{235}U and 13 lb (6.1 kg) of ^{239}Pu. The TNT equivalent yield of the Hiroshima bomb was 13.5 kt, and that of the Nagasaki bomb was 22 kt. Calculating the efficiency of the Hiroshima bomb is difficult since it was composed of uranium that varied in enrichment of ^{235}U from 50 percent to the modern norm of 93 percent. Modern fission bombs yield a TNT equivalent of anywhere from 1 kt to 150 kt.

See also: CHAIN REACTION; FALLOUT; FISSION; FUSION; FUSION BOMB; NUCLEAR BOMB, BUILDING OF

Bibliography

RHODES, R. *The Making of the Atomic Bomb* (Simon & Schuster, New York, 1986).

SERBER, R. *The Los Alamos Primer: The First Lectures on How to Build an Atomic Bomb* (University of California Press, Berkeley, CA, 1992).

WOLFSON, R. *Nuclear Choices: A Citizen's Guide to Nuclear Technology* (MIT Press, Cambridge, MA, 1991).

STANLEY GOLDBERG

FLAVOR

Flavor is the generic name for the labels that distinguish the six quarks and the six leptons. The six quark flavors are up, down, charm, strange, top, and bottom. Three of these, up, charm, and top have an electric charge of +2/3 (in units where the proton has a charge of +1); these are sometimes called generically up-type quarks. The other three—down, strange, and bottom quarks—have a charge of −1/3 and are generically referred to as down-type quarks. Leptons also come in two sets of three, those with a charge of −1 (the electron, the muon, and the tau) and those with a charge of zero (the electron neutrino, the mu neutrino, and the tau neutrino).

For all the particles except the neutrinos the feature that distinguishes the three similarly charged particles is mass. Each flavor has a distinct mass. As shown in Table 1, the masses for the six quarks and for the three charged leptons are very different. All three neutrinos have very small masses, possibly even zero.

All the more massive quark and lepton types are unstable. They decay via charged weak interactions, that is, by processes involving an intermediate W^+ or W^- boson. In strong and electromagnetic interaction processes, which occur much more rapidly than weak processes with a similar energy release, flavor never changes. Similarly production or annihilation of a quark and antiquark always involves matched flavor and antiflavor in strong or electromagnetic cases, but when a W boson is involved there *must* be different flavors to maintain electric charge conservation.

If all three flavors of neutrino are massless, then how do we distinguish one from the next? The answer lies in the fact that the pattern of weak decays for leptons is very simple. In every lepton process involving emission, absorption, production, or decay of a W, the lepton converts between a given flavor of charged lepton and the matching flavor of neutrino. Thus, if a W^- decays to produce an electron, the accompanying neutrino is an electron-type antineutrino, whereas when the charged lepton is a mu then the accompanying antineutrino is a muon type. It has been verified in experiments that the neutrinos produced in association with muons do not behave in the same way as would an electron-type neutrino. In fact, the Nobel Prize in 1988 was awarded to Leon Lederman, Melvin Schwartz, and Jack Steinberger for the first experimental demonstration of this remarkable fact. A similar demonstration for tau neutrinos has not yet been achieved.

This property results in three lepton flavor conservation laws. For example, the electron flavor conservation law says that in any process the number of electrons, plus the number of electron-type neutrinos, minus the number of their antiparticles (positrons plus electron-type antineutrinos) does not change. A similar rule applies for each generation. One consequence is that three neutrino types are stable particles, as is the electron, which is the lightest electrically charged particle of any type and hence is forbidden to decay by conservation of electric charge (along with conservation of energy).

If neutrinos actually have small masses, then these three separate conservation laws may not be exact. Experimentally all we know is that the masses are smaller than the values shown in the accompanying table. To date there is no convincing evidence, either for neutrino masses or for violations of the three lepton flavor conservation laws. Some physicists speculate that such violations do occur, chiefly because they would help explain some data on the production of neutrinos by the Sun that do not match expectations.

Table 1   Standard Model Matter Particles

Grouping	Particle Symbol	Particle Name	Antiparticle Symbol	Antiparticle Name	Mass* (GeV/c^2)
U-type quarks	u	up or u-quark	$\bar{u}$	anti-u or u-bar	$(5 \pm 3) \times 10^{-3}$
Electric charge $= \frac{2}{3}$	c	charm or c-quark	$\bar{c}$	anticharm or c-bar	1.5 ± 0.4
($-\frac{2}{3}$ for anti)	t	top or t-quark	$\bar{t}$	antitop or t-bar	176 ± 20
Down-type quarks	d	down or d-quark	$\bar{d}$	antidown or d-bar	$(1 \pm 0.5) \times 10^{-2}$
Electric charge $-\frac{1}{3}$	s	strange or s-quark	$\bar{s}$	antistrange or s-bar	$(2 \pm 1) \times 10^{-1}$
($+\frac{1}{3}$ for anti)	b	bottom or b-quark	$\bar{b}$	antibottom or b-bar	4.3 ± 0.2
Charged leptons	e^-	electron or e-minus	e^+	positron or e-plus	5.1×10^{-4}
Electric charge $= -1$	μ^-	muon or mu-minus	μ^+	muon or mu-minus	0.106
($+1$ for anti)	τ^-	tau or tau-minus	τ^+	tau or tau-plus	1.784
Neutral leptons or neutrinos	ν_e	electron-neutrino or nu-e	$\bar{\nu}_e$	anti-electron-neutrino or nu-e-bar	$< 5 \times 10^{-9}$
Electric charge $= 0$	ν_μ	muon-neutrino or nu-mu	$\bar{\nu}_\mu$	nu-mu-bar	$< 3 \times 10^{-4}$
	ν_τ	tau-neutrino or nu-tau	$\bar{\nu}_\tau$	nu-tau-bar	$< 3 \times 10^{-2}$

*Since quarks cannot be isolated, the definition of their masses has some technical subtlety. The ranges given here are taken from the Particle Data Group's best estimates; they do not represent standard statistical errors.

Since all the quarks do indeed have masses, the pattern of weak decays of the more massive quark flavors is much more complicated. We call the least massive quark of each charge the first generation quarks, the next in mass belongs to the second generation, and the highest mass flavor of each charge is the third generation member. All possible transitions occur, but with varying strength, depending on whether the quarks connected by the W process are in the same generation or in different ones. If mass differences allow it, a decay within the same generation is favored. Decays that cross one generation are suppressed, while decays such as b to $u + W$, which cross two generations, are even less likely.

The peculiar set of names for the various quark flavors got its start because of the odd pattern of quark weak decays. K mesons are now understood as particles made from one quark and one antiquark, with one strange quark or antiquark along with one from the first generation. When these particles were first discovered they were much longer-lived than physicists expected for a meson in their mass range, and so people said these particles were strange. The name stuck, even after the long lifetimes were understood. When Murray Gell-Man first suggested the quark model he called the third quark—which was introduced to explain K mesons and other such strangely long-lived particles—the strange quark.

The names up and down are a bit more arcane. Nuclear physicists long ago realized that, as far as the strong interactions are concerned, the proton and the neutron are almost identical particles. The electromagnetic interaction certainly makes them different, and they do also have slightly different masses. However, it turns out that as far as strong nuclear forces go, we can treat them as identical except for a two-valued label. Now, the other two-valued quantity we know is the spin of a fermion. In this case we can think of the two choices as a spin angular momentum ($\bar{h}/2$) pointing up or pointing down. By analogy these became the names of the two lightest quark flavors, which are the ones which make up the proton (uud) and neutron (udd). However, remember that this up and down are only labels to distinguish the different quark charges, they have no relation to any directional quantity in real physical space. Actually Gell-Mann originally called these two quarks p and n, using lowercase letters for the quarks and uppercase P and N for the proton and the neutron. People found this notation too easily led to confusion, and the usage u and d along with the names up and down became standard for these quarks.

With three flavors called up, down, and strange, what should we name the next flavor to be discovered? The name charm was a piece of whimsy, first used by James D. Bjorken and Sheldon Glashow who speculated that quarks should follow the same pattern as leptons, and thus that the existence of a muon neutrino as well as a muon suggested there should be a fourth quark flavor. The names top and bottom for the third generation of quark flavors are an echo of the up and down names. In fact, originally some authors speculating on the existence of the third generation named these two quarks "truth" and "beauty", but that was too much even for particle physicists. Nowadays those names are seldom heard, though beauty is still occasionally used for the *b* quark name instead of bottom.

See also: ANTIMATTER; BOSON, *W*; BOSON, *Z*; COLOR CHARGE; ELEMENTARY PARTICLES; LEPTON; LEPTON, TAU; NEUTRINO, HISTORY OF; POSITRON; QUARK; QUARK, BOTTOM; QUARK, CHARM; QUARK, DOWN; QUARK, STRANGE; QUARK, TOP; QUARK, UP; SPIN

Bibliography

CLOSE, F.; MARTEN, M.; and SUTTON, C. *The Particle Explosion* (Oxford University Press, New York, 1987).

HALZEN, F.; and MARTIN, A. D. *Quarks and Leptons: An Introductory Course in Modern Particle Physics* (Wiley, New York, 1984).

HELEN R. QUINN

FLUID DYNAMICS

Fluid dynamics is the study of the motion of liquids and gases. Hydrodynamics is the study of liquids, whereas aerodynamics studies the mechanics of air (and other gases). These studies overlap and the above terms are often applied interchangeably. Fluid dynamics has a wide range of application in science and engineering, including atmospheric studies, oceanography, transportation (as in automobiles and airplanes), turbines, pumps, fans, blood flow in living organisms, building design, nuclear physics, and in sports such as baseball and golf.

Fluids, which include liquids and gases (and therefore plasmas) are substances that deform continuously and without limit when acted on by an external force (shear stress), no matter how small. Solids, which are not fluids, also deform with shear stress, but offer a large resistance and at some point, a limit of deformation beyond which the solid breaks. Another distinguishing characteristic is that fluids flow even with small applied forces, while solids tend to require much larger forces.

Sometimes the distinction between solid and fluid is arbitrary. Materials like glass and glaciers flow like liquids over very long periods of time. A granular solid like salt behaves like a fluid when poured; toothpaste has both solid and fluid properties, depending on the magnitude of applied pressure. Microscopically the phases distinguish themselves in the molecular spacing and intermolecular forces. In solids, the molecules are closely spaced and have strong binding forces. In liquids, the molecules are generally farther apart and interact with weak cohesive forces. The molecules are farther apart in gases than in liquids, and there is little or no binding force between the molecules. Another distinguishing feature of fluids is that molecules in a fluid lack any kind of long range order (pattern) and are constantly changing position and orientation, even when the fluid is at rest (macroscopically).

One macroscopic distinction between liquids and gases is compressibility, meaning that the density changes as pressure is applied to the system. Compressibility is a direct result of the microscopic properties, namely the intermolecular spacing and forces. Liquids are nearly incompressible. This is generally not true for gases, although they are often treated as incompressible for the sake of ease in computation. For example, when air flows less than 50 m/s the assumption of incompressibility generally gives acceptable results.

Fluid mechanics ignores the atomic nature of matter by treating fluids as a continuous medium. This assumption is quite valid so long as the smallest "piece" of fluid contains a very large number of atoms. In such a medium, any microscopic fluctuations in the fluid properties are invisible on the measurable scale. This is typically the case in observed phenomena and with usual measuring instruments. Thus it is the macroscopic quantities such as pressure, density, and flow velocity that characterize the dynamics of a fluid.

Theoretical studies in fluid mechanics require the application of Newton's laws and conservation of energy and momentum. Since a typical fluid possesses a mathematically unmanageable number of particles, models, approximations, and simplifications must be made. One of the primary simplifications in fluid dynamics occurs in the distinction between nonviscous and viscous flow. The measure of resistance to the shear stress is called the viscosity, which is an internal friction in a fluid resulting from particles of different velocities exchanging momentum. For example, honey is more viscous than water because honey requires a larger force than water to create or maintain a certain amount of flow. In fluid dynamics viscosity is an energy dissipating force and is analogous to sliding friction.

Ideal Fluids

The least complicated theoretical analysis is seen in the study of an ideal or Eulerian fluid. An ideal fluid is continuous, incompressible, nonviscous, and does not exchange heat with its surroundings. This ideal fluid model came about in 1755 when Leonhard Euler first developed the calculus of three-dimensional vector fields, used to study velocity fields. The concept of a velocity field results from velocity vectors that have time-dependent components in each direction (x, y, and z). Thus the velocity in the x direction is $v_x = v_x(x, y, z, t)$, with similar terms for v_y and v_z. The acceleration $\mathbf{a} = d\mathbf{v}/dt$ of a fluid element with mass dm requires a special derivative since this element is not a single, discrete particle as it is in mechanics. The substantial or material derivative has the following form:

$$\mathbf{a} = \frac{D\mathbf{v}}{Dt} = v_x \frac{\partial \mathbf{v}}{\partial \mathbf{x}} + v_y \frac{\partial \mathbf{v}}{\partial y} + v_z \frac{\partial \mathbf{v}}{\partial z} + \frac{\partial \mathbf{v}}{\partial t}$$

$$= (\mathbf{v} \cdot \mathrm{grad})\mathbf{v} + \frac{\partial \mathbf{v}}{\partial t},$$

where the capital D's are a reminder of the special derivative. The right-most expression is a condensed form of the middle term. The gradient operator (grad) is a mathematical function common in vector calculus that describes how the velocity components change in the region surrounding the fluid element. Applying Newton's second law, $dF = dm \times a$, the final equation of motion for the fluid, known as Euler's equation, is

$$-\mathrm{grad}(p) = \rho\left(\frac{\partial \mathbf{v}}{\partial t} + (\mathbf{v} \cdot \mathrm{grad})\mathbf{v}\right),$$

where p is the pressure, ρ is the density, and $\mathbf{v}$ is the velocity. Careful inspection of this equation reveals a force term (p) equaling a mass (ρ) term multiplied by an acceleration $\mathbf{Dv}/Dt$. This equation is of fundamental importance in fluid dynamics.

The velocity field can be illustrated using streamlines, providing a visual representation of a moving fluid. Streamlines are imaginary lines tangent to the velocity vector of a tiny element of fluid at all points in the fluid. If the flow is steady, so that the fluid velocity is always the same value at a given point in the fluid, then the streamlines represent the trajectory of a moving "particle" of fluid. In this scenario, no two streamlines may cross and no particles may pass from one streamline to another. Such flow is called laminar, because the fluid moves in layers, each layer having a higher or lower speed than those adjacent to it. Figure 1 shows a laminar streamline pattern.

Using conservation of mass, the Euler relation leads to the continuity equation for fluids (which remains valid even when the fluid is not ideal):

$$\rho_1 v_1 A_1 = \rho_2 v_2 A_2 = \mathrm{constant},$$

where A represents the cross sectional area of flow and the subscripts 1 and 2 represent different points in the fluid stream. For incompressible flow, $\rho_1 = \rho_2$ and $vA = $ constant. The equation demands that the mass of fluid entering a region per time must equal the mass of fluid leaving that region per time. The

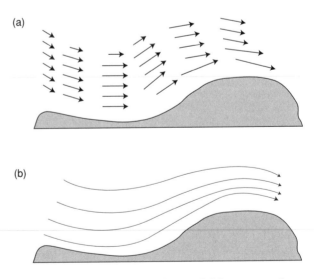

(a)

(b)

Figure 1 (a) Laminar velocity field past an obstacle. (b) Streamlined flow past an obstacle.

continuity equation requires that fluids flow faster through regions of smaller area. Examples of this include rapids that form in rivers where the water is shallow or constrained by close banks and water that forms a narrow stream as it speeds up when poured from a pitcher. For the case of steady (time-independent) flow Euler's Equation leads to a conservation of energy relation known as Bernoulli's principle, derived by Daniel Bernoulli in 1738. Bernoulli's equation is

$$\frac{1}{2}\rho v_1^2 + \rho g z_1 + p_1 = \frac{1}{2}\rho v_2^2 + \rho g z_2 + p_2 = \text{constant},$$

where z represents the vertical height and g is the gravitational constant 9.8 m/s^2. Bernoulli's equation provides a relationship among kinetic energies, gravitational potential energies, and pressures at two arbitrary locations in the fluid stream (1 and 2). The equation resembles the conservation of mechanical energy equation for a single particle, with the addition of the pressure terms. It also shows how a fluid may exchange speed and/or height for pressure. Bernoulli's equation is not valid if any energy adding or depleting mechanism is present, such as pumps, turbine blades, or viscosity. In many applications, changes in z (and thus the potential energy term) are negligible or absent and lead to a special case of Bernoulli's principle, the venturi effect. The venturi effect states that when the velocity of a fluid increases, its pressure decreases and vice versa.

Many examples and applications of the venturi effect are common such as the lift on an airplane wing, the "sucking" effect of tornadoes, and design of automotive carburetors. A few examples are presented here to demonstrate its scope and usefulness.

• When two naval vessels travel side by side, the water flowing between them is constrained and by the continuity equation must move faster than the water on the outer sides. The high speed water between the ships results in a lower pressure. The ships are drawn together and, to prevent collision, an outward force from the rudder is necessary.

• A Venturi flowmeter measures fluid speeds. As shown in Fig. 2 a fluid (assumed incompressible with density ρ) flows through a pipe, which has a constriction in it. The fluid flows faster in the constricted region, as a result of the continuity relation. This results in a lower pressure, which draws the liquid (with density $\rho_0 > \rho$) upward in

the U-shaped tube (called a manometer), which is connected to each region. Knowledge of the densities, cross-sectional areas of each region, and the difference in height (h) of the manometer liquid is sufficient to determine the speed of the fluid in the pipe.

• Bernoulli's equation also explains the curve on a spinning sphere, such as a baseball. Because of a retarding effect of the surface of the ball, which must be slightly rough, air flow past the ball has different speeds, depending on the orientation of the spin. The difference in speed causes a pressure difference, which gives rise to a net force (centripetal), causing the ball to curve.

Although greatly simplified, the ideal fluid model allows for a wide range of understanding and application of fluid dynamics through the application of Euler's equation or the fundamental relationships known as Bernoulli's and the continuity equations.

Viscous Fluids

To truly understand the dynamics of real fluids one must include viscosity. The Navier–Stokes equation (developed by Cloude Louis Marie Henri Navier in 1822 and independently by George Stokes in 1845) governs the motion of a viscous fluid and reduces to Euler's equation when viscosity is zero. Viscosity plays the role of friction changing kinetic energy to heat and causing changes in the trajectory of fluid particles from the ideal situation. The retarding force on an object moving through a viscous fluid is called viscous drag. The presence of the viscosity term in the Navier–Stokes equation makes it nonlinear, so very few complete solutions exist. Nev-

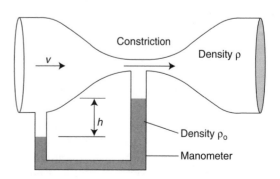

Figure 2   Venturi flowmeter.

ertheless, as discussed above, many problems involving real, viscous fluids can be solved by ignoring viscous effects. This is possible because in many practical, real-world situations, the bulk of a fluid flows as if no viscosity were present.

There is, however, an important region of the fluid in which viscosity plays a vital role, no matter how insignificant it is elsewhere. The boundary layer is the region immediately adjacent to an object either placed in the flow of the fluid or moving through the fluid. The boundary layer is a comparatively thin film of fluid that is attached to the surface of the object. At the fluid–surface interface the fluid has zero velocity relative to the surface. As one moves perpendicularly away from the surface the fluid velocity quickly increases until it reaches the velocity found in the mainstream fluid (which is essentially undisturbed by the presence of the object). Ludwig Prandtl formulated this boundary layer concept in 1904, making a major breakthrough in theoretical fluid dynamics beyond the Euler model of ideal fluids.

Fluid properties such as velocity, density, and viscosity, and object properties such as the size and geometry determine the thickness and the flow characteristics of the boundary layer. The Reynolds number (Re), introduced by Osborne Reynolds, combines these quantities in such a way as to describe whether the flow is laminar or turbulent:

$$\mathrm{Re} = \frac{\rho v D}{\eta},$$

where η is the viscosity and D is some dimension of the object (e.g., the diameter if it is a sphere). For low Reynolds numbers (Re < 2000) the flow remains smooth and laminar. This typically occurs for very viscous fluids or for very low velocities. As the Reynolds number increases (Re > 3000), the nonlinear aspects of Navier–Stokes become important, which causes turbulence to develop downstream from the object. Turbulent (nonlaminar) flow involves random, rapidly changing velocities along with rotational motion. This chaotic, rotational motion of the fluid extracts more energy from the system, causing greater retarding forces than are present with only viscous drag. To reduce this energy loss, objects may be streamlined so that turbulence is reduced or eliminated. No simple pattern exists for turbulent flow and no simple theory explains it.

See also: BERNOULLI'S PRINCIPLE; CAVITATION; HYDRODYNAMICS; NAVIER–STOKES EQUATION; NONLINEAR PHYSICS; PLASMA; QUANTUM FLUID; REYNOLDS NUMBER; SHOCK WAVE; TURBULENT FLOW; VISCOSITY; VORTEX

Bibliography

ADAIR, R. K. "The Physics of Baseball." *Phys. Today* **48** (5), 26–31 (1995).

BRIGGS, L. J. "Effect of Spin and Speed on the Lateral Deflection (Curve) of a Baseball; and the Magnus Effect for Smooth Spheres." *Am. J. Phys.* **52**, 589–596 (1959).

FEYNMAN, R. P.; LEIGHTON, R. B.; and SANDS, M. *The Feynman Lectures on Physics*, Vol. 2 (Addison-Wesley, Reading, MA, 1964).

FISHBANE, P. M.; GASIOROWICZ, S.; and THORNTON, S. T. *Physics for Scientists and Engineers* (Prentice Hall, Englewood Cliffs, NJ, 1993).

TIPLER, P. A. *Physics,* 3rd ed. (Worth, New York, 1982).

TOKATY, G. A. *A History and Philosophy of Fluidmechanics* (G. T. Foulis, Oxfordshire, Eng., 1973).

KENNETH D. HAHN

FLUID STATICS

A fluid at rest can be specified by its shape, volume, and density. Since an ideal fluid cannot offer resistance to shear, its internal force is always directed perpendicular to any surface that comes in contact with it. The pressure p of a fluid can be defined in terms of the force F exerted by the fluid at right angles to a surface of area S as $p = F/S$.

The density ρ (mass/volume) of a fluid is a function of temperature and pressure. In gases, since they expand (almost linearly) with temperature (K), and contract with the application of pressure, ρ is a function of T and p. Since the volume of liquid does not vary appreciably with T and p, for a commonly encountered range of temperature and pressure, ρ can be approximated by a constant.

The variation of pressure in a fluid at rest can be calculated by considering the net force on a thin disk of fluid at a depth y measured from the bottom of the vessel (or ocean, etc.) The differential equation resulting from this mechanical construction is

$$d\,p/d\,y = -\rho y. \qquad (1)$$

This can be solved with appropriate boundary conditions to yield pressure as a function of elevation y measured from the bottom.

For most liquids, it is a good approximation to assume that ρ is constant. Then Eq. (1) can yield an expression for a pressure difference between two levels 1 and 2 with height y_1 and y_2, respectively, measured from the bottom:

$$p_2 - p_1 = -\rho g(y_2 - y_1) \qquad (2)$$

It is usually more convenient to measure depth from the top surface of a liquid. Let us take y_2 to be the height of the top surface as measured from the bottom. Then $y_2 - y_1$ corresponds to the point of interest as measured from the top surface, that is, it is the depth h of this point. We now have a simpler form of Eq. (2) as

$$p = p_0 + \rho g h, \qquad (3)$$

where p_0 corresponds to the atmospheric pressure at the top surface of the liquid.

One of the consequences of the relation given in Eq. (3) is that a change Δp_0 of the external pressure results in the change Δp of the internal pressure of the incompressible fluid. This is known as Pascal's principle or Pascal's law, which states that any change in pressure applied to an enclosed fluid is transmitted unaltered to the rest of the body of fluid. A hydraulic press is an application of this principle.

Figure 1 is a schematic drawing of a hydraulic press, which is a machine that delivers a large force by applying a smaller force. In a U-tube arrangement, the tube is filled with a fluid and one arm on the left has smaller cross-sectional area A than the one on the right with area B. ($B/A > 1$.) When a force f is applied to the sealed cover plate at A in the downward direction, it creates a pressure change of f/A, which is transmitted to the rest of the U-tube. At B the pressure is still f/A but the upward force delivered at B is $f(B/A)$. The force f is multiplied by a factor B/A.

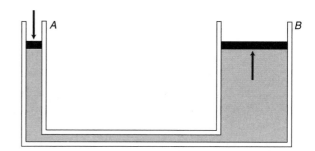

Figure 1 Schematic of a hydraulic press.

Another aspect of static fluid is the buoyancy with which it supports an object partially or totally immersed in it.

When the fluid is compressible and the fluid density ρ becomes a function of y or p in Eq. (1), we no longer have the solution shown in Eq. (2). This will be the case with gases such as air. We can obtain the variation of pressure with altitude of the atmosphere of the earth. We need to make some simplifying assumptions that are not very realistic. One is that the temperature of the air does not change with altitude and the second is that the gravitational acceleration does not vary with altitude. With these assumptions, and assuming the air to approximate ideal gas, ρ is proportional to p. The solution to Eq. (1) can now be written as

$$p = p_0 \exp(-g\,y\rho_0/p_0), \qquad (4)$$

where ρ_0 and p_0 are the values at sea level that are known. With the available values substituted for g, ρ_0, and p_0, one can obtain a usable expression for p as

$$p = p_0 \exp(-1.15 \times 10^5\,y), \qquad (5)$$

where y is measured in meters.

See also: BUOYANT FORCE; DENSITY; PASCAL'S PRINCIPLE; PRESSURE

Bibliography

GIANCOLI, D. C. *Physics for Scientists and Engineers*, 2nd ed. (Prentice Hall, Englewood Cliffs, NJ, 1984).

SEARS, F. W.; ZEMANSKY, M. W.; and YOUNG, H. D. *College Physics* (Addison-Wesley, Reading, MA, 1985).

CARL T. TOMIZUKA

FLUORESCENCE

An atom or molecule that has been activated by the absorption of a discrete amount of energy will always have some finite probability for releasing this energy, or some fraction of it, in the form of light. Activation can be by the absorption of light or by impact with other particles; it can occur even during the formation process of the atom or molecule. The light that is emitted is referred to as luminescence or, more specifically, photoluminescence if the activation is by light, sonoluminescence if the activation is by sound, radioluminescence if the activation is by collisions with particles that emanate from the radioactive decay of unstable nuclei, and chemiluminescence when activation is by chemical reaction. In all of these cases, the major mechanism responsible for the luminescence involves a process whereby a more energetic state of a specific atom or molecule (an upper state) makes a transition to a less energetic state (a lower state) by the emission of a quantum of light (i.e., a photon). This mechanism for the production of the light is called fluorescence. Depending on the difference in energy of the upper and lower states, the light that is emitted in fluorescence can range from the very short wavelengths of gamma rays and x rays, through the ultraviolet, visible, and infrared regions of the spectrum, to the much longer wavelengths of microwaves and radio waves. In each of these regions there exist important applications of the fluorescence process.

Fluorescence can always occur spontaneously in the sense that the transition requires (after activation) no further stimulation. In this case, the photon that is emitted may emerge at any time subsequent to the activation. The average time required for emission is called the radiative time and depends critically on the nature of the upper and lower states and on the energy of the emitted photon. Transitions between upper and lower states that are most favorable for the emission of light are referred to as allowed transitions, and for these the radiative times range from femtoseconds for soft x rays to nanoseconds for visible light to milliseconds for infrared light. For so-called forbidden transitions, these times are significantly longer. For a certain class of forbidden transitions, the fluorescence is sometimes referred to as a phosphorescence.

When the upper state of the atom or molecule has the possibility to lose its energy by processes other than fluorescence, such as through collisions with other species and/or, for molecules, by breaking or rearranging chemical bonds in processes referred to as photochemistry, the probability for fluorescence will be less than unity. In such cases, one defines a fluorescence quantum yield to be the probability that the atom or molecule emits the photon. This is determined by measuring the fraction of activated species that fluoresce.

In addition to spontaneous fluorescence, it is possible to stimulate the upper state to emit its photon by having present another photon with an energy equal to that of the energy difference between the upper and lower states. The emitted photon is then found to have not only the same energy as the stimulating photon, but also to be moving in the same direction and with the same polarization (i.e., direction of its electric field). This so-called stimulated fluorescence is the basis for the laser.

See also: EMISSION; ENERGY, ACTIVATION; LASER; LUMINESCENCE; PHOTON; POLARIZATION

Bibliography

BIRKS, J. B., ed. *Organic Molecular Photophysics,* Vol. 2 (Wiley, London, 1973).

PRINGSHEIM, P. *Fluorescence and Phosphorescence* (Interscience, New York, 1949).

STEPANOV, B. I., and GRIBKOVSKII, V . P. *Theory of Luminescence* (Iliffe, London, 1968).

SANFORD LIPSKY

FLUX

See ELECTRIC FLUX; MAGNETIC FLUX

FLY'S EYE EXPERIMENT

The Fly's Eye experiment makes use of the Fly's Eye detector to study cosmic rays above 10^{17} eV. The Fly's Eye detector has two arrays of telescopes separated by a distance of 3.5 km. One array has sixty-seven telescopes, and the other has thirty-six telescopes. The detector takes on the name "Fly's Eye" because these telescopes are arranged to view the entire sky, similar to the eyes on a fly. Each telescope is made up of a 1.5-m-diameter mirror with fourteen photomultipliers at its focal plane to record the spatial and temporal patterns of light created by distant cosmic ray tracks.

As a high-energy cosmic ray enters the atmosphere, it collides with atomic nuclei, and its energy is used to create a tightly bunched bundle of hadrons. Particles in this bundle interact with air molecules as they travel through the atmosphere. By the time this bundle strikes the ground, most of the particles have evolved into photons and leptons and are confined in a circular disk that is several meters thick and 100 m in diameter. As the disk travels through the atmosphere, charged particles in it excite the air nitrogen molecules and cause them to fluoresce. Particle cascades as far as 30 km away can be observed by the Fly's Eye detector as glowing spots that move with the speed of light. This unique observational ability of the Fly's Eye detector makes it the world's largest calorimetric detector; its range of detection extends over an area of about 1,000 km^2.

Measurements obtained by this detector suggest that the flux of cosmic rays above 10^{17} eV varies inversely as the 3.2 power of the energy and is dominated by heavy elements such as iron nuclei. Above 3×10^{18} eV, the flux varies inversely as the 2.6 power of the energy and is dominated by light nuclei, such as nuclei of hydrogen atoms. Flux of cosmic rays above 8×10^{19} eV is extremely small. These findings support the conclusion that below 3×10^{18} eV, cosmic rays are produced in the Milky Way. Above that energy, cosmic rays are made outside of our galaxy. The greatly reduced flux above 8×10^{19} eV supports the extragalactic origin of cosmic rays in this energy region, since any rays produced in distant galaxies would be attenuated by the microwave background radiation created at the time of the big bang.

A 3.2×10^{20} eV (5.1×10^8 ergs) event is the highest energy event recorded so far. Given the current understanding of the magnetic field strength in between galaxies, the arrival direction of this particle is expected to point in the general direction of the source that produced it. So far, searches carried out to locate possible sources of this particle failed to find viable candidates. A high-resolution Fly's Eye detector is being built to solve the mystery of the origin of high energy cosmic rays.

See also: COSMIC RAY

Bibliography

BALTRISAITIS, R. M., et al. "The Utah Fly's Eye Detector." *Nuclear Instruments and Methods A* **240,** 410 (1985).
GAISSER, T. K. *Cosmic Rays and Particle Physics* (Cambridge University Press, Cambridge, Eng., 1990).

EUGENE C. LOH

FORCE

Force is the interaction between particles in the universe. Examples include the gravitational force between the Sun and Earth and the electrical force between electrons and protons.

In Newtonian mechanics, force is considered to be the cause of changes in motion of an object. It follows, then, that if there is no change in motion of an object, there is no net force acting on the object. This is the essence contained in Newton's first law. This law claims that an object in a uniform state of motion, which could be represented by an object at rest or moving at constant velocity, continues in that state of motion unless acted upon by a force. This law is exemplified by a spacecraft moving through empty space, far away from any sources of gravitation. Such a spacecraft does not need to continuously fire its rocket engines in order to continue moving. Once it is given an initial velocity, it will continue to move with the engine off, according to Newton's first law.

Most forces can be modeled phenomenologically as a push or a pull. For example, the pull of the earth on a person's body results in the everyday experience of weight. Forces belong to the category of physical quantities known as vectors. The simple interpretation of a vector is a quantity with both magnitude and direction. The direction of a force is certainly important to its complete description. For example, in lifting a heavy object, it is important to apply a force in the upward direction. A force on the object in the downward direction will give a completely different result.

Because of the vector nature of forces, it is possible that a number of forces can simultaneously act on an object with no resulting change in motion of the object. In this case, the net force must be zero, that is, the force vectors must cancel each other, and we say that the object is in translational equilibrium. This equilibrium could be static, as in the case of a chandelier hanging at rest under the action of gravity pulling downward and the tension in the chain from which it hangs pulling upward. The equilibrium could also be dynamic, such as in the case of an automobile moving at constant velocity. In this case, the gravitational force, the normal force from the roadway, the friction force from the road, and the air resistance all cancel so that the net force is zero.

If an object experiences a net force, the result is a change in the motion, which appears as a change in the momentum of the object. For an object of fixed mass, a change in momentum is characterized by a change in velocity. The rate of change of the velocity is acceleration. The acceleration of an object resulting from a net force on the object depends on the amount of opposition that the object offers to having its motion changed. This opposition is measured by the property that is known as mass.

The exact relation between force, mass, and acceleration is found in one form of Newton's second law:

$$\mathbf{F}_{net} = m\mathbf{a}.$$

The notation on the left suggests that if a combination of several forces is acting, it is the net result of the combination that determines the change in motion. This equation indicates that, for a given mass, a larger net force will result in a proportionally larger acceleration. This is a familiar everyday experience—if one pushes harder on an object, the motion changes more rapidly. On the other hand, if we imagine providing the same acceleration to a variety of objects, the equation indicates that the force required is proportional to the mass of the object. This is also a familiar experience—the more massive an object is, the harder it is to change its motion.

The fact that forces represent interactions between entities leads to another of Newton's laws, Newton's third law. This law demonstrates the symmetry of the interaction—each interacting object exerts an equal but oppositely directed force on the other. A common statement of Newton's third law is this: If object A exerts a force on object B, then B exerts an equal force in the opposite direction on A. For example, imagine a child standing on skates

next to a wall. As the child pushes horizontally on the wall, the wall pushes on the child with an equal force in the opposite direction. As a result, the motion of the child changes—he or she moves away from the wall.

In one model of forces, the interaction is considered to act directly between particles, due to special properties of the particles, such as mass or electric charge. This is generally known as action-at-a-distance, and it raises the question as to how the particles communicate their presence to one another. Another model, field theory, addresses this question by picturing one particle, because of its special property, establishing an altered space around itself. The altered space is called a field (gravitational field, electrical field, etc.). The second particle interacts with the field, due to its similar special property, resulting in the force. Yet another model, quantum field theory, describes the interacting particles as communicating by emitting and absorbing exchange particles with each other. For example, electrically charged particles are imagined to exchange photons. This exchange of photons results in the electromagnetic force. Each of these descriptions has a corresponding mathematical formulation that predicts the behavior of interacting particles under certain conditions.

As mentioned above, the interaction between particles (or between a particle and a field) depends on special properties of the particles. For most of the twentieth century, four fundamental interactions were identified: gravitational, electromagnetic, strong, and weak.

Gravitational force is the attractive interaction acting between particles that have the property of energy. The most common form of energy that results in significant effects due to the gravitational force is mass-energy, so we often state that gravitation is an interaction between particles with mass. It is an inherently weak force, requiring planetoid-sized collections of mass for observable results. It is a long-range force, falling off as an inverse square of the distance between spherical distributions of mass such as planets. In quantum field theory, the exchange particles are called gravitons.

Electromagnetic force is the attractive or repulsive interaction acting between particles that have the property of electric charge. It is a much stronger force than the gravitational force, resulting in significant effects on microscopic particles, such as electrons and protons, as well as on macroscopic objects, such as coils of wire in motors. It is a long-range force, falling off as an inverse square of the distance between point sources. The interaction between stationary electric charges is called the electric force, while that between moving electric charges is the magnetic force. Electric charges are labeled as positive or negative. Electric forces between like charges are repulsive, while unlike charges result in an attractive force. Moving electric charges result in regions of space that are labeled as north or south magnetic poles. Similar to the electric interaction, like magnetic poles repel, while unlike magnetic poles result in an attractive force. In quantum field theory, the exchange particles are photons.

Strong force is the attractive force acting between particles that have the property of color charge, which is a property possessed by quarks. For this reason, the strong force is often called the color force. The primary role of the strong force is to hold quarks together in a particle belonging to the category called hadrons. A residual strong force will also act between color-neutral hadrons, just as a van der Waals force acts between electric charge-neutral dipoles. This results in the attractive force that holds the protons together in the nucleus, balancing their mutual repulsion due to the electromagnetic force. The residual strong force is very short-range, acting only over a separation distance on the order of 10^{-15} m. In quantum field theory, the exchange particles for the strong force (color force) are called gluons.

Weak force is the most difficult of the four interactions listed here to envision as a push or a pull. It is much more instructive to consider this as an interaction, rather than a force. The most evident role of the weak force is in its effect on nuclear particles, resulting in the transmutation of neutrons into protons and vice versa. The observable effect of this transmutation is the ensuing beta decay of radioactive nuclei. The weak force is short-range, acting over distances small compared to a nucleon, thus qualifying more as a force *within* a particle rather than *between* particles. In quantum field theory, the exchange particles are called W (with either positive or negative electric charge) and Z (electrically neutral).

During the early part of the nineteenth century, a list of fundamental forces would have treated electricity and magnetism separately. It was not clear at that time that they were different manifestations of the same fundamental force. With the work of Hans Christian Oersted, Michael Faraday, and James Clerk Maxwell, these two forces were unified into the electromagnetic force. In the latter part of the twentieth century, physicists made concerted efforts to further unify the forces, with the ultimate goal of demonstrating that all forces are manifestations of one single interaction. With this approach, at high enough energies, the four forces listed above would theoretically become indistinguishable. At lower energies, however, the interactions develop unique qualities and appear to be different types of forces. Success has been made in theoretically unifying the electromagnetic and weak forces into the electroweak interaction.

A category of forces that does not appear explicitly in the discussion above is that of contact forces. These include the force that a hand applies to a book sliding across a table, the normal force from the floor upward on a standing person, the tension in a string, the friction force of a hockey puck sliding across ice, and so on. In reality, these forces have the electromagnetic force as a basis. When a hand pushes a book across a table, it is an electrical repulsion between the atoms of the hand and those of the book that causes the change in motion. The normal force from the floor is also an electrical repulsion between the atoms of the floor and the atoms on the bottom of the feet. The tension in a string is due to the electrical attraction among the atoms making up the string. The friction between a sliding hockey puck and the ice is due to momentary electromagnetic interactions that occur between the atoms of the puck and those of the ice as they pass by each other.

On a macroscopic level, then, the only forces that affect us directly in our everyday experience are the gravitational force (in our interaction with the earth), and the electromagnetic force (in all other experiences involving forces). Macroscopic objects interact because of the forces among the microscopic particles from which they are composed. The strong and weak forces affect microscopic processes all around us, but we do not interact with them in terms of observable, everyday phenomena.

A distinction is often made between conservative forces and non-conservative, or dissipative, forces. A conservative force is one that allows for transformations between kinetic and potential energies in a system such that the sum of the kinetic and potential energies is constant. For example, in the case of a ball thrown in the air in the absence of air resistance, the ball-earth system will have the same total of its kinetic and gravitational potential energies throughout the motion—the gravitational force is conservative. The electrical force is another example of a conservative force—as an electron is accelerated in an electron gun in a television set, the sum of kinetic and electric potential energies is constant.

On the other hand, let us consider the air resistance on the ball thrown into the air. The effect of air resistance is to increase the internal energy of the ball and the air—there is an increase in temperature. This energy comes at the expense of the total of kinetic and potential energies, whose sum will decrease due to the air resistance. Thus, air resistance is a non-conservative force—it results in a transformation to internal energy. In a similar way, the friction force between a sliding object and the surface on which it slides results in an increase in temperature of the book and the surface. The kinetic energy associated with its motion is being transformed to internal energy. Friction is another example of a non-conservative force.

Forces acting between particles are inherently conservative—the fundamental forces discussed above are conservative if we imagine them acting only between a pair of particles. If we include the interaction of forces with the surrounding matter, however, we introduce the possibility of non-conservative forces. In the cases of air resistance and friction mentioned previously, the object of interest interacts with the surrounding matter, resulting in the possibility for energy associated with the object to transform to internal energy in the surrounding matter.

See also: CENTRIFUGAL FORCE; CENTRIPETAL FORCE; ELECTROMAGNETIC FORCE; ELECTROMOTIVE FORCE; ELECTROSTATIC ATTRACTION AND REPULSION; FIFTH FORCE; GRAVITATIONAL FORCE LAW; INTERACTION, ELECTROMAGNETIC; INTERACTION, ELECTROWEAK; INTERACTION, FUNDAMENTAL; INTERACTION, STRONG; INTERACTION, WEAK; NEWTONIAN MECHANICS; NEWTON'S LAWS; NORMAL FORCE; NUCLEAR FORCE; VAN DER WAALS FORCE

Bibliography

HALLIDAY, D., RESNICK, R., and WALKER, J. *Fundamentals of Physics* (Wiley, New York, 1993).

HECHT, E. *Physics* (Brooks/Cole, Pacific Grove, CA, 1994).

HESSE, M. B. *Forces and Fields* (Thomas Nelson and Sons, London, 1961).

JEWETT, J. W. *Physics Begins with an M . . . Mysteries, Magic, and Myth* (Allyn & Bacon, Boston, 1994).

KAUFMAN, W. J. *Particles and Fields: Readings from Scientific American* (W. H. Freeman, San Francisco, 1980).

JOHN W. JEWETT JR.

FOUCAULT PENDULUM

See PENDULUM, FOUCAULT

FOURIER SERIES AND FOURIER TRANSFORM

Any complex sound (vibration or periodic disturbance) can be broken down into its pure sine wave components. Likewise, any complex sound can be created from the sum of pure sine waves having frequencies that are multiples of the fundamental frequency. This very remarkable theorem by Jean Baptiste Joseph Fourier says that any periodic curve in time or space (continuous, periodic, single-valued function in space or time), however complicated, may be built up by adding together a series of sinusoidal functions whose amplitudes a, frequencies ω, and relative phases are judiciously selected.

An astounding result of this theorem is the fact that the complex sound from a violin playing the note A (440 Hz) can be created exactly by a large number of sine wave generators, each one playing a particular frequency, loudness, and phase. Often, only a rather small number of waves will be needed to duplicate the complex sound to a satisfactory approximation. The more extreme the variations in the amplitudes and slopes (as, for example, in the sound of a single drum beat), the more terms are needed in the Fourier series to adequately represent it.

Any oscillating function $f(t)$ can be exactly reproduced by a sum of the form

$$
\begin{aligned}
f(t) = {} & a_0\cos 0\omega t + a_1 \cos 1\omega t + a_2\cos 2\omega t \\
& + a_3\cos 3\omega t + \ldots + a_n\cos n\omega t + \ldots \\
& + b_0\sin 0\omega t + b_1\sin 1\omega t + b_2\sin 2\omega t \\
& + b_3\sin 3\omega t + \ldots + b_n\sin n\omega t + \ldots,
\end{aligned}
$$

for some frequency ω. This infinite sum of sine and cosine terms can be made to represent any oscillating function by choosing the appropriate amplitudes, frequencies, and phases. In more compact form we can write

$$f(t) = a_0 + \sum_{n=1}^{\infty} a_n \cos n\omega_0 t + \sum_{n=1}^{\infty} b_n \sin n\omega_0 t,$$

where $\omega_0 = 2\pi f_0 = 2\pi/T_0$ and $\omega_n = n\omega_0$ (time varying), and

$$f(x) = a_0 + \sum_{n=1}^{\infty} a_n \cos n k_0 x + \sum_{n=1}^{\infty} b_n \sin n k_0 x,$$

where $k_0 = 2\pi\sigma_0 = 2\pi/\lambda_0$ and $k_n = nk_0$ (space varying). The a_0, a_n, and b_n are amplitude coefficients.

Mathematicians prior to Fourier used series of this type to solve certain problems where it was evident that such a series should exist. They were surprised when Fourier showed that any arbitrary periodic function could also be expressed this way. For example, consider the following time-varying function:

$$y(t) = 2 + 3 \sin 2\omega t + 5 \sin 3\omega t + 4 \cos 5\omega t.$$

Compare this with the Fourier series

$$y(t) = a_0 + a_1 \sin 1\omega_1 t + a_2 \sin 2\omega_2 t + a_3 \sin 3\omega_3 t$$

$$+ \ldots b_1 \cos 1\omega_1 t + b_2 \cos 2\omega_2 t + b_3 \cos 3\omega_3 t + \ldots.$$

By inspection we see that it is made up of the following coefficients:

$$a_0 = 2,$$

$$[b_2 = 3, \; \omega_2 = 2\omega_0]$$

$$[b_3 = 5, \; \omega_3 = 3\omega_0]$$

$$[a_5 = 4, \; \omega_5 = 5\omega_0].$$

The resultant curve $y(t)$ is shown in Fig. 1a, and its "frequency distribution," that is, the frequencies and their amplitudes needed to make up the curve, is shown in Fig. 1b.

The complex wave $y(t)$ has been synthesized from the three harmonic functions and a constant. By inspection we analyzed this special case $f(t)$ to determine the amplitudes and frequencies needed to make this function. Fourier showed how to analyze a general $y(t)$ and get the amplitudes and frequencies that define $y(t)$. For any function of period T we can fix the fundamental frequency ω_0 by $T = 2\pi/\omega_0$. Then, to get the values a_n, b_n, $\omega_n = n\omega_0$, cor-

responding to the frequencies, we must first integrate both sides of $f(t)$ with respect to t over one full period T. We get

$$\int_0^T f(t)\,dt = \int_0^T a_0\,dt + \int_0^T \sum a_n \cos n\omega_0 t\,dt$$

$$+ \int_0^T \sum b_n \sin n\omega_0 t\,dt$$

$$= \int_0^T a_0\,dt + 0 + 0 = a_0 T,$$

since the integrals of any sine or cosine over one full period are zero. Our formula for the first constant a_0 is

$$a_0 T = \int_0^T f(t)\,dt$$

(a)

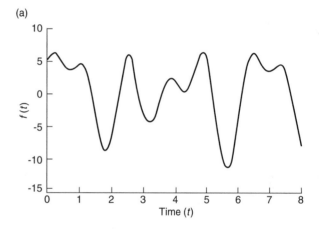

(b)

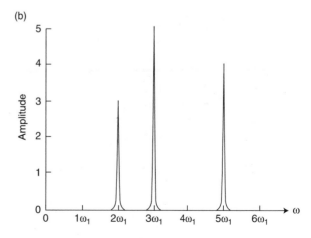

Figure 1 (a) The complex wave $y(t)$, and (b) its amplitude-frequency distribution.

or

$$a_0 = \frac{1}{T}\int_0^T f(t)\ dt.$$

The other coefficients are found as follows. To determine a_n, multiply both sides of $f(t)$ by cos $m\omega t\ dt$, where the m are integers like the n, and integrate over one full period, say from $t = 0$ to $t = T$. Then we get

$$\int_0^T f(t)\cos\ m\omega t\ dt = \int_0^T a_0\cos\ m\omega t\ dt$$
$$+ \int_0^T \Sigma\ a_n\cos\ n\omega_0 t \cos\ m\omega t\ dt$$
$$+ \int_0^T \Sigma\ b_n\sin\ n\omega_0 t \cos\ m\omega t\ dt.$$

In this rather strange integral only certain terms give nonzero contributions—remarkably, only those for which $n = m$. We get

$$\int_0^T f(t)\cos\ m\omega t\ dt = 0 + a_n \int_0^T \cos^2\ n\omega_0 t\ dt$$
$$= a_n \int_0^T [(1 + \cos\ 2n\omega_0 t)/2]\,dt$$
$$= a_n T/2.$$

This gives us our second formula:

$$a_n = (2T)\int_0^T f(t)\ \cos\ n\omega_0 t\ dt.$$

Here we used the following orthogonality conditions for sine and cosine functions:

$$\int_0^T \sin\ n\omega_0 t \sin\ m\omega_0 t\ dt = \begin{cases} 0, & n \neq m \\ T/2, & n = m \end{cases}$$

$$\int_0^T \sin\ n\omega_0 t \cos\ m\omega_0 t\ dt = 0, \quad \text{all } n \text{ and } m.$$

(These can be readily derived using trignometric identities.) Similarly, to find b_n, multiply both sides of $f(t)$ by sin $m\omega_0 t\ dt$ and integrate over one full period and get

$$b_n = (2/T)\int_0^T f(t)\ \sin\ n\omega_0 t\ dt.$$

We now have the recipe to calculate all the coefficients. As an example, we calculate b_n for the function $y(t)$ that was created earlier:

$$b_n = (2/T)[\int_0^T 2\ \sin\ n\omega_0 t + 3\sin\ 2\omega_0 t \sin\ n\omega_0 t$$
$$+ \int_0^T 5\sin\ 3\omega_0 t \sin\ n\omega_0 t$$
$$+ \int_0^T 4\cos\ 5\omega_0 t \sin\ n\omega_0 t]\ dt.$$

Note here that n is an integer that runs from zero to infinity, while the function contains only certain integers in the argument. We get nonzero values for the integrals only when $n = m$. We get $[b_2 = 3, \omega_2 = 2\omega_0]$ and $[b_3 = 5, \omega_3 = 3\omega_0]$ with all other $b_n = 0$.

Likewise, we calculate the a_n by multiplying $f(t)$ by cos $m\omega_0 t$ and integrating over one full period. Since only the cosine term survives, we get

$$a_n = (2/T)\int_0^T 4\ \cos\ 5\omega_0 t \cos\ n\omega_0 t\ dt.$$

This gives $[a_5 = 4, \omega_5 = 5\omega_0]$ and zero for all other values of n. Likewise for a_0 we get

$$a_0 = (1/T)\int_0^T f(t)\ dt = (1/T)\int_0^T 2\ dt = 2.$$

This gives $[a_0 = 2]$. These calculated values match exactly with those of our initial $y(t)$.

Periodic signals that arise in nature are not necessarily sinusodial in form, although these do frequently occur. Sinusoidal examples are the complex musical tones from musical instruments, water waves, and the short and long wavelength ripples that describe sand dunes. The Fourier expansion can represent even those functions that are not periodic or cannot be expressed in closed analytic form. There are restrictions. The functions must be made of sections or pieces of analytic curves, not necessarily joined at the ends. For simplicity, here we restrict our examples to simple periodic cases.

For example, a square wave shown in Fig. 2a can be represented as a function of infinite sums of sines and cosines. Such an alternating voltage function can arise by rapidly switching battery terminals many times per second. Fourier shows how to calculate the waves that make the square wave.

First we must define the function on the single periodic interval from $t = 0$ to $t = T$:

$$f(t) = \begin{cases} + V_0, & 0 \le t \le T/2 \\ - V_0, & T/2 \le t \le T. \end{cases}$$

Here $\omega = 2\pi/T$. Now we calculate the Fourier coefficients for an expansion of this function. We get

$$a_0 = (1/T)\left[\int_0^{\frac{T}{2}} V_0 \, dt + \int_{\frac{T}{2}}^T -V_0 \, dt\}\right] = 0.$$

(a)

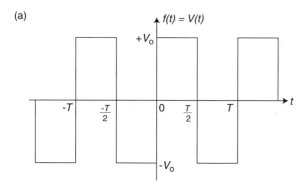

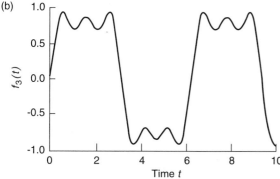

(c)

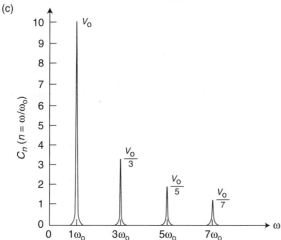

Figure 2 (a) A square wave, (b) the sum of the first three waves that make up a square wave, and (c) the amplitude-frequency distribution for a square wave.

This says the DC or constant term is zero, and

$$a_n = (2/T)\int_0^{\frac{T}{2}} [V_0 \cos 2\pi nt/T] \, dt$$

$$+ (2/T)\int_{\frac{T}{2}}^T [-V_0 \cos 2\pi nt/T] \, dt$$

$$= [V_0/\pi n](2\sin \pi n = \sin 2\pi n) = 0$$

for all n. This says there are no cosine terms, and

$$b_n = (2/T)\int_0^{\frac{T}{2}} [V_0 \sin 2\pi nt/T] \, dt$$

$$+ (2/T)\int_{\frac{T}{2}}^T [-V_0 \sin 2\pi nt/T] \, dt$$

$$= [V_0/\pi n](2 - 2\cos \pi n) = 2V_0/\pi n)(1 - \cos \pi n).$$

When n is an even integer, all $b_n = 0$. When n is odd, $b_n = 4V_0/\pi n$. Thus the Fourier series for the square wave is

$$f(t) = (4V_0/\pi)[1/1 \sin 1\omega_0 t + 1/3 \sin 3\omega_0 t$$

$$+ 1/5 \sin 5\omega_0 t + 1/7 \sin 7\omega_0 t \ldots.$$

This can be written in a more compact form:

$$f(t) = (4V_0/\pi)\sum_{n=0}^{\infty} [\sin (2n - 1) \omega_0 t]/(2n - 1).$$

All a_n are zero for all n. The b_n are zero only when the n are even numbers $(2, 4, 6 \ldots)$. We say the even harmonics are zero. Since b_n exists only for odd n, this series contains only odd harmonics. Their amplitudes decrease as $1/n$. Signal generators set to these frequencies and periods would produce an electrical signal on an oscilloscope that will approximate the square wave more accurately, as successively more terms of the series are added. The sum of the first three terms are shown in Fig. 2b. The amplitude-frequency distribution is shown in Fig. 2c.

An important and practical acoustical problem is a plucked guitar string of length L that is initially pulled aside a distance h from its center (Fig. 3a). Just before release, its shape is a "triangle." When it

609

is released it will vibrate with a shape predicted by the Fourier amplitudes and frequencies that describe its initial shape. Here we use the space (x) representation of the Fourier series. The periodicity occurs over the length L, and the wavelengths λ allowed are $2L/1$, $2L/2$, $2L/3$, $2L/4 \cdots 2L/N$. Recall the parameter $k_0 = 2\pi/\lambda_0$ for the space representation is analogous to $\omega_0 = 2\pi/T_0$ for the time representation.

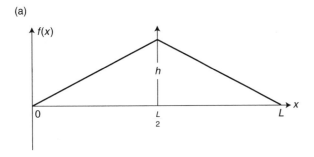

(a)

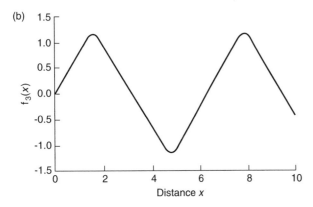

(b)

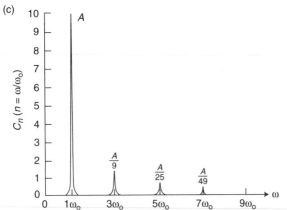

(c)

Figure 3 (a) The triangle shape of a plucked string just before release, (b) the sum of the first three sine waves that make up a triangle wave, and (c) the amplitude-frequency distribution for a triangle wave.

First we must define the function. Looking at Fig. 2b, the geometry of the string can be defined by

$$f(x) = \begin{cases} 2hx/L, & 0 \leq x \leq L/2 \\ 2h(L-x)/L, & L/2 \leq x \leq L. \end{cases}$$

Also by inspection, $f(x) = 0$ at all times at the end points: $x = 0$ and $x = L$. This says that all cosine terms (i.e., the b_n) must be zero. We need calculate only the a_n:

$$a_n = (2/L)\int_0^{\frac{L}{2}} f(x) \sin nk_0x \, dx$$

$$+ (2/L)\int_{\frac{L}{2}}^{L} f(x) \sin nk_0x \, dx$$

$$a_n = (2/L)\int_0^{\frac{L}{2}} 2h/x \sin nk_0x \, dx$$

$$+ (2/L)\int_{\frac{L}{2}}^{L} 2h(L-x)/L \sin nk_0x \, dx$$

$$a_n = [8h/\pi^2 n^2] \sin \pi n/2.$$

Since the $b_n = 0$, we get

$$f(x) = (8h/\pi^2)[0 + 1/1 \sin 1k_0x - 1/9 \sin 3k_0x$$

$$+ 1/25 \sin 5k_0x].$$

Again in this case only odd harmonics exist, and here we find that their amplitudes decrease as $1/n^2$. The alternating negative sign indicates that every other sine term is shifted by π, that is, one half wave. The sum of the first three terms is shown in Fig. 3b. Note how its shape already approximates the plucked string of Fig. 3a. Its amplitude-frequency distribution is shown in Fig. 3c.

Although it appears at first glance that the Fourier expansion is suited only to the analysis of periodic functions, we can get a Fourier representation even for nonperiodic functions. We can do this by using the simple trick of taking the limit in which the period T becomes arbitrarily large ($T \rightarrow \infty$). The sum over discrete frequencies $n\omega_0 = 2\pi n/T$ then becomes an integral over all possible ω. Let us demonstrate this, starting from the Fourier sum for a function of period T:

$$f(t) = \sum_{n=-\infty}^{\infty} c_n \exp(in\omega_0 t)$$

with

$$c_n = 1/T \int_{-\frac{T}{2}}^{\frac{T}{2}} f(t) \exp(-in\omega_0 t) \, dt$$

and $\omega = 2\pi/T$. Here we have used the shorthand $c_n = a_n + ib_n$ and the definition $e^{i\theta} = \cos\theta + i\sin\theta$ to combine the two sums into a single expression.

Now the sum over n includes frequencies that are spaced by an amount $\delta\omega = 2\pi/T$. As T gets larger this spacing shrinks, so eventually we can replace the sum by an integral over all values of ω. To see this, let us insert a factor $1 = \delta\omega(T/2\pi)$ into the Fourier sum

$$f(t) = \sum c_n \exp(in\omega_0 t) = \sum (\delta\omega)(T/2\pi)\exp(i\omega t)$$

$$\rightarrow (\text{as } T \rightarrow \infty) \rightarrow \int d\omega \, g(\omega)\exp(i\omega t),$$

(a)

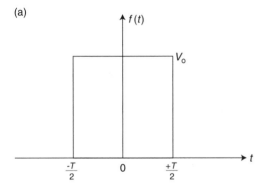

(b)

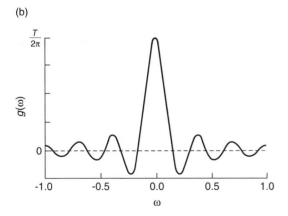

Figure 4 (a) A rectangular voltage pulse, and (b) its amplitude-frequency distribution.

where the previous formula for c_n provides the definition

$$g(\omega) = \lim (T/2\pi) c\omega$$

$$= 1/2\pi \int f(t) \exp(i\omega t) \, dt.$$

Thus the discrete set of Fourier coefficients c_n are replaced by the continuous function $g(\omega)$, which is known as the Fourier transform of the original function $f(t)$. Given either of the related pair of functions $f(t)$ and $g(\omega)$, we have full information on both the time and the frequency dependence of the physical process in question by using the Fourier transform to generate the other member of the pair. This turns out to be a very powerful tool in many areas of physics.

As an example, let us apply these integrals to find what frequencies and amplitudes are needed to build a (nonperiodic) function such as a step voltage pulse of amplitude V_0 and duration τ in Fig. 4a. First we must define the function. From Fig. 4a we see that

$$f(t) = \begin{cases} 0, & -\infty < t \leq -T/2 \\ V_0, & -T/2 \leq t \leq T/2 \\ 0, & +T/2 \leq t < +\infty. \end{cases}$$

To get all the frequencies needed to create the pulse, we calculate

$$g(\omega) = \int f(t) \exp(-i\omega t) \, dt$$

$$= \int V_0 \exp(-i\omega t) \, dt = V_0 \left[\exp(-i\omega t)\right]/i\omega \big|$$

$$= V_0[\exp(+i\omega\tau/2) - \exp(-i\omega\tau/2)]/i\omega$$

$$= V_0[2 \sin \omega\tau/2]/\omega = V_0\tau [\sinh \omega\tau/2].$$

When $\omega = 0$, $g(\omega) = V_0\tau$. The amplitude-frequency distribution is shown in Fig. 4b. When the infinite number of frequencies ω, with the proper amplitudes, given by the distribution $g(\omega)$ are added, they will produce the step function $f(t)$ shown in Fig. 4a.

The net result of Fourier analysis is that we need use nothing more complicated than sinusoidal currents and voltages to study the reaction of electrical (and mechanical) circuits to complex waves. We can mathematically break the complex signal into its separate sinusoidal components, compute what the effect each of these will have in the circuit, and finally add the effects.

See also: MOTION, PERIODIC; OSCILLATION; SOUND; WAVELENGTH

Bibliography

FRANKLIN, P. *An Introduction to Fourier Methods and the Laplace Transformation* (Dover, New York, 1949).
GASKILL, J. D. *Linear Systems, Fourier Transforms, and Optics* (Wiley, New York, 1978).
STEWARD, E. G. *Fourier Optics: An Introduction* (Wiley, New York, 1983).

WILLIAM BICKEL

FRACTAL

Fractal is defined in an abstract language by mathematicians as any geometrical figure (a point set) whose Hausdorff dimension is larger than its topological dimension. Hausdorff dimension is a way to quantify the degree of complexity of zigzag lines. Topological dimension is a measure of "connectedness" of a figure, that is, whether one needs to cut at a point or along a line to divide a geometrical figure in two parts at an arbitrary location.

In a more qualitative manner, fractal is defined by Benoit B. Mandelbrot as a shape made of parts similar to the whole in some way. Some coastlines such as the one of western Britain, as noted by Mandelbrot, or the complicated one of Norway with fjords, observed by Jens Feder, exhibit this fractal property. When one magnifies a segment of the coastline, it looks as complicated as the overall view of it. As one continues in turn to magnify a portion of this magnified coastline, it again looks as complicated and jagged as the original pattern. There are many natural and artificially generated patterns that exhibit this characteristic behavior.

One method to quantify the fractal behavior of a pattern is to measure its fractal dimension. This is related to the Hausdorff (or Hausdorff–Besicovitch) dimension. As one tries to measure the "true length" L_0 of a complex curve like a coastline with fjords, river mouths, and so on, one might begin by using a divider (a device like a compass but with both legs equipped with a sharp needle point) that spans a short length δ. We "walk" the divider along this curve and count the number of steps $N(\delta)$ to cover the entire length of the curve. Let us call the approximate length of the curve obtained by this measurement $L(\delta)$, obviously a function of the step δ as:

$$L(\delta) = N(\delta)\delta.$$

To improve the measurement, we make the step δ shorter and shorter. We should expect:

$$\lim_{\delta \to 0} L(\delta) = \lim_{\delta \to 0} N(\delta)\delta = L_0.$$

This is not the case with fractals. As we make the steps shorter and shorter, we find that we do not appear to reach a limiting value for $L(\delta)$. Instead, we find that the expression

$$L(\delta) = a \cdot \delta^{1-D} \tag{1}$$

describes the behavior reasonably well. For example, Mandelbrot applied this approximation to the coastline of Britain by making a log-log plot of Eq. (1). The slope of the plot is $1 - D$, and the value he obtained was $D = 1.3$. This D is called the fractal dimension. Ordinary (nonfractal) curves should give exactly $D = 1$, as a in this case would be the true length L_0. Mandelbrot also measured D for a circle by using the same procedure and obtained $D = 1$ as expected. Hausdorff–Besicovitch dimension is defined along the same line of reasoning, but it uses a mathematical concept of "measure" along with general concepts of set theory.

Fractal Dimension

There are a large number of recipes for generating fractals. One of them is the triadic Koch curve (Fig. 1). This process begins with a simple line segment of length 1 ($n = 0$), to which a generator operation is applied. The generator for the Koch curve is an operation of replacing a given line segment with four segments, each equal in length to one-third the length of the original segment, to create a tent-like triangle in the middle ($n = 1$). The total length of this figure is 4/3. The next step ($n = 2$) is to apply

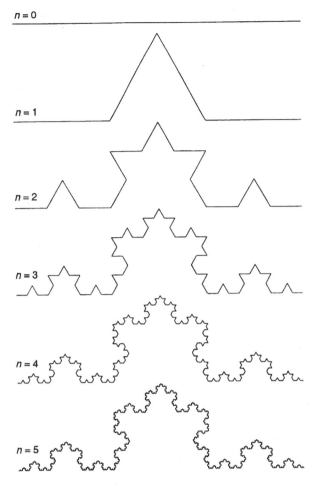

$n = 0$

$n = 1$

$n = 2$

$n = 3$

$n = 4$

$n = 5$

Figure 1 Construction of the triadic Koch curve.

the same generator to each one of the four shorter ($1/3$ each) line segments to create a more complicated figure made up of $4^2 = 16$ segments, each of length $\delta = 3^{-2} = 1/9$ of the original length. Then $L(\delta) = L(1/9) = 4^2\delta = (4/3)^2 = 16/9$. As we continue to the nth step, we will have

$$L(\delta) = (4/3)^n \ (= 4^n \, \delta), \qquad (2)$$

where the line segment for each successive step is

$$\delta = 3^{-n} \qquad (3)$$

To rewrite this expression in the form $L(\delta) = \delta^{1-D}$, take a logarithm of Eq. (3) to find n as

$$n = -(\ln \delta)/\ln 3. \qquad (4)$$

Now Eq. (2) can be rewritten as

$$
\begin{aligned}
L(\delta) &= (4/3)^n \\
&= e^{n(\ln 4 \, - \, \ln 3)} \\
&= e^{-\ln \delta(\ln 4 \, - \, \ln 3)/\ln 3} \\
&= e^{\ln \delta(1 \, - \, \ln 4/\ln 3)} \\
&= \delta^{1 \, - \, D}, \qquad (5)
\end{aligned}
$$

where $D = \ln 4/\ln 3 \approx 1.262$. This is the fractal dimension of the triadic Koch curve when the generator is applied successively until $n \rightarrow \infty$. Figures such as the one with $n = 2, 3, \ldots$ and so on, are justifiably called prefractals.

Fractals and Physics

There are many different types of fractals and correspondingly many ways to view the properties of fractals, such as scaling, which has to do with the question of whether or not we recover the same set of points when we change the length scale. A figure that is invariant to scaling is called self-similar. Along with the similarity, a method for quantifying the similarity is introduced into the theory of fractals as similarity dimension. This aspect of the theory of scaling is related to one of the significant recent developments in thermodynamics, that is, the theory of critical phenomena.

One of the many examples of fractals in nature is associated with the diffusion-limited aggregation process (or DLA), such as the pattern of growth of ice crystals on a glass plate or of dendrite in metals. Many fractal patterns develop as a result of the processes where the rate limiting factor is the diffusion of atoms and molecules in liquids, solid surfaces, or in the bulk of solids.

See also: CRITICAL PHENOMENA

Bibliography

BARNSLEY, M. F. *Fractals Everywhere* (Academic Press, San Diego, CA, 1988).

BUNDE, A., and HAVIN, S., eds. *Fractals and Disordered Systems* (Springer Verlag, New York, 1991).

FEDER, J. *Fractals* (Plenum Press, New York, 1988).

MANDELBROT, B. *Fractal Geometry of Nature* (W. H. Freeman, New York, 1982).

CARL T. TOMIZUKA

FRAME OF REFERENCE

A frame of reference is essential for quantitative measurements of physical events. It consists of a spatial coordinate system and a clock that an observer uses to measure the position and time of a physical event in the frame. Typically, a three-dimensional coordinate system and a set of standardized clocks are used, but a frame of reference can be defined more generally in terms of a set of physical bodies with clocks moving according to arbitrary laws.

An observer maps a trajectory for a body by measuring its position at each moment in time in a given reference frame. This trajectory represents a law of motion in that frame that can be described in terms of an equation of motion. The motion may appear to be different when viewed by an observer in another reference frame located at a different point in space, or moving relative to the original frame. If we know how the relative positions and orientations of the two frames change with time, we can define a transformation that converts the measured coordinates of the trajectory from one frame to the other. We say that the laws of motion are invariant under this transformation if the same equation of motion applies in both frames.

Light travels so quickly that it appears to be instantaneous. Isaac Newton assumed that the speed of light was infinite, so that a single clock can define the time for all frames of reference, regardless of their positions or relative motion. Frames of reference that share the same definition of time are called Newtonian frames.

Inertial frames of reference have different positions and relative velocities for their spatial coordinate systems. In the absence of external forces, a body is stationary in its center-of-mass reference frame where the spatial coordinate axis is at the body's center of mass. Newton's laws describe the motion of a body in an inertial reference frame. In some other inertial frame, moving at a constant relative velocity v, the body itself appears to move with a velocity v as described by Newton's laws in the absence of an applied force. A Galilean transformation converts the velocity and position of the body as seen from one Newtonian inertial frame to another. A force F acting on a body with inertial mass m produces the same acceleration, $a = F/m$, in all inertial frames.

The calculation of trajectories is often simplified by transforming to a convenient frame. In scattering experiments (e.g., billiard balls on a pool table), a projectile is fired at a stationary target, and both projectile and target move after the collision. The resulting positions and velocities of target and projectile are measured in the laboratory frame in which the target is initially at rest. The calculations are easier to perform in the center-of-mass frame, where both the target and projectile initially move toward each other. Scientists generally perform the calculations in the center-of-mass frame and use a Galilean transformation to convert the results to the laboratory frame for comparison with the experimental measurements.

Accelerated reference frames are noninertial frames. Newton's laws do not describe the motion of bodies in accelerated frames, since a body at rest in such a frame requires a force acting on it to keep it accelerating with the frame. A rotating reference frame is a special example of an accelerated frame. A body at a fixed point in a rotating frame moves in a circle when viewed from an inertial frame, where it experiences an acceleration toward the axis of rotation due to the centripetal force. Conversely, a body at rest in an inertial frame appears to move in a circle when viewed from a rotating frame. We define an effective force, the centrifugal force, to account for this observed circular motion in the rotating frame. In addition, if a body in a rotating frame moves further away from the axis of rotation, it increases its circular velocity as seen in an inertial frame, and so appears to be accelerating. The Coriolis force is defined to account for this apparent acceleration. Newton's laws are modified to describe motion in a rotating frame of reference by adding the centrifugal and Coriolis forces to the applied force. This approach is used extensively in calculations of motion on Earth, for example; in meteorological calculations for weather predictions; and in flight path calculations for long-range missiles.

Newton's assumption that a single clock can be used in all inertial frames followed from assuming that the speed of light is infinite. Although this is often a very good approximation, the speed of light is really finite, so the equations of motion require modification when velocities approach the speed of light. James Clerk Maxwell derived a set of equations to describe electromagnetic interactions that directly involves the speed of light in a vacuum. These equations are not invariant under Galilean transformations, but Hendrick Antoon Lorentz showed that they are invariant under Lorentz transformations. These transformations show that position and time are not independent and are both changed when transforming from one frame to another. This four-dimensional frame of reference is called Minkowski spacetime.

Albert Einstein used these ideas in his special theory of relativity. He assumed that the speed of light is constant in all inertial frames and that all inertial frames have the same laws of nature (the principle of relativity). He showed from these assumptions that Lorentz transformations, not Galilean transformations, provide the correct method for converting space and time coordinates from one frame to another. In practice, the difference between the two is very small when the velocity differences between frames are much smaller than the speed of light, so that Newton's laws are generally valid. At very high velocities, however, relativistic corrections to Newton's laws become very important. This has been verified in experiments with elementary particles.

Einstein's general theory of relativity introduced the idea of curved spacetime. Einstein assumed that the laws of physics could be described locally in terms of inertial frames of reference related by Lorentz transformations. Over long enough distances, however, the space is not flat, as assumed in standard inertial frames. Einstein postulated that space is curved by the presence of bodies with inertial mass, and that this curvature of spacetime is responsible for the force of gravity.

See also: Center-of-Mass System; Centrifugal Force; Centripetal Force; Coriolis Force; Frame of Reference, Inertial; Frame of Reference, Rotating; Galilean Transformation; Lorentz Transformation; Newton's Laws; Relativity, Special Theory of; Spacetime

Bibliography

Kittel, C.; Knight, W. D.; and Ruderman, M. A. *Berkeley Physics Course,* Vol. 1: *Mechanics* (McGraw-Hill, New York, 1965).

Levich, B. G. *Theoretical Physics* (North-Holland, Amsterdam, 1971).

Resnick, R., and Halliday, D. *Physics,* 3rd ed. (Wiley, New York, 1977).

Taylor, E. F., and Wheeler, J. A. *Spacetime Physics* (W. H. Freeman, San Francisco, 1966).

Philip A. Sterne

FRAME OF REFERENCE, INERTIAL

Galileo Galilei made the discovery, and Isaac Newton took it as his first law of mechanics: A free particle remains at rest or travels with constant speed along a straight line. But a straight line relative to what? To describe motions, and physical events in general, we need a reference frame. For example, if we choose a frame rigidly attached to the earth, the various points of the earth remain at rest in this frame, while the stars trace out vast circles in the course of each day. Among all possible reference frames there is one class that plays a special role in classical mechanics (and also in special relativity), namely the class of inertial frames. Their defining property is that Newton's first law holds in them; their geometry is assumed to be euclidean, and they must also be endowed with a suitable time-keeping system. One can picture an inertial frame as an imaginary cubical lattice of unit rulers, filling all space and having identical standard clocks attached to each lattice point. These clocks can be synchronized, for example, by shooting equal bullets from equal guns in all directions from the spatial origin at time $t = 0$. As a lattice clock is passed by one of these bullets, it is set to read $t = r/u$, where r is its distance from the origin and u the speed of the bullets. In principle, any event can then be assigned the (x,y,z) coordinates of the nearest lattice point and the time t read on the nearest clock. In practice, of course, a more indirect method must be used.

Under the assumption of Newtonian kinematics it is easy to see that the existence of one inertial frame implies the existence of infinitely many others, all moving uniformly (i.e., with constant speed and without rotation) relative to each other. To Newton, the frame of the fixed stars was the basic inertial frame that determined all the others.

Newton's laws of mechanics all have the remarkable property of being equally valid in every inertial frame. This is the famous Newtonian relativity principle. Anyone who has flown in a fast airplane has directly experienced this; as long as the plane flies uniformly relative to the fixed stars (which it usually does), the laws of mechanics work on board just as they do on Earth. Given enough space, we could easily play billiards or ping-pong in the plane without in the least modifying our usual style.

Albert Einstein was greatly impressed with this property of the Newtonian laws, but he also believed in the essential unity of all physics. Thus, he postulated that *all* physical laws are equally valid in all inertial frames and, in fact, that in them they find their simplest expression. This is called Einstein's relativity principle. However, prompted by discoveries in electrodynamics, Einstein also postulated one additional property of inertial frames: In *every* inertial frame, he said, light travels rectilinearly with the *same* constant speed c in all directions. On the basis of Newtonian kinematics this is quite absurd. For example, if a light signal travels down the length of an airplane at speed c, and the plane travels relative to the fixed stars at speed v, we expect that signal to travel at speed $v + c$ relative to the stars and not at speed c. Nevertheless, Einstein discovered that one can have a different kinematics—relativistic kinematics—less intuitive but every bit as self-consistent as Newton's, in which one particular speed (and for this he picked the speed of light) is invariant from inertial frame to inertial frame. Einstein's kinematics, together with its profound implications for the rest of physics when coupled with his relativity principle, is called special relativity. This has become one of the best-verified theories in all of physics.

One of its most startling ingredients is the relativity of time. In Newton's theory, as two inertial frames pass through each other, their clocks will always agree, provided that the zero settings are suitably matched. Therefore, Newton's time is *absolute*. In Einstein's theory, the clocks of two such frames agree only on one ever-shifting plane at right angles to their relative velocity; on each side of this plane the clocks of the forward moving frame lag more and more behind the clocks of the other frame. Thus, time in Einstein's theory is *relative*. Also, length has become relative, in the sense that the lattice of each inertial frame, when observed at a single moment from a second inertial frame, appears shrunk in the direction of the relative velocity of the two frames.

In spite of their utility in classical mechanics and special relativity, however, there are problems with the very concept of inertial frames. Philosophically, inertial frames offend scientific thinking inasmuch as they act but cannot be acted on. Physically, their defining property cannot really be used to locate them, since gravity is everywhere so that a truly "free" particle does not exist. And cosmologically, though the universe of galaxies must decelerate after the big bang owing to the mutual gravitational attraction of the galaxies, would we not expect, simply by symmetry, each galactic center to be at rest in an inertial frame?

The final blow against extended inertial frames was delivered by Einstein's general relativity, which, however, also showed a way out. The blow was that, according to general relativity, gravitating matter curves space so that no extended euclidean (i.e., flat) space like an inertial frame can exist in the real world. The way out was Einstein's concept of a local inertial frame. Newton had already shown that gravity is effectively eliminated, as far as mechanics is concerned, in a freely falling nonrotating cabin in a gravitational field. Particles in the cabin are effectively free and Newton's first law is really verified. A good example of this is provided by astronauts floating freely in their space capsules. Appealing to the unity of physics once more, Einstein, in his equivalence principle, extended the mechanical result to the whole of physics. Such freely falling cabins, now called local inertial frames, are to take the place of the earlier infinite inertial frames. It is in them that the laws of physics are valid in their special-relativistic form. Thus, whereas special relativity is an ideal theory of physics in an ideally flat world without gravity, and with infinite inertial frames, its relevant reference frames in the real world are the local inertial frames. The relevant inertial reference frames for modern Newtonian mechanics are larger but not infinite. Roughly speaking, they can embrace a gravitationally bound system, such as a single galaxy, or a bound cluster of galaxies, but no more.

See also: FRAME OF REFERENCE; FRAME OF REFERENCE, ROTATING; GALILEAN TRANSFORMATION; LORENTZ TRANSFORMATION; NEWTONIAN MECHANICS; NEWTON'S LAWS; RELATIVITY, GENERAL THEORY OF; RELATIVITY, SPECIAL THEORY OF

Bibliography

RINDLER, W. *Essential Relativity,* 2nd ed. (Springer-Verlag, New York, 1977).

WOLFGANG RINDLER

FRAME OF REFERENCE, ROTATING

A rotating reference frame is understood to be a frame that is rotating as a rigid body about an axis (or a set of axes) fixed in a specified inertial reference frame. Any rotating reference frame is a non-inertial frame. Nevertheless, a rotating reference frame is often a useful tool for describing motion of objects that are spinning, circling, or moving along spirals.

The rotating reference frame is suitable for analyzing measurements made on any rotating platform, and in relating them to observations of the same phenomena made in an inertial reference frame. Earth as a whole is a rotating laboratory since it spins around its axis and also moves on an elliptical path with the Sun at one of the focus points. The effect of the elliptical motion of the earth on observations made in earth-bound experiments can be almost always ignored since they are extremely small and occur on a long time scale (as compared with typical experimental time scales). In most laboratory experiments, the noninertial effects associated with the spinning of the earth, such as the Coriolis effect, are likewise hard to discern. In large "outdoors" systems, however, these noninertial effects are often anything but subtle. Circulation of large air masses is affected by the Coriolis effect, causing weather patterns and tornadoes to rotate in the counterclockwise direction in the Northern Hemisphere, and in the clockwise direction in the Southern Hemisphere. Projectiles are deflected on the earth due to the Coriolis effect, making long-range artillery inaccurate unless aiming is corrected. The Foucault pendulum, on display in many science museums, is a startling and more peaceful way to demonstrate the effect.

Satellites, cyclotrons, and centrifuges are examples of devices for which a rotating coordinate system may be convenient, simplifying their mathematical analysis. To understand the mechanism of a cream separator, for example, it is easier to analyze the radial diffusion of cream in a reference frame in which the liquid is not rotating, rather than to watch its complicated swirling motion. In a rotating frame of reference, however, even simple actions can sometimes appear confounding. Thus a ball thrown in the air by a child on a merry-go-round moves along an unexpected path as viewed on the carousel, but its motion looks simpler (parabolic) to the bystanders. A table-tennis match on a rotating platform would be a devilishly challenging game!

Translations and Rotations

The distinction between a purely translational, straight-line motion of a small object (point parti-

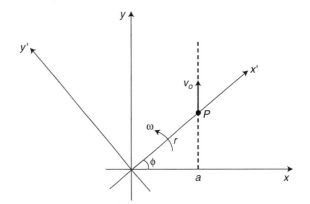

Figure 1   Motion along a straight line.

cle) in a fixed inertial frame, and a motion that involves rotation, is subtle. If a particle moves along a straight line on a plane, it is in effect also rotating around any point on that plane lying outside the line. Conversely, a particle following a circular path can be considered as moving instantaneously along straight (tangent) lines.

It is instructive to consider in some detail the following two examples. In Fig. 1, starting at $t = 0$, point P moves with a constant speed v_0 along the straight line $x = a$. Consequently, P rotates about the origin of the XY reference frame with an angle given by $\phi = \arctan(v_0 t / a)$; that is, with a time-varying angular speed,

$$\omega(t) = \frac{d\phi}{dt} = \frac{a v_0}{a^2 + v_0^2 t^2}. \tag{1}$$

Similarly, one can express the radius of rotation for P, which is r, as a function of time $r(t) = (a^2 + v_0^2 t^2)^{\frac{1}{2}}$, and the radial velocity as

$$v(t) = \frac{dr}{dt} = \frac{v_0^2 t}{r(t)}. \tag{2}$$

Thus the movement of point P can be described as a translational motion in terms of (Cartesian) variables $x = a$, $y = v_0 t$, or as a rotational motion with angular variables $\phi(t)$, $r(t)$; that is, a rotation at a decreasing rate and a lengthening radius. Yet another way to describe it is to say that in the frame $X'Y'$, rotating with the angular speed ω as given in Eq. (1) relative to XY, point P moves along the straight line $y' = 0$ with the speed $v(t)$ given in Eq. (2).

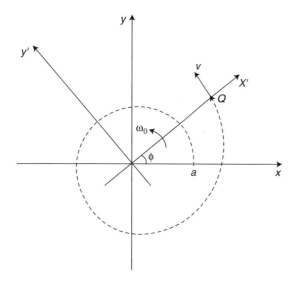

Figure 2   Motion along a spiral.

A converse example is given in Fig. 2, where point Q moves in the XY frame with the constant angular speed ω_0 and the constant radial speed v_0, spiraling away from the origin, $\phi(t) = \omega_0 t$, $r(t) = a + v_0 t$. The description of this motion in terms of the Cartesian coordinates $x = r \cos \phi$, $y = r \sin \phi$ is cumbersome. On the other hand, in the $X'Y'$ coordinate system rotating with the angular speed ω_0 relative to XY, point Q moves with the constant speed v_0 along the straight line $y' = 0$ or the X' axis.

Whether translational or rotational description is used, or whether the reference frame used is inertial (fixed) or noninertial (rotating), is often a matter of convenience. In choosing a reference frame, however, one must keep in mind that physical laws, such as Newton's second law of motion, require modification when they are written in a noninertial frame of reference.

Transformations

Observations made in a fixed, inertial frame of reference can be related to those made in a rotating frame using transformations of coordinate systems. In planar cases, such as those illustrated in Figs. 1 and 2, coordinates $\mathbf{x}' = (x', y')$ of any point in the rotating system can be expressed in terms of $\mathbf{x} = (x, y)$ coordinates of the inertial frame through

$$x' = x \cos \phi - y' \sin \phi,$$
$$y' = x \sin \phi + y \cos \phi. \tag{3}$$

The angle ϕ above is, in general, a function of time $\phi = \phi(t)$. For $\phi = $ constant, these equations describe the relationship between two inertial frames, one of which has been turned by the angle ϕ relative to another. In the example given with Fig. 2, in which the prime system rotates with the constant angular speed ω_0 relative to the unprimed system, $\phi = \omega_0 t$.

General rotation in three dimensions is conveniently and traditionally represented through the three Euler angles ϕ, θ, ψ, defined in Fig. 3. Any rotation S is decomposed into a sequence of three rotations of the right-handed coordinate system: first through angle ϕ about the z axis, next through angle θ about the (new) x axis, and finally through angle ψ about the (twice-rotated) z axis again. The resulting rotation can be symbolically represented as

$$S(\phi,\theta,\psi) = R(\psi,3) R(\theta,1) R(\phi,3), \tag{4}$$

where the numbers following the angles indicate the axis about which the rotation is to be performed. Thus $R(\phi, 3)$ represents the three-dimensional rotation about the third or z axis, which is given by the transformations in Eq. (3) augmented by the equation $z' = z$. (Mathematically, S and R are 3×3 matrices—the three rows of $R(\phi,3)$, for example, are given by the following: $\cos \phi$, $\sin \phi$, 0; $-\sin \phi$, $\cos \phi$, 0; 0, 0, 1.) The right-hand side of Eq. (4) should be read from right to left. Except for special cases, the order in which individual rotations are executed will affect the final "rotated" state. The Euler angles in Eq. (4) can be functions of time, so the expression $S(t) = S(\phi(t), \theta(t), \psi(t))$ represents a general time-dependent rotation in three dimensions. In the

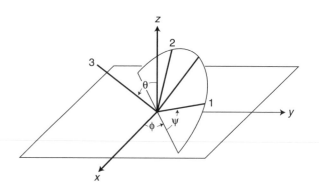

Figure 3   Euler's angles.

example given with Fig. 2, $\phi(t) = \omega_o t$, $\theta(t) \equiv \psi(t) \equiv 0$, and $S(t) = R(\omega_o t, 3)$.

Noninertial Effects

An inertial frame of reference is, by definition, a frame in which Newton's laws of motion are obeyed. Any frame of reference rotating with respect to some inertial frame is a noninertial frame of reference. In a rotating frame, Newton's laws must be modified.

This is best illustrated by returning again to the rotation about the z axis with a constant angular speed ω_o. If $\mathbf{i}_o = (1,0,0)$, $\mathbf{j}_o = (0,1,0)$, and $\mathbf{k}_o = (0,0,1)$ are unit vectors in the inertial frame along the x, y, and z axis, respectively, then the transformation $R(\omega_o t, 3)$ gives the unit vectors in the rotating frame as $\mathbf{i}(t) = (\cos \omega_o t, -\sin \omega_o t, 0)$, $\mathbf{j}(t) = (\sin \omega_o t, \cos \omega_o t, 0)$, and $\mathbf{k} = (0,0,1)$.

The position of a particle in the rotating frame can be represented as $\mathbf{r}(t) = x\mathbf{i}(t) + y\mathbf{j}(t) + z\mathbf{k}(t)$. Differentiating this expression twice with respect to time, and noting that $d\mathbf{i}/dt = -\omega_o \mathbf{j}$ and $d\mathbf{j}/dt = \omega_o \mathbf{i}$, gives

$$\mathbf{a}_{\mathrm{rot}} = \frac{d^2 \mathbf{r}}{dt^2} = a_x \mathbf{i} + a_y \mathbf{j} + a_z \mathbf{k}$$
$$+ 2\omega_o(v_y \mathbf{i} - v_x \mathbf{j}) - \omega_o^2(x\mathbf{i} + y\mathbf{j}), \quad (5)$$

where $v_x = dx/dt$, $a_x = d^2x/dt^2$, and so on. The last two terms of Eq. (5) are sometimes called Coriolis acceleration and centrifugal acceleration, respectively.

For a particle of mass m that is subject to force $\mathbf{F}$ in the inertial frame, the effective force observed in the rotating frame of reference is given by $\mathbf{F}_{\mathrm{rot}} = \mathbf{F} - \mathbf{F}_1 - \mathbf{F}_2$, where, according to Eq. (5), $\mathbf{F}_1 = -2\omega_o(v_y \mathbf{i} - v_x \mathbf{j})$ and $\mathbf{F}_2 = \omega_o^2(x\mathbf{i} + y\mathbf{j})$. With this notation, Newton's second law in the rotating system takes the form $\mathbf{F}_{\mathrm{rot}} = m\mathbf{a}_{\mathrm{rot}}$. In calling $\mathbf{F}_1$ and $\mathbf{F}_2$ the "Coriolis force" and "centrifugal force," one must be mindful that these are not forces in the ordinary sense. These additional terms, sometimes referred to as "fictitious forces," are manifestations of the noninertial frame of reference and are not due to positions or motions of any other bodies as observed in an inertial frame.

See also: ACCELERATION, ANGULAR; CENTRIFUGE; CORIOLIS FORCE; CYCLOTRON; FRAME OF REFERENCE, INER-TIAL; MOTION, CIRCULAR; MOTION, ROTATIONAL; PENDULUM, FOUCAULT; VELOCITY, ANGULAR

Bibliography

HALLIDAY, D., and RESNICK, R. *Fundamentals of Physics*, 4th ed. (Wiley, New York 1992).

REICHERT, J. F. *A Modern Introduction to Mechanics* (Prentice Hall, Englewood Cliffs, NJ, 1991).

SYMON, K. R. *Mechanics*, 3rd ed. (Addison-Wesley, Reading, MA, 1971).

ANDRZEJ HERCZYŃSKI

FRANCK–CONDON PRINCIPLE

The Franck–Condon principle arises in the approximate treatment of the stimulated emission and absorption of light by molecules. It is derived within the framework of the "Fermi golden rule" (a first-order perturbation expression for the transition rate from molecular state i to molecular state f, induced by light of frequency ω), the dipole coupling approximation for the interaction of the radiation with the electrons and nuclei of the molecule, and the Born–Oppenheimer (BO) approximation for separating the electronic and nuclear degrees of freedom. The first two approximations lead to the transition rate expression

$$\omega_{fi} = \frac{|A_0|^2 \pi e^2 \omega^2}{2\hbar c^2} \left| \langle f | \left(-\sum_j \mathbf{r}_{ej} + \sum_j Z_j \mathbf{R}_{N/j} \right) | i \rangle \right|^2, \quad (1)$$

where A_0 determines the intensity of the radiation, Z_j is the magnitude of the charge on nucleus j, $\mathbf{r}_{ej}$ is the position of the jth electron, $\mathbf{R}_{N/j}$ is the position of the jth nucleus, and the states $|i\rangle$ and $\langle f|$ describe the complete dynamics of the isolated molecule, including all electronic coordinates and spins, and nuclear coordinates and spins, in a laboratory center-of-mass frame. (We shall suppress writing the spin dependence in the following discussion, but it must be included in a detailed treatment.)

The BO approximation for separating the nuclear and electronic degrees of freedom is based on the large time-scale difference for nuclear and electron motions associated with the very small electron mass, compared to the masses of the nuclei. As a consequence, at any particular nuclear configuration, the electrons are able to "explore" all of the configuration space available to them, producing the interferences responsible for energy quantization. Mathematically, this is imposed on solutions of the total system Schrödinger equation by writing the solutions in the product form

$$\psi_i(\{\mathbf{q}_{ej}\}, \{\mathbf{Q}_{Nj}\}) = \chi_{\nu_i, J_i}(\{\mathbf{Q}_{Nj}\}) \phi_i(\{\mathbf{q}_{ej}\}, \{\mathbf{Q}_{Nj}\}), \quad (2)$$

where $\{\mathbf{Q}_{Nj}\}$ denotes all of the nuclear degrees of freedom (including spins), $\{\mathbf{q}_{ej}\}$ denotes all of the electron degrees of freedom (including spins), ν_i denotes the collection of vibrational quantum numbers associated with the nuclear internal state of the total state i, and J_i denotes the collection of angular momentum quantum numbers. Note that i also includes the spin state quantum numbers. Strictly speaking, a single product as in Eq. (2) is not sufficient and a sum of similar terms is required. This is because the BO electronic states are calculated in a center-of-mass frame that rotates with the nuclei. As a result, the BO states are functions only of the internal relative nuclear coordinates, and do not depend on the orientation angles of the nuclear framework. But the transition dipole matrix element must be calculated with complete wave functions, including nuclear orientation. This requires that the BO wave functions must be rotated from the molecular to the lab oriented frame, introducing an additional dependence on the Euler angles of the nuclear framework. The nuclear wave function already includes these angles in its rotational wave function. Again, to reduce notational complexity, we shall use the single product form, Eq. (2), but understand that the $\mathbf{Q}$ in ϕ_i, ϕ_f, χ_i and χ_f includes the Euler angles of the nuclear framework as well as the internal relative nuclear coordinates. In order to reduce the notational complexity, we abbreviate the electronic and nuclear coordinates by $\mathbf{q}$ and $\mathbf{Q}$, respectively, and when only spatial variables are involved, we shall denote them by $\mathbf{r}$ for electrons and $\mathbf{R}$ for nuclei. To illustrate, in the simple case of a diatomic molecule, there is a single vibrational quantum number ν_i, and the nuclear rotational state is labeled by j_i, m_i, where $\hbar\sqrt{j_i(j_i + 1)}$ gives the magnitude of the nuclear rotational angular momentum, and $\hbar m_i$ gives its z component. The spin quantum number is suppressed.

The quantity $\langle f | [-\sum_j \mathbf{r}_j + \sum_j Z_j \mathbf{R}_j] | i \rangle$ is called the transition dipole matrix element (denoted by μ_{fi}), and we note that in the second term, there is *no* dependence on the electronic degrees of freedom, except through the initial and final states. This term therefore vanishes unless the two electronic states are the same, due to the orthogonality of the electronic part of the wave functions. Thus, this term generates the pure vibrational-rotational part of the molecular spectrum, and it is of no interest here. We assume different electronic states i and f, so the matrix element becomes

$$\mu_{f,\nu_f,J_f;i,\nu_i,J_i} = -\int d\mathbf{Q} \int d\mathbf{q} \chi^*_{\nu_f J_f}(\mathbf{Q}) \phi^*_f(\mathbf{q},\mathbf{Q})$$
$$\sum_j \mathbf{r}_j \chi_{\nu_i J_i}(\mathbf{Q}) \phi_i(\mathbf{q}, \mathbf{Q}). \quad (3)$$

The dipole operator does not depend on spin so the first result is that no change occurs in the electronic and nuclear spin states (there are no "intermultiplet" transitions). In addition, even though the dipole operator in Eq. (3) does not depend on the nuclear coordinates, the nuclear wave functions are *not* orthogonal because they are eigenstates of *different* BO Hamiltonians. That is, they are nuclear eigenstates for motion on different BO potential surfaces. In addition, there is a dependence on nuclear position in the electronic wave functions that is not eliminated when the integral over electron coordinates is carried out. The dependence on the internal relative nuclear coordinates Q_{rel} is weak and is commonly neglected in the first approximation. The dependence on the nuclear framework Euler angles $\hat{Q}$ *cannot* be neglected.

We define the matrix $M_{fi}(\mathbf{Q})$ by

$$M_{fi}(\mathbf{Q}) = \int d\mathbf{q} \phi^*_f(\mathbf{q}, \mathbf{Q}) \sum_j \mathbf{r}_j \phi_i(\mathbf{q}, \mathbf{Q}), \quad (4)$$

and express the dipole transition matrix as

$$\mu_{f,\nu_f,J_f;i,\nu_i,J_i} = -\int d\mathbf{Q} \, \chi^*_{\nu_f J_f}(\mathbf{Q}) M_{fi}(\mathbf{Q}) \chi_{\nu_i J_i}(\mathbf{Q}). \quad (5)$$

M_{fi} can be expanded about the relative internal nuclear equilibrium position (potential minimum) of either state i or state f; if only the leading term is retained, one obtains

$$M_{fi}(\mathbf{Q}) \approx M_{fi}(Q_{\text{rel,equil}}, \hat{Q}), \qquad (6)$$

and

$$\boldsymbol{\mu}_{f,\nu_f,J_f;i,\nu_i,J_i} = -\int dQ_{\text{rel}}\chi^*_{\nu_f J_f}(Q_{\text{rel}})\chi_{\nu_i J_i}(Q_{\text{rel}})$$

$$\int d\hat{Q} Y^*_{J_f}(\hat{Q}) Y_{J_i}(\hat{Q}) M_{fi}(Q_{\text{rel,equil}}, \hat{Q}). \quad (7)$$

The integral involving only internal relative nuclear wave functions $\chi_{\nu,J}(Q_{\text{rel}})$ is referred to as the "Franck–Condon overlap integral" and it is the ultimate basis of the Franck–Condon principle. It follows that, to the degree that the Fermi golden rule, the dipole approximation, and the BO separation hold, *the transition rate for electronic transitions is governed by the degree of overlap of the initial and final nuclear wave functions.* Physically, this implies that the transition occurs "vertically" (i.e., at fixed nuclear positions) since the upper and lower vibrational states are evaluated at the same Q_{rel}. This results from the BO assumption that the nuclei were stationary compared to the electrons. Only final vibrational states that have sufficient amplitude in the regions where the initial vibrational state is nonzero will have large overlaps with it (assuming there is not a lot of cancellation due to oscillations in the vibrational wave functions). Since the rotational energy levels are very closely spaced compared to the electronic and vibrational energy spacings, it is the dependence of the overlap integral on ν_f and ν_i that determines the details in the electronic spectrum.

In the simple case of a diatomic molecule, the Q_{rel}-integral is over the inter-nuclear distance R:

$$\boldsymbol{\mu}_{f,\nu_f,J_f;i,\nu_i,J_i} = -M_{fJ_f,iJ_i}(R_{\text{equil}})$$

$$\int_0^\infty dR R^2 \chi^*_{\nu_f J_f}(R)\chi_{\nu_i J_i}(R). \qquad (8)$$

If the BO diatomic potentials governing the vibration have their minima at about the same point, R_{equil}, and have similar shapes (case A in Fig. 1), we expect the Franck–Condon overlap integral to be largest for $\nu_f = \nu_i$. The upper state has a somewhat larger equilibrium position when the electron is excited from a bonding to an antibonding orbital (case B), and in that case, one generally finds that the maximum overlap occurs at those final vibrational states which have their inner classical turning points

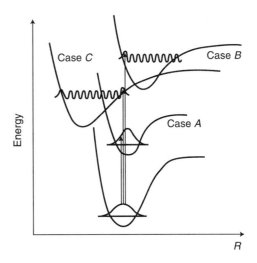

Figure 1 Franck–Condon overlap integrals.

in the region where the lower vibrational state is large. This is a reflection of the fact that high vibrational levels tend to oscillate rapidly except near the turning points, where they tend to accumulate amplitude. If the electron is excited from an antibonding orbital to a bonding one (case C), then the higher vibrational state will have its minimum at a smaller radial distance than the lower state. The transition will then be dominated by final vibrational states with outer turning points that are at the same distances (if any) as the region where the lower vibrational state is significant.

Similar expressions govern photodissociation, except that the final state is a scattering rather than a bound state. It remains true that transitions having the largest Franck–Condon overlaps are those where the vertical transition ends up in the neighborhood of a classical turning point of the scattering state. Predissociation occurs when the excited scattering state has a barrier (e.g., a centrifugal barrier) that requires the nuclei to tunnel in order to separate.

See also: DEGREE OF FREEDOM; DIPOLE MOMENT; EMISSION; EXCITED STATE; GROUND STATE; SCHRÖDINGER EQUATION; SPIN; SPIN, ELECTRON; WAVE FUNCTION

Bibliography

DAVYDOV, A S. *Quantum Mechanics* (Addison-Wesley, Reading, MA, 1965).

SCHATZ, G. C., and RATNER, M. A. *Quantum Mechanics in Chemistry* (Prentice Hall, Englewood Cliffs, NJ, 1993).

DONALD J. KOURI

FRANKLIN, BENJAMIN

b. Boston, Massachusetts, January 17, 1706;
d. Philadelphia, Pennsylvania, April 17, 1790;
electricity.

In his *Autobiography,* Benjamin Franklin notes that his paternal ancestors lived in the village of Ecton in Northamptonshire, England, for more than 300 years as farmers and tradesmen. About 1682, to avoid religious persecution, his father Josiah emigrated to New England. Benjamin's mother Adiah, the second wife of Josiah, was a daughter of Peter Folger, one of the first settlers of New England. Benjamin was born the youngest of ten sons and the fifteenth child among seventeen of Josiah, the eighth of Adiah. Benjamin attended school only two years but was an avid, life-long student of literature, philosophy, science, and languages. At the age of ten he began assisting his father in his business of candle and soap making. At twelve he was apprenticed to his half brother James, a printer. When James launched a weekly newspaper, Benjamin at sixteen began contributing anonymous articles in which he critiqued many aspects of society under the guise of a poor young widow, "Silence Dogood." Resentful of beatings by his brother, Benjamin at seventeen ran off to Philadelphia. Thus began his extraordinary odyssey.

By dint of industry and thrift, Franklin prospered as a printer. In 1729 he bought a fledgling newspaper, *The Pennsylvania Gazette;* over the next decade his skill as editor and chief reporter enabled it to become the most widely read paper in the colonies. In 1730 he formed a common-law marriage with Deborah Read; their family included his son William (mother unknown), their son Francis, and their daughter Sally. From 1732 to 1757 Franklin published *Poor Richard's Almanack;* it was extremely popular and fostered his reputation for wit and wisdom. He also became a bookseller, established a circulating library, organized a debating club that in 1743 developed into the American Philosophical Society, helped to found in 1751 an academy that later became the University of Pennsylvania, and promoted many other civic projects, including a fire company, police force, hospital, and paving and lighting streets. In 1748 Franklin retired from business, turning his print shop over to a partner, and thereafter devoted himself chiefly to scientific research and civic affairs. Among many diplomatic and political roles, Franklin served for years in England as agent for several of the colonies, in France as minister, as a delegate to the Continental Congress, as postmaster general, as a member of the committee to draft the Declaration of Independence, and a delegate to the Constitutional Convention. Franklin crossed the Atlantic eight times, a voyage that took four to six weeks. He lived for a total of twenty five years in London (1724–1726, 1757–1762, and 1764–1773) and Paris (1776–1785), although in America his home remained in Philadelphia until his death.

Franklin was greatly esteemed in his day as a scientist as well as a sage. In the early eighteenth century, electricity was a greater mystery than gravity had been a century earlier. Franklin, almost entirely self-educated and far from any center of learning, solved that mystery. He devised, executed, and correctly interpreted a series of simple, compelling experiments and formulated lucid explanations. Among his several major discoveries, foremost was his concept of electricity as a single fluid, manifest as a positive or negative charge, depending on whether the fluid was present in excess or deficit relative to the neutral condition. He also explained the distinction between insulators and conductors, the role of grounding, the operation of a capacitor such as the Leyden jar, all concepts involved in his most celebrated result, the elucidation and taming of lightning. Franklin introduced into the electrical lexicon terms such as charge, plus or minus, positive or negative, armature, battery, and conductor. His book, *Experiments and Observations on Electricity, made at Philadelphia in America,* consisting of letters he had sent to a colleague in England, was a sensation in Europe; it went through five editions in English (1751–1774) and was translated into French, German, and Italian. It was read not only by scholars but by the literate public, including the clergy and aristocracy. Contemporary scientists often likened Franklin to Isaac Newton.

The link that Franklin made between leaping sparks and lightning bolts was indeed comparable to that Newton made between falling apples and the moon's orbit. Before Franklin, lightning was considered a supernatural phenomenon. If a house was struck by lightning, the fire brigade would douse neighboring structures, but only pray over the struck one, not wanting to intrude on divine punishment. Such views also led to the custom of storing munitions in churches. In 1767 lightning detonated tons of powder in a Venice church; 3,000 people were killed and a large part of the city destroyed.

Despite ample demonstrations of the efficacy of Franklin's lightning rods, he had to weather thunderous attacks for his audacity in stealing a prerogative of the Almighty. He remained unruffled, writing in 1753, "Surely the Thunder of Heaven is no more supernatural than the Rain, Hail or Sunshine of Heaven, against the Inconveniences of which we guard by Roofs & Shades without Scruple."

Like so much else he did, the scope of Franklin's scientific work was remarkable. He wrote major papers on population growth and on meteorology, devised experiments on heat conduction and evaporation, measured ocean temperatures and charted the Gulf Stream, studied bioluminescence and the stilling of water waves by a surface layer of oil, invented a flexible catheter, constructed bifocal eyeglasses, and promoted an efficient wood-burning stove. He also advanced arguments in favor of the wave theory of light.

Many high honors were bestowed on Franklin, but perhaps even more telling were what might be termed low honors. His celebrity was immense, particularly in France. His image appeared everywhere on medallions and banners, snuff boxes and inkwells, often with the motto *Eripuit celeo fulmen sceptrumque tyrannis* (He snatched lightning from the sky and the scepter from tyrants). Louis XVI became so annoyed by this veneration that he gave his favorite mistress a chamber pot with a Franklin medallion at the bottom of the bowl. Franklin's popularity aided significantly his efforts to accelerate the vital flow of arms and funds supplied by the French to support the American Revolution.

While Franklin was always alert for practical applications, in his work on electricity and most of his other scientific studies, his style was that of an explorer, eager for adventure and insight rather than profit or utility. During the several years when he was chiefly occupied with his electrical studies, from about 1745 to 1752, Franklin often confessed apologetically to friends that he had become obsessed with his experiments. He called them philosophical amusements, which he pursued despite what seemed then an almost total lack of prospective applications. Three years before he conceived of the lightning rod, Franklin averred that electricity at least "may help to keep a vain man humble."

See also: BATTERY; CAPACITANCE; CAPACITOR; CHARGE; CHARGE, ELECTRONIC; CONDUCTOR; ELECTRICAL CONDUCTIVITY; ELECTRICITY; INSULATOR; LIGHTNING; NEWTON, ISAAC

Bibliography

CLARK, R. W. *Benjamin Franklin* (Random House, New York, 1983).

COHEN, I. B. *Benjamin Franklin's Science* (Harvard University Press, Cambridge, MA, 1990).

COHEN, I. B. *Science and the Founding Fathers* (W. W. Norton, New York, 1995).

FRANKLIN, B. *Experiments and Observations on Electricity, made at Philadelphia in America*, republished 1774 ed., with critical and historical introduction by I. B. Cohen (Harvard University Press, Cambridge, MA, 1941).

FRANKLIN, B. *Writings*, edited by J. A. L. LeMay (Library of America, New York, 1987).

TANFORD, C. *Ben Franklin Stilled the Waves* (Duke University Press, Durham, NC, 1989).

DUDLEY HERSCHBACH

FRAUD

See ERROR AND FRAUD

FRAUNHOFER DIFFRACTION

See DIFFRACTION, FRAUNHOFER

FRAUNHOFER LINES

White light from the Sun can be dispersed (spread out into the colors of the rainbow) by a prism or a diffraction grating into a continuous distribution of colors from red to violet. The Fraunhofer lines are narrow dark lines that appear in the dispersed solar spectrum. Although William Wollaston is credited with first discovering these dark lines, it was Joseph von Fraunhofer who made a detailed study of the phenomenon during the early nineteenth century.

The experiments by Wollaston and Fraunhofer represent the first observation of an absorption spectrum. When light from a source having a continuous emission spectrum strikes an atom or molecule, part of that spectrum is absorbed. Light is absorbed when an electron is induced to change from one energy state to another. Since the energy levels and, therefore, the absorption lines are characteristic of that atom or molecule, one can use absorption spectra to identify which elements or compounds comprise a sample. In order to understand why narrow lines appear instead of all the light being absorbed, we need to use quantum theory.

According to quantum theory, light can be thought of as consisting of massless particles called photons. Each photon carries with it an energy equal to $h\nu$, where h is Planck's constant, and ν is the frequency of the light. Light is absorbed by atoms and molecules in small, discrete steps called quanta. When a photon strikes an atom or molecule, the photon's energy may, or may not, be absorbed. Quantum theory states that a photon may be absorbed only if its energy is exactly equal to the difference between two energy levels within an atom or molecule. Thus, the absorption lines correspond exactly to differences in atomic or molecular energy levels.

The Sun's overall spectral distribution is that of a "blackbody" at a temperature of roughly 5,000 K. The atoms in the Sun's atmosphere, however, are cooler so they can absorb light and are responsible for most of the Fraunhofer lines. Additional lines come from absorption by gases in Earth's atmosphere.

Fraunhofer rediscovered the lines during the course of his work in developing optical instruments. He was well known for the quality of his optical instruments and his theories on diffraction. The invention of the ruling engine in the latter part of the eighteenth century greatly simplified Fraunhofer's manufacture of diffraction gratings. As a result, diffraction gratings are often used as a substitute for prisms.

The diffraction grating disperses light into its constituent colors much like a prism. Thus, Fraunhofer could easily check the quality of his work by shining light from a lamp through a prism or diffraction grating, producing light of a single color. This monochromatic light could then be sent through optical elements to check for chromatic aberrations. When Fraunhofer used sunlight instead of a lamp, he discovered numerous lines imbedded in the continuous spectrum. He cataloged more than 500 lines.

The origin of these lines remained a mystery for many years. It was not until Gustav Kirchhoff performed his work, in collaboration with Robert Bunsen, in absorption spectroscopy that an explanation appeared. He noted the similarity between the absorption spectrum of atoms, particularly sodium, and that of the Fraunhofer lines. Although most lines were readily attributed to the known elements, certain lines could not be. These lines were due to absorption by the then-undiscovered-element helium. In the Sun, helium is the product of the nuclear reaction that fuses hydrogen atoms together. Helium does not occur in great abundance on Earth, coming mainly from radioactive alpha decay.

See also: GRATING, DIFFRACTION; LIGHT; PHOTON; QUANTUM THEORY, ORIGINS OF

Bibliography

HERZBERGER, M. *Modern Geometrical Optics* (Interscience, New York, 1958).

TRAVERS, B., ed. *World of Scientific Discovery* (Gale, Detroit, 1994).

JOHN E. MATHIS
ROBERT N. COMPTON

FREE BODY DIAGRAM

The free body diagram is a convenient construction for understanding and solving problems involving forces in mechanics. When dealing with macroscopic objects (neither large galaxies nor tiny atoms) it is helpful to divide forces into two categories: contact and field. Contact forces are those exerted on something by actually touching it—the push of a parent's hand on the back of a child swinging, the support force of a bookcase acting on a book. The prime example of a field force is the gravitational force, which acts without physical contact.

Force is a central concept in Newtonian mechanics. Newton's second law, relates force to the acceleration of an object:

$$\Sigma \mathbf{F} = m\mathbf{a}, \qquad (1)$$

(a)

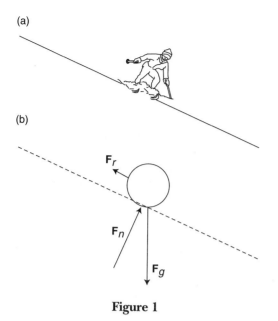

(b)

Figure 1

are stationary. To analyze the forces, consider one block at a time and draw a free body diagram for each. In Fig. 2b this is done for the block on the right-hand side. $\mathbf{F}_g$ and $\mathbf{F}_N$ are the same as before, the new feature is the tension $\mathbf{T}$, a contact force provided by the cord attached to the two blocks. In this case the sum of all the forces acting on the block will be zero because it is not moving.

As a final example, consider the planet Jupiter and its thirteen satellites. To study the dynamics of this very complex system, one could start by sketching a free body diagram for each of the fourteen objects. Concentrating on Ganymede, the largest of the satellites, the free body diagram would change with time and would consist of vectors, representing gravitational forces, pointing from Ganymede to the instantaneous positions of the other satellites, to Jupiter, to the Sun, and perhaps to other planets. The lengths of these vectors follow from Newton's law of gravity, and the sum of all of them would have to be computed for each instant to find the changing acceleration of Ganymede.

The free body diagram helps to isolate a chosen object and thus makes it easier to find and analyze the forces acting on it. This step is essential before Newton's second law can be applied.

where $\Sigma\mathbf{F}$ is the sum of all vector forces acting *on* the object, m is the mass of the object, and its acceleration is $\mathbf{a}$.

To apply the second law it is first necessary to identify the forces present in a given situation. Figure 1a shows a skier sliding down a straight slope. Suppose we wish to calculate the skier's acceleration through the application of Newton's second law. The first step is to analyze the forces acting on the skier. This is done by drawing another diagram, a free body diagram, in which the skier is represented by, say, a circle and all the forces are drawn with as much detail as possible as in Fig. 1b. $\mathbf{F}_g$ is the gravitational force, always perpendicular to Earth's surface. $\mathbf{F}_r$ is the resistance force due to the friction between the skis and the snow and also air resistance. It opposes the motion and so is directed opposite to the velocity of the skier. $\mathbf{F}_N$ is the support force provided by the snow on the skier. All of these together are added vectorially and constitute the $\Sigma\mathbf{F}$ on the left-hand side of Eq. (1). For a 110 lb (50 kg) skier on a 30 percent slope with a coefficient of kinetic friction of $\mu = 0.05$, $\mathbf{F}_g = 490$ N, $\mathbf{F}_N = 469$ N, $\mathbf{F}_r = 23.5$ N. The vector sum of all the forces is 117 N, pointing down the slope. The skier's acceleration is 2.34 m/s^2 down the mountain.

Consider, now the system in Fig. 2a—two blocks on inclined surfaces connected by a cord. Suppose that the surfaces are polished so that the frictional forces are negligible, and, further, that the blocks

(a)

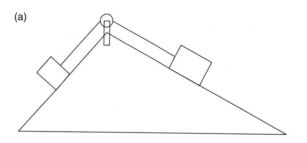

(b)

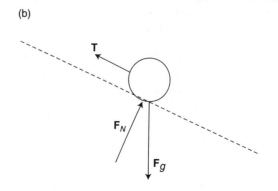

Figure 2

See also: ACCELERATION; FORCE; NEWTONIAN MECHANICS; NEWTON'S LAWS; VECTOR

Bibliography

HALLIDAY, D., RESNICK, R., and WALKER, J. *Fundamentals of Physics*, 4th ed. (Wiley, New York, 1993).

P. L. ALTICK

FREE ENERGY

See ENERGY, FREE

FREE FALL

An object is said to be freely falling if the only (significant) external force acting on it is gravity. In physics, a freely falling object is not necessarily moving downward; a thrown baseball freely falls after it leaves the thrower's hand even if it is moving horizontally or even upward. On the other hand, a feather, even if dropped from rest, does *not* freely fall because the force of air friction on the feather is comparable to the gravitational force acting on the feather.

One of the most curious characteristics of gravity is that two objects freely falling near each other always fall with the same acceleration, even though one might naively expect the more massive object to fall faster. The explanation offered by Newtonian mechanics is that while the gravitational force acting on an object is proportional to its mass ($F_g = mg$, where g is the constant of proportionality), the force needed to give that object a certain acceleration a is also proportional to its mass ($F_{net} = ma$). If gravity acts alone on the object, then $F_{net} = F_g$, implying that $mg = ma$, or $a = g$, independent of the object's mass.

The magnitude of the acceleration g experienced by all freely falling objects at a given location near a gravitating body quantifies the strength of the body's gravitational field at that location. Near the surface of the earth, $g = 9.8$ m/s^2 (g slightly decreases with altitude, but varies by less than 1 percent within about 30 km of the earth's surface).

In situations where a freely falling object's acceleration g can be considered constant, the downward component of the object's velocity increases linearly with time. Near the earth, this component increases by 9.8 m/s (32 ft/s = 22 mph) every second. In a coordinate system whose z axis is defined to point vertically upward, we can express this idea mathematically as follows:

$$v_z = v_{z0} - gt, \qquad (1)$$

where v_z and v_{z0} are the object's vertical velocity components at time t and time $t = 0$ respectively (negative velocities imply downward motion here). It is possible to prove that this implies that the vertical displacement of a freely falling object during the time interval from 0 to t is

$$z = z_0 + v_{z0}t - \tfrac{1}{2}gt^2, \qquad (2)$$

where z and z_0 are the object's vertical positions at time t and 0 (negative values imply downward displacements here).

Since the force of gravity only acts in the vertical direction, gravity does not affect the horizontal com-

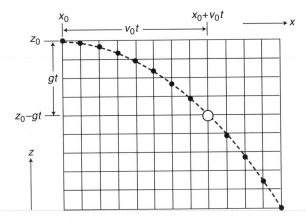

Figure 1 When g is constant, the path of a freely falling object is a parabola. The object shown was launched at time $t = 0$ with zero vertical velocity and a horizontal velocity of v_0.

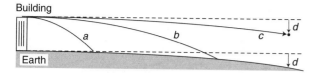

Figure 2 If we launch an object horizontally from the top of a tall building, it will fall along a parabolic trajectory (a) to the ground. If we launch it with a greater horizontal speed, its parabola (b) becomes longer and shallower. If its horizontal speed is sufficiently large (c), the distance d that it falls in a given time becomes the same as the distance that the earth's surface drops from under the object due to the earth's curvature. In such a case, the object will never reach the ground; it will freely fall endlessly *around* the earth.

ponent of a freely falling object's velocity. Galileo Galilei was the first to recognize that an object whose downward velocity component constantly increases while its horizontal velocity component remains constant will follow a parabolic path (see Fig. 1).

If an object's horizontal speed is large enough (about 8 km/s near the earth's surface), the distance the object falls away from a horizontal line in a given time is about the same as the distance that the earth's surface moves away from that line due to the earth's curvature (see Fig. 2). Such a falling object will thus never reach the ground; it instead falls endlessly *around* the earth. We say that an object freely falling around a gravitating body orbits that body. Isaac Newton's greatest triumph was to demonstrate that one can explain everything known about the motions of the planets using the hypothesis that they are freely and endlessly falling around the Sun.

Objects in an orbiting space shuttle appear to be weightless not because there are no gravitational forces acting on them (at the shuttle's greatest altitude, g is still greater than 8 m/s²), but because all objects fall with the same acceleration. An object placed at rest in midair in the shuttle cabin appears to float because both it and the shuttle are freely falling together toward the earth with exactly the same acceleration.

See also: GRAVITATIONAL ATTRACTION; GRAVITATIONAL CONSTANT; GRAVITATIONAL FORCE LAW; NEWTONIAN MECHANICS; NEWTON'S LAWS

Bibliography

FRAUTSCHI, S. C.; OLENICK, R. P.; APOSTOL, T. M.; and GOODSTEIN, D. *The Mechanical Universe* (Cambridge University Press, Cambridge, Eng., 1986).

FRENCH, A. P. *Newtonian Mechanics,* (W. W. Norton, New York, 1971).

THOMAS A. MOORE

FREEZING POINT

The freezing point of a pure substance is the temperature at which the substance undergoes a change of phase from liquid to solid. It is identical to the melting point. The heat of fusion is the amount of thermal energy absorbed by the solid (usually per gram or per mole) as it undergoes melting under normal atmospheric pressure. It is therefore the enthalpy change, represented as ΔH_f. (It is also known as the latent heat, carried over from eighteenth-century terminology.) On freezing, the same amount of thermal energy is given off to the surroundings.

The freezing point depends on the pressure, but the effect is small (compared, for example, with the changes in boiling point with pressure) because the volume changes on freezing are small. The change in freezing point with pressure is given by the Clapeyron equation

$$\frac{\Delta T}{\Delta P} = \frac{T \Delta V}{\Delta H},$$

where T is the temperature (in kelvins), ΔV is the difference in volume (e.g., in cubic meters per mole), and ΔH is the difference in enthalpy (or the heat of fusion, e.g., in joules per mole) between solid and liquid. The pressure P is measured in pascals (1 Pa = 1 N/m² and 1 atm = 1.01325×10^5 Pa).

For example, a pressure increase of 10^7 Pa (somewhat larger than would be expected for a man on an ice skate) would decrease the freezing point of water slightly less than 1°C. Atmospheric pressure decreases the freezing point of water about 0.008°C below the triple point (pressure of 4.579 torr, or 610.5 Pa).

Freezing points are also lowered by impurities, whether the impurities have higher or lower melting points than the primary substance. (The only exception would be an impurity that is fully miscible with the solid phase of the primary substance and has a higher freezing point.) For example, salt water has a lower freezing point than fresh water, and air dissolved in water in equilibrium with the atmosphere lowers the freezing point of the water approximately 0.002°C. (Hence, the "normal" freezing point of water, called 0°C or 273.15 K, is 0.008°C + 0.002°C = 0.01°C below the triple point of water, defined to be 273.16 K.)

Measurement of the freezing point is a sensitive test of purity. Even when the true freezing point is not known, a cooling curve (temperature versus time) in the vicinity of the freezing point provides information on the degree of purity. A pure substance should give a flat cooling curve during the time it takes to freeze (apart from an expected dip at the beginning, caused by supercooling, before the first crystals appear). If there are impurities, a downward slope is expected for the cooling curve because the composition of the liquid phase changes as the major component freezes out, leaving a higher impurity level and lower freezing point.

See also: ENTHALPY; PHASE, CHANGE OF; THERMODYNAMICS; TRIPLE POINT; WATER

Bibliography

BAUMAN, R. P. *Modern Thermodynamics with Statistical Mechanics* (Macmillan, New York, 1992).

ROBERT P. BAUMAN

FREQUENCY, NATURAL

Some objects, if disturbed slightly, oscillate about a stable equilibrium position. The time for one complete oscillation is called the period, and the reciprocal of the period is the frequency. If the period T is measured in seconds, then the frequency $f = 1/T$, equal to the number of oscillations per second, is measured in hertz (Hz). If the disturbance is not excessive (anharmonic oscillator), the period is independent of the nature of the disturbance and depends only upon the properties of the oscillating body.

A string, tautly stretched between two fixed points, is an example of a body with a natural frequency. A violin string vibrates when disturbed by a bow, a piano wire when struck by a hammer, and a harp wire when plucked by a finger. A stringed instrument's sounds result from the wire's vibration that induces sound waves through the air and causes the eardrum to resonate. The frequency of the eardrum and sound wave equal the natural frequency of the instrument's vibrating wire.

The simplest vibration of a stretched string is a sinusoidally periodic motion for which half of a sine wave fits exactly between the string's fixed ends, and the wavelength is, therefore, twice the distance between the end points (see Fig. 1, where $n = 1$). Since the wavelength equals the product of the wave speed v and the period T, the fundamental natural frequency f_1 satisfies

or

$$2L = vT = \frac{v}{f_1},$$

$$f_1 = \frac{v}{2L}.$$

A string's characteristics, namely, its linear mass density ρ (kg/s) and tension T (N), determine the

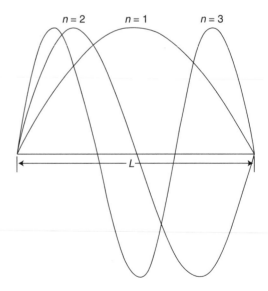

Figure 1 Sinusoidally periodic motion.

wave speed $v = \sqrt{T/\rho}$ (m/s). The magnitude of the disturbing force influences only the amplitude.

In addition to the fundamental frequency f_1, a string may vibrate with overtones. A sinusoidal oscillation for which an integral number of wavelengths fits exactly between the fixed end points produces an overtone. Hence, the nth natural overtone frequency satisfies

$$\frac{2L}{n} = \frac{v}{f_n}$$

or

$$f_n = nf_1,$$

where n is an integer. In a tuned musical instrument, the natural frequencies of the strings are overtones of each other and produce, as Pythagoras discovered, pleasing tones. For example, the fundamental frequency of high C is exactly twice the fundamental frequency of middle C.

When disturbed, a string does not usually vibrate sinusoidally with a natural frequency equal to the fundamental frequency or one of its overtones. The oscillation is typically a superposition of sound waves for which the dominant contribution is the fundamental frequency. A violin differs from a cello because the proportions of overtones are different.

The natural frequency phenomenon also occurs in two dimensions for stretched membranes (drum) and in three dimensions for a tuning fork. Solids vibrate with different natural frequencies along different body axes. For example, a tuning fork struck on its top does not produce the clear note that is made by striking it correctly on its side.

When disturbed slightly, a pendulum or swing oscillates with a natural frequency $f = \sqrt{g/l}$, where l denotes the pendulum's length and g is gravity's acceleration. Because the pendulum frequency is independent of the disturbing force, it provides the physical mechanism for a clock.

A mass m attached to a spring obeying Hooke's law (spring constant k) oscillates with a natural frequency

$$f = \frac{1}{2\pi} \sqrt{\frac{k}{m}}.$$

More generally, the natural frequencies of small oscillations for any body about a stable equilibrium position are called the normal mode frequencies.

Electromagnetic waves in a cavity are a nonmaterial example of natural frequency for which an integral number of half-wavelengths must fit exactly within the cavity's walls. An alternating current circuit, composed of an inductor (L henrys) and a capacitor (C farads), oscillates with a natural frequency

$$f = \frac{1}{2\pi\sqrt{LC}}.$$

Every atom has natural frequencies determined by a quantization condition: The circumference of an orbiting atomic electron must equal an integral number of de Broglie wavelengths. Similar precise oscillations of electrons in ammonia or cesium molecules are the physical basis for the atomic clock.

See also: Circuit, AC; Electromagnetic Wave; Hooke's Law; Oscillation; Oscillator; Oscillator, Anharmonic; Pendulum; Wavelength

Bibliography

Goldstein, H. *Classical Mechanics,* 2nd ed. (Addison-Wesley, Reading, MA, 1980).

Whittaker, E. T. *A Treatise on the Analytical Dynamics of Particles and Rigid Bodies* (Cambridge University Press, Cambridge, Eng., 1988).

George Rosensteel

FRESNEL, AUGUSTIN-JEAN

b. Broglie, France, May 10, 1788; *d.* Ville-d'Avray, France, July 14, 1827; *optics.*

Fresnel was the second of four sons born to Jacques Fresnel, an architect-builder, and Augustine Mérimée, daughter of the overseer of a chateau upon which Jacques carried out improvements. When the disorders of the French Revolution closed down the architect's projects, the Fresnels retired to a Norman village, where the children received along with their "three R's" stern lessons in religion, ethics, and conservative politics. In 1804, placing high in a nationwide mathematics examination, the future physicist won admission to the *École Polytechnique* in Paris. A creation of the revolution, the Polytechnique drew its teaching staff from among the leading scientists of France and gave its elite students a rigorous grounding in advanced mathematics and the physical sciences. After two years of instruction here Fresnel took an additional three years of technical training in the School of Bridges and Roads, which qualified him for government service as a civil engineer. In breaks from his professional responsibilities he pursued scientific research. In whatever he did Fresnel labored earnestly and unsparingly, driven by a strong sense of duty, a compulsion to achieve something important, and somber forebodings of an early grave. In fact, he died of tuberculosis at age thirty-nine.

In 1814, while engaged in building roads in the provinces, Fresnel first fixed his attention on the problem of the nature of light. None of the various inquiries to which he had previously devoted his spare hours had borne fruit. Time and again he had learned from correspondents in Paris that his "discoveries" were already common knowledge. Frustrated, he turned to optics with renewed hopes of original findings and boldly initiated a line of research aimed at demonstrating, in opposition to prevailing scientific beliefs, that light consists of waves and not corpuscles. Although the corpuscular hypothesis had been contested before, most recently by Thomas Young, it stood as the established view, enjoying the sanction of Isaac Newton and providing reasonable explanations for the rectilinear propagation of light, the laws of reflection and refraction, as well as most other optical phenomena. That the effects associated with heat, electricity, magnetism, and certain other physical agencies also found a handy explanation when referred to hypothetical fluids further enhanced its credibility. Dissatisfied with this explanatory scheme of multiple weightless fluids, Fresnel hypothesized that vibrations in a single universal fluid (the ether) might be responsible for many of the diverse phenomena nature presents. In pursuit of this idea he set about to find evidence that light is vibratory. His research was hardly begun when he learned of Napoleon's return from Elba. Distraught at the reappearance of the exiled emperor, the state engineer deserted his post and volunteered for a royalist army mobilized to turn back the invader. When Napoleon triumphed,

Fresnel was suspended from his job and put under police surveillance.

With Napoleon's final exile and the Second Restoration of the Bourbons, Fresnel was reactivated into the Corps of Bridges and Roads, but meanwhile his suspension and a grant of leave allowed him to pass some time in Paris, where he got caught up on recent work in optics, including that of Young, and to return home to Normandy, where he launched into a study of diffraction. Diffraction attracted him because the narrow bands of color observable inside and outside the shadow of the diffracter (a thin object like a hair) gave indications that the propagation of light is not strictly rectilinear, a point favoring the wave hypothesis. Fresnel obtained positive support for waves when, borrowing an idea from acoustics, he used the principle of interference to derive a simple formula that successfully predicted band positions. Thus far, his theory of diffraction was rather crude, but after recasting it in terms of wavelets (Huygens's principle) and carrying out a difficult mathematical analysis using differential and integral calculus, he brought it to perfection. When the paper presenting the new theory won the physics prize of the Paris Academy of Sciences in 1819, the wave theory of light took an important step forward.

By this time Fresnel had been assigned to Paris, where he benefited from collaboration with his mentor and friend François Arago. Elucidating the complexities of polarized light was the leading focus of the collaboration and presented the greatest challenge for the wave theory. By showing experimentally that under certain circumstances light has an asymmetric aspect, polarization raised a serious difficulty for the wave hypothesis, since, unlike corpuscles, waves—at least longitudinal waves—could not be different on different sides. This problem was much on Fresnel's mind when he and Arago began experiments with polarized light. Initially they found that in circumstances where ordinary light produces interference bands, rays polarized in mutually perpendicular planes have no effect on one another. Not knowing how to interpret this puzzling result, Fresnel pressed on with his research, studying in turn chromatic polarization, the reflection of polarized light, and the rotation of the plane of polarization associated with the passage of light through quartz and certain liquids. While yielding many valuable results, these investigations advanced the case for waves in only the limited sense that the principle of interference received new confirmation.

But Fresnel still did not understand what polization was and how the wave hypothesis could accommodate it. The problem came down to knowing what kind of waves constitute light. It seemed that they had to be longitudinal because transverse waves are propagated only in solid media, and the properties of a solid could hardly be ascribed to space. Yet Fresnel was reluctant to dismiss transverse waves altogether. Vibrations at right angles to the direction of propagation clearly embraced the possibility of a lateral asymmetry. By 1821 Fresnel reached the bold conclusion that light waves must be transverse. As a rationale he proposed a mechanical hypothesis that retained a fluid medium for the luminous vibrations, but it was a weak effort, and his real justification for transverse waves was the testimony of his experiments with polarized light, which consistently pointed to forces acting at right angles to the rays. Polarized light, he could now say, had its basis in ether vibrations executed in a definite, fixed plane at right angles to the direction of the wave, while ordinary light consisted of systems of waves polarized in all directions. Other physicists were loath to accept the idea that light waves are transverse, and even Arago, who faithfully supported Fresnel in everything else, deserted him here.

Fresnel effectively answered his critics with a successful application of the concept of transverse waves to double refraction. In an 1821 article, Fresnel suggested that the two rays of double refraction correspond to perpendicular components of the vibrations of the ray incident on the doubling crystal. From this simple idea he rapidly traversed an arduous course to a full-blown mathematical theory of double refraction that met every test of experiment. This was Fresnel's last major contribution to wave optics, though in his later years he made a beginning toward a mathematical theory of dispersion. By about 1830 the wave theory triumphed over the corpuscular theory. Despite Young's priority in introducing the principle of interference, Fresnel made the more substantial contribution to this conceptual revolution by his rigorous, systematic development of wave optics.

A model physicist, Fresnel brought to his research an ingenious mind, deft hands, and the discipline of an excellent scientific education. He was equally proficient in experiment and mathematics and combined the two effectively. Characteristically, he initiated his investigations with experiments and proceeded, by mathematical analysis, to theory. He set as his goal mathematical theories from which precise consequences could be deduced and tested by further experiments. His theoretical bent did not mean, however, that he was above using his talents for practical purposes. The echelon lens (now known as the Fresnel lens), widely used in lighthouses of the period, was his invention. After 1824 his scientific work slackened as responsibilities on the Lighthouse Commission absorbed more and more of his energy. To the problems encountered here—the improvement of the lenses and the design, construction, and placement of lighthouses—he brought the same inventiveness, concentration, and perseverance displayed in his scientific work.

See also: DIFFRACTION; DIFFRACTION, FRESNEL; DOUBLE-SLIT EXPERIMENT; ETHER HYPOTHESIS; INTERFERENCE; LIGHT; LIGHT, WAVE THEORY OF; OPTICS; POLARIZED LIGHT; YOUNG, THOMAS

Bibliography

BUCKWALD, J. Z. *The Rise of the Wave Theory of Light* (University of Chicago Press, Chicago, 1989).

KIPNIS, N. *History of the Principle of Interference of Light* (Basel, Boston, 1991).

SILLIMAN, R. H. "Fresnel and the Emergence of Physics as a Discipline." *Historical Studies in the Physical Sciences* **4**, 137–162 (1975).

ROBERT H. SILLIMAN

FRESNEL DIFFRACTION

See DIFFRACTION, FRESNEL

FRICTION

Friction occurs at boundaries between, and within, solids, liquids, and gases. For fluids (liquids and gases), the analysis involves viscosity. The discussion herein will be limited to surface interactions between bulk solids in which the surfaces are not too rough, so that motion perpendicular to the surfaces may be neglected in comparison to the motion parallel to the surfaces. The frictional force is then nearly independent of whether the surface is "rough" or "smooth."

Friction always acts to prevent slippage of adjacent surfaces. The frictional force is parallel to the sliding surfaces. It may be in the direction of the motion or opposed to the motion. For example, we could not walk and a car or bicycle could not move without friction; but neither we nor the vehicles could stop without friction.

Measurements of forces lead to a simple mathematical model for friction. First, as we would expect, the frictional force depends on the pressure forcing the surfaces together. It also depends on the total area of contact. However, because the pressure P is equal to the normal (perpendicular) force f_N divided by the area of contact, or $P = f_N/A$, the product of these two factors gives

$$f_{\text{friction}} \propto PA = \frac{f_N}{A} A = f_N,$$

independent of area. This is expressed by Coulomb's law of friction. For surfaces in motion,

$$f_{\text{friction}} = \mu_k f_N,$$

where μ_k is a constant called the coefficient of kinetic friction. For surfaces at rest, the expression is better written

$$f_{\text{friction}} \leq \mu_s f_N$$

with μ_s the coefficient of static friction, generally larger than μ_k. For surfaces at rest, the frictional force cannot exceed $\mu_s f_N$, but it also cannot exceed the force applied parallel to the surface. There is no frictional force for undisturbed objects at rest. Typical values of the coefficients fall between 0.2 and 0.9, but values as low as 0.03 and well above 1 have been reported.

Guillaume Amontons described friction in 1699 and Charles Augustin de Coulomb summarized the information then known in *The Theory of Simple Machines,* published in 1779. These rules have proved extremely serviceable, even though they have long been recognized as approximations. For example, it is easily shown that the coefficient of static friction is

equal to the tangent of the angle of repose, the largest angle of an incline before an object begins to slip down the incline. Yet when an object begins to slip, it may move erratically, exhibiting stick-and-slip behavior.

Even surfaces that appear smooth to the eye or to the touch are rough at the atomic level. Cleaved crystals may have extremely flat surfaces, but usually with some steps. Polished surfaces may be quite flat overall, but with hills (called asperities) scattered randomly over the surface. Except in experiments conducted with extreme caution under high vacuum, there is also a surface coating that may include gases, oil, oxides, water, and other contaminants. If these extraneous materials are excluded, frictional properties are quite different from usual measured properties. Very flat surfaces in close contact exhibit extremely high friction.

Two extreme views of friction, at the atomic level, may be roughly described as physical and chemical models. In the physical model, which has been compared to inverting the Swiss Alps over the Austrian Alps, the cause of friction is penetration of the peaks of one surface into the valleys of the other. Motion is then only possible by slight motion of the surfaces in the perpendicular direction, or by cleavage of the peaks of one or both surfaces, or by plastic deformation of the surfaces.

The other extreme view interprets friction as arising from adhesion, or bonding, between the peaks of one surface and the peaks of the other surface. Detailed studies indicate the chemical model of bonding and tearing is more nearly correct, but it cannot be the entire story, as shown by the tendency of a hard surface to plow grooves into a softer surface.

In microscopic modeling, it is often argued that the actual area of contact is much smaller than the apparent area of contact, and the actual area varies with the normal force applied. The amount of contact also depends strongly on the nature of the impurities present on the surface, and on the time of contact. It is this effect that seems to be primarily responsible for the difference between μ_s and μ_k, and for the increase in μ_s observed when surfaces are in contact for long periods of time. We take advantage of this dependence by adding oil or other lubricants to decrease friction in machinery.

We also minimize the importance of friction by leverage. In the wheel and axle, as in a wagon, the frictional force acts at the surface of a small shaft. Because of leverage, this friction has an effect smaller, by a ratio of lengths, than the dragging of a nonrotating wheel on the surface of the road. Friction is further reduced by substituting rolling friction for sliding friction, through the use of ball-bearing supports.

Rolling friction is very small for hard wheels on a hard surface. Hard wheels on a soft surface deform the surface, so that the wheel must go uphill to move forward. Soft wheels on a hard surface are flattened on the bottom, leading to a similarly increased force to roll the wheel forward. In addition, there is a subtle difference in direction and location of the surface forces, such that the net forces are typically neither parallel to the direction of motion nor pointed toward the center of rotation, so that a frictional drag force does not cause an increase in angular speed of the wheel.

As, in practice, static coefficients of friction are greater than kinetic coefficients, the resistance to skid (parallel or perpendicular to the direction of motion) of a rolling wheel is greater than for a skidding wheel. This is crucially important for automobiles. As may be shown easily with toy cars, when the front wheels lock before the rear wheels, there is loss of steering but usually no severe spin. If the rear wheels lock first, the rear wheels quickly slide forward and the skid becomes rear-end first. Antilock brakes pump the brake system to avoid locking either front or rear axles, thus retaining steering control, avoiding skids, and perhaps taking some advantage of the higher static coefficient of friction.

See also: COULOMB, CHARLES AUGUSTIN; NORMAL FORCE

Bibliography

BOWDEN, F. P., and TABOR, D. *Friction: An Introduction to Tribology* (Doubleday, New York, 1973).
RABINOWITZ, E. "Resource Letter F-1 on Friction." *Am. J. Phys.* **31,** 897–900 (1963).

ROBERT P. BAUMAN

FUNDAMENTAL INTERACTION

See INTERACTION, FUNDAMENTAL

FUNDAMENTAL PARTICLES

See ELEMENTARY PARTICLES

FUSION

Nuclear fission and fusion are two processes, involving the nuclei of atoms, that can result in a significant release of energy. Nuclear fission, first discovered by Otto Hahn and Fritz Strassman in 1939, is the splitting of a heavy nucleus (such as uranium-235) into two lighter nuclei of comparable mass.

In contrast to nuclear fission, fusion is the process by which two nuclei combine or "fuse" into a heavier element. Fusion is the mechanism that powers stars such as our Sun, and it is the primary source of energy in the observable universe. For example, energy production in the Sun and similar stars occurs through a series of fusion reactions beginning with the fusion of two protons (appropriately named the proton-proton cycle). The reactions are represented by

$$^1\text{H} + {}^1\text{H} \rightarrow {}^2\text{H} + \beta + \nu + Q^1,$$

$$^1\text{H} + {}^2\text{H} \rightarrow {}^3\text{He} + \gamma + Q^2,$$

$$^3\text{He} + {}^3\text{He} \rightarrow {}^4\text{He} + {}^1\text{H} + {}^1\text{H} + Q^3,$$

where Q is the energy release in the reaction, and ν, β, and γ are the neutrinos, electrons, and gamma rays, respectively. The overall energy release of the cycle ($Q^1 + Q^2 + Q^3$) is approximately 25 MeV.

Although fission and fusion are opposite processes, it is possible to understand why each can release significant amounts of energy by referring to Fig. 1, the binding energy curve, which shows the average binding energy per nucleon (protons and neutrons—the constituents of atomic nuclei—are commonly referred to as nucleons) as a function of the number of nucleons. The binding energy of a nucleus is the energy that must be supplied to the nucleus to separate it into its constituent protons and neutrons.

To quantify binding energy, we can apply Albert Einstein's famous law resulting from special relativity

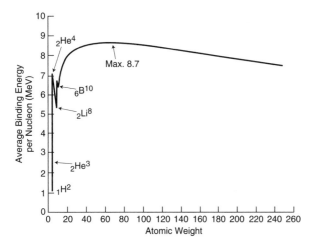

Figure 1 The binding energy curve showing the relationship between average binding energy per nucleon and the number of nucleons.

concerning mass and energy, $E = mc^2$. Thus, the binding energy is approximately the mass of the nucleus subtracted from the sum of the masses of the individual constituent nucleons multiplied by the square of the speed of light. The average binding energy per nucleon is simply this value divided by the number of nucleons. As shown in the binding energy curve, very light nuclei (such as hydrogen) and very heavy nuclei (such as uranium) have the least average binding energy per nucleon. This explains why the fusion of light nuclei and the fission of heavy nuclei both generate significant amounts of energy; in both cases, the reaction products are nuclei that have more tightly bound configurations. According to Einstein's law, these less-massive reaction products are accompanied by a significant release of energy. The binding energy curve also shows why the fission or fusion of iron and its neighboring elements, the most stable nuclides found in nature, would not produce energy.

Conditions Required for Fusion

Nuclear fusion requires extraordinary conditions of temperature and pressure. The difficulty in fusing two nuclei directly arises from the strong electrostatic (coulomb) repulsion that even the lightest similarly charged nuclei (proton) encounter as they approach one another. To fuse, the two nuclei must overcome the long-range coulomb force (which falls off slowly as the inverse square of the separation dis-

tance) and get within a separation less than 10^{-14} cm. At this distance, the strong but very short-range attractive nuclear force becomes operative and causes the two nuclei to fuse.`

To overcome this repulsive force, a significant number of the nuclei must be moving at great speed. Because the average kinetic energy of a particle is equal to $3/2kT$, where k is Boltzmann's constant $(1.38 \times 10^{-23}$ J/K) and T is the Kelvin temperature, this requirement translates into temperatures in excess of tens of millions of degrees Kelvin. For such large values, it is convenient to express the levels of temperature in the units of kilo-electron-volts (1 keV is equivalent to 11.6×10^6 K). At these temperatures, most of the electrons are no longer bound to the positively charged nucleus, and the medium is now in the plasma state, a hot gas composed of positively charged ions (the atomic nuclei) and free electrons. Although plasmas are not abundant on Earth, they are commonly referred to as the fourth state of matter and make up more than 99 percent of the luminous universe. For example, at the center of the Sun, the temperature of the plasma is calculated to be nearly 2 keV. This plasma predominantly consists of fully ionized hydrogen (bare protons and free electrons).

Thus, to achieve fusion, matter must be in the plasma state and at very high temperatures. Though not appropriate for all conditions, for the purpose of this discussion, plasmas can be described as fluids with electrically charged constituents. Plasmas thus respond to externally applied electric and magnetic fields; they conduct current (and therefore can generate their own magnetic fields); and they can support numerous wave-like phenomena because of the long-range coulomb force between the constituent particles.

Although high temperatures are necessary for fusion, achieving significant energy release also requires that the hot plasma be sustained for a sufficiently long time at a high enough density to allow many of the nuclei to react. Because fusion is a binary process involving two nuclei, the number of interactions per second per unit volume is proportional to the density of particles (ions/cm³). The number of fusion events per unit volume is thus simply proportional to $n\tau$, where τ is the lifetime of the hot plasma, commonly referred to as the confinement time, and n is the number of plasma ions per unit volume. The product $n\tau$, whose significance was first recognized by John D. Lawson, is another critical parameter for fusion. Depending on the particular fusion reaction, large values of $n\tau$ must be obtained simultaneously with the high temperatures previously described if a significant amount of energy from fusion is to be obtained. For example, the fusion of the two heavy isotopes of hydrogen, deuterium (^{2}H) and tritium (^{3}H),

$$^2H + {}^3H \rightarrow {}^4He + n + Q \text{ (17.6 MeV)}$$

requires $n\tau$ of approximately 10^{14} ions·cm^{-3}·s (at a temperature of 5–10 keV) if the fusion energy release is to equal the energy required to assemble and heat the plasma.

Fusion Energy

As previously discussed, extreme temperature and pressure (which is simply proportional to the product of the ion density and temperature) conditions are required to achieve fusion. However, the need for advanced energy sources to meet future world demands and the attractive features of fusion as an energy source have led to a quest by the advanced industrial countries to develop terrestrial fusion reactors. During the twenty-first century, the world will need to reduce the relevance of fossil fuels and replace them as the primary worldwide energy source, due to both their decreasing availability and their impact on the earth's ecosystem. Fusion, fission, and advanced solar energy are the only identified alternatives to fossil fuels.

Energy production by fusion has several significant advantages. For example, the fuel used in a reactor—deuterium (D) and tritium (T)—is plentiful and easily available. Deuterium is a stable isotope of hydrogen with a fractional abundance of 1.5×10^{-4}. The oceans contain 100 trillion tonnes of deuterium. Tritium, being radioactive with a half-life of 12.3 years, is not abundantly found in nature. However, it can readily be transmitted from lithium by the energetic neutron produced in D-T fusion via the ^{7}Li(n, ^{4}He) ^{3}H reaction. Lithium is abundant in the earth's crust with estimated reserves exceeding 250 billion tonnes. In addition to the availability of fuel, the fusion of deuterium and tritium requires the least stressing plasma conditions, although other fuel cycles are possible. For both of these reasons, D-T fusion is the leading candidate for fusion energy.

Not only is the potential supply of fusion fuel plentiful, the energy release per fusion event is enormous—typically a million times greater than from burning fossil fuels and eight times that of fission. This means that small amounts of fuel are required to power the reactor. For example, a billion-watt power plant operating for one year would require less than 0.2 tonnes of fusion fuel, compared to more than 2×10^6 tonnes of coal or oil. Fusion fuel is therefore virtually unlimited. Furthermore, fusion produces neither combustion products (such as acid rain) nor greenhouse gases, does not involve large-scale mining and transportation, and the fuel is not localized in specific geographical regions but is distributed more or less uniformly over the earth.

Fusion also has distinct advantages over fission reactors. Fusion reactors are inherently safe—the nature of the process makes a "run-away" and meltdown reactor scenario impossible. Also, in contrast to the actinides (which are the highly radioactive elements) produced directly by fission, the by-products of the fusion reaction are by themselves not radioactive. D-T fusion produces helium and a neutron. The fusion neutron can induce radioactivity by nuclear transmutation of materials that constitute the reactor; however, prudent engineering and advanced materials can minimize this radioactivity and produce levels typically 10^{-3} times that produced in fission reactors. This radioactivity is also short-lived with half-lives measured in decades rather than the thousands of years for fission waste.

Despite all of the desirable properties of fusion, after much research, effective reactors have not yet been demonstrated and are still decades away. Nonetheless, significant progress has been made, and two distinct and very credible pathways to fusion energy have been identified and developed: magnetic fusion and inertial confinement fusion, each of which are delineated by the manner in which the hot plasma is confined.

Magnetic Fusion

Magnetic fusion takes advantage of the fact that the fusion plasma responds to electric and magnetic fields. It relies on a clever arrangement of magnetic fields to confine a relatively low-density plasma. Present technology (such as the strength of materials) limits the maximum confining magnetic field to about 10 T (10^5 G), approximately 200,000 times greater than the earth's natural field. The energy density, or pressure associated with a magnetic field

$\bar{B}$, is equal to $\bar{B}^2/8\pi$. For 10 T, this pressure is approximately 390 atm. To be confined, the plasma pressure must be less than this magnetic pressure. Experiments and theory show the ratio of the plasma pressure to magnetic pressure (commonly referred to as β) to be in the range of 10 percent for stable confinement. Given the 5–10 keV fusion temperatures, this translates into plasma densities of $\sim 10^{14}$ ions/cm^3. To achieve the Lawson number discussed above, the plasma must be confined for approximately 1 s.

This concept is further delineated by the different geometries of the confining magnetic fields and the techniques by which they are generated. While many variations have been studied, by far the most developed and successful approach is the tokamak invented by Russian scientists Igor Tamm and Andrei Sakharov and American scientist Lyman Spitzer.

The tokamak is shaped like a torus (approximately doughnut-shaped) and the magnetic field is produced by currents flowing in external coils (the toroidal field) and induced in the fusion plasma (the poloidal field). The current in the plasma (typical values are 20 million A) is induced by the solenoidal coil in the center of the torus. The plasma acts as a one-turn, secondary winding of a transformer. The resulting magnetic lines of force are closed (i.e., each line meets itself), and the plasma particles spiral along the field as they travel around the torus.

In addition to being confined, the plasma must be heated to the temperatures required for fusion. As in any conductor, the plasma has a finite resistance; thus the plasma is heated by the flowing current. This ohmic or resistive heating is only effective to temperatures of 1 or 2 keV because the plasma resistance decreases as the plasma gets hot. Additional methods of heating the plasma are required and several techniques have been successfully developed. They can be divided into two principal classes: (1) heating by injecting energetic neutral particles and (2) heating by radio frequency (frequency of billion cycles per second) electromagnetic waves.

In the first method, neutral deuterium atoms with energies of about 120 keV are injected into the plasma across the magnetic field (they cross the magnetic field because they have no net charge). Once in the plasma, they are quickly ionized and transfer their energy to the fusion plasma by collisions with the plasma ions. Today this process is well understood, and the largest tokamaks routinely employ neutral beam heating powers of more than 30 MW.

Plasma ions trapped in a magnetic field spiral around the field lines with a characteristic frequency called the ion cyclotron frequency. This frequency depends on the strength of the magnetic field and the mass of the ion. When electromagnetic waves of this frequency are launched into the plasma, a resonance interaction takes place between the waves and particles, and energy can be efficiently transferred into the plasma. This is the process of radio-frequency heating, which is also employed in modern tokamaks. A similar process called electron cyclotron heating also can occur with electrons.

In addition to these external methods of heating, the plasma can also be significantly heated by trapping the energetic particles produced in the fusion process. For example, in D-T fusion, about 20 percent of the energy released per reaction (3.5 MeV) goes into the alpha particle (^{4}He), which is trapped by the confining fields. The alpha particle heats the plasma in a manner similar to both of the methods discussed above. When significant fusion takes place, this process can be the dominant heating mechanism for the plasma. In principle, once fusion occurs, all external heating processes can be turned off and the alpha heating alone can sustain the plasma. The plasma is then said to ignite. The parameter Q, the ratio of fusion power to the external power input to the plasma, is a measure of the performance of a particular device. The value of Q at plasma ignition is infinity. Detailed reactor studies have shown that a Q value of approximately 20 will be required for economic power production for tokamaks.

The majority of magnetic fusion research has been carried out with tokamaks. The largest machines are the Joint European Torus (JET) in England, the Tokamak Fusion Test Reactor (TFTR) at Princeton University, and the JT-60 Tokamak in Japan. Figure 2 shows the TFTR, which has generated more than 10 MW of fusion power with a Q of approximately 0.3. This represents the best values achieved by any fusion device to date, and culminates the rapid progress made in magnetic research since the mid-1970s with a series of increasingly capable tokamaks that use both deuterium and deuterium-tritium fuel.

The next step in the magnetic fusion program is to construct a facility that will have a Q greater than 10 and will produce more than a billion watts of fusion power. This proposed tokamak, to be built by a

Figure 2   The Tokamak Fusion Test Reactor (TFTF) at Princeton University.

consortium of nations (the United States, Japan, Russia, and the European community), has been named ITER for International Tokamak Experimental Reactor. The ITER will be a major step in the development of fusion energy.

Inertial Confinement Fusion

The other principal approach to fusion is called inertial confinement fusion (ICF). Unlike magnetic fusion, ICF uses no confining fields to contain the plasma; instead, it creates enormous densities and pressures by imploding a small spherical capsule, or target, containing fusion fuel. Figure 3 illustrates the principles of ICF. An intense source of power from a driver, usually a laser or particle beam, irradiates a 1-mm-diameter target, which has a typical mass of 1 mg. The driver energy can be focused directly onto a fusion capsule (this is called direct drive), or it can be first directed to a "converter" that converts the drive energy into x rays, which then impinge on the fusion target[9] (this is called indirect drive).

The irradiance on the target exceeds 100 TW/cm^2, vaporizing the surface and creating a hot plasma that rapidly expands outward. To conserve momentum, the remaining target is forced inward at a typical velocity of 300 km/s. The target behaves as a spherical rocket moving at 0.1 percent of the speed of light. When the target stagnates, this kinetic energy is converted into enormous pressure and density. Pressures exceeding 200 billion atm and densities greater than 10^{25} ions/cm^3 can be achieved. At these densities, the Lawson criteria give a confinement time of 10^{-11} s. The fusion reaction rate, which is proportional to the plasma density, proceeds so rapidly that the fusion fuel is consumed before the plasma disassembles. As we will show

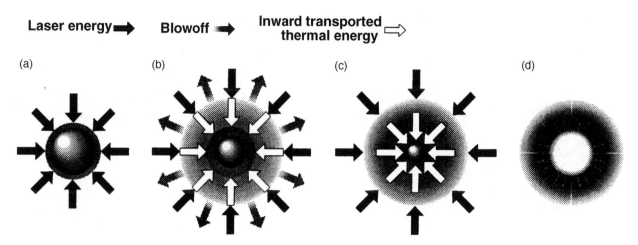

Figure 3 Principles of inertial confinement fusion (ICF): (a) laser beams rapidly heat the surface of the fusion target, forming a surrounding plasma envelop; (b) fuel is compressed by the rocket-like blowoff of the hot surface material; (c) during the final part of the laser pulse, the fuel core reaches 20 times the density of lead and ignites at 100,000,000°C; and (d) thermonuclear burn spreads rapidly through the compressed fuel, yielding many times the input energy.

below, such ICF targets can generate energy significantly greater than the energy input from the driver. The target output divided by the incident driver energy is called the target gain G.

To make power economically, the driver must not require a large fraction of the target output energy. It can be shown that the cost of electricity is dependent on the product of the target gain G and the driver efficiency η. This product must be approximately 10 for economic power production. Driver efficiencies for ICF power plants from 5 percent to greater than 20 percent are thought to be feasible, requiring target gains in the range of 50 to 200.

Furthermore, as shown in Fig. 3, ICF is inherently a pulsed process, requiring targets to be irradiated 5 to 10 times per second to produce power. For example, if the fusion energy is converted into electricity with an efficiency of 30 percent, irradiating ten 300-MJ-yield targets per second would produce 1 GW of electrical power.

In magnetic fusion, the principal difficulty is containing a low-density, hot plasma for seconds and longer. In contrast, the critical challenge in the ICF process is the spherical implosion of a target to high-density conditions. The underlying principle of ICF can be understood by simple energetics and by examining the efficiency in which an inertially confined target undergoes fusion.

If all of the fuel in a milligram of deuterium and tritium underwent fusion, approximately 400×10^6

J of energy would be released. If the fuel were a sphere at solid density (approximately 5×10^{22} ions/cm^3), it would have a radius of 1 mm. If all of the energy of the driver were used by the target, approximately 1×10^6 J could heat the fuel to fusion temperatures of 10 keV. However, the amount of fuel that would undergo fusion is simply the ratio of the fusion reaction time (the inverse of the reaction rate) to the inertial confinement time. This latter time is proportional to the fuel radius divided by the speed of sound in the plasma (the average velocity of the plasma ions), which scales as the square root of the fuel temperature. For a 1-mm-radius target at normal density and at 10 keV, this "burn-up" efficiency is 1×10^{-3}. Such a target would clearly be uninteresting for power production. However, if the target were imploded to approximately 1/20 of its normal size and to a corresponding density 8×10^3 higher, the confinement time would decrease proportionately, but the fusion reaction time would decrease by nearly 1×10^4. The "burn-up" efficiency would then be nearly 40 percent, and the target would produce more than 150 times the energy used to heat it.

An additional complexity arise because a typical target implosion only has an efficiency of about 5 percent (meaning that only about 5 percent of the absorbed driver energy ultimately resides in the compressed fuel). Thus, to achieve high energy gain, the driver pulse must be delivered in a com-

02-10-1284-4798A   9-85

Figure 4 The world's largest ICF facility, the Nova laser at Lawrence Livermore National Laboratory, in which ten laser beamlines simultaneously focus on a fusion target.

plex temporal manner so as to compress the fuel and only heat the central region to fusion temperatures. If properly applied, compressing the fusion fuel takes only 1 percent of the energy per mass required to heat the fuel to fusion temperatures. Fusion then initiates in this central "hot spot" and the alpha particles deposit their energy in the dense fuel, heating the remaining fuel to fusion temperatures. Ignition in the ICF context is therefore similar to that described above for magnetic fusion using both of these features. Sophisticated computer simulations predict overall gains of 50 to 100, and that power generation is possible with drivers that have efficiencies greater than 10 percent.

High-peak-power and high-peak-energy drivers have been built to study the physics of ICF, and significant progress has been achieved. The major facilities include the U.S. Nova and Omega lasers and the Particle Beam Ion Accelerator, the French Phebus laser, Russian Iskra 5 laser, and Japanese Gekko 12 laser. Much of the physics of how energy is coupled to a target and the details of implosion have

been elucidated and confirmed by numerous sophisticated experiments and computer calculations.

At present, the highest energy ICF facility is the Nova laser, shown in Fig. 4. It can irradiate a target with a peak power and energy of nearly 50 GW and 50,000 J of laser light at a wavelength of 0.35 μm. Experiments at Nova, Omega, and Gekko 12 have successfully imploded targets to the temperatures and densities required for fusion. However, even the Nova laser is not powerful enough to achieve these conditions simultaneously. As a result, only fusion power in excess of 4 GW with an energy gain of 0.5 to 1 percent have been achieved.

With the success of experiments and the advances in laser technology, the U.S. Department of Energy is now planning to build the National Ignition Facility. The facility will irradiate a variety of targets with 500 GW and 1.8 MW of ultraviolet laser energy to demonstrate fusion ignition and energy gain in the laboratory. This facility, which will explore both the direct and indirect target concepts, will establish the physics basis for developing ICF for energy production.

See also: CYCLOTRON RESONANCE; FISSION; FUSION BOMB; FUSION POWER; NUCLEAR BINDING ENERGY; NUCLEAR PHYSICS; REACTOR, NUCLEAR; TOKAMAK

Bibliography

BETHE, H. A.; BACHER, R. F.; and LIVINGSTON, M. S. *Seminal Articles on Nuclear Physics, 1936–1937* (American Institute of Physics, New York, 1986).

CONN, R. W. "The Engineering of Magnetic Fusion Reactors." *Sci. Am.* **249** (4), 60–72 (1983).

CORDEY, J. G.; GOLDSTON, R. J.; and PARKER, R. R. "Progress Toward a Tokamak Fusion Reactor." *Phys. Today* **45** (1), 22–30 (1992).

FICKETT, A. P.; GELLINGS, C. W.; and LOVINS, A. B. "Efficient Use of Electricity." *Sci. Am.* **263** (3), 64–74 (1990).

FURTH, H. P. "Progress Toward a Tokamak Fusion Reactor." *Sci. Am.* **241** (2), 50–61 (1979).

HOGAN, W.; BANGERTER, R.; and KULCINSKI, G. "Energy from Inertial Fusion." *Phys. Today* **45** (9), 42–50 (1992).

LINDL, J. D.; MCCRORY, R. L.; CAMPBELL, E. M. "Progress Toward Ignition Burn and Propagation in Inertial Confinement Fusion." *Phys. Today* **45** (9), 32–40 (1992).

WILHELMSSON, H. *Plasma Physics, Nonlinear Theory and Experiments* (Plenum, New York, 1977).

E. MICHAEL CAMPBELL

FUSION BOMB

The explosive energy in a fission bomb is produced by a chain reaction splitting nuclei of heavy fissile isotopes such as uranium-235 (^{235}U) or plutonium-239 (^{239}Pu). The binding energy per nucleon of the fragment isotopes is always higher than the binding energy per nucleon of the original fissile nucleus. It is that excess energy that is released when the nucleus fissions. For a given geometry, however, the fission process imposes unavoidable limitations on the amount of energy released in a fission bomb. The fissionable material in the unexploded bomb must not constitute a critical mass. After detonation, as the chain reaction proceeds, the pressure inside the core causes a rapid expansion of fissionable material rendering the mass, as a whole, subcritical. The reaction shuts down.

This limitation is not a factor in a fusion bomb. The fusion bomb exploits the merging, or fusing, of two hydrogen isotope nuclei, deuterium (^{2}H) and tritium (^{3}H), to form an isotope of Helium (^{4}He):

$$^{2}H + {}^{3}H \rightarrow {}^{4}He + {}^{1}n + energy \qquad (1)$$

The binding energy per nucleon for the nuclei of light isotopes such as ^{2}H and ^{3}H is less than the binding energy per nucleon of the fused product, ^{4}He. The more deuterium and tritium available, the more powerful the bomb. Because such a bomb relies on the fusing of the nuclei of two hydrogen isotopes, the device is almost always called a hydrogen bomb.

In order for the fusion reaction to take place, the participating nuclei must have high enough kinetic energy to overcome the electrostatic repulsion between the protons in the deuterium and tritium. This allows the more powerful, attractive, but short-range, nuclear forces to prevail. The kinetic energy required corresponds to a temperature of about $10^{8\circ}$ K, easily attainable in the explosion of a fission bomb. The high temperature, however, while necessary, is not sufficient. Unless extremely high pressure is applied to the deuterium-tritium mixture, the result is a so-called boosted fission bomb, where the pressure supplied by implosion is of too brief a duration to provide sustained fusion burning of the deuterium-tritium mixture.

Solving the problem of creating both high temperature and high pressure was the chief roadblock to the invention of the fusion bomb. In 1951, after several false starts, the mathematician Stanislaw Ulam and the physicist Edward Teller collaborated in solving this puzzle. While some of the details are still highly classified, the major features of the hydrogen bomb are well known.

The components of a hydrogen bomb can be divided into two major parts: the fission primary and fusion secondary. The primary is nothing more than a boosted fission bomb, which, in modern devices, is about the size of a soccer ball. The energy of the exploding primary is divided between electromagnetic radiation, chiefly in the form of x rays, which propagate at the speed of light ahead of the blast, and heat waves, which propagate at the speed of sound.

While the fireball is just beginning to expand and the blast wave has hardly started, the x rays have already been scattered by an outer casing of ^{238}U into the dense, rigid, plastic foam that surrounds the secondary. These x rays contain about 3 percent of the total energy of the primary (for a 40-kt primary that would amount to more than 1 kt TNT equivalent). This flood of x-ray energy instantly vaporizes the plastic foam, turning it into a very hot, high-pressure plasma. Under this intense and uniform pressure, the secondary is compressed to a fraction of its initial volume.

The cylindrical secondary is composed of the following components: An outer shell of ^{238}U, which, initially, scatters the onslaught of x rays from the primary back into the surrounding plastic. The next layer is composed of lithium deuteride (^{6}Li^{2}H). At the very center of the secondary is a cylinder of ^{239}Pu, sometimes referred to as "the sparkplug." Under the enormous pressure of the high temperature plasma, the stick of ^{239}Pu fissions. Some of the resulting neutrons convert ^{6}Li$^+$ ions into tritium, which, because of the high temperature and pressure, fuses with deuterium ions. The neutrons, thus liberated [see Eq. (1)], manufacture more tritium from the lithium ions adding to the fusion reaction. At these temperatures, the shower of liberated neutrons have enough energy to fission ordinary ^{238}U, which is immune to fission by neutrons created in "ordinary" fission bombs. In fact, the fissioning of the ^{238}U shell produces about half the total energy and most of the radioactive fallout of the fusion bomb—a bomb more aptly described as a fission-fusion-fission bomb. All these processes in the secondary occur in a time less than the time required for the blast wave from the initial primary detonation to reach the secondary.

The very first thermonuclear device test, code named "Mike," was detonated by the United States in November 1952 on the tiny island Elugalab, adjacent to Eniwetok in the Marshall Islands. It was expected to yield a TNT equivalent of a couple of megatons, but actually produced 10.4 megatons equivalent of energy. It created a 3-mile-diameter fireball. (The Trinity test bomb in 1945 had generated a fireball only one-twelfth as large.) The bomb used liquid tritium and deuterium, which required cryogenic refrigeration equipment. The total weight of the device was 65 tons. The explosion vaporized Elugalab and left a crater 1/2 mile deep and 2 miles in diameter. The largest hydrogen bomb detonation was a test by the Soviet Union in 1961, which yielded 58 megatons TNT equivalent, a device of such frightening proportions that it is said to have induced the Soviet nuclear pioneer, Igor Kurchatov, to abandon further work on nuclear weapons.

The first deliverable U.S. H bomb, the Mark 17, completed in 1954, was rated at 15 megatons TNT equivalent. It was 25 ft long, had a diameter of almost 5 ft, and weighed 22 tons. It was designed to be carried in the B-36 bomber. Since that time, bombs and missile warheads of similar yields have become significantly smaller and lighter.

See also: FALLOUT; FISSION; FISSION BOMB; FUSION; FUSION POWER; NUCLEAR BOMB, BUILDING OF

Bibliography

MORLAND, H. *The Secret That Exploded* (Random House, New York, 1981).
RHODES, R. *Dark Sun: The Making of the Hydrogen Bomb* (Simon & Schuster, New York, 1995).
WOLFSON, R. *Nuclear Choices: A Citizen's Guide to Nuclear Technology* (MIT Press, Cambridge, MA, 1991).

STANLEY GOLDBERG

FUSION POWER

Nuclear fusion is the process by which two light nuclei fuse together to form a single heavier nucleus. This normally occurs only at temperatures above about 10 million degrees centigrade, when the kinetic energy of these light nuclei can overcome the electrostatic repulsion between their positive charges. Nuclear fusion results in the release of a relatively large amount of energy per reaction, which is the ultimate source of power for the Sun and stars, as well as for the hydrogen bomb. Recent laboratory experiments on controlled nuclear fusion are close to achieving net power production in devices such as tokamaks. Nuclear fusion is the opposite of nuclear fission, in which a single heavy nucleus splits into two lighter nuclei.

In principle, all nuclei below the atomic number of iron can undergo nuclear fusion to form more stable nuclei, but the energy release per fusion reaction is much higher for the lightest elements, due to their much lower binding energy per nucleon. The physics of nuclear fusion in the Sun involves complex and relatively slow processes such as the proton-proton chain and carbon-nitrogen cycle, which eventually convert hydrogen into helium. The detailed mechanics of solar fusion are not yet completely understood; for example, the measured solar neutrino flux is much less than calculated from the expected fusion rates.

The main reactions used in experiments with controlled (and uncontrolled) fusion on Earth are:

$$D + D = He^3(0.82 \text{ MeV}) + n(2.45 \text{ MeV}) \quad (1)$$

$$D + D = T(1.01 \text{ MeV}) + H(3.02 \text{ MeV}) \quad (2)$$

$$D + T = He^4(3.5 \text{ MeV}) + n(14.1 \text{ MeV}) \quad (3)$$

where H, D, and T represent hydrogen and its isotopes deuterium (D) and tritium (T), and energies are expressed in millions of electron volts (MeV). The fusion products include neutrons (n) and alpha particles (He^4).

At the very high temperatures necessary for thermonuclear fusion, all matter is ionized into a plasma in which the electrons are no longer bound to the nuclei. In the core of the Sun and stars this plasma has a temperature of about $T = 15$ million degrees centigrade and a density of about $n = 10^{30}$ atoms/m^3, which is more than 100 times the density of liquid water. In tokamak fusion plasmas the temperatures are up to $T = 400$ million degrees centigrade, but the typical densities of $n = 10^{20}$ atoms/m^3 are much lower. The Sun's plasma is confined by its immense gravitational forces, whereas the tokamak plasma is confined by high magnetic fields. In another approach, the fusion plasmas in inertial confinement fusion experiments (and hydrogen bombs) are confined for only a short time until the fuel explodes.

The fusion reaction with the largest cross section is D + T, as represented in Eq. (3). In that case, the requirements for a self-sustained fusion reaction are approximately

$$n\tau \geq 6 \times 10^{20} \text{ m}^{-3} \cdot \text{s}$$

$$T \geq 100 \text{ million degrees centigrade,}$$

where τ is the energy confinement time for the plasma. When these conditions are satisfied, the energy loss from the plasma is balanced by the energy input from the 3.5 MeV alpha particles, assuming they thermalize within the plasma. The 14 MeV neutrons escape from the plasma, and their energy can potentially be used to generate electrical power. The development of nuclear fusion as a practical power source remains one of the greatest challenges for the twenty-first century.

See also: FISSION; FUSION; FUSION BOMB; PLASMA; SUN; TOKAMAK

Bibliography

BACHALL, J. N. "The Solar Neutrino Problem." *Sci. Am.* **262** (5), 54–61 (1990).

CONN, R. W.; CHUYANOV, V. A.; INOUE, N.; SWEETMAN, D. R. "The International Thermonuclear Experimental Reactor." *Sci. Am.* **266** (4), 102–110 (1992).

GROSS, R. A. *Fusion Energy* (Wiley-Interscience, New York, 1984).

STEWART J. ZWEBEN

G

GALAXIES AND GALACTIC STRUCTURE

The study of galaxies started when Galileo aimed his telescope at the Milky Way and realized that its diffuse light is resolved into a large number of stars. Charles Messier, William Herschel, and William Parsons (the Earl of Rosse) were among the early observers who cataloged numerous diffuse objects (nebulae) distributed over the sky. The development of astronomical photography at the end of the nineteenth century, together with reflecting telescopes with relatively fast focal ratios, enabled the fainter outer parts of galaxies and the prevalence of spiral patterns to be revealed. The so-called spiral nebulae posed a particular problem, since what we now know to be unresolved star light could be interpreted as a vortex of gas forming protostars. In April 1920, Heber Curtis and Harlow Shapley met at the National Academy of Sciences to debate the size of the Milky Way and the distances to the spiral nebulae. These issues were not settled until 1924 when Edwin Hubble discovered Cepheid variable stars in the Andromeda nebula, showing that the spiral nebulae indeed lie outside the Milky Way and that the universe contains many island universes (galaxies).

Galaxies are composed of luminous material in the form of stars and gas and nonluminous material of unknown nature, held together by gravity. Astronomers determine the masses of galaxies from their gravitational influence on smaller systems, or from the orbits of their constituent stars and gas. A typical galaxy, like the Milky Way, has a mass of about 200 billion solar masses, a diameter of 30 Kpc, and rotates once per few hundred million years.

Properties of Galaxies

Galaxies show a large variety of shapes and sizes. Galaxies of regular shape are commonly divided into two main classes: spiral galaxies and elliptical galaxies. Further, spiral galaxies can be separated into ordinary spirals and barred spirals. There is another type of galaxy known as a lenticular, which shows morphological properties of both elliptical and spiral galaxies. Galaxies that show little or no symmetry are known as irregulars. There are different classification systems envisaged by several astronomers that bear upon the appearance of galaxies on photographic prints: Hubble, de Vaucouleurs, Morgan, van den Bergh, and so on. The Hubble classification of galaxies that is commonly used includes elliptical galaxies, ordinary spiral galaxies, barred spiral galaxies, lenticular galaxies, and irregular galaxies.

We would expect the shapes of galaxies to depend on fundamental physical parameters, such as their total mass and angular momentum. Normal galaxies emit energy most strongly by starlight, as opposed to thermal emission from interstellar dust or synchrotron radiation from an active nucleus. Consequently, the collective properties of stars, such as their distribution of ages, can be expected to be important. We do not fully understand the physical processes that create these complex forms, but it is plausible that either stars formed first from a protogalactic gas cloud, which later collapsed without dissipation, or the gas first collapsed onto a plane dissipatively and formed stars later. The former process naturally accounts for the origin of galactic bulges, whose stars have large random velocities and occupy a spherical volume around the galaxy's center. The latter process could explain the origin of galactic disks, whose stars have ordered circular motions and are confined to a plane.

The distances separating stars are so large that galaxies are considered to be collisionless systems of particles. This means that the orbits of stars do not change dramatically since their birth, and the overall structure of galaxies may be persistent with time in the absence of a gravitational encounter with another massive galaxy. Thus the appearance of galaxies contains information related to the conditions of their formation, as long as the light is argued to trace the underlying mass distribution. In fact, if massive stars are present, the morphology can be strongly affected since such stars are very luminous, yet they trace only a fraction of the mass. Another factor that complicates the connection between the distribution of visible light and the distribution of mass is absorption by interstellar dust, which depends on the wavelength of observation, the amount and distribution of the dust, and the angle of inclination of the main plane of the galaxy with respect to our line of sight.

Elliptical Galaxies

Elliptical galaxies vary in shape from spherical to relatively oblate and in some cases prolate or triaxial. The projected contours of equal surface brightness are well approximated by ellipses. The intensity of their image profile is centrally concentrated, decreasing smoothly with increasing radius. Elliptical galaxies lack detectable disks. Elliptical galaxies are subclassified by their ellipticity $e = 10(a \cdot b)/a$, where a and b are the major and minor axis diameters of the galaxy. A spherical elliptical galaxy will have an ellipticity $e = 0$ and thus labeled E0. The galaxies with the highest ellipticities found have $e = 7$.

Normal ellipticals are almost devoid of gas and dust. Since star formation requires gas and dust, we infer that ellipticals are no longer actively forming stars and that their stellar population is old, as supported by their red colors. Moreover, the observed large broadening of the atomic lines, interpreted as a relatively large stellar velocity dispersion, is not characteristic of recent star-forming events.

Elliptical galaxies many times larger than normal ellipticals have been found at the center of galaxy clusters. These giant ellipticals, designated as cD, are the most luminous galaxies among rich clusters of galaxies. They have complex or multiple nuclei that are believed to arise by gravitationally capturing smaller galaxies (a process known as galactic cannibalism).

Spiral Galaxies

Spiral galaxies, like the Milky Way, are composed of a disk and a central bulge. The predominant disk component contains spiral arms where active star formation takes place. The spiral arms contain the youngest, brightest, hottest stars as well as gas and dust, which are essential in the process of star formation. Due to the ongoing star formation, the spiral arms show blue colors, dust absorption, and gas line emission. The disks rotate and this is what gives them their characteristic, flattened shapes. The bulges and halos of spiral galaxies are similar to elliptical galaxies in their ages, chemical abundance, and the kinematics of the constituent stars. Barred spirals differ from normal spirals in that their bulge is elongated, forming a bar from which the spiral arms unwind.

Spiral galaxies contain a mixture of stellar ages: young stars, known as Population I stars, populate the spiral arms and old stars, known as Population II stars, fill the bulge and halo. Population I stars follow the rotation of the disk in an ordered fashion. They also present high quantities of chemical elements heavier than hydrogen and helium: oxygen, nitrogen, carbon, and so forth. In contrast, Population II stars have a lower abundance of elements heavier than hydrogen and helium and random, large spatial velocities allowing them to populate the entire galaxy from the bulge out to the sparse halo.

The velocity field of stars and gas, determined from the Doppler velocities measured as a function of radial position for disk galaxies inclined to our line of sight, is found to be in uniform circular motion. Emission from neutral hydrogen appears at a wavelength of 21 cm, and can often be traced to larger radial distances than the light from stars, thus affording an important dynamical probe for the mass distribution. These 21-cm-Doppler measurements show constant velocities of rotation as far out in radius as the detection limit, which means that the enclosed mass is increasing proportional to the radius. Consequently, astronomers conclude that there must be a considerable amount of nonluminous material extending well beyond the visible regions of the galaxy.

Lenticular galaxies, designated by SO, look like spirals without any spiral structure. Like spirals, lenticulars can be decomposed in bulge and disk and the disks can also be represented by exponential brightness profiles.

Irregular Galaxies

Irregular galaxies are galaxies with morphologies distinct from either giant ellipticals or spirals like the Magellanic Clouds. Irregulars are almost always dwarf galaxies, galaxies with low surface brightness, and low-mass galaxies. Since there are few, if any, dwarf galaxies with spiral structure, evidently coherent spiral density waves cannot be maintained for disk masses lower than some limit. Although some dwarf galaxies show elliptical shapes, they do not seem to be related to giant ellipticals by a simple scaling of their physical properties. Roughly half the dwarf galaxies show large amounts of gas and ongoing star formation, whereas the other half shows little or no gas.

Even though dwarf galaxies are very abundant, accounting for 80 percent of the galaxies in the Local Group, they are not known to contribute much to the luminosity density of the universe. They are nonetheless interesting because they have apparently not processed their supply of interstellar gas into new generations of stars as efficiently as spirals and ellipticals have, and therefore they may provide important clues to the chemical evolution in galaxies.

See also: ACTIVE GALACTIC NUCLEUS; DARK MATTER; GRAVITATIONAL LENSING; INTERSTELLAR AND INTERGALACTIC MEDIUM; STARS AND STELLAR STRUCTURE

Bibliography

BINNEY, J., and TREMAINE, S. *Galactic Dynamics* (Princeton University Press, Princeton, NJ, 1987).

HUBBLE, E. P. *The Realm of the Nebulae* (Yale University Press, New Haven, CT, 1936).

MIHALAS, D., and BINNEY, J. *Galactic Astronomy,* 2nd ed. (San Francisco, Freeman, 1981).

SANDAGE, A. *The Hubble Atlas of Galaxies* (Carnegie Institution, Washington, DC, 1961).

SHU, F. H. *The Physical Universe* (University Science Books, Mill Valley, CA, 1982).

MARIANNE TAKAMIYA

GALILEAN TRANSFORMATION

Two inertial reference frames S and S', with S' moving at constant velocity $\mathbf{V}$ with respect to S, are shown in Fig. 1. For example, S may be fixed with respect to the ground and S' fixed in a car moving along a straight highway at constant speed. Let an observer in each frame independently measure the location and the time of occurrence of the same single pointlike event, such as the turning on of a tiny lamp. Each observer has a suitable array of meter sticks and clocks with which to carry out these measurements.

For convenience we choose the corresponding axes of the two reference frames to be parallel, the X and the X' axes to coincide, and the relative motion to occur along the common XX' axis. We also calibrate the clocks so that they read $t = t' = 0$ when the two frames coincide.

Observers S and S' record in their notebooks the space and time coordinates of the event as x, y, z, t and x', y', z', t', respectively. Question: If you are

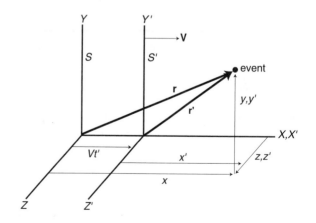

Figure 1

given either set of measurements, how can you find the other set?

From Fig. 1 we can speculate that the connections between the two sets of measurements are

$$x = x' + Vt' \qquad y = y' \qquad z = z' \qquad t = t' \quad (1)$$

Given the measurements of S', you can use these Galilean transformation equations to find the corresponding measurements of S.

To go the other way, that is, to find the measurements in frame S' given those in frame S, simply exchange primed and unprimed quantities in Eq. (1) and reverse the sign of V.

Guided again by Fig. 1, we can write the first three components of Eq. (1) in vector form as

$$\mathbf{r} = \mathbf{r}' + \mathbf{V}t'. \quad (2)$$

Here $\mathbf{r}$ and $\mathbf{r}'$ are the position vectors that locate the event in frames S and S', respectively.

Suppose that the event in Fig. 1 is the passage of a moving particle through a specified point. Recalling that $\mathbf{v} = d\mathbf{r}/dt$ and bearing in mind that $\mathbf{V}$ is a constant, we can differentiate Eq. (2) and find the relation between the velocities of this particle as measured by S and by S'. This Galilean transformation equation for velocities is then

$$\mathbf{v} = \mathbf{v}' + \mathbf{V}. \quad (3)$$

The acceleration of the moving particle is defined by $\mathbf{a} = d\mathbf{v}/dt$. Differentiating Eq. (3) with respect to time yields the Galilean transformation equation for accelerations:

$$\mathbf{a} = \mathbf{a}'. \quad (4)$$

Thus the acceleration of a moving particle has the same value in all inertial reference frames. In formal language we say that the acceleration of a particle is *invariant with respect to a Galilean transformation.* Other quantities that are similarly invariant are the force acting on the particle, its mass when the particle is at rest, and its electric charge.

Consider now a rod lying along the XX' axis of Fig. 1 and at rest in frame S. Let each observer measure the positions (x_1, x_2) and (x_1', x_2') of the two ends

of the rod, noting also the times of measurement (t_1, t_2) and (t_1', t_2'). From Eq. (1), the *differences* between the measurements of these event pairs are related by

$$(x_2 - x_1) = (x_2' - x_1') + V(t_2' - t_1') \quad (5a)$$

and

$$(t_2 - t_1) = (t_2' - t_1'). \quad (5b)$$

Here $(x_2 - x_1)$ is simply the length L of the (stationary) rod as seen by S. For S', however, $(x_2' - x_1')$ will be the length L' of the (moving) rod *only* if S' measures the two end positions at the same time. That is, we must have $t_2' = t_1'$. Making it so in Eq. (5a) leads to $L = L'$. Thus the Galilean transformation equations predict that the measured length of a rod moving at constant velocity is not changed by its motion.

The Galilean equations also predict that time intervals measured by a clock are unchanged by the motion of the clock [see Eq. (5b)]. In particular, if $t_2' - t_1' = 0$, then $t_2 - t_1$ must also equal zero. Thus, if two events are simultaneous in one inertial frame, the Galilean transformation equations predict that they will be simultaneous in any other such frame.

Question: Are the predictions of the Galilean transformation equations correct? Experimentation shows that these equations, which are based on Newton's views of the nature of space and time, hold where the laws of Newtonian mechanics predict correct results. This vast realm is that of our daily experience of moving objects such as baseballs, cars, orbiting satellites, and planets. Newtonian mechanics, for example, was completely adequate to control the motion of the space probe Voyager 2, which reached Neptune on August 25, 1989, within five minutes of its scheduled arrival time after twelve years en route.

For speeds approaching that of light, however, the Galilean transformation equations fail and come to be seen as a low-speed limiting case of the Lorentz transformation equations, which are based on Albert Einstein's views of the nature of space and time. These equations predict that moving rigid rods *do* change their measured lengths (they shrink) and moving clocks *do* change their measured readings (they slow down), but these effects are normally too small to measure at speeds much less than that of light. Newtonian mechanics also fails at high speeds and comes to be seen as an important limiting case of Einstein's special theory of relativity. We must be guided by the laws of relativistic mechanics in dealing with fast-moving particles such as electrons, which can easily reach speeds extremely close to that of light. A relativistic treatment is also necessary for slow-moving objects if we want to make extremely precise measurements. Thus, when an atomic clock (stable to about one part in 10^{13}) is transported between distant cities, relativistic corrections must be applied to its readings.

See also: LORENTZ TRANSFORMATION; NEWTONIAN MECHANICS; RELATIVITY, SPECIAL THEORY OF; RELATIVITY, SPECIAL THEORY OF, ORIGINS OF

DAVID HALLIDAY

GALILEI, GALILEO

b. Pisa, Italy, February 15, 1564; *d.* Arcetri, Italy, January 8, 1642; *mechanics, cosmology.*

Galileo was the eldest son of Vincenzio Galilei and Giulia Ammannati. He had two brothers, Michelangelo, who became musician to the Grand Duke of Baviera, and Benedetto, who died in infancy, and three sisters, Virginia, Anna, and Livia, and possibly a fourth, Lena. His father was a musician and the author of an influential treatise on music that shows a gift for polemics that his son was to develop in his own writings. Galileo was enrolled at the University of Pisa in 1581 but left in 1585 before taking a degree. He studied Euclid and Archimedes and was soon able to tutor students in Florence and Siena. In 1586 he composed a short treatise, *The Little Balance,* in which he reconstructed the reasoning that he believed had led Archimedes to devise a way of detecting whether the goldsmith, who had fashioned King Hero's crown, had substituted a baser metal for gold. Galileo used his newly won insight to construct a hydrostatical balance that is the first instance of his interest in applied sciences. He was appointed to the Chair of Mathematics in Pisa in 1589. His notebooks (published only in the twentieth century) show the influence of the professors at the Jesuit College in Rome. The popular notion of Galileo as a hard-headed and thorough-going experimentalist owes much to his first biographer,

Vincenzo Viviani, who claimed that Galileo ascended the Leaning Tower of Pisa sometime between 1589 and 1592 to refute Aristotle by showing that bodies fall at the same speed regardless of their weight. In his treatise, *On Motion,* written around 1591, Galileo makes frequent mention of towers, but he does not state that all bodies fall at the same speed but that their speed is proportional to the differences between their specific gravity and the density of the medium through which they descend. In other words, the young Galileo reached the erroneous conclusion that different-size bodies of the same material fall at the same rate while same-size bodies of different materials do not.

In 1592 Galileo was appointed to the Chair of Mathematics in Padua. He lectured on prescribed topics such as Euclid's *Elements,* the *Sphere* of Sacrobosco, Ptolemy's *Almagest,* and the Pseudo-Aristotelian *Mechanics.* Medical students made up the majority of his audience. Galileo gave private lectures on fortifications and military engineering to young noblemen, and he manufactured and sold a mathematical instrument, the military and geometrical compass or sector.

The first indication of Galileo's commitment to Copernicus appeared in a letter that he wrote to his former colleague at Pisa, Jacopo Mazzoni, in May 1597. In August of the same year he received a copy of Kepler's *Mysterium Cosmographicum,* in which the heliocentric theory was vindicated on mathematical and symbolic grounds. After reading the preface, Galileo wrote to Kepler to voice his approval of the view that Earth is in motion and to express his fear of making his position known to the public at large.

Galileo never married but from his common-law wife, Marina Gamba, he had three children: Virginia (born August 13, 1600), Livia (born August 18, 1601), and Vincenzo (born August 21, 1606). For all his children Galileo carefully cast horoscopes.

Around 1602 Galileo began making experiments with falling bodies in conjunction with his study of the motion of pendulums and the problem of the brachistochrone, namely the curve between two points along which a body moves in the shortest time. He first expressed the correct law of freely falling bodies (*s* is proportional to t^2) in a letter to Paolo Sarpi in 1604, but he claimed to have derived it from the assumption that speed is proportional to distance. (He only realized later that speed is proportional to the square root of the distance.)

In July 1609, when Galileo was in Venice, he heard that a Dutchman had invented a device to make distant objects appear nearer, and he immediately attempted to construct such an instrument himself. Others were at work on similar devices, but by the end of August 1609 Galileo had produced a nine-power telescope that was better than those of his rivals. He returned to Venice where he gave a demonstration of his spying-glass from the top of the Campanile of San Marco. The practical value for sighting ships at a distance impressed the Venetian authorities who confirmed Galileo's appointment for life and raised his salary from 520 to 1,000 florins, an unprecedented sum for a professor of mathematics. Galileo never quite mastered the optics of his combination of a plano-convex objective and a plano-concave eyepiece (our opera glass), but he succeeded in producing a thirty-power telescope, which he turned to the sky in 1610. What he saw is reported in the *Sidereus Nuncius,* which appeared in March 1610. The work was to revolutionize astronomy. The Moon was revealed to be covered with mountains (Galileo was even able to make a rough estimate of their height), the Milky Way dissolved into a multitude of starlets, new stars appeared as if out of nowhere, and, more spectacular still, four satellites were found orbiting Jupiter. This was a particularly important discovery since, if Jupiter revolved around a central body with four attendant planets, it could no longer be objected that Earth could not orbit around the Sun with its moon. Jupiter's satellites did not win the day for Copernicanism, but they removed a major obstacle to having it seriously entertained by astronomers.

In July 1610 Galileo was appointed Mathematician and Philosopher of the Grand Duke of Tuscany. Shortly thereafter he discovered that Venus has phases like the Moon, which was of great significance since it proved that Venus went around the Sun. In the summer of 1611 Galileo debated the cause of floating bodies with Peripatetic philosophers. Galileo maintained, along Archimedean lines, that the cause of floating was the relative density, against the view of his Aristotelian opponents that it was the shape. In the autumn of that year, Christoph Scheiner, a Jesuit who taught at the University of Ingoldstadt in Germany, announced that he had discovered spots on the Sun. Galileo took him to task for suggesting that these spots were small satellites orbiting around the Sun and insisted that he had observed the sunspots before Scheiner. In an age conscious of priorities, this was a claim that did not endear him to his rival. It also opened a breach between Galileo and the Jesuits.

In December 1613, theological objections were raised at a dinner at the Court of the Grand Duke in Pisa. Galileo was absent, but his disciple Castelli defended Galileo's views when he was asked his opinion by Christina of Lorraine, the mother of Cosimo II. Galileo felt that the matter was important enough to write a long letter to Castelli, dated December 21, 1613, which he expanded into the *Letter to the Grand Duchess Christina,* his most detailed pronouncement on the relations between science and scripture. Borrowing the *bon mot* of Cardinal Cesare Baronio, "the intention of the Holy Spirit is to teach us how to go to heaven, not how the heavens go," Galileo developed the view that God speaks through the Book of Nature as well as the Book of Scripture. In 1616 the Copernican theory was examined by the Holy Office, the Roman Congregation charged with the defense of Catholic orthodoxy. The result was a ban on the book by Copernicus and all other works containing the same teaching. Galileo himself was not mentioned but an unsigned memorandum found in the proceedings states that Galileo was enjoined not only to relinquish the theory that Earth moved but not to discuss it. (The authenticity of this document has been queried.)

Galileo returned to Florence at the end of May 1616 and turned his mind to a non-controversial topic: the determination of longitudes at sea. He hoped that accurate tables of the periods of revolutions of the satellites of Jupiter would make it possible for seamen to know their location merely by observing the satellites through a telescope, but the tables were never accurate enough for the method to be useful. In the Autumn of 1618 great excitement was generated over the appearance, in rapid succession, of three comets. Galileo thought that comets were merely optical phenomena caused by refraction in the atmosphere, and he criticized the account of Orazio Grassi, the professor of mathematics at the Roman College, who claimed the comets were celestial bodies beyond the sphere of the Moon.

What changed Galileo's Copernican fortune was the election of a Florentine, Urban VIII, to the Roman Pontificate in 1623. Galileo felt that he could now write about Earth's motion. In January 1630 his long awaited *Dialogue on the Two Chief World Systems* was completed. It is divided into four days. In the First Day, Galileo criticizes the Aristotelian division of the universe into two sharply distinct regions, the terrestrial and the celestial. He does this by attacking the apparently natural distinction between rectilinear and circular motion upon which Aristotle rested his case and by pointing out similarities between Earth and the Moon. In the Second Day, Galileo argues that the motion of Earth would be imperceptible to its inhabitants, and that the diurnal rotation of Earth on its axis is simpler than the daily revolution of all the planets and stars postulated by Ptolemy. In the Third Day, the annual revolution of Earth around the Sun is said to offer a simpler interpretation of the apparent stations and retrogressions of the planets. The Fourth Day makes the ingenious but erroneous claim that the tides are evidence for Earth's motion. The *Dialogue* also contains the correct law of falling bodies and a discussion of the principles of the relativity of motion and the conservation of motion, but Galileo assumed that inertial motion was circular rather than rectilinear.

The *Dialogue* went to press in June 1631. The publisher decided to print a thousand copies, a large edition for the time, and the work was not completed until February 21, 1632. It only reached Rome at the end of March or early April. In the summer of 1632, Urban VIII ordered a Preliminary Commission to investigate the licensing of the *Dialogue*. In the file on Galileo in the Holy Office, the commission found the unsigned memorandum of 1616 that enjoined him not to hold, teach or defend in any way that Earth moves. The Commissioners considered the injunction genuine and concluded that Galileo had contravened a formal order of the Holy Office. In the light of this discovery, Galileo was summoned to Rome, arriving, after much delay, on February 13, 1633. Despite his vigorous denial, Galileo was judged by the Holy Office to have contravened the orders of the Church. On the morning of June 22, 1633, he was taken to a hall in the convent of Santa Maria Sopra Minerva in Rome and was made to kneel while the sentence was read; it condemned him to imprisonment. Still kneeling, Galileo formally abjured his error. He was allowed to leave for Siena and later, in 1634, to return to Florence, where he was confined to his house in Arcetri.

Galileo sought comfort in work, and within two years he completed the *Discourses on Two New Sciences,* the book to which his lasting fame as a scientist is attached. The first of these two new sciences is a novel mathematical treatment of the structure of matter and the strength of materials. Galileo showed that there is a limit to the size of objects made of the same material and maintaining the same proportions. The second science is natural motion, which

was discussed, for the first time, in the light of the times-squared law of freely falling bodies and the independent composition of motions. These laws enabled Galileo to show that the path of projectiles is a parabola. When he cast about for a publisher, he came up against a new problem: The Church had issued a general prohibition against printing or reprinting any of his books. Galileo's manuscript was sent to Louis Elzevir in Holland, where it appeared in 1638. Although Galileo became blind in that year, he never succeeded in obtaining the pardon he longed for, and he had to remain under house arrest until his death on January 8, 1642, five weeks before his seventy-eighth birthday.

See also: COPERNICAN REVOLUTION; COPERNICUS, NICOLAUS; FREE FALL; KEPLER, JOHANNES; KEPLER'S LAWS; NEWTON, ISAAC; NEWTONIAN MECHANICS; NEWTON'S LAWS

Bibliography

BAGIOLI, M. *Galileo Courtier* (Chicago University Press, Chicago, 1993).

CLAVELIN, M. *The Natural Philosophy of Galileo* (Harvard University Press, Cambridge, MA, 1974).

DRAKE, S. *Galileo at Work: His Scientific Biography* (Chicago University Press, Chicago, 1978).

DRAKE, S. *Galileo* (Oxford University Press, Oxford, Eng., 1980).

DRAKE, S. *Galileo: Pioneer Scientist* (University of Toronto Press, Toronto, 1990).

FANTOLI, A. *Galileo: For Copernicanism and for the Church* (Vatican Observatory Publications, Rome, 1994).

FINOCCHIARO, M. A. *The Galileo Affair: A Documentary History* (University of California Press, Berkeley, 1989).

KOYRÉ, A. *Galileo Studies* (Humanities Press, Atlantic Highlands, NJ, 1978).

REDONDI, P. *Galileo Heretic* (Princeton University Press, Princeton, NJ, 1987).

SHEA, W. R. *Galileo's Intellectual Revolution* (Science History Publications, New York, 1972).

WILLIAM R. SHEA

GALVANOMETER

The galvanometer is a very sensitive device used to measure electrical current passing through it; that

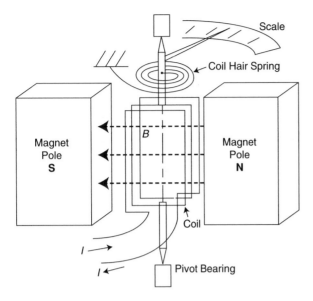

Figure 1 Galvanometer movement. A coil that carries current *I* rotates in the magnetic field *B* from a permanent magnet.

is, it is a very sensitive ammeter. The analog galvanometer is usually of the d'Arsonval type. The d'Arsonval movement consists of a rectangular coil of wire that is suspended in the magnetic field from a permanent magnet (see Fig. 1). The current to be measured is passed through this coil, and because of forces on the coil from the magnetic field, the coil rotates through an angle. Due to a restoring (opposing) torque from the suspension mechanism, the coil rotates through an angle that is proportional to the current flowing. A pointer or "needle" that is attached to the coil then moves over a scale. This scale may be calibrated so that the number that the needle points to indicates the current that is flowing.

The most important application of the galvanometer is in analog voltmeters and ammeters, where it is the essential electrical mechanism. A galvanometer is also used with a Wheatstone bridge for precision measurement of resistance and with a precision potentiometer for precision measurement of voltage differences. The ballistic galvanometer is a type of galvanometer that is especially designed for the measurement of the total charge that flows in a brief pulse of current. However, digital meters (which use electronic sensing circuits and do not use a moving pointer to indicate a value) have largely replaced the Wheatstone bridge and precision potentiomenter.

In a highly sensitive galvanometer, the coil is suspended in the magnetic field by fine wires; these become the axis of rotation of the coil, carry the current to and from the coil, and supply the restoring torque because they are twisted as the coil rotates. Usually a small mirror, rather than a pointer, is attached to the coil in this case. Light reflected from the mirror falls on a scale marked in millimeters. The sensitivity to current (called simply the sensitivity) of this type of galvanometer can be 10^{-6} A/mm to 10^{-10} A/mm. Unfortunately, vibrations can interfere with the operation of the galvanometer and can even damage it.

More rugged galvanometers use small shafts on the ends of the coil, which mount into hard metal or jeweled bearings. The shafts become the axis of rotation in this case. A fine coiled hair spring is mounted between the shaft and the frame supporting the shafts. If the coil is turned, a restoring torque by the hair spring tends to return the coil to its equilibrium position. The sensitivity of this type of meter is usually quoted as the amount of current that causes the needle to move to the end of its range on the scale, or full scale current. A typical value for this type of galvanometer is 50 μA for full scale, although sensitivities are commonly found from 1 μA to 1 mA full scale.

The wires of the coil that are perpendicular to the magnetic field (*B*) each experience a force when current passes through them that is proportional to the current. These wires are on opposite sides of the coil, and the forces act to turn the coil in the same sense, so that there is a torque about the center axis of the coil. This torque is then proportional to the current *I* that is passing through the coil. The coil comes to equilibrium when the torque from the wire suspension (or coil spring) is large enough to counterbalance the torque due to the current flowing in the coil. Then the angle θ through which the coil rotates is directly proportional to the current *I* flowing through the coil.

See also: AMMETER; POTENTIOMETER; VOLTMETER

Bibliography

BELL, D. *Electronic Instrumentation and Measurements* (Reston Publishing, Reston, VA, 1983).

CHIANG, H. H. *Electrical and Electronic Instrumentation* (Wiley, New York, 1984).

DENNIS BARNAAL

GAMMA DECAY

See DECAY, GAMMA

GAMMA RAY BURSTERS

In 1963 the United States launched the first of the Vela satellites to monitor the Soviet Union's compliance with the Limited Test Ban Treaty, which banned nuclear tests in and beyond Earth's atmosphere. To the surprise of U.S. researchers, the satellite detected many explosions, later found not to be coming from anywhere in the solar system but from distant space. In 1973 these puzzling observations of bursts of gamma rays were finally made public in a paper by Ray Klebesadel, Ian Strong, and Randy Olson, and have been defying our understanding since. Many satellites have followed after the Vela satellites trying to unravel this puzzle: KONUS, Geminga, and the Gamma Ray Observatory, to name a few.

It is clear that some very energetic event in space emits bursts of gamma rays, which are very high energy photons (energies above 100,000 eV) or very high frequency light (frequencies above 10^{19} Hz). We can observe these only in satellites since Earth's atmosphere prevents most of them from reaching detectors on Earth's surface. The bursts last from milliseconds to minutes and show a richness of time variations throughout the event. These events come to us from many different directions, but we still do not know how distant they are. Two main possibilities are under debate at the present: a galactic origin, meaning that gamma ray bursts originate in our own galaxy, the Milky Way; or a cosmological origin, meaning that they may come from different galaxies throughout the universe. A galactic origin implies a less energetic source than the cosmological one, since the further away the stronger the source of this radiation must be.

Proposed models for gamma ray bursts in our galaxy usually involve an energetic explosion around a neutron star or a black hole, while cosmological models involve the collision of two neutron stars or black holes, among many even more exotic models. The extreme amount of energy necessary to emit such bursts across the universe is a tough hurdle for models of cosmological origin. The energy necessary at cosmological distances (10^{51} ergs) is a reasonable fraction of the total energy necessary to keep a neutron star together.

The Gamma Ray Observatory has been able to detect about one of these events a day and has enough directional information to give a sense that the sources are close to an isotropic distribution in the sky, which favors somewhat the cosmological origin for the events. Future space observatories will help determine the origin of these events when the pointing resolution improves and correlations between the direction of the bursts and known objects can be found.

Observations have found that these events are not all the same. Three sources of somewhat less energetic gamma ray bursts repeat and are now called Soft Gamma Ray Repeaters. Counterparts for these events have been identified in radio and x-ray observations. The counterparts place the source for the events in supernova remnants, indicating that the sources are likely to be neutron stars.

See also: BLACK HOLE; COSMOLOGY; GAMMA RAY OBSERVATORY; MILKY WAY; NEUTRON STAR

Bibliography

FISHMAN, G.; BRAINERD, J.; and HURLEY, K. *Gamma-Ray Bursts*, AIP Conference Proceedings No. 307 (AIP Press, New York, 1994).

ANGELA V. OLINTO

GAMMA RAY OBSERVATORY

The Gamma Ray Observatory (GRO), launched on April 5, 1991, is one of NASA's four Great Observatories. The GRO is designed to observe the universe as it appears in gamma rays, the most energetic form of electromagnetic radiation, with energies in excess of 10,000 eV. The GRO can detect gamma rays over a broad energy spectrum and is the first such satellite to be able to survey the entire sky in gamma rays. Earth's atmosphere effectively screens out gamma rays; thus, the GRO permits a new view of the energetic universe that is impossible to see from the ground.

There are four instruments aboard the GRO. The Burst and Transient Source Experiment (BATSE) is designed to study transients such as gamma ray bursts and solar flares. The Oriented Scintillation Spectrometer Experiment (OSSE) examines low-energy gamma rays, while the Compton Telescope (COMPTEL) images sources of medium-energy gamma rays. Finally, the Energetic Gamma Ray Experiment Telescope (EGRET) studies the highest-energy gamma rays. Gamma rays cannot be focused using lenses and mirrors that work with visible light. Instead, the four instruments aboard the GRO use high-energy particle detectors, such as sodium iodide crystals, that are induced to emit pulses of visible light whenever a gamma ray enters them. EGRET detects electron-positron pairs that are created when energetic gamma rays enter a spark chamber consisting of electrified wires. The pairs create sparks that are seen in the chamber and that reveal the path of the incoming gamma ray.

BATSE has detected more than 1,100 gamma ray bursts. These mysterious phenomena, which last anywhere from a fraction of a second up to a few minutes, occur about once per day. Afterwards, no new or peculiar counterparts to the bursts remain. BATSE has found that gamma ray bursts occur very uniformly across the sky; however, the number of faint bursts drops off much faster than would be expected if the sources of gamma rays are uniformly distributed in space. These two observations imply that the distribution of burst sources is approximately spherical and extends only to a finite distance. Some theorists think that this distance extends to the edge of the observable universe and that the bursts are cosmological, while others suggest that the bursts are galactic and originate in a large halo around our own galaxy.

The GRO has also observed high-energy sources called active galactic nuclei (AGNs), which some theorists believe are supermassive black holes drawing in surrounding material. Radio-loud AGNs, called blazars, reside in elliptical galaxies and have been studied at the highest energies by EGRET, while radio-quiet AGNs, called Seyfert galaxies, are found in active spiral galaxies and have been examined by OSSE and COMPTEL. The GRO has also studied gamma rays emitted from the decay of radioactive nuclei, such as cobalt-56 and aluminum-26, that are produced by supernova explosions of massive stars in our galaxy. Finally, the GRO has also observed gamma rays emanating from radio pulsars, such as the Crab and Vela pulsars, as well as from new ones such as the Circinus and Geminga pulsars.

See also: ACTIVE GALACTIC NUCLEUS; ELECTROMAGNETIC RADIATION; GAMMA RAY BURSTERS; PULSAR

Bibliography

GEHRELS, N.; FICHTEL, C. E.; FISHMAN, G. J.; KURFESS, J. D.; and SCHONFELDER, V. "The Compton Gamma Ray Observatory." *Sci. Am.* **269** (6), 68–77 (1993).

JEAN M. QUASHNOCK

GAS

Gas is one of the three classical states of matter; the other two are solid and liquid. Classically, a solid retains its shape even without being constrained by a container, while a liquid will flow so that, in equilibrium, it will fill its container up to the same level throughout, essentially retaining its volume. A gas, however will expand to fill its container; that is, a gas is several orders of magnitude more compressible than a liquid.

The distinction between a solid and a liquid is not always as clear as it would seem. One can argue, for example, whether taffy is a solid or a liquid; it flows like a liquid if you wait long enough but feels like a solid when you take off the wrapper. The physicist resolves the argument by defining a crystalline solid as a structure in which the atoms are arranged in a regular (i.e., periodic) array exhibiting long-range order (i.e., positional correlations exist over ranges that are long compared with the interatomic distance or, more properly, with the unit-cell size). A liquid exhibits short-range order, and gas exhibits no positional correlations for distances much greater than the molecular size.

What usually makes it easy to distinguish a liquid from a gas (both are called fluids) is that most substances have a recognizable phase transition between the liquid state and the gaseous state called vaporization (or evaporation). The reverse phase transition, from the gas to the liquid, is called condensation. In general, the density of the liquid is greater than that of the gas at a phase transition (i.e., the phase transition is first order). The pressure of the gas as a function of the temperature of the phase transition is called the vapor-pressure curve. At any low enough temperature, a liquid with a free surface has a gas above it at the corresponding vapor pressure. For example, the fuel tank of a car contains vaporized gasoline even if there are just a few drops of liquid fuel left in the bottom.

For every substance that has a liquid-gas phase transition, however, there is a temperature above which the liquid and the gas become indistinguishable; that is, above this critical temperature no pressure is enough to liquefy the substance.

Ideal Gas

When a gas is at a pressure much less than the pressure that makes it condense at that temperature, then its density is also low. In this case, the gas molecules are mostly far away from each other, where "far" means at distances that are large compared to the size of a gas molecule. Condensation into the liquid state occurs because the attraction between the gas molecules tends to make them stick together. The ideal gas model discussed below assumes that molecular size is so small compared to a typical distance between molecules that it is neglected completely. It may seem strange that a physical theory that predicts behavior so close to the real observed behavior of gases assumes that the gas molecules have zero size. However, taking molecular size and molecular attraction into account gives a better model and a theory (van der Waals theory) that predicts realistic gas laws, even close to condensation density.

The ideal gas law was published by Robert Boyle in England in 1660 and again sixteen years later by Edme Mariotte in France. It says that at constant temperature the volume of an enclosed gas is inversely proportional to its pressure. In other words, if you let the gas expand so as to double its volume, its pressure will have halved. It has been said that Boyle used to sit by the great vats in a brewery as a child and watch the bubbles rise in the beer. As they rose to the lower pressures nearer the surface, he noted the bubbles grew. In his early thirties Boyle built a laboratory, which included barometers for measuring pressure. Careful volume measurements using glass tubes filled with mercury led him to the finding that at a given temperature

$$(\text{pressure}) \times (\text{volume}) = \text{constant},$$

where the "constant" depends on the temperature of the gas. It was not until well into the nineteenth century that the proportionality relation between this constant and the *absolute temperature* (also called Kelvin temperature) was clarified.

For a mass M of a gas having molecular weight m, this relation may be written

$$pV = (M/m)RT.$$

The constant R is called the universal gas constant and is expressed in units consistent with the units of the absolute temperature T. The temperature of melting ice is 273.15 K. With $M = m$ and $p = 1$ atm (i.e., 1.013×10^5 N/m^2), the volume of an "ideal" gas at this temperature is 22.4 L. Thus we have

$$R = 8.3145 \text{ J·K}^{-1}\text{·mol}^{-1}.$$

One mole is one molecular weight, and there are a thousand liters in one cubic meter. Kinetic theory relates this experimental number to the number of molecules in one mole of gas (i.e., Avogadro's number). That it should indeed be a universal constant was discovered by Joseph Louis Gay-Lussac in France in 1802, almost a century before its value was determined.

Kinetic Theory

Assume that a macroscopic volume of gas contains a large number of identical molecules distributed uniformly throughout a container with the distance between molecules large compared with the molecular diameter. Assume also that there are no external forces and no intermolecular forces except collision forces, allowing the molecules to move in perpetual random motion of straight lines between collisions (which are perfectly elastic) with other molecules and with the walls of the container. Under these conditions, kinetic theory shows that

$$pV = \tfrac{2}{3}N(\tfrac{1}{2}m\overline{v}^2),$$

where p is the pressure of the gas, V is the volume occupied by the gas, N is the number of molecules contained in the volume, and $\tfrac{1}{2}m\overline{v}^2$ is the average kinetic energy per molecule. This equation can then be rewritten as the ideal gas law:

$$pV = Nk_BT,$$

so that k_BT is equal to two-thirds times the average kinetic energy per molecule. The variable k_B is known as the Boltzmann constant and now has an established value of 1.3807×10^{-23} J·K^{-1}. Since the Avogadro number (N_A, the number of molecules in a mole) is equal to the ratio of the universal gas constant to the Boltzmann constant, we are able to calculate the Avogadro number:

$$N_A = \frac{R}{k_B}$$

$$= \frac{8.3145 \text{ J·K}^{-1}\text{·mol}^{-1}}{1.3807 \times 10^{-23} \text{ J·K}^{-1}}$$

$$= 6.02 \times 10^{23} \text{ mol}^{-1}.$$

See also: AVOGADRO NUMBER; BOLTZMANN CONSTANT; BOYLE'S LAW; COLLISION; GAS CONSTANT; IDEAL GAS LAW; KINETIC THEORY; LIQUID; MOMENTUM

Bibliography

GOLDEN, S. *Elements of the Theory of Gases* (Addison-Wesley, Reading, MA, 1964).

STEFAN MACHLUP

GAS CONSTANT

The gas laws of Robert Boyle and J. A. C. Charles are combined to yield

$$PV = aT,$$

where P is the pressure of the gas within a container having a volume V, T is the absolute (Kelvin) temperature of the gas, and a is a constant to be determined by experiment.

The evaluation of the constant a comes about through a series of experiments. First, the factor a is a different constant for each gas studied. The various values of a converge on a value that is nearly the same for all gases (a *universal* gas constant) if multiplied by $1/n$, where n is the number of moles of gas in the sample. The equation given above becomes the ideal gas law,

$$\frac{PV}{n} = \left(\frac{a}{n}\right)T = RT,$$

where $R = a/n$ is the universal gas constant. For the purpose of evaluation of R, the ideal gas law is rewritten as

$$\frac{PV}{nT} = R.$$

The experiment to evaluate R holds the value of temperature at an arbitrary value, such as the ice point (normal melting point) of 273.15 K. The values of P, V, and n are measured, and R is calculated, but there are still slight systematic variations in the calculated values of R, even for the same gas, that are not due to random experimental errors.

The difficulty with the experiment as described is that the experiment is performed on *real* gases, but the evaluation of R is based on a formula for an *ideal* gas. One of the assumptions used to derive the ideal gas law from first principles is that the gas particles do not interact with each other. Noninteraction can be experimentally achieved by means of smaller and smaller pressures, or equivalently, smaller and smaller molar gas densities n/V.

James F. Schooley cites data to evaluate R using the helium-4 gas isotope and methane gas at 273.15 K (see Fig. 1). In the limit of zero molar gas density, the two experimental gases converge on the value of R equal to

$$R = \lim_{n/v \to 0} \frac{PV}{nT} = 0.0820568 \text{ l·atm·K}^{-1}\text{·mol}^{-1},$$

or in SI units,

$$R = 8.31441 \text{ J·K}^{-1}\text{·mol}^{-1}.$$

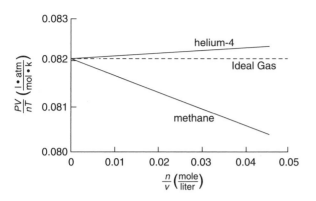

Figure 1 Data used to evaluate R using helium-4 gas isotope and methane gas at 273.15 K.

The graph of the theoretical ideal gas, shown for comparison, is a horizontal line indicating that R is not a function of n/V. In other words, gas particles in an ideal gas do not interact with each other. As a practical matter, however, systematic errors were introduced by the presence of water adsorbed on the surface of the gas container, which could not be accounted for in an error analysis.

Much greater accuracy is achieved by using an acoustic interferometer to measure the speed of sound in a gas. The gas constant R is obtained by plotting the square of the speed of sound as a function of pressure, again, in the limit of zero pressure. At the National Physical Laboratory in England, an argon gas acoustic interferometer operating at the triple point of water, namely 273.16 K, was used to determine a new value of the gas constant R equal to 8.314510 J·K^{-1}·mol^{-1} with an uncertainty of 8.4 parts per million.

Boltzmann's constant k_B, which is equal to the gas constant R divided by Avogadro's number N_A, is the only other physical constant related to the gas constant.

See also: GAS; IDEAL GAS LAW

Bibliography

COHEN, E. R., and TAYLOR, B. N. "The Fundamental Physical Constants." *Phys. Today* **49** (8), BG9–13 (1996).

SCHOOLEY, J. F. *Thermometry* (CRC Press, Boca Raton, FL, 1986).

ZEMANSKY, M. W., and DITTMAN, R. H. *Heat and Thermodynamics* (McGraw-Hill, New York, 1996).

RICHARD H. DITTMAN

GAUGE BOSON

See BOSON, GAUGE

GAUGE INVARIANCE

Gauge invariance is a central concept in modern theoretical physics. What is involved can be illustrated, in everyday terms, by the example of a topographic map, in which altitude is indicated by contours of equal elevation. Pick a point A and a higher point B on the map; to determine the elevation gain from A to B, and therefore the work done in lifting an object (or taking a hike) from A to B, we read off the elevation of B and the elevation of A and subtract. This difference between elevations is the same irrespective of the zero point with respect to which elevation contours are measured. By convention, elevations are measured with respect to sea level, but a contour map would be equally valid for determining elevation gains if the contours were measured with respect to the center of the earth or any other reference point. The elevation contours on the map are measures of what physicists term a gauge potential (in fact, they are directly proportional to the potential energy of an object in the earth's gravitational field), and the elevation gain from A to B is a physical quantity. The change from a contour map computed with origin at sea level, to a second contour map computed with any other choice of elevation zero, is what physicists term a gauge transformation, and the fact that the elevation gain from A to B is invariant with respect to this transformation is a statement of gauge invariance. In order to give all the information conveyed by a contour map without introducing an arbitrary choice of zero elevation contour, which physicists term a choice of gauge, one would have to give a table of elevation differences for all pairs of points A and B on the map. This system, which would require the hiker to carry an encyclopedia rather than a single sheet of paper, is clearly a very cumbersome way to convey the same information that is efficiently contained in the elevation contours.

With this example in mind, we can now define gauge invariance for physical systems in a general way. Gauge invariance is a property of physical systems in which the most concise description of physical force fields is given by auxiliary quantities called potentials, with the force fields uniquely specified by giving the potentials, but not the other way round. In a gauge invariant system, many different potentials can lead to the same force fields and the same physical effects; the transformation between potentials leading to the same fields is called a gauge transformation, and the invariance of physical effects to such transformations is called gauge invariance. The term "gauge invariance" was first used in 1919 by the mathematician Hermann Weyl in a now-obsolete theory of gravitation and electromagnetism, but it was soon seen to apply in a much wider context.

Consider now electrostatics for a system of charged conductors. Around the conductors, draw surfaces of equal electrical potential, which are just like the elevation contours on the topographic map. Pick two points A and B; the difference between the potential $\phi(B)$ at B and the potential $\phi(A)$ at A, which is the analog of elevation gain, is a physical quantity determining the work W done in transporting a particle of charge Q from A to B according to the formula

$$W = Q[\phi(B) - \phi(A)].$$

The work or energy difference computed from this formula is clearly independent of where we place the zero point of potential energy. A transformation between potentials computed with two different zero potential points is a gauge transformation, and mathematically takes the form

$$\phi(X) \rightarrow \phi(X) + C,$$

with C a constant and with X a general point. Applied with X taken first as A and then as B, the constant C cancels out in the difference, giving`

$$\phi(B) - \phi(A) \rightarrow \phi(B) - \phi(A),$$

so the difference of potentials, and hence the work W computed from it, are both gauge invariant. More generally, in an electrostatic system the force on a charge is given by $\mathbf{F} = Q\mathbf{E,}$ with $\mathbf{E}$ the electric field at the point where the charge is located. Since the electric field is given by the negative of the gradient of the potential,

$$\mathbf{E} = -\boldsymbol{\nabla}\phi,$$

and since the gradient of a constant C is zero, the force and electric field are also invariant under gauge transformations of the potential.

Electrodynamics, which governs charges in motion, is described by the Maxwell Equations, which are differential equations for the electric and magnetic fields. In this case, two potentials are required, the scalar potential ϕ introduced above together with a vector potential $\mathbf{A}$. The electric field $\mathbf{E}$ and the magnetic field $\mathbf{B}$ are computed from first derivatives of the potentials with respect to space and time, and there are transformations of the potentials—the gauge transformations of electrodynamics—which leave the fields invariant. Thus, the potentials of electrodynamics depend on an arbitrary choice of gauge, but the field strengths, which are the physical quantities, are gauge invariant. In the most general case, fixing the gauge requires specifying not just one constant, as in the topographical map and electrostatics examples, but a function that can depend on both the spatial coordinates and the time.

The gauge invariance of electrodynamics is associated with a conservation law, the conservation of electric charge. (This is a specific example of an important theorem proved by Emmy Noether, which shows that conserved quantities in physical systems are in general associated with invariances.) Electric charge conservation is in turn related to the fact that electrostatics is governed by an inverse square force law. Since the force exerted by a point charge Q on a test charge is proportional to the electric field of the point charge, the inverse square force law is equivalent to the statement that the electric field of a point charge at a distance R from the charge varies as Q/R^2, and that the potential varies correspondingly as Q/R. Since the surface area of a sphere of radius R is $4\pi R^2$, the product of the electric field at radius R times the surface area of a sphere through R is the constant $4\pi Q$; this conserved total flux measures the electric charge Q of the point charge source of the electric field.

Massive particles obey Yukawa-like equations rather than the Maxwell equations, and have a potential law that varies with distance as

$$\frac{e^{-MR}}{R},$$

with M the particle mass. This looks like an electrostatic potential (which varies as $1/R$) only when the mass M is exactly zero, so the photon (the force carrier of the electromagnetic field) behaves as a particle of exactly zero mass. For a Yukawa particle with nonzero mass M, the analog of the total flux vanishes as the sphere radius approaches infinity because of the rapidly decreasing exponential factor; charge is screened, not conserved, and the resulting forces are of short range. The analog of the Maxwell equations obeyed by massive particles contains the mass M in a term that does not obey gauge invariance. Hence, gauge invariance, the inverse square force law, and masslessness of the photon are all associated features. The theory of massless photons interacting with charged particles, including quantum effects, is called quantum electrodynamics, and is the first gauge theory to be well-understood. It governs all of the technological applications of electromagnetism that have transformed modern life.

Theories of the weak and strong interactions are also gauge theories of a related type, in which (unlike electrodynamics) the result of two successive gauge transformations depends on the order in which the transformations are performed. In the strong interaction theory, quantum chromodynamics, the gauge symmetry is unbroken, leading to novel phenomena such as quark confinement. In the electroweak theory, which unifies the electromagnetic and weak interactions, the gauge symmetry is partially broken. The unbroken symmetry corresponds to the massless photon and quantum electrodynamics; the broken sector corresponds to massive analogs of the photon (called weak gauge bosons), which carry the short range weak forces.

Albert Einstein's general relativity is also a kind of gauge theory, which is why Hermann Weyl first introduced the concept of gauge invariance in the context of attempts to unify gravitation and electromagnetism. It is also why physicists still believe that the principle of gauge invariance will ultimately lead to the unification of all of the forces in a single elegant theory.

See also: BOSON, GAUGE; EINSTEIN, ALBERT; GAUGE THEORIES; GRAND UNIFIED THEORY; MAXWELL'S EQUATIONS; QUANTUM CHROMODYNAMICS; QUANTUM ELEC-

TRODYNAMICS; RELATIVITY, GENERAL THEORY OF; THEORETICAL PHYSICS; YUKAWA, HIDEKI

Bibliography

CHENG, T.-P., and LI, L.-F. *Gauge Theory of Elementary Particle Physics* (Clarendon Press, Oxford, Eng., 1984).

YANG, C. N. "Gauge Fields" in *Selected Lectures: Hawaii Topical Conference in Particle Physics,* Vol. 1, edited by S. Pakvasa and S. F. Tuan (World Scientific, Singapore, 1982).

STEPHEN L. ADLER

GAUGE THEORIES

Gauge theories are the basis for the standard model of elementary particle forces, which gives a remarkably successful quantitative account of the electromagnetic, strong, and weak interactions of elementary particles, verified by experiments up to the highest energies attained in currently operating accelerators. Gauge theories are so named because they are constructed using the principle of gauge invariance, according to which a concise mathematical description of physical effects is obtained by using gauge potentials, which are themselves not uniquely physically determined. To uniquely specify the gauge potentials, the observer must make a choice of gauge by specifying one or more arbitrary functions of the space-time coordinates. All physical effects are then invariant under the change from one such arbitrary choice of gauge to another, which is called a gauge transformation.

The oldest known gauge theory, which serves as a prototype for the construction of recent generalizations, is the theory of electromagnetism. In electrodynamics, the electromagnetic force fields associated with moving charged particles are most simply described by scalar and vector potentials, which can be calculated (for a specified choice of gauge) from the particle motions. When a charged particle moves, its contribution to the potentials changes, and this in turn alters the force fields that act on other charged particles that are present, thus influencing their motions. In the classical description of this phenomenon, an electromagnetic wave propagates out from the initial charged particle and influences the motions of other charged particles as it encounters them. In quantum mechanics, wave motions have a particle aspect (and vice versa), and this leads to an alternative description of the same phenomenon. In the quantum mechanical description, a spin-1 boson force carrier called a photon, which is the elementary quantum of the electromagnetic field, is exchanged one or more times between the initially moving charged particle and every other charged particle which it influences. Because of the exact gauge invariance of electromagnetism, the photon is an exactly massless particle, and this corresponds to the fact that electromagnetism is a long-range force. (In our everyday lives, the long-range nature of electromagnetism is evident in the working of a compass in Earth's magnetic field, in the operation of motors and dynamos, and in the propagation of radiant energy from the Sun and the distant stars to Earth.) The photon is customarily denoted by the symbol γ in the physics literature because the "gamma rays" of early twentieth-century radioactivity are now understood to be energetic photons.

The gauge theories describing the weak and strong interactions are generalizations of the gauge theory of electromagnetism called non-Abelian or Yang–Mills gauge theories. In electromagnetism, the result of two successive gauge transformations is independent of the order in which these transformations are carried out. By contrast, in non-Abelian gauge theories, the result of two successive gauge transformations depends on the order in which they are performed, in other words, gauge transformations are noncommutative. This leads to physical behavior that is markedly different from that familiar in electromagnetism. For example, we have seen that electromagnetism is a long-range force; by contrast, non-Abelian gauge theories can lead to short-range forces of two different types. In the gauge theory of weak interactions, the non-Abelian gauge symmetry is partially spontaneously broken and charges that couple to the corresponding parts of the gauge field are screened. In the gauge theory of strong interactions, the non-Abelian gauge symmetry remains unbroken, and charges that couple to the gauge field are confined. Let us now examine in more detail these two types of Yang–Mills theory behavior.

The gauge theory of the weak interactions is actually unified with the gauge theory of electromagnetism in a larger structure called the electroweak gauge theory. This theory has four spin-1 force-carrying gauge fields, which obey highly symmetric equations of motion with a non-Abelian gauge invariance. These four gauge fields are coupled to the fields for a pair of spin-0 particles called the Higgs particles, and the potentials of the theory are structured so that the minimum potential state of the Higgs particles, called the ground state of the theory, is asymmetrical and so spontaneously breaks the non-Abelian gauge symmetry. This asymmetry results in large masses for the particles associated with three of the spin-1 gauge fields. These particles are customarily called *weak* gauge bosons because the forces which they mediate are the short-range weak interactions such as those observed in beta decay radioactivity. The corresponding weak charges to which they couple are said to be screened because the associated force fields do not extend out to large distances. However, even in the asymmetrical ground state of the electroweak theory, the fourth electroweak gauge field remains massless and retains the simpler gauge invariance (called an Abelian gauge invariance) characteristic of electromagnetism. It has the photon as its associated particle and gives rise to the long-range electromagnetic force.

In the physics literature, the weak bosons are customarily denoted W^+ and W^-, which are electrically charged and have masses equal to 85.5 proton masses, and Z^0, which is electrically neutral and has a mass equal to 97.2 proton masses. The corresponding force ranges, which are inversely proportional to these masses, are of the order of 10^{-16} cm. The Higgs particles are believed to be even heavier than the weak bosons, and their discovery is a principal goal of future high energy accelerator experiments.

The gauge theory of the strong interactions is called quantum chromodynamics because its eight force-carrying spin-1 gauge fields couple to charges called color charges. These eight gauge fields obey highly symmetric equations of motion with an exact, unbroken, non-Abelian gauge invariance; the associated gauge bosons are called gluons and are denoted by the symbol g. The color charges to which the gluons couple are carried by both the quarks, which are the constituent particles from which all strongly interacting particles are constructed, and by the gluons themselves. Because the gluons' fields are strongly self-coupled, their lines of force or flux lines tend to clump together like taffy. When a quark and an antiquark, bound together in a strongly interacting particle called a meson, are pulled apart, the flux lines running between them bunch up into a narrow flux tube or "bag" of approximately constant cross-sectional area. Thus the ratio of flux to area, which is a measure of the color force field, is nearly constant along the line joining the quark to the antiquark. To pull the quark and antiquark apart by a distance R, one has to supply an amount of energy equal to the force field times R. Since the force field is constant, the energy needed to pull a quark and antiquark apart grows in direct proportion to their distance of separation. Therefore it is not possible to obtain an isolated quark by separating a bound quark and antiquark infinitely far apart, since this would require supplying an infinite amount of energy, so color charges are said to be confined. A related property of quantum chromodynamics is that in violent, small distance collisions, the color forces act very weakly; this feature, called asymptotic freedom, played an important role historically in helping physicists decipher the structure of the strong interactions.

In actual high energy physics experiments, one never encounters flux tubes that are very long. The reason is that as soon as the energy in the flux tube is great enough to create a quark-antiquark pair out of the vacuum, this process occurs and the flux tube splits into two short flux tubes, one linking the original quark to the antiquark that emerged from the vacuum, and the other linking the quark that emerged from the vacuum to the original antiquark. The original meson is seen to split into two mesons. In collisions occurring at very high energies, this flux tube splitting process occurs multiple times, leading to the copious production of strongly interacting particles observed in accelerator experiments.

The standard gauge theory model of electromagnetic, weak, and strong forces contains nearly twenty empirically determined parameters, in the form of masses and coupling constants. For this reason, and because it does not include gravitation, it is believed not to be a complete theory of elementary particle forces. Attempts to achieve further unification, for example through grand unified theories that link the strong force with the electroweak force, also involve gauge theories.

See also: BOSON, GAUGE; BOSON, HIGGS; BOSON, *W*; BOSON, *Z*; COLOR CHARGE; ELECTROMAGNETISM; GAUGE INVARIANCE; GRAND UNIFIED THEORY; INTERACTION, ELECTROWEAK; INTERACTION, STRONG; INTERACTION, WEAK; PHOTON; QUANTUM CHROMODYNAMICS; QUANTUM ELECTRODYNAMICS; QUARK; QUARK CONFINEMENT; SPIN; SYMMETRY BREAKING, SPONTANEOUS

Bibliography

CHENG, T.-P., and LI, L.-F. *Gauge Theory of Elementary Particle Physics* (Clarendon Press, Oxford, Eng., 1984).

STEPHEN L. ADLER

GAUSS'S LAW

Generally, the term "Gauss's law" is used to denote one of a set of four equations, known as Maxwell's equations, which summarize the laws governing electric and magnetic fields. Named after Karl Frederich Gauss the term as generally used describes a connection between the properties of the electric field **E** and the charges that create the electric field. The law is based on a theorem from vector calculus know as Gauss's theorem, or the divergence theorem. In some cases the term "Gauss's law" refers to two of Maxwell's equations: the one noted above and a second one also based on the divergence theorem but involving the magnetic field rather than the electric field.

Gauss's law for the electric field is closely related to Coulomb's law; that is, the law describing the force between two point charges. The two are, however, not equivalent. When doing a problem using Gauss's law one must typically make some assumption regarding the directional properties of the electric fields. These directional properties are governed by a second of Maxwell's equations. Coulomb's law on the other hand has the directional properties as part of the law.

Maxwell's equations can be written in differential or integral form and, therefore, so can Gauss's law. In the differential form Gauss's law is

$$\mathbf{V} \times \mathbf{E} = \frac{\rho}{\varepsilon_0},$$

where **E** is the electric field at some point, ρ is the charge density at that point, and ε_0 is a constant known as the permittivity of free space and is equal to 8.85×10^{-12} F/m. The differential form establishes a local characteristic of the electric field since it states what happens at a point. The integral form describes a global property of the electric field since it describes a property of the electric field over the surface of some closed volume, known as a Gaussian volume. The integral form is

$$\oint \mathbf{E} \cdot d\mathbf{A} = \frac{q_{\text{in}}}{\varepsilon_0},$$

where on the left-hand side we integrate $\mathbf{E} \cdot d\mathbf{A}$ over the surface of a closed volume, and on the right-hand side q_{in} denotes the algebraic sum (i.e., taking into account pluses and minuses) of the charge inside (hence the subscript "in") the closed Gaussian volume. Both equations are written using the *Système International d'Unités* (SI). Using other systems of units will change the constants that appear in the equations.

Generally the integral form is used and an expression for the electric field is sought based on a specific distribution of charge. One begins by choosing a Gaussian volume that is an imaginary volume of any desired shape and size. The choice of a specific shape and size is dependent on the exact problem under consideration and is a key factor in the use of Gauss's law. After selection of the Gaussian volume one considers infinitesimal areas dA of the surface of the Gaussian volume. The integrand involves a dot product of two vectors $\mathbf{E}$ and $d\mathbf{A}$. Thus far dA has only a magnitude, that of the infinitesimal area, and no direction. Its direction is defined to be perpendicular to the area dA and pointing outward from the Gaussian volume (see Fig. 1). The integration of $\mathbf{E} \cdot d\mathbf{A}$ is done over the surface of the Gaussian volume. The integral is a surface integral and the small circle appearing at the midpoint of the in-

tegral sign denotes that this surface integral must be done over a closed surface (i.e., a surface that encloses a volume), which is what one has automatically due to the fact that one is dealing with a Gaussian volume. The circle on the integral sign is not used by all authors. The left-hand side of the equation is said to be the flux of the electric field through the surface of the Gaussian volume. The evaluation of the right-hand side of Gauss's law is generally very simple. All one needs to do is find the amount of charge contained inside the Gaussian volume and then divide by ε_0.

While Gauss's law is valid in all circumstances, the successful use of Gauss's law often requires that the charge distribution have a great deal of symmetry. Three types of symmetry can occur in various Gauss's law problems: spherical, cylindrical, or planer symmetry. Another common and important use of Gauss's law is in the establishing of a number of properties of conductors, such as the location of any excess charge on a conductor and the electric field in and near the surface of a conductor.

For the magnetic field case the differential form of Gauss's law is

$$\mathbf{V} \times \mathbf{B} = 0$$

and its integral form is

$$\oint \mathbf{B} \cdot d\mathbf{A} = 0.$$

This equation states that the flux of the magnetic field through the surface of any Gaussian volume is zero. The physical consequence of this law is that there is no magnetic analog to electric charge, or, in other words, that magnetic monopoles do not exist. Because of this fact, Gauss's law for magnetic fields cannot be used as it is for the electric field case to calculate the magnetic field. To do this type of calculation in the magnetic field case one typically uses Ampère's law, another of Maxwell's equations.

See also: AMPÈRE'S LAW; CHARGE; CONDUCTOR; COULOMB'S LAW; FIELD, ELECTRIC; FIELD, MAGNETIC; MAXWELL'S EQUATIONS; VECTOR

Bibliography

HALLIDAY, D., and RESNICK, R. *Physics,* 4th ed. (Wiley, New York, 1992).

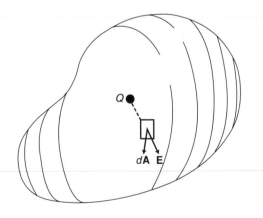

Figure 1 A Gaussian volume enclosing a point charge Q.

JACKSON, J. D. *Classical Electrodynamics,* 2nd ed. (Wiley, New York, 1975).

LORRAIN, P.; CORSON, D.; and LORRAIN, F. *Electromagnetic Fields and Waves,* 3rd ed. (W. H. Freeman, San Francisco, 1988).

PURCELL, E. M. *Berkeley Physics Course,* Vol. 2: *Electricity and Magnetism,* 2nd ed. (McGraw-Hill, New York, 1985).

SEARS, F. W.; ZEMANSKY, M. W.; and YOUNG, H. D. *University Physics,* 7th ed. (Addison-Wesley, Reading, MA, 1989).

JAMES L. MONROE

GEIGER COUNTER

A Geiger counter, also called Geiger–Müller (GM) counter, is a radiation detector and counter. The counter displays a running total of the combined number of gamma, beta, and alpha particles entering and interacting inside the detector. The detector is a Geiger tube.

The Geiger tube is typically a cylinder filled with an inert gas, usually Argon. Its diameter ranges from 1 to 10 cm, and its length is 2 to 10 times its diameter. One end face of the cylinder is often made of a thin mica or mylar foil to allow most of the betas (electrons) and alphas (helium nuclei) into the detector. Gammas (photons) including x rays more easily penetrate into the Geiger tube. Most Geiger counters are capable of measuring fluxes up to about 1,000 particles per second.

A Geiger counter is delineated from other gas radiation detectors by its higher electric field. The electric field in the Geiger tube is created by placing a positive voltage (typically 500 V) on a thin tungsten wire stretched along the center of the cylinder (see Fig. 1). The cylinder is maintained at ground (zero voltage) and is usually made of metal or glass. If made of glass, it is sprayed with a conducting coating. The positive electrode (the tungsten wire) is called the anode and the cylinder is the cathode.

When radiation interacts with the argon gas, it ionizes the atoms into free electrons and positively charged ions. The electric field accelerates the free electrons toward the tungsten anode. Each electron knocks additional electrons free by collisions with other argon atoms. This leads to an avalanche of electrons toward the tungsten wire. Because of the large electric field, the avalanche spreads, causing

the whole Geiger tube to discharge. The electrons from the discharge are collected in 10^{-6} s or less, thereby creating a short current pulse. An electronic circuit counts these pulses coming from the tungsten wire. The size of each pulse is nearly the same for all events, independent of the type of radiation.

The count rate of a Geiger counter is limited by the transit time of the positive ions, which take about 10^{-4} s to reach the surface of the cathode. Positive argon ions will release radiation when they interact with the cathode, thereby resulting in a second discharge. To quench these secondary discharges, the electronics that receive the current pulse can be designed to make the voltage on the tungsten wire dip to a low value until all of the positive ions are collected on the cathode. This is the slow, external quenching technique.

To allow the Geiger counter to handle higher counting rates, a fast, self-quenching method can be employed. In this method a small amount of organic gas or halogen gas is added to the argon gas. This quenching gas must have a lower ionization potential than argon. Because of this lower ionization potential, an argon ion colliding with a quenching gas molecule is likely to capture an electron from it and become neutral. The transfer of energy and ionization from the argon to the quenching gas is good because the quenching gas is more likely to lose its energy by dissociation rather than emit new radiation. Furthermore, the quenching gas has broad and intense absorption bands, meaning that it can absorb gamma radiation from the argon ions over a wide range of wavelengths. The quenching gas allows the Geiger tube to recover from the discharge more quickly, thereby allowing the detector to achieve measurements of higher counting rates. A

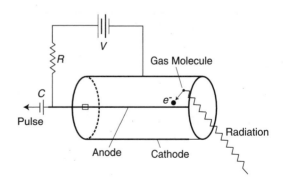

Figure 1 Diagram of the construction of a Geiger counter.

problem with the quenching gas is that it is slowly depleted during the use of the Geiger counter. A Geiger tube containing an organic quenching gas has a usable life of about 10^9 counts.

The voltage applied to the tungsten filament has an acceptable range of values often stretching over 200 V. This is called the Geiger region. If the voltage is too high, the Geiger tube will discharge continuously. If the voltage is too low, the electron avalanche is more localized so that the current pulse becomes too small to be counted by the Geiger counter.

See also: IONIZATION CHAMBER

Bibliography

LAPP, R. E., and ANDREWS, H. L. *Nuclear Radiation Physics,* 3rd ed. (Prentice Hall, Englewood Cliffs, 1963).
TSOULFANIDIS, N. *Measurement and Detection of Radiation* (McGraw-Hill, New York, 1983).

WILLIAM R. WHARTON

GENERATOR

See ELECTRIC GENERATOR

GEODE

A geode is a semispherical to globular rock mass with a central cavity lined by crystals growing from the outer wall into the center. Geodes range widely in size, from almost microscopic to more than 50 cm. Small geodes, typically less than 3 cm, are termed "vugs."

Most geodes begin as open cavities in rocks, some as the open interior of fossil shells or gas cavities within volcanic rocks. The cavities then slowly become partially filled by crystallization from water-rich solutions present at low temperatures and shallow depths. Commonly, the most inner crystals are only the last of many crystalline layers deposited from solutions that have filled the cavity at various points in time.

Quartz, the most abundant polymorph of silicon dioxide, is a common mineral in geodes and may range from clear to showy colored varieties, such as amethyst. Agates, which are banded, filled cavities of fine-grained quartz (termed "chalcedony"), are a gaily colored end product of crystallization within a former geode. Calcite, the most abundant polymorph of calcium carbonate, is also a common mineral of geodes. A large number of other minerals may crystallize in geodes, such as barite (barium sulfate), celestite (strontium carbonate), anglesite (lead sulfate), and sphalerite (zinc sulfide).

It is not uncommon for the crystals of a geode to be different from the ones that make up the rock or vein in which they are found. For example, barite, celestite, and quartz commonly occur in geodes in limestones that are mainly calcite.

The intrinsic beauty of geodes is enhanced by cutting through them with a diamond saw followed by polishing. It is in this prepared form, which may even involve artificial staining, that geodes are sold commercially or displayed in museums.

See also: CRYSTAL; CRYSTAL STRUCTURE

Bibliography

SANBORN, W. B. *Oddities of the Mineral World* (Van Nostrand Reinhold, New York, 1976).
WHITE, J. S. *The Smithsonian Treasury: Minerals and Gems* (Smithsonian Institution Press, Washington, DC, 1991).

WILLIAM G. MELSON

GEOPHYSICS

Geophysics is the branch of physics that addresses the application of physical principles and methods to the problems and processes of the planet Earth, from its inner core to its outer environs in space. Applied geophysics or geophysical surveys use physical tools for mineral and oil exploration, while research or experimental geophysics studies the solid Earth, its fluids, its sister planets in the solar system, and its magnetic fields.

The Disciplines of Geophysics

Because of the diversity of interests involved in geophysics, it appears sometimes that the only thing that unites geophysicists is the common use of physics and mathematics. Thus, while it is possible to call both an oceanographer and a space physicist "geophysicists," it is quite clear that these two scientists apply their skills to the study of very different media. This theme of unity in diversity will become more apparent when the integrative aspects of geophysics is discussed.

Applied geophysics. Sometimes called exploration geophysics, applied geophysics is goal-oriented and practical in aim: the discovery of new deposits of oil, gas, and minerals for the use of society. The area of application here is the solid crust of the earth, the upper few tens of kilometers that cover the more fluid, or at least pliable, mantle.

Electrical, magnetic, radioactive, elastic, and gravitational properties of the rocks and minerals that make up the crust are used, alone or in combination, to determine the likely presence of unusual concentrations of such minerals as copper, iron, phosphate, or hydrocarbons.

Seismology. The study of earthquakes and related phenomena such as tsunamis has occupied keen observers of nature for ages. Ancient Chinese philosophers ascribed the violent shaking of the earth's surface accompanying earthquakes to forces deep below the surface. The catastrophic fires that accompanied the Lisbon earthquake in the eighteenth century prompted a major section of Voltaire's *Candide.* In 1992, a giant tidal wave, or tsunami, generated by an undersea earthquake, swept several hundred Indonesian villagers into the sea.

Seismology is concerned with earthquakes, their causes, and their prediction. In addition, seismologists use the seismic waves traversing the earth's layers during an earthquake to glean a glimpse of the structure of these layers, their thickness, their composition, and how they influence the surface features of continents and oceans. The speed of an elastic seismic wave increases or decreases as it encounters more- or less-dense material. From this simple observation, seismologists have been able to decipher the deep structure of the earth—to penetrate some of the layers of "Chinese boxes" that make up the earth.

Tectonophysics. One of the major advances in understanding the behavior of this planet we call Earth took place in the past 30 years with the development of the theory of plate tectonics. Starting with observations made by submarines during World War II, and later by ocean-going research ships, patterns of magnetic lineaments found on the ocean floor were correlated with the age of rocks there. The dense basaltic rock making up the ocean floor in the Atlantic, for example, was found to be younger near the major crustal crack called the Mid-Atlantic Ridge. This pattern was repeated symmetrically on the other side of the ridge. More recently, geophysicists found that fresh lava is being produced at or near similar ridges traversing all the oceans of the earth.

These observations led to the idea that the crust and solid part of the underlying mantle formed a series of interlocking plates, large and small, that grind against each other in a continuous ever-shifting dance of birth and death. These observations also led to the rapid growth of the discipline of tectonophysics, which deals not only with the deformation of these plates but with continental drift, seafloor spreading, the down-warping in the trenches adjacent to island arcs, and the structural history of continents. What produced the series of great faults, such as the San Andreas Fault in California, that circumscribe the Pacific Ocean? Why did the great supercontinent Pangaea break up some 3 billion years ago? Tectonophysicists attempt to answer these questions by narrowing the uncertainties about the behavior of rocks under great stresses. They try to explain the changes that have taken place over the nearly 4 to 5 billion years since the earth became an individual planet.

Geodesy. As more and more knowledge accumulates about movement of plates during the geologic past, and as it becomes clear that this movement continues today, more precise methods of measuring the rates of physical change become critical. Happily this need is met by developments in an ancient discipline: geodesy.

Geodesy is the study of the shape, size, and gravitational field of the earth. Until recently, geodesists were either theorists occupied with explaining the variations of the gravitational field from place to place (for example, the force of gravitational pull is greater at the poles than it is at the equator), or engineers using precision instruments to survey a construction site or lay the foundations for a road.

Two technologies provided a far more precise tool with great applications. Satellites with exact, predetermined orbits have become ubiquitous. When powerful laser beams are placed in these satellites, pinpoint determination of changes on the surface of the earth becomes possible. In fact, it is now possible to say with some certainty at what rate part of California is slipping out to sea along the San Andreas Fault. Or, in another application of this satellite technology, to determine if a bulge on the surface near Los Angeles is growing or standing still, whether and where the surface of the sea is rising or lowering, and how the plates are interacting with each other.

Geomagnetism. Even in ancient times, natural magnetic minerals such as magnetite were known and used, whether in navigation or in games. And in the early days of the modern development of science, the magnetic field of the earth was scrutinized intensely. What produced it? How did it work? And later: Why did the poles of that field migrate across the globe over geologic time? How did it relate to the charged particles coming in from the Sun to form the magnetosphere enveloping the earth?

In the geologic past, it was found that magnetic minerals were trapped in rocks of different kinds. This phenomenon is called paleomagnetism, and its study reveals many of the past patterns of plate movement and seafloor spreading.

In the atmosphere, and the space beyond, the earth's magnetic field extends outward, controlling the vast amounts of electrons and protons streaming down from the Sun. Some of these particles are thus trapped in huge radiation belts circling the earth. Others reach the ionosphere, only to collide with molecules of atmospheric gases, which become ionized and energized. Such ionization may disrupt radio communications or produce light phenomena, most visibly as auroras.

Volcanology. Volcanic eruptions are the result of superheated molten rock, or lava, forcing the solid layer of rock above to break asunder, producing either an explosion of lava or a combination of gases and volcanic ash that can travel many miles in minutes. In some cases, both of these scenarios play themselves out at the same site at different times.

Another form of volcanism is not associated with major explosions but rather takes place over long periods of time and over huge surface areas. Lava seeps out along small cracks or vents and forms over time a large volcanic plateau.

Examples of the first type of explosive volcanism are abundant in the plate margin areas of the crust, especially near the subduction zones of the plates. Thus we find Mount St. Helens and its sister volcanoes in the northwestern United States clustered near the Pacific Ocean subduction area. Similarly, the volcanoes of Japan are near the edge of the Pacific Plate.

The second type of explosive volcanism is usually found in the interiors of continents. One of the most dramatic examples of this is the Deccan Plateau of India.

To study volcanoes and volcanism, a scientist must be familiar with petrology (the study of rocks) and geochemistry (the study of the chemical constituents of rocks and fluids). In other words, a volcanologist strives to answer fundamental questions about the composition of the deep layers of the earth, the distribution of elements, and heat in the interior.

Hydrology. The study of water in all of its forms, hydrology is becoming a critical science for our survival. Water resources are not only increasingly scarce but are frequently affected by human activity in ways that make them polluted or otherwise unusable. In other cases, disputes make water practically unusable.

Whether in the form of water vapor in the atmosphere, ice in glaciers, raging torrents in rivers, or pore water in groundwater, water is an essential commodity. Water-related political disputes among nations (or even among states within the same nation) are increasing.

In the meantime, pollution—directly into a stream or indirectly by seepage underground—is rendering important tracts of rivers and aquifers poisoned and unfit for consumption. Overdrawing major reservoirs of groundwater, without adequate replenishment by rain, makes these reservoirs potentially barren for a long time to come.

A hydrologist's task, therefore, is to delineate these problems and to try to solve them. It is not surprising, then, to find common ground between science and engineering in the study of water resources and the hydrologic cycle.

Ocean sciences. Water is also found in the oceans, covering most of the earth's surface. Oceanography can be divided into several subdisciplines: physical oceanography (the physics of the oceans and interactions with the atmosphere), geological oceanography (the study of the ocean floors and sedimentation), biological oceanography, and chemical oceanography.

To carry out studies of the oceans, their currents, waves, properties, and dynamics, various platforms have been developed. Among these are specialized submersibles, fixed platforms, buoys, satellites, and underwater laboratories with or without divers.

Because of the increasing importance of ocean studies for resources or as part of the global-change picture, the demand for oceanographers of various kinds is growing rapidly.

Atmospheric sciences. The gaseous envelope that surrounds the earth in a life-sustaining cocoon has become the focus of increasing scientific interest. The answers being developed by these studies may mean the difference between a gradual deterioration in the quality of human life or finding the answers to some of the most intractable problems facing us today. Is the greenhouse effect real, and, if real, how serious is it? Global warming: What does it involve? Is there a pattern? And what is its relationship to the ozone hole?

No less important, can scientists get to the point of doing some real long-range weather prediction? Can thunderstorms be harnessed? Atmospheric geophysics combines physics and mathematics to provide some of the answers.

Planetology. The Apollo missions to the Moon have provided excitement and some solid scientific evidence on the nature of the Moon's surface, the rocks composing it, and how it formed. Other spacecraft have mapped parts of Mars, Venus, Mercury, Jupiter, and all the planets except for Pluto. Some missions have come close to comets and asteroids. In fact, an encounter between a comet and the surface of Jupiter has been recorded "live" by sophisticated tracking instruments. Other studies, closer to home, have concentrated on the study of the debris coming to the surface of the earth: meteorites.

Planetologists compare the planets of the solar system, their natural satellites, and other bodies found in the vicinity, such as asteroids and interplanetary dust. Planetologists interact with other scientists as they try to explain the origins of the solar system and its composition. In most cases, they were originally geochemists, paleontologists, petrologists, space physicists, or biologists before becoming planetologists. While their work is of great scientific significance, there is little promise of economic benefit in planetology.

Space–planetary physics. The Sun emits a spectrum of particles and rays: x rays, gamma rays, protons, electrons, cosmic rays, light, radio waves, and so on. Most of this material is trapped in the earth's upper atmosphere at high altitudes. It is studied by satellite observations or by computing the time radio beams reflected from the earth take to return.

The configurations of both the magnetosphere and the ionosphere are studied in detail. Air glow is analyzed and so is the Van Allen Radiation Belt. The components of the solar wind are related to the shape and structure of the magnetospheres of the planets and the interplanetary magnetic field. Many experiments explore the nature of space plasmas and how they behave under extreme conditions, such as those encountered near solar flares or the magnetic sheath that surrounds spacecraft.

Geophysicists and Professionals

Who are the practitioners of geophysics? Who are the men and women who tackle the problems outlined above? What are their qualifications?

Most geophysicists have a Ph.D. in science, usually with a heavy concentration on physics and mathematics but sometimes with undergraduate degrees in chemistry and even biology. But it is this fundamental grounding in the principles of physics and the ability to use mathematics to formulate physical problems that distinguishes geophysicists from other physical scientists, such as geologists, who also study the earth.

Many geophysicists work as researchers at universities and various influential laboratories. Others are employed by government agencies or industries. Opportunities for work exist in the laboratory, field, or office. Field work could mean trekking around a desert, walking in space, peering through a submarine window, or collecting ice samples from the South Pole.

Integration of Disciplines

The unity of the field of geophysics is best expressed by the planet Earth metaphor, for it is the earth with its lithosphere, hydrosphere, biosphere, and atmosphere that forms the focus of geophysical studies. Even space physicists and planetologists are concerned with comparisons between Earth and its sister planets or with interactions between Earth and space plasma.

Thus, from an organizational point of view, we find that geophysical societies have grown around the idea of unity among diversity that characterizes the science. The most prominent societies of this type are the International Union of Geophysics and Geodesy, which meets once every four years, and the American Geophysical Union, which is also international in scope despite its name.

This integration also is manifested in the many interfaces found among the disciplines outlined above. The atmosphere and the ocean are clearly intertwined. Winds erode rocks of the continents. The water found on continents is related in one grand hydrologic cycle to that found in the air and in the ocean. The magnetosphere of the earth interacts with that of the other neighboring planets. Paleomagnetism is related to plate tectonics and the latter is related to earthquakes and volcanoes. Gravity is related to all the components of the earth. Cycles run within cycles.

Planet Earth in Its Future Journey

People study science to gain an understanding of natural forces that affect their lives. The study of geophysics poses some direct challenges to these aims.

Human society, on the eve of the new millennium, is challenged by artificially induced changes to the environment; growing scarcity of natural resources; ancient, unpredictable catastrophes; a rapid increase in population; and new revolutions in our understanding of the earth. Plate tectonics has unified the many different strands of knowledge about the lithosphere. The concept of global change has provided a conceptual framework for the fluid envelope surrounding our planet. Comparative planetology has made it clear that similar processes appear to operate in the solar system.

After nearly 5 billion years, our planet is mature. Life has evolved in myriad ways and continues to evolve. The human segment of this life has mastered all the environmental niches, from the extreme cold of Antarctica to the desolation of Death Valley. In the process, human society faces difficulties adjusting to and coping with growth.

Geophysics in all its fields, but above all in its synoptic application of fundamental principles, will play a very crucial role in developing an understanding of these challenges and in devising solutions to the problems.

See also: ATMOSPHERIC PHYSICS; EARTHQUAKE; SEISMOLOGY; VOLCANO

Bibliography

AGU Handbook (American Geophysical Union, Washington, DC, 1992).

BATES, C. C.; GASKELL, T. F.; and RICE, R. B.; eds. *Geophysics and the Affairs of Man* (Pergamon Press, Oxford, 1982).

Careers in Geophysics (American Geophysical Union, Washington, DC, 1986).

PRESS, F., ed. *Planet Earth: Readings from Scientific American* (W. H. Freeman, San Francisco, 1974).

G. N. RASSAM

g-FACTOR

The gyromagnetic ratio, or *g*-factor (often represented by the letter *g* with an appropriate subscript), is a term closely related to the angular momentum and the magnetic moment of particles. Just as there are different particles and different motions leading to angular momenta, there are different *g*-factors as well. The different *g*-factors are the orbital *g*-factor, the spin *g*-factor, the Landé *g*-factor, the proton *g*-factor, and the neutron *g*-factor.

The orbital *g*-factor, g_1, is the ratio of the orbital magnetic dipole moment, μ_1, expressed in units of the Bohr magneton, μ_B, to its orbital angular momentum L, expressed in units of $\hbar$. Therefore, g_1 is given by

$$g_1 = \frac{\mu_1/\mu_B}{L/\hbar}.$$

Although g_1 has a value equal to one, in general, μ_1 may be expressed as

$$\mu_1 = \frac{g_1\mu_B}{\hbar}L = g_1\mu_B\sqrt{l(l+1)},$$

where *l* is the orbital quantum number.

The spin *g*-factor, g_s, has a value very close to 2.0; it has been measured with a high degree of precision to be 2.00232. A relation connecting g_s, μ_B, the magnetic dipole moment of the electron μ_s, and the spin angular momentum, *S* can be obtained as

$$g_s = -\frac{\mu_s/\mu_B}{S/\hbar}$$

The Landé *g*-factor, *g* or g_j, is given by

$$g_j = 1 + \frac{[j(j+1) + s(s+1) - l(l+1)]}{2j(j+1)},$$

where j is the total angular momentum quantum number and s is the spin quantum number. Clearly, g_j can have different values, depending on the values of l, s, and j.

The g-factors corresponding to the nuclear particles are even more intriguing. If the protons and neutrons are treated as point particles and one then proceeds to use the successful Dirac theory, one will deduce that the proton g-factor, g_p, should have a value very close to two; whereas the neutron, which has no charge, should have a g-factor, g_n, equal to zero. The experimental values for g_p and g_n are very close to 5.6 and -3.82, respectively. Thus, there is compelling reason to anticipate additional hidden structures in these nuclear particles. Recent developments have given further credence to the presence of the constituent particles of the nucleons—quarks.

See also: BOHR MAGNETON; MAGNETIC MOMENT; MOMENTUM; QUARKS; SPIN

Bibliography

BREHM, J. J., and MULLIN, W. J. *Introduction to the Structure of Matter* (Wiley, New York, 1989).

EISBERG, R., and RESNICK, R. *Quantum Physics of Atoms, Molecules, Solids, Nuclei, and Particles*, 2nd ed. (Wiley, New York, 1985).

SANKOORIKAL L. VARGHESE

GIBBS, JOSIAH WILLARD

b. New Haven, Connecticut, February 11, 1839;
d. New Haven, Connecticut, April 28, 1903;
thermodynamics, statistical mechanics.

Gibbs was the only son of the Professor of Sacred Literature at Yale College of the same name. Gibbs's health, like that of his unmarried sister, was precarious, and he lived with her, his married sister, and brother-in-law in the family house. His ordered existence centered upon Yale as student, tutor, and faculty member. He entered Yale in 1854, graduating in 1858. Atypically, he continued at Yale in engineering

and received his Ph.D. in 1863 with a dissertation on gear design. For three years he was an unpaid tutor at Yale. He then took a postdoctoral tour of Europe. Gibbs attended semester-long courses at the Collége de France and at the universities of Berlin and Heidelberg. The courses were given by the leading mathematicians and experimental physicists of Europe. At the same time, he completed an intense reading program in the works of other British, German, and French mathematical physicists and mathematicians.

Returning to New Haven in 1869, Gibbs resumed his unpaid position until appointed Professor of Mathematical Physics in 1871. However, Yale only offered Gibbs a salary after Johns Hopkins tried to lure him to Baltimore in 1879. The faculty at Johns Hopkins knew of him because of his growing reputation in Europe from his research in thermodynamics. Gibbs rarely traveled to meetings and did not join the American Physical Society, but he always sent reprints of his work to all the important physicists in Europe and the United States.

Gibbs's research interests went from engineering to mechanics and then to thermodynamics, vector analysis and the elctromagnetic theory of light, and statistical mechanics. His output was modest in quantity but always of superior quality. His publications share the characteristic of being based on only general principles: his mechanics on Hamiltonian equations, his thermodynamics only on its two basic laws. Unlike others, he avoided assumptions about molecules or their interactions. He changed thermodynamics to exclude molecular considerations and named "statistical mechanics" for the study of ensembles of mechanical systems that obey Hamiltonian dynamics. In his research, the implications of these basic laws are developed at the most general level possible, usually in geometrical form, consistent with reaching physically significant not mathematically interesting results. Gibbs always moved from the simplest to the most general, complex case.

With this approach Gibbs transformed thermodynamics. In 1873 the law of conservation of energy was understood, but there were three interpretations of the second law and misunderstandings about the relationship between energy and entropy. In his first thermodynamics paper, Gibbs stated both laws clearly, identified entropy as a state function, and investigated the thermodynamic properties of a homogeneous substance. He also developed new thermodynamic diagrams to display these properties. In his second paper, Gibbs considered a substance existing in three physical states. He mapped the thermodynamic equilibrium of these states and showed where they could coexist and identified the critical point. In his third paper, Gibbs generalized the notion of heterogenity to include chemical mixtures. He wrote the first law as the change in internal energy as a function of changes in entropy, work done, and the sum of the masses in the mixture multiplied by the "thermodynamic potential" for each mass. Thermodynamic potential could be used for changes in the density, magnetic, or electrical properties. Focusing on changes in mass, Gibbs developed his phase rule that established whether a particular chemical mixture was stable, and if not the direction in which that chemical mixture would move to reestablish equilibrium. This was an old problem for chemists and critical for the growing chemical industry in Europe and the United States. However, Gibbs did not bring this to the attention of chemists. This was done by James Clerk Maxwell in Britain. In Europe, Gibbs's work was the foundation for the careers of Wilhelm Ostwald and J. H. van't Hoff who also trained the first generations of physical chemists, many of whom returned to the United States and established departments in this field in the newly established American universities.

Gibbs then turned to vector analysis and in 1879 taught one of the first courses in the subject. His work was derived from William Rowan Hamilton's quarternions and William Kingdon Clifford's mechanics. However, Gibbs modified both, and his work was closer to that of Hermann G. Grassmann. In his publications, he ran afoul of Peter Guthrie Tait, who fought to keep Hamilton's quarternions in their pure form against the tide of vectors. Gibbs countered by demonstrating their use in astronomy and their possibilities for electromagnetism.

In his final publication, Gibbs fused his earlier interest in mechanics and thermodynamics. He considered the behavior of ensembles of mechanical systems, defined only as obeying Hamilton's mechanics, that shared different characteristics. His microcanonical ensemble was of systems that shared the same total energy but not the same positions or velocities. Generalizing, he constructed his canonical ensemble, where the systems contained only the same number of particles but did not share the same values of total energy. He demonstrated that such an ensemble behaved like a thermodynamical system. Finally he considered his grand ensemble of systems containing different numbers of particles. Early proponents of quantum theory, trying to emphasize the need for quanta in physics, faulted Gibbs's statistical mechanics. All his ensembles depended on the equipartition theorem, that the average energy of each degree of freedom of motion in every mechanical system was the same. This did not seem to apply to molecules of real gases. He also assumed the ergodic hypothesis, that all his systems eventually passed through every possible configuration consistent with its total energy. Finally there was a real paradox. The entropy of a grand ensemble of mechanical systems depended upon whether the particles making up the systems were distinguishable or indistinguishable from those alike in all other respects in the other systems in the ensemble. For Gibbs this was a technical problem; for his critics it was a matter of the properties of real molecules. Later work in quantum statistical mechanics, that postulated a statistical distribution rather than deriving it, demonstrated that Gibbs's approach was very profitable.

Gibbs's research was appreciated in Europe. He was elected to the major physical and mathematical societies there. Although he was a member of the important scientific societies Gibbs made less impression in the United States, where physicists were dedicated to experiment and less aware of the power of physical theory.

See also: ENSEMBLE; PHASE RULE; STATISTICAL MECHANICS; THERMODYNAMICS, HISTORY OF

Bibliography

BUMSTEAD, H. A., and VAN NAME, R. G., eds. *The Scientific Papers of J. Willard Gibbs* (Dover reprint, New York, [1906] 1961).

GIBBS, J. W. *Elementary Principles of Statistical Mechanics Developed with Special Reference to the Rotational Foundations of Thermodynamics* (Dover reprint, New York, [1902] 1960).

KLEIN, M. J. "The Physics of J. W. Gibbs." *Phys. Today* **43**, 40–48 (1990).

WHEELER, L. P. *Josiah Willard Gibbs: The History of a Great Mind* (Yale University Press, New Haven, CT, 1952).

ELIZABETH GARBER

GRAND UNIFIED THEORY

The simplest grand unified theories (GUTs) are beautiful but speculative syntheses of the basic

Table 1 "Known" Forces

Force	Range	Strength	Particle	Mass[a]
Electromagnetic	∞	1/137	Photon	0
Weak	10^{-16} cm	$\approx 1/30$	W and Z	$\approx 100\ m_P$
Strong	10^{-13} cm	≈ 1	Gluons	0 or $\approx m_P$
Gravity	∞	$\approx 10^{-38} \dfrac{E^2}{m_\mathrm{P}^2 C^4}$	Graviton?	0

[a]Where m_P stands for proton mass.

forces between elementary particles in a single symmetrical structure. To see how they work, we must first examine the forces that they unify. What we know about the world at very short distances is summarized (or caricatured) in Table 1 and Table 2.

There are "force carrier particles" associated with each type of force, as shown in Table 1. For example, the electrical force between charged particles is related to the fact that when a charged particle is accelerated, it emits electromagnetic radiation. But in the quantum mechanical theory, the radiation is seen to consist of photons, particles of light. Table 2 shows one "family" of matter particles. For reasons that we do not understand, there are two other families with similar properties. But all of the matter that we are familiar with (because it is stable) is built from the first family.

Gravity is listed separately in Table 2 because if we were only interested in the physics of individual particles we would not know about it at all. It is only because we have some experience with huge collections of particles put together into planets and stars that we know about gravity. It plays no detectable role in the interactions between individual particles in any experiment we can do, or even plausibly imagine, because its effects are to small compared to those of the other three interactions. GUTs do not attempt to include gravitation interactions.

The particle physics forces look complicated at long distances, because the strong and weak forces are short range, and so have very different behavior from electricity and magnetism. By looking at the world at shorter distances, we have learned that the three particle physics forces have much in common. At collision energies large compared to 100 times the proton mass energy, which we can almost produce today, in particle accelerators near Geneva, Chicago, and San Francisco, all the force particles and all the matter particles produced are moving at nearly the speed of light all the time, and they appear very similar indeed.

This world is completely beyond direct human experience. Common sense will not work here at all. Common sense is the distillation of what people absorb from their experience of the world. This does not include experience with velocities near the speed of light or with the tiny spin of a single electron. For example, to the particle physicist, there is little difference between different dimensional measures (e.g., time, length, mass, and energy). They are all measuring essentially the same thing. Because of relativity and quantum mechanics, the question is always, "How small?" Once you get down to the small distances at which particle physicists work, relativity and quantum mechanics are crucial in everything. All particles are moving at very nearly the speed of light, and all processes are as quantum mechanical as they can be. For this reason, particle physicists usually use units in which the speed of light c and the fundamental quantum of angular momentum h are chosen to be equal to 1, because we are always in the relativistic and quantum mechanical regime.

In quantum mechanics, many quantities are quantized, which means they occur only in integer multiples of a certain basic value or quantum. One such quantity is electric charge. When you first learn

Table 2 Forces For Various Particle Types

Force	Electron	Neutrino	u Quark	d Quark
Electromagnetic	$\times$		$\times$	$\times$
Weak	$\times$	$\times$	$\times$	$\times$
Strong			$\times$	$\times$
Gravity	$\times$	$\times$	$\times$	$\times$

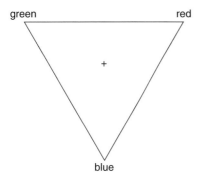

Figure 1

about it in the classical physics of electricity and magnetism, there is no reason to expect that electric charge comes in discrete chunks. However, that is the way the world works. The electricity we are interested in results from the movement of electrons. All electrons look the same. In particular, that means that each electron has exactly the same electric charge, and that if we have a bunch of electrons, we can find the total electric charge of the bunch just by counting them and multiplying by the charge of a single electron.

Another important property of this world is that all interactions, even when particles are created or destroyed, leave the total charge unchanged. This is charge conservation. Thus any state of any quantity of matter can be given a label that describes its total electrical charge and this label will not change no matter how that state evolves, as long as it remains isolated from any other matter. Not only is charge conserved, but as far as we know, all charges in the universe are related to the charge of the electron. For example, the electron charge is equal to minus the proton charge. Experimentally, we know that this is true to an incredibly high accuracy, because atoms with equal numbers of protons and electrons appear neutral.

The weak and strong forces can also be described in terms of quantized charges, at least at very short distances. The strong force is somewhat simpler. The charges associated with the strong force are called "color" charges. Each type of quark comes in three different quantum mechanical states that are called "colors." (While the metaphor is obscure, some "colorless" particle states like protons are composed of one quark of each color—hence, the analogy with the primary colors: red, blue, and green.) The color charges are labels for these quantum me-

chanical states. The theory of the interactions of these colored quarks is known as quantum chromodynamics (QCD).

The crucial fact about QCD is the "threeness" of the quarks. The three colors of quarks are entirely equivalent and can be represented as the corners of an equilateral triangle, as shown in Fig. 1, where the vertices of the triangle are balanced around the point in the center, which corresponds to something with 0 color charge. Strong interaction processes correspond to the possible transformations of the triangle into itself. For example, there is a process that takes a red quark into a green one. The corresponding force particle, the gluon, mediates an interaction in which a red quark becomes a green quark, and the opposite transformation takes place for another nearby quark, that is,

$$u_{\text{red}} + d_{\text{green}} \rightarrow u_{\text{green}} + d_{\text{red}}.$$

This structure has a name that comes not from physics but from mathematics. It is called $SU(3)$, where $SU(n)$ is the group of $n \times n$ unitary matrices with unit determinant, which specify rotations among n complex numbers. The 3 denotes the three color charges. These charges, together with the force particles, the gluons, which are emitted and absorbed as quark color is changed, are responsible for the strong force in the same way that electric charge and the photon are responsible for the electrical forces. The big difference between electrical forces and strong forces comes from the fact that the gluons carry color charge (as well as anticolor charge), and hence they also interact with one another, while photons are themselves electrically neutral. For example, in the interaction described

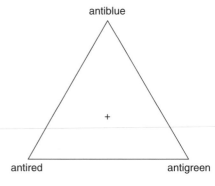

Figure 2

above, a gluon emmited by the u quark and then absorbed by the d must carry (red + antigreen) color.

The threefold symmetry of the quark triangle is important. The mathematics of this symmetry can be used to build a consistent, relativistic, quantum mechanical description of the interaction. The symmetry is not an end in itself, but it is crucial because it determines the dynamics.

The antiquark has the opposite charges shown in Fig. 2. In a sense, the antiquark color charges can be obtained from those of the quarks. They are just the vector sums of the charges of the three pairs of different quarks, so that antiblue = green + red, as shown in Fig. 3.

There is something special about the connection between the electric charges of the quarks and the color charges. The u quark has charge 2/3 (in units of the positron charge, which is minus the electron charge). The d quark has charge $-1/3$. Put a d quark and a positron together, and the color charges shown in Fig. 4 result. The positron sits in the middle because it is colorless. Now displace things in the direction perpendicular to the page a distance proportional to their electric charge. After a while, we end up with a regular tetrahedron, balanced about its central point, because each of the three colors of the d quark move 1/3 as much as the positron and in the opposite direction. If there were only d quarks and antiquarks and electrons and positrons in the world, and no weak interactions, the color interactions and the electromagnetic interactions could be combined into a single, unified dynamics based on the fourfold symmetry, called $SU(4)$, with this regular tetrahedron as the

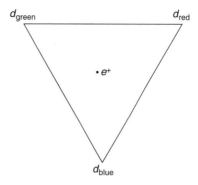

Figure 4

basic object. This works because the sum of the charges of the four particles is zero, so that the tetrahedron remains balanced—a requirement for the symmetry.

There is an elegant thing about embedding electricity and magnetism into a theory of this kind. The electric charges are now quantized automatically, because of the symmetry structure. Furthermore, the charge of the electron necessarily has the right relation to the charges of colored quarks so that atoms can be neutral. One way of seeing this is to realize that everything in the world has to be built out of the multiples of the basic tetrahedron.

Of course, even if no other particles existed, there would be something very wrong with this theory. In our world, the color $SU(3)$ interactions and the electromagnetic interactions look very different. Something must happen to break the unified $SU(4)$ apart again into $SU(3)$ and electromagnetism, without destroying the essential symmetry between them. Particle physicists have a partial picture of this kind of effect, which we also see at work in the weak interaction. This picture is called the Higgs mechanism. It will appear next, in the context of the weak interaction symmetry $SU(2)$.

In discussing the weak interaction, we will ignore something absolutely crucial—the idea of spin—for simplicity. (The discussion below becomes more accurate if one interprets everything as applying only to the "left-handed" particles—those whose angular momentum along their direction of motion is $-\frac{1}{2}\hbar$.) The weak interactions are based on a twofold, or $SU(2)$ symmetry, in which particles are organized into pairs or doublets. The lightest matter particles, the u and d quarks, and the electron (e^-) and its neutrino (ν), are organized into doublets as follows:

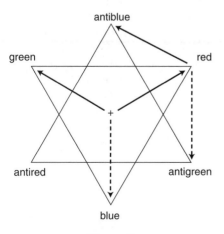

Figure 3

This is the first family—the others are organized similarly. The key interaction is one in which the particle at the top of one doublet changes identity and becomes the particle at the bottom, while the particle at the bottom of the other doublet likewise changes identity to the particle at the top. For example, this interaction can induce scattering processes like

$$\nu_e + d \rightarrow e^- + u.$$

Notice that electric charge is conserved in the process, even though the particles at the top and bottom of each doublet have different charge, because the difference is the same for both. This interaction is associated with a W particle in the same way that electric forces are associated with photons and strong forces with gluon exchange. The charge difference of 1 unit between the top and bottom of the doublets is related to the electric charge of the W, which is the same as that of the positron.

The trouble with this is that the W looks very unlike the photon, and the weak interactions very unlike electromagnetism. The photon has no mass. The W has a mass 100 times that of the proton. Electromagnetic forces have a very long range, falling off only like a power of the separation. Weak forces have a range of only about 10^{-15} cm. At longer distances, they are negligible. The explanation of this difference is the Higgs mechanism, which generates the W and Z masses. In fact, at distances smaller than 10^{-16} cm, the W (and its neutral partner, the Z) come apart into a massless or photonlike piece and something else. That something else is an entirely new kind of matter, unlike anything else we have ever seen. We know from our theoretical description of the weak interactions and from the observed properties of the W and Z particles how it combines with the photonlike parts. This process of combination is the Higgs mechanism. (This process is called spontaneous symmetry breaking. There are a number of theories about what the new kind of matter might be. One predicts the so-called Higgs bosons. These can be tested at even higher energy accelerators.)

The fact that the top and bottom of the $SU(2)$ doublets do not have the same electric charges means that electricity and magnetism and the weak interactions are inextricably bound up together by the Higgs mechanism. This partial unification of weak and electromagnetic interactions was understood in the 1960s and the early 1970s through the efforts of many physicists. For their efforts in the area, Sheldon Glashow, Abdus Salam, and Steven Weinberg received the Nobel Prize for physics in 1979.

All the pieces of the standard model of particle physics have now been discussed, and we can see how they fit together into the simplest GUT. Just add the neutrino to the tetrahedron of the three colors of d quark and the positron. The neutrino has no color and no electric charge, so it sits at the middle of the tetrahedron. But it is part of a doublet with the positron under the weak interaction $SU(2)$. To describe the $SU(2)$ charge, we need another dimension, just as we had to go from two dimensions to three to get from the $SU(3)$ triangle to the $SU(4)$ tetrahedron. Now another dimension beyond three is a little hard to visualize, but the idea is the same as going from the triangle to the tetrahedron. Just move the neutrino in one direction in the new dimension, and move the tetrahedron $1/4$ as much in the opposite direction. For the right displacement, the most symmetrical four dimensional object (called a 4-simplex by mathematicians) is obtained. Then it is easy to check that all of the charges, $SU(2)$, $SU(3)$, and electric are contained in the charges of the $SU(5)$ that describes this new four-dimensional object.

This leaves some of the particles unaccounted for: the u quarks, the u and d antiquarks, and the electron. That is ten particles all together because of the three colors for each quark or antiquark. Just as we could understand the three antiquark charges in $SU(3)$ as equivalent to the sums of two different quark charges, so in $SU(5)$ we find a set of ten objects whose properties are those of the ten distinct combinations of three different elements of the basic five charges of the 4-simplex, as shown in the following equation:

$$10 = \frac{5 \times 4 \times 3}{3 \times 2 \times 1}$$

Further, we find that these ten combinations have precisely the correct color, electromagnetic, and weak properties to be identified as the u quarks, the u and d antiquarks, and the electron. Everything fits with nothing left over. This is the $SU(5)$ theory. It is the simplest GUT, but there are a number of generalizations almost as economical, based on other symmetries. The beautiful fit of the matter particles into $SU(5)$ is one of the triumphs of GUTs.

The extra force particles associated with the extra possible transitions in $SU(5)$ should produce some fascinating new physics. Because quarks, antiquarks, and electrons all appear together in the ten, a process like the following is possible:

$$d_{\text{red}} + u_{\text{blue}} \rightarrow \overline{u}_{\text{green}} + e^-,$$

where $\overline{u}_{\text{green}}$ denotes the antiparticle of a green up quark. If this process occurs inside a proton, the proton can decay into a positron and a neutral particle called a pion, which itself quickly decays into two photons. Proton decay has been looked for experimentally, and the rate of proton decay from the process shown above is now known to be at least about 10^{32} years. Thus this process must be very weak if it exists at all. This means that the force particle that mediates this transition, called X (after W) must be very heavy.

But in fact, we can compute the X mass, approximately. At energies above the X mass, all the forces in $SU(5)$ have the same strength. However, the strength of the forces are measured by dimensionless "coupling constants" that depend on the energy scale. The $SU(3)$ QCD coupling gets stronger as one comes down from high energy to the energies that can be probed, while the $SU(2)$ electroweak coupling and the electromagnetic coupling change much less. By measuring the couplings at low energies and extrapolating them up to very high energies, we can predict the X mass in the $SU(5)$ theory (or any particular GUT). This analysis also gives a consistency check on the theory, because the three couplings must come together at the same energy scale. The simplest $SU(5)$ model fails at this point. The couplings do not quite come together, and the mass of the X is predicted to be too small to be consistent with the measured lower limit on the proton lifetime. However, a number of simple modifications of the model can give predictions consistent with the observations.

One of the most interesting possible modifications is supersymmetric $SU(5)$, in which the particles of the model all have partners with different spin. Supersymmetry is a hypothetical symmetry that re-lates the properties of the observed particles with those of (as yet unobserved) superpartners. In supersymmetric $SU(5)$, the three couplings come together correctly, but at an even larger energy scale than in the nonsupersymmetric theory. Many particle theorists find this theory quite compelling.

Ultimately, though, it is nature that must decide. If supersymmetric GUTs describe the world at very small distance, experimenters should soon find some of the predicted superpartners, and the next generation proton decay experiment might see proton decay.

See also: BOSON, HIGGS; BOSON, W; BOSON, Z; COLOR CHARGE; INTERACTION, ELECTROMAGNETIC; INTERACTION, STRONG; INTERACTION, WEAK; QUANTUM CHROMODYNAMICS; QUARK; SUPERSYMMETRY; SYMMETRY; SYMMETRY BREAKING, SPONTANEOUS

Bibliography

CREASE, R. P., and MANN, C. C. *The Second Creation: Makers of the Revolution in Twentieth-Century Physics* (Macmillan, New York, 1986).

DAVIES, P., ed. *The New Physics* (Cambridge University Press, Cambridge, Eng., 1989).

DIMOPOULOS, S., and GEORGI, H. "Softly Broken Supersymmety and $SU(5)$." *Nucl. Phys.* **B193**, 150 (1981).

GEORGI, H., and GLASHOW, S. L. "Unity of All Elementary Particle Forces." *Phys. Rev. Lett.* **32**, 438 (1974).

GEORGI, H.; QUINN, H.; and WEINBERG, S. "Hierarchy of Interactions in Unified Gauge Theories." *Phys. Rev. Lett.* **33**, 451 (1974).

PATI, J., and SALAM, A. "Lepton Number as the Fourth Color." *Phys. Rev. D* **10**, 275 (1974).

HOWARD GEORGI

GRATING, DIFFRACTION

A diffraction grating is a transparent or a reflecting surface with many parallel lines or grooves ruled on it. Sizes range from a few millimeters square to about 150×300 mm for gratings used with astronomical telescopes. Groove spacings range from 7/mm to 150,000/mm. A grating with fewer than 200 grooves/mm is called an echelle.

When light strikes a grating, its various wavelengths (colors) are diffracted at different angles, and can be separated from each other. A spectrum is produced, with the greatest deviation for the longest wavelength. The grating equation is

$$m = d(\sin \theta_1 + \sin \theta_2),$$

where m is the integral order of interference (difference in optical path length between two successive grooves), and θ_1 and θ_2 are the angles of incidence and of diffraction.

The width of the grating determines the resolving power—how close in wavelength two lines can be, yet still be seen as separate. The larger the grating, the larger the resolving power.

The groove shape determines in what order the maximum amount of light is diffracted. A sawtooth groove shape used at the proper angle, the blaze angle, can diffract almost all of the light into a single order. The wavelength diffracted in the first order at the blaze angle is called the blaze wavelength.

Initially gratings were ruled on plane and concave reflecting metal surfaces by precise mechanical engines using a diamond stylus. Transmission grating replicas were made by coating the ruled surface with transparent glue that, when dry, was peeled off and cemented onto a glass substrate. Plane gratings require focusing optics for their use in spectroscopy, whereas concave gratings require no auxiliary optics. Whenever possible, plane gratings are used because of their superior accuracy of spacing and uniform groove shape.

Now gratings are ruled in aluminum and other coatings deposited on fused silica substrates. They are also produced by holographic methods because of their uniformity of spacing and groove shape, and the ability to tailor the spacing and blaze to the task at hand. It is also possible to fabricate high-efficiency, multilayer dielectric diffraction gratings.

For many applications a grating spectrometer is the instrument of choice because of its simplicity and ease of use, compared with a prism spectrometer, Fabry–Pérot interferometer, or Fourier transform spectrometer.

Concave gratings make use of the properties of the Rowland circle. When a slit, concave grating and detectors are placed on a circle of diameter equal to the radius of curvature of the grating, all wavelengths are in focus at the detectors.

A small but typical laboratory spectrometer has a plane grating 50 mm square ruled with 600 grooves/mm, and uses a collimating mirror and a camera mirror of 250 mm focal lengths. The blaze wavelength is 800 nm with an efficiency of 65 percent, and the resolving power is 30,000 in the first order. A large research instrument might have a plane grating with 250 mm of ruled width, 300 grooves/mm, a blaze angle of 64°, mirrors of 3,000 mm focal length, and a resolving power of 750,000 when used at the blaze angle for observing red light.

See also: DIFFRACTION; DIFFRACTION, FRAUNHOFER; GRATING, TRANSMISSION; HOLOGRAPHY; INTERFERENCE; RESOLVING POWER

Bibliography

DAVIS, S. P. *Diffraction Grating Spectrographs* (Holt, Rinehart and Winston, New York, 1970).

HUTLEY, M. C. *Diffraction Gratings* (Academic Press, San Diego, CA, 1982).

JENKINS, F. A., and WHITE, H. E. *Fundamentals of Optics* (McGraw-Hill, New York, 1976).

SUMNER P. DAVIS

GRATING, TRANSMISSION

Light displays interference effects due to its inherent wave nature. When light falls upon a set of apertures, or slits, each slit can be considered a source of light. Diffraction results from interference of the wave fronts originating from each slit. The positions at which constructive interference occurs depend on the wavelength of the light. If more than one wavelength, or color, is present in the incident light, then the light is dispersed according to its wavelength, as occurs in a prism.

An array of a large number of equally spaced, parallel slits, all of the same width, is called a diffraction grating. The slits in a diffraction grating are referred to as lines or rulings. Diffraction gratings are of two types: transmission gratings and reflection gratings. In transmission gratings, the incident light passes through the grating. In spectroscopy, diffraction gratings can be used as the dispersing element in spectrometers for measurement of the wavelengths present in light.

The diffraction grating was first conceived in 1786 by the American astronomer David Ritten-

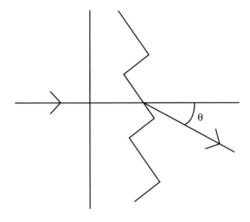

Figure 1 Profile of a blazed transmission grating.

house, who laid hairs across two fine screws to produce the parallel lines necessary for the grating. In 1821 the German physicist Joseph von Fraunhofer, who is credited with the invention of the modern diffraction grating, produced a transmission grating similar to Rittenhouse's by winding fine wire around a set of screws. Fraunhofer also produced the first glass transmission gratings by using a diamond to rule glass plates with up to 8,000 lines per inch. In the late nineteenth century, the precision with which gratings could be produced was vastly increased by the invention of the ruling machine by Henry A. Rowland at Johns Hopkins University.

Later developments in diffraction grating design included grooves ruled at an angle, producing a saw-toothed profile said to be "blazed" (see Fig. 1). Blazed gratings have the advantage of directing more intensity into higher diffracted orders.

For light incident perpendicular to the plane of a grating (i.e., normal incidence), the angle θ at which a given wavelength λ of light is diffracted is given by the grating equation

$$m\lambda = d \sin \theta,$$

where m is the order of diffraction and d is the grating constant (the distance between each line).

The resolution of the grating, the ability to disperse one wavelength from another, is enhanced by increasing either the number of lines N in the grating or the order of diffraction that is observed. A measure of the resolution is given by the resolving power R, which is found to be given by $R = mN$.

See also: DIFFRACTION, FRAUNHOFER; GRATING, DIFFRACTION; RESOLVING POWER

Bibliography

BARNES, R. M., and JARRELL, R. F. "Gratings and Grating Instruments" in *Analytical Emission Spectroscopy*, edited by E. L. Grove (Marcel Dekker, New York, 1971).
HUTLEY, M. C. *Diffraction Gratings* (Academic Press, London, 1982).
KIRKBRIGHT, G. F., and SARGENT, M. *Atomic Absorption and Fluorescence Spectroscopy* (Academic Press, London, 1974).

JEFFERSON L. SHINPAUGH

GRAVITATIONAL ASSIST

Deepspace missions sometimes use close swingbys of the planets to explore the vast reaches of the solar system. These planetary swingbys are referred to as gravity-assist swingbys and allow planetary spacecraft to gain or lose velocity so that they reach their final destination. While use of this technique has become more commonplace in recent years, the concept of gravity assist was known much earlier in this century.

Two of the earliest pioneers, Yu. V. Kondratyuk and Fridrikh A. Tsander, documented their ideas on gravity assist between 1918 and 1925. Kondratyuk proposed the "use of a satellite for flight in the solar system when it is required to gather velocity and the return from this flight when it is required to absorb energy." The gravity-assist concept was initially documented in western literature in the 1950s by V. A. Firshoff (1954), D. F. Lawden (1954), and G. A. Crocco (1956).

The gravity-assist concept is applied in interplanetary travel by controlling the spacecraft to swing by a planet at a specified altitude. The gravitational attraction of the planet causes the spacecraft trajectory to bend during its swingby. This results in a spacecraft velocity gain or loss with respect to the Sun. Figure 1 illustrates a gravity-assist Earth swingby. The trajectory of the spacecraft with respect to Earth is a hyperbola; the spacecraft approaches and departs along the asymptotes of this hyperbola with a constant speed, called the V_∞. The spacecraft achieves its maximum velocity with re-

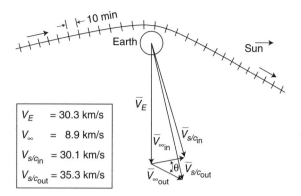

V_E	= 30.3 km/s
V_∞	= 8.9 km/s
$V_{s/c_{in}}$	= 30.1 km/s
$V_{s/c_{out}}$	= 35.3 km/s

Figure 1 Illustration of gravity-assist swingby.

spect to Earth at closest approach. The angle between the incoming and outgoing V_∞s is referred to as the bend angle of the flyby (θ). The velocity vector of Earth with respect to the Sun at the time of the flyby is indicated by V_E. Adding the incoming and outgoing V_∞s to the velocity of Earth yields the velocity vectors of the spacecraft with respect to the Sun before and after the flyby. The effect of the hyperbolic flyby in an Earth-centered reference frame is simply to rotate the V_∞ through an angle equal to the bend angle; there is no net energy change for the spacecraft trajectory with respect to Earth as a result of the flyby. However, the rotation of the Earth-centered V_∞ has the effect of increasing the magnitude of the velocity vector in a Sun-centered reference frame. The amount of bending and the magnitude of the spacecraft velocity change depends on the swingby closest approach altitude. The spacecraft velocity increases for spacecraft swingbys across the trailing hemisphere of the planet and decreases for swingbys across the leading hemisphere of the planet.

The change in spacecraft velocity is due to an exchange of energy between the spacecraft and planet. So if the spacecraft is speeded up from the gravity assist, the planet is slowed. However, due to the relatively large mass of the planet, the change in planet velocity is infinitesimal while the change in spacecraft velocity is significant. The spacecraft velocity change achieved from the planetary swingby means that less propellant is required to be carried onboard to provide a velocity change.

Several planetary missions have used gravity-assist swingbys to reach their target destination. The first planetary mission to employ this concept was Mariner 10, which was launched in 1973. Mariner 10 used a Venus gravity-assist to reduce its velocity to enable a swingby of Mercury. The gravity-assist from the Mercury swingby was then used to adjust the spacecraft orbital period around the Sun so that the spacecraft returned to Mercury for two additional swingbys.

The next usage of planetary gravity assist was with the Pioneer 11 mission to Jupiter. This mission was initially intended to only be a Jupiter flyby mission. However, the close swingby was able to provide a near 180 degree gravity-assist turn that enabled Pioneer 11 to fly by Saturn five years later on the opposite side of the solar system.

Both Voyager 1 and 2, launched in 1977, used a gravity-assist swingby of Jupiter to reach Saturn. Voyager 2 continued on to Uranus and Neptune, using the gravity assist of each planetary encounter to target the spacecraft to the next planet.

Gravity-assist planetary swingbys may also be used to change the plane of the spacecraft trajectory by flying over the poles of the planet. An example is the Ulysses mission, where the spacecraft was initially launched in 1990 to Jupiter. At Jupiter the spacecraft flew by the north pole, kicking the spacecraft out of the ecliptic plane of the solar system. This enabled Ulysses to later fly over the poles of the Sun.

Several current and planned missions make extensive use of planetary swingbys not only to reach their target planet, but also to allow extensive trajectory modifications after the arrival at the destination planet. Galileo was launched in 1989 and performed gravity-assist swingbys at Venus and Earth (twice) to enable a Jupiter arrival in 1995. A close swingby of the satellite Io will reduce the spacecraft velocity at Jupiter to assist in the capture of the spacecraft by the gravity field of Jupiter. Once in orbit at Jupiter there will be ten close swingbys with the other Galilean satellites.

The Cassini mission to Saturn will be launched in 1997. Its interplanetary trajectory will include swingbys of Venus (twice), Earth, and Jupiter to provide a Saturn arrival in 2004. Once in orbit, the Cassini spacecraft will use over thirty swingbys of the moon Titan to modify its trajectory and allow a thorough exploration of the Saturnian system.

See also: GRAVITATIONAL ATTRACTION; PLANETARY SYSTEMS; SPACE TRAVEL; TRAJECTORY; VELOCITY

Bibliography

CESARONE, R J. "A Gravity Assist Primer." *AIAA Student Journal* **27** (1), 16–22 (1989).

DIEHL, R. E., and D'AMARIO, L. A. "Interplanetary Spacecraft Dynamics" in *The Astronomy and Astrophysics Encyclopedia,* edited by S. P. Maran (Van Nostrand Reinhold, New York, 1992).

KOHLHASE, C. E. *The Voyager Neptune Travel Guide* (Jet Propulsion Laboratory Publication 89–24, Pasadena, CA, 1989).

NOCK, K. T., and UPHOFF, C. W. "Satellite Aided Orbit Capture." *American Astronautical Society Paper 79–165* (1979).

ROGER E. DIEHL

GRAVITATIONAL ATTRACTION

Of the four fundamental interactions, gravitation is by far the most obvious in everyday life (though ironically it is also by far the weakest). As a result, people have pondered the mystery of gravitational attraction for millennia, inventing a variety of models to explain its nature.

Gravitational Attraction Before Newton

In the fourth century B.C.E., the Greek philosopher Aristotle offered one of the first comprehensive models of motion and the causes of motion. In Aristotle's model, terrestrial objects are mixtures of four elements: earth, water, air, and fire. The natural state of earth (and to a lesser extent, water) was to be at rest as close as possible to the center of the earth. An object made primarily of these elements displaced away from the earth would seek to return to its natural place and thus be attracted downward. On the other hand, the natural state of fire (and to a lesser extent, air) was to move upward away from the earth. Celestial bodies were thought to be constructed of a fifth element whose natural state was to move endlessly in perfect circles. Thus, in Aristotle's model, gravity was the downward-seeking tendency of two of the five elements.

European scholars generally accepted this understanding of gravitational attraction for nearly two millennia. In the early seventeenth century, acceptance of Aristotle's model began to crumble, particularly after 1609, when Galileo Galilei and others turned the newly invented telescope on the heavens and saw that the Moon appeared to be constructed of rock just like the earth.

Newton's Laws and Universal Gravitation

In 1687 Isaac Newton published a book entitled *Philosophiae Naturalis Principia Mathematica* (Mathematical Principles of Natural Philosophy) that offered for the first time a comprehensive mathematical theory of motion, a universal model of gravitation, and a complete explanation of celestial motion. Newton's work revolutionized physics, in a real sense marking its birth as a mathematical science.

Three crucial insights made Newton's triumph possible. The first was his recognition that objects naturally move in straight lines and that forces act to change the direction of that motion, the direction of the change being the same as the direction of the force. Newton's assertion that straight-line motion (instead of circular motion) was natural even for celestial bodies was a bold break with the Aristotelian model, but it enabled him to see that the planets would have to be continually attracted toward the Sun to follow a circular orbit around it.

The second insight was that gravitation is universal. Newton claimed late in life that observing a falling apple led him in his youth to reflect on what would happen if Earth's gravity extended all the way out to the Moon and to begin doing some calculations based on this idea. Well before 1687, he also knew that the moons of Jupiter and Saturn seemed to obey the same mathematical laws as the planets orbiting the Sun. Even so, it is a giant leap to Newton's grand statement that every particle in the universe exerts on every other particle an attractive gravitational force directed along the line connecting the objects and whose magnitude is

$$F_{\text{grav}} = G \frac{Mm}{r^2}, \tag{1}$$

where M and m are the masses of the objects involved, r is the distance between them, and G is a universal constant (whose measured value is 6.67×10^{-11} N·m²/kg²).

This law applies only to point particles. Any object massive enough (like Earth) to exert a significant gravitational force, however, will hardly be a point particle. Newton's third insight was that a spherical object both exerts and responds to gravitational forces as if its mass were concentrated at its center. Newton's mathematical proof of this assertion enabled him to treat large objects like Earth and the Sun as if they were point particles.

Newton was able to show that everything that was known at the time about falling objects and about the motion of the planets followed from these principles. (Figure 1 illustrates how these ideas can be applied to projectile and planetary motion, respectively.) In the centuries that followed, this model continued to prove enormously successful. When astronomers began to make measurements of planetary orbits fine enough to display the gravitational influences of the planets on each other, they found that the model accurately described these effects. Astronomers used the model to predict both the existence and approximate position of Neptune on the basis of its effects on other planets, leading to its discovery in 1846. In recent years, astronomers have used the model to infer the existence of unseen dark matter in nearby galaxies, at least two planets orbiting a distant pulsar, and a giant black hole at the center of the M81 galaxy.

General Relativity

The special theory of relativity, published by Albert Einstein in 1905, is based on the assumption that the laws of physics are the same in all inertial reference frames. Einstein immediately began an effort to generalize the theory to cover both inertial and noninertial (accelerating) frames. By 1907 he had realized that such a theory would necessarily involve a treatment of gravity, since observers in accelerating reference frames experience forces opposite to their acceleration that would look to them like gravitational forces. However, it took Einstein eight more years to finish his general theory of relativity, partly because his efforts kept nudging him toward a different theory than he originally imagined.

General relativity is founded on the observation, first made by Galileo, that all objects fall with the same acceleration. (Recent experiments have shown that the ratio of the gravitational accelerations of aluminum and gold is *one* to at least eleven decimal places.) This idea means that two small objects, launched from the same point with the same initial speed and direction, will follow the same trajectory in a gravitational field, independent of their mass, composition, or other characteristics. This suggests that the trajectory has nothing to do with the objects, but is rather a property of the spacetime through which they move.

Moreover, if all objects fall with the same acceleration, then in a freely falling reference frame (such as an orbiting space shuttle), all objects will appear to have zero gravitational acceleration. In such a frame, an object at rest will remain at rest, and objects in motion will remain in motion in a straight line, as if there were no gravity.

In a reference frame at rest on the earth, objects' trajectories appear bent toward the earth by gravity. According to Einstein, however, this is an artifact of choosing an inappropriate reference frame. The gravitational force that we feel in a frame at rest on the ground is akin to the rearward force that we seem to feel when we are in an accelerating car: Both "forces" can be made to vanish by choosing to work in a different reference frame.

Closer investigation shows, however, that the effects of gravity are not completely eliminated in a freely falling reference frame. Two objects placed side-by-side and initially at rest in a freely falling

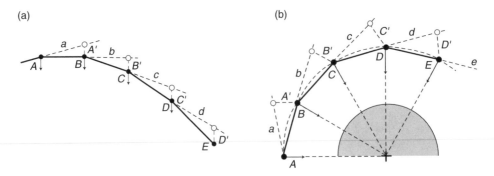

Figure 1 Gravitational attraction applied to (a) projectile and (b) planetary motion.

reference frame will each accelerate toward the center of the earth (as will the frame itself), but since the direction of the center of the earth is slightly different for the two objects, they will appear to have a small relative acceleration toward each other. Thus, their trajectories through spacetime, while initially parallel, will begin to curve toward each other as time passes. Such tidal effects cannot be erased by choosing an appropriate reference frame; they represent the aspect of gravity that is real (frame independent).

In general relativity, all these ideas are linked as follows. The geodesic hypothesis asserts that a free object in space will follow a unique path called a geodesic through spacetime, a path that depends only on spacetime and not on the nature of the object. A geodesic on a curved surface is the straightest possible path on that surface; similarly, a geodesic in spacetime is (as revealed in a freely falling frame) the straightest possible trajectory through that spacetime. The tidal effects of gravity, however, mean that initially parallel trajectories eventually curve toward each other. This indicates that spacetime is curved; if spacetime were flat (Euclidean), initially parallel geodesics (straight lines) would remain parallel. In 1915 Einstein found a set of equations (now called the Einstein field equations) that successfully linked the frame-independent tidal effects of gravity (i.e., the curvature of spacetime) to the presence of matter and energy. John A. Wheeler nicely summarized the complete theory this way: "Space tells matter how to move, matter tells space how to curve" (Misner, Thorne, and Wheeler, 1973, p. 5).

Thus, according to general relativity, there really is no attractive gravitational force between two objects. Rather, a massive object causes the surrounding spacetime to become curved, and nearby objects simply follow the straightest paths that they can through this curved spacetime. Depending on an object's initial position and velocity near the earth, its geodesic might accelerate directly toward the earth, follow a parabolic path in space that bends toward the earth, or follow an elliptical orbit around Earth. In a frame at rest on Earth's surface, an object appears to be attracted toward the earth because that is the way its geodesic goes; to hold the object at rest, we have to exert an upward force on it to accelerate it away from its natural downward geodesic.

While the predictions of general relativity and Newton's theory are in many circumstances indistinguishable, Einstein showed that general relativity accounted for a 40-arcsecond-per-century shift in the orientation of Mercury's orbit that Newton's model could not explain. In 1919 photographs taken during a total eclipse verified that light from stars behind the Sun was bent by the curved spacetime around it by exactly the amount predicted by general relativity. It was not until the 1970s that it became possible to test the predictions of general relativity in detail. Since then the theory has survived a number of stringent tests, including ongoing observations of a binary pulsar system that show the system to be radiating energy in the form of gravitational waves at exactly the rate predicted by the Einstein field equations. The Global Positioning System (GPS) has to take account of general relativitistic effects on radio signals and clocks to achieve its goal of being able to locate a radio receiver on Earth's surface within a few meters.

Quantum Gravity

During the late 1940s, Julian Schwinger, Richard Feynman, and Shinichiro Tomonaga developed a very successful quantum theory of the electromagnetic field called quantum electrodynamics (QED). This theory for the first time offered a comprehensive theory of the electromagnetic field that was consistent with both special relativity and quantum mechanics. In the early 1970s, Sheldon Glashow, Abdus Salam, and Steven Weinberg extended the theory to include the weak nuclear force, and in the late 1970s a number of physicists developed analogous quantum field theory (quantum chromodynamics, or QCD) to describe the strong nuclear force. These theories cover three of the four known fundamental interactions, leaving only gravity out.

In all of these theories, subatomic particles are imagined to exert forces on each other by exchanging field particles that represent quantum excitations of the field in question (in the case of the electromagnetic field, these field particles are photons). If you imagine two people throwing a basketball back and forth, you may be able to see that the recoil from throwing and catching the basketball will tend to push the two people apart. It is harder to see how exchanging objects like this could give rise to an attractive force (maybe imagine exchanging boomerangs instead of basketballs), but in quantum field theory this particle exchange can lead to both repulsive and attractive forces. The particle-exchange concept here is not the basic idea on which we build the theory, but rather Feynman's visual image for expressing the meaning of the equations of quantum field theory, which are the theory's real basis.

In a straightforward gravitational quantum field theory, particles would exert gravitational forces by exchanging gravitons, which the incredible weakness of the gravitational interaction would make virtually undetectible. However, the approach for constructing a quantum field that works so well for the other interactions yields nonsense when applied to gravity. In all quantum field theories, the equations in some circumstances yield infinite results, but in the three successful quantum field theories, these infinities can be effectively canceled out using a process called renormalization. In the case of gravity, though, the infinities cannot be canceled this way.

The failure of the straightforward approach has spawned a number of more speculative approaches to the problem of developing a quantum theory of gravity, including supersymmetry theories and string theories. To date, none of these models has proven to be unambiguously successful; finding a workable quantum theory of gravitation thus remains an unsolved problem.

See also: EINSTEIN, ALBERT; FIELD, GRAVITATIONAL; GRAVITATIONAL FORCE LAW; GRAVITATIONAL WAVE; GRAVITON; NEWTON, ISAAC; NEWTONIAN MECHANICS; NEWTON'S LAWS; QUANTUM CHROMODYNAMICS; QUANTUM ELECTRODYNAMICS; QUANTUM FIELD THEORY; RELATIVITY, GENERAL THEORY OF; SPACETIME; SUPERSTRING; SUPERSYMMETRY

Bibliography

COHEN, I. B. *The Birth of a New Physics,* rev. ed. (W. W. Norton, New York, 1985).

ELLIS, G. F. R., and WILLIAMS, R. M. *Flat and Curved Space-Times* (Clarendon Press, Oxford, Eng., 1988).

HORGAN, J. "Particle Metaphysics." *Sci. Am.* **270** (2), 97–106 (1994).

MISNER, C. W.; THORNE, K. S.; and WHEELER, J. A. *Gravitation* (W. H. Freeman, San Francisco, 1973).

THOMAS A. MOORE

GRAVITATIONAL CONSTANT

The gravitational constant, usually denoted *G,* is a fundamental constant of nature that determines the strength of gravitational interactions between massive bodies. It is often referred to as Newton's constant, because it first appeared in Newton's universal law of gravitation,

$$F_{12} = G \frac{m_1 m_2}{r_{12}^2}. \tag{1}$$

In this equation, F_{12} is the magnitude of the force between two objects of gravitational masses m_1 and m_2 separated by a distance r_{12}. The force is an attractive one that pulls the objects toward one another. The gravitational constant G is given by, in MKS units,

$$G = 6.67259 \pm .00085 \times 10^{-11}\,\mathrm{N{\cdot}m^2/kg^2}. \tag{2}$$

Einstein's theory of general relativity gives corrections to Newton's law of gravitation. These are small in many circumstances (e.g., for planetary orbits or for falling objects near the earth's surface). It remains true in Einstein's theory that the strength of gravity is parameterized by the single constant G.

It follows from Newton's law of gravitation that on the surface of the earth an object of mass m is subject to a gravitational force

$$F = mg, \tag{3}$$

where

$$g = \frac{Gm_e}{r_e} = 9.80665\ \mathrm{m/s^2}, \tag{4}$$

with m_e and r_e here being the mass and radius of the earth, respectively. Newton's second law of motion ($F = ma$) and the equality of gravitational and inertial mass then imply that a freely falling body just above the surface of the earth will undergo an acceleration g independent of its mass.

See also: ACCELERATION; GRAVITATIONAL ATTRACTION; GRAVITATIONAL FORCE LAW; NEWTON'S LAWS; RELATIVITY, GENERAL THEORY OF

Bibliography

HEWITT, P. G., *Conceptual Physics,* 6th ed. (Scott, Foresman, Glenview, IL, 1989).

YOUNG, H. D. *University Physics: Extended Version for Modern Physics,* 8th ed. (Addison-Wesley, Reading, MA, 1992).

ANDREW STROMINGER

GRAVITATIONAL FIELD

See FIELD, GRAVITATIONAL

GRAVITATIONAL FORCE LAW

The gravitational force law was discovered by Isaac Newton. As he reflected on this and other scientific achievements, he wrote: "If I have seen farther than others, it is because I have stood on the shoulders of giants." One of the giants to whom he was indebted was Johannes Kepler, whose empirical laws of planetary motion were crucial to Newton's discovery of gravitation.

If a planet moves in an elliptical orbit around the Sun, Newton's first law of motion (the law of inertia) implied, the planet will move under the influence of an external force. Newton's challenge was to work out what kind of force would guide a planet in its elliptical orbit. First, he was able to show that the force exerted on a planet would have to be directed toward the Sun if it were to move in accordance with Kepler's second law of planetary motion. Such a force is called a central force. (For a simple example of a central force, consider a stone tied to a string. When it is whirled around, the force is directed along the string toward the center of the circle. The resulting motion in this case, however, is a circle because the force is constant.)

With the direction of the force thus established, Newton considered the next obvious question: How does the force vary with the distance? From Kepler's observation that the farther away a planet is, the longer it takes to make a complete revolution around the Sun, Newton concluded that the force would have to grow weaker with the distance. In fact, he was able to show mathematically that if the force varied

inversely as the square of the distance, a planet would move in an elliptical orbit around the Sun. He thus discovered that the gravitational force between the Sun and the planets varied as the inverse square of the distance of separation.

Newton also sought to know the character of the force between Earth and the Moon: Is this force also an inverse square relation? As a way of testing this hypothesis, he compared the falling motion of an apple to the falling motion of the Moon as it went around Earth. The Moon was known to be 60 times farther from Earth's center than a falling apple. If Earth's force were indeed inverse-square, an apple would fall at the rate $(60)^2 = 3{,}600$ times greater than the Moon. From Galileo's study of falling motion, it was known that an apple would fall 16 ft, or 192 in., in the first second. The Moon would then fall only 0.05 in. (that is, 192 in./3,600) in 1 s. Newton was able to show that in one second the Moon did, in fact, fall about 0.05 in. from a straight path. The implication of this conclusion was, of course, revolutionary. Not only did Earth's gravitational force reach to the Moon, but the force that kept the Moon in its orbit around Earth was the same force that caused an apple to fall. Newton thus discovered the universality of the gravitational force; gravity operated not only terrestrially but celestially as well.

Pulling all these ideas together, Newton arrived at his law of universal gravitation: Every body in the universe attracts every other body with a force directly proportional to the product of their masses and inversely proportional to the square of their distance of separation. The mathematical expression of the law of universal gravitation is

$$F = G\frac{M_1 M_2}{R^2},$$

where G is the universal gravitational constant, M_1 and M_2 denote the masses of the bodies, and R is the distance of separation between the bodies.

Gravity is the only known force that applies to every object in the universe. The gravitational constant G is the same whether the force acts between stars or between Earth and an apple. It does not distinguish the terrestrial from the celestial. Given the law of gravitation, for example, Kepler's empirical laws of planetary motion and the falling motion of an apple can now be described in a single, unified way; they no longer refer to seemingly separate and unrelated phenomena.

Insights provided by the law of gravitation are deeper than empirical laws. For example, Kepler's third law of planetary motion states that the square of the period T of revolution of a planet is proportional to the cube of the semimajor axis r of its orbit:

$$T^2 \propto r^3.$$

Newton's law of gravitation, by contrast, yields the proportionality constant, enabling Kepler's law to be written as

$$T^2 = \frac{4\pi^2}{GM}\, r^3,$$

where M is the mass of the Sun. Notice that this formula leads to the determination of the mass of the Sun. Indeed, a similar formula applies to the orbiting motion of any object—from artificial satellites to galaxies. Newton's law of gravitation expressed in this form thus provides a powerful technique for determining masses of celestial objects, including the mass of Earth.

Such is the power and utility of Newton's law of gravitation. It is not difficult to see why it is regarded as the "greatest generalization ever achieved by the human mind."

See also: GRAVITATIONAL ATTRACTION; GRAVITATIONAL CONSTANT; INVERSE SQUARE LAW; KEPLER, JOHANNES; KEPLER'S LAWS; NEWTON, ISAAC; NEWTON'S LAWS

Bibliography

COHEN, I. B. *The Newtonian Revolution* (Cambridge University Press, Cambridge, Eng., 1980).

CONSOLMAGNO, G. J., and SCHAEFER, M. W. *Worlds Apart: A Textbook in Planetary Sciences* (Prentice Hall, Englewood Cliffs, NJ, 1994).

WESTFALL, R. S. *Never At Rest: A Biography of Isaac Newton* (Cambridge University Press, Cambridge, Eng., 1980).

SUNG KYU KIM

GRAVITATIONAL LENSING

Gravitational lensing is an astrophysical phenomenon that occurs when a massive object, such as a galaxy or cluster of galaxies, lies along our line of sight to a background object. With such an alignment the gravitational field of the foreground object alters the path of the radiation from the background source. The measurement of the deflection of light by the Sun during the solar eclipse of 1919 prompted calculations of the magnitude of these gravitational lensing effects, and it was predicted that they should be observable, particularly in an extragalactic context. In 1979 the effects of gravitational lensing were for the first time convincingly demonstrated in the spectra of a distant quasar, and since then many other observations of gravitational lensing have been reported. Current research in the field is directed toward the discovery of more lens systems and toward using these systems as a tool to attack various problems in astrophysics.

Gravitational lensing has been observed in many forms. When the mass distribution in the lens is sufficiently compact in opatial extent and the lens and source are closely aligned, "strong lensing" occurs

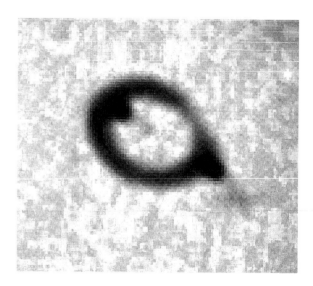

Figure 1 Gray-scale representation of a radio image of an Einstein ring gravitational lens. The two bright spots are the double image of the nucleus of the background galaxy; the ring is an Einstein ring image of a radio lobe associated with the background galaxy. This image was constructed by the author from data acquired with the National Radio Astronomy Observatory's Very Large Array Telescope.

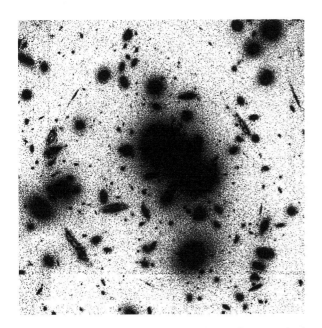

Figure 2 Gray-scale representation of an optical image of a cluster of galaxies. Several of the background galaxies surrounding the center of the cluster are distorted into small arcs. This image, provided by E. L. Turner, was acquired with the Hubble Space Telescope.

and two or more separate images of the background source are produced. If the mass distribution is circularly symmetric and the alignment is perfect, then the multiple images appear in the form of an apparent ring of emission that surrounds the lens. This configuration is known as an Einstein ring. Figure 1 shows a radio image of an Einstein ring produced by the close alignment of two galaxies. For gravitational lensing of extragalactic objects by galaxies the angular diameter of the ring, or the typical separation between multiple images, is a fraction of an arc second, easily observable with modern telescopes. The multiple images of strong lens systems provide observational constraints in the inference of the mass distribution responsible for the lensing, providing a new way to measure astronomical mass distributions. Dynamical studies of the orbits of stars in galaxies have led to the conclusion that galaxies contain significant amounts of dark matter; analysis of several gravitational lens systems has confirmed these results.

Gravitational lensing is observable not only when multiple images of the background source are pro-

duced, but also when the background source is merely distorted (weak lensing). For a circular or nearly circular background source, this distortion appears in the form of stretching the images into arcs that surround the mass distribution. An example of weak lensing is shown in Fig. 2, where a cluster of galaxies distorts the images of a background population of galaxies. As in the case of strong lensing, the pattern of arcs surrounding galaxy clusters are being used to infer the mass distribution in clusters, providing evidence that a smooth distribution of matter, in addition to the matter associated with the visible galaxies, must be present.

Another observable consequence of gravitational lensing is microlensing, the temporary brightening of a background source that occurs when a lens of small mass passes through the line of sight. This differs from the phenomena described above in that the angular separation of the images is too small to be resolved, but the focusing of the light rays by the lens produces a measurable effect. Microlensing provides astronomers with a means to detect small mass concentrations that are not themselves luminous. Surveys for microlensing are under way that have the goal of determining whether the dark matter in galaxies is in the form of dark stellar or substellar objects. The success or failure of these searches will provide important information on the nature of the dark matter in galaxies.

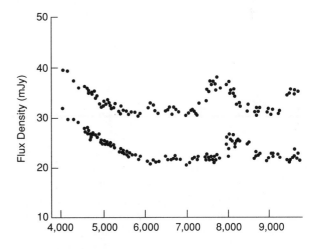

Figure 3 Radio flux as a function of time for two images of a gravitationally lensed quasar. The time delay between images appears as the horizontal offset in the location of features in the records.

Finally, gravitational lensing effects are detected in observations of the temporal behavior of lensed quasars and active galactic nuclei. These objects, believed to be at the centers of high redshift galaxies, are often variable in one or more bands of the electromagnetic spectrum. The different images of the background source are associated with different paths of propagation from the source to the observer. In general, these paths do not have the same length, and changes in the flux measured in the images will be observed in all images, but at different times. Figure 3 shows the flux measured at radio wavelengths in a two-image system. These data provide a measurement of the difference in propagation time associated with the pair of images, which in turn provides an independent measure of the cosmological distance scale.

See also: ACTIVE GALACTIC NUCLEUS; DARK MATTER; ELECTROMAGNETIC SPECTRUM; GALAXIES and GALACTIC STRUCTURE; QUASAR

Bibliography

BLANDFORD, R. D.; KOCHANEK, C. S.; KOVNER, I.; and NARAYAN, R. "Gravitational Lens Optics." *Science* **245,** 797 (1988).

BLANDFORD, R. D., and NARAYAN, R. "Cosmological Applications of Gravitational Lensing." *Annu. Rev. Astron. Astrophys.* **30,** 311 (1992).

TURNER, E. L. "Gravitational Lenses." *Sci. Am.* **259** (July), 54 (1988).

J. N. HEWITT

GRAVITATIONAL WAVE

Electrically charged particles exert forces on each other and these forces depend on the position of each particle. If one of those charges suddenly moves, the change in the force on other particles is not instantaneous. Rather, electromagnetic waves carry information about the change, and they carry energy and pressure as well. As required by Albert Einstein's special theory of relativity, the waves must move at a certain fundamental speed, $c \approx 3 \times 10^8$ m/s.

We know that gravitation must have the same features. A change in position of a source of gravity, say, the Sun, must not instantaneously change the gravitational force of the Sun on Earth. Rather, the change in force is carried to Earth by waves that also travel at c and carry energy and pressure.

The action of an electromagnetic wave is transverse. As an electromagnetic wave passes a charged particle, the charged particle undergoes accelerations in the plane perpendicular to the direction in which the wave is advancing. The wave accelerates the particle in one direction, then, one half-cycle later, in the opposite direction. The accelerations caused by a gravitational wave also are transverse but are a bit more difficult to understand. The effects of the wave are called "tidal" distortions, and they are similar to the gravitational distortions caused by the Sun and Moon on Earth. If a gravitational wave traveling vertically strikes an object in a laboratory, the object will be stretched, say, in the north–south direction, and shortened in the east–west direction. One half-cycle later the directions of elongation and shortening will be exchanged.

Our understanding of gravitational waves is fairly recent. In Isaac Newton's classical theory of gravity there are no waves; the action of gravity is instantaneous. This is incompatible with the well-verified special theory of relativity, which demands that no influence can propagate faster than c. Einstein developed his theory of gravity—general relativity—to be compatible with the requirements of special relativity. For years the mathematical subtleties of general relativity caused confusion about the physical meaning of gravitational waves and whether, for example, they carry energy.

We now know that gravitational waves have very real physical properties, play a role in astrophysical processes, and can, in principle, be detected. All of these physical aspects of the waves are unfamiliar because gravity is by far the weakest of all physical interactions. The electrical force between a proton and an electron is more than 10^{40} times greater than the gravitational force. Despite this weakness, gravitational forces can be important because all bits of matter generate the same kind of gravity. There are no positive and negative charges as there are with electrical forces.

Electromagnetic waves are typically generated by a small amount of electric charge being rapidly shaken back and forth through short distances, in an atom, an antenna, the hot filament of a light bulb, and so on. Large-scale motions of matter are not effective in generating electromagnetic waves because there is an approximate equality between the positive and negative charges in any large collection of particles. Exactly the opposite is true in the case of gravity. Its inherent weakness makes small-scale motions ineffective, but in large-scale motions all bits of matter contribute together.

To generate gravitational waves we would want to shake the largest possible mass through the largest possible distance. Such mechanical motions cannot be carried out nearly as rapidly as electronic motions, so gravitational waves cannot be generated with high frequencies and short wavelength. One of the consequences of the necessarily low frequency is that gravitational waves almost always can be considered "classical" waves. Quantum mechanics requires that gravitational radiation be composed of fundamental excitations that are called "gravitons" (similar to the "photons" of electromagnetic radiation). At low frequencies an individual graviton has very low energy, and only the wavelike nature of the collection of gravitons needs to be considered.

Gravity is too weak for its waves to be generated with reasonable strength by laboratory equipment. If we could shake a 1-ton mass through distances of 100 m at a frequency of 10,000 Hz, we would generate around 10^{-16} W in gravitational waves. Only the enormous mass motions of astrophysical events can produce significant gravitational wave energy. The astrophysical events that are most interesting as possible sources of waves are the oscillations of the remnants of supernova explosions and the spiraling of a pair of neutron stars into each other. Black hole processes (the collision of black holes, the infall of stars into black holes) are potentially very powerful and interesting sources of waves, but it is uncertain how commonly such events occur.

Some effects of astrophysical gravitational waves almost certainly have been observed. In work that won them the 1993 Nobel Prize for physics, Joseph Taylor and Russell Hulse discovered a binary pair of neutron stars, one of which, PSR1913 + 16, was a pulsar. By timing the arrival of the radio pulses from this system, they and others calculated the rate at which gravitational wave energy would be radiated according to Einstein's theory. In a separate calculation they determined the rate at which the system was losing energy by measuring the speed at which its orbit was shrinking. The two rates of energy loss were in excellent agreement.

This "observation" of gravitational waves is indirect, but scientists are hoping to make direct observations perhaps by the end of the century. Besides verifying basic physical theory, gravitational

wave detectors could function as a new class of "telescope" to study large scale astrophysical motions. All other telescopes (optical, radio, x ray, and so on) detect radiation from small-scale particle motions.

The detectors require apparently impossible sensitivity. Gravitational waves are expected to arrive at Earth with strengths to cause distortions no more than one part in 10^{21}. Even if the entire Earth were the detector, the distortion would be only a tiny fraction of the size of an atom. With such weak interactions the waves will travel through Earth with a negligible loss of energy. Earth and any other objects are almost totally transparent to the waves.

Two basic designs are being used for the detectors of the tiny distortions. The earliest detectors were massive, suspended, usually cylindrical bars. When a gravitational wave falls on one of these bar detectors, the time-varying distortion (much less than the diameter of a nuclear particle) sets the bar ringing at its natural frequency, and its ringing motion is detected electronically. The alternative is to use laser beams to monitor the positions of suspended test masses. A number of detectors of both bar and laser design are now operating at a variety of sites around the world, but none has yet achieved the sensitivity to detect the expected waves, and indeed no waves have yet been observed. Scientists are hoping that this will change in the next few years with the development of better detecting technology and larger detectors, especially the 4-km LIGOs (laser interferometric gravitational wave observatories) in the United States.

See also: Black Hole; Electromagnetic Wave; Graviton; Neutron Star; Newton, Isaac; Quantum Gravity; Quantum Mechanics; Relativity, General Theory of; Relativity, Specific Theory of; Specific Gravity

Bibliography

Davies, P. C. W. *The Search for Gravity Waves* (Cambridge University Press, Cambridge, Eng., 1980).

Ruthen, R. "Catching the Wave." *Sci. Am.* **266** (3), 90–101 (1992).

Schwarzschild, B. "Hulse and Taylor Win Nobel Prize for Discovering Binary Pulsar." *Phys. Today* **46** (12), 17–19 (1993).

Travis, J. "LIGO: A $250 Million Gamble." *Science* **260** (5108) 612–614 (1993).

Richard H. Price

GRAVITON

The graviton is a massless elementary particle presumed to be associated with the gravitational field. It has never been observed, but there are strong theoretical reasons to believe that the graviton exists. In Einstein's classical theory of general relativity, disturbances in the gravitational field travel at the speed of light. These disturbances are carried by gravitational waves, much as disturbances in an electromagnetic field are carried by electromagnetic waves. In a quantum theory according to Bohr's complementarity principle, all forms of matter and energy have dual descriptions as waves and as particles. For example, a quantum state of the electromagnetic field may be described either as a collection of photons or as a superposition of electromagnetic waves (more precisely, in a Fock basis of photons or as a linear superposition of coherent states). The photon description is more useful for example in the photoelectric effect, while the wave description is more appropriate for the emission of radio antenna.

Applying the complementarity principle to gravity, one concludes that there should exist a massless particle, the graviton, dual to gravitational waves. However, while the gravitational wave description has proven useful as an example in studying the energy emission of the binary pulsar PSR1913, there is no observed physical phenomena, analgous to the photoelectric effect, in which a description in terms of gravitons is appropriate. In principle the hydrogen atom could decay from an excited state to its ground state by emission of a graviton rather than a photon. However, the probability of such an occurrence is about one out of 10^{50}. Therefore, direct observation of gravitons is unlikely.

While electromagnetic waves are emitted from sources with oscillating electric dipole moments, gravitational waves arise from oscillating quadrapole mass sources. In the particle description, this leads to one unit of spin for the photon and two for the graviton. It is not known how to describe in full generality quantum mechanical interactions for relativistic particles with spins greater than one.

Thus, while it is likely that the graviton exists in some form, it is an elusive object that is virtually impossible to detect and whose properties are poorly understood.

See also: COMPLEMENTARITY PRINCIPLE; ELECTROMAGNETIC WAVE; EXCITED STATE; GRAVITATIONAL WAVE; GROUND STATE; PHOTON; QUANTUM GRAVITY; WAVE–PARTICLE DUALITY

Bibliography

WEINBERG, S. *Gravitation and Cosmology: Principles and Applications of the General Theory of Relativity* (Wiley, New York, 1972).

ANDREW STROMINGER

GRAVITY

See CENTER OF GRAVITY; GRAVITATIONAL ATTRACTION; QUANTUM GRAVITY; SPECIFIC GRAVITY

GREAT ATTRACTOR

The Great Attractor is one of the largest structures detected in the universe. With a mass equivalent to half a million galaxies, it stretches over hundreds of millions of light-years. Even though its center lies roughly 200 million light-years from Earth, the large mass of the Great Attractor is pulling the Milky Way galaxy and most of the surrounding galaxies toward it. This information led to discovery of the Great Attractor. In the mid-1980s, seven astronomers—David Burstein, Roger Davies, Alan Dressler, Sandra Faber, Donald Lynden-Bell, Roberto Terlevich, and Gary Wegner—observed an almost uniform flow of over 400 galaxies. The Seven Samurai, as they became known in the astronomy community, postulated that there must be some large mass pulling on all of these galaxies. A careful mathematical analysis determined the most likely value of the mass of the Great Attractor and its distance from Earth.

Given the indirect way in which the Great Attractor was discovered, it is not surprising that many astronomers were reluctant to believe in its existence. Since it was discovered, there have been various arguments put forth both in favor of it and against it. The case is not closed yet, but detailed investigations have tended to confirm at least the rough idea that there is an unusually large number of galaxies in the region of the Great Attractor.

Evidence for the Great Attractor

The Seven Samurai set out to map the velocities of a set of 400 elliptical galaxies. If the universe is completely uniform, then the big bang model of cosmology predicts that each galaxy should be moving away from Earth at a velocity proportional to its distance from Earth. Thus galaxies close to Earth should be receding relatively slowly and faraway galaxies should be moving away rapidly. The ratio between the recession velocity and the distance of a galaxy from Earth is called the Hubble constant. If the universe were uniform, this ratio would be the same for all galaxies observed; hence the term "Hubble constant."

As early as 1976, a group headed by Vera Rubin found that the ratio of galaxy velocities to distances was not exactly constant. Some velocities were slightly smaller than predicted and others were slightly larger. These deviations are called peculiar velocities. It is the peculiar velocities of the elliptical galaxies that the Seven Samurai were interested in. They felt that the deviations seen must have been the result of observers not having probed large enough regions of the universe. On very large scales, it was thought, the universe should be very close to uniform, and thus peculiar velocities should be very small. In fact, the Seven Samurai found just the opposite.

The sample of galaxies studied by the Seven Samurai consisted of those both near and far. The average distance, greater than 100 million light-years, was large enough to lead the group to expect negligibly small peculiar velocities. The average galaxy, however, had a peculiar velocity of order 600 km/s. Even more astounding, most of these galaxies were moving in the same direction.

By carefully dividing the galaxies into different regions in the sky, the Seven Samurai saw that those in one region were moving faster than anywhere else. They came to interpret this as due to a large concentration of mass, the Great Attractor. The Great Attractor pulls on objects close to it more strongly than on those far away. Thus galaxies close to it have larger velocities than those far away.

The Seven Samurai inferred that there must be a large mass concentration in the region of the Great Attractor. Immediately thereafter, Ofer Lahav compiled a map of the galaxies in the region and showed that there are indeed many more galaxies than average in this region. Lahav's map provided more concrete evidence that the Great Attractor exists. It is somewhat surprising that until then, this mass concentration had not been observed. One reason for the relatively late discovery of the Great Attractor is its size; it extends so far that it is necessary to map a sizable region before understanding how exceptional the Great Attractor is. Part of the ignorance also stemmed from the fact that the Great Attractor lies near the galactic plane. That is, starlight from the Milky Way galaxy obscures a sizable fraction of the Great Attractor.

Although the Great Attractor is roughly 200 million light-years from the Milky Way galaxy, it was detected by observing galaxies within a radius of about 100 million light-years. Thus, the observed galaxies are mostly flowing toward the Great Attractor and away from Earth. Initially no galaxies on the far side of the Great Attractor were observed flowing back toward it, and therefore toward Earth. If this "back side infall" were to be observed unambiguously, it would silence critics of the Great Attractor hypothesis. Unfortunately, the observational situation remains unclear. A number of detections of back-side infall have been made; two of the Seven Samurai, Dressler and Faber, in fact used different techniques and probed deeper than the original survey to detect infall. However, because these surveys probe so deeply, there are a number of uncertainties that are difficult to correct for.

Implications of the Great Attractor

Cosmologists work within the framework of a homogeneous universe. Relatively small irregularities, such as galaxies and even clusters of galaxies, often can be accommodated within this framework. However, extremely large structures such as the Great Attractor tend to pose difficulties for cosmological models. There was a great deal of debate as to whether various cosmological theories could produce a structure as large as the Great Attractor. This problem has been alleviated through the discovery by the Cosmic Background Explorer (COBE) satellite of inhomogeneities in the microwave background. These inhomogeneities are larger than those expected at the time of the discovery of the Great Attractor. Any successful model must account for these larger inhomogeneities; it has thus become more likely that a structure as inhomogeneous as the Great Attractor could have been produced.

Nonetheless, the Great Attractor continues to be an astounding example of a large inhomogeneous region in a universe that is proving to be much more complex and interesting than anyone had imagined.

See also: BIG BANG THEORY; COSMIC BACKGROUND EXPLORER SATELLITE; COSMOLOGY; GALAXIES and GALACTIC STRUCTURE; HUBBLE CONSTANT; MILKY WAY

Bibliography

ANDERSEN, P. H. "Ripples in the Universal Hubble flow." *Phys. Today* **10,** 17–19 (1987).

DRESSLER, A. *Voyage to the Great Attractor* (A. Knopf, New York, 1994).

LYNDEN-BELL, D.; FABER, S. B.; BURSTEIN, D.; DAVIES, R. L.; DRESSLER, A.; TERLEVICH, R. J.; and WEGNER, G. "Spectroscopy and Photometry of Elliptical Galaxies: V. Galaxy Streaming Toward the New Supergalactic Center." *Astrophys. J.* **326,** 19–49 (1988).

SCOTT DODELSON

GREAT WALL

The Great Wall is one of the largest structures observed in the universe. It consists of thousands of galaxies stretched out in a sheetlike structure. The galaxies cover a region with a length of over 500 million light-years, a width over 200 million light-years, but a compact vertical length of only about 20 million light-years. The entire complex resides some 200 million light-years from Earth.

The discovery by Margaret Geller, John Huchra, and collaborators in 1989 led many astrophysicists to reappraise what is known about structure in the universe. Perhaps most importantly, the mapmaking survey that uncovered the structure [the Center for Astrophysics (CFA) survey] produced dramatic pictures which killed off the notion that the universe is smooth and homogeneous. On the contrary, the inhomogeneities in the CFA survey—such as the Great Wall—were as large as could be detected given the survey's size. It is therefore possible that the Great Wall is even larger than we now suppose! The universe as seen by the CFA survey, and others like it, appears to have large empty voids, surrounded by galaxies arranged in sharp sheetlike structures.

The Great Wall is the largest and most impressive of these.

The CFA Survey

By simply observing the position of a galaxy in the sky, one does not know its distance from Earth. Two galaxies can appear very close together in the sky but one could be much farther away than the other. For a long time, this fundamental fact limited the understanding of structure in the universe.

In order to supplement the two-dimensional information gained by observing a galaxy's position, astronomers must measure a galaxy's redshift. The wavelength of light emitted by an object moving away from us is longer than it would be if the object were at rest. The color of light is related to its wavelength, so a receding object appears redder than a similar object at rest. By measuring this redshift, astronomers can accurately measure how fast a given galaxy is receding from us. According to Hubble's law, galaxies are moving from us in direct proportion to their distance from us. Thus, by measuring a galaxy's redshift, astronomers are in fact measuring its distance from us.

The CFA survey revisited a portion of the sky mapped in two dimensions by Fritz Zwicky and collaborators. The map was made by looking for all galaxies brighter than some fixed lower limit. Some 30,000 galaxies were observed in the northern hemisphere in this way. Geller and Huchra set out to measure redshifts for a sizable fraction of these. Their strategy was to measure in "strips." They would get the redshifts for all galaxies (brighter than the limit) in a region of the sky that spanned 135° in longitude but only six degrees in latitude. The longitude span is equivalent to the region of our globe spanned by Europe and Asia; at the depths observed by the survey, this range in longitude corresponds to a length of over 500 million light-years. This surveying strategy allowed the CFA team to see the results of the survey before all the strips were finished. Most of the redshifts were taken with the 1.5-m telescope at Mount Hopkins.

The Great Wall

The results from the first slice of the CFA survey, released in 1986, were shocking. The team placed a dot at a given distance and a given longitude if a galaxy was observed in the thin slice at that position.

Instead of seeing dots uniformly spread throughout the region, one observes long filaments of connected points separating regions with no galaxies at all. It is an accident, of course, that the dots denoting galaxies resemble the body of a human being.

One feature of the stick figure is slightly misleading. The torso, which extends from the center of the diagram down toward Earth at the bottom, consists of galaxies that in reality are very close together. They are members of the Coma cluster. Since they are traveling very fast due to the strong gravity of Coma, they have very large redshifts that are not of cosmological origin. It is impossible though to separate out the contribution of the redshift coming from the cosmological expansion and the part coming from the gravity of Coma. So an accurate determination of their distance cannot be made. This effect is only a problem when observing features parallel to the viewer's line of sight. The "arms" of the stick figure do not suffer from this illusion; thus the long stretch of galaxies from one end of the figure to the other is a real physical structure that spans the full range of the survey, some 500 million light-years. After the first slice was completed, it was not possible to comment on the width of that structure since the slice covered only 20 million light-years (corresponding to six degrees). In subsequent years, however, adjoining slices were released and the same structure appeared in each. Thus it is not a long string, but rather a "Great Wall." Its extent is at least as large as the range of the survey that discovered it; that is to say, it is at least 500 million light-years long and 200 light-years wide. As can be seen from the figure, it has very little extent in the other direction (along our line of sight); it has a thickness less than one-tenth the size of its other dimensions.

The Future

The technology necessary to measure redshifts has improved dramatically in recent years. At the time the Great Wall was discovered in 1989, the CFA survey contained redshifts for over 6,000 galaxies. This is ten times more redshifts than were available from the combined efforts of all astronomers in 1956, and the progress continues. Within a few years after the Great Wall was discovered, the number of redshifts in the CFA survey had more than doubled. There have been a variety of other surveys that have mapped different parts of the sky with similar numbers of redshifts. By the turn of the century, with the advent of surveys such as the Sloan Digital Sky Survey, there will be on the order of one million galax-

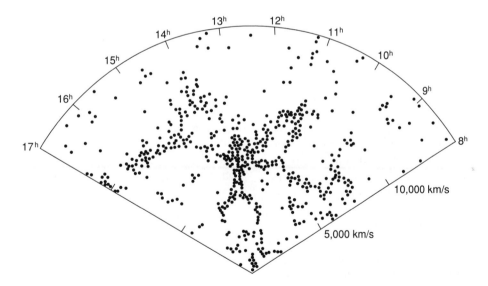

Figure 1 The first CFA slice. Each "hour" represents fifteen degrees in longitude. The distance from us can be obtained by dividing the velocities (e.g., 10,000 km/sec) by the Hubble constant. The "head" of the human figure lies several hundred million light-years from Earth. The extended "arms" of the figure compose the Great Wall (de Lapparent, Geller, and Huchra, 1986, with author permission).

ies with accurate positions and redshifts. The structure measured in the CFA survey is but one example of what can hope to learn in the coming years.

See also: COSMOLOGY; GALAXIES AND GALACTIC STRUCTURE; HUBBLE CONSTANT; REDSHIFT; UNIVERSE

Bibliography

GELLER, M. J., and HUCHRA, J. P. "Mapping the Universe." *Science* **246,** 897–903 (1989).
DE LAPPARENT, V.; GELLER, M. J.; and HUCHRA, J. P. "A Slice of the Universe." *Astrophys. J.* **302,** L1–L5 (1986).
SCHWARZSCHILD, B. "Gigantic Structure Challenge Standard View of Cosmic Evolution." *Phys. Today* **6,** 20–23 (1990).

SCOTT DODELSON

GRO

See GAMMA RAY OBSERVATORY

GROUND STATE

In quantum mechanics, states of definite energy are called stationary states or allowed states. The stationary state for a particular system with the lowest possible energy is called the ground state of that system. The ground state may be a bound state or an unbound one, depending on the system. When the ground state is bound, there are generally one or more excited states, which may also be bound or unbound, and there is a finite energy difference between the ground state and the first (or lowest) excited state. In many bound-state situations, the potential energy is taken to be zero at infinite separation, and the bound states have negative total energies. The most negative of these energies corresponds to the ground state.

Certain elementary systems are of particular importance. A free particle is unbound, and can have any positive energy whatsoever. Its ground state has zero energy. A particle of mass m in an infinite potential well of length L ("particle in a box") is bound, and the energies of its various states are given by

$$E_n = \frac{n^2\pi^2\hbar^2}{2mL^2},$$

where n can be any nonzero integer. The ground state of the particle in the box is the state with $n = 1$. A particle executing simple harmonic motion with frequency ω is bound, and the energies of its various states are given by

$$E_n = (n + \tfrac{1}{2})\hbar\omega,$$

where n is any integer including zero. The simple harmonic oscillator in the ground state has the quantum number $n = 0$.

In the elementary quantum-mechanical treatment of the hydrogen-like atom without spin, the energies of the atom's bound states are given by

$$E_n = -\frac{\mu Z^2 e^4}{2(4\pi\varepsilon_0)^2\hbar^2}$$

where μ is the reduced electron mass, Z is the atomic number, and e is the electron charge. The quantum number, n, can have any integer value except zero. In this system, the ground state has quantum number $n = 1$.

See also: ENERGY LEVELS, EXCITED STATE; QUANTUM MECHANICS; QUANTUM NUMBER; STATIONARY STATE

Bibliography

ANDERSON, E. *Modern Physics and Quantum Mechanics* (Saunders, Philadelphia, 1971).
EISBERG, R., and RESNICK, R. *Quantum Physics of Atoms, Molecules, Solids, Nuclei, and Particles,* 2nd ed. (Wiley, New York, 1985).
GRIFFITHS, D. *Introduction to Quantum Mechanics* (Prentice Hall, Englewood Cliffs, NJ, 1995).

JAMES R. HUDDLE

GROUP VELOCITY

In wave phenomena, the group velocity is the rate at which a signal, a pulse of energy, a wave packet, or the modulation of a carrier wave propagates through space. Energy and momentum can be transported by disturbances that are strictly periodic in time and space, but the transmission of specific information requires a more complex wave form that is finite in its spatial extent and in its temporal duration. Since the basic equations governing wave motion are linear, the Fourier integral theorem allows any such wave form to be constructed from a linear superposition of strictly periodic sinusoidal waves having different frequencies ν, and wavelengths λ. Accordingly, a one-dimensional transverse pulse, such as the one shown in Fig. 1, can be represented by the function

$$\psi(x,t) = \frac{1}{\sqrt{2\pi}} \int_{-\infty}^{+\infty} A(k)\, e^{i(kx - \omega t)}\, dk,$$

where the sinusoidal waves described by the exponential are characterized by wave numbers $k = \lambda/2\pi$ and angular frequencies $\omega = 2\pi\nu$, and the spectral composition of the wave packet they form is described by the set of amplitudes $A(k)$. The velocity of propagation v_p, of each of the constituent waves is called a phase velocity and is given by $v_p = \omega/k$. If all of the constituent waves have the same phase velocity, as is true for the passage of light through a vacuum, for example, their behavior is said to be dispersion free. In this case, ω is linearly proportional to k, the proportionality constant being the common phase velocity v_p. More frequently, media through which waves travel are dispersive, and ω is a more complicated function of k.

In general, when a wave packet propagates through a dispersive medium, its shape becomes distorted after a period of time, since its constituent waves have different phase velocities. However, if the dependence of ω on k is only slightly nonlinear, the spectral distribution function $A(k)$ is sharply peaked around some value k_0, and $\omega(k)$ can be approximated by

$$\omega(k) \approx \omega_0 + \frac{d\omega}{dk}\bigg|_0 (k - k_0).$$

Substitution of this expression into the integral form for $\psi(x,t)$ followed by a small amount of algebra leads to the result

$$\psi(x,t) \approx \psi[x - (d\omega/dk)\,\big|_0 t, 0]\, e^{i[k_0(d\omega/dk)\,|_0 - \omega_0]t},$$

which shows that, aside from an overall phase factor, the wave packet propagates through space undistorted in shape with a group velocity given by $v_g = (d\omega/dk)\,\big|_0$.

The forgoing discussion applies to all wave phenomena: mechanical, electromagnetic, and quantum mechanical. Several cases warrant special

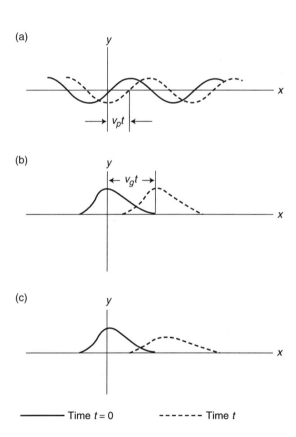

Figure 1 Snapshots of one-dimensional transverse disturbances traveling along the $+x$ direction: (a) a strictly periodic sinusoidal wave with a phase velocity v_p; (b) a pulse with a group velocity v_g propagating undistorted in a dispersion free medium; and (c) a pulse propagating in a highly dispersive medium, displaying distortion and the absence of a well-defined group velocity.

mention. For non-dispersive media, the group and phase velocities are equal ($v_g = v_p$). For light waves propagating through a dispersive medium of index of refraction $n(k)$, ω and k are related by the expression $\omega(k) = ck/n(k)$, where c is the speed of light in vacuum. The phase velocity of any constituent wave is then given by $v_p = c/n(k)$, and the group velocity of any wave packet can be expressed as $v_g = c/[n(\omega) + \omega(dn/d\omega)]$. For nonrelativistic matter waves, the energy of a quantum mechanical wave packet describing a free particle of mass m is $\hbar\omega$, while its momentum is $\hbar k$, where $2\pi\hbar$ is Planck's constant. In this case, the relationship between ω and k is given by $\omega = \hbar k^2/2m$. Then the phase velocity of any constituent wave is $v_p = \hbar k/2m$, while the group velocity of any wave packet is $v_g = \hbar k/m$. Therefore, for matter waves, the group velocity and the phase velocity of a free particle are related by the expression $v_g = 2v_p$. For all wave phenomena, c serves as an upper limit on the group velocity, in accordance with the principles of special relativity.

See also: BROGLIE WAVELENGTH, DE; DISPERSION; ELECTROMAGNETIC WAVE; FOURIER SERIES AND FOURIER TRANSFORM; PHASE VELOCITY; WAVE MOTION; WAVE PACKET; WAVE SPEED

Bibliography

ANDERSON, E. E. *Modern Physics and Quantum Mechanics* (Saunders, Philadelphia, 1971).

HALLIDAY, D.; RESNICK, R.; and KRANE, K. S. *Physics*, 4th ed. (Wiley, New York, 1992).

HECHT, E., and ZAJAC, A. *Optics*, 2nd ed. (Addison-Wesley, Reading, MA, 1987).

JACKSON, J. D. *Classical Electrodynamics*, 2nd ed. (Wiley, New York, 1975).

MICHAEL S. LUBELL

H

HAAS–VAN ALPHEN EFFECT, DE

The de Haas–van Alphen (DH-VA) effect, discovered in Bi in 1930, is an oscillatory dependence of the magnetization on the applied magnetic field strength. This effect is detectable in most other elemental metals and ordered alloys at low temperatures and strong magnetic fields. It is a diamagnetic property since it is due to the orbital motion of conduction electrons in the magnetic field. Related magneto-oscillatory behavior is noticeable in other properties such as the electrical resistivity (Shubnikov–de Haas effect). While of fundamental interest, the DH-VA effect also provides a powerful tool for investigating the electronic structure in metals.

Central to an understanding of the DH-VA effect are the quantized magnetic energy levels called Landau levels. These levels have the energy E_n for the motion of conduction electrons perpendicular to the field with

$$E_n = (n + \gamma)\hbar\omega_c,$$

where $n = 0, 1, 2 \ldots$, γ is a constant, and $\hbar$ is Planck's constant divided by 2π. The cyclotron frequency $\omega_c = eB/m$ where B is the magnetic flux density, e is the electronic charge, and m is called the cyclotron effective mass. The value of m depends on the electron's interactions within the metal. At a temperature of absolute zero, electrons fill all of the Landau levels with energy less than a certain value, the Fermi energy E_F, leaving the higher levels empty.

The DH-VA oscillations appear as a result of Landau levels passing through E_F in succession with increasing (or decreasing) B. Since this occurs periodically in $1/B$, the oscillatory magnetization M_1 can be written as

$$M_1 = M_0 \sin(2\pi F/B)$$

neglecting higher harmonics. The amplitude M_0 is a slowly increasing function of B at a fixed temperature and F is termed the DH-VA frequency. The value of F is important because it is used to measure the Fermi surface (FS).

The FS is the constant energy surface of conduction electrons in wave vector or momentum space at the maximum energy E_F. The extremal FS cross-sectional area A perpendicular to the field is given by $A = (2\pi e/\hbar)F$. Measurements of F for all different directions of the field relative to the crystal axes permit the shape and volume of the FS to be determined. The FS volume is accurately related to the concentration of conduction electrons.

Other FS parameters (e.g., the conduction-electron effective mass and magnetic moment) are obtained from the DH-VA amplitude. The amplitude, however, decreases rapidly with increasing temperature because thermal motion blurs the FS. Temperatures in the liquid helium range (the order of 4 K) and fields greater than about 1 T are usually required for detectable amplitudes.

The DH-VA and related oscillatory effects characterize the main methods for measuring the Fermi surface parameters accurately. This detailed information permits the electronic structure in metals to be determined over a wide energy range. Recent DH-VA studies in high-temperature, oxide superconductors help advance our understanding of the electronic properties in this class of potentially important materials.

See also: CYCLOTRON RESONANCE; DIAMAGNETISM; FERMI SURFACE.

Bibliography

KITTEL, C. *Introduction to Solid-State Physics,* 6th ed. (Wiley, New York, 1986).

MacKINTOSH, A. R. "The Fermi Surface of Metals." *Sci. Am.* **209** (1), 110–120 (1963).

SHOENBERG, D. *Magnetic Oscillations in Metals* (Cambridge University Press, Cambridge, Eng., 1984)

J. J. VUILLEMIN

HADRON

Hadrons are a class of subatomic particles. Two types of hadrons—the proton and neutron—are constituents of the atomic nucleus. The others are created in high-energy collisions such as those from cosmic rays or those at particle accelerators.

A hadron is a particle made of strongly interacting constituents—quarks (and/or gluons). The so-called strong force causes quarks to bind together tightly to form particles such as protons and neutrons. In fact, this force is so strong that quarks are never found separately but only inside hadrons. Although quarks carry strong charge, hadrons have zero strong charge.

Hadrons are subdivided into two categories: the mesons and baryons. Mesons are color-neutral particles made from a quark and an antiquark. Examples include the pion ($\pi^+ = u\bar{d}$), the kaon ($K^- = s\bar{u}$), and the B meson ($B^0 = d\bar{b}$). Baryons are color-neutral

hadrons made from three quarks, one of each of the three possible quark color charges. For every baryon type made of three quarks, there is a corresponding antibaryon type made of the corresponding three antiquarks. Protons and neutrons are baryons.

While all quarks have fundamental strong interactions, all hadrons have residual strong interactions with other hadrons due to their strongly interacting constituents (the quarks), just as electrically-neutral atoms have residual electromagnetic interactions due to their electrically-interacting constituents. This residual interaction between the hadrons due to their color-charged constituents is responsible for the strong nuclear force, which is the force that binds protons and neutrons together to form atomic nuclei.

The size of the hadrons is fixed by the competition between the potential energy of the interaction between the quarks and the kinetic energy that occurs, via the Heisenberg uncertainty principle, when the constituents are confined to a small area. Since the kinetic energy of a confined object grows as the region shrinks while the potential energy decreases as the region shrinks, there is a preferred size that has the minimum total energy.

The kinetic energy provides (via $E = mc^2$) an important part of the mass of the hadron; for the proton and the neutron it is the dominant contribution to their mass. This explains, for example, why the proton mass is much larger than the sum of the masses of the quarks that it contains. (A similar competition between kinetic and potential energy determines the size of atomic nuclei and of atoms, but in both these cases the forces are less strong, so the objects are large enough that the kinetic energy is small and hence does not play a major role in the mass of the composite object.)

Hadrons can be formed from any combination of quark flavors except the top, which decays too rapidly to ever form a composite particle. All mesons and all baryons except the proton are unstable and decay by strong, electromagnetic, or weak interactions. Understanding the possible hadron types, their complex set of possible decays, and the relative rates of each allowed decay is a major part of particle physics and provides numerous tests of the theory called the standard model. The *Review of Particle Physics,* a compilation of particle knowledge published every other year, contains more than 300 pages of listings of particle types and their known decays.

One simple rule for classifying decays is that a weak decay occurs whenever the flavor of a quark is changed or whenever a quark disappears on meeting an antiquark of a different flavor. Electromagnetic decays can be recognized by the fact that they produce photons, whereas strong decays always result only in some less massive hadrons.

The proton is stable within the standard model and by experiment is known to have a half-life of greater than 10^{31} years. Models that go beyond the standard model, called grand unified theories because they unify the three particle interaction types into a single picture, introduce additional extremely weak (i.e., rare) interactions that cause proton decay.

Neutrons also can become stable when bound inside an atomic nucleus. The patterns of stable and unstable nuclei are a major subject of study in nuclear physics.

See also: COLOR CHARGE; DECAY, NUCLEAR; FLAVOR; INTERACTION, STRONG; NEUTRON; PROTON; QUARK; UNCERTAINTY PRINCIPLE

Bibliography

CLOSE, F. E., MARTEN, M. and SUTTON, C. *The Particle Explosion* (Oxford University Press, New York, 1994).
WEINBERG, S. *The Discovery of Subatomic Particles* (W. H. Freeman, New York, 1990).

R. MICHAEL BARNETT

HALL EFFECT

When a conducting rod carrying a current is placed in a transverse magnetic field, an electric potential appears across the rod in a direction perpendicular to both the current and the magnetic field. This is the Hall effect.

Suppose the rod is along the x axis as in Fig. 1, with width w in the y direction and depth d in the z direction. The magnetic field **B** is in the z direction (perpendicular to the page), and the current is in the x direction, along the rod. If there is a density n of carriers of charge q moving with a drift velocity v_x, the current is given by $I = (nwd)qv_x$. This current is either due to positive charges moving to the right or negative charges moving to the left. The magnetic force on particles moving with velocity **v** is given by

$\mathbf{F_B} = q\mathbf{v} \times \mathbf{B}$. In this case the force is in the $-y$ direction, no matter which sign the charges have. The carriers pile upon the bottom of the rod, and an opposite charge appears on the top of the rod. The charges build up until they produce a transverse electric field $E_y = v_xB$, such that the electric force qE_y on the carriers exactly cancels the magnetic force. This means there is no net transverse force on the carriers, so there is no transverse current. The Hall electric potential $V_H = wE_y$ across the sample is then given by

$$V_H = (1/nqd)IB.$$

The factor $R = 1/nq$ is called the Hall constant, and it gives information about the charge carriers in a conductor. R is much larger for semiconductors than metals, because the density of carriers is much smaller. The sign of the Hall constant tells the sign of the charge on the carriers. It is well known that in a metal it is the electrons that carry the current, so it should be no surprise that in copper and many other metals the Hall constant is negative. In semiconductors that have excess electrons (n-type), the Hall constant is negative, but in p-type semiconductors the conduction is by holes, which act as positive charge carriers, so the Hall constant is positive. Also in some metals such as berylium R is positive, which shows that conduction is by holes.

In 1980 the quantum Hall effect was discovered by Klaus von Klitzing. In semiconductors it is possible to produce narrow conducting layers in which the carrier motion is two dimensional, for example, near the surface of a metal-oxide-semiconductor transister. Applying a large magnetic field perpendicular to the layer, he measured the Hall resistance, $R_H = V_H/I$, which according to our theory should be B/nqd. When he held the current fixed

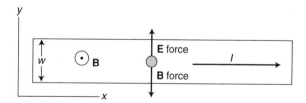

Figure 1 Experimental setup for Hall effect, showing a charge carrier with magnetic and electric forces in balance.

and varied the number of electrons in the layer, von Klitzing found that there were steps in R_H at values h/e^2p, where p is an integer. This is remarkable because it depends only on the fundamental constants e and h. The effect depends on the quantization of the electron motion in a magnetic field. The electrons move in circular orbits perpendicular to the field with the cyclotron frequency, $f_c = eB/2\pi m$. According to quantum theory, these orbits correspond to energy levels with spacing hf_c. Each level can hold a large number of electrons—eB/h per unit area. A step in R_H occurs each time a quantum level is completely filled.

See also: CHARGE; CONDUCTION; CONDUCTOR; ELECTRICAL CONDUCTIVITY; ELECTRON; ELECTRON, CONDUCTION; ENERGY; ENERGY LEVELS; FIELD, MAGNETIC; FORCE; HOLES IN SOLIDS; MOTION; SEMICONDUCTOR; VELOCITY

Bibliography

HALLIDAY, D., and RESNICK, R. *Fundamentals of Physics,* 4th ed. (Wiley, New York, 1992).

KITTEL, C. *Introduction to Solid-State Physics,* 6th ed. (Wiley, New York, 1986).

LAURA M. ROTH

HARMONIC MOTION

See MOTION, HARMONIC

HARMONICS, SPHERICAL

Spherical harmonics, $Y_{lm}(\theta,\phi)$, represent a set of simultaneous solutions in spherical coordinates to a pair of differential equations that physically represent the square of the orbital angular momentum $\mathbf{L}^2$ and the z component of the angular momentum L_z, that is,

$$\mathbf{L}^2 Y_{lm} = l(l+1)\hbar^2 Y_{lm} \qquad (1)$$

and

$$L_z Y_{lm} = m\hbar Y_{lm}, \qquad (2)$$

where

$$\mathbf{L}^2 = -\hbar^2 \left[\frac{1}{\sin\theta} \frac{\partial}{\partial\theta} \left(\sin\theta \frac{\partial}{\partial\theta} \right) + \frac{1}{\sin^2\theta} \frac{\partial^2}{\partial\phi^2} \right] \qquad (3)$$

and

$$L_z = -i\hbar \frac{\partial}{\partial\phi}. \qquad (4)$$

The orbital angular momentum l takes values 0, 1, 2, ..., and for each l, the magnetic subquantum number m takes values 0, ± 1, ± 2, ..., $\pm l$. If the potential energy V of a quantum mechanical system is spherically symmetric, or equivalently, if the system is subject to a central force $\mathbf{F} = -\hat{\mathbf{r}}(dV(r)/dr)$, the angular part of the spatial wave function of the Schrödinger equation is always given by the spherical harmonics regardless of the detailed behavior of the radial potential $V(r)$. Thus, these functions play a special role in quantum mechanics.

Following the standard phase convention, the spherical harmonics satisfy the symmetry relations

$$Y^*_{lm}(\theta,\phi) = (-1)^m Y_{l-m}(\theta,\phi) \qquad (5)$$

and

$$Y_{lm}(\pi - \theta, \pi + \phi) = (-1)^l Y_{lm}(\theta,\phi). \qquad (6)$$

It is also related to the associated Legendre function $P_l^m(\cos\theta)$ by the expression

$$Y_{lm}(\theta,\phi) = (-1)^m \left[\frac{(2l+1)(l-m)!}{4\pi(l+m)!} \right]^{1/2} e^{im\phi} P_l^m(\cos\theta), \qquad (7)$$

and thus is given explicitily by

$$Y_{lm}(\theta,\phi) = \frac{(-1)^{l+m}}{2^l l!}\left[\frac{(2l+1)(1-m)!}{4\pi(l+m)!}\right]^{1/2}$$

$$e^{im\phi}(\sin\theta)^m\left[\frac{\partial}{\partial(\cos\theta)}\right]^{l+m}(\sin\theta)^{2l}. \quad (8)$$

When $m = 0$, it reduces to a Legendre polynomial $P_l(\cos\theta)$, i.e.,

$$Y_{l0}(\theta,\phi) = \left[\frac{2l+1}{4\pi}\right]^{1/2}P_l(\cos\theta). \quad (9)$$

In spectroscopic applications, spherical harmonics are often replaced by the more convenient tensor operators $C_q^{(k)}$ defined as

$$C_q^{(k)} = \left[\frac{4\pi}{2k+1}\right]^{1/2}Y_{kq}(\theta,\phi). \quad (10)$$

The Cartesian coordinates (x,y,z) can be expressed in terms of the radius r and spherical harmonics by

$$x = \frac{r}{\sqrt{2}}(C_{-1}^{(1)} - C_1^{(1)}),$$

$$y = \frac{ir}{\sqrt{2}}(C_{-1}^{(1)} + C_1^{(1)}), \quad (11)$$

$$z = rC_0^{(1)}.$$

For a multielectron system, the Coulomb interaction between two electrons separated by a distance $r_{12} = |\mathbf{r}_1 - \mathbf{r}_2|$ can be expanded in spherical harmonics, that is,

$$\frac{1}{r_{12}} = \sum_{l=0}^{\infty}\left(\frac{r_<^l}{r_>^{l+1}}\right)P_l(\cos\theta), \quad (12)$$

where θ is the angle between $\mathbf{r}_1$ and $\mathbf{r}_2$, $r_<$ ($r_>$) is the smaller (greater) of $|\mathbf{r}_1|$ and $|\mathbf{r}_2|$, and

$$P_l(\cos\theta) = \frac{4\pi}{2l+1}\sum_{m=-l}^{l}Y_{lm}^*(\theta_1,\phi_1)Y_{lm}(\theta_2,\phi_2). \quad (13)$$

Or, in terms of the tensor operator $C_m^{(l)}$,

$$P_l(\cos\theta) = \sum_{m=-l}^{l}C_m^{(1)*}(\theta_1,\phi_1)\,C_m^{(1)}(\theta_2,\phi_2). \quad (14)$$

Equations (13) and (14) are known as the addition theorem of spherical harmonics.

Certain properties of spherical harmonics are very useful to know when combining angular momenta. The product of two spherical harmonics that are functions of the same angles are given by

$$Y_{l_1 m_1}Y_{l_2 m_2} = \sum_{lm}\left[\frac{(2l_1+1)(2l_2+1)(2l+1)}{4\pi}\right]^{1/2}Y_{lm}^*$$

$$\begin{pmatrix} l_1 & l_2 & l \\ m_1 & m_2 & m \end{pmatrix}\begin{pmatrix} l_1 & l_2 & l \\ 0 & 0 & 0 \end{pmatrix}, \quad (15)$$

where $m_1 + m_2 + m = 0$, $l = (l_1 + l_2)$, $(l_1 + l_2 - 2)$, ..., $|l_1 - l_2|$, and the $3 - j$ symbol of Wigner is defined in terms of the Clebsch–Gordan coefficients $(l_1 m_1 l_2 m_2 | l_1 l_2 l - m)$ by

$$\begin{pmatrix} l_1 & l_2 & l \\ m_1 & m_2 & m \end{pmatrix} =$$

$$(-1)^{l_1 - l_2 - m}\left(\frac{1}{2l+1}\right)^{1/2}(l_1 m_1 l_2 m_2 | l_1 l_2 l - m). \quad (16)$$

See also: COORDINATE SYSTEM, CARTESIAN; COORDINATE SYSTEM, SPHERICAL

Bibliography

EDMONDS, A. R. *Angular Momentum in Quantum Mechanics* (Princeton University Press, Princeton, NJ, 1957).

RACAH, G. "The Theory of Complex Spectra." *Phys. Rev.* **62**, 438–462 (1942).

ROSE, M. E. *Elementary Theory of Angular Momentum* (Wiley, New York, 1957).

TU-NAN CHANG

HEALTH PHYSICS

Health physics, or radiological health, is a specialty that deals with the protection of individuals and population groups against the harmful effects of ionizing radiation while permitting, and even encouraging, the application of radiation for the benefit of humankind.

For fifty years after the discovery of x rays (in 1895, by Wilhelm Conrad Röntgen) and of radioactivity (in 1896, by Antoine Henri Becquerel), radiation was used in ways that would seem outrageous to modern health physicists. Doctors, who irradiated patients for diagnostic or therapeutic reasons, often failed to take even rudimentary precautions for themselves. Researchers carried radium sources in their pockets. After all, what harm could be done by something that could not be seen, felt, tasted, or heard? Despite published reports describing harmful effects of excessive exposure to radiation, protection efforts evolved slowly and the effort remained associated primarily with medical practice.

The health physics profession was born in 1942 when its first groups were organized under the direction of Robert S. Stone, the head of the Health Division of the Manhattan Engineering District (MED). The MED was an organization developing the atomic bomb, and the groups were named "health physics" partly because of the need for secrecy and partly because many physicists were working in the Health Division. The need for radiation protection services grew rapidly as the uses of ionizing radiation expanded in the postwar era. In 1955, the Health Physics Society (HPS) was formed; it has steadily grown over the years to its current level of about 6,500 members. Professional organizations such as the National Council on Radiation Protection and Measurements, (NCRP), the International Commission on Radiation Protection (ICRP), and the International Commission on Radiological Units (ICRU) maintain voluntary radiation protection standards. Several government agencies regulate the use of radiation sources in the United States. The U.S. Nuclear Regulatory Commission is responsible for the basic federal radiation protection standard, Title 10, Code of Federal Regulations, Part 20 (10 CFR, Part 20), "Standards for Protection Against Radiation." The U.S. Department of Energy (DOE) establishes radiation protection standards for operations conducted within government-owned facilities. The Center for Devices and Radiological Health (CDRH) of the Public Health Service (PHS) provides assistance to the states regarding radiation safety in medicine and in environmental protection. The Conference of Radiation Control Program Directors (CRCPD) was formed in 1968 as an independent professional organization to coordinate radiation protection standards in state and local programs. The Environmental Protection Agency (EPA) develops overall radiation exposure standards that other agencies incorporate into their regulations. The National Institute of Standards and Technology (NIST) plays a supporting role by providing basic data and calibrations for measurements of radioactivity, radiation exposure, and doses.

The American Board of Health Physics (ABHP) was formed in 1959 as an independent organization to offer professional certification to qualified individuals. Upon passing an examination, the applicant is registered as a certified health physicist (CHP). By 1994, the ABHP had granted 1,353 certificates. Certification is open to graduates in health physics, environmental science, industrial hygiene, environmental engineering, nuclear engineering, radiation technology, and occupational medicine programs with at least a Master of Science degree and relevant professional experience. Employment opportunities exist in the industrial, military, research/university, medical, and government (state and federal regulatory agencies) sectors of the economy. The typical salary starts at about $30,000 per year for entry-level positions; senior level, experienced professionals earn up to $90,000 per year.

The goals of radiation protection revolve around two types of radiation-induced biological effects. The first type is nonstochastic effects, which occur soon after irradiation and thus can be definitely linked to radiation exposure. These effects include temporary and permanent hair loss, skin redness, skin shedding, and ulceration, progressing to wasting and death of skin. A threshold exists for nonstochastic effects; that is, a certain level of exposure is required for the effects to occur. The second type, stochastic effects, is best described as random damage to cells that increases the chance of developing cancer later in life. These effects manifest themselves as a statistical increase in the number of cancer occurrences within the group of people exposed to increased amounts of radiation. Currently, it is believed that no threshold exists for stochastic effects; any exposure to ionizing radiation, no matter how small, carries a certain risk of developing cancer later in life.

Thus, at first thought, every effort to minimize levels of radiation exposure should be worthwhile. However, the risk is usually very small and similar to other risks occurring in everyday life (like taking an airplane trip). In addition, the population is exposed to a certain background radiation from cosmic rays and terrestrial sources. Since this radiation has always been present in the human environment, the added risk associated with any additional radiation exposure should be judged against the risk from background radiation exposures. Any society, no matter how wealthy, has only limited resources to spend on public health. If these resources are spent unwisely, the public's health deteriorates. In radiation protection, this concept is represented by the so-called ALARA principle, which calls for implementation of protection measures that keep the radiation exposure levels well below the threshold of nonstochastic effects, and make the stochastic risk of radiation exposure as low as reasonably achievable.

The duties of a health physicist are to monitor implemented radiation protection measures and to enforce the ALARA principle. Personnel monitoring is used to record radiation exposure of individual workers by means of a variety of personal dose monitors, and portable monitoring is used for radiation detection and exposure control. The health physicist must keep personnel and the environment under constant surveillance. If control measures are found to be ineffective, or if they break down, the health physicist must be able to evaluate the degree of hazard and to make recommendations regarding remedial action.

Professional aspects of health physics include measurements of different types of radiation and radioactive materials; establishment of quantitative relationships between radiation exposure and biological damage; transport of radioactivity through the environment; and design of radiologically safe equipment, processes, and environments.

Health physics professionals must understand complex relationships between people and the physical, chemical, biological, and even social components of the environment. To perform effectively, they must also be knowledgeable in major interdisciplinary areas that include basic physical, life, and earth sciences as well as applied areas such as toxicology, industrial hygiene, medicine, public health, and engineering.

See also: NUCLEAR MEDICINE; RADIATION PHYSICS

Bibliography

CEMBER, H. *Introduction to Health Physics,* 2nd ed. (Pergamon Press, New York, 1989).

TURNER, J. E. *Atoms, Radiation, and Radiation Protection* (Pergamon Press, New York, 1986).

KATHREN, R. L., and ZIEMER, P. L., eds. *Health Physics* (Pergamon Press, New York, 1980).

DRAGANIĆ, I. G.; DRAGANIĆ, Z. D.; and ADLOFF, J. P. *Radiation and Radioactivity on Earth and Beyond* (CRC Press, Boca Raton, FL, 1989).

WLAD T. SOBOL

HEAT

In physics and technology, we recognize three distinct meanings of heat, each related to energy. Terminology is not standardized, so only the context permits the reader to distinguish between these three, quite different, meanings.

One interpretation of heat is as temperature, or "hotness." Although this usage is presently frowned upon in physics, it remains in colloquial language (e.g., heat and humidity) and in special technical meanings, such as color temperature (red heat versus white heat) and heat treatment (raising to a prescribed temperature).

A second interpretation of heat is as the portion of internal energy that changes with change of temperature. Internal energy includes nuclear energy (usually very large, but constant), chemical energy (large and constant unless a chemical reaction occurs), and electron energy (large, but nearly independent of temperature). It also includes strain energy (reversible and irreversible) and thermal energy (or heat). Thermal energy consists of several types of molecular energy, including translational kinetic energy, rotational energy, and usually the kinetic and potential energy components of vibrational energy. It may also include the potential energy of phase transitions.

The law of equipartition of energy tells us that the thermal energy stored in these several molecular modes, or degrees of freedom, is exchanged freely between them, seeking a common level. That common level of average thermal energy in each mode (at normal temperatures) is $\frac{1}{2}kT$ (k, the Boltzmann constant, is 1.38×10^{-23} J/K), and is thus proportional to the temperature.

The third interpretation of heat, nearly always represented as Q (or q) in thermodynamics, we call thermal energy transfer. Because thermal energy and thermal energy transfer are a form of energy and an amount of energy transferred, respectively, each has units of energy, which may be the joule (SI unit), calorie, foot-pound, Btu, and so forth. The similarity of name and units contributes to confusion between these two distinct quantities.

Antoine Lavoisier was the first to propose a complete theory tying together the three meanings of heat, incorporating ideas of Joseph Black. Lavoisier suggested that a fluid, *calorique* (caloric, or heat, in English), was responsible for the sensation of temperature. Adding caloric to a body (i.e., thermal energy transfer Q) increased the amount of caloric (thermal energy) in the body and caused a higher temperature. The caloric capacity or specific caloric (heat capacity or specific heat) was a measure of how much caloric must be added for a given temperature rise. Caloric could become "latent" under certain conditions, but cutting metal or freezing water released the latent caloric, which then again produced a sensible temperature effect.

Soon afterward, Benjamin Thompson, Count Rumford, demonstrated that Lavoisier's theory was not consistent with experiment. In particular, he showed that any arbitrary amount of thermal energy could be obtained from brass being bored, and the amount was greater when less brass was cut (a dull cutting tool). Furthermore, the brass chips had the same heat capacity as the original uncut brass. Nevertheless, caloric theory has strongly influenced modern terminology and thought.

It was not until the middle of the nineteenth century that the efforts of Robert Mayer and Hermann von Helmholtz, in Germany, and especially the experimental efforts of James Prescott Joule, an English brewer, led to an understanding of energy and its relationship to the three meanings of heat. Temperature is a measure of the average amount of thermal energy in each of the (classical) modes of energy storage, as discussed above. However, the thermal energy of a system may be increased or decreased in many ways, in addition to the possible transfer of thermal energy (Q). Work done on a system increases the energy of the system, which may therefore increase the amount of thermal energy. Thermal energy may be produced internally, by conversion of chemical, or nuclear, or strain energy into molecular motions, or by conversion of kinetic energy of an object into thermal energy when it strikes a rigid surface.

Joule was the first to carry out accurate measurements to find the conversion factor between traditional units of mechanical energy (joule or foot-pound) and thermal energy (the calorie or Btu). He added energy to mercury by stirring (i.e., doing work W; $Q = 0$) and measured the temperature rise. His result is commonly described as the mechanical equivalent of heat.

Energy may be transferred between a system and its surroundings in many ways. Thermal energy is transferred as a consequence of a temperature difference; $Q > 0$ indicates energy added to the system, and $Q < 0$ indicates thermal energy transferred from the system to the surroundings. Energy may be transferred as work W when a force acts through a distance, which may change the kinetic energy, the rotational energy, the strain energy, or the thermal energy (or, for electrical work, the chemical energy) of the system. Energy may also be added by transfer of material, including convection or addition of fuel, and by other modes that cannot be classified as Q or W. Such energy transfers are often incorporated into the first-law equation, which is related to the first law of thermodynamics (conservation of energy) but is less general. The first-law equation may be written

$$\Delta E = Q + W + \sum_i X_i$$

which shows that the change in the total energy of the system E is equal to the thermal energy transferred to the system plus the work done on the system plus energy added by other modes of transfer.

For any change of state, between given initial and final states, ΔE is fixed (because E is a state function), but either Q or W may have arbitrary values, subject to the condition imposed on their sum by the first-law equation. The values of Q and W depend on the path, that is, on the details of the process, and are often not operationally defined. Thus they cannot be known when irreversible steps are involved in the process.

Because Q and W represent mutually exclusive modes of energy transfer, it is not possible, in any process, to "convert" Q to W (heat to work) or vice versa. Also, when the process has been completed, there is no longer any Q or W (in contrast to the thermal energy and mechanical energy, which persist). Neither Q nor W are a measure of any quantity that may be changed; hence the symbol ΔQ should be avoided as misleading. [$Q = \Delta E$, so $\Delta Q = \Delta(\Delta E)$.

Only in caloric theory would ΔQ represent a change in caloric in the system.] Similarly, an infinitesimal amount of thermal energy transfer is not a change in Q and therefore not a mathematical differential (definite or indefinite). To avoid the misleading symbol dQ, notations such as $\overline{d}Q$, δQ, and simply q, have been employed.

When a net amount of energy is transferred to a system of constant mass without change of motion of the system, there is usually an increase in temperature of the system ΔT. (Exceptions may occur when there are internal changes such as chemical reactions or, more commonly, a change of phase of the system.) If the energy transfers are (apart from work of expansion against the atmosphere) entirely thermal energy transfers, Q, caused by and measured by temperature differences, the measurements are called calorimetry.

The heat capacity is defined, for energy transferred as thermal energy, as $C = Q/\Delta T$. The value depends on the process conditions and is not defined if a phase change occurs and $\Delta T = 0$. If the volume of the system is held constant, the value is labeled C_v; if pressure is constant C_p. It can be shown that $C_v = \Delta E/\Delta T$ and $C_p = \Delta H/\Delta T$, where H is enthalpy; $H \equiv E + PV$. The ratio of the heat capacity of a substance to the heat capacity of water, or by extension the heat capacity in calories per gram, is called the specific heat of the substance, but the term is sometimes further extended to any value of heat capacity per unit mass.

To avoid the uncertainty in process conditions, it is generally accepted practice to assume, unless otherwise specified, that processes for which Q is measured are carried out reversibly at constant pressure (usually atmospheric pressure) and constant temperature, with no work done except against the atmosphere. (A reversible process is one that may be reversed at any point by an infinitesimal decrease in driving influence or increase in opposing influence.) These "heats" are then equal to the enthalpy changes ΔH and are so listed in tabulations. Also, the "heat" quantities are typically defined such that they are likely to be positive.

Examples include heat of solution, which is the amount of thermal energy given off by a solution to return to the original temperature (usually room temperature) when a solute is added; heat of fusion, which is the amount of thermal energy absorbed by a substance when it melts; heat of vaporization, which is the amount of thermal energy absorbed by a liquid as it vaporizes; and heat of sublimation, the amount of thermal energy absorbed by a solid as it vaporizes, or sublimes. For historical reasons, heats of fusion, vaporization, and sublimation are often called latent heats, a reminder that there is no temperature (i.e., heat) effect evident when the thermal energy is transferred.

"Heat" is also a verb, with ill-defined meaning. It may indicate increasing the temperature of an object by adding thermal energy, as in heating water to the boiling point. It may mean increasing the temperature of an object by adding energy in another form, as in doing work on a gas by compressing the gas. It may mean increasing the temperature of a system without addition of energy, by a conversion of energy to thermal energy within the system, as in overheating of an athlete or a car. Or it may mean the transfer of thermal energy to a system without change of temperature of the system, as when heating melting ice or boiling water on a stove. The term "heat," like the noun, is convenient because it permits reference to a group of processes without requiring specification of the process, but it is therefore not a satisfactory technical term.

See also: CALORIMETRY; CHEMICAL PHYSICS; ELECTRICAL CONDUCTIVITY; ENERGY, CONSERVATION OF; ENERGY, KINETIC; ENERGY, POTENTIAL; EQUIPARTITION THEOREM; FUSION; HEAT, CALORIC THEORY OF; HEAT, MECHANICAL EQUIVALENT OF; HEAT CAPACITY; HEAT PUMP; HEAT TRANSFER; KINETIC THEORY; MOLECULAR SPEED; PHASE; PHASE, CHANGE OF; PHASE TRANSITION; SPECIFIC HEAT; TEMPERATURE; THERMODYNAMICS; THERMOELECTRIC EFFECT

Bibliography

BAUMAN, R. P. *Modern Thermodynamics with Statistical Mechanics* (Macmillan, New York, 1992).

BRUSH, S. G. *The Kind of Motion We Call Heat: A History of the Kinetic Theory of Gases in the 19th Century* (North-Holland, Amsterdam, 1976).

ROBERT P. BAUMAN

HEAT, CALORIC THEORY OF

The term "calorique" used to denote the subtle, weightless, and highly elastic fluid of heat first appeared in print in the *Méthode de Nomenclature Chimique* (1787) in which Antoine Lavoisier, Guyton de Morveau, Claude Louis Berthollet, and Antoine François de Fourcroy laid the foundations of the modern language of chemistry. Thereafter, as the new chemistry and the material theory of heat gained ground, both "calorique" and its English equivalent "caloric" came quickly into common use.

Caloric was one of a number of supposedly weightless or imponderable fluids (the fluids of electricity and magnetism were others) that were increasingly invoked, from the mid-eighteenth century, to account for physical phenomena. However, the immediate roots of the caloric theory lie in discussions of the nature of heat that were conducted in the 1770s in Scotland (by Joseph Black and certain of his pupils) and Paris (at the time when Lavoisier was formulating his new theory of combustion). The theory replaced three views that had divided physicists and chemists earlier in the eighteenth century. One of these was that heat was a vibration of the particles of ordinary, ponderable matter; the second was that it resulted from the motion of a material "fire" that was present in all substances; the third, combining elements of the first two and taught by the influential Herman Boerhaave at Leiden, involved the motion of both ordinary matter and of "fire." In the caloric theory, by contrast, the mere quantity of caloric in a body, rather than any kind of motion, determined temperature. The strength of the theory lay in its explanatory power in areas of inquiry that, at the time, were attracting unprecedented interest. The emerging concept of latent heat, for example, lent itself well to the idea that a "matter of heat" combined, in a quasi-chemical way, with ordinary matter, overcoming the natural forces of cohesion and, thereby, causing a change of state. Even more convincing was the explanation of the heats of chemical reaction. In the caloric theory, it was supposed that the capacity of the products of an exothermic reaction to contain heat was less than that of the substances entering the reaction, so that "excess" heat was expelled, causing a rise in temperature. Similar explanations were used to account for the heat liberated in the percussion of a solid and in the adiabatic compression of a gas; in both of these cases, it was supposed that a reduction in volume led to a reduction in "heat capacity" and, hence, to the liberation of heat.

Despite the ability of the caloric theory to provide plausible qualitative explanations of phenomena of great current interest, it was always recognized that it had weaknesses. One, picked on by Benjamin Thompson, Count Rumford, in 1798, was its unconvincing explanation of the heat of friction. In this case, unlike that of percussion, it was hard to see how mere rubbing could cause a decrease in the volume of a body. Rumford built on this familiar charge against the caloric theory with an investigation of the heat produced in the boring of a cannon. His observation that heat would go on being produced in unlimited quantities so long as boring continued was unremarkable. But a more original part of his case rested on his measurement of the specific heat of the metal chips produced in the boring, which he showed to be the same as that of the bulk metal. According to one version of the caloric theory, that of the Scot William Irvine, the specific heat of the chips should have been lower, since, for Irvine, specific heat was a measure of the capacity of a body to contain heat. If this assumption were made, Rumford's argument against the caloric theory became persuasive. But the fact that not all calorists followed Irvine was just one of several reasons why contemporaries regarded it as inconclusive. Certainly, the later view of John Tyndall, who believed that Rumford's work had served to "annihilate" the caloric theory, is wholly unfounded.

It is a mark of the failure of Rumford's challenge that the caloric theory was more widely accepted between 1800 and 1815 than at any other time in its history. It gained acceptance not only with the new chemistry of Lavoisier but also through the work of John Dalton, whose exposition of the theory in his *New System of Chemical Philosophy* (1810) incorporated caloric in an elaborate view of gas structure, and of Pierre-Simon Laplace, whose physics of short-range forces depended on the existence of caloric and other imponderable fluids. However, by the time one of the most detailed accounts of the theory, that of Laplace in book XII of the fifth volume of his *Traité de Mécanique Céleste*, was published in 1823, the fortunes of the theory were beginning to wane. The growth in support for the wave theory of light, following the work of Thomas Young and Augustin Jean Fresnel, and the gathering evidence of the similarities between radiant heat and light, helped to resurrect the old belief that heat was motion: on this view, temperature was determined by the motion of particles of ordinary matter, while radiant heat became a vibration in the ether that was supposed to pervade the universe.

The growing challenge to caloric after 1815 ushered in a period of agnosticism in which most physicists and chemists were reluctant to commit themselves on the nature of heat. This was not regarded as a significant impediment, since much of the best work in this period did not depend on the truth of any particular theory. Joseph Fourier's *Traité Analytique de la Chaleur* (1822) was a particularly distinguished demonstration that thermal physics could be advanced by an author who eschewed any attempt at explanation. And even when Sadi Carnot used the language of the materiality of heat and the typical calorist principle of the conservation of heat in his *Réflexions sur la Puissance Motrice du Feu* (1824), he was careful to stress the uncertainty of the caloric theory. By the time of his death in 1832, Carnot had rejected caloric, and some two decades later Rudolf Clausius and William Thomson showed how the essential argument of the *Réflexions* could be adapted to the new principle of the conservation of energy.

Once it was accepted that, in an ideal Carnot engine, some heat had to be consumed in order to produce work and that the heat lost was related to the work by the mechanical equivalent of heat, belief in caloric was no longer tenable. Such support as the theory still had in the mid-nineteenth century quickly disappeared. By then, however, calorists were few in number, and, almost without exception, physicists looked back on the mathematical treatments of caloric by Laplace and, more recently, by Siméon Poisson (in his *Théorie Mathématique de la Chaleur* of 1835) as relics of a theory that had long outlived its usefulness.

See also: CARNOT, NICOLAS-LÉONARD-SADI; CLAUSIUS, RUDOLF JULIUS EMMANUEL; HEAT, MECHANICAL EQUIVALENT OF; KELVIN, LORD; LAPLACE, PIERRE-SIMON; THERMODYNAMICS, HISTORY OF

Bibliography

CANTOR, G. J. N., and HODGE, M. J. S., eds. *Conceptions of Ether: Studies in the History of Ether Theories, 1740–1900* (Cambridge University Press, Cambridge, Eng., 1981).

FOX, R. *The Caloric Theory of Gases from Lavoisier to Regnault* (Clarendon, Oxford, Eng., 1971).

GUERLAC, H. "Chemistry as a Branch of Physics: Laplace's Collaboration with Lavoisier." *Historical Studies in the Physical Sciences* **7**, 193–276 (1976).

MCKIE, D., and HEATHCOTE, N. H. DE V. *The Discovery of Specific and Latent Heats* (Edward Arnold, London, 1935).

ROBERT FOX

HEAT, MECHANICAL EQUIVALENT OF

Julius R. Mayer, a German physician, introduced the concept that heat is a form of energy. In 1842 he wrote, "The warming of a given weight of water from 0° to 1° degree Centigrade corresponds to the fall of an equal weight from a height of about 365 meters." It was James P. Joule (the son of an English brewer), however, who performed the careful experimental work that gave certainty to the fact that heat was indeed a form of energy, and therefore a unit of heat, expressed in the British thermal unit (BTU), could be related to the accepted unit of energy, which in Joule's day was the foot-pound. Joule devised an apparatus that, in a sense followed Mayer's prescription. Using a carefully insulated copper vessel, he built a device that allowed a set of brass paddles to be turned in a quantity of water as weights fell at a constant rate through a known distance. The viscous resistance of the water was sufficient to provide the balancing force. In his 1850 paper in the *Philosophical Transactions of the Royal Society* (Vol. 140, p. 61), he gave details not only of the equipment but of the particular set of experiments he carried out. After dropping the weights twenty times through a distance of 63 in., the temperature of the water rose, an effect due strictly to the work done on the water through friction (see Fig. 1). The temperature rises were on the order of one-half degree Fahrenheit. Then, taking into account the specific heats of water, copper, and brass, Joule calculated the me-

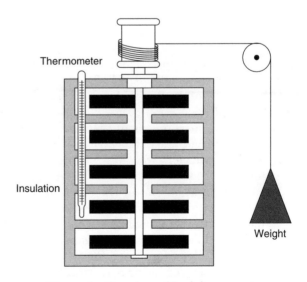

Figure 1 Drawing of Joule's apparatus.

chanical equivalent of heat as "that quantity of heat capable of increasing the temperature of a pound of water . . . by 1° Fahr [i.e., one BTU], requires for its evolution the expenditure of a mechanical force [i.e., energy] represented by the fall of 772 lbs through a space of one foot." The accepted value today makes one BTU equivalent to 778.26 ft lb.

Joule performed further experiments, one using mercury rather than water in his equipment, and a second where he used the sliding friction of cast iron on itself, obtaining 774 and 775 ft lb, respectively, for the mechanical equivalent of 1 BTU.

Mayer's investigations into the equivalence of heat as energy began during his tenure as ship's doctor when, in Java, he noticed the bright red color of his patients' blood, which he attributed to a high level of oxygen in the bloodstream. This physiological difference from cold-climate, much darker, blood set his mind to work. Out of his musings he considered heat related to the kinetic energy of the constituent atoms of the material. This concept was similar to the conclusions reached by Benjamin Thompson, who was also known as Count Rumford, when he observed the high temperatures of the shavings produced during the drilling out of a cannon's bore. Mayer was convinced that when energy was changed from a kinetic form to either a potential or a heat form, the total amount of energy remained a constant—that is, energy is conserved. Unfortunately his writings and ideas were not given much credence until near the end of his lifetime; eventually he was awarded the Copley Medal of the Royal Society of London in 1871. Similarly, because he had no formal training, Joule's early efforts for publication of his work on the mechanical equivalent of heat bore scant attention until he was "discovered" by William Thompson (Lord Kelvin) with whom he then began to collaborate. Like Mayer, Joule received the Copley Medal (1866). He was also elected president of the British Association for the Advancement of Science.

An interesting anecdote about Joule, which may be true, is that he took one of his carefully made and calibrated thermometers with him on his honeymoon. Joule's purpose was to measure the temperatures at the top and bottom of the waterfall near their vacation site. It is not recorded whether he was able to measure any change and thereby get another value for the mechanical equivalent of heat.

A relatively simple experiment to measure such a change in temperature may be performed by placing some lead shot in a meter-long mailing tube, closing both ends with corks. Hold the apparatus vertically and measure the temperature of the shot (drill a small hole in one cork, push a thermometer through so the bulb is buried in the shot, remove the thermometer, and close up the hole with a small cork). Raise the tube vertically the length of the tube. Invert the tube quickly so the shot can fall vertically. Raise the tube and invert twenty to forty times. The gravitational potential energy gained by the shot through your work on it is transformed into kinetic energy as it falls. Then the external kinetic energy of motion of the shot is transformed into internal kinetic energy—or heat—increasing its temperature.

Heating water by the passage of electric current through a resistor embedded in the water is a modern technique for measuring the mechanical equivalent of heat, even though one of Joule's early experiments was done precisely this way.

Modern measurements for energy are usually given in joules (approximately 0.1 kg dropped 1 m or, exactly, 1 N of force acting through a distance of 1 m) or electron volts (one electron dropped through a potential difference of $1 \text{ V} = 1.6 \times 10^{-19}$ J). Heat is measured in calories (the energy needed to raise the temperature of 1 g of water 1°C) or BTUs. Incidentally, the "calorie" listed on a food package is actually a *kilo*calorie, the energy needed to raise the temperature of 1 kg of water 1°C. The present-day mechanical equivalent of heat is given by the relation 4.185 cal and is equivalent to 1 J.

See also: ENERGY; ENERGY, KINETIC; ENERGY, POTENTIAL; HEAT; HEAT, CALORIC THEORY OF; KELVIN, LORD; TEMPERATURE

Bibliography

MAGIE, W. F. *A Source Book in Physics* (Harvard University Press, Cambridge, MA, 1969).

WEAVER, J. H. *The World of Physics* (Simon & Schuster, New York, 1987).

EDWARD J. FINN

HEAT, SPECIFIC

See SPECIFIC HEAT

HEAT CAPACITY

The heat capacity of a substance is a measure of the heat input required to increase the temperature of the sample. In the absence of phase transitions, a heat input ΔQ to a sample results in a temperature change ΔT of the sample. The average heat capacity of the sample over ΔT is defined as the ratio $\Delta Q/\Delta T$, and in the limit as ΔT goes to zero, this ratio is the heat capacity of the sample.

The specific heat is defined as the heat capacity per unit mass, while the molar heat capacity is the heat capacity per mol. The term "mol" refers to the number of atoms in the solid, where one mole is equivalent to Avogadro's number of atoms. At temperatures close to room temperature, the molar heat capacity of most solids is close to 25 J/K·mol. This value can be explained by assuming that all of the atoms in the solid vibrate like classical harmonic oscillators, with energies $3k_B T$ (25T J). In fact, the room temperature value of the heat capacity of a solid generally reflects the number of atoms in the sample, while values of the specific heat correspond to the molar density of samples. For instance, the molar heat capacities at room temperature of Al and Hf are similar, 24.2 J/mol·K and 25.7 J/K·mol. The specific heat of Al is 898 J/K·kg, more than six times greater than the specific heat of Hf, which is 144 J/K·kg. It is not surprising that Al is often used in the construction of thermal baths in calorimeters, as this material can absorb more heat without changing temperature beyond a set limit.

The heat capacities of solids generally approach zero value at temperatures approaching absolute zero. This observation reflects the true quantum nature of the oscillations of the atoms in a solid. Instead of a continuous energy spectrum ($3k_B T$, with all $T \geq 0$), only discrete, quantized values of the vibration energy of atoms occur, $E_n = \hbar\omega(n + \frac{1}{2})$, where n is any integer greater than or equal to zero and $\hbar$ is Planck's constant. Thus, this theory provides a minimum excitation energy for the vibrating atoms in a solid, and as the average energy (i.e., temperature) of the atoms in a solid decreases, it becomes less and less likely that an individual oscillator will be excited ($n \neq 0$). Thus, quantum mechanics predicts that the heat capacity of a solid goes to zero at low temperatures, in agreement with experiment. Albert Einstein first advanced this theory, and his prediction, along with a later, more sophisticated

model due to Peter Debye, shows good agreement with experimental heat capacity data.

Heat capacity measurements are a probe of the energy states of substances. They contribute to an understanding of the physical phenomena associated with these states, such as superconductivity or magnetism. Heat capacity measurements are sensitive indicators of phase transitions, such as melting or changes in magnetic order.

See also: QUANTUM MECHANICS; SPECIFIC HEAT; SPECIFIC HEAT, EINSTEIN THEORY OF; THERMODYNAMICS

Bibliography

CEZAIRLIYAN, A. *Specific Heat of Solids* (Hemisphere Publishing, New York, 1988).

KITTEL, C. *Introduction to Solid-State Physics*, 6th ed. (Wiley, New York, 1986).

PITZER, K. S., and BREWER, L. *Thermodynamics*, 2nd ed. (McGraw-Hill, New York, 1961).

ZEMANSKY, M. W. *Heat and Thermodynamics*, 4th ed. (McGraw-Hill, New York, 1957).

ERIC J. COTTS

HEAT ENGINE

Mechanical energy may be quantitatively converted to thermal energy, for example by friction or by compression of a gas, and thermal energy may be quantitatively converted to mechanical energy by a process such as isothermal (constant temperature) or reversible adiabatic expansion of a gas. (An adiabatic process involves no transfer of thermal energy; a reversible process is one that may be reversed at any point by an infinitesimal decrease in driving influence or increase in opposing influence.) However, it is not possible to convert thermal energy totally to mechanical energy in a process at constant temperature of system and surroundings, or in a cyclic process—a process in which the system returns to its original state.

A heat engine is a device for converting thermal energy to mechanical energy [or heat to work, in the vernacular, although heat (Q) and work (W) are mutually exclusive modes of energy transfer, so it is not literally possible to convert heat to work].

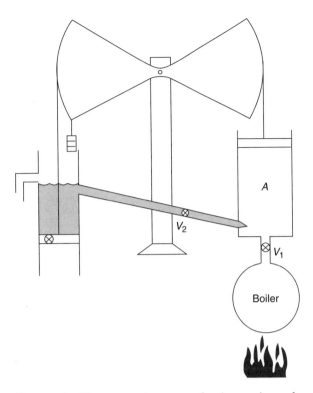

Figure 1 Newcomen's atmospheric engine, designed in 1705, fed steam at atmospheric pressure into the cylinder *A*, raising the piston and lowering the pump piston. Cool water sprayed into the cylinder condensed the steam, lowering the pressure and bringing down the piston, which raised the pump piston and water.

Heat engines were developed at least 2,000 years ago, but practical engines are largely a product of the eighteenth century and later. Efforts to understand the operation of heat engines provided the primary incentive for the development of the science of thermodynamics (i.e., heat and work).

Thomas Savery's steam pump (1698) allowed steam from a boiler to enter a chamber that was subsequently cooled, producing a vacuum that could lift water from a well or mine. It was soon replaced by Thomas Newcomen's engine (1705), which functioned similarly except that the expansion-condensation chamber was a cylinder with a piston connected by a lever to a piston that lifted water (see Fig. 1). The design was substantially improved by James Watt between 1763 and 1782. He added a condensation chamber separate from the expansion cylinder, keeping the hot parts of the engine hot and the cold parts (relatively) cold, thereby improv-

ing the efficiency of the engine. Watt also introduced the double-acting steam engine, in which steam first drove the piston in one direction, then was introduced at the opposite end of the cylinder to drive the piston back (see Fig. 2). A mechanism that transformed the linear motion of the piston to circular motion increased the applicability of the engines.

Regardless of the details of the cyclic process, to produce mechanical energy, thermal energy must be put into the system at some "high" temperature (typically heating of a boiler or combustion of gasoline), and some fraction of this energy must be discharged, at some lower temperature, into cooler surroundings. The principle was analyzed by Nicholas-Léonard-Sadi Carnot, who described a heat engine operating through a reversible isothermal expansion at the high temperature, an adiabatic reversible expansion, an isothermal compression at the low temperature, and adiabatic compression. This is called the Carnot cycle.

The Stirling engine (1816) employs a noncondensable gas, such as air. The Stirling cycle involves (1) an isothermal expansion at the upper temperature, with absorption of thermal energy; (2) an isochoric (constant volume) transfer through a regenerator, giving up thermal energy to the regenerator; (3) an isothermal compression at the lower

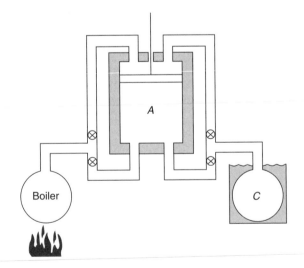

Figure 2 Watt's double-acting engine (1785). Steam from the boiler is alternately fed to the top or bottom of the well-insulated cylinder *A*, while the opposite end is connected to the cool condenser *C*.

temperature, giving off thermal energy; and (4) an isochoric transfer through the regenerator, absorbing thermal energy from the regenerator. The cycle has the same theoretical efficiency as the Carnot cycle, but it involves nonlinear linkages between two pistons.

The Otto cycle approximates the behavior of internal combustion engines, such as the gasoline engine. An adiabatic compression step is followed by heating at constant volume (combustion step), then by an adiabatic expansion and by cooling at constant volume.

The Diesel cycle is a variation in which air is compressed adiabatically to reach a high temperature, then fuel is injected. The fuel ignites spontaneously (without a spark), causing expansion during the power stroke. Exhaust gases are then replaced with fresh air. The heating step (combustion) ideally occurs at constant pressure. Diesel engines work best in larger sizes, with high compression ratios.

The Rankine cycle, for condensable vapors, involves four steps: (1) compression of a liquid; (2) warming and vaporizing the liquid; (3) adiabatic expansion and cooling of the vapor; and (4) condensation of the vapor.

A heat engine operated in reverse moves thermal energy from a lower temperature to a higher temperature, requiring work input to the engine. They are then called heat pumps, or refrigerators. The operating cycle of most heat pumps approximates a reversed Rankine cycle.

The efficiency of any heat engine is the net mechanical energy output (work, W), divided by the thermal energy input (Q_H) at the upper temperature. The maximum possible efficiency of an engine operating between T_H and T_C is

$$\varepsilon = \frac{W}{Q_H} = \frac{T_H - T_C}{T_H} = 1 - \frac{\Delta T}{T_H},$$

which is known as Carnot's theorem. The temperatures must be expressed on an absolute scale, usually the Kelvin scale. The Carnot efficiency is the efficiency for any reversible heat engine operating between the two fixed temperatures, T_H and T_C. It is independent of the working fluid.

The highest theoretical efficiencies are limited to those cycles in which the thermal energy transfers occur only at T_H and at T_C. In contrast, the Otto cycle requires the addition of thermal energy while the temperature is changing, and removal while the

temperature is dropping. Even Carnot efficiencies generally appear to be low. For example, an engine operating between room temperature (25°C = 298 K) and boiling water temperature (100°C = 373 K) has a Carnot efficiency of

$$\varepsilon = \frac{\Delta T}{T_H} = \frac{75}{373} = 27\%,$$

and the actual efficiency would be less.

The Carnot efficiency could only be achieved if the process were carried out reversibly, and thus infinitely slowly, with no power output; the work done per unit of time would be zero. To obtain thermal energy transfer (heat flow), there must be a difference between the temperatures of the heat reservoirs, T_H and T_C, and the high and low temperatures of the system, T'_H and T'_C. Approximate analyses of real engines have been carried out by Curzon and Ahlborn and others, taking into account the actual temperature range over which thermal energy is transferred. They found that a satisfactory approximation, for many purposes, is the equation

$$\varepsilon = \frac{W}{Q_H} = \frac{\sqrt{T_H} - \sqrt{T_C}}{\sqrt{T_H}}.$$

For example, the highest practical efficiency for any heat engine operating between 25°C and 100°C would be approximately

$$\varepsilon = \frac{\sqrt{373} - \sqrt{298}}{\sqrt{373}} = 11\%.$$

Higher energy conversion efficiencies may be obtained if there is no step in which the energy appears in randomized form, as thermal energy. Electrochemical cells and fuel cells, electric motors, and photoelectric devices are not limited to Carnot efficiencies.

In practice, the operation of many heat engines is noncyclic, with material injected at one point and discarded at another. This may affect the measured efficiencies.

See also: ADIABATIC PROCESS; CARNOT CYCLE; ENGINE, EFFICIENCY OF; HEAT; HEAT PUMP; HEAT TRANSFER; ISOCHORIC PROCESS; ISOTHERMAL PROCESS; REFRIGERATION; THERMODYNAMICS

Bibliography

BAUMAN, R. P. *Modern Thermodynamics with Statistical Mechanics* (Macmillan, New York, 1992).

JONES, J. B., and HAWKINS, G. A. *Engineering Thermodynamics* (Wiley, New York, 1960).

SANDFORT, J. F. *Heat Engines* (Doubleday, New York, 1962).

ROBERT P. BAUMAN

HEAT PUMP

Thermal energy flows naturally from high-temperature regions to low-temperature regions; for example, from the red-hot end of an iron rod to the cool end, or from a warm room in winter to the cold out-of-doors. Thermal energy from the room passes through the walls to the exterior, from warm to cold. A heat pump forces thermal energy to flow in the opposite direction, the "unnatural" direction: it pumps thermal energy from cold to hot.

According to this cold-to-hot definition, refrigerators and air conditioners qualify as "heat pumps" because they pump thermal energy from the cold interior of the refrigerator or room (shaded area in Fig. 1) to the warm exterior. Although they heat the exterior, the real purpose is to keep the interior cool. Commonly, however, the term "heat pump" applies to those devices that pump thermal energy in order to heat a space rather than to cool it. For example, a residential heat pump transfers thermal energy from the cold air outside to the interior of the house in order to warm the house: now, the shaded area in Fig. 1 is the outside of the house. In the summer, it operates in the opposite direction so it becomes an ordinary air conditioner, cooling the inside and heating the outside.

Heat pumps are an efficient way to use electrical energy for heating. A fixed amount of electrical energy produces more heating if used to run a heat pump than if the same energy is used to heat with resistance heaters such as baseboard strip heaters. Baseboard strip heaters convert electrical energy directly into thermal energy by sending electrical current through a resistive wire made of nichrome or a similar material. The electrical energy is completely converted into thermal energy, causing the wire to heat up. Since all of the electrical energy goes into heating the house, in a sense baseboard heaters are 100 percent efficient. However, heat pumps produce several times as much heating for the same amount of electrical energy, the exact amount depending on the interior and exterior temperatures.

Heat pumps are a form of "heat engine," a term that includes refrigerators, car engines, and any device that moves thermal energy between a low temperature T_L and and a high temperature T_H, and does work or has work done on it. A car engine and its cooling system transfers thermal energy from the hot gases in the cylinders to the cool atmosphere, and the engine creates work to move the car. In a refrigerator an electric motor does work to operate the compressor, which moves thermal energy from inside the refrigerator to the warm room air.

A common measure of the performance of any heat engine is the ratio

$$\frac{\text{desired energy output}}{\text{required energy input}}.$$

For a heat engine like a car engine, where the goal is to do work W and the energy input is the energy of the fuel, efficiency is defined as

$$\varepsilon = \frac{W}{Q_H},$$

where Q_H is the thermal energy extracted from the high-temperature region. For the car engine, this is the hot cylinder gases, whose thermal energy comes from the fuel. Higher efficiencies are desirable be-

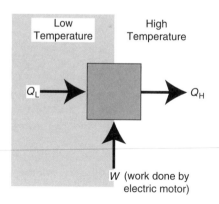

Figure 1 Heat pump.

cause more work is done for a given amount of energy input.

For heat pumps, performance is measured by the coefficient of performance (COP), which equals the ratio of the heat Q_H pumped into the warm room (the desired effect) to the work-energy W needed to operate the heat pump:

$$\text{COP} = \frac{Q_H}{W}.$$

Large COPs are desirable because they mean that more heat is pumped into the room for a given amount of work expended.

The first law of thermodynamics (conservation of energy) requires that $Q_H = Q_L + W$ in Fig. 1. This law, along with the second law of thermodynamics (the law of entropy), means that the COP cannot be larger than

$$\text{COP}_{max} = \frac{T_H}{T_H - T_L},$$

where the temperatures are absolute temperatures such as Kelvin (K).

When the two temperatures are nearly the same, the denominator is small and COP_{max} is large. For a room temperature of 20°C (68°F or 293 K) and an outside temperature of freezing (0°C = 273 K), $\text{COP}_{max} = 15$. For these conditions, it is theoretically possible to have 15 times as much heating effect from a heat pump as from an electrical resistance heater for the same amount of electrical energy. Due to frictional and other losses, the actual COP is smaller, typically in the range of 2.5 to 3.5.

As the outside temperature decreases, the COP also decreases, so that at 0°F (−18°C) COP_{max} is 8, and the actual COP is about 2. This is still good compared with resistance heating, but is not especially favorable when compared with the cost of gas heat. Therefore, heat pumps are used more in the southern United States where outside temperatures are higher during the winter. Heat pump systems often have auxiliary resistance heaters to handle extremely cold conditions, when their heat output decreases.

If the compressor is inside the house, part of the thermal energy produced by a heat pump comes from the energy used by the motor that runs the compressor. This energy ends up as thermal energy that can be used to heat the house, and this equals the thermal energy output of a resistance heater for the same amount of electrical energy. But, in addition, there is thermal energy in the air outside the house, and the heat pump uses the electrical current to pump outside thermal energy into the room. Therefore, a heat pump can easily out-perform resistance heating in terms of energy consumption. In many cases, the compressor is in a separate unit outside the house so the "compressor heat" dissipated into the outside air and is unavailable to heat the house.

Water-assisted heat pumps are designed to maintain a large COP even on very cold winter days. One type uses well water as the low-temperature source and relies on the fact that the temperature of well water does not vary much even during cold weather. By running the heat pump between well water at 50°F (10°C = 283 K) and the inside of the house at 68°F (20°C = 293 K), COP_{max} is 29.3, although COPs of real heat pumps at these temperatures are only 4 to 5.

See also: CONSERVATION LAWS; ENTHALPY; ENTROPY; HEAT; THERMODYNAMICS

Bibliography

GOLDSTEIN, M., and GOLDSTEIN, I. F. *The Refrigerator and the Universe* (Harvard University Press, Cambridge, MA, 1993).

PITA, E. G. *Air Conditioning Principles and Systems: An Energy Approach* (Wiley, New York, 1981).

LAWRENCE A. COLEMAN

HEAT TRANSFER

Energy that is transferred from one location to another due solely to a difference in temperature is referred to as heat. There are three mechanisms by which heat is transferred: Conduction, Convection, and radiation.

Conduction

If one end of a rod is held in a flame, after some time a rise in temperature will be observed at the other end of the rod due to the conduction of heat through the rod. In general, conduction is the slowest of the mechanisms of heat transfer. On the atomic scale, the oscillation of the atoms at the end of the rod placed in the flame increase in amplitude when the temperature of the rod rises. Part of the energy of these atoms is transferred to their cooler neighboring atoms down the rod by the forces between the atoms, increasing the oscillations of the neighboring atoms. The same process transfers energy to the next atoms along the rod and so on until the energy reaches the other end of the rod. Although there is no net movement of material in the rod, the heat is transferred from one end to the other.

Experiments have shown that the factors governing conduction of heat are the cross-sectional area A, the length L, and the temperature difference between the ends of the rod. The rate at which heat (Q) is transferred through the rod (in joules/second = watts) is given by

$$\frac{\Delta Q}{\Delta t} = kA\frac{T_{\text{hot}} - T_{\text{cold}}}{L}.$$

The constant of proportionality k is the thermal conductivity of the material. Materials with high thermal conductivities conduct heat rapidly. Most metals are good conductors of heat. Cooking pots are often made with copper bottoms because the thermal conductivity of copper is nearly twice that of aluminum and nearly eight times that of steel. Materials with low thermal conductivities, such as polyurethane foam and fiberglass, conduct heat poorly and are often used as insulation. Thermal insulators such as these have a low thermal conductivity in great part because of the low conductivity of the dead air spaces contained within them. An interesting example of insulation is the protective tile used on the exterior of the space shuttle. These tiles have such an unusually low thermal conductivity and low heat capacity that a tile glowing hot in the center can be held by the edges using bare hands.

A useful quantity related to thermal conductivity is the R value of insulation, defined by

$$R = \frac{L}{k},$$

where L is the thickness and k is the thermal conductivity of the insulation. R values are useful in the calculation of heat load to buildings because the R value of two layers of insulation is simply the sum of the R values of each layer. Finding the effective thermal conductivity of the two layers is a more difficult calculation.

Convection

Convection is the transport of heat by the mass motion of material within a liquid or gas. It is the second fastest mechanism of heat transport. The basis for convection is that warmer fluid expands, becoming less dense, and rises; cooler fluid, being more dense, falls. This motion sets up convection currents that transport heat throughout all regions of the fluid. Convection is easily observed in a pot of heating water by placing a few drops of food coloring into the water; the food coloring will be observed to follow the convection currents in the water. If the water were unable to circulate in convection currents and one was forced to rely solely on conduction to heat a pot of water, a much longer time would be required to boil water.

Atmospheric convection plays an important role in determining weather patterns, while convection in the oceans transports heat over long distances. The land next to the ocean gets hot in the day and cool at night, while the ocean remains at an almost constant temperature. When the heated air over the land rises in the daytime or the cooled air over the land falls at night, convection currents are created. These are the sea breezes familiar to anyone who has visited the beach; very rarely will the air be completely still at the seashore. The heat transported by the Gulf Stream from tropical latitudes to Great Britain is responsible for the more moderate climate there than at similar latitudes elsewhere.

Forced convection of the body's blood supply carries heat from the interior of the body to its surface. Controlling this blood flow regulates the body's temperature. Air next to the body is warmed and rises, setting up convection currents that help carry away body heat. This effect is enhanced by any wind that is present (the familiar wind-chill factor.) Note that evaporation of perspiration is also very important in transferring heat from the skin to the air. In an automobile, forced convection of the coolant carries heat out of the engine to the radiator to help cool the engine.

Radiation

When people step from shade to direct sunlight, they feel warmth from the sunlight. This warmth is due to the transport of heat by radiation from the Sun to their skin. The heat takes the form of electromagnetic waves such as infrared light, visible light, and ultraviolet light. This energy is transported through the vacuum of space and does not require the motion of any material or atoms (as opposed to conduction and convection.) Radiation is the fastest method of heat transport. All objects are continuously radiating and absorbing heat in the form of electromagnetic waves. In the absence of other factors, an object cools if it is radiating more heat than it absorbs and warms if it radiates less heat than it absorbs. Thus, the surface of Earth warms in the daytime when more energy is received from the Sun than is radiated away by Earth, and cools at night when energy is radiated by Earth and none is absorbed from the Sun. Cloudy nights are generally warmer than clear nights because some of Earth's radiation is absorbed by clouds and partially reradiated back to the surface.

The rate at which heat (Q) is radiated by an object with surface area A, at temperature T (in kelvins), is described by the Stefan–Boltzmann law:

$$\frac{\Delta Q}{\Delta t} = \sigma e A T^4.$$

The constant e is the emissivity of the object, which is affected by the roughness or smoothness of the surface and by the color of the object. $\sigma = 5.6705 \times 10^{-8}$ W/m^{2}K^4 is a fundamental physical constant called the Stefan–Boltzmann constant. The wavelength of the emitted radiation is dependent on the temperature, reaching the visible portion of the spectrum when very hot.

See also: CONDUCTION; CONVECTION; ELECTROMAGNETIC WAVE; HEAT CAPACITY; STEFAN–BOLTZMANN LAW

Bibliography

HEWITT, P. G. *Conceptual Physics,* 7th ed. (HarperCollins, New York, 1993).

JONES, E. R., and CHILDERS, R. L. *Contemporary College Physics,* 2nd ed. (Addison-Wesley, Reading, MA, 1993).

OHANIAN, H. C. *Principles of Physics,* (W. W. Norton, New York, 1994).

YOUNG, H. D. *University Physics,* 8th ed. (Addison-Wesley, Reading, MA, 1992).

JOHN C. RILEY

HEISENBERG, WERNER KARL

b. Würzburg, Germany, December 5, 1901; *d.* Munich, Germany, February 1, 1976; *quantum theory, nuclear physics.*

Heisenberg was the younger son of August and Anna Wecklein Heisenberg. The family belonged to the academic upper middle class of Wilhelmian Germany. Heisenberg's father, an authority on the Byzantine empire, taught classical languages at a Würzburg gymnasium (high school) and became professor of Greek philology at the University of Munich. His mother was the daughter of a Munich gymnasium principal. His brother, Erwin, became a chemist in Berlin. In 1937 Heisenberg married Elisabeth Schumacher, the daughter of a noted Berlin professor of economics. They had seven children.

Heisenberg attended primary schools in Würzburg and Munich before entering his grandfather's gymnasium in 1911. Distinguishing himself in mathematics and classical piano, he graduated at the top of his class and entered the University of Munich in 1920. There he studied theoretical atomic physics and hydrodynamics with Arnold Sommerfeld. After receiving his doctorate in the record time of three years, Heisenberg served as a postdoctoral assistant to Max Born in Göttingen and Niels Bohr in Copenhagen. In 1927 he was appointed Professor of Theoretical Physics at the University of Leipzig, at that time Germany's youngest full professor. In 1942 Heisenberg assumed directorship of the Kaiser Wilhelm Institute for Physics in Berlin, a government-sponsored research institute. He remained with the institute thereafter.

Heisenberg is best known for his contributions to quantum mechanics, the new physics of the atom and its interactions with light and other atoms. As a leading member of the small group of mainly European young men who created quantum mechanics during the 1920s, Heisenberg made the initial breakthrough in the field. He relied on laboratory data about the atom to reinterpret the equations of the motion of electrons in atoms as quantum expressions. This led to the use of mathematical entities known as matrices; his new quantum mechanics is still called matrix mechanics. The uniting of matrix mechanics with the alternative wave mechanics developed by Erwin Schrödinger, as well as the introduction of electron spin, resulted in modern quantum mechanics. Heisenberg received the 1932 Nobel Prize for physics for his work on quantum mechanics and its applications to hydrogen molecules.

However successful, the new quantum mechanics of atomic events still required an interpretation or a set of rules for linking the everyday world of the laboratory with the strange world of the atom. In 1927 Heisenberg presented the uncertainty, or indeterminacy, principle that, together with Bohr's complementarity principle, formed the so-called Copenhagen interpretation of quantum mechanics. Although debated ever since, it remains the dominant interpretation.

As elaborated by Bohr, the Copenhagen interpretation relies upon the wave–particle duality—the notion that, under certain circumstances, particles can behave as waves and waves as particles. Quantum objects exist as both waves and particles until observed in a laboratory. The act of observation involves choosing one side of the duality, thereby disturbing nature in such a way that Heisenberg's uncertainty principle comes into play. According to this principle, the position and speed (or momentum) of a quantum object at a given instant cannot be measured simultaneously with absolute precision. The more precise the measurement of one variable, the more imprecise, or uncertain, is the other. The same held for the variables of energy and time. Among the many consequences of this principle is the renunciation of strict causality, the exact determination of the future on the basis of the present. Since we cannot measure all of the mechanical variables of a quantum object at a given time with absolute precision, we cannot determine its future motion with absolute certainty; we can make only statistical predictions about its probable future motions. As Born showed, the probabilities can be derived from the wave functions of Schrödinger's wave mechanics. A number of physicists, most notably Albert Einstein, objected to the introduction of probabilities and statistics into the foundations of physics. Although Einstein insisted that nature is not statistical, which he expressed by his famous statement that God "does not play dice," he did not succeed in refuting the new physics.

Following the completion of quantum mechanics, Heisenberg worked with Wolfgang Pauli and others to obtain a relativistic form of quantum mechanics for application to high-energy particles and fields. After the discovery of the neutron in 1932, Heisenberg developed the first neutron-proton theory of the nucleus. With the discovery of nuclear forces, Heisenberg intensified his work on relativistic quantum field theories. To the end of his life he searched for a unified theory of elementary particles based upon a unification of all forces, or fields.

Coming of age at the end of World War I, Heisenberg directly experienced the social upheavals of the postwar period in Germany. He became an ardent follower of the German youth movement, developing an unbreakable attachment to Germany. After Adolf Hitler's rise to power in Germany, Heisenberg suffered many indignities but convinced himself that he could help German physics best by remaining at his post. His reactions to the Nazis were typical of the non-Jewish cultural elite of Germany. After the discovery of nuclear fission and the outbreak of World War II, Heisenberg became a leader in German development of nuclear fission. Moving to Berlin in 1942, he headed the main German effort aimed at attaining a sustained chain reaction and, at first, an atomic bomb. The project did not achieve either goal. Although never a Nazi, he traveled as a German cultural representative to Nazi-occupied countries and maintained direct access to important government figures. He was greatly criticized after the war for his wartime activities. His aims and motives are still the subject of debate. After the war, Heisenberg again became a leading figure in German science policy, attempting to re-establish international relations and successfully arguing for a West German nuclear reactor program and against West German access to nuclear weapons. He also traveled widely, speaking frequently on the philosophical and cultural significance of quantum mechanics.

See also: BORN, MAX; COMPLEMENTARITY PRINCIPLE; MATRIX MECHANICS; QUANTUM MECHANICS, CREATION OF; UNCERTAINTY PRINCIPLE; WAVE–PARTICLE DUALITY, HISTORY OF

Bibliography

BEYERCHEN, A. *Scientists Under Hitler: Politics and the Physics Community* (Yale University Press, New Haven, CT, 1977).

CASSIDY, D. C. "Heisenberg, Uncertainty, and the Quantum Revolution." *Sci. Am.* **266,** 106–112 (1992).

CASSIDY, D. C. *Uncertainty: The Life and Science of Werner Heisenberg* (W. H. Freeman, New York, 1992).

HEISENBERG, W. *Physics and Philosophy: The Revolution in Modern Science* (Harper & Row, New York, 1958).

HEISENBERG, W. *The Physicist's Conception of Nature,* trans. by A. J. Pomerans (Greenwood, Westport, CT, 1970).

HEISENBERG, W. *Physics and Beyond: Encounters and Conversations,* trans. by A. J. Pomerans (Harper & Row, New York, 1971).

HEISENBERG, W. *Across the Frontiers,* trans. by P. Heath (Harper & Row, New York, 1974).

JAMMER, M. *The Conceptual Development of Quantum Mechanics* (McGraw-Hill, New York, 1966).

WALKER, M. *German National Socialism and the Quest for Nuclear Power, 1939–1949* (Cambridge University Press, New York, 1989).

DAVID C. CASSIDY

HEISENBERG UNCERTAINTY PRINCIPLE

See UNCERTAINTY PRINCIPLE

HELMHOLTZ, HERMANN L. F. VON

b. Potsdam, Germany, August 31, 1821; *d.* Berlin, Germany, September 8, 1894; *thermodynamics, hydrodynamics, electrodynamics.*

Helmholtz was born just outside Berlin, the son of a secondary school teacher. He studied medicine and served briefly as an army doctor, then taught anatomy and physiology at the universities of Königsberg, Bonn, and Heidelberg. Easily the most versatile scientist of his age, he formulated the principle of the conservation of energy, invented the ophthalmoscope, laid the foundations of modern physiological optics and acoustics, and produced many elegant popular lectures on scientific and philosophical topics. In 1871, amid great political fanfare, Prussia appointed the former physiologist to the prestigious chair of physics at the University of Berlin. In 1887, as the doyen of German natural science, he was named president of the newly created Imperial Institute for Physics and Technology.

Throughout his career Helmholtz sought the underlying principles that unify and govern natural phenomena. He wrote on hydrodynamic theory and on the energetics of reaction chemistry, and near the end of his career worked to found scientific laws on the principle of least action and reconcile the second law of thermodynamics with the laws of mechanics. The roots of all these contributions, however, lay in his epic formulation of the principle of the conservation of energy in 1847.

Physicists had traditionally described the living force (*vis viva*) of a system of moving bodies as the sum of the components' individual masses times the square of their velocities. Because perpetual motion is impossible, the living force can be increased only if work is done on the system. However, engineering tradition agreed that in real systems living force could be *lost,* through friction or inelastic collisions. Another, more mathematical tradition of analytic mechanics described mechanical systems through mathematical functions known as potentials. Potentials were functions of the spatial position of the objects in the system, and they were known to possess a fixed value for every state of the system, no matter how the state had moved or changed in reaching that state. Sadi Carnot had used this remarkable property indirectly in his famous analysis of heat engines working through closed cycles. In general, however, this tradition attributed no physical significance to the potential, only an abstract, mathematical usefulness.

Helmholtz synthesized these approaches in 1847. He wrote that changes in the living force of a system must be compensated by changes in a mathematical entity possessing the properties both of the work and the potential. He called that entity the tension force in the system. The tension force represented for Helmholtz the total amount of force that is stored in a system and is available to do work in increasing the living force. Living forces and tension forces are interconvertible, but their sum must be constant. Within a few years physicists had renamed Helmholtz's tension force the potential energy of a system; the living force became the kinetic energy; and their sum was the total energy of the system.

Helmholtz recognized that most of the phenomena that interest physicists cannot be analyzed in practice directly in terms of the motions of their ultimate parts. He devoted most of his 1847 paper to discussing how his conservation principle might be applied to apparently non-mechanical phenomena such as optics, frictional electricity, electromagnetism, electric circuits containing voltaic batteries or thermoelectric sources, and heat and the expansion of gases. He defended the mechanical theory of heat, arguing that living force apparently lost through friction or inelastic collisions is actually conserved as the vibratory kinetic energy of molecules (free heat) or is converted to tension force. He also discussed the existing experimental data, much of it from James Joule, pointing to fixed conversion equivalents among different forms of energy. Helmholtz also applied energy conservation to living organisms. He argued that animals convert the tension force stored chemically in foodstuffs to heat (in the form of body heat) and work (as muscular exertion) and that they have no other particularly vital source of energy. Physiological issues like this, in particular the origin of the body heat, seem to have motivated Helmholtz's initial interest in the physics of energy conservation.

Energetics guided all Helmholtz's science, including his later contributions to the development of electrodynamics. In the nineteenth century, continental physicists pursued action-at-a-distance approaches to these effects. They sought electrodynamic forces analogous to the force of gravitation, by which a moving charge carrier acts directly across space to influence a similar entity. German physicist Wilhelm Weber derived such a force law in 1846; however, the force that he postulated varied not only with the distance but with the relative velocities and accelerations of the bodies between which the electrodynamic effects acted. In England, Michael Faraday and James Clerk Maxwell conceived electrodynamic phenomena in a very different way. Their field approach envisioned electrodynamic action as propagated across space through contiguous effects in the medium.

Helmholtz introduced Maxwell's field theory to the continent in a form that continental physicists could more readily understand. He believed that forces which vary with velocity cannot conserve energy; therefore, he attacked Weber's theory and sought a more abstract approach to electrodynamics. Helmholtz derived a generalized potential law, from which he claimed that all the various contending theories of electrodynamics, including those of Maxwell and Weber, could be deduced as special cases, depending on the assumptions one made about the medium. Only experiment could decide

which special case was actually realized in nature, Helmholtz taught, but he came more and more to favor the Maxwellian limit, in which action-at-a-distance effects became negligible. In his laboratory, Helmholtz set his students to experimental tests of the Maxwellian alternative. In 1888 one of them, Heinrich Hertz, demonstrated the existence of the transverse electromagnetic waves that could be predicted from Maxwell's theory or from Helmholtz's generalized potential. Ironically, Helmholtz's success in popularizing Maxwell's theory on the continent caused his own approach to electrodynamics to be quickly forgotten. Here as elsewhere, however, Helmholtz's insistence on abstraction, synthesis, and ultimate principles powerfully influenced the classical physics of the nineteenth century.

See also: ELECTROMAGNETISM, DISCOVERY OF; ENERGY, CONSERVATION OF; ENERGY, KINETIC; ENERGY, POTENTIAL; HYDRODYNAMICS; LEAST-ACTION PRINCIPLE; LIGHT, ELECTROMAGNETIC THEORY OF; THERMODYNAMICS, HISTORY OF

Bibliography

CAHAN, D., ed. *Hermann von Helmholtz and the Foundations of Nineteenth-Century Science* (University of California Press, Berkeley, 1993).

HELMHOLTZ, H. VON *Selected Writings of Hermann von Helmholtz,* edited by R. Kahl (Wesleyan University Press, Middletown, CT, 1971).

KOENIGSBERGER, L. *Hermann von Helmholtz,* trans. by F. A. Welby (Dover, New York, 1965).

R. STEVEN TURNER

HISTORY

See PHYSICS, HISTORY OF

HOLES IN SOLIDS

According to the band theory of electrons in solids, a completely full band cannot carry current, but a partially filled band can. If a band is nearly full, the vacant electron orbitals near the maximum energy of the band behave as if they were positive charge carriers. These vacant energy states, called holes, are important for conduction in some semiconductors (e.g., Ge and Si) and some metals (e.g., Al, Bi, Cr, and Fe). Doping semiconductors with atoms that have a lower number of valence electrons than the host material enhances hole conduction.

The Band Theory

In the free electron model of transport, the periodic potential due to the ion cores of a solid are ignored, and conduction electrons are able to fill all energy states. If the periodic potential due to the ionic cores is included in the quantum-mechanical description of electronic transport, gaps develop between allowed energy bands at the Brillouin Zone boundaries.

The total wave vector (or electron momentum) for a filled energy band is zero. If an electron with wave vector k_e is promoted from the (full) valence band to the conduction band, conservation of momentum dictates that the net wave vector of the valence band with one missing electron is $-k_e$. Alternatively, the collective behavior of the valence band can be described as if it contains a hole with a wave vector $-k_e$. The equation of motion for such a hole in an applied electromagnetic field is identical to that for a particle with charge $+e$.

Transport Properties

The mobility of electrons and holes is defined to be the magnitude of their average drift velocity per unit applied electric field. The total conductivity σ of a solid in which both electrons and holes carry current is given by

$$\sigma = ne\mu_n + pe\mu_p,$$

where e is the charge of an electron, n and p are the concentrations of electrons and holes, μ_n and μ_p are the respective mobilities.

In materials with only one type of carrier, the Hall effect can be used to determine both the sign and concentration of the carriers. The Hall effect occurs when a conductor carrying current is placed in a transverse magnetic field. An electromotive force (the Hall emf) is created that is perpendicular to both the current and magnetic field. By measuring the magnitudes Hall emf E_H, the current density j, and the magnetic field strength B, the carrier concentration n is calculated as follows:

$$ne = \frac{jB}{E_H}.$$

The sign of the carriers is determined by the sign of the Hall emf.

Doping in Semiconductors

Doping semiconductors with a material having fewer electrons in the valence shell can enhance hole conduction. For example, a gallium atom (three valence electrons) acts as an acceptor in a germanium (four valence electrons) matrix. The gallium atom tries to form four covalent bonds with neighboring germanium atoms. In this process a valence electron is transferred from a germanium atom to the gallium. Thus, a hole is created that is free to move about the lattice, while the transferred electron remains localized to the gallium atom. If the concentration of holes is sufficiently high, they will carry a majority of the current. Semiconductors in which hole conduction dominates are called p-type semiconductors, because the majority of carriers are positively charged.

See also: ACCEPTOR; BRILLOUIN ZONE; CONDUCTION; DOPING; ELECTRICAL CONDUCTIVITY; ELECTRON; ELECTRON, CONDUCTION; SEMICONDUCTOR; SOLID; TRANSPORT PROPERTIES

Bibliography

ASHCROFT, N. W., and MERMIN, N. D. *Solid-State Physics* (Holt, Rinehart and Winston, New York, 1976).

KEER, H. V. *Principles of the Solid State* (Wiley Eastern Limited, New Delhi, India, 1993).

KITTEL, C. *Introduction to Solid-State Physics,* 6th ed. (Wiley, New York, 1986).

KEVIN AYLESWORTH

HOLOGRAPHY

The word "holography" is derived from Greek roots that literally mean "entire picture." The act of doing holography is to make holograms. There are many different types of holograms, but they all share one common distinguishing feature, they recreate truly three-dimensional images of original objects.

The field of holography was originally discovered by Dennis Gabor in 1947. He was awarded the Nobel Prize for physics in 1971. In the early 1960s, Emmett N. Leith and Juris Upatnieks of the United States and Yu. Denisyuk of Russia independently discovered additional methods in using laser light to make holograms.

What is a Hologram?

A hologram is a recording on a light-sensitive medium (e.g., photographic emulsion) of interference patterns formed between two or more beams of light derived from the same laser.

In a common scheme for making a hologram, a laser beam is split in two by using a partially reflecting flat mirror. One of the beams is spread out by a lens or curved mirror and directed onto the emulsion coated on a glass plate. This is called the reference beam (R). The remainder of the light is spread out to illuminate the three-dimensional object being recorded. The light scattered by the object toward the plate is called the object beam (O). Because all the light is from the same laser, the two beams are mutually coherent and form distinct interference patterns.

To understand this process better, first consider the simplest of all objects, a point in space. Figure 1a shows two beams, situated far from the plate and interfering at 90° with respect to each other. The interference pattern is precisely the same as that from a Young's double slit with very wide fringe separations. The exposed and developed hologram is a diffraction grating consisting of $d = \lambda$, where d is the distance between fringes and λ is the wavelength of the light.

Figure 1b shows how to reconstruct the wave fronts of O. Laser light from R illuminates the hologram. The diffracted light, according to the equation $m\lambda = d \sin \theta$, yields $\theta = 90°$ for $m = 1$. There is no room for higher orders. Thus, all the diffracted light form a virtual image of O. If R were directed backward (R') as shown in Fig. 1c, O is recon-

structed backward also. A screen placed in the location O' will show the read image of the object.

If we replace the point O with a three-dimensional object illuminated by laser light, the new object beam O consists of a large collection of point sources representing the scattering points of the object. The recording on the plate now consists of a superposition of gratings. When illuminated by R or R', a virtual or real image of the object can be observed, respectively.

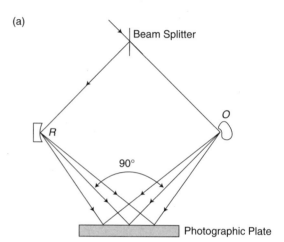

(a)

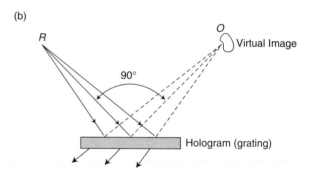

(b)

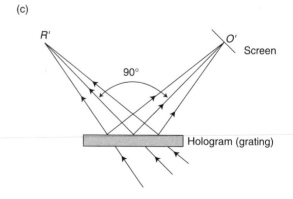

(c)

Figure 1

The above recording is called a transmission hologram. It has the property that any small area on it is capable of recreating a complete picture of the object. This can be understood by realizing that any area on a hologram is simply a *smaller* hologram.

Figure 2 represents the general interference pattern between two point sources of light, R and O, on a plane containing the sources. The lines represent the locations of interference maxima; halfway between them are the interference minima. The perpendicular bisector of RO is the zeroth-order interference, the loci of points that have the same optical path to R and O. In three-dimensional space, the pattern is a figure of revolution with RO as the axis. It is a family of hyperboloids with foci R and O.

In region A sufficiently far from R and O, the pattern is precisely that of the Young's fringes, as discussed before. Region B consists of waves moving in opposite directions, forming standing waves. The antinodes along the line joining R and O are separated by $\lambda/2$. Region C represents fringes of a Michelson interferometer.

Generally, the distance RO is many thousand wavelengths of light, and the patterns are microscopic and beyond visual observation. Figure 2 is a special case in which R and O are only a few wavelengths apart for simplicity.

If a hologram is made by placing the plate in region B, parallel to the zeroth order of Fig. 2, it records the standing wave pattern. The emulsion of the plates is usually about 10λ thick; thus it records up to twenty hyperboloidal planes. This remarkable "white-light reflection" hologram can be viewed with a point source of incandescent light from R. Because it is a "volume" hologram, it performs Bragg diffraction and selects the same wavelength λ from the white light and reconstructs the wave front of O.

Making Simple Holograms

Figure 3 shows a simple system for making transmission holograms. The light from a 1- to 5-milliwatt HeNe laser is spread by a front-surfaced curved mirror. Some light arrives at the plate directly and serves as the reference beam, where R is the focal point of the mirror. Another part of the light illuminates the object and is scattered onto the plate as the object beam. The plate and object are supported by a steel plate on top of an inflated rubber tube, which absorbs mechanical vibrations from beneath. All components are held down by magnets or glue.

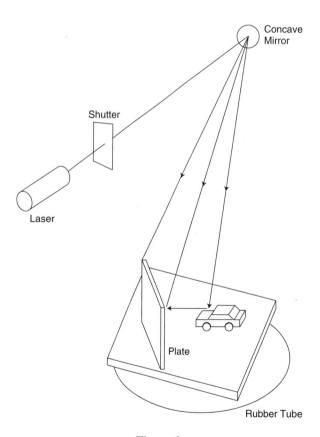

Figure 3

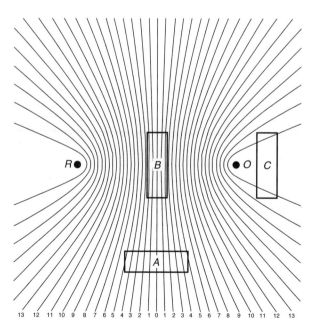

Figure 2

This is necessary because during the exposure, which may be several seconds long, any relative movement between the object and the plate will smear the microscopic interference patterns being recorded, resulting in failure. In general, much more complicated optical arrangements are necessary to illuminate large scenes in more artistic or useful ways.

Using a ruby laser, which can emit more than 1 J of light energy in less than 20 ns, a hologram can be made of deep moving objects, such as live people. The brief exposure allows the recording of interference patterns without smearing. Obviously, great precautions must be taken to protect the eyes of the subjects.

Figure 4 shows a simple setup for recording a white-light reflection hologram. Notice that the object is in contact with the plate and on the opposite side of R.

Because it is a volume hologram, the separation between Bragg planes determines the reconstructed wavelength. If the processed emulsion is prevented from expanding or shrinking, the reconstructed

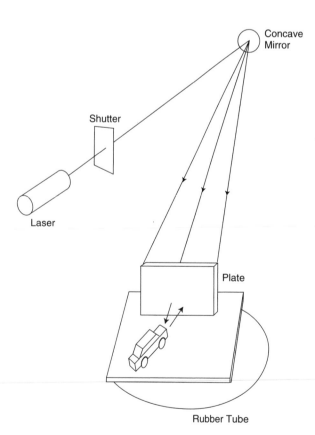

Figure 4

image using white light will have the same color as the laser light used in the recording. Therefore, using an emulsion that is sensitized for red, green, and blue wavelengths can result in amazingly realistic images in full color.

There are many hybrid variations on the transmission and reflection holograms. The most common types are used for security purposes, such as those on credit cards. These holograms are actually transmission holograms embossed on a reflective surface and usually sacrifice the vertical parallax.

Another popular form is the integral hologram. This type of hologram is synthesized from a series of two-dimensional images (i.e., photographs or computer graphics).

Applications

Beside the obvious uses of holography in art and image reproduction along with its incidental niche in security, three major areas of application have emerged.

Information storage. When R' is addressed onto 1 mm^2 of a transmission hologram, an image containing about eight megabits of information emerges onto the projection screen (photographic paper) at the speed of light. At a distance of 30 cm between the hologram and the screen, this is tantamount to down-loading one megabyte of information per nanosecond! Theoretically, each cube having a volume of λ^3 can store one bit of information holographically. Thus, a 1 cm^3 hologram can hold one terabyte (10^{12} bytes) of information, which can be transmitted through space without wires (free space interconnect). This capability is now being demonstrated in the laboratory. Using exotic synthetic crystals, this holographic memory system can read and write. Thus, holograms will play a leading role in the forthcoming optical computer.

Holographic optical elements (HOEs). Holograms can be made so that they work like lenses and/or mirrors. Furthermore, holographic optics can be designed by computers (CDH) to perform operations impossible for standard optics. Already, electro-optic chips, in which electronics are mixed with microscopic lasers, optical fibers, and CDHs, are being actively studied. The massively parallel processing capability inherent in optics will help solve the present input–output congestion in electronic computers.

Outside the computer, HOEs have already been successfully used in scanning, head-up display (a clear window through which a driver can read necessary information), and a variety of medical instruments.

Interferometry. Just as Albert A. Michelson was able to use interferometry to define the length of a meter with the highest accuracy (at the time), so can holographic interferometry (holometry) be used for high-accuracy measurements in three dimensions. This is achieved by looking through a hologram and simultaneously viewing the virtual image that has been stored and the real object. The virtual image interferes, in real time, with light scattered by the real object, showing the differences through the live fringes.

See also: COHERENCE; DIFFRACTION; GRATING, DIFFRACTION; GRATING, TRANSMISSION; IMAGE, OPTICAL; IMAGE, VIRTUAL; INTERFERENCE; INTERFEROMETRY; MICHELSON–MORLEY EXPERIMENT

Bibliography

JEONG, T. H. *Laser Holography: Experiments You Can Do* (HoloInfo Center, Lake Forest, IL, 1995).
SAXBY, G. *Practical Holography* (Prentice Hall, Englewood Cliffs, NJ, 1994).

TUNG HON JEONG

HOOKE'S LAW

Robert Hooke of England began his scientific career as Robert Boyle's assistant but soon rose to prominence as a scientist. His irascible personality and virulent attacks on Isaac Newton obscured his fame and hurt his reputation greatly. Nonetheless, Hooke was one of the great scientists of the seventeenth century. He was one of the founders of the Royal Society of London, and his own experimental work did much to establish its reputation.

In 1678 Hooke announced that he had made a discovery concerning springs. It was a simple law, accurate over a wide range, destined to play an important role in physics and engineering. If a vertically mounted spring is supported rigidly at its upper end and weights are added to its lower end, then, as Hooke discovered, the distance that the spring is stretched is found to be directly proportional to the weight (force of gravity) applied to the spring. In general, when a helical spring is either stretched or compressed, it will return to its undeformed size and shape after the deforming force is removed. But while the spring is being deformed, it exerts a restoring force F on the object deforming the spring, given by the formula

$$F = -kx,$$

where x is the displacement of one end of the spring measured with respect to the other end and k stands for a proportionality constant. The minus sign indicates that the vectors F and x point in opposite directions. In other words, if the spring is stretched and x is positive, the restoring force produced by the spring is negative, or in the opposite direction.

The proportionality constant k is known variously as the spring constant, the stiffness the spring, or the modulus of elasticity. The value of k depends upon the type of material in the spring, its size, and the way it was manufactured. The greater the value of k, the stiffer the spring.

Hooke's law holds for a steel spring with remarkable accuracy over a wide range of stretches. The important characteristic of the spring is that it be elastic, which means that it returns to its original size upon the removal of the stretching force. At the yield point a spring has been stretched beyond its elastic limit and will not spring back to its original size, but will be permanently lengthened somewhat. If a spring is stretched much beyond the yield point, then it will reach its breaking point.

There is behavior similar to Hooke's law in many cases of stretching, compression, twisting, and bending, for example,

1. a straight wire being pulled: stretch of wire $\propto$ tension;
2. a rod being stretched or compressed: change of length $\propto$ force;
3. a rod being twisted: angle of twist $\propto$ torque; and
4. a beam being bent: sag of beam $\propto$ load.

These examples show the principle that a generalized deformation is proportional to the generalized deforming force for an elastic body.

See also: ELASTICITY; ELASTIC MODULI AND CONSTANTS

Bibliography

GOULD, P. L. *Introduction to Linear Elasticity* (Springer-Verlag, New York, 1994).

TIPLER, P. A. *Physics for Scientists and Engineers* (Worth, New York, 1991).

RICHARD H. DITTMAN

H-THEOREM

In 1872 Ludwig Boltzmann introduced the *H*-theorem in his study of the thermal equilibrium of dilute gases. From a microscopic dynamical point of view, the theorem purported to explain the approach to thermal equilibrium of a dilute gas of molecules. In so doing, a kinetic theory explanation for the second law of thermodynamics was achieved. If correct, this means the time reversibility of Newtonian dynamics used for the microscopic description of the dynamics of the gas molecules is reconciled with the time irreversibility of the macroscopic thermodynamic description of gases, in particular with the second law of thermodynamics. The correctness of Boltzmann's theorem was hotly debated during the thirty years following its introduction, and Boltzmann's suicide in 1906 has been attributed to bouts of deep depression that were exacerbated by this debate.

Boltzmann did not introduce the symbol *H* until 1896. In 1872 he used the symbol *E*, no doubt from the German word for entropy, *entropie*. The second law of thermodynamics implies that the entropy evolves in time to a maximum. However, Boltzmann's *E*-function evolves in time to a minimum according to his *E*-theorem (later called the *H*-theorem). Boltzmann was confused by this discrepancy and did not see that it was simply a matter of a minus sign and that his *E*-function (later caller the *H*-function) provided a statistical interpretation for the entropy. Initially, he took a strictly deterministic view of the time evolution of the *H*-function. Only in his 1896 treatise of the gas theory did Boltzmann introduce the idea of molecular disorder and connect the idea of entropy with a statistical or combinatorial point of view. It was Max Planck who took up these connections shortly after 1896 and by 1910 the combinatorial, that is statistical, definition of entropy achieved a central position in statistical mechanics. Ironically, Boltzmann's tombstone contains the simple epitaph: $S = k \ln W$, an expression first written by Planck and capturing the statistical meaning of entropy [here S denotes entropy, k is Boltzmann's constant and $\ln W$ is the natural logarithm of W, the probability (from the German word for probability, *wahrscheinlichkeit*)].

In Boltzmann's treatment, the *H*-theorem describes the time evolution of the *H*-function, which is defined in terms of the distribution function $f(\mathbf{r},\mathbf{p},t)$ for an N molecule gas. The number, $f(\mathbf{r},\mathbf{p},t)\,(dr)^3(dp)^3$, is the number of molecules in the six-dimensional differential volume $(dr)^3(dp)^3$ centered at the coordinate $\mathbf{r}$ and the momentum $\mathbf{p}$ (the multiple integral of $f(\mathbf{r},\mathbf{p},t)$ over volume and momenta and equals the total number of molecules, N). H is defined by the integral:

$$H = \int (dr)^3 (dp)^3 f(\mathbf{r},\mathbf{p},t) \ln [f(\mathbf{r},\mathbf{p},t)]. \quad (1)$$

Boltzmann argued that $f(\mathbf{r},\mathbf{p},t)$ satisfies the Boltzmann equation, which is based on the microscopic Newtonian dynamics of molecular collisions. From Boltzmann's equation for $f(\mathbf{r},\mathbf{p},t)$, it follows rigorously that $(d/dt)\,H < 0$, except at equilibrium for which $(d/dt)\,H = 0$. These statements imply the *H*-theorem: The *H*-function decreases monotonically with time and achieves its minimum at equilibrium.

Boltzmann further showed that at equilibrium, $f(\mathbf{r},\mathbf{p},t)$ is the Maxwell–Boltzmann distribution, that is, the canonical distribution of classical statistical mechanics. In this way, Boltzmann had seemingly derived the second law of thermodynamics from the kinetic theory of gases. The catch to this apparently major achievement is to be found in the Boltzmann equation for $f(\mathbf{r},\mathbf{p},t)$. This equation does not follow rigorously from Newtonian mechanics but instead depends upon an assumption called the *stosszahlansatz*. The propagation of this property through time is called the propagation of molecular chaos and is the problematic point in the development. Only for a highly simplified version of the Boltzmann equation, the McKane model, has the propagation of chaos rigorously been proved.

The basic idea of proving an *H*-theorem for any form of microscopic dynamics has been generalized to contexts other than that of dilute gases. Many of these alternative descriptions take the form of a master equation that describes the time evolution of a probability distribution. Given the master equation, the *H*-theorem follows rigorously. However, the master equations are phenomenological in origin and have not been derived rigorously from Newtonian dynamics. Both in this broader context and in the original dilute gas context of Boltzmann, the truth of the *H*-theorem depends on the truth of the arithmetic inequality $(x - y)\ln(y/x) < 0$, which is true for any x not equal to y.

Josiah Willard Gibbs introduced an *H*-function defined over the entire 6*N*-dimensional phase space for *N* molecules. In place of Eq. (1), he wrote

$$H = \int (dr)^{3N}(dp)^{3N} \rho \ln(\rho) \qquad (2)$$

in which ρ denotes the Liouville distribution in 6*N*-dimensional phase space. It follows rigorously from deterministic Newtonian mechanics that this quantity is a time invariant, unlike Boltzmann's *H*. This apparent conflict reflects, once again, the underlying antagonism between microscopic time reversibility and macroscopic time irreversibility. The work of Boltzmann's student Paul Ehrenfest, and later of Ehrenfest's student George Uhlenbeck, has made it clear that the resolution to this mystery lies in the contraction of the description manifested here by the difference between *f* and ρ. Saying that *f* represents a coarse graining of ρ is one expression for this difference. In the context of quantum mechanics, this difference is exhibited by density matrices vis-a-vis reduced density matrices. With these modern perspectives, we now see that any contraction of the description, or any coarse graining, will convert a Gibbs *H*-theorem into a Boltzmann *H*-theorem. Together with the statistical interpretation of entropy, these ideas explain the origin of the second law of thermodynamics.

In 1948 Claude Shannon generalized these ideas to include a theory of information. Without the minus sign necessary for the entropy interpretation, Boltzmann's *H*-function is interpreted instead as information *I* [the natural logarithm (ln) must be converted to base 2 logarithms, i.e., bits]. This quantitative definition of information *I* is identical with Leon Brillouin's notion of negentropy. The extension of Boltzmann's ideas implies that the second law of thermodynamics also governs information flow in microelectronic devices. Another, nearly simultaneous, extension by Erwin Schrödinger, also using the idea of negentropy, implies that the second law of thermodynamics governs biochemical energy transductions. Thus, the impact of the *H*-theorem on our understanding of nature's diverse manifestations has been enormous.

See also: BOLTZMANN, LUDWIG; ENTROPY; GIBBS, JOSIAH WILLARD; MAXWELL–BOLTZMANN STATISTICS; PHASE SPACE; STATISTICAL MECHANICS; THERMODYNAMICS

Bibliography

KUBO, R.; TODA, M.; and HASHITSUME, N. *Statistical Physics II: Nonequilibrium Statistical Mechanics* (Springer-Verlag, Berlin, 1985).

KUHN, T. S. *Black-Body Theory and the Quantum Discontinuity* (Oxford University Press, New York, 1978).

RONALD F. FOX

HUBBLE, EDWIN POWELL

b. Marshfield, Missouri, November 20, 1889; *d.* San Marino, California, September 28, 1953; *cosmology.*

Hubble was the third of seven children. In 1898 his father transferred to the Chicago agency of his fire insurance firm and moved the family first to Evanston, and then to Wheaton, just outside Chicago. At Wheaton High School, Hubble was a star athlete as well as a scholar, and he continued his athletic accomplishments at the University of Chicago. Hubble was two years younger than most of his classmates, but he was 6′ 3″ tall and very well coordinated. He starred on the basketball team that won the Big Ten title his senior year, and he often placed in Big Ten dual track meets in both the shot put and the high jump. He was also elected vice president of his senior class. In 1910 Hubble went to Oxford University as a Rhodes Scholar, a high honor awarded to a single outstanding student-athlete-leader in each state.

After three years at Oxford studying law, traveling through Europe during vacations, and competing in athletics (high jump, broad jump, shot put, hammer throw, running events, and water polo), Hubble returned to the family home, now in Louisville, Kentucky. For a year he taught physics and Spanish at a high school across the river, in New Albany, Indiana, and coached the basketball team to an undefeated regular season and a trip to the state championship, where the team won its first two games before being eliminated. Hubble also passed the Kentucky bar exam, but he did not practice law.

In 1914 Hubble returned to the University of Chicago and the Yerkes Observatory as a graduate student in astronomy. He hoped to finish his doctoral dissertation on a photographic investigation of faint nebulae (wispy patches of light barely visible in the heavens with a good telescope) and take up a position at the Mount Wilson Observatory in southern California in the summer of 1917. In April, however, the United States declared war on Germany. Hubble rushed through his dissertation, took his final oral exam, and reported to the army for duty three days later. He served in France and made the rank of major before the war ended. In 1919 he finally joined the Mount Wilson Observatory.

In 1924 Hubble married the sister-in-law of a colleague from the Lick Observatory in northern California, but the Hubbles had no children.

During World War II Hubble was chief of ballistics and director of the Supersonic Wind Tunnels Laboratory at the Army Proving Grounds in Aberdeen, Maryland, and was awarded the Medal for Merit for his wartime work. Except for this period, he worked all his life at Mount Wilson.

Hubble's scientific achievements made him the foremost astronomer of the twentieth century and one of the most influential scientists of all time in changing our understanding of the universe. He was on the cover of *Time* magazine in 1948 and the Hubble Space Telescope is named after him. Perhaps the best indication of his fame, though, is the renaming of the Wheaton high school in 1991 after Hubble rather than the school's star football player, the legendary Red Grange—the "Galloping Ghost" at the University of Illinois, All-American halfback in 1923 and 1924, All-American quarterback in 1925, member of the college Football Hall of Fame, player for the Chicago Bears, and a charter member of the Professional Football Hall of Fame.

What Hubble did to make himself even more famous than the Galloping Ghost was, first, demonstrate conclusively after centuries of fruitless speculation by other astronomers that spiral nebulas (faint patches of light in the sky) are independent galaxies at great distances beyond our own galaxy. Using the new 100-in. telescope at Mount Wilson, Hubble found Cepheid variable stars in spiral nebulas. These stars are a useful indicator of distances because their brightness correlates with the duration of time of their change from maximum to minimum and back to maximum brightness. Hubble used the relationship established for Cepheids in our galaxy between absolute luminosity (apparent brightness measured at a standard distance from the object) and period (the time it takes for the star to vary in brightness from maximum to faintest and back to maximum). He measured periods for Cepheids in spiral nebulas and then assumed they had the absolute luminosity corresponding to Cepheids of the same period in our galaxy. (Knowing the period, he could read off the expected absolute luminosity from a graph of the period–luminosity relation.) Finally, with an estimated absolute luminosity, he calculated how far away the Cepheids (and the nebulas they were embedded in) had to be for their absolute luminosity to be diminished to the apparent luminosity that he measured (luminosity is reduced in inverse proportion to the distance squared).

Proving in the mid-1920s that spiral nebulas are galaxies, a great accomplishment in itself, was only the starting point for Hubble. Step by step, he determined distances to ever more distant galaxies, first using Cepheid variables, and later calculating an average absolute luminosity for galaxies at known distances and then using this in conjunction with measured apparent luminosities of galaxies so distant that Cepheids were not detectable in them to estimate distances for these galaxies (again calculating distance from the difference between estimated absolute luminosity and measured apparent luminosity). A few velocities had been measured by other astronomers, including Vesto M. Slipher at the Lowell Observatory in Arizona, from the Doppler effect (a redshift in the spectrum of light from an object moving away from the observer). At the Mount Wilson Observatory, Hubble directed a program (largely carried out by his colleague Milton Humason) of measuring velocities for the galaxies whose distances he also was determining. By 1935 they had determined velocities and distances for more than 100 galaxies. Hubble's work throughout the 1930s demonstrated that more distant galaxies are moving away from us at greater velocities (his famous velocity-distance relation, which can, in turn, be used to estimate distances for even more distant galaxies from their measured velocities, or Doppler redshifts). Scientists, including Albert Einstein, had assumed the universe to be static; Hubble showed that it is expanding.

World War II interrupted Hubble's work on cosmology, and his life ended in 1953, soon after completion of the 200-in. telescope on Palomar Mountain and too soon for conclusive answers from the research program he planned. Although subsequent investigations generally have confirmed Hubble's relativistic, expanding model of the universe, evaluation of his work should not be based solely on this fact. Instead, Hubble's cosmology should be appreciated more for the assumptions it overthrew, for the vistas it opened, and for its being one of the great accomplishments of human intellect.

See also: COSMOLOGY; DOPPLER EFFECT; GALAXIES AND GALACTIC STRUCTURE; HUBBLE SPACE TELESCOPE; REDSHIFT; UNIVERSE, EXPANSION OF, DISCOVERY OF

Bibliography

CHRISTIANSON, G. *Edwin Hubble: Mariner of the Nebulae* (Farrar, Straus & Giroux, New York, 1995).

HETHERINGTON, N. "Hubble's Cosmology." *Am. Sci.* **78,** 142–151 (1990).

HETHERINGTON, N. *The Edwin Hubble Papers: Previously Unpublished Manuscripts on the Extragalactic Nature of Spiral Nebulae* (Pachart, Tucson, AZ, 1990).

HETHERINGTON, N. *Hubble's Cosmology: A Guided Study of Selected Texts* (Pachart, Tucson, AZ, 1995).

HUBBLE, E. *The Realm of the Nebulae* (Yale University Press, New Haven, CT, 1936).

OSTERBROCK, D.; BRASHEAR, R.; and GWINN, J. "Self-Made Cosmologist: The Education of Edwin Hubble" in *Evolution of the Universe of Galaxies, Edwin Hubble Centennial Symposium,* edited by R. Kron (Astronomical Society of the Pacific, San Francisco, 1991).

NORRISS S. HETHERINGTON

HUBBLE CONSTANT

In 1929 American astronomer Edwin Powell Hubble showed that other galaxies in the universe are moving away from Earth with a speed of recession that is proportional to their distance. This relationship, called Hubble's law, is expressed as

$$v = H_0 d,$$

where v is the speed of the galaxy, d is its distance, and H_0 is Hubble's constant (denoted as K by Hubble). The simplest and accepted interpretation is that the universe is expanding, as predicted by big bang cosmological models. The Hubble constant quantifies the rate at which the universe is expanding.

The Hubble constant sets the scale of the universe—its age and size—but in fact, is not constant; it slowly decreases with time. The time back to the big bang is equal to $1/H_0$ times a numerical factor that is less than 1 (assuming no cosmological constant) and depends upon the amount of matter in the universe (2/3, for the critical density universe). The time back to the bang is less than $1/H_0$ because the universe expanded more rapidly in the past and has slowed due to the self-gravitational attraction of the matter in the universe. The size of the observable universe is equal to c/H_0 times another numerical factor (2, for the critical density universe), and c is the speed of light.

In principle the Hubble constant is simple to measure: it is given by the ratio of the speed of any distant galaxy divided by its distance. The speed can be easily and accurately gotten by measuring the Doppler shift of atomic spectral features detected in the light from the galaxy; for receding objects the Doppler shifts are toward long wavelengths and are known as red shifts. Distances are much more difficult. They require standard candles: given an object (e.g., bright star) of known intrinsic brightness, the distance can be determined by the analog of the inverse-square law in the expanding universe. Finding and calibrating standard candles is the problem. To span the great distances in the universe, a succession of intertwined standard candles, known as the cosmic distance ladder, is used.

Errors in the distance ladder have plagued cosmologists since the discovery of the expansion. Hubble's original determination was 550 km·s^{-1}·Mpc^{-1}, corresponding to a Hubble time $1/H_0 = 1.8$ Gyr, which is less than the age of the solar system. Many errors have been uncovered since, and accurate distance determinations made since the 1980s have led to the present range, 40 km·s^{-1}·Mpc^{-1} to 90 km·s^{-1}·Mpc^{-1}, corresponding to a Hubble time between 25 Gyr and 11 Gyr. For the critical density universe, which is favored by ideas coming from the study of the early universe, such as inflation, this range implies a time back to the bang between 16 Gyr and 7 Gyr. Unless the value of the Hubble constant is close to the lower end, the time back to the bang is too small to be consistent with the ages of the oldest stars, dated at between 13 Gyr and 19 Gyr.

In 1994 Canadian astronomer Wendy Freedman and her collaborators achieved an important milestone on the road to a definitive measurement of the Hubble constant when they succeeded in using the Hubble space telescope to detect Cepheid variable stars in the Virgo cluster, some 16 Mpc away. These very bright, pulsating stars are reliable standard candles. Within five years the Hubble Space Telescope team hopes to study enough Cepheids to determine H_0 to a precision of 10 percent. When this is done, not only will the scale of the universe finally be accurately determined, but also ideas such as inflation will be tested.

See also: BIG BANG THEORY; COSMOLOGICAL CONSTANT; COSMOLOGY; COSMOLOGY, INFLATIONARY; HUBBLE, EDWIN POWELL; UNIVERSE, EXPANSION OF

Bibliography

FREEDMAN, W., et al. "Distance to the Virgo Cluster Galaxy M100 from Hubble Space Telescope Observations of Cepheids." *Nature* **371**, 757–767 (1994).

FUKUGITA, M.; HOGAN, C. J.; and PEEBLES, P. J. E. "The Cosmic Distance Scale and the Hubble Constant." *Nature* **366**, 309–312 (1993).

WEINBERG, S. *First Three Minutes* (Basic Books, New York, 1972).

MICHAEL S. TURNER

HUBBLE SPACE TELESCOPE

The Hubble Space Telescope (HST) is in many respects the most powerful optical telescope ever built. It is not the largest—its 2.4-m primary mirror is small compared to the Keck telescope in Hawaii, which has a 10-m diameter. However, by orbiting 300 miles above the surface of the earth, it avoids the disturbing effects of the atmosphere. This permits it to see much finer detail than ground-based telescopes and also to observe wavelengths such as the ultraviolet that do not reach the earth's surface.

The HST is named for the American astronomer Edwin Hubble, who in the 1920s made two of astronomy's fundamental discoveries. He provided evidence that the faint spiral nebulas were actually distant galaxies—that the universe is filled with galaxies—and that virtually all the galaxies were moving away from us (i.e., the universe is expanding). The latter discovery led to the concept of the big bang as the start of the expansion. It is fitting that two of the key projects undertaken by the HST are the study of the evolution of galaxies and study of the big bang.

The HST is approximately 14 m long, 5 m in diameter, and weighs 11,500 kg. It is designed to use the full capacity of the space shuttle, which put it into orbit April 25, 1990. The solar panels, which unfurled in orbit, are 10 m long and were contributed by the European Space Agency. Light that travels down the telescope tube and strikes the primary mirror is reflected back up to a small secondary mirror centered in the tube. This returns light back toward the primary and through a hole in its center, an optical design called the Ritchy—Cretien form of Cassegrain (see Fig. 1). Behind the hole are the four main scientific instruments: two cameras and two spectrographs. Both cameras can make visual and ultraviolet images; one of the spectrographs is entirely devoted to ultraviolet observations. The cameras are designed to provide much higher resolution than can be obtained from the ground. The spectrographs have very small entrance apertures, allowing the HST's high resolution to enable spectroscopy of individual objects in crowded fields such as the centers of globular star clusters, observations that are impossible from the ground. The spectrographs also can obtain much higher signal-to-noise ratios and higher spectral resolution than previous orbiting ultraviolet telescopes, allowing measurement of weak spectral features never before seen.

Shortly after launch, it was discovered that the HST's primary mirror suffered from spherical aberration, an optical defect that caused images to contain about 15 percent focused light, with the rest spread out in a haze. The defect was eventually traced to faulty test equipment used in manufacturing the mirror years before launch. Although intensive computer processing could eliminate most of the haze from images, and spectroscopic observations could still be made, the telescope's ability to image faint objects such as those at the edge of the universe was ruined. In December 1993, space shuttle astronauts made more than ten major repairs to the telescope, including the installation of new gyroscopes, solar panels, and small, very precise corrective mirrors that returned the telescope to the originally designed optical performance. Installing the corrective optics required removing one of the HST's original five instruments, a high-speed photometer. Resolution is now close to the limit im-

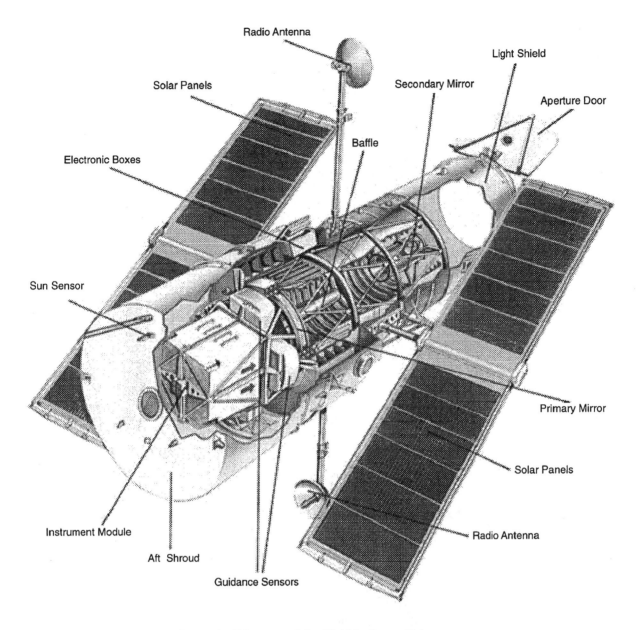

Figure 1 Diagram of the Hubble Space Telescope.

posed by the wave properties of light, nearly 0.05 arc sec, or the angle subtended by one-fourth at a distance of 100 km. The tracking accuracy, controlled by the gyroscopes and automatic trackers that lock onto stars previously selected as guide stars, is typically 0.004 arc sec.

Use of the HST requires careful preparatory work. Before launch, the entire sky was mapped and nearly 20 million guide stars were identified—by far the most comprehensive star catalog ever made. An individual astronomer (any in the world are eligi-

ble) applies to use the HST about one year in advance, specifying the scientific question to be investigated and the proposed observations. About 800 applications are typically received each year. Six to eight panels of astronomers representing different fields of expertise spend a week ranking the proposals and allocating telescope usage. A medium-sized approved program might be given twenty-five hours of telescope use.

Many significant discoveries were made by the HST during its first four years of operation. Most made use of the unprecedented resolution, or both high resolution and high precision spectroscopy. Examples of the former include probing the cores of globular star clusters and giant elliptical galaxies. Adding spectra to measure doppler shifts have shown very high velocities right at the core of several elliptical galaxies. This is strong circumstantial evidence that a black hole of about 10^9 solar masses is located there. In the Orion nebula, a young star-forming region 1,500 light-years from the Sun, imaging has shown evidence of disks of material around many of the stars. These may well be examples of solar systems in formation. By looking at the very distant universe, the HST looks backward in time. Since light has taken about 10 billion years to reach Earth from the most distant objects, the objects are seen as they were 10 billion years ago, when they were young. These observations are providing evidence of many more irregular galaxy forms and collisions between galaxies in the early universe than are seen today. The scale of the universe—how far away the galaxies are and when the big bang occurred—is being calibrated by one of the HST's large key projects, a search for Cepheid variable stars in moderately distant galaxies. Cepheids pulsate with periods of days or weeks, and the pulsation period is related to their intrinsic luminosity. Measuring the period and apparent brightness gives the distance. Edwin Hubble's identification of Cepheids in the Andromeda Galaxy in 1924 was the first proof that galaxies were far beyond the Milky Way. Study of these stars in more distant galaxies by the HST seems to be building evidence that the universe is somewhat smaller and younger than previously thought. Another key prediction of the big bang theory is that only the first three elements—hydrogen, helium, and a trace of lithium—should have been produced primordially. The HST observations have helped show that this is indeed correct, and that the other elements have built up with time from supernovas during the history of our galaxy.

With the ability to be serviced by the space shuttle, the HST, which is scheduled to be given new instruments in the future, such as a more powerful spectrograph and an infrared camera, should continue to operate for more than a decade.

See also: ABERRATION, SPHERICAL; BIG BANG THEORY; HUBBLE, EDWIN POWELL; HUBBLE CONSTANT; UNIVERSE; UNIVERSE, EXPANSION OF; UNIVERSE, EXPANSION OF, DISCOVERY OF

Bibliography

FIENBERG, R. T. "HST: Astronomy's Discovery Machine." *Sky and Telescope* **79** (4), 366–372 (1990).

FIENBERG, R. T. "Hubble's Agony and Ecstasy." *Sky and Telescope* **81** (1), 14–20 (1991).

SHAW, J. M., and SHEAHEN, T. P. "Correcting Hubble Vision." *Sci. Teach.* **61** (9), 16–19 (1994).

DOUGLAS DUNCAN

HURRICANE

A hurricane is the most intense phase of a tropical cyclone, a warm-centered low pressure system that forms and grows as it moves across specific areas of tropical and sometimes subtropical waters. In its early stages, a tropical cyclone is most easily seen as an organized pattern of clouds and low-level winds.

A tropical cyclone is classified according to its near surface sustained wind speed. This measurement refers to the instantaneous wind speed measured at 10 m, averaged over 1 min. Gusts may be 10 to 40 percent greater than the sustained speed. (In practice, measurements may be made at different heights and then extrapolated to 10 m.) For a tropical disturbance, the maximum sustained wind speed is less than 20 kn (where 1 kn = 0.5144 m/s). A tropical depression has a maximum wind speed between 20 and 34 kn, while a tropical storm has maximum winds between 34 and 63 kn. A tropical cyclone that is classified as a hurricane has reached a wind speed level of 64 kn or greater.

Once a storm has achieved hurricane status, it is further categorized by its intensity. In the United States, the Saffir–Simpson scale is commonly used for this purpose. This scale was formulated to allow characterization of the intensity of Atlantic hurricanes. It expresses hurricane intensity in terms of (usually estimated) maximum near surface sustained wind speed and depth of storm surge (an abnormal rise in the sea along a shore in response to the strong winds, the lowered pressure around the center of the storm, and the topography of the ocean bottom). Strictly, the Saffir–Simpson scale applies only along the East Coast and the Gulf Coast of the United States. A Category 1 (weak hurricane) produces winds of 65 to 82 kn. Damage is usually light to moderate, and the storm surge is usually 1.2 to 1.5 m above normal water levels. A Category 2 (moderate hurricane) produces winds of 83 to 95 kn. Major structural damage to mobile homes is very likely. Some harm may be done to roofing, windows, and doors, but there will be few major problems with buildings. High water in the storm surge will be 1.8 to 2.4 m above normal water levels. A Category 3 (strong hurricane) is associated with winds of 96 to 113 kn. Some damage occurs to roofs, windows, and doors. Structural harm occurs to some small homes and utility buildings; mobile homes are destroyed. A few wall failures on large buildings are likely. The surge will be 2.7 to 3.7 m above normal water levels. A Category 4 (very strong hurricane) is characterized by winds of 114 to 135 kn. There will be extensive damage to roofing, windows, and doors. Complete failure of roof structures on many small houses will occur, and mobile homes will be destroyed. Some walls of large buildings will fail. Storm surge will be 3.9 to 5.5 m above normal water level. A Category 5 (devastating hurricane) possesses winds greater than 135 kn. Very severe and extensive roof, window, and door damage should be expected, with complete failure of roof structures on most homes and many industrial buildings. Some large buildings will suffer complete structural failure, while some smaller ones will be overturned and sometimes blown away. Storm surge will be greater than 5.5 m above normal water level.

Defined on the basis of stronger than normal winds, hurricanes are roughly circular systems, with diameters typically between 400 and 500 km. Seen from above, a typical hurricane is capped by a circular shield of cirrus cloud. Opaque around the center, the shield thins to become translucent near the outer edge of the storm. Often a clear "eye" is seen near the center of the shield, though sometimes the eye cannot be seen because it is hidden under cloud. The cirrus shield is surrounded by and overlies long inward-spiraling cloud bands formed of lines of heavy showers.

The heavy rain that falls in hurricanes comes from two main regions of deep, vigorous cumulonimbus clouds: the annular wall of cloud surrounding the eye (the eyewall), and the curving bands of cloud that spiral in towards the heart of the hurricane (the spiral rain bands). These regions of heavy rainfall are a consequence of the continuous flow of warm, moist air towards the central low pressure minimum; this occurs in a surface layer of two to three kilometers thickness. In the eyewall and the rain bands, this moist surface air is transported upwards. As it rises, water vapor condenses, producing cloud and rain drops and releasing enormous amounts of latent heat. Visually, one sees very deep, powerful cumulonimbus clouds forming the eyewall and the rain bands. At the top of the troposphere, the outflow from the hurricane is marked by the cirrus cloud shield.

Tropical cyclones are thermodynamic engines driven by a cycle of evaporation of warm water from the ocean surface followed by condensation of water vapor and the release of latent heat in cumulonimbus clouds. Warm sea surface temperatures and strong sea-level winds lead to high evaporation rates into the inflowing air. A strong hurricane tracking across a warm tropical ocean evaporates massive quantities of water into the near-surface airflow. The enormity of the latent heat of evaporation due to the passage of a hurricane across a tropical ocean produces an extensive and substantial cooling of the sea surface.

The eye is a small region of cloudless sky and light winds; its diameter is typically 20 km or less, but it can be up to 40 km across. The sky is often clear because the air in this central column is sinking from great heights to near the surface. This air is being strongly warmed as it subsides and so tends to be too warm and too dry for any cloud to form.

The central eye and the surrounding annulus of very heavy rain and intense winds covers only a small fraction of the area affected by a typical hurricane. Outside this central region, the weather is dominated by strong gusty winds and periods of heavy rain showers that fall from the inward-spiraling rain bands. These bands can be many hundreds of kilometers long and many tens of kilometers wide. Between the rain bands, the weather conditions are still windy but tend to be cloudfree.

The central, minimum sea-level pressure in a hurricane is typically about 50 hPa lower than the pressure at the edge of the storm; in extreme cases, this difference can approach 100 hPa. It is this large horizontal pressure gradient across the sea surface which makes the wind blow so strongly.

Peak surface winds are found beneath the eyewall cloud, in a circular annulus surrounding the eye. This narrow zone of strong winds produces dramatic and abrupt changes in the weather when the eye crosses a particular spot. In the case where the center of the eye passes directly overhead, its approach is heralded by the band of devastating winds circulating counterclockwise (in the Northern Hemisphere) around the center of low pressure. The high wind drops rapidly to near-calm conditions and nearly clear skies as the center passes by, only to be quickly followed by equally devastating winds blowing from a direction opposite to those experienced earlier.

The occurrence of tropical disturbances is highly localized and strongly seasonal. Tropical cyclones form only over the open sea; this ocean/land split reflects the necessary requirement of a (relatively smooth) warm water surface to serve as an energy source. Further, they form only in preferred regions like the West Atlantic, Western North Pacific, and South Indian Oceans; over a twenty-year period (1958–1977), no tropical cyclones were observed to form over the entire South East Pacific and South Atlantic Oceans. The Northern Hemisphere's oceans see approximately twice as many cyclones as do the Southern's, even though the area of tropical and subtropical ocean is larger south of the Equator. Further, there are distinct differences between the Atlantic and Pacific Oceans. The North West Pacific experiences about 3 times as many tropical cyclones as the North Atlantic does in a typical year.

On average, the majority of cyclones in the Northern Hemisphere occur from July to October, with a peak of activity during August and September. On the other hand, the Southern Hemisphere experiences most such systems from December to March, with a peak during January and February. Sensitivity to location and the strong seasonality reflects the requirement that the sea surface be sufficiently warm before a tropical system can develop. The warmest sea surface temperatures occur in September and February in the Northern and Southern Hemispheres, respectively—these months and those that immediately precede them are the times of peak tropical cyclone activity. The critical value of sea surface temperature for hurricane development appears to be about 27°C. Tropical systems cannot form or persist in regions where sea surface temperatures are cooler than this value.

Another factor which limits where hurricanes can form is latitude. A distinct feature of these disturbances is the rotating wind pattern. The rotation is derived from Earth's rotation through the Coriolis effect. Such organized circulations cannot occur within about 5° latitude of the Equator. Hence, hurricanes are restricted to forming in regions poleward of 5° latitude, but equatorward of the 27°C sea surface isotherm.

Atmospheric features also play a crucial role in hurricane development. One important factor is the difference in the wind direction and speed between the sea surface and the top of the troposphere. If this difference is too great, a cyclone will not form even over a favorably warm ocean. Similarly, if the air in the middle of the troposphere is too dry, the deep convective clouds that are the essential components of a hurricane will tend to evaporate when growing up into relatively dry air.

See also: ATMOSPHERIC PHYSICS; CORIOLIS FORCE; TORNADO

Bibliography

ANTHES, R. A. *Tropical Cyclones, Their Evolution, Structure, and Effects,* Meteorological Monograph 19(41) (American Meteorological Society, Boston, MA, 1982).

GRAY, W. "Global View of the Origin of Tropical Disturbances and Storms." *Monthly Weather Review* **96,** 669–700 (1968).

HAWKINS, H. F., and IMBEMBO, S. M. "The Structure of a Small, Intense Hurricane: Inez 1966." *Monthly Weather Review* **104,** 418–442 (1976).

JOHN T. SNOW

HUYGENS, CHRISTIAAN

b. The Hague, Netherlands, April 14, 1629; *d.* The Hague, Netherlands, July 8, 1695; *mechanics, optics.*

Huygens was one of the major figures of the Scientific Revolution. His basic approach to physics was remarkably modern. He usually chose a well-defined problem, provided a mathematical solution, and used a hypothetico-deductive method to test and expand his results.

His grandfather was secretary to William the Silent, the leader of the Dutch revolt against Spanish rule. His father, uncle, and older brother all served William's successors and the new Dutch republic. Primarily educated at home, Huygens did attend the mathematics lectures of Frans van Schooten at the University of Leiden and studied law at the College of Orange in Breda. Freed by a changing political situation to pursue science instead of the family profession, he went to Paris to join the Académie Royale des Sciences of Louis XIV. He was also the first foreign member of the Royal Society of London.

When he was seventeen he proved mathematically that projectiles follow a parabolic path, only afterwards learning about Galileo's work on the subject. A decade later he created the first accurate, fully functioning pendulum clock. In 1659, while determining the constant of gravitational acceleration (g), he discovered the isochronism of the cycloid. That is, if a body falls along an inverted cycloid, the time that it takes to reach the bottom is independent of the height from which it begins. Pursuing this discovery further, he invented the mathematical theory of evolutes and an ideally perfect cycloidal pendulum clock. Jakob and Johann Bernoulli and others built on this exemplar of mathematical physics to create the field of rational mechanics.

In the 1650s Huygens also analyzed collisions of moving bodies using his hypothesis that all motion is relative. He believed that every physical event was merely the manifestation of matter in motion and collision. Thus, in later years he rejected the Newtonian explanation of gravity in favor of a modified vortex theory that never successfully integrated with his mathematical studies of centrifugal force and motion in a resisting medium.

His interest in optics also started in his twenties, when he and his older brother ground their own lenses for telescopes. He correctly hypothesized that Saturn had a ring and discovered its largest moon, Titan, with one of their telescopes. In addition, he began a geometrical study of the optical properties of lenses, especially aberration, that he continued to work on throughout his life but, characteristically, never published. In 1690 he did publish a pair of treatises on gravity and light. The treatise on light includes an introduction to his hypothetico-deductive method and an explanation of the double refraction of Iceland spar by means of a pulse, or wave, theory of light. He considered each point on a wave front to be the center of a weak secondary wave; the resulting new wave front is the curve (envelope) which is tangent to all those secondary waves (the Huygens principle). He contended that the speed of light is finite and (in contrast to Newton's particle theory) slower in a dense than in a rare medium.

In mathematics, he created a new geometrical technique for approximating pi, wrote the first published treatise on probability, and participated in the mathematical debates that led to the formulation of calculus. In technology, he invented the spiral spring watch and participated in the development of the air pump and microscope. In cosmology, he speculated that there are planets around other suns, populated with people who would understand geometry because it is basic to all of nature.

See also: COLLISION; LENS; LIGHT, SPEED OF; LIGHT, WAVE THEORY OF; NEWTONIAN MECHANICS; OPTICS

Bibliography

ARNOL'D, V. I. *Huygens and Barrow, Newton and Hooke: Pioneers in mathematical analysis and catastrophe theory from evolvents to quasicrystals,* trans. by E. J. F. Primrose (Birkhauser, Basel, 1990).

BELL, A. E. *Christian Huygens and the Development of Science in the Seventeenth Century* (Arnold, London, 1947).

BOS, H. J. M.; RUDWICK, M. J. S.; SNELDERS, H. A. M.; and VISSER, R. P. W. *Studies on Christiaan Huygens: Invited Papers from the Symposium on the Life and Work of Christiaan Huygens, Amsterdam, 22–25 August 1979* (Swets & Zeitlinger, Lisse, 1980).

BURCH, C. B. "Huygens' Pulse Models as a Bridge Between Phenomena and Huygens' Mechanical Foundations." *Janus* **68,** 53–64 (1981).

STRUIK, D. J. *The Land of Stevin and Huygens: A sketch of science and technology in the Dutch Republic during the Golden Century* (Reidel, Dordrecht, 1981).

YODER, J. G. *Unrolling Time: Christiaan Huygens and the Mathematization of Nature* (Cambridge University Press, Cambridge, Eng., 1988).

JOELLA G. YODER

HYDRODYNAMICS

Hydrodynamics (aerodynamics, fluid mechanics, fluid dynamics) is the study of the flow properties of any deformable continuous medium, from the viscous cooling volcanic lava to the atmosphere and oceans of Earth, to the convection and general circulation within the Sun and the streaming of the tenuous gas in interplanetary and interstellar space. Hydrodynamics is part of everyday life in the air and water around us and in the industries that support us. For instance, oceanography and meteorology treat the fluid motions initiated by the uneven heating by the Sun and strongly deflected by the rotation of Earth. Hydrodynamics is fundamental to the design of aircraft and ships, where it is essential to maximize thrust and lift and minimized drag. The design of steam and gas turbines and jet engines is based on optimizing the shape of the turbine nozzles and blades to extract the maximum work from the propellant (water, air, steam, mercury vapor, etc.). The essential heat exchange in air conditioners, refrigerators, and internal combustion engines is based on convective transport of heat in forced flows of water and air. Hydraulics, with widespread applications in machinery of all kinds, including aircraft, ships, and automobiles, is the special branch of hydrodynamics that deals with the transmission of fluid pressure in enclosed quasi-static bodies of fluid.

Given the central role of fluid motions in our world, it is hardly surprising that hydrodynamics has been applied in one form or another since ancient times. Discounting swimming with our natural appendages, it would appear that the first real hydrodynamic devices were canoes, paddles, and then sails, followed by water wheels driven by rivers to hoist water into irrigation ditches or to take over the back breaking task of grinding grain for human consumption.

Now the rapid motion of air or water in even so simple a duct as a smooth straight pipe or canal can be exceedingly complex. The complexity arises from the fact that neighboring portions of a flow get out of step with each other, jostling, deforming, and intermixing as the fluid moves along the channel. The phenomenon of chaotic entangling, mixing, and jostling is called turbulence. It is opposed only by the viscosity of the fluid, which resists all deformations of the fluid and guarantees that the flow sticks to the walls of the channel (except in extreme cases of air flowing over a smooth surface at velocities approaching or exceeding the speed of sound). The water in a pipe, for example, sticks to the smoothest wall, so that the speed of the flow varies from zero at the wall to some maximum at the center of the pipe. Thus, even a small deflection of fluid brings fast and slow fluid into collision, thereby initiating turbulence.

If the fluid motion is slow, the viscosity suppresses the internal mixing of the fluid and constrains the motion to the minimum smooth laminar flow representing the general passage of the fluid. But if the fluid flow is fast enough, the jostling and colliding get the upper hand and the fluid motion develops swirls and eddies in addition to the general flow. As a result of the seemingly unlimited forms and variations of fluid motions, the science of hydrodynamics is as vital and alive today as it was 100 or 150 years ago when the modern scientific (experimental and theoretical) study of hydrodynamics was getting underway. New and pressing problems, such as terrestrial weather and climate, the development of more efficient and faster aircraft, the convection that delivers heat to the surface of the Sun (at slightly variable rates), or the optimization of the flame front and outflow in a rocket engine, to mention but a few of the many applications, continually motivate investigations of ever more complex aspects of the hydrodynamics. Even the many simple facets of traditional, century-old hydrodynamics have yet to be fully discovered and described.

The liquid iron and nickel core of Earth and the hot ionized gases of stars and galaxies conduct electricity, so that over those large dimensions magnetic fields are carried along with the fluid motion, distorting and swirling with the fluid to produce new field configurations and introducing magnetic stresses that alter the flow. Thus, the hydrodynamics of planetary liquid metal cores, stellar convective zones, stellar winds, and interstellar gas is an active subject that is essential to understanding the universal appearance of magnetic fields and magnetic activity in the astronomical universe.

To take a closer look at the dynamics of a fluid, the first point to be understood is that any non-rigid (i.e., non-solid) material viewed on a scale large compared to the spacing of the individual atoms can be well approximated as a continuous fluid. Hence, the dynamical behavior of the material is described by the hydrodynamical form of Newton's equation of motion. An essential concept, then, is the pressure of the fluid, defined simply as the force per unit area exerted by each small region of fluid on the contiguous regions of fluid. So if the pressure increases in some direction, the contiguous regions push back harder from that direction and the region of fluid is accelerated in the direction opposite to the pressure increase. The force per unit volume is equal to the change in pressure p per unit length in that direction, so Newton's equation becomes

$$\rho \frac{d\mathbf{v}}{dt} = -\boldsymbol{\nabla} p$$

for a fluid coasting freely under its own inertia. Here ρ is the density of the fluid, $\mathbf{v}$ is the velocity of a moving element of fluid, $d\mathbf{v}/dt$ denotes the rate of change of the velocity with time (i.e., the acceleration) and ∇p (the gradient of p) represents both the direction of increase of p and the change of p per unit distance in that direction (i.e., $-\nabla p$ represents the vector force per unit volume, just as the vector $\mathbf{v}$ represents the direction and magnitude $v = |\mathbf{v}|$ of the flow and is described by the three components (v_x, v_y, v_z) in the orthogonal x, y, and z directions).

In the tenuous gases in the interplanetary and interstellar space the atoms rarely collide, so the pressure is defined as the rate at which momentum is transported by the atoms streaming freely from one region into the next. Momentum transport and force are equivalent in Newtonian mechanics, so the equation of motion is the same in both the collisional and the collisionless gas. Irrespective of collisions, the pressure is given by NkT for N atoms per unit volume at a temperature T, where k is the Boltzmann constant (equal to 1.38×10^{-23} J·K^{-1}, or 1.38×10^{-16} ergs·K^{-1}).

A word of explanation of $d\mathbf{v}/dt$ is in order here. If one follows a particular small region of fluid in a larger flow, $d\mathbf{v}/dt$ is just the acceleration vector $\mathbf{a}$ of that region. The concept of following the individual elements of fluid is referred to as the Lagrangian formulation of hydrodynamics. However, it is sometimes desirable to think of the spatial pattern of the fluid streaming, without following the individual fluid elements. In that case $\mathbf{v}$ is a continuous function of position $\mathbf{r}$ or x_i and time t, written $\mathbf{v}(\mathbf{r}, t)$. The acceleration $d\mathbf{v}/dt$ then has two parts. One part is $\partial \mathbf{v}/\partial t$, representing the time rate of change of the velocity $\mathbf{v}$ observed at any fixed point in space. The other is the acceleration of each small region of fluid as the region moves at the speed $\mathbf{v}$ to where $\mathbf{v}$ has a different value. The faster the fluid streams across a variation in $\mathbf{v}$, the greater the acceleration, so this contribution is described by $(\mathbf{v} \cdot \nabla)\mathbf{v}$. The equation of motion is then written

$$\rho\left[\frac{\partial \mathbf{v}}{\partial t} + (\mathbf{v} \cdot \nabla)\mathbf{v}\right] = -\nabla p$$

and is called the Euler equation.

A hydrodynamic flow is conveniently represented by the stream lines, defined by the instantaneous direction of $\mathbf{v}(\mathbf{r},t)$ throughout the region and given formally by the family of solutions of the two differential equations

$$\frac{dx}{v_x} = \frac{dy}{v_y} = \frac{dz}{v_z}.$$

If there are additional forces exerted on the fluid, they are added to the right-hand side of the equation of motion. Thus, for instance, a gravitational acceleration of magnitude and direction $\mathbf{g}$, introduces a force $\rho\mathbf{g}$, so that the Euler equation is modified to

$$\rho\left[\frac{\partial \mathbf{v}}{\partial t} + (\mathbf{v} \cdot \nabla)\mathbf{v}\right] = -\nabla p + \rho\mathbf{g}.$$

The viscosity μ of a fluid plays an important role in several ways, including the suppression of turbulence. In a fluid of uniform density, the viscosity consists entirely of the resistance to shear. The velocity shear is defined most simply for a unidirectional flow in which the speed v varies perpendicular to the flow at a rate denoted by v/L. The viscosity μ is defined as the ratio of the force F per unit area (required to maintain the shear) to the shear rate, that is, $F = \mu v/L$. In most fluids (e.g., gas, water, mercury, etc.) the viscosity is essentially independent of v/L. But in slurries and polymerized fluids, μ may depend very much on v/L, such fluids being referred to as non-Newtonian.

The effect of the viscosity in a continuous flow of homogeneous fluid is to push each small element of fluid toward the mean velocity of the surrounding elements of fluid. The difference between the velocity at any point and the mean velocity on a small sphere centered on that point is proportional by the Laplacian $\nabla^2 = \partial^2/\partial x^2 + \partial^2/\partial y^2 + \partial^2/\partial z^2$ operating on $\mathbf{v}$. It can be shown, then, that the viscosity introduces a volume force $\mu\nabla^2\mathbf{v}$, so that

$$\rho\left[\frac{\partial \mathbf{v}}{\partial t} + (\mathbf{v} \cdot \nabla)\mathbf{v}\right] = -\nabla p + \rho\mathbf{g} + \mu\nabla^2\mathbf{v}.$$

The equation of motion including the viscosity is known as the Navier–Stokes equation. It provides a complete mathematical description of the flow of an incompressible fluid, including the chaotic turbulent state. It should not come as a surprise, then, that the general mathematical solution of the Navier–Stokes is intractable. The many known analytical solutions to the Navier–Stokes equation do not include cases of turbulence, although they embrace many non-chaotic fluctuating flows.

It was pointed out earlier that a flow becomes turbulent when the inertia of neighboring regions of fluid dominates the viscous forces between the regions. The inertial forces are characterized by $\rho(\mathbf{v} \cdot \nabla)\mathbf{v}$ on the left hand side. Recalling that ∇ implies the change per unit length, it follows that in a flow of characteristic scale L the inertial force term $\rho(\mathbf{v} \cdot \nabla)\mathbf{v}$ is of the general order of $\rho v^2/L$, while the viscous term $\mu\nabla^2\mathbf{v}$ is of the order of $\mu v/L^2$. The ratio of the inertial to viscous forces is called the Reynolds number, N_R, equal to $\rho v L/\mu$. Thus, for instance, if L is set equal to the radius of a pipe with smooth walls, the flow becomes turbulent when the Reynolds number exceeds approximately 10^3.

In high-speed flows, or large-scale flows in a stratified atmosphere, or in sound waves, the density of the fluid varies with the time and with position. The relation between the density change and the velocity $\mathbf{v}$ is easily described from the principle that matter is conserved (i.e., matter is neither created nor destroyed). The net flow of mass out of a region (per unit volume of the region) is described by the divergence $\nabla \cdot (\rho\mathbf{v})$ of the mass flux $\rho\mathbf{v}$, with the result that ρ declines at the rate

$$\frac{\partial \rho}{\partial t} = -\nabla \cdot (\rho\mathbf{v}),$$

often called the equation of continuity. If ρ is uniform, then $\nabla \cdot \mathbf{v} = 0$.

The pressure of a compressible gas depends upon both the density and the temperature through the equation of state, $p \cong R\rho T = NkT$, where R is a constant. For adiabatic changes, p varies in proportion to ρ^γ, or $T^{\gamma/(\gamma-1)}$ ($T \sim \rho^{\gamma-1}$), where γ is the ratio of the specific heats, C_p/C_v for constant pressure and volume, equal to 1.40 in air and 1.67 in helium.

The viscosity opposes the compression or expansion associated with changes in ρ, so that a second viscosity coefficient appears with a term $\nabla(\nabla \cdot \mathbf{v})$ in the Navier–Stokes equation. When working with an incompressible fluid, it is convenient to eliminate the pressure p, taking advantage of the fact that the curl of a gradient is identically zero. Then we define the vorticity as the curl of $\mathbf{v}$, written $\nabla \times \mathbf{v}$ and note the vector identity $(\mathbf{v} \cdot \nabla)\mathbf{v} = (\nabla \times \mathbf{v}) \times \mathbf{v} + \nabla\frac{1}{2}v^2$. The curl of the Navier–Stokes equation becomes the vorticity equation

$$\frac{\partial \omega}{\partial t} = \nabla \times (\mathbf{v} \times \omega) + \nu\nabla^2\omega,$$

where $\nu \equiv \mu/\rho$ is the kinematic viscosity and $\omega = \nabla \times \mathbf{v}$ is the vorticity. This relation exhibits some of the many remarkable properties of hydrodynamic flows. For instance, insofar as ν is negligible, the vorticity ω is transported bodily with the flow $\mathbf{v}$. The field lines of ω, called the vortex lines, can be considered as moving precisely with the fluid. Writing the vorticity equation in the Lagrangian form

$$\frac{d\omega}{dt} = \frac{\partial \omega}{\partial t} + (\mathbf{v} \cdot \nabla)\omega = (\omega \cdot \nabla)\mathbf{v}$$

it follows that the vorticity is increased by the stretching of the vortex lines in the non-uniform flow $\mathbf{v}$, at the rate $(\omega \cdot \nabla)\mathbf{v}$ at which $\mathbf{v}$ increases in the direction ω of the vortex lines. The vortex lines continually lengthen in an irregular flow (just as a drop of ink would be drawn into a ribbon) and are increasingly kinked and looped on ever smaller scales, representing the formation of ever smaller swirls and eddies. The creation of smaller eddies extends down to the scale where the Reynolds number of the individual eddy falls to a value of the order 10^2 and viscosity suppresses any smaller eddies.

The viscosity converts the kinetic energy of the motion into thermal motion of the atoms and molecules of the fluid. The mean rate of dissipation of kinetic energy and, hence, the heating rate within a volume of fluid with static boundaries, can be shown to be $\mu\omega^2$ ergs·cm^{-3}·s^{-2}, which is usually largest in the smallest eddies in a turbulent flow.

Numerical simulation of fluid motions has opened a new line of investigation that complements laboratory experiments. It must be kept in mind that, besides demonstrating a flow, it is essential to understand the mechanics of the flow in terms of basic principles. The problem of understanding becomes increasingly difficult with increasing Reynolds numbers and chaos in the flow, so that the theory of turbulence turns toward probability distributions, rather than working with the immense sea of information necessary to provide a detailed description. Even the contemporary super computer cannot provide a simulation of strong turbulence, because the number of degrees of freedom of a hydrodynamic flow with Reynolds number N_R in n dimensions is of the general order of $(0 \cdot 1\, N_R)^n$. The best that can be done in the study of large-scale flows in nature (e.g., winds in the terrestrial atmosphere, ocean currents, convection in the Sun, etc.), with Reynolds numbers ranging upward from 10^5 to 10^{12} or more, is to declare that the effect of the turbulence on the mean large-scale flow is only to add an effective eddy viscosity ν_e to the kinematic viscosity ν of the fluid, with ν_e of the order of the mean velocity difference $v(L)$ across some appropriate scale L, multiplied by the scale L, so that $\nu_e \sim 0 \cdot 1 L|\nabla v(L)|$. Thus, the effective Reynolds number vL/ν_e for the mean flow is reduced to some value such that the mean flow is only weakly chaotic at most and so can be simulated by existing numerical methods. The concept is imprecise at best, and its weakness is obscured by the euphemism that the turbulence has been suitably parameterized. Even with so grand a simplification, the numerical simulation of a system as complicated as the vertically stratified terrestrial atmosphere can be treated only on a coarse grid. So it will be years before a precise global simulation of the terrestrial atmosphere is possible.

See also: FLUID DYNAMICS; FLUID STATICS; NAVIER–STOKES EQUATION; REYNOLDS NUMBER; TURBULENCE; TURBULENT FLOW; VISCOSITY

Bibliography

BATCHELOR, G. K. *An Introduction to Fluid Dynamics* (Cambridge University Press, Cambridge, Eng., 1967).

LANDAU L. D., and LIFSCHITZ, E. M. *Fluid Mechanics* (Pergamon Press, London, 1959).

E. N. PARKER

HYDROGEN BOMB

See FUSION BOMB

HYDROGEN BOND

The properties of conglomerates of molecules such as water or ammonia cannot be understood by looking at an individual molecule itself. The interaction between the individual molecules often has dramatic consequences for their properties. Examples of interactions between molecules include the van der Waals force and the hydrogen bond.

The hydrogen bond was first mentioned in the chemical literature in 1912. It was introduced to explain the properties of certain ammonia derivatives. However, the nature of this chemical bond could not be investigated until quantum mechanics was created in 1925.

The hydrogen bond requires, as its name already suggests, the participation of a hydrogen atom. The other bond partners are typically highly electronegative elements of small atomic radius like oxygen, nitrogen, and fluorine. The electronegativity of an atom characterizes its ability to attract the bond electrons in a covalent bond. For example, the electronegativity of oxygen (3.5) is much bigger than that of hydrogen (2.1). To explain the picture of the hydrogen bond that emerges from quantum mechanics, we consider a specific example, the hydrogen bond formed between two water molecules. The electrons in the O–H bond of a water molecule are strongly attracted by the electronegative oxygen atom, so the electron density at the hydrogen atom is reduced. The positive charge of the hydrogen nucleus is therefore not so well shielded by electrons. A free-electron pair of the oxygen atom in a second water molecule is attracted by this deshielded hydrogen nucleus, leading to the formation of a hydrogen bond. Two unique features of the hydrogen atom are necessary to stabilize this type of bonding: the hydrogen atom has no inner electrons that would repel the free-electron pair, and the small size of the hydrogen atom, which allows the free-electron pair to get close to the attracting nucleus. On the other hand, the small size of the hydrogen atom limits the number of the hydrogen-bonded partners to one. Furthermore, the deshielded hydrogen nucleus can effectively attract only electron pairs that are well localized in space. Thus only electronegative atoms that possess such electron pairs can have an effective interaction with the hydrogen nucleus. Usually, as in the above example, the distance between the hydrogen atom and the hydrogen-bonded atom is larger than the distance between the hydrogen atom and its covalent bond partner. The hydrogen bond may be so strong that the bond distances become equal, as in the case of the FHF^- ion in solid KHF_2. This compound involves fluorine, the most electronegative of all elements (electronegativity of 4). In this case it is not possible to distinguish between the covalent bond and the hydrogen bond.

The energy needed to break a hydrogen bond is typically of the order of 5,000 cal/mol, compared with 50,000–100,000 cal/mol typical for regular types of bonds (1 kcal = 4,186 J). The low binding energy of the hydrogen bond explains why it can be broken at room temperature.

Important physical properties of compounds are affected by the formation of hydrogen bonds. The hydrogen bond explains the unusual high melting and boiling points of water and other hydrides, such as NH_3 and HF, formed with elements of the second row of the periodic table. Without the stabilizing effect of the hydrogen bonds, the melting and boiling points of water would be $-100°C$ and $-80°C$, respectively. The crystal structure of ice is an effect of the hydrogen bond. Each of the two free-electron pairs of the oxygen atom in a water molecule participates in a hydrogen bond, so that each oxygen atom is surrounded by four hydrogen atoms. In liquid water (and other first row hydrides) a large number of molecules are held together by hydrogen bonds. The individual water molecules form small dipoles with the oxygen as the negative part and the hydrogens as the positive part of these dipoles. The formation of hydrogen bonds leads to a lining up of these dipoles that results in a large dielectric constant for water. In much the same way, the hydrogen bond affects the properties of other compounds like alcohols and carboxylic acids.

Despite the weakness of the hydrogen bond, it plays an important role in the structure of biological molecules. The large number of hydrogen bonds in these structures assures their stability. Proteins are made up of polypeptide chains, which can be connected to themselves and to other polypeptide chains via hydrogen bonds. For example, silk fiber has a β-keratin structure in which individual polypeptides are held together by hydrogen bonds to form planar structures. The genetic information of living things is stored in nucleic acids such as deoxyribonucleic acid (DNA). The two strains of DNA are another important example of a biological structure held together by hydrogen bonds.

See also: COVALENT BOND; DIELECTRIC CONSTANT; IONIC BOND; MOLECULE; QUANTUM MECHANICS, CREATION OF; WATER

Bibliography

COULSON, C. A. *Valence,* 2nd ed. (Oxford University Press, New York, 1961).

GREENWOOD, N. N., and EARNSHAW, A. *Chemistry of the Elements* (Pergamon Press, New York, 1984).

MASTERTON, W. L.; SLOWINSKI, E. J.; and STANITSKI, C. L. *Chemical Principles with Qualitative Analysis,* 6th ed. (Saunders, Philadelphia, 1986).

PAULING, L. *The Nature of the Chemical Bond,* 2nd ed. (Cornell University Press, Ithaca, NY, 1948).

STREITWIESER, A., JR., and HEATHCOCK, C. H. *Introduction to Organic Chemistry,* 3rd ed. (Macmillan, New York, 1985).

MATTHIAS ERNZERHOF

HYDROMETER

A hydrometer is an instrument used to measure the density or specific gravity of liquids. It consists of a glass bulb, weighted at the bottom, and attached to a narrow uniform stem that is marked with a calibrated scale. When placed in a liquid, a hydrometer floats, its weight being such that only the stem protrudes from the liquid. The reading on the scale, at the surface of the liquid, is a measure of the density of the liquid.

How the Hydrometer Works

Hydrometers work because any object immersed in a fluid experiences a buoyant force equal to the weight of the fluid that the object displaces (Archimedes' principle). The weight of the fluid displaced is equal to the product of the density of the fluid, the volume displaced, and the acceleration due to gravity, g. Figure 1 shows identical hydrometers floating in two different liquids: water, with density ρ_w, and another with a lower density ρ. In each case the weight of the hydrometer is balanced by the buoyant force. More of the less-dense liquid must be displaced to balance the weight of the hydrometer, so the stem of the hydrometer on the right is submerged an additional distance h. To find the relationship between h and ρ, carry out the following steps, using the notation given in the diagram. Set the two buoyant forces equal to each other ($g\rho_w V_w = g\rho V$); use that equation to write an expression for the change in volume $V - V_w$ and equate that to the change in volume of the submerged part of the stem, $h \times A$. The result is

$$h = -\frac{V_w}{A}\left(1 - \frac{\rho_w}{\rho}\right)$$

This shows that the relationship between h and ρ is a nonlinear one, but that a graph of h versus $(1/\rho)$ will give a straight line. The graph can be used to calibrate homemade hydrometers, using experimental data for h in liquids of known density.

Uses

The uses of hydrometers include testing products such as petroleum, milk, and alcoholic beverages; measuring the concentrations of solutions; checking the state of charge of lead-acid storage batteries; and serving as detectors for control devices. For precise results, liquids tested must be at the temperature for which the hydrometer was calibrated. A label on the hydrometer such as "Sp. Gr. (60/60)°F," for example, indicates that the scale readings give the specific gravity of a liquid at 60°F relative to the specific gravity of water at 60°F. For very precise measurements, the effects of surface tension and surface contamination must be accounted for.

Scales

The stem of the hydrometer can be marked off to show density, specific gravity, or percent concentration directly. In precision hydrometers these are nonlinear scales, due to the nonlinear relationship between h and ρ. Linear scales, for specialized uses, include the API scale, devised by the American Petroleum Institute for measuring the density of petroleum products, and the Baumé scale, for making solutions of specific concentrations. Reference handbooks contain tables and equations that can be

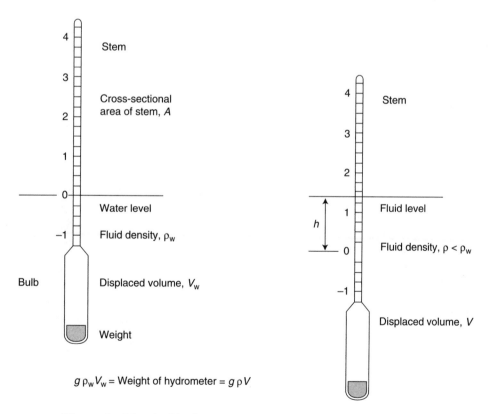

Figure 1   Identical hydrometers floating in two different liquids.

used to convert these linear scale readings to density or specific gravity.

See also: ARCHIMEDES' PRINCIPLE; BUOYANT FORCE; DENSITY; SPECIFIC GRAVITY

Bibliography

BAUMEISTER, T., and MARKS, L. S., eds. *Standard Handbook for Mechanical Engineers,* 7th ed. (McGraw-Hill, New York, 1967).

HOLZBOCK, W. G. *Instruments for Measurement and Control,* 2nd ed. (Reinhold, New York, 1962).

SEARS, F. W.; ZEMANSKY, M. W.; AND YOUNG, H. D. *College Physics,* 7th ed. (Addison-Wesley, Reading, MA, 1991).

WEAST, R. C., ed. *CRC Handbook of Chemistry and Physics,* 65th ed. (CRC Press, Boca Raton, FL, 1984).

CHARLES E. TAYLOR

HYPERFINE STRUCTURE

Hyperfine structure (hfs) was observed shortly after the development of interferometers first by Albert Michelson in 1891 and then by Charles Fabry and Alfred Pérot in 1897. The name arises from the fact that in many atomic spectra a single spectral line, with increasingly high resolution, becomes a group of closely spaced lines and upon finer resolution these are resolved into still more lines. The first group is called fine structure and the next hfs. The latter are separated by energies of order 10^{-3} to 1 cm^{-1} corresponding to 10^{-7} to 10^{-4} eV, depending upon the atom. In 1924 Wolfgang Pauli suggested that this structure was due to the interaction of the electrons with the nuclear magnetic field. He pointed out that if there is a nonvanishing nuclear moment it must be directed along the nuclear angular moment. Then the different lines in the hfs multiplet would result from different orientations of the nuclear angular momentum relative to the total angular momentum of the atom. The total angular momentum vector **F** is composed of the total electronic angular momentum **J** and the nuclear angular momentum **I** vectors:

$$\mathbf{F} = \mathbf{I} + \mathbf{J} \tag{1}$$

and is a constant of motion of the atom. The magnitudes of both **I** and **J** are fixed within a hfs multiplet, and it is only their relative orientation, and therefore the magnitude of **F**, that varies. In 1930 Enrico Fermi first applied the quantum theory to this problem. He found that the angular momentum addition rules lead to the fact that the number of lines in a multiplet is the smaller of $(2j + 1)$ or $(2i + 1)$ where j and i are the angular momentum quantum numbers corresponding to **J** and **I,** respectively. The same theory leads to the hfs interval rules, which say that the lines are ordered according to f, the quantum number of **F,** and that they are separated by Cf. Here C is a constant, varying from atom to atom, and f is the quantum number of the upper line. This is the normal hfs interval rule. The inverted interval rule describes a multiplet with the same structure but with the largest f having the lowest energy. The constant C is proportional to the magnitude of the nuclear magnetic dipole moment, and hfs measurements were used in the early days of nuclear physics to determine this moment and the nuclear angular momentum for many different atoms.

The hyperfine separation of the ground state of atomic hydrogen is one of the most accurately measured numbers in all of physics. It was measured in 1963 by Norman Ramsey and is known to twelve significant figures with an error of ±3 in the twelfth figure. This, the most common isotope of hydrogen, has a nucleus which is a single proton and only one electron with zero orbital angular momentum in the ground state. Both particles have an intrinsic angular momentum (spin) of one-half so the two angular momentum vector can, quantum mechanically, only be oriented either parallel ($f = 1$) or antiparallel ($f = 0$). The $f = 1$ level lies higher so this is an example of normal hfs. It is the separation of these two levels, about $1,420 \text{ mHz} = 5.873 \times 10^{-6}$ eV that Ramsey measured. Application of the Dirac theory of the electron gives agreement to about 0.1 percent with the measured number. A better agreement requires a quantum field theoretical description of the electron and the radiation.

Hydrogen is the most common element of the universe and makes up the bulk of the matter in stars and interstellar space. This hyperfine emission is one of the more common forms of radiation and is used to determine the temperature and bulk velocity relative to the earth. The former is due to the thermal broadening of the line and the latter is due to its Doppler shift.

See also: FERMI, ENRICO; GROUND STATE; INTERFEROMETRY; INTERFEROMETER, FABRY-PÉROT; NUCLEAR MOMENT; PAULI, WOLFGANG; QUANTUM FIELD THEORY; QUANTUM NUMBER

Bibliography

BRANSDEN, B. H., and JOACHAIN, C. J. *Physics of Atoms and Molecules* (Longman, New York, 1983).

KOPFERMANN, H., and SCHNEIDER, E. E. *Nuclear Moments* (Academic Press, New York, 1958).

MARVIN H. MITTLEMAN

HYSTERESIS

Hysteresis is the nonreversibility of a physical system during the application of an external variable owing to some kind of "memory" attributed to the system. The memory is usually due to a frictional effect which inhibits the change being forced by the external variable. The amount of hysteresis depends upon the history of the system and occurs in a wide variety of physical systems.

The most common and historical example of hysteresis is seen in the behavior of a ferromagnet under the influence of an externally applied magnetic field. The features are illustrated by the "ferromagnetic curve" or "hysteresis loop," which can be described as follows. A ferromagnet can have a magnetization, M, equal to zero when the small, ferromagnetically ordered domains within it are randomly oriented. The application of an external field, $\mathbf{H}$ (in units of A/m) can overcome the resistance of the domains to reorient, thereby aligning them along the applied field. This results in the familiar macroscopic magnetization, $\mathbf{M}$, of a "magnet." As the field is increased the alignment of the domains will eventually be complete and the magnetization will reach a maximum. This is shown in Fig. 1 as the curve which begins at zero and maximizes to the right. The total internal field in the magnet is called the magnetic induction field, $\mathbf{B}$ (in units of Tesla), and is the sum of the magnetization produced by aligning the domains and the applied field. Normally the applied field can be ignored compared to the magnetization. In Fig. 1, μ_0 is called the permeability of free space. Reducing the external field to zero does not reduce the magnetization to zero, again because the domains tend to remain aligned leaving a remnant field shown as $\mathbf{B}_r$ in Fig. 1. The applied field in the reverse direction that is necessary to bring the magnetization to zero is called the coercive field, $\mathbf{H}_c$. Applying an increasingly larger field in the reverse direction will eventually saturate the magnetization in that direction. The cycle can be completed by again reversing the field—applying the field in the original direction—until the former positive magnetization is obtained. The entire cycle will produce a curve with an area $\int \mathbf{B} \cdot d\mathbf{H}$ which represents the work done per unit volume during the cycle. The larger the coercive field or the larger the area under this magnetization curve, the "harder" the magnet is said to be. Soft magnets have a small area within the magnetization curve.

Analogously, another common system that shows hysteresis is that of ferroelectric compounds. Here the work per unit volume done during a cycle is the area under the applied electric field versus the polarization.

Hysteresis is seen in certain structural phase transitions where the work done overcomes the chemical forces of transforming from one ordered structure to another. Macroscopic physical systems can show hysteresis in the form of mechanical "D-denting" similar to the common oil-canning in which the work done is force times distance. There are examples of hysteresis in superconducting materials, especially evident in the current versus voltage cycle of the appropriate geometry of a Josephson

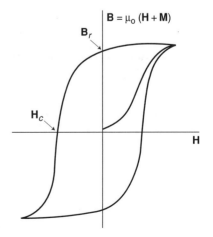

Figure 1 Sample hysteresis loop.

junction. Also because the "magnetization" of a superconductor is negative, superconductors show an inverse or backward magnetization curve.

See also: ANTIFERROMAGNETISM; FERROMAGNETISM; MAGNETIC MATERIAL; MAGNETIZATION; PHASE TRANSITION; SUPERCONDUCTIVITY

Bibliography

BLAKEMORE, J. S. *Solid-State Physics* (Saunders, Philadelphia, 1974).

LYNN, J. W., ed. *High-Temperature Superconductivity* (Springer-Verlag, New York, 1990).

JOHN E. DRUMHELLER

I

ICE

Ice is water (H_2O) in the solid state. The melting point T_m of ice at one atmosphere pressure, $P = 1$ atm $= 760$ torr $= 9.81 \times 10^4$ Pa, is, by definition of the Celsius temperature scale, $T_m = 0°C = 273.15$ K. At that temperature and pressure, ice has a density $\rho = 0.917$ g/cm^3 = 917 kg·m^{-3}. Ice is less dense than liquid water which, at $T = 0°C$ and $P = 1$ atm, has a higher density of $\rho = 1.000$ g/cm^3 = 1000 kg·m^{-3}. As a consequence, ice floats in water such that about 10 percent of the ice volume extends above the surface of the water. The specific volume V being the inverse density ($V = 1/\rho$) of ice is $V = 1.0908$ cm^3/g $= 1.0908 \times 10^{-3}$ m^3·kg^{-1}. The heat of fusion necessary to melt ice is $Q = 3.34 \times 10^5$ J·kg^{-1} or 6,010 J/mol.

Ice has two exceptional properties where it differs from (almost) all other solids materials. (Other materials with such exceptions are gallium, bismuth, and germanium.) The exceptional properties concern the phase transition between ice and liquid water, that is, the melting when going from the solid to the liquid phase or, reversely, the freezing when going from the liquid to the solid phase.

The first exceptional property of ice is its shrinking in volume, by approximately 10 percent, when melting to liquid water. (Compare the density ρ and specific volume V of ice and of water given above.) This exceptional property is called the anomalous volume contraction of ice upon melting. Since most people live in places where the average temperature is above the melting point of ice, the reverse effect of that anomaly, namely, the anomalous volume expansion of water upon freezing, is more familiar. Examples from everyday life are bursting water pipes or car radiators when the temperature drops below freezing.

The second exceptional property of ice is a lowering of the melting point T_m with increasing pressure. Ice skating is an everyday phenomenon where this property is involved. The high pressure just below the edge of an ice skate causes the melting of a thin film of ice. This means that the skate actually slides on a thin film of liquid water which, after passage of the skate, immediately refreezes. Another example of the same effect is the motion of glaciers which actually slide downhill on a thin film of water at the interface between the glacier and the rock beneath. The exact lowering of the melting point of ice is 0.0075°C for each atmosphere increase in pressure or, in short, $dT_m/dP = -7.5 \times 10^{-3}$ °C/atm $= -74$ K·GPa^{-1}.

The two exceptional properties of ice are thermodynamically related by the Clausius–Clapeyron equation, which relates properties at a phase transition, $dT_m/dP = (V_l - V_s)\ T_m/Q$. Here V_l and V_s are the specific volumes of the liquid and solid phase of water, respectively, and T_m is the melting point on the absolute temperature scale. Since T_m (in kelvin units) and the heat of fusion Q are positive quantities, the anomalous contraction of ice upon melting, $(V_l - V_s) < 0$, necessitates, by the Clausius–Clapeyron equation, the anomalous lowering of the melting point of ice with increasing pressure, $dT_m/dP < 0$.

The anomalous properties of ice are easy to understand, on a conceptual level, by considering the positions (configuration) of the water molecules in ice and in liquid water. The configuration of water molecules is, to a large extent, a consequence of the peculiar, heartlike shape of each water molecule. When the large oxygen atom and the two small hydrogen atoms form a H_2O molecule, the H atoms are located at an angle of $104°$ from the central O atom rather than in a straight line. The nonalignment of the three atoms gives the H_2O molecule a polar character which, in turn, gives rise to the large dielectric constant of water. The shape of the H_2O molecules and their size, or volume V_{H_2O}, can be considered to be the same, regardless whether the molecules are in water vapor, in liquid water, or in ice. It is only the spacing and the orientation of the H_2O molecules that determine the actual phase.

In liquid water, the H_2O molecules are close together, subject to short-distance thermal motion, but randomly oriented. However, when water freezes to ice, the H_2O molecules arrange themselves in a hexagonal, three-dimensional ring structure. A consequence of the hexagonal arrangement of molecules shows up on the macroscopic scale by the six-cornered shape of snow flakes. On the microscopic scale, the structure is such that always six H_2O molecules surround a tiny void. (The void is a very small region of vacuum, that is, of empty space. This is to say that the void is *not* a tiny air bubble.) The size of such a void is roughly half the size of a single H_2O molecule. More accurately, the volume of the void V_0 is about 60 percent of the volume of a H_2O molecule, $V_0 \approx 3/5 \, V_{H_2O}$. The ratio between the volume of a void and the volume of six surrounding H_2O molecules is then $V_0/(6 \, V_{H_2O}) = (3/5 \, V_{H_2O})/(6 \, V_{H_2O}) = 1/10$. Thus the 10 percent volume expansion of freezing water is a direct consequence of the hexagonal arrangement of the H_2O molecules during ice formation requiring additional space of $1/10 \, V_{H_2O}$ for each molecule's contribution to a void. When, in the reverse process, ice melts to water, the hexagonal arrangement of the H_2O molecules collapses, causing the voids to disappear. This explains, conceptually, the 10 percent volume contraction of ice upon melting.

The hexagonal structure of ice is present for pressure up to about 2,000 atm. Under higher pressure, ice undergoes phase transitions to other structures, also called modifications. Besides the hexagonal structure, ten other modifications of ice are known to date.

See also: BUOYANT FORCE; DENSITY; DIELECTRIC CONSTANT; HYDROGEN BOND; PHASE TRANSITION; PRESSURE; TEMPERATURE SCALE, CELSIUS; TEMPERATURE SCALE, KELVIN

Bibliography

HEWITT, P. G. *Conceptual Physics*, 6th ed. (Scott, Foresman, Glenview, IL, 1989).

SEARS, F. W., and SALINGER, G. L. *Thermodynamics, Kinetic Theory, and Statistical Thermodynamics*, 3rd ed. (Addison-Wesley, Reading, MA, 1975).

MANFRED BUCHER

IDEAL GAS LAW

The ideal gas law is a simple equation of state describing how a gas responds to macroscopic changes in pressure, temperature, or volume. It is an accurate empirical expression derived from the laws of Robert Boyle and Jacques Charles. Boyle's law observes that for gas at a constant temperature, the volume is inversely proportional to pressure. This is complemented by Charles's law that volume is directly proportional to temperature at constant pressure. If the gas is composed of a single kind of molecule (or atom, if it is a monatomic gas like sodium), the ideal gas law is stated in either of two ways:

$$PV = nRT$$

or

$$PV = Nk_BT,$$

where P, V, T, n, and N are thermodynamic variables of pressure, volume, temperature, the number of moles of the gas, and the number of molecules. R, which is equal to $8.314510 \, \text{J} \cdot \text{K}^{-1} \cdot \text{mol}^{-1}$, is the universal gas constant, and k_B, which equals $1.380658 \times 10^{-23} \, \text{J} \cdot \text{K}^{-1}$, is the Boltzmann constant. Note that the two equations above are equivalent since the ratio, $N/n = R/k_B$, is the Avogadro number N_A, which equals 6.022×10^{23} molecules per mole.

Behavior of real gases approaches that of an ideal gas at high temperature and low density. Most gases at room temperature and atmospheric pressure act like ideal gases. For high density or low temperature, intermolecular interactions complicate the response of the gas, leading to relations such as the van der Waals equation:

$$P = \frac{RT}{v - b} - \frac{a}{v^2},$$

where v is a molar volume (V/n) and a and b are empirical constants characteristic of the particular gas.

To gain a deeper understanding of why gases behave this way, we need a clearer picture of how macroscopic quantities such as temperature and pressure derive from interactions at the microscopic level. This is the point at which explanations of an ideal gas and the more general equation of state diverge. A large-scale coherent response for an ideal gas assumes infrequent interactions occur among individual molecules and that temperature and pressure reflect only global averages of molecular motion. For simplicity, we assume all molecules in the gas are composed of the same material, that is, a pure gas. We assume that the number of molecules is large and the average distance between molecules is much larger than the size of each molecule. This reduces the complexity to the consideration of moving point masses in a large volume. We assume these point masses interact only through elastic collisions, that is, they conserve momentum and energy and do not change the shape of the molecules. Further, we assume the particles move randomly according to Newton's laws of motion. To limit external effects, we assume the gas is in thermal equilibrium with the walls of the container, that is, the total kinetic energy transferred to the walls by collisions is the same as the total kinetic energy received from the walls. These approximations provide the basis for defining averaged macroscopic quantities.

A molecule with velocity components v_x, v_y, v_z and kinetic energy $\epsilon = 0.5 \, mv^2 = 0.5 \, m(v_x^2 + v_y^2 + v_z^2)$ strikes a wall elastically, that is, the momentum in the

x direction changes from $p_x = mv_x$ to $p_x = -mv_x$, preserving the energy component, $0.5 \, mv_x^2$. The change in momentum is equivalent to the average force exerted by the molecule on a wall over a time Δt. For a cubical box of length d, the molecule travels to the far wall and returns, exerting the force F on this wall every time period $\Delta t = 2d/v_x$. The force exerted by a single particle is $F = (-2mv_x)/\Delta t = (-mv_x^2/d)$. The average force exerted by N particles is $-Nm \langle v_x^2 \rangle / d$ where $\langle v_x^2 \rangle$ is the average particle speed in the x direction. The pressure felt by this wall is the average force per unit area, $(Nm \langle v_x^2 \rangle / d)/d^2$; d^3 is recognized as the volume V of the cube. Since there is no preferred direction of motion for any particle, $\langle v_x^2 \rangle = \langle v_y^2 \rangle = \langle v_z^2 \rangle = \langle v^2 \rangle / 3$. The average pressure can then be expressed as $P = (2/3) N \langle \epsilon \rangle / V$ or $(2/3)(N/V) \langle 0.5 \, mv^2 \rangle$. Rearranging and comparing with the ideal gas law, we see

$$PV = Nk_BT = N(2/3) \langle 0.5 \, mv^2 \rangle.$$

That is, temperature is directly proportional to the average kinetic energy of the molecules.

In the above example, the particles were free to move in the x, y, or z dimensions. We say the particles had three translational degrees of freedom. Since there was no preferred direction of motion, there was equipartition of energy among the three dimensions; that is,

$$k_BT/2 = (1/3)\langle 0.5 \, mv^2 \rangle = (m/2)\langle v_x^2 \rangle$$

$$= (m/2)\langle v_y^2 \rangle = (m/2)\langle v_z^2 \rangle.$$

In applying the ideal gas law there are four important limits: isothermal, isobaric, isovolumetric, and adiabatic changes; that is, constant temperature, constant pressure, constant volume, and constant mass and energy, respectively. Adiabatic changes indicate that the system is isolated from external influence, so the mass and energy are conserved within the system. Adiabatic changes are the only variations that involve more than two variables. The first law of thermodynamics describes internal energy transfer, that is, an increase in heat, $\Delta U = Q = nC_V\Delta T$, resulting directly from the work expended by the system, $-P\Delta V$. Here, C_V is the molar heat capacity of the gas at constant volume. At constant pressure, the molar heat capacity is $P\Delta V \mid P =$

$nC_P\Delta T$. Conservation of mass within the system requires that $\Delta n = 0$. Changes in the variables of the ideal gas law require

$$P\Delta V + V\Delta P = nR\Delta T = (R/C_V)(-P\Delta V)$$

using the first law. This could also be written $nC_P\Delta T - nC_V\Delta T = nR\Delta T$, so $R = C_P - C_V$, or writing $\gamma = C_P/C_V$, $R = (\gamma - 1)C_V$. Dividing by PV, the equation above becomes

$$\frac{\Delta V}{V} + \frac{\Delta P}{P} = \frac{(1 - \gamma)\Delta V}{V}.$$

This integrates to $\gamma\ln V + \ln P = $ constant, or

$$PV^\gamma = \text{constant},$$

the law for an ideal and adiabatic gas. This can also be expressed as $TV^{(\gamma - 1)} = $ constant.

Often, it is more convenient to work with a variable density $\rho = N/V$, giving an ideal gas law expressed as $P = \rho k_B T$. This equation is often preferred in atmospheric or ionospheric applications where the volume is not a measurable quantity. It has the advantage of reducing the equation from four variables to three. Differentiating the equation leads to the following general expression:

$$\frac{dP}{d\rho} = \frac{P}{\rho} + \frac{\rho d(k_B T)}{d\rho}.$$

The adiabatic limit is expressed as $P = C\rho^\gamma$ and $T = C_1\rho(\gamma - 1)$, where C and C_1 are the constants.

The ideal gas law played a significant role in the development of thermodynamics. It lead to the concept of an absolute temperature scale (Kelvin) associated with the total kinetic energy of the gas, thus tying the microphysics of individual particles to thermal properties on the macroscale. This is well demonstrated by following the cycles of the Carnot engine for an ideal gas.

See also: BOYLE'S LAW; CHARLES'S LAW; EQUILIBRIUM; EQUIPARTITION THEOREM; KINETIC THEORY; THERMODYNAMICS

Bibliography

ANDERSON, H. L., and COHEN, E. R. "Thermodynamics" in *A Physicist's Desk Reference*, 2nd ed., edited by H. L. Anderson (American Institute of Physics, New York, 1989).

ALICE L. NEWMAN

IDEALIZATION

See APPROXIMATION AND IDEALIZATION

IMAGE, OPTICAL

An optical image of an object is produced when the rays of light emanating from the object are refracted by a lens or reflected by a mirror. The refraction or reflection causes the light rays to change their direction of propagation. The optical image is formed at the location in space where the light rays from a point on the object converge, or from where they appear to emanate. When the rays actually converge, the image is known as a real image and can be seen on a screen placed at the location of the image; when they only appear to emanate from a point, the image is called a virtual image and cannot be seen on a screen. While the image looks like the object, many of its properties can differ. The location of the image is generally different from the location of the object. The size of the image can be smaller or larger than the size of the object. And the image can be erect (right-side up) or inverted (up-side down). In addition, optical images can be created by a system of more than one optical element (lenses and/or mirrors) and the optical image resulting from the first optical element becomes the object for the second one, and so on.

Optical elements have a focal point and the distance from the optical element to this point is known as the focal length f. Parallel light rays can be considered to be emitted from an object an infinite distance away. The focal point of an optical element is defined as the point where parallel rays that strike the element converge, or appear to originate. In the case of a lens, if the focal point is on the same side of the lens as the parallel light rays, the focal length is defined to be negative; on the other side, it is positive. A lens with a positive focal length is called a convex or coverging lens; a negative focal length is known as concave or diverging. Similar considerations apply for spherical mirrors. A concave mirror, however, has a positive focal length while a for a convex mirror, it is negative.

Defining the distance of the object from an optical element as o, and the image distance as i, these distances are related to the focal length by

$$\frac{1}{i} + \frac{1}{o} = \frac{1}{f},$$

which is known as the lens equation. For an object an infinite distance away, this equation results in $i = f$, as expected from the definition of focal point. If i is positive, then the image is real and on the opposite side of a lens from the object; if i is negative, the image is virtual and appears to be on the same side of the lens as the object. The magnification M of an optical element is defined as the ratio of the object size to the image size. From geometrical considerations,

$$M = -\frac{i}{o},$$

and the image is erect or inverted depending upon whether M is positive or negative.

See also: LENS; MIRROR, PLANE; OPTICS; REFLECTION; REFRACTION

Bibliography

FISHBANE, P. M.; GASIOROWICZ, S.; and THORNTON, S. T. *Physics for Scientists and Engineers* (Prentice Hall, Englewood Cliffs, NJ, 1993).

HECHT, E. *Optics*, 2nd ed. (McGraw-Hill, New York, 1974).

SERWAY, R. A., and FAUGHN, J. S. *College Physics*, 2nd ed. (Saunders, Philadelphia, 1989).

STEVEN T. MANSON

IMAGE, VIRTUAL

Lenses, mirrors, and optical devices containing combinations of lenses and mirrors form two primary types of optical images: real and virtual. In a real image, a bundle of light rays originating at each point on an object is brought to focus at a corresponding point on the image. Such an image can be projected on a screen as with a slide projector. The cornea and lens of the human eye work in conjunction to project a real image on the eye's retina, just as a camera lens projects a real image on the film plane. In each of these examples, light rays actually pass through the real image.

In a virtual image, on the other hand, a bundle of light rays originating at each point on the object only appears to come from a corresponding point on the image. Since light rays do not actually pass through the image, a virtual image cannot be projected on a screen. A virtual image can, however, be viewed by the human eye. As a matter of fact, most optical devices used in conjunction with the human eye, such as binoculars, telescopes, and microscopes, produce virtual images.

A concave spherical mirror is commonly used as a shaving or cosmetic mirror to produce a virtual image of the face behind the mirror. If the object (a face in this case) is positioned between the mirror M and its focal point F, lights rays leaving a point O on the object are reflected according to the law of reflection so as to appear to come from a point I on the image. The entire virtual image shown by the dashed arrow in Fig. 1 is upright and magnified compared to the original object. Although the light rays that enter the viewer's eye E only appear to

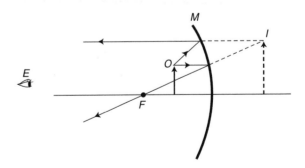

Figure 1 Concave spherical mirror.

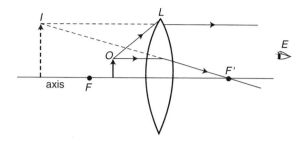

Figure 2   Converging lens.

come from the virtual image, the result is the same as if they actually originated at the image.

A converging lens used as a magnifying glass provides another example of virtual images. If a small object is positioned between a lens L and its focal point F, light rays leaving an object point O will be refracted according to the law of refraction so as to appear to originate at an image point I. The entire image shown by the dashed arrow in Fig. 2 is upright and magnified just as was the case for the concave mirror. Even though the light rays that enter the viewer's eye E only appear to come from the image, again the resulting view is identical were they to originate at the image. In viewing optical images, the human eye does not distinguish between virtual and real images.

Finally, holography provides a means of producing both real and virtual images by means of the diffraction of light rays. An advantage of viewing a virtual holographic image such as that produced on a bankcard is that it appears three-dimensional.

See also: HOLOGRAPHY; IMAGE, OPTICAL; LENS; LIGHT; REFLECTION; REFRACTION

Bibliography

HECHT, E. *Optics,* 2nd ed. (Wiley, New York, 1987).
HEWITT, P. G. *Conceptual Physics,* 7th ed. (HarperCollins, New York, 1993).
TIPLER, P. A. *Physics,* 3rd. ed. (Worth, New York, 1991).

ROBERT MORRISS

IMPULSE

Impulse is a measure of the cumulative effect of a force that acts over a period of time. Usually the

force acts for a brief period of time, such as when a blow is struck or in a collision. For example, suppose that a constant force of magnitude F_0 is applied in a fixed direction to a body B for a duration of time T, beginning at an instant of time t_1. Then the body experiences an impulse I given by the product $I = F_0 \times T$. This is illustrated in Fig. 1, where the force is plotted as a function of time:

The impulse is equal to the shaded area under the horizontal line representing the force. Therefore the impulse would be the same if the force F_0 were doubled, while the time T were halved, and so on. The units of impulse are units of force times units of time, such as Newton-seconds in the SI system of units.

Since the force acting on a body is equal to the rate of change of its linear momentum (this in Newton's third law of motion), then for a constant force the momentum of the body will increase linearly with time. Thus if the body B has a linear momentum p_1 at time t_1, then at time $t_1 + T$ its momentum will be $p_2 = p_1 + F_0 \times T$. This illustrates the principle that "the impulse on a body is equal to the change in its linear momentum." That is, $I = p_2 - p_1$, where $I = F_0 \times T$ for the example just given.

Generally a force on a body will not be a constant in either magnitude or direction. For the sake of simplicity, suppose that the force F applied to a body changes its magnitude with time, that is, $F(t)$, but acts in a fixed direction. This might be the force of a foot on a soccer ball that is being kicked. Figure 2 illustrates how a plot of such a force might look.

The impulse of the foot on the ball is again given by the area under the curve $F(t)$, except that the

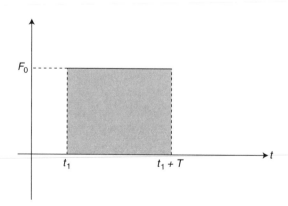

Figure 1   Illustration of impulse with force plotted as a function of time.

force now is not represented by a horizontal line, as in Fig. 1, but by the bell-shaped curve of Fig. 2. Mathematically the impulse is expressed as the integral

$$I = \int_{t_1}^{t_2} F(t)\, dt. \tag{1}$$

Note that Eq. (1) gives the result $I = F_0\,(t_2 - t_1) = F_0 T$ if $F(t)$ is a constant, F_0, over the period of time from t_1 to t_2.

Newton's third law of motion (for the body in question, say the soccer ball) is expressed mathematically as

$$\frac{dp}{dt} = F(t), \tag{2}$$

where dp/dt is the rate of change of the body's linear momentum. Integrating both sides of Eq. (2) with respect to time, we obtain

$$\int_{t_1}^{t_2} F(t)\, dt = \int_{t_1}^{t_2} \frac{dp}{dt}\, dt, \tag{3}$$

or

$$I = \int_{t_1}^{t_2} dp(t) = p(t_2) - p(t_1), \tag{4}$$

which is just the statement that "the impulse on a body is equal to the change in its linear momentum" for the more general case of Fig. 2. Note that this statement follows from Newton's third law of motion, Eq. (2).

Most generally, a force on a body may change in magnitude and direction so that Newton's Eq. (2) must be written in terms of vectors, namely as

$$\mathbf{F}(t) = \frac{d\mathbf{p}}{dt}, \tag{5}$$

from which it follows that

$$\mathbf{I} = \int_{t_1}^{t_2} \mathbf{F}(t)\, dt = \mathbf{p}(t_2) - \mathbf{p}(t_1), \tag{6}$$

where the impulse is now also a vector. Often when people speak of an impulse in such general circumstances as Eq. (6), they mean the magnitude $|\mathbf{I}|$ of the impulse vector.

Recall that the linear momentum of a body of rest mass m and moving with velocity $\mathbf{v}$ is given by $\mathbf{p} = m\mathbf{v}$, provided that the magnitude of the velocity v is negligible compared to the speed of light, c (i.e., $v \ll c$). Otherwise the momentum is $\mathbf{p} = m\mathbf{v}(1 - v^2/c^2)^{-1/2}$.

For a body rotating about a fixed axis, Newton's equation of motion can be written in the "angular form"

$$T(t) = \frac{dL}{dt}, \tag{7}$$

where L is the angular momentum of the body about its axis of rotation, and T is the applied torque. For example, the body might be a fly-wheel and the torque may be due to a force that is struck on the wheel at a fixed perpendicular distance D from the axis. The torque in this case is given by $T = F \times D$. Integrating both sides of Eq. (7) with respect to time gives, in complete analogy with the steps of Eqs. (2) to (4), the following result:

$$A = \int_{t_1}^{t_2} T(t)\, dt = L(t_2) - L(t_1). \tag{8}$$

This equation states that "the angular impulse A applied to the body (the fly-wheel here) is equal to the change in angular momentum of the body."

See also: FORCE; MOMENTUM; NEWTON'S LAWS; VECTOR

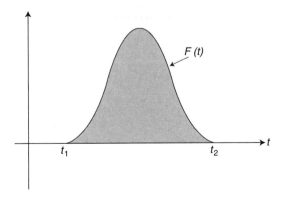

Figure 2 Sample plot of force magnitude as it changes with time.

Bibliography

FETTER, A. L., and WALECKA, J. D. *Theoretical Mechanics of Particles and Continua* (McGraw-Hill, New York, 1980).

SEARS, F. W.; ZEMANSKY, M. W.; and YOUNG, H. D. *University Physics,* 7th ed. (Addison-Wesley, Reading, MA, 1987).

J. W. DAREWYCH

INDEX OF REFRACTION

See REFRACTION, INDEX OF

INDUCTANCE

Current in an electric circuit produces a magnetic field in the space around its wires. If the current varies with time, the changing magnetic field will induce an electromotive force (EMF) in the circuit itself, and in other nearby circuits as well. The generic term "inductance" refers to either self-inductance or mutual inductance. In SI units, inductance is measured in henrys.

Circuits that include components with long wires wrapped around ferromagnetic materials can have inductances of several henrys or more, and circuits with direct, short current paths have inductances of less than 10^{-8} H. In fast-switching computer circuits and ultrahigh-frequency communications circuits, even these tiny, unavoidable inductances will affect circuit behavior.

The concept of magnetic flux linkage is central to understanding both self- and mutual inductance: The magnetic field caused by a current is most easily visualized by constructing a family of curves, called magnetic field lines, tangent at each point to the magnetic field vector. Magnetic field lines loop back on themselves to form closed curves encircling the current path. The density of these field lines can be made equal to the magnetic field intensity at every point in space. The net number of magnetic field lines threading through a circuit path in one particular direction is called the magnetic flux linking the circuit.

Applying Faraday's law of electromagnetic induction, the EMF induced in a circuit is equal to the rate of change of the magnetic flux linking the circuit.

If only one circuit is present, the magnetic field and the magnetic flux linking the circuit are proportional to the current, so the induced EMF is proportional to the rate of change of the current.

Where the predominant source of the magnetic flux linking a circuit is the current in *another* circuit, the induced EMF will depend on the rate of change of that other current.

When two or more circuits are in close proximity, they will interact through mutual induction, even though there are no conducting connections between them. Also, each of the circuits has its own self-induction. The self- and mutual induced EMFs are additive, because magnetic fields obey the law of superposition.

The direction of the induced EMF in a circuit is given by Lenz's Law: The induced EMF is in the direction that would change the current so as to oppose a change in the magnetic flux linking the circuit. Thus self-induced EMF is often called back EMF.

See also: CIRCUIT, AC; ELECTROMAGNETIC INDUCTION, FARADAY'S LAW OF; ELECTROMOTIVE FORCE; FIELD, MAGNETIC; FIELD LINES; INDUCTANCE, MUTUAL; LENZ'S LAW; SELF-INDUCTANCE

Bibliography

FEYNMAN, R. P.; LEIGHTON, R. B.; and SANDS, M. *The Feynman Lectures on Physics,* Vol. 2 (Addison-Wesley, Reading, MA, 1964).

TIPLER, P. A. *Physics for Scientists and Engineers,* 3rd ed. (Worth, New York, 1991).

DAVID A. DOBSON

INDUCTANCE, MUTUAL

The mutual inductance M of a pair of circuits is the proportionality constant between the induced EMF in either one of the circuits (caused by a changing current in the other circuit) and the rate of change of that other current:

$$\text{EMF}_1 = M \, di_2/dt \qquad (1)$$

$$EMF_2 = M \, di_1/dt, \qquad (2)$$

M has the same value in both equations. In SI units, EMF is given in volts, di/dt is given in Amperes per second, and M is given in henrys.

Because of mutual induction, energy can be transferred between electric circuits that are not connected together by electrical conductors at any point. The energy transfer between these circuits is mediated by electromagnetic fields.

Based on an energy argument, an upper limit on the value of M for any two circuits is given by

$$M^2 < L_1 L_2 \qquad (3)$$

where L_1 and L_2 are the self-inductances of the two circuits. In a situation where circuits 1 and 2 each include a coil, and these coils are wound on a common ferromagnetic core, nearly all the magnetic flux due to a current in either coil links every turn of both coils. As a result, M approaches the maximum value permitted by the inequality above. The ratio of EMFs induced in the two coils is equal to the ratio of the number of turns on the coils:

$$EMF_2/EMF_1 = M/L_1 = L_2/M = N_2/N_1. \qquad (4)$$

This result, known as the transformer equation, is of profound importance in the conversion of alternating current electric power from one voltage to another for safe, efficient transmission and distribution.

There is no lower limit on the magnitude of M. For example, if the center of one coil lies on the axis of the other coil and the two coils have their axes mutually perpendicular, the net linkage of magnetic flux from one coil with the other will be zero, and M will be zero. Rotating either coil so they are no longer perpendicular causes some flux linkage and increases the magnitude of M. This strategy for varying inductance was employed in some early radio receivers to tune in a station.

The sign of M depends on the sign conventions used for the currents and the relative orientation of the coils. It is determined using Lenz's law.

The magnetic field produced when a round coaxial cable carries equal currents in opposite directions on its inner and outer conductors is confined entirely inside the cable. No magnetic flux from current in the cable itself links an external circuit. The M values in Eqs. (1) and (2) are always equal, so they must both be zero. Thus, we can be sure that no EMF will be induced in the cable by currents in any circuits outside the cable, even though they may be nearby and produce magnetic fields that penetrate the coaxial cable. This is the principle by which low-voltage signals in coaxial cables are protected from contamination by induced voltage noise, a technique of great importance in audio frequency circuits.

See also: CIRCUIT, AC; COAXIAL CABLE; ELECTROMAGNETIC INDUCTION; ELECTROMAGNETIC INDUCTION, FARADAY'S LAW OF; FIELD, MAGNETIC; INDUCTANCE; INDUCTOR; LENZ'S LAW; MAGNETIC FLUX; SELF-INDUCTANCE

Bibliography

TIPLER, P. A. *Physics for Scientists and Engineers,* 3rd ed. (Worth, New York, 1991).

FEYNMAN, R. P.; LEIGHTON, R. B.; and SANDS, M. *The Feynman Lectures on Physics,* Vol. 2 (Addison-Wesley, Reading, MA, 1964).

DAVID A. DOBSON

INDUCTION

See ELECTROMAGNETIC INDUCTION

INDUCTOR

An inductor is an electrical device with two leads (like the resistor and the capacitor); it is an essential component in many AC circuits. It consists simply of a coil of wire. For a capacitor the voltage is proportional to the charge; for a resistor the voltage is proportional to the current; for an inductor the voltage is proportional to the rate of change of the current:

$$V = -L(dI/dt). \qquad (1)$$

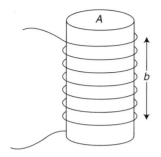

Figure 1 Solenoid.

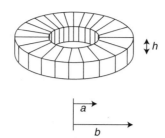

Figure 2 Toroid.

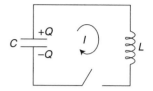

Figure 3 *LC* tank.

The proportionality factor L is called the inductance; it depends on the geometry of the coil and is measured in henrys (a henry is a volt-second-per-ampere). For an air-core solenoid (a helix of n turns with cross-sectional area A and length b (see Fig. 1),

$$L = \mu_0 n^2 A / b, \qquad (2)$$

where μ_0 is the permeability of free space; for a toroid (n turns wrapped around a "donut" of rectangular cross section, with inner radius a, outer radius b, and height h (see Fig. 2),

$$L = (\mu_0 n^2 h / 2\pi) \ln(b/a). \qquad (3)$$

The fundamental principle underlying the behavior of an inductor is Faraday's law. When current flows in the coil, it sets up a magnetic field; if the current changes, the magnetic field also changes, producing an electric field. It is this induced electric field that generates the voltage V between the two terminals of the conductor. According to Lenz's law, the voltage is in such a direction as to oppose the change in current that caused it—hence the minus sign in Eq. (1) and the term "back EMF."

Because inductors oppose changes in current, they function as a kind of inertia in electric circuits (rather like mass, which opposes changes in velocity for mechanical systems). This is seen most dramatically in the *LC* tank circuit, in which a charged capacitor is connected to an inductor (see Fig. 3). The moment the switch is closed, the capacitor begins to discharge, driving current through the inductor. But, when the discharge is complete, the inductor will not let the current stop flowing; it "overshoots," charging up the capacitor in the opposite direction, and then the whole process repeats itself. In principle, the charge will slosh back and forth forever, with frequency

$$f = 1/(2\pi\sqrt{LC}), \qquad (4)$$

where C is the capacitance (in practice there is always some stray resistance in the circuit that eventually damps out the oscillations). The heart of a radio or television is an *LC* tank, in which either the inductance or (more commonly) the capacitance is adjustable, so that f can be "tuned" to respond to a particular broadcast frequency. This is a resonant system, and it takes only a tiny driving signal from the antenna to stimulate large oscillations in the tank.

See also: CAPACITANCE; CAPACITOR; CIRCUIT, AC; ELECTROMAGNETIC INDUCTION; ELECTROMAGNETIC INDUCTION, FARADAY'S LAW OF; INDUCTANCE; INDUCTANCE, MUTUAL; LENZ'S LAW; RADIO WAVE; RESONANCE; SOLENOID

Bibliography

PURCELL, E. M. *Electricity and Magnetism*, 2nd ed. (McGraw-Hill, New York, 1985).

DAVID GRIFFITHS

INDUSTRY, PHYSICISTS IN

The industrial workplace has become a critically important element in the career path for scientists. Recent figures indicate that approximately 40 percent of Ph.D. physicists are employed in industry, a sharp increase from the 30 percent level only 10 years earlier (approximately 1980). The industrial world entered by those trained in physics also has changed: worldwide competition has forced a shortened research-to-product cycle; in many instances, industrial laboratories are faced with decreasing resources with which to accomplish this task, and physicists must be more nimble and team-oriented when solving problems in such an environment. As recently as 10 years ago, a physicist could look forward to a long career investigating fundamental scientific problems in an industrial laboratory interested in long-term programs. Young scientists gravitated toward well-known institutions such as AT&T Bell Laboratories or the IBM Research Center to be involved with such research and to be associated with Nobel Prize–winning scientists and corporate research fellows.

As the twentieth century draws to a close, however, the role of the physicist in industry is changing. The clear delineation between basic research (how things work) and applied research (making things work) has been blurred. Physicists must now exhibit a broader range of interests and skills to succeed. They must be prepared to take on such nontraditional roles, for instance, as engineering and product analyses, manufacturing, and even customer interaction. In addition, an industrial physicist will find a much closer interaction between industrial and university laboratories, which creates exciting opportunities to work with both academic and nonacademic researchers on a variety of problems. So, although the traditional view of corporate research is changing, industrial careers still present exciting and fast-moving technical work.

Physics in the Industrial Environment

In the latter half of the twentieth century an increasing number of industries have come to depend on advanced-technology developments to ensure product designs that meet the needs of the competitive marketplace. These advanced technologies in turn often stem from basic discoveries or developments in physics. With broad training in science, and with their acquired physics-based knowledge and measurement skills, physicists are in a unique position to contribute to the needs of these industries in such ways as:

1. An increasingly sophisticated product development approach that incorporates computer-aided modeling and simulation.
2. The exploitation of new and advanced materials and material properties (e.g., ceramics, superconductors, nonlinear optics, and so on).
3. The use of advanced "smart" materials and sensors (e.g., fiberoptics, semiconductor lasers) as integral parts of the product design to monitor and control the state of the product.
4. The employment of increasingly sophisticated large scale "systems" (e.g., satellite-based systems, such as communication networks and the Global Positioning System).

Among the many contributions physicists have made to industrial product development, perhaps none is more important than the work that made possible the "information age"—the invention of the transistor. Brought together as a team at Bell Telephone Laboratories at the end of World War II, physicists John Bardeen, William Shockley, and Walter Brattain conducted research that would earn them the 1956 Nobel Prize in physics. Their story provides important insight into the world of the industrial physicist, including the link between basic and applied research, the importance of teamwork and corporate financial support, and (an element that is so often overlooked) the pure drama and excitement of unlocking the mysteries of nature.

The research that eventually led to the invention of the transistor began as part of Bell's push to develop new communications devices for the emerging postwar markets. Bolstered by solid-state advancements in radar during the war, the management at Bell Laboratories became convinced that the future of communications lay in a basic understanding of the solid state. In particular, some success had been achieved in the use of semiconductive rectifiers for microwave signal detection. But a stumbling block to further advancements was that these developments were largely of an empirical nature. The management at Bell Laboratories recognized the potential of these small devices in the large-scale communications systems then being envisioned; however, they also realized that no real progress toward that goal would be made without an understanding of the basic principles on which these devices operated. Thus the direction from management to Bardeen, Shockley, and Brattain was "to achieve an understanding of the physics underlying the electrical properties of the semiconductive state." In the course of that effort, the transistor was born.

In both theoretical and experimental efforts, the three physicists were able to overcome the problems in semiconductor reliability that had plagued many earlier efforts elsewhere, by making use of the considerable background and expertise in device development then available at Bell Laboratories. However, their initial experimental devices, while incorporating some of the then-new ideas on the quantum behavior of electrons in solids, yielded little or no current amplification.

Still convinced of the soundness of their theoretical approach, Bardeen then made the crucial conjecture that perhaps their models for electron flow within the device needed to be broadened to encompass surface quantum effects at the semiconductor/metal contact point of the rectifier junction. A series of revamped experiments by Shockley, Brattain, and R. B. Gibney, a physical chemist, confirmed the existence of a substantial surface-state shielding effect. Armed with these results, a new design for the point-contact device, reflecting these findings, was conceived by Shockley. This time a large amplification effect was seen, the first observation of transistor action. Shortly following these discoveries, Shockley made the bold suggestion of utilizing a sequence of semiconductive layers within the crystal. Thus was the junction transistor born, and with it the solid-state revolution in electronics that followed.

A detailed treatment by these physicists of the story underlying their theoretical and experimental efforts to understand the physics of semiconductor device behavior may be found in their Nobel Lectures. In his Nobel Lecture, Shockley notes of being asked frequently if experiments he had planned were "pure or applied research?" To this he replied that to him it was "more important to know if the experiment will yield new and probably enduring knowledge about nature," but he also believed that it would in fact take a combination of pure and applied research to "confer the greatest benefit on mankind" that was sought in Alfred Nobel's will.

The Physicist's View of Industry

After being trained to apply analytical skills in problem solving, the physicist approaches industry with much the same view as that often associated with academia. He or she may plan intellectual pursuits, unencumbered by the reality of resource limitations or the internal competition for these funds. The newly hired physics researcher's perception of industry may be characterized by expectations of enlightened company management, abundant company funding for new ideas, freedom of "choice" in research projects, complete state-of-the-art facilities, encouragement for attendance at technical meetings and membership in professional societies, a high-quality technical staff, and a support staff to perform administrative functions.

For many physicists arriving from academia, individual research projects leading to peer-reviewed publications and presentations at technical meetings may head the list. The relationship of such work to the success or growth of the company was once of secondary importance in the mind of the researcher. In the changing, highly competitive environment, this view may no longer be true, but scientific work must still be of the highest standard and a number of industries and sponsors still evaluate people with respect to scientific publications and peer reviews.

From an industry view, on the other hand, the important "real world" issues are somewhat different and include such considerations as overall corporate strategy, costs and profits for the stockholders, priority funding, carefully scrutinized budgets, and control of capital investments. As noted earlier, modern industry must move quickly and creatively to remain competitive in the worldwide arena. Although research remains an extremely important aspect of overall strategic planning, this environment places difficult boundary conditions on the technology manager. He must weave technology advances into new products and manufacturing techniques, while facing a broad range of requests for limited funds. In addition, these new products must represent the company's strategic goals and please upper management and shareholders alike.

This real world environment fosters the need for physicists with interests and skills previously deemed of lesser importance than technical skills. The ability to be a team player (not preoccupied with who is named as first author) counts for a great deal when potential promotions are being evaluated. It has been noted that, in many ways, the number of publications is not as important as leadership in the company.

The choice of career paths in industry—of research, management, or a combination thereof—is also important for the physicist. In larger firms especially, dual paths are available that allow a scientist to choose paths of either research or management. The research ladder is typified by a progression from research scientist, to senior scientist, to principal scientist, to executive scientist, staff scientist, or science fellow. The management ladder is typified by a progression from research scientist to branch head, to department head, to vice president.

The research ladder allows increasing autonomy in research programs, prestige within the company, and recognition among peers at other organizations. The managerial ladder may mean more in financial compensation but not necessarily more prestige. Smaller corporations, by necessity, may allow a physicist to progress upward on both ladders as responsibilities in both areas increase.

The New Industrial Physicist

A critical look at the new environment, industry requirements, hiring practices, and strategic goals was presented at the American Institute of Physics Corporate Associates Meeting held at AT&T Bell Laboratories in October 1994. It was noted there, for example, that traditional physicist roles, which were to ensure against technical surprises by competitors, investigate fundamental phenomena, produce hardware technology, and create instrumentation technology for other fields of science, have been expanded in the new environment. Physicists now also are expected to participate in applied research, investigate ways to lower costs of technology, investigate ways to lower costs of manufacturing and of products, and use a mixture of skills in physics and engineering to bring products to market rapidly.

Imbedded in the myriad new roles is a requirement for a broader mix of skills on the part of the physicist. In general, industry is no longer looking for the old academic type, that is, the lone wolf pursuing knowledge for its own sake. Because teamwork is an essential part of modern industry, communication and people skills are high on the list of traits necessary for the modern industrial physicist. This includes the ability to communicate not just internally, but also to customers, to the public, and to the media as well. Cross-over skills are also deemed important for selected promotional and leadership activities in industry, where a technical background can bring additional insight and analyses. This opens opportunities for the physicist in such areas as patent issues, marketing, and regulatory affairs, in addition to technical management.

A Look at the Future

Physicists in industry will spend more time in "technology exploitation" for commercial products and in research for a "clear and present benefit" as their work becomes more applied. The workplace trends indicate that people move around more, less time is spent at each company; no more "tenure"— scientists will work hard throughout their career; and global competition enhances the view of best quality and best cost, so scientists must increasingly compete with others inside a company.

Entering and contributing in the technical environment of modern industry requires physicists who have excellent technical skills, are creative in solving problems and seeing the "big picture," have good communication skills (both internal and external), have good "people skills" (i.e., they can work with people of different backgrounds, work on teams, and praise others with less concern for who gets credit), have the ability to lead by example, are flexible and have a desire to continue learning (two experiences are much greater than one experience repeated), and have an alignment of personal and corporate goals.

This last item is becoming more important as research is being redefined for the agile industries of the future. Previously, research was generally divided into either basic or applied. Now the emphasis is on "strategic research," where the goal is defined first, then the research effort (both basic and applied) needed to achieve that goal is carried out.

The modern industrial physicist, tuned to corporate needs and goals, can be at home in this challenging and fast-moving new environment.

See also: BASIC, APPLIED, AND INDUSTRIAL PHYSICS

Bibliography

BARDEEN, J. "To a Solid State." *Science* **5** (9), 143–145 (1984).

GEPPERT, L. "Industrial R & D: The New Priorities." *IEEE Spectrum* **31** (9), 30–41 (1994).

LUBKIN, G., and GOODWIN, I. "Physics Round Table: Reinventing Our Future." *Phys. Today* **47** (3), 30–39 (1994).

HOLDEN, C.; MOFFAT, A.; KAISER, J.; SELVIN, P.; and FOX, K. "Careers '95: The Future of the Ph.D." *Science* **270** (5233) 121–146 (1995).

WILSON, A. H. "The Theory of Electronic Semiconductors." *Proc. R. Soc. London* **A133**, 458 (1931).

WILSON, A. H. "The Theory of Electronic Semiconductors–II." *Proc. R. Soc. London* **A134**, 277 (1931).

WILSON, A. H. "A Note on the Theory of Rectification." *Proc. R. Soc. London* **A136,** 487 (1932).

PAUL V. MARRONE

RAYMOND F. MISSERT

JACK LOTSOF

INERTIA, MOMENT OF

The moment of inertia is a fundamental concept in rotational motion. It is a measure of how much effort is required to change the speed of rotation of an object. The moment of inertia, customarily denoted by I, is an element of the rotational equation of motion, which states that torque is equal to the moment of inertia times the angular acceleration ($\Gamma = I\alpha$).

Mathematical Properties

Although a single small mass has the simplest calculable moment of inertia, the inertia concept is essential for large (extended) objects, which are analyzed as collections of small masses.

The moment of inertia for a single small mass. Suppose a small mass m is made to follow a circle of radius r at a constant speed v (see Fig. 1). This is accomplished by a "centripetal" force, $\mathbf{F}_c$, directed toward the center of rotation and having the magnitude $mv^2 r$. If in addition to $\mathbf{F}_c$ a force $\mathbf{F}_{tan}$ is applied at right angles to $\mathbf{F}_c$, then Newton's second law of motion states that the tangential force equals the mass times the acceleration ($\mathbf{F}_{tan} = m\mathbf{a}_{tan}$). Multiplying both sides of this equation by the radius of the circle gives $r\mathbf{F}_{tan} = rm\mathbf{a}_{tan}$. The left-hand side of this equation is the torque (denoted here as Γ) acting on the mass. Therefore, we have $\Gamma = rm\mathbf{a}_{tan}$. Torque

has the physical dimensions of force times distance and common units of measure for it are foot-pounds or newton-meters. Since the tangential acceleration $\mathbf{a}_{tan}$ is equal to $r\alpha$, where α is the angular acceleration, we can perform a substitution and obtain $\Gamma = mr^2\alpha = I\alpha$, where $I = mr^2$ is the moment of inertia.

A rigid body and forced rotation. An object is a rigid body if the distances between all of its constituent parts are unchanging. It could be a collection of small individual masses or a solid, continuous mass distribution as in Fig. 2. Rigidity requires that all parts of the body must have the same rotation period and the same angular acceleration. The rigidity also automatically provides structurally the radial force $\mathbf{F}_c$. Then the sum of all the torques acting on the object equals the sum of the individual I's for each small mass times their common α. Moments of inertia for some simple objects are shown in Fig. 3.

Perpendicular axis theorem. If a rigid body is thin and planar, its moment of inertia about an axis perpendicular to that plane is equal to the sum of the two moments of inertia about any pair of perpendicular axes in that plane and through the perpendicu-

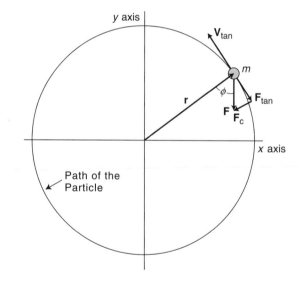

Figure 1 Mechanics of a small mass m restricted to circular motion in a plane. The total force on the mass is $\mathbf{F} = \mathbf{F}_c + \mathbf{F}_{tan}$. The centripetal force is $\mathbf{F}_c = m(\mathbf{V}_{tan})^2/r$, and the tangential force is $\mathbf{F}_{tan} = m\mathbf{a}_{tan} = \mathbf{F}\sin\phi$. The tangential force causes a clockwise angular acceleration, which in this case is a slowing of $\mathbf{V}_{tan}$.

(a)

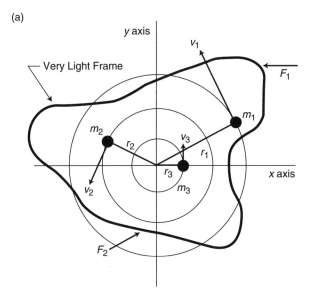

(b)

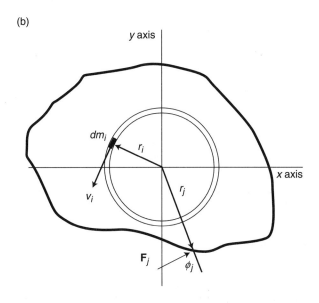

Figure 2 (a) A rigid body consisting of three nonidentical masses attached to a thin, flat, negligibly light frame. Each mass has a tangential velocity proportional to its radius. The moment of inertia about the z axis is $I = m_1 r_1^2 + m_2 r_2^2 + m_3 r_3^2 = \Sigma_i m_i r_i^2$. Counterclockwise torque is produced by externally applied forces $\mathbf{F}_1$ and $\mathbf{F}_2$. (b) A continuous mass distribution consisting of differentially small mass elements dm_i. Each individual mass element has a moment of inertia $dI_i = dm_i r_i^2$, and the total moment of inertia is given by $I = \Sigma_i dI_i = \Sigma_i dm_i r_i^2$ including all masses. The total torque is given by $\Gamma = \Sigma_j \mathbf{F}_j \sin \phi_j$.

lar axis. For example, the moment of inertia of a thin uniform circular disk (as shown in Fig. 3) about an axis lying in its plane and passing through the center is $MR^2/4$.

Parallel-axis theorem. The moment of inertia about an axis parallel to an axis through the center of mass, but displaced a distance τ from it, is $I = I_{cm} + M\tau^2$, where $M = \Sigma_i m_i$ is the total mass. As an example, $I = 5MR^2/2 + MR^2/2 = 7MR^2/5$ for a rolling sphere.

The general case of rotation in three dimensions. Although I can be defined for any space axis, two particular cases are of special importance: (1) three mutually perpendicular axes that pass through the center of mass, and (2) a physically forced axis (called a constrained axis), in particular a space-fixed axle.

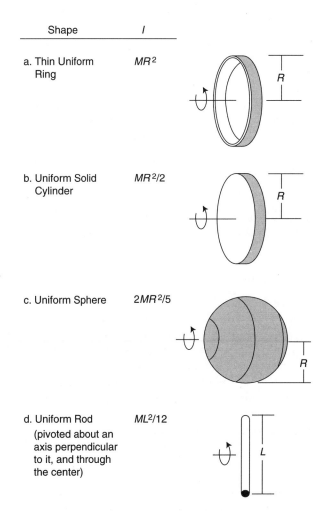

Shape	I	
a. Thin Uniform Ring	MR^2	
b. Uniform Solid Cylinder	$MR^2/2$	
c. Uniform Sphere	$2MR^2/5$	
d. Uniform Rod (pivoted about an axis perpendicular to it, and through the center)	$ML^2/12$	

Figure 3 Some common shapes and their moments of inertia about the axes shown.

History

An empirical prescientific understanding of moment of inertia must be one of the older skills of civilization, since it dates from the invention of the wheel. Although the scaling of wheel diameter and the design of the inner support structure are dominated by material strength and environmental factors, the question of acceleration might have entered the minds of the wheelwrights. Effects of the moment of inertia appeared in windmills and water-powered mill wheels during the middle ages. The earliest fully scientific appreciation of the importance of moment of inertia is attributable to Galileo Galilei. In 1581 he noted the length dependence and amplitude independence of the periods of church chandeliers. Galileo again encountered the concept of moment of inertia when he rolled cylinders down an inclined plane in order to decrease the rate of gravitational fall. Among several others claiming to be inventors of the pendulum clock, Christiaan Huygens prevailed because of his superior escapement, pioneering analytic understanding of I, and publication of its principle in *Horologium* (1658). He held the first working pendulum clock patent. His analysis of I and rotational motion extended even to the products of inertia, which enter into motion that is not confined to principal axes. The full mathematical equations of a rigid body came later from Leonhard Euler, who originated the term "products of inertia." Further analysis is due, among others, to Joseph Louis Lagrange, Augustin Louis Cauchy, and Louis Poinsot.

Applications

The moment of inertia is a basic parameter in reciprocating and turbine engines and also in their drivetrains. Flywheels with large I have always been required in reciprocating engines for moderating the cyclic accelerations. Since there is a natural precession of a principal axis about L for a free body, it is particularly important for constrained objects rotating at high speed to have a principal axis precisely aligned with the constraining axle (dynamic balancing). If this is not done, strong periodic forces are experienced on the axle, which can be especially troublesome if near a natural resonant frequency. Scientific analysis for optimized sport performance makes use of computers programmed with the Euler equations. The time-varying moment of inertia is analyzed through replay and spatial digitization of

videotapes of athletic performances. Platform diving, high jumping, figure skating, and gymnastics are only the most spectacular beneficiaries of this technique. Analysis and control of the orientation of free spacecraft and artificial satellites by working with their moments of inertia is essential to their successful operation. Proper placement of payloads and fuel tanks will achieve optimal principal axes relative to the onboard thrusters, or angular-momentum-reservoir gyros, or viscous damping elements. The optimal torques from the control elements can then achieve or maintain spacecraft orientation or stability by operating on the moments of inertia.

A dominating feature of diatomic and polyatomic molecules is their spectrum of rotational energy states, which are discretely separated due to quantum mechanics. The energy separation of these states is inversely proportional to their moments of inertia. An increase of these moments with increasing rotation rate is observed and is due to centrifugal stretching (just as the earth has an equatorial bulge due to its rotation). Some nuclei of atoms have mass distributions that are ellipsoidal and may have different Is along their principal axes. The I for these nuclei is less than a rigid body calculation and is similar to a viscous flow of the constituent protons and neutrons.

See also: CENTER OF MASS; CENTRIPETAL FORCE; GALILEI, GALILEO; HUYGENS, CHRISTIAAN; INERTIAL MASS; MOTION, ROTATIONAL; PENDULUM; QUANTUM MECHANICS; RIGID BODY; SPECTROSCOPY, MICROWAVE; TENSOR

Bibliography

BELL, A. E. *Christiaan Huygens and the Development of Science in the Seventeenth Century* (Edward Arnold, London, 1947).

FOWLES, G. R. *Analytical Mechanics,* 2nd ed. (Holt, Rinehart and Winston, New York, 1970).

KAPLAN, M. H. *Modern Spacecraft Dynamics and Control* (Wiley, New York, 1976).

R. A. KENEFICK

INERTIAL MASS

There are two quite different types of mass in Newtonian mechanics: inertial mass (m_I) and gravitational mass (m_G). The former measures a body's resistance to acceleration; the latter is the gravitational analog of electric charge.

The reason a truck is harder to set in motion than a ping-pong ball is its greater inertial mass. Inertial mass is that which enters Newton's second law: $\mathbf{F} = m_I\mathbf{a}$, that is, (vector) force equals mass times (vector) acceleration. Consider a given force, say a spring attached to a wall and extended to a given length. By attaching various objects to the free end of that spring, letting go, and measuring the initial accelerations, one can get the ratios of their inertial masses. Alternatively, inertial mass is that which is multiplied by the (vector) velocity to give the (vector) momentum: $\mathbf{p} = m_I\mathbf{v}$. Since momentum is conserved in all collisions, collisions between particles can also be used to determine the ratios of their inertial masses.

Gravitational mass, on the other hand, is that which determines the force $\mathbf{F} = m_G\mathbf{g}$ that a body experiences in a given gravitational field $\mathbf{g}$, say that at the surface of the earth. Hence it corresponds to the *weight* of the body. The amazing thing is that these two types of mass are *equal* if we adopt suitable units. But what has gravity to do with inertia? This equality is a genuine puzzle in Newton's theory and must be accepted as an axiom. Modern experiments have established its validity to an incredible accuracy of one part in 10^{12}. Newton tested it by observing that the periods of simple pendulums depended only on their length and not on the bob. A lighter bob is pulled less by Earth's gravity, but it also needs proportionately less force to reverse the motion in a given time.

An important manifestation of the equality of the two types of mass is what has been called Galileo's principle, an extension of what Galileo Galilei actually found in his famous free-fall experiments from the Leaning Tower of Pisa: Given any gravitational field due to arbitrarily moving sources (think, for example, of the solar system), the motion of a test particle in this field depends only on where and when and with what velocity it is released. A truck and a ping-pong ball, identically released, will travel side by side forever. Why? The force in the field $\mathbf{g}$ is $m_G\mathbf{g}$, which must equal $m_I\mathbf{a}$ by Newton's second law, so that $\mathbf{a} = \mathbf{g}$ no matter what the mass. But $\mathbf{a}$ fully determines the motion via the differential equation $d^2\mathbf{x}/dt^2 = \mathbf{a}$ together with our initial conditions.

In general relativity, which is the modern theory of gravity, gravitating sources curve spacetime and test particles follow geodesics (i.e., the straightest possible lines in curved spacetime). Such a geodesic is determined by an initial direction $dx:dy:dz:dt$ in spacetime. But that is equivalent to knowing an initial velocity (dx/dt, dy/dt, dz/dt) in space. The mystery of why all particles follow the same motion in a gravitational field is thereby resolved; it is simply a generalization of Galileo's original law of inertia (Newton's first law) according to which particles in the *absence* of gravity move straight in space and time.

In special relativity, Newton's second law, $\mathbf{F} = m_I\mathbf{a}$, is no longer valid; force and acceleration are not even necessarily in the same direction. But momentum conservation is still strictly valid, and inertial mass can still be defined by $\mathbf{p} = m_I\mathbf{v}$. It is now a *variable* quantity, increasing with the speed of the particle according to the law $m_I = m_0(1 - v^2/c^2)^{-1/2}$, where m_0 is the so-called rest mass and c is the speed of light. But above all, m_I is now recognized as an exact measure of the particle's total energy E (internal and kinetic) according to Einstein's famous formula $E = m_I c^2$.

See also: GRAVITATIONAL ATTRACTION; GRAVITATIONAL FORCE LAW; INERTIA, MOMENT OF; NEWTONIAN MECHANICS; NEWTON'S LAWS; RELATIVITY, GENERAL THEORY OF; RELATIVITY, SPECIAL THEORY OF; WEIGHT

Bibliography

RINDLER, W. *Essential Relativity*, 2nd ed. (Springer-Verlag, New York, 1977).

WOLFGANG RINDLER

INFLATIONARY COSMOLOGY

See COSMOLOGY, INFLATIONARY

INFRARED

The infrared (IR) frequencies of the electromagnetic spectrum lie below the red part of the visible spectrum—hence, infra-red. Equivalently in wavelength, the infrared lies to the long wavelength side of the visible. The visible spectrum is defined simply by the sensitivity of the human eye and covers roughly a factor of 2 in frequency. The infrared region ranges from a high frequency of about 4×10^{14} Hz, where it meets the red, to a low of about 1×10^{11} Hz, where it meets the microwave region. In wavelength, the IR spans the range from ~0.75 μm to ~1 mm or a factor of more than 1,000, a very much broader range than the visible. (Sometimes the wavelength region from 0.1 mm to 1 mm is assigned to the submillimeter microwave band.)

Much of the importance of the infrared spectral region can be understood from the properties of blackbody radiation (better described as "hot body" radiation). Max Planck in 1912 gave the complete description of blackbody radiation in the Planck radiation law. However, some years earlier Wilhelm Wien had shown that the wavelength, λ_m, of the maximum intensity in the electromagnetic radiation which every solid body emits is inversely proportional to the absolute temperature of the body. Thus, $\lambda_m (\mu m) = 2,898 / T$ (K). The temperature of the Sun's outer atmosphere is about 5,700 K and therefore produces maximum light intensity at $2,898/5,700 = 0.51$ μm (510 nm), exactly where the human eye is most sensitive. By contrast, our body temperature is about 37 °C or 310 K. Therefore our bodies radiate in the infrared with highest intensity at a wavelength of $\lambda = 2,898/310 = 9.3$ μm. This wavelength is in the mid-infrared.

Recent development of sensitive semiconductor detectors of infrared radiation has led to the development of infrared imaging or "night" vision devices which image the IR from warm bodies. These IR cameras are very useful, for example, for imaging buildings in the winter to reveal heat leaks through windows and poorly insulated walls and satellite imaging of crop conditions. In the Persian Gulf War, these devices were spectacularly successful in revealing the location of tanks and personnel hidden under thin camouflage. The temperature differences with the surrounding ground showed up nicely as a brightness contrast in the IR.

The fact that H_2O in vapor or droplet form, CO_2, CH_4, and all triatomic and larger molecules have regions of strong absorption in the infrared leads to the "greenhouse effect" in Earth's atmosphere. Greenhouse warming comes about because these gases are transparent to most of the (short) wavelengths of solar radiation which then warms Earth's surface. However, the long-wavelength infrared radiation from Earth's surface back into space is partly trapped by the "greenhouse" gases leading to a gradual warming of Earth as concentrations of CO_2 and CH_4 increase.

All molecules, except for the homonuclear diatomics such as O_2, N_2, and H_2, have strong infrared absorption lines which arise from stretching and bending vibrational modes of the molecules. Molecules with dipole moments such as HCl, CO, H_2O and OH^- also have absorption lines due to rotational transitions. These characteristic absorption lines can serve as the basis for identification of gases in the atmospheres of Earth and other planets, and in interstellar gas clouds. This is useful, for example, in trace element monitoring of pollutants in Earth's atmosphere, for measurements of the temperature of the methane in Jupiter's atmosphere, and for the temperature and density of gases such as CO, HCOOH, and H_2O in interstellar gas clouds.

Infrared spectroscopy by absorption requires a light source, usually a hot body, which produces a broad spectrum in the infrared. The IR light transmitted through the gas will show the characteristic absorption lines of the gas when it is dispersed by a prism or a diffraction grating spectrometer [or by a Fourier-transform infrared (FTIR) spectrometer, which uses the principles of a Michelson interferometer]. For astrophysics, the light source is a background star; for laboratory work it is usually a glowing filament or more recently a tunable laser. Unfortunately IR spectroscopy is complicated by the fact that glass windows strongly absorb in most of the IR region. Similarly, Earth's atmosphere transmits poorly. Therefore, special windows made from pure germanium or KBr, for example, must be used and the spectrometer must be purged of water vapor. Interstellar spectroscopy is almost entirely satellite based for the same reason.

The emission of IR radiation is complementary to absorption and may also be used for identification of molecular species when the molecules are excited by light or collisions with other particles. This sometimes occurs in interstellar gas clouds and in gases excited by electrical discharges. In extreme cases the excitation may be strong enough to produce laser action in the infrared just as in the visible.

See also: ATMOSPHERIC PHYSICS; ELECTROMAGNETIC RADIATION; ELECTROMAGNETIC SPECTRUM; FOURIER SERIES AND FOURIER TRANSFORM; INTERFEROMETRY; MOLECULE; RADIATION, BLACKBODY; RADIATION PHYSICS; SEMICONDUCTOR; SPECTROSCOPY; TEMPERATURE; WAVELENGTH

Bibliography

COLTHRUP, N. B.; DALY, L. H.; and WIBERLEY, S. E. *Introduction to Infrared and Raman Spectroscopy,* 3rd ed. (Academic Press, Boston, 1990).

GANS, P. *Vibrating Molecules: An Introduction to the Interpretation of Infrared and Raman Spectra* (Chapman and Hall, London, 1971).

ALVIN D. COMPAAN

INFRASONICS

Infrasonic waves are sound waves with frequencies below the lowest frequency that the human ear can hear (about 25 Hz). These frequencies are felt rather than heard, particularly at high intensities. Infrasonic waves have the same speed of propagation in air as do audible sounds at the same temperature, 343.9 m/s at 20°C. The speed varies with temperature (T) according to the relation $331.7 + 0.61T$ m/s. Infrasonic, like audible and ultrasonic sound waves, obey normal physical laws such as reflection, refraction, interference, and diffraction. The wavelengths of infrasonic waves are all greater than 14 m (45 ft), so that they are scattered very little by trees and buildings and consequently do not cast acoustical shadows. Normal microphones are not able to detect infrasounds.

Earthquakes and seismic waves are elastic waves in the crust of the earth that occur at infrasonic frequencies. Low frequency sound waves also accompany tornadoes and volcano eruptions. Elephants can communicate hundreds of miles across the desert using infrasonic waves. The audible sounds that humans associate with "elephant talk" are in fact high-frequency overtones of fundamental frequencies below 20 Hz. Blue whales transmit infrasound through the sea up to distances of tens of miles.

Rocket boosters, large ship propellers, and turbojet aircraft engines often produce intense frequencies between 5 and 25 Hz, as do chemical and nuclear explosions and lightning. Extremely intense levels (greater than 130 dB) of infrasound between 5 and 10 Hz are often present near the engine rooms of large ships.

Research studies on the effects of high intensity infrasound on humans have shown numerous physiological effects such as fatigue, drowsiness, nausea, dizziness, and pain. Infrasonic frequencies of about 7 Hz have been found to impede normal brain activity. This has been associated with the 7-Hz α wave of an electroencephalogram. The human jaw appears to have a resonant frequency at this same frequency so that, when subjected to high intensity infrasound of about 7 Hz, speaking is impossible. Whole-body exposure tests to infrasound by NASA using frequencies of 15 to 25 Hz at 130 dB caused severe middle-ear pain, respiratory difficulties, and headaches in the subjects. Infrasounds accompany everyday noises. They are produced in vehicles at high speeds and have been associated with causing drowsiness in truck and bus drivers.

The human body is continually being subjected to low-level infrasonic waves when walking and running. At normal walking speed, 2 paces per second (2 m/s or 4 mph), the body is subjected to an infrasonic vibration of 2 Hz. When running the frequency increases to about 6 Hz.

Very low infrasound with frequencies of 1 Hz or less travel through the atmosphere for thousands of miles without any appreciable absorption so that they can traverse the earth several times before decaying.

See also: SEISMIC WAVE; SOUND; ULTRASONICS

Bibliography

TABULEVICH, V. N. *Microseismic and Infrasound Waves* (Springer-Verlag, New York, 1992).

TEMPEST, W. *Infrasound and Low-Frequency Vibrations* (Academic Press, New York, 1976).

JOHN ASKILL

INSULATOR

An electrical insulator resists the flow of electricity. Application of a voltage difference across a good insulator results in negligible electrical current. In comparison, a conductor allows current to flow readily. Controling the flow of current in electrical wiring and electronic circuits requires both insulators and conductors. For example, wires typically consist of a current-carrying metallic core sheathed in an insulating coating.

Resistivity is the measure of a material's effectiveness in resisting current flow. Materials with resistivities higher than 10^8 Ω m are usually considered to be good insulators; these include glass, rubber, and many plastics. Resistivities as high as 10^{16} Ω m can be achieved in exceptional insulating materials. Normal conductors may have resistivities as low as 10^{-8} Ω m. The enormous variation in room-temperature resistivity is one of the largest for any physical attribute of matter.

The charge carriers responsible for current in most conductors are electrons, moving relatively freely in a metal. In insulators, electrons cannot move freely. When atoms of simple metals combine to form a solid, the outer valence electrons become free for conduction. In an ideal insulator, all electrons stay tightly bound to the atoms, so that there are no electrons that can be readily moved through the material for conduction.

A more complete understanding of insulators and conductors requires consideration of electronic band structure. The electrons in an isolated atom possess discrete energies, a consequence of quantum mechanics. These discrete levels evolve into bands of allowed energies when the atoms condense into a solid. Forbidden regions separate the allowed bands, as shown schematically in Fig. 1. The electrons in the solid fill in the bands, from lower to higher energy.

The distinction between an insulator and a conductor lies in how the electrons fill in the allowed bands. For a simple metal, the highest band containing electrons will be only half full. The thermal energy (at ordinary temperatures) will be sufficient to generate conduction electrons—electron states of slightly higher energy are available in the incompletely filled band.

In comparison, the highest energy band containing electrons is completely full in a good insulator. The thermal energy of the electrons is not sufficient for promotion from this band, known as the valence band, to the next band with available energy states, known as the conduction band. The gap between the valence and conduction band, known as the band gap, is at least several electron volts (eV) wide in an insulator—thermal electron energies are 100 times smaller.

The distinction between an insulator and a semiconductor is one of degree. Although both have completely filled valence bands at 0 K, the band gap of a semiconductor is smaller than an insulator. For narrower band gaps, thermal energy is more capable of promoting electrons into the conduction band.

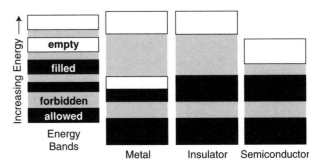

Figure 1   Bands of allowed energy in a solid.

Insulators play a critical role in many aspects of technology, from large scale to the microscopic. Electrical power transfer relies on high-voltage transmssion lines for which insulators are required to prevent losses to ground. High-voltage transformers rely on special insulating oils. Dielectric materials enhance the charge storage of a capacitor. Even bits of information in a computer memory required thin insulators in the form of oxide layers in increasingly smaller transistor circuits.

See also: CONDUCTOR; ELECTRICAL RESISTANCE; ENERGY LEVELS; RESISTOR; TRANSFORMER; TRANSISTOR

Bibliography

EISBERG, R., and RESNICK, R. *Quantum Physics of Atoms, Molecules, Solids, Nuclei, and Particles* (Wiley, New York, 1985).

KITTEL, C. *Introduction to Solid-State Physics,* 6th ed. (Wiley, New York, 1986).

SHUGG, W. T. *Handbook of Electrical and Electronic Insulating Materials* (Van Nostrand Reinhold, New York, 1986).

TIPLER, P. A. *Physics for Scientists and Engineers,* 3rd ed. (Worth, New York, 1991).

GEORGE H. WATSON

INTERACTION

In everyday usage the word "interaction" suggests two things that affect each other in any way. We can talk about the interaction between two people, whether it is a telephone conversation or a hug. We can discuss the interactions between two nations, or between a bat and a ball. In physics, Newton's law tells us that for every action there is an equal and opposite reaction, that is to say, if *A* exerts a force on *B*, then *B* exerts an equal and opposite force on *A*. Interaction, in physics, describes both of these equal and opposite forces.

When we study any system in science, one of the first questions we ask about it is the nature of the interactions between the various parts of the system. By breaking a system down into component parts and separating the interactions between parts from the processes within the parts, we can simplify problems and separate the study of processes that occur at different spatial and time scales. As we go to simple physical systems, such as atoms or nuclei, the interactions between components also become simpler; they are the forces each particle feels because of the presence of the other particles.

The words "force" and "interaction" thus appear almost interchangeable in physics usage, though force has a more specific definition. However, the word "interaction" also takes on a new meaning beyond its everyday usage; it is used to denote the underlying physical mechanism that causes the forces. For example, we say the electromagnetic interaction between two charged particles leads to a force between them. Equally, we say that these forces are the effect on each particle of the electric and magnetic force fields due to the other particle. The force fields define the magnitude and direction of the force that would be felt by a specified charge at any location. The force field transmits the effect of a charge to any other charge. The long-range interaction between two separated charges is mediated by the force field.

It is traditional in physics to divide forces into two classes: dissipative and conservative. This distinction is based on whether the action of this force involves dissipation of energy into some surrounding medium or no such dissipation. Dissipative forces (such as friction, elastic restoring forces, etc.) thus involve the presence of some bulk medium, whereas the conservative forces can act on isolated particles. Only this latter kind of force can be described in terms of a force field (or the related potential energy function), and only for these do we think in terms of some underlying fundamental interaction that is the source of the force field. Force fields are due to particular properties of matter. For example, electric charges create surrounding electric and magnetic force fields (the latter only if they are moving), and mass-energy (in fact, any form of energy) is the source of gravitational force fields.

There are four well-known fundamental interaction types: strong, electromagnetic, weak, and gravitational. For each of these one can define related force fields. The dissipative forces in bulk matter, be it solid, liquid, or gas, are all residual effects of the underlying electromagnetic structure of the electrically neutral matter, whereas the fundamental electromagnetic interactions are the interactions between charged particles and/or electric currents. This distinction, between fundamental interactions between individual particles and residual effects of these interactions due to the structure of complex objects, parallels the distinction between conservative and dissipative forces.

Once scientists began to think of the fundamental interactions in terms of force fields, they recognized that these fields can have effects that are quite unlike the classical notions of force. One can have traveling waves of the force field, such as radio waves or light, both of which are traveling electromagnetic waves. In quantum theory this goes a step further. There is a particle type associated with the excitations of the traveling force field waves. In the case of electromagnetism, the particles are called photons; these are quanta of electromagnetic radiation. The emission or absorption of photons is thus another effect of the electromagnetic interaction. The transitions of atoms from one quantum state to another with the associated photon emission is just as much an effect of the electromagnetic interaction as is the force between two separated charges. All radioactive transitions or decays are consequences of the fundamental interactions. Thus, physicists today use a broader meaning of interaction than its original everyday one. It describes any effects from the force field and its excitations, including processes such as the decay of an isolated particle.

See also: FIELD; FORCE; INTERACTION, ELECTROMAGNETIC; INTERACTION, ELECTROWEAK; INTERACTION, FUNDAMENTAL; INTERACTION, STRONG; INTERACTION, WEAK; NEWTON'S LAWS

HELEN R. QUINN

INTERACTION, ELECTROMAGNETIC

The theory of electromagnetic interactions encompasses all electric and magnetic effects and all electromagnetic radiation of any wavelength. The "classical" theory of these effects was based on the famous work of James Clerk Maxwell, who published a paper in 1864 entitled "A Dynamical Theory of the Electromagnetic Field." The four equations, known today as Maxwell's equations, were brought together in this work, which built on the previous studies of many others, including André-Marie Ampère, Charles Augustin Coulomb, Michael Faraday, Carl Friedrich Gauss, and Hermann von Helmholtz, among others. Maxwell's work introduced the concept of electric and magnetic fields to explain the remote or "action at a distance" effect of one charge or magnet on another.

These fields are quantities that span space without requiring any supporting medium. They cause forces on any electric charge or magnet placed within them. Conversely, any electric charge or current and any magnetized material causes electric and magnetic fields. The pattern of the fields depends on the size and distribution of the magnetism, currents, and charges. All the forces between charges, currents, and magnets can be understood and calculated precisely from knowledge of their relative positions and motions by calculating the electromagnetic fields created by each source. One can add the fields from several separate sources to find the resulting force on any other charge or magnet placed at any point in space.

The strength and direction of the forces depend on the strength and direction of the magnetic and electric fields. These force fields are continuous vector quantities, with a magnitude and a direction at each point in space. Furthermore, electric and magnetic fields can be mapped by "lines of flux," which are parallel to the direction of the field at any point in space. The density of the lines of flux in any region is proportional to the field strength. A remarkable thing about this map of the field is that electric lines of flux terminate only on electrical charges. This is expressed mathematically in Maxwell's equations by Gauss's Law $\nabla \cdot \mathbf{E} = 4\pi\rho$, where $\mathbf{E}$ is the electric field strength and ρ is the charge density in appropriate units. Magnetic flux lines, which never terminate, are densest near the poles of a magnet; they appear to emerge from one pole and return to the opposite pole, but by continuing the lines through the region of the source magnet itself, one sees that they are in reality closed loops of magnetic flux. Any electric current creates closed loops of magnetic flux surrounding the wire. This can be readily observed by bringing a compass close to a wire carrying a current.

A wide range of physical phenomena can be understood as electromagnetic waves, that is, electromagnetic fields that vary in time and space in a wave-like fashion. These waves can carry energy and information from one place to another. Radio waves, microwaves, x rays, gamma rays, and light itself are electromagnetic waves; they differ only in their wavelengths. It took the work of many physicists over some years after Maxwell's initial formulation of his equations to understand that radiation and absorption of electromagnetic waves is in fact described by these equations.

The forces on an electric charge due to a particular electric field depend on the amount of the charge. The electric force is parallel to the electric field direction. The magnetic force on a charge depends on the amount of the charge but also on the velocity of the charge (and thus the electric current). The magnetic force on an electric charge is perpendicular to both the direction of the magnetic field lines and to the direction of motion of the charge.

The fact that magnetic effects depend on the velocity of a charge should immediately catch the attention of the alert reader. Perhaps, too, this reader has been concerned by the fact that a static charge creates an electric field while a moving charge, that is, an electric current, creates a magnetic field in addition. The understanding of how the physics of Maxwell's equations is indeed independent of the frame of reference was developed in the years from about 1900 to 1907, chiefly from work by Jules-Henri Poincaré, Hendrick Lorentz, and Albert Einstein. This work revealed an inextricable interrelationship between the electric and magnetic effects, and, furthermore, led to Einstein's crucial assertion that the speed of light is the same for all observers. The definition of what is an electric field and what is a magnetic field depends on the frame of reference of the observer. An observer moving with the same steady speed as an electric charge will find that the charge creates only an electric field. Another observer, relative to whom the charge is moving, will detect both electric and magnetic fields due to that charge. However, both observers will agree as to the forces on any other charge due to these fields.

The classical theory of electromagnetism was one of the great successes of nineteenth-century physics. In conjunction with the theoretical progress in understanding electromagnetism, a practical understanding of electricity and magnetism developed in that period. This was the foundation for the design of devices for the generation of electric power and of ways to use this power in factories and homes. The ubiquitous electric motors we nowadays take so much for granted provided and continue to provide the motive power for many appliances and tools. Factories of the twentieth century look very different from those of the nineteenth, chiefly because multiple electric motors replaced huge steam turbines powering webs of belt-driven pulleys. Electric lighting, too, has become standard and has possibly changed our lives more than any other single invention. Like the use of electric power, the use of electromagnetic waves for communication, in radio and then television signals, has certainly changed the world.

In the twentieth century a new development of physics, the quantum theory, was first developed in the context of electromagnetic interactions. The quantum theory of electromagnetism, known as quantum electrodynamics (QED), is built on Maxwell's electromagnetic fields and provides a theory of the quantum processes of emission and absorption of electromagnetic radiation as photons, the quantum version of electromagnetic waves. Any electromagnetic wave is the sum of many photons. When the relevant number of photons is large enough, the classical theory correctly describes the behavior.

The quantum theory, however, successfully treats many electromagnetic phenomena that were not explained in Maxwell's theory. One example is the photoelectric effect, in which electrons are ejected from a metal surface by light (or, in quantum language, by photons). The fact that the energy of the electrons emitted from a given surface depends only on the color and not the intensity of the light striking on the surface could not be understood prior to the quantum theory. Einstein's explanation of the photoelectric effect in terms of photons was a key step in the development of quantum theory. The photon energy is fixed by the photon wavelength, which in turn is fixed by the color of the light. Each electron is ejected by absorbing a single photon and thus aquires a definite additional energy. If this energy is great enough to overcome the attraction of the surface material, the electron will escape. Its escape energy can thus be any value up to the energy of the photon it absorbed.

In quantum theory the electromagnetic attraction between the positive electric charge of the atomic nucleus and the negative electric charge of the electrons results in a set of possible states for electrons in any given atom. The characteristic spectral lines of atoms (e.g., the yellow of a sodium vapor lamp and the blue of a mercury one) can then be understood as due to the emission of photons from an atom in which an electron makes a transition from an excited or high-energy state to a lower energy or less-excited state in the same atom. This was also explained in quantum theory. The difference between the energy of the electron in the two states determines the energy of the photon. This in turn determines its wavelength and thus the color of the emitted light.

All the complexities of chemistry and the structure of materials are due to the properties of electrons in atoms and the interactions between atoms due to their electromagnetic substructure. With the exception of gravity, all the many forces we experience on a day-to-day basis are due to electromagnetism and to the electrical substructure of atoms. The fundamental electromagnetic interactions between charged particles and electromagnetic fields lead to interactions between charged particles. The residual effects of the electrical substructure of electrically neutral matter are responsible for the rigidity of a chair that supports us, for the friction that allows us to walk across the floor, for the pull of a spring, and the tug of the wind in the branches of a tree. All these forces occur due to the interaction of the electrons in the surface of one atom with those in the surface of another and are thus fundamentally electromagnetic in nature.

See also: CHARGE; CHARGE, ELECTRONIC; EINSTEIN, ALBERT; ELECTROMAGNETISM; FIELD, ELECTRIC; FIELD, MAGNETIC; GAUSS'S LAW; LORENTZ, HENDRIK ANTOON; MAXWELL'S EQUATIONS; PHOTOELECTRIC EFFECT; QUANTUM ELECTRODYNAMICS; QUANTUM THEORY, ORIGINS OF

Bibliography

FEYNMAN, R. P. *QED: The Strange Theory of Light and Matter* (Princeton University Press, Princeton, NJ, 1988).
WHITTAKER, E. T. *A History of the Theories of Aether and Electricity,* 2 vols. (Nelson, London, 1951, 1953).

HELEN R. QUINN

INTERACTION, ELECTROWEAK

The term "electroweak interaction" describes the modern theory of two interactions: electromagnetic and weak. In the mid-1800s James Clerk Maxwell's work unified two apparently quite different types of phenomena, electricity and magnetism, into a single theory of electromagnetism. Nearly one hundred years later, the developers of electroweak theory made another giant conceptual leap, unifying Maxwell's electromagnetism (or rather its quantum field theory descendant, quantum electrodynamics, or QED) with the theory of the weak interactions.

This entry will first briefly review the nature of these two interactions, which are described in more detail elsewhere in this encyclopedia. Then it will give a nontechnical description of the way they are combined in the modern electroweak theory.

QED, or quantum electrodynamics, deals with electric and magnetic properties and the interaction between particles due to their electric charges and magnetic moments. These interactions occur via the exchange of photons. Photons are massless particles that are the quantum excitations of the electromagnetic fields and have one unit of intrinsic angular momentum or spin. (Spin, like all angular momenta in quantum theories, is measured in units of $\hbar$, Planck's constant divided by 2π).

Weak interactions are the processes responsible for the class of radioactive decays known at beta decay. They occur much less readily than electromagnetic processes with a similar energy release. The archetypal beta decay process is a transition in which a neutron disappears, producing instead a proton, an electron, and an anti-electron type neutrino. Similar weak-interaction-mediated decay processes are responsible for the fact that all the more massive types of quark or lepton are unstable. For example, a muon decays to produce a muon-type neutrino, an electron, and an anti-electron-type neutrino; a K meson, which contains a strange (s) quark (or antiquark), decays to produce a collection of pi mesons, which contain no s quarks but only u- and d-type quarks.

The electroweak theory was developed in the period from 1961 to 1967, primarily through the work of Sheldon Glashow, Steven Weinberg, and Abdus Salam, who were awarded the Nobel Prize in 1979 for this work. Further technical developments in the early 1970s, especially the work of Gerard 't Hooft

and Martinus Veltmann, were crucial to the experimental tests that led to broad acceptance of this theory. They introduced techniques to perform the calculations needed to determine what outcome of experiments this theory would predict. Experimental work since that time has resoundingly confirmed the predicted outcomes.

The electroweak theory introduces particles that act as the mediators of weak interactions in the same way that the photon mediates electromagnetic processes. These particles, the W and Z bosons, again like the photon, carry one unit of intrinsic angular momentum or spin. The strength of the interaction of these particles is comparable to that of the photon. However, unlike the photon, the W and Z particles are massive. This causes the beta-decay weak interactions to occur at rates much lower than electromagnetic decays (which produce photons) with comparable energy release. The mass of the exchanged particle also leads to an interaction probability that falls off much more rapidly with distance than in the electromagnetic case.

In the electroweak theory the electromagnetic and weak interactions have a deep relationship. What happens is somewhat peculiar. We start out with a mathematical theory that appears to represent four massless spin-1 particles. Three of these, with electric charges +1, 0, and −1, are related by a symmetry (technically they form a multiplet under the group $SU(2)$). The symmetry requires these three to have the same interaction strength, which we write as g. The fourth particle, which also has electric charge zero, has a different interaction strength, g'. In addition to these spin-1 particles the theory appears to contain a set of four spin-0 particles. The electric charges match the four spin-1 states. The most peculiar thing about the theory at this stage is that the parameter that would normally represent the square of the mass of these spinless particles is negative.

In such a theory a phenomenon called spontaneous symmetry breaking occurs. The naive interpretation of the mathematical theory turns out to be wrong, but one can find a new interpretation of the ground state and the excitations of the theory that makes much better physical sense in that it contains only particles that have masses such that $m^2 \geq 0$. A quite satisfactory set of particles emerges.

We find two charged, massive spin-1 particles, which are the W^+ and W^-, the particles that mediate beta decay and other similar weak processes. The quantum excitations that form these massive particles include both the original massless spin-1 charged states and the spin-0 charged states. (There are only two distinct polarization states for a massless spin-1 particle, but three such states for a massive spin-1 particle, so in effect the original spinless particles have become the third spin-polarization state of the massive W bosons.)

The four neutral (charge zero) states (two spin 1, two spin 0) also get mixed together, but here the story is even more complicated. The net result is one massive spin-1 particle, the Z boson; one massless spin-1 particle, the photon; and one massive spin-0 particle, the yet-to-be-observed Higgs boson. Notice that once again we have lost a spin-0 particle and gained a mass for one spin-1 particle. The Z boson mediates a new type of weak interaction, only discovered after it had been predicted by this theory. An example of this interaction is a process in which a neutrino is deflected by exchanging momentum with a proton via a mediating Z boson. Since neutrinos are not detected, all that one sees in an experiment is a proton that suddenly acquires momentum and energy. The large mass of the Z makes the probability of such interactions rare. With an intense neutrino beam, and a large enough number of protons in the volume instrumented for detection of recoil motions, the effect is observable. It occurs at the rate predicted by this theory.

Both the Z boson and the photon are mixtures of the two original neutral spin-1 states. This is why this theory is one where the weak and electromagnetic interactions are related. It also results in a different mass and interaction strength for the Z boson than those of the W bosons.

A number of well-measured quantities depend only on the two interaction-strength parameters g and g' and one additional parameter, v, that sets the scale of all particle fundamental masses. Below we give approximate formulas that show how physically measurable quantities are related by these parameters. More precise calculations determine the relationships between measurements and parameters more accurately. The approximate forms given here are sufficient to show how the weak and electromagnetic parameters are intertwined.

The weak beta decay strength is $G_{Fermi} = g^2/8M_W^2 = 1/(4v^2)$. Thus measurement of beta decay rate thus sets the scale v, which is found to be approximately $270\,\text{GeV}/c^2$. The usual electromagnetic coupling strength e, (or fine structure constant α), is given by

$$\alpha = \frac{e^2}{4\pi} = \frac{g^2 g'^2}{(g^2 + g'^2)}.$$

The ratio of g' to g is given by the measured strength of the Z-mediated weak process compared to the W-mediated process, so that ratio, together with the measured value of $\alpha \approx 1/137$, fixes the values of both these parameters. The masses of the W and Z bosons in this electroweak theory are given by $M_W^2 = g^2 v^2/2$, $M_Z^2 = (g^2 + g'^2)v^2/2$.

The theory thus predicted the masses of the W and Z bosons, particles which, at the time, had never been observed. Subsequent experiments confirmed that these predictions were correct. Carlo Rubbia and Simon van der Meer were awarded the 1984 Nobel Prize for their development of the facility that could produce and detect the W and Z bosons and their leadership of the discovery experiment.

The electroweak theory has now been tested in numerous experiments; results again and again have matched the predictions of this theory. Together with quantum chromodynamics (QCD), which is the quark and gluon theory of the strong interactions, it is now referred to as the standard model of particle processes. Particle physicists like to speculate on grand unified theories, which unify the electroweak theory with QCD, in much the same way as the electroweak theory unifies electromagnetic and weak phenomena, but so far there is no evidence to confirm (or to contradict) these speculations. An even more ambitious goal is a theory that also include quantum gravity in the unification. To date it appears that the approach known as string theory is the most promising development in this direction.

See also: BOSON, HIGGS; BOSON, W; BOSON, Z; GRAND UNIFIED THEORY; LEPTON; MUON; PHOTON; PLANCK CONSTANT; QUANTUM CHROMODYNAMICS; QUANTUM ELECTRODYNAMICS; QUARK

Bibliography

CLINE, D. B.; RUBBIA, C.; and VAN DER MEER, S. "The Search for Intermediate Vector Bosons." *Sci. Am.* **244** (March), 48 (1982).

WEINBERG, S. "Unified Theories of Elementary Particles." *Sci. Am.* **231** (July), 50 (1974).

HELEN R. QUINN

INTERACTION, FUNDAMENTAL

Fundamental interactions are the properties of the fundamental building blocks of matter that are responsible for forces between these objects or decays of one such object into some others. There are four known classes of fundamental interactions in physics: gravitation, electromagnetism, and the strong and weak nuclear interactions.

All of the fundamental interactions in nature can be understood in terms of force fields similar to electric and magnetic fields. These fields lead to equal and opposite forces between pairs of objects; they also lead to particle decay and other such transitions between different quantum states of a system. For example, the decay of an excited atom to its stable ground state by the emission of a photon is a process mediated by the electromagnetic interaction.

In the case of gravitation we are all well aware from our everyday experience that there is a force field surrounding Earth that pulls any object toward Earth. Furthermore, we find that the force field is conservative, by which we mean that if we use energy to raise an object to some height, then, in a vacuum where no other forces come into play, we can recover exactly the same amount of energy by letting the object fall.

Every massive object, no matter what its substructure, is surrounded by a gravitational force field. The force between any two massive objects is given by the form first stated by Isaac Newton,

$$F_{12} = \frac{Gm_1 m_2}{r_{12}^2},$$

where m_1 and m_2 are the masses of the two objects and r_{12} is the distance between them. The constant G is universal, which means that precise experiments show that its value does not depend on the type of matter involved. Both the existence of a conservative force field that causes forces at a distance between objects and this universality are indications that we are dealing with a fundamental property of nature.

A more accurate theory of universal gravitation, Einstein's theory of general relativity, replaces the masses of the object by the energy—mass being just one form of energy with the famous relationship $E = mc^2$. Even massless objects such as photons experience gravitational forces due to their energy.

The first observation of this was the famous 1919 eclipse experiments that detected the bending of the path of light from a distant star due to the gravitational pull of the Sun.

A force field can transmit energy by forming traveling waves. In the electromagnetic case we have many names for these waves, depending on the wavelength which determines how we sense them—x rays, light, microwaves, and radio waves are all traveling electromagnetic field excitations. Similarly, traveling gravitational waves can carry away gravitational energy. There is evidence for this effect in collapsing binary star systems.

All the fundamental interactions apart from gravitation are now understood in terms of the quantum theory of the force fields involved. In quantum field theories there is a particle associated with the quantum excitations of each force field. In the case of an electromagnetic field the particle excitation is called a photon. So far there is no entirely satisfactory quantum field theory for gravitation, but scientists have invented a name for the particle that, in such a theory, would carry gravitational interactions. It is called a graviton.

The idea of force-at-a-distance between electric charges and/or currents is also quite familiar. Magnets and current-carrying wires exert a force at a distance on other magnetic objects. Static electric charges likewise exert forces on other charges without coming in contact with them. These effects can be described as due to electric and magnetic fields, and are generically called electromagnetic interactions. Once again the interaction is universal; the force between charges depends only on the amount and spatial distribution of charges, not on any other properties of the charged matter. Electromagnetic force fields are also conservative; energy expended in moving a charge within such a force field can be recovered as kinetic energy when the charge is released and allowed to move freely in the force field. Thus one can define electrical and magnetic potential energy, in much the same way as one defines gravitational potential energy, as the energy stored up by doing work against the force field.

Apart from gravity, all the forces we experience in everyday life have to do with the rigidity, elasticity, and surface properties of matter. All these properties are due to the electromagnetic interaction, which is responsible for the binding of negatively charged electrons to positively charged nuclei to form atoms, and for the residual forces between the electrically neutral atoms that arrange them into molecules, crystals, solids, liquids, and so on. Forces between two matter objects arise from the changes in the internal energy of an object when its structure is deformed because of the close approach of another object. Forces of friction, the pull of a spring, the rigidity of a chair or a table top, all arise from the electrical substructure of the material involved.

All of chemistry too is due to residual electrical effects between atoms. In matter the region occupied by the electrons of one atom often overlaps that of another atom and hence electron rearrangements and shared electrons can occur. More distant atoms affect one another because the electrical force fields outside an atom do not vanish even though the atom is electrically neutral. The negative and positive charges would have to be located at precisely the same place for their fields to cancel exactly. The understanding of chemistry and of solid state physics is not simple; the variety of properties that follow from the electron structure of matter is incredibly rich and complex.

We have no everyday direct experience of the remaining two interactions. However, without them we would not exist, and the types of stable matter in the universe would be very different. The strong interaction provides the force that holds quarks together to form protons and neutrons, as well as the forces between protons and neutrons that cause them to form complex atomic nuclei. The radioactive decays of nuclei also involve another quite different interaction, known as the weak interaction.

Historically the term "strong force" (or "strong nuclear force") was used for the interaction between protons and neutrons that causes them to form the nucleus. We now understand that this interaction is not itself a fundamental interaction. The entire field of particle physics grew out of attempts to understand the underlying mechanism for nuclear forces. Many other kinds of particles were discovered as scientists sought to study proton and neutron interactions. These discoveries led eventually to the theory of quarks and their interactions, which particle physicists now call the standard model. In this theory protons and neutrons, as well as the many other particles, are composite objects made from even smaller particles called quarks. The fundamental strong interaction is now recognized to be the interaction between quarks rather than that between protons and/or neutrons.

Just as particles with electric charges produce and feel electromagnetic fields, so particles that produce and feel the strong force field have a property called color charge, and strong force fields are also known as color force fields. This name has nothing whatever to do with the usual meaning of the word color. The quantum excitations of the strong force field are called gluons because their effect is to "glue" the quarks (and the gluons themselves) into composite particles which are neutral with regard to color charge. These particles are called hadrons.

Neutrons and protons are the examples of hadrons found in ordinary matter. The binding of neutrons and protons to form the nucleus of an atom is a residual effect of the strong interactions of the quarks inside the hadrons, just as the binding of electrically neutral atoms to form molecules is a residual effect of their electrical substructure.

Nuclei are for the most part very stable and do not change as we move matter, so strong forces play a hidden, though important, role in our lives. Nuclear fission and fusion are residual strong interaction effects processes in which nuclei do change. In fission a massive nucleus breaks apart into two or more less massive ones. In fusion two very low mass nuclei combine to form a more massive one. Both processes only occur if the sum of the product masses is less than the total mass of the initiating particle(s). The lost mass energy appears as kinetic energy of the products.

The weak interaction does not play any important role in the forces between the particles in a nucleus, so the term "weak nuclear force" is something of a misnomer. For the typical separation of two protons in a nucleus we find the weak interaction provides a force between them that is only about 10^{-7} times as strong as the electrical one. The gravitational attraction between them is tiny, approximately 10^{-36} times smaller than their electrical repulsion. The residual strong force between them is, however, about 20 times stronger than the electrical repulsion, so the nucleus is held very tightly together.

What then is the role in nature of the weak interaction? W and Z bosons are the quantum excitation of the weak field in the same sense that the photon is the quantum excitation of the electromagnetic field. The W boson plays a very special role: only by emitting or absorbing a W can the flavor type of a quark or a lepton be changed. Each of the more massive types of quark or lepton decays by emitting a W boson, thereby making a transition to a less massive quark or neutrino type, respectively. The W itself then very rapidly decays to produce either an additional light quark and antiquark, or a neutrino and a less massive charged lepton. The Z boson mediates non-flavor-changing weak effects. Except in very special circumstances, for example in electron-positron collisions at an accelerator tuned to enhance Z production, Z effects are tiny and difficult to observe.

If it were not for the weak interaction all of the six known flavors of quark (u = up, d = down, s = strange, c = charm, b = bottom, and t = top) and all the charged leptons (electron, muon, or tau) would be stable objects. There would be many more stable types of matter in the world around us than the neutrons and protons and electrons that comprise all atoms!

Even neutrons are not stable, isolated neutrons having a mean-life of about fourteen and a half minutes. They decay to give a proton, an electron, and an antielectron-type neutrino. This process, known as beta decay, is a characteristic weak decay process. A down quark inside the neutron has emitted a (virtual) W boson and thus made a transition into an up quark, thereby converting the neutron to a proton. The W immediately decays to produce the electron and its matching antineutrino.

The neutrons inside many types of nuclei are stable. This can be understood as a consequence of the Pauli exclusion principle, which applies to neutrons and protons (and any other fermions) just as it does to electrons. Inside a nucleus a neutron is stable if it cannot lower its energy by becoming a proton (and emitting an electron and an antineutrino). This occurs when all the allowed lower energy states for protons in that nucleus are already occupied. In fact, in some nuclei the population of neutron and proton states may make it energetically favorable for a proton to become a neutron, which it can do by either capturing an electron or emitting a positron (it also emits a neutrino). To see where weak interactions occur in the radioactive decay chain of a heavy nucleus, look for transitions in which the total number of protons increases or decreases by one, with a corresponding change in the number of neutrons.

All the many particles discovered in cosmic rays and accelerator experiments are unstable, and many have extremely short half-lives. One way to see that there are four distinct fundamental interaction processes is to look at the half-lives. For a given energy release, strong interaction processes occur fastest, typically with half-lives of the order of 10^{-24} s. Processes that produce one or more photons, which signals an electromagnetic decay, have half-lives in the range of 10^{-22} s. Processes involving change of quark flavor and/or neutrinos, either of which signal a weak decay, occur much more slowly, with half-lives of order 10^{-18} s, or even longer. Gravitational interactions are the weakest of all, so weak they play essentially no role in fundamental particle processes. However, they become significant when there are large numbers of particles, since the gravitational effects of all the particles add up, unlike the strong or electromagnetic case where neutral combinations can be formed.

While the four well-understood fundamental forces explain all observed forces and particle decays they may not be the only fundamental forces in nature. There may be other extremely weak forces whose effects we have yet to observe. In the theory of matter known as the standard model, it is found that these four types of interactions are not sufficient to explain all the properties of matter. Some further type of interaction is needed to explain where the masses of all the particles come from. It is not yet understood exactly what form this new interaction takes.

The simplest way to achieve a theory that includes masses for the particles is to introduce a particle known as the Higgs boson that mediates a new type of interaction between fundamental particles and gives them their masses. We have no evidence that this is the answer chosen by nature. Other theories with more complicated additional interactions have been proposed as alternate explanations of particle mass. Only by further experimentation at very high energy accelerators will we be able to tell which of these ideas is correct. We do know that the theory of the four interactions alone is not a complete theory of nature. It would predict that all fundamental particles have zero mass. This is not at all what we observe.

As well as the additional interaction type needed to explain particle masses, physicists speculate that there are other additional interactions that are like the known weak interactions, but even weaker in their effects at everyday energies. These interactions lead to one type of effect that can be searched for, namely the decay of protons. The present experimental information is that the half-life of the proton is greater than about 10^{32} years, so any interaction that causes it to decay must indeed be very weak.

The reason to introduce these interactions, for which there is not as yet a shred of experimental evidence, is that they occur very naturally whenever we try to write a theory in which the strong, electromagnetic, and weak interactions are unified. The mathematical properties of these three very different interactions are so similar that it is very tempting to believe that they are just different low-energy manifestations of a single unified interaction. This idea, which goes by the name of grand unified theory, is very popular with particle physicists because, mathematically, the unified theory is much simpler than having three separate theories for these three types of interactions.

Eventually physicists would like to find a theory that unifies all the types of forces, including gravity. At the present time, no one has succeeded in writing a consistent theory that includes a quantum theory of gravity as well as all the properties of matter and the other interactions. Einstein's dream, a unified theory of all interactions, remains only a dream.

None of the additional interactions discussed above are the so-called fifth force widely publicized in the early 1990s. That was based on certain apparent anomalies in the gravitational force field. All the claimed anomalies can be explained by changing the assumed nearby matter density profile. Currently there is no convincing evidence for gravitational anomalies and so no need for this so-called fifth force to be introduced. The gravitational constant has been observed to be very precisely the same independent of the type of matter. This makes it extremely difficult to construct theories of additional gravitation-like forces that are not already ruled out by experimental data.

See also: BOSON, GAUGE; BOSON, HIGGS; BOSON, NAMBU–GOLDSTONE; BOSON, W; BOSON, Z; INTERACTION; INTERACTION, ELECTROMAGNETIC; INTERACTION, ELECTROWEAK; INTERACTION, STRONG; INTERACTION, WEAK; PARTICLE PHYSICS; PHOTON; QUANTUM FIELD THEORY; QUARK, BOTTOM; QUARK, CHARM; QUARK, DOWN; QUARK, STRANGE; QUARK, TOP; QUARK, UP

Bibliography

CLOSE, F. E. *The Cosmic Onion: Quarks and the Nature of the Universe* (Heinemann, London, 1983).

ROLNICK, W. B. *The Fundamental Particles and Their Interactions* (Addison-Wesley, Reading, MA, 1993).

HELEN R. QUINN

INTERACTION, STRONG

Among the elementary particles are hadrons, which are the subnuclear particles that make up the atomic nucleus and account for the forces that bind the nucleus together. They consist of baryons, such as the proton and neutron, and mesons, such as the pion. Baryons are spin-$\frac{1}{2}$ fermions and play a role in the nucleus similar to electrons in the atom. They are the constituent particles out of which the nucleus is built. Mesons are integer-spin bosons and are analogous to photons or electromagnetic quanta in the atom. Just as the electromagnetic forces that bind the atom can be understood in terms of the continuous exchange of photons between electrically charged atomic constituents, the nuclear force that binds nuclei can be understood as arising from the exchange of mesons, principally the pi, rho, and omega mesons.

The nuclear force is much stronger than the electromagnetic force, thus the term "strong interaction." For example, the nuclear attraction between protons and neutrons at nuclear distances (10^{-13} cm) is roughly 100 times larger than the electric coulomb forces between protons. This difference in the strength of the strong and electromagnetic forces originates in differences in the interactions between electrons and photons, on the one hand, and nucleons and mesons, on the other hand. In either case, the basic process is the emission of a boson by a fermion when it is disturbed. For example, an accelerated electron emits a photon with a probability of about 1 percent. The emitted photons constitute the electromagnetic radiation created when charges are accelerated. Similarly, a sufficiently disturbed proton can emit a neutral pi meson, but this occurs with a much larger probability. The large probability manifests itself in a much stronger force induced by the exchange of pions between nucleons.

The strong force, unlike the electromagnetic force, is very short range and restricted to nucleons spatially separated by no more than about 10^{-13} cm.

The electromagnetic force, by contrast, can extend over arbitrarily large distances, though it slowly decreases with distance. This is why, in the ordinary macroscopic world, electromagnetic forces are important and nuclear forces are not. Ultimately, the ranges of the forces are controlled by the properties of the exchanged bosons, in particular, their masses.

The photon, being massless, is easily emitted and yields the long-range force. The pion has a mass of about 140 MeV (about one-seventh of the proton mass) and cannot travel as far to create a long-range effect. The precise connection between the range of a force r and the mass of the changed boson m, is $r = \hbar/(mc)$, where $\hbar$ is Planck's constant, and c is the speed of light. Thus, the pion with a mass of 140 MeV yields a range of order 10^{-13} cm.

According to the modern theory of hadrons, baryons and mesons are not fundamental indivisible objects. They consist of smaller constituents called quarks. Quarks are fermions and come in a variety of types known as up, down, strange, charm, bottom, and top, but only the up and down quarks play a significant role in the structure of protons, neutrons, and pi mesons. The remaining quark types are constituents of more exotic and short-lived hadrons that are extensively studied in high-energy particle collisions.

The proton, neutron, and other baryons are each composed of three quarks. For example, the proton is two up quarks and a down quark. The neutron is two downs and an up. Antiprotons and antineutrons are similarly constructed out of antiquarks.

Mesons, on the other hand, are made out of a single quark and a single antiquark. For example, a positively charged pi meson is an up quark and an antidown quark. A negative pion is a down and an antiup. A neutral pion is a quantum superposition of up, antiup and down, antidown.

Once again, the familiar pattern of forces mediated by boson exchange recurs, this time to bind the quarks into hadrons. The fundamental boson, similar in some respects to the photon, is called the gluon. A disturbed quark can emit a gluon, and gluon exchange induces a very strong force between quarks and between quarks and antiquarks. The theory of quarks and their interaction with gluons is called quantum chromodynamics (QCD). QCD is the underlying explanation of the strong nuclear force that in some ways resembles the residual van der Waals force between neutral atoms that are responsible for molecular forces.

Molecules are composed of electrically neutral atoms. Nevertheless, the atomic forces that cause molecules to form are electromagnetic in origin. In the same way, hadrons are neutral with respect to color charge. (The term "color" has no more to do with the ordinary concept of color than quark flavor has to do with the taste of quarks. It is just a whimsical term for the source of the gluon field.) The ordinary strong force between nucleons (protons and neutrons) is analogous to the forces between atoms.

The color charge of a quark and its relationship to gluon emission are analogous to the corresponding electromagnetic concepts but involve the more abstract mathematical concepts of group theory. Each kind of quark comes in three colors, which are arbitrarily labeled red, green, and blue. Likewise, a gluon is described by a color and an anticolor label. Thus, there are red-antiblue gluons, green-antigreen gluons, and so forth. A red quark can emit a red-antiblue gluon and become a blue quark. This feature of QCD makes it more mathematically complex than quantum electrodynamics (QED).

Quark confinement is one of the consequences of this increased complexity. Quarks cannot appear in arbitrary combinations. For example, a single isolated quark is not allowed by the theory. Similarly, two quarks are not allowed, though a quark and antiquark pair is allowed. Three quarks are allowed and form an ordinary baryon. The precise rules are subtle, but, roughly, only objects in which the colors appropriately cancel to form color-neutral objects are allowed. Thus, a quark with its antiquark is allowed. According to the mathematical rules of color counting, a quark of each type—that is, one red, one green, and one blue—is also a color-neutral object.

The theory of QCD has achieved a great deal of success in explaining the qualitative properties and behavior of hadrons. It may also be used to calculate many of the properties of hadrons when they collide at high energies. However, due to the strength of QCD interactions, it is not yet possible to use QCD to make high-precision calculations of hadron properties.

See also: COLOR CHARGE; ELECTROMAGNETIC FORCE; FLAVOR; HADRON; NUCLEAR FORCE; QUANTUM CHROMODYNAMICS; QUARK CONFINEMENT

Bibliography

FRITZSCH, H. *Quarks: The Stuff of Matter* (Basic Books, New York, 1983).

GOTTFRIED, K., and WEISSKOPF, V. F. *Concepts of Particle Physics* (Oxford University Press, Oxford, Eng., 1984).

LEONARD SUSSKIND

INTERACTION, WEAK

The weak interaction is a short-range interaction, so it affects only isolated particles or nucleons in a nucleus where they are less than a Fermi (10^{-13} cm) apart. The contribution of the weak interaction to nuclear energies is only approximately 10^{-12} ergs, a very small fraction of typical nuclear binding energies. The importance of the weak interaction lies in the fact that it is the only interaction that can change one type of quark or lepton into another. Therefore, it is responsible for the instability of all the more massive fundamental particles and for certain crucial steps in the chain processes that produce energy from stars and radioactive decays.

The first known weak interaction was beta decay. In this process, a neutron changes to a proton and emits an electron and an antineutrino:

$$n \rightarrow p + e^- + \bar{\nu}_e.$$

The free neutron has a lifetime of 17.3 m. The emitted electrons (originally named the beta radiation) have a continuous spectrum rather than one that is simply a peak at a single energy. If only a proton and an electron were produced, because of conservation of energy and momentum, one would expect each to have a predetermined energy. Thus, the range of electron energies produced in beta decay at first suggested that conservation of energy might not be true for this decay. In 1930 Wolfgang Pauli proposed instead that the range of electron energies could be explained by the existence of a new particle (now called a neutrino) that had only weak interactions. At first his suggestion was considered even more radical than a lack of conservation of energy, but it proved to be the correct explanation. It is the neutrino that carries off some of the energy released in beta decay. This particle is electrically neutral and has zero mass as far as is known. It does not participate in either the strong or the electromagnetic interactions; it feels only weak and gravitational interactions. Neutrinos hitting Earth from the Sun or elsewhere will, with a high probability, pass completely through the earth and out the other side without interacting.

The neutrinos produced in beta decay are now known to be antielectron-type neutrinos, $\bar{\nu}_e$. There are also two other types of neutrinos: $\nu_\mu (\bar{\nu}_\mu)$ the muon-type neutrinos, and $\nu_\tau (\bar{\nu}_\tau)$, the tau-type neutrinos. All of these neutrinos have spin-$\frac{1}{2}$ and are fermions. Because they interact only via weak interactions, neutrinos are very difficult to observe experimentally. Now, experiments with neutrinos at high-energy particle accelerators are routine. However, one needs a very large volume detector and a very intense source of neutrinos to observe even a few neutrino-induced processes.

Since the neutron has a mass of 939.56563(28) MeV/c^2, which is slightly larger than the sum of the proton mass of 938.27231(28) MeV/c^2 and the electron mass of 0.51099906(15) MeV/c^2, the free neutron undergoes beta decay (β^-), but the free proton cannot emit a positron, (β^+). However, for nucleons bound in a nucleus, one can have either β^- decay and β^+ decay or neither, depending on the binding energies. These processes are characterized by the participation of four fermions:

$$(Z,A) \rightarrow (Z + 1,A) + e^- + \bar{\nu}_e \; (\beta^- \text{ decay})$$

$$(Z,A) \rightarrow (Z - 1,A) + e^+ + \nu_e \; (\beta^+ \text{ decay}),$$

where Z is the atomic number of the nucleus, and A is its atomic weight.

The principle that determines which decays are possible is the conservation of energy and mass, as given in Albert Einstein's equation $E = mc^2$. If there is more rest energy in the initial state than in the final state, the process will occur; otherwise, it does not occur. The Pauli exclusion principle, which says that no two identical particles can be in the same state, also plays a role here. The Pauli principle applied separately to the neutrons and the protons in a nucleus explains why there are cases where a nucleus with one less neutron and one additional proton can be more massive than the original nucleus, even though a proton is less massive than a neutron. In this case, β^- decay is forbidden and the neutrons in such a nucleus are stable. β^+ emission, in which a proton decays inside a nucleus to become a neutron, is similarly explained. Energy conservation together with the Pauli exclusion principle determine the various radioactive decay chains of nuclei as well as the possibility of fusion processes.

The weak interaction is the only one of the four fundamental interactions that can change one element into another without changing the atomic weight. Do not confuse the term atomic weight, which is the total number of neutrons plus protons in a nucleus, with the mass of the nucleus. A gives only an approximate mass estimate for a nucleus. For example,

$$^{37}\text{Cl} \rightarrow {}^{37}\text{Ar} + e^- + \bar{\nu}_e \;(\beta^- \text{ decay})$$

and

$$^{11}\text{Cl} \rightarrow {}^{11}\text{B} + e^+ + \nu_e \;(\beta^+ \text{ decay}).$$

In each of these decays the product nucleus is less massive than the original, although they share the same "atomic weight." Neutrinos can also initiate weak processes in matter, for example,

$$\bar{\nu}_e + p \rightarrow n + e^+.$$

This is how neutrinos were first observed by Clyde Cowan and Frederick Reines in 1953. The source of the antineutrinos was a nuclear reactor, and the target was a large vat containing a hydrogenous substance with Cd added. Any produced positron meets an electron and annihilates in a time of about 10^{-9} s. The photons produced by such annihilations were observed. The produced neutron was captured significantly later (10^{-5} s) by the Cd, and gamma rays emitted in the capture reaction were also seen. Only the neutrino-initiated process $\bar{\nu}_e + p \rightarrow n + e^+$. could explain this sequence of events. Furthermore, the rate at which such events were seen was as expected from the theory of weak interactions.

The weak interaction grows in strength as the energy increases. It becomes comparable in strength to the electromagnetic interaction at approximately 100 GeV, where the two interactions get unified into what is now called the electroweak interaction. At energies large compared to 100 GeV, weak and electromagnetic process are mathematically very similar; all can be viewed as aspects of a unified electroweak theory.

This theory of the weak interaction has evolved from the original Fermi theory in 1930, to the theory of Robert Marshak, E. C. G. Sudarshan, Richard Feynman, and Murray Gell-Mann in 1959, to the present gauge theory of the electroweak interaction in which electromagnetism and the weak interaction are unified. The electromagnetic interaction is mediated by the exchange of virtual photons, while the weak interaction is mediated by the exchange of the vector bosons (W^+ and W^- for the charged weak interaction, and the Z^0 for the neutral weak interaction). The large masses of the gauge bosons ($M_W = 80.22(26)$ GeV/c^2 and $M_Z = 91.187(7)$ GeV/c^2 account for the fact that the weak interaction has a very short range, much less than one Fermi (10^{-13} cm), which is approximately the size of the neutron or proton.

In this modern theory, beta decays are described in terms of quark processes. A neutron contains one up-type quark and two down-type quarks, whereas a proton contains two up type and one down type. In the beta decay process a down quark emits a (virtual) W^- boson and becomes an up type. The W^- almost immediately disappears to produce the electron and antielectron-type neutrino. The electroweak theory was developed by Sheldon L. Glashow, Steven Weinberg, and Abdus Salam in the 1970s, for which they received the 1979 Nobel Prize for physics. The theory predicted the existence of W bosons and of an additional type of weak interaction process mediated by an uncharged massive particle, the Z boson.

Since the W boson is much more massive than a neutron, it is clearly impossible (by conservation of energy) to actually produce real W in the beta decay of a neutron. That is the meaning of the word "virtual" in the description of beta decay above. A virtual particle is one that occurs at an intermediate stage in a quantum process but cannot be directly observed. However, real W and Z particles can be produced in high-energy collision experiments. The Nobel Prize for physics in 1984 was awarded to Carlo Rubbia and Simon van der Meer for their leadership in building the experiment and the accelerator facility that discovered the first direct evidence for the W and Z particles. Subsequent experiments have not only confirmed the existence of these bosons but have provided many precision tests of relationships predicted by this electroweak theory.

Weak decay involving W bosons is the only process in which one fundamental quark or lepton can be transformed into a different one. In beta decay, a down quark becomes an up quark by emitting a W^- boson. Similarly, every more massive quark type can decay to any less massive one of different charge by emitting a (possibly virtual) W boson with the appropriate charge. The possible final states are then governed by the available energy. The W^- can decay to produce any quark with charge $-1/3$ (down, strange, or bottom) together with any antiquark of a charge $-2/3$ (i.e., the antiparticle of a charge $+2/3$ quark, up, charm, or top), provided sufficient energy is available. Other possible products from W^- decay are any lepton (electron muon or tau together with the matching antineutrino). The pattern and relative probabilities of these decays can all be fit by the Glashow, Weinberg, Salam electroweak theory. W^+ bosons have a similar set of possible decays; simply replace particles by antiparticles (and vice versa, antiparticles by particles) in the list of W^+ decays above. Thus, it is the weak interaction that is responsible for the instability of all the more massive types of quarks and leptons and, hence, for the fact that these fundamental particles are not found as the basic constituents of ordinary matter. Only up and down quarks (and electrons) are found in stable matter. More than 150 different types of unstable particles that contain more massive types of quarks are known, and their weak decay patterns are all understood in terms of quark decays.

The least massive charged mesons, called pions, also decay by a weak interaction process, but in this case, rather than being emitted in the decay of a single quark, the virtual W is produced by annihilation of the quark and antiquark within the meson. The virtual W then decays to produce a muon and a muon-type neutrino (or antineutrino, depending on the muon charge).

Just as the more massive quark types are unstable because of weak decays, so too are the more massive charged leptons, the muon and the tau. The muon, a particle just like an electron but more massive, decays via the weak interaction as follows:

$$\mu^- \to e^- + \bar{\nu}_e + \nu_\mu$$

$$\mu^+ \to e^+ + \nu_e + \bar{\nu}_\mu.$$

Its lifetime is $2.19703(4) \times 10^{-6}$ s (approximately 2.2 μs). This decay illustrates a property of lepton decays alluded to above—the fact that there is a direct relationship between each charged lepton type and a distinct neutrino type. Notice in the decay above that when we start with a negatively charged muon the final state contains a muon-type neutrino, but when we start with the antimuon (positively charged) the final state contains an antineutino of muon type. We can define a quantity called lepton number for each lepton type, which is conserved in all processes. Electron number, for example, is de-

fined as $+1$ for an electron (e^-) or for an electron-type neutrino (ν_e), and -1 for a positron (e^+) or for an antielectron-type neutrino ($\bar{\nu}_e$); all other particles have zero electron number. Similar definitions apply for muon number and tau number.

The tau lepton has a mass $m_\tau = 1{,}777.1(5)$ MeV/c^2. It was discovered in 1975 by Martin Perl at the Stanford Linear Accelerator Center (SLAC). It decays via the weak interaction into the electron and the muon:

$$\tau^- \to e^- + \bar{\nu}_e + \nu_\tau$$

$$\tau^- \to \mu^- + \bar{\nu}_\mu + \nu_\tau$$

and

$$\tau^+ \to e^+ + \nu_e + \bar{\nu}_\tau$$

$$\tau^+ \to \mu^+ + \nu_\mu + \bar{\nu}_\tau.$$

However, the tau is massive enough to decay also into strongly interacting particles or hadrons; for example,

$$\tau^- \to K^- \nu_\tau$$

$$\tau^+ \to K^+ \bar{\nu}_\tau$$

$$\tau^- \to \pi^- \pi^0 \nu_\tau$$

$$\tau^+ \to \pi^+ \pi^0 \bar{\nu}_\tau.$$

The tau lifetime is $295.6(3.1) \times 10^{-15}$ s. Each of these hadronic decays of the tau can be understood in terms of quarks. For example, a K^+ meson is a combination of an up quark and an antistrange quark produced from a virtual W^+ boson.

As far as is known, the neutrinos are massless. The experimental bounds are

$$m_{\nu_e} < 5.8 \text{ eV}/c^2,$$

$$m_{\nu_\mu} < 0.27 \text{ MeV}/c^2,$$

and

$$m_{\nu_\tau} < 31 \text{ MeV}/c^2.$$

All of these bounds are consistent with massless neutrinos. If neutrinos actually have small masses, then it is probable that the three separate lepton number conservation laws are not exact, though the sum of all three together remains an exactly conserved quantity.

Fundamental particle interactions have certain discrete invariances, which relate the rate for a process to that for some particular similar process. For example, in either strong interactions or electromagnetic interactions, it is found that any process has exactly the same rate as its mirror image process. This is encoded in particle physics as a symmetry called parity. Parity is a combination of mirror reflection and a rotation about an axis perpendicular to the mirror, so that all position vectors (relative to the center of the mirror) are reversed ($x \rightarrow -x$). A second relationship, called charge conjugation invariance (C), says that the rate for any strong or electromagnetic process is identical to the rate for a second process that has all particles replaced by their antiparticles (and vice versa). Until 1956 it was believed that the weak interaction, like the other interactions (strong, electromagnetic), were invariant under P and C transformations. Then T. D. Lee and Chen Ning Yang realized that parity invariance had not been checked for the weak interaction. They suggested an experiment in which this could be checked. The experiment performed by Chien-Shuing Wu in 1957 consisted of the beta decay of cobalt-60 into nickel-60:

$$\mathrm{Co}^{60} \rightarrow \mathrm{Ni}^{60} + e^- + \nu_e.$$

The Co^{60} nucleus has an intrinsic angular momentum called spin. The spin of the cobalt-60 nuclei were aligned with an external magnetic field $\mathbf{B}$ and the emitted electrons were detected. The electrons were found to be emitted only in the downward (opposite $\mathbf{B}$) direction; no electrons are emitted upward. If P invariance applied, one would expect equal numbers of electrons emitted up and down. Since no electrons were emitted upward, to everyone's surprise, it was found that P is maximally violated in weak interactions. Lee and Yang were awarded the Nobel Prize for their work. The result implies that neutrinos are left-handed and the antineutrinos are right-handed. One describes this fact by saying that the neutrinos have helicity -1 and the antineutrinos have helicity $+1$, where helicity is defined as the projection of a particle's spin along the direction of its motion divided by the magnitude of the total spin of the particle. There is no evidence so far for the existence of right-handed neutrinos or left-handed antineutrinos in nature, but if neutrino masses are not exactly zero, then these particles may also exist.

For some time it was thought that a weaker invariance was still true. The combination of the two operations, parity and charge conjugation (CP), appeared to give related rates. For example, in the weak decays

$$\pi^+ \rightarrow \mu^+ + \nu_\mu$$

$$\pi^- \rightarrow \mu^- + \bar{\nu}_\mu,$$

the μ^+ are emitted polarized, with their spin opposite to the direction to that in which they travel. This must happen in order to conserve angular momentum, since there are only left-handed neutrinos and the pion spin is zero. If P is applied, one obtains a right-handed neutrino and a μ^+ with spin parallel to its travel direction. However, if C is applied as well as P, the process in the second equation is obtained, with a μ^- that has spin parallel to its travel. This does occur in nature ($\pi^- \rightarrow \mu^- + \bar{\nu}_\mu$), since the antineutrino is right-handed. The rates for these two processes are the same. Thus, the combination PC (or CP, the order of the two operations does not matter) was considered to be conserved in weak interactions. In 1964, however, Val L. Fitch and James W. Cronin observed a small (0.1%) CP or PC violation effect in K decays. They received the Nobel Prize in 1980 for their work. High-energy physicists are still trying to understand the CP violation.

The weak interaction is important for producing elements and energy in stellar interiors, including the Sun. The proton-proton chain (important below 15×10^6 K) is the major mechanism for energy release from the Sun. It includes the following steps:

$$p + p \rightarrow {}_1\mathrm{H}^2 + e^+ + \nu_e$$

$${}_1\mathrm{H}^2 + p \rightarrow {}_2\mathrm{He}^3 + \gamma$$

$${}_2\mathrm{He}^3 + {}_2\mathrm{He}^3 \rightarrow {}_2\mathrm{He}^4 + 2p.$$

The carbon-nitrogen cycle (important above 15×10^6 K) is

$${}_6\mathrm{C}^{12} + p \rightarrow {}_7\mathrm{N}^{13} + \gamma$$

$$_7N^{13} \rightarrow {}_6C^{13} + e^+ + \nu_e$$

$$_6C^{13} + p \rightarrow {}_7N^{14} + \gamma$$

$$_7N^{14} + p \rightarrow {}_8O^{15} + \gamma$$

$$_8O^{15} \rightarrow {}_7N^{15} + e^+ + \nu_e$$

$$_7N^{15} + p \rightarrow {}_6C^{12} + {}_2He^4.$$

In each of these chains the step that produces a neutrino is a weak interaction process (charged) mediated by the W boson. It is thought that all heavy elements were produced in such chains of processes in the interior of stars, and so the weak interaction is essential to the presence of these elements in the universe.

Neutrinos from the Sun (solar neutrinos) were observed in the 1970s by Raymond Davis and colleagues. However, the number observed is less than the number expected from theory. The theory involves not only weak interaction theory but also a detailed model of the core of the Sun. More recent experiments, sensitive to different neutrino energies, also find discrepancies from the predictions based on standard models of the Sun. This mismatch of calculation and experiment is called the solar neutrino problem. There are four major assumptions in the theoretical model:

1. The pp-solar-cycle is the dominant energy source of the Sun:

$$4p + 2e^- \rightarrow {}_2He^4 + 2\nu_e$$

2. The Sun is in quasi-equilibrium.
3. The neutrinos travel without change or interaction from production in the solar core to their detection at the earth.
4. The neutrino detection rates are understood.

One possible explanation is that the three types of neutrinos are not exactly massless and distinct and that there are small effects that mix up the three neutrino types as they travel through the Sun. This explanation has not yet been confirmed, but experiments are being designed that can test whether neutrinos have masses in the range required to explain solar neutrino rates.

See also: CHARGE CONJUGATION; *CPT* THEOREM; INTERACTION, ELECTROWEAK; LEPTON; LEPTON, TAU; NEUTRINO; NEUTRINO, HISTORY OF; NEUTRINO, SOLAR; PARITY

Bibliography

CLINE, D. B., RUBBIA, C., and VAN DER MEER, S. "The Search for Intermediate Vector Bosons." *Sci. Am.* **244** (March), 48 (1982).

QUIGG, C. "Elementary Particles and Forces." *Sci. Am.* **250** (April), 84 (1985).

MARK A. SAMUEL

HELEN R. QUINN

INTERFERENCE

In physics the term "interference" is used in a number of different ways, for instance to describe the difficulty of distinguishing a desired signal in the presence of another signal or background noise. This article is concerned solely with the most common use of the term by physicists, namely, to characterize the unintuitive effects that can be observed when two or more waves (with whose elementary properties the reader is assumed to be familiar) simultaneously traverse the same points in space. The simplest and best known illustration of interference is provided by a pair of narrow slits in an opaque screen on which shines a monochromatic (for instance, red) light source; a sheet of paper behind the slits will exhibit alternate bright and dark bands of red in the region where the light waves reaching the paper from the two slits overlap, even though the illumination on the paper from either slit alone shows no trace of any such variation in light intensity. In other words, the waves emanating from the individual slits appear to be "interfering" with each other; correspondingly the alternate bright and dark bands usually are termed "interference fringes," with the interference termed "constructive" at the bright bands and "destructive" at the dark. Illuminating with red light a screen containing a large number of narrow slits (e.g., 10,000) carefully equally spaced a few wavelengths apart (the wavelength of red light is about 7×10^{-7} m) produces even more dramatic unintuitive interference effects. In particular, when such a screen, known as a diffraction grating, is illuminated from a direction perpendicular to the grating, the grating appears to deflect (or as physicists say, diffract) the incident light into just a few widely separated special directions, as can be demonstrated using a lens to focus the transmitted light onto a sheet of paper. In these circumstances just a few widely spaced very bright bands will be observed, located at those very special points on the paper where all the individual waves that leave the individual slits and make their way through the lens are able to interfere constructively; at all other points the paper is essentially completely dark, corresponding to essentially completely destructive interference of those individual waves leaving the slits.

Paradoxically, we are able to construct uncomplicated (yet quantitatively very accurate) explanations of such interference effects only because in nature waves—though capable of interference in the sense just described—propagate independently of (i.e., without interference by) other waves, to a very good first approximation at least. In a chamber quartet recital, for example, the sound waves from the violin traverse the auditorium in a fashion that does not depend on whether or not the pianist is playing. The electromagnetic wave transmission from the television Channel 4 antenna reaches our sets quite unaffected by the signal from Channel 13 or by any of the much longer wavelength signals from AM radio stations in the area, and continues to be unaffected when Channel 13 shuts down or increases its power. The only known simple means of accounting for this just-described independent propagation of waves is the assumption that if $Y_1(\mathbf{r}, t)$ denotes the amplitude of a lone wave 1 traversing points $\mathbf{r}$ in space as a function of time t, and if $Y_2(\mathbf{r}, t)$ is the corresponding amplitude of a second lone wave 2, then the wave amplitude $Y(\mathbf{r}, t)$ experienced at space points $\mathbf{r}$ and instants t when both waves are simultaneously present must be just the sum of the two independently propagating individual wave amplitudes, that is,

$$Y(\mathbf{r},t) = Y_1(\mathbf{r},t) + Y_2(\mathbf{r},t). \qquad (1)$$

This rule for wave propagation, known as the "principle of superposition," is more than an assumption, however; rather it is a consequence of the fact that to a very good first approximation waves of all types in all media obey a special differential equation, the "wave equation," which happens to be "linear." The linearity implies that $Y(\mathbf{r}, t)$ given by Eq. (1) indeed is a solution of the wave equation if Y_1 and Y_2 each are solutions; if the sum on the right-hand side of Eq. (1) were not a solution of the wave equation, $Y(\mathbf{r}, t)$ given by Eq. (1) could not possibly represent the actual wave amplitude resulting from the simultaneously propagating wave amplitudes Y_1 and Y_2.

It is a quite general result of wave theory that the energy being carried by the wave $Y(\mathbf{r}, t)$ in the vicinity of $\mathbf{r}$ at time t is proportional to $[Y(\mathbf{r}, t)]^2$. One of the simplest solutions $Y(\mathbf{r}, t)$ of the wave equation can be visualized as the progression past points $\mathbf{r}$ of a periodic alternation of wave amplitude maxima $Y = A$ and minima $Y = -A$ (akin to the crests and troughs of ocean water waves relative to the undisturbed ocean); in this solution, which is known to represent the propagation of a monochromatic light wave or a pure tone sound wave, A^2 is the measure of the average (over a wave cycle) of the wave energy. For two such simultaneously propagating waves, therefore, the superposition principle Eq. (1) provides the justification for associating destructive interference at a point $\mathbf{r}$ with the simultaneous arrival at $\mathbf{r}$ of a Y_1 wave crest and a Y_2 wave trough, corresponding (momentarily at least) to a net wave energy at $\mathbf{r}$ whose measure $(A_1 - A_2)^2$ is comparatively small. Constructive interference at $\mathbf{r}$ is similarly associated with the simultaneous arrival at $\mathbf{r}$ of the crests of both Y_1 and Y_2.

In order that the interference at $\mathbf{r}$ actually be observable, the crest-trough confluences (destructive interference) or crest-crest confluences (constructive interference) discussed in the preceding paragraph must be persistent, implying that the wave amplitudes of the interfering waves at $\mathbf{r}$ must be coherent, that is, they must vary relative to each other in some well-defined unchanging way, for some sufficiently extended period of time. For light waves, this necessary coherence generally cannot be achieved unless the interfering waves ultimately arise from the same source, as, for example, in the double-slit interference experiment described earlier, wherein the slits were illuminated by the same red light source; because the frequency of red light is about 5×10^{14} Hz (cycles per second), for destructive interference to persist for even a microsecond when each slit is illuminated by a different red light source imposes the requirement (presently achievable only with the most carefully controlled light sources) that the two independent light sources vary in an unchanging way relative to each other for 500 million cycles. Sound waves from independent sources normally also do not interfere;

there are no seats in a concert hall where the sound reaching the ear from two orchestra violins playing the same notes is less loud than the sound from either violin alone. Because sound frequencies are so much lower than light frequencies, however, interference from different sound sources can be observed with much less difficulty than with light waves. When two audible pure tones whose frequencies f_1 and f_2 differ by only a few hertz simultaneously arrive at the ear, a waxing and waning (or beating) of the sound intensity at a frequency equal to $f_1 - f_2$ can be heard as the crests of the two pure tones at the ear slowly go in and out of coincidence; piano tuners do fine tuning by listening for the beats between the fundamental of a given key and the second harmonic of the key one octave down.

For light waves the superposition principle was incorporated into Huygens's principle of wave propagation by Christiaan Huygens as early as 1678, long before it was realized that waves obeyed a wave equation, in the course of a continuing controversy about whether light propagated as a wave or as a stream of particles ("corpuscles" was the term employed by Isaac Newton, who did not favor the wave theory of light. Since light in free space, for example, the light reaching Earth from the Sun, appears to travel in straight line "rays," the supposition that the light consists of a stream of particles is natural. Moreover, Newton's corpuscular theory could account for the observed trajectories of light rays reflected from a smooth surface or bent by traversal through a smooth interface between two media. According to Huygens's principle, however, light propagation can be visualized as the advance of a succession of wave fronts, with each new wave front formed from the superposition and consequent interference of spherical wavelets sent forth from every point on the previous wave front. It then is simple to show that the pattern of constructive and destructive interference among these wavelets causes plane wave fronts to advance via new plane wave fronts; similarly, spherical wave fronts advance via new spherical wave fronts. In short, the purely wave-theoretic Huygens's principle, without any reference whatsoever to particles, also explains why light appears to travel along rays, the rays being the lines drawn perpendicular to the succession of advancing wave fronts. Furthermore, Huygens's principle—which today can be rigorously derived from the wave equation—also can account for the observed behavior of light rays on reflection or refraction.

Newton's corpuscular theory of light was not abandoned by the community of physicists until after about 1862, when Jean Bernard Léon Foucault reported his experimental finding that the velocity of light in water was less than in air. The corpuscular theory could explain the refraction of light rays between air and water only by assuming that the velocity of light in water was greater than in air; the wave theory, as formulated using Huygens's principle, required that the velocity of light in water be less than in air, the result Foucault found. James Clerk Maxwell's purely theoretic derivation of the magnitude of the velocity of light, published in 1873, provided the final clincher for the thesis that light propagates as a wave, indeed an electromagnetic wave. The corpuscular theory of light began to fall out of favor as early as 1801, however, when Thomas Young first observed double-slit interference fringes; the notion that the two streams of particles arriving from the two slits can somehow cancel each other at some points so as to produce dark areas on the sheet seems obviously unreasonable, as if two bullet holes in a target could somehow merge to produce no bullet hole at all. Even stronger evidence for the wave theory of light and the utility of the interference concept came from the work of Augustin Fresnel in about 1820. Fresnel realized that the constructive and destructive interference among wavelets that causes light to apparently travel along rays requires the participation of wavelets from all points on the oncoming wave front; thus, for example, interposing a screen that blocks a portion of the oncoming wave front results in light propagation that cannot be visualized in a simple corpuscular ray fashion. On this basis Fresnel was able to quantitatively account for the diffraction (bending) of light into the shadow zone behind a semi-infinite plane screen. Fresnel's theory shows that this bending is "proportional to" the wavelength in the incident wave, thereby offering a simple explanation of the observation that sound waves generally penetrate around the corner of a building, whereas light waves do not (compare the 1-ft wavelength of a 1,000 Hz sound wave in air with the wavelength of red light stated earlier). Fresnel diffraction theory also quantitatively predicts the pattern of interference fringes which characteristically are observed at the edge of any shadow produced by a sharp edged opaque barrier. An especially convincing demonstration of the validity of Huygens's principle and the interference concept is the bright spot observed at the center of the shadow of a circular disk (e.g., a penny) oriented perpendicular to the incident light wave; the bright spot results from the necessarily constructive (as can be shown) net interference of all the secondary wavelets reaching the symmetrically situated point at the shadow's center. The presence of this bright spot is so unintuitive that Siméon-Denis Poisson, who first recognized that Fresnel's theory predicted the existence of such a spot, regarded this (at the time not yet observed) result as an incontrovertible refutation of the wave theory of light.

Despite these just-described successes of the wave theory of light and the interference concept, it now is known—and has been known ever since Arthur Holly Compton's x-ray scattering experiments in 1923—that electromagnetic waves, including visible light, do manifest particle-like properties, despite the apparently obvious unreasonableness of the belief that particles can destructively interfere. Even more remarkably, the elementary particles (namely electrons, protons, and neutrons) comprising ordinary matter have been shown to have wave-like properties, including the ability to interfere. Electron diffraction, seemingly quantitatively explainable only on the same wave interference basis that accounts for the red light diffraction grating observations discussed earlier, was observed by Clinton Joseph Davisson and Lester H. Germer in 1927; since that date interference also has been observed with beams of neutrons, helium atoms, and even diatomic sodium molecules. Further pursuit of these dual wave-particle properties of electromagnetic radiation and of material particles is beyond the scope of this article.

It is essential to remember that the discovery of particle-like properties associated with light waves in no way diminishes the essential role of interference, through the principle of superposition and Huygens's principle, in quantitatively accounting for light propagation. Indeed, as is so commonly the case with scientific advances, recognition of the importance of interference in light propagation has stimulated important applications of interference to technology as well as to new pure science. The locations of interference fringes can be very sensitive to the wavelengths and to the relative displacements of the sources from which the interfering coherent waves emanate. Thereby, interferometers are able to very accurately measure not only wavelengths, but also object dimensions, distances, and velocities. Illustrations of the use of interferometers include the demonstration by Albert A. Michelson and Edward W. Morley that the velocity of light measured by an observer on Earth does not depend on the direction of propagation of the light relative to Earth's velocity; Michelson's measurements of stellar diameters; modern measurements of atmospheric emissions (which are performed with commercially available interferometers); the construction of "interference filters" that permit the transmission of light in a selected band of frequencies only (atmospheric sensing measurements also have employed commercially available interference filters); the testing of optical components, for example, a telescope mirror, to be certain it has been ground to the designed shape and smoothness; studies of fluid flow; studies of stressed rotating vibrating objects, for example, turbine blade assemblies; detection of strains in transparent materials; and verifying the alignments of the photolithographic "masks" used in fabricating integrated circuits, for example, computer chips. One of the most ingenious applications of interference is holography, invented by Dennis Gabor in 1948, wherein a three-dimensional image of an object can be constructed by illuminating a photographic film that has recorded the interference pattern between the light reflected from the object and the original light source; holography has important applications in industrial design, for example, of musical instruments. Interference has had far fewer technological applications with sound waves than with light, but standing acoustic waves, generated by interference between a sound wave and its reflection, have been used to measure velocity and absorption in various materials, at audible and ultrasonic frequencies.

See also: COHERENCE; COMPTON EFFECT; DAVISSON–GERMER EXPERIMENT; DIFFRACTION, FRESNEL; DOUBLE-SLIT EXPERIMENT; GRATING, DIFFRACTION; HOLOGRAPHY; INTERFEROMETRY; LIGHT, WAVE THEORY OF; MAXWELL'S EQUATIONS; MICHELSON–MORLEY EXPERIMENT; WAVE–PARTICLE DUALITY, HISTORY OF

Bibliography

BUCHWALD, J. Z. *The Rise of the Wave Theory of Light* (University of Chicago Press, Chicago, 1989).

HALLIDAY, D.; RESNICK, R.; and KRANE, K. S. *Physics,* 4th ed. (Wiley, New York, 1992).

HARIHARAN, P. *Basics of Interferometry* (Academic Press, Sand Diego, 1992).

EDWARD GERJUOY

INTERFEROMETER, FABRY–PÉROT

For almost one hundred years, the Fabry–Pérot interferometer (named for Charles Fabry and Alfred Pérot) has been widely used for studying the structure of spectral lines. It has seen important applications for research in atomic physics, astronomy, astrophysics, and nuclear structure. It is now being used as a resonant cavity for lasers. In its most commonly used form the interferometer produces circular fringes, appearing as a number of sharp and bright concentric rings on a dark background, which can be seen with a telescope or projected on a screen. The fringes result from the interference between multiple beams. The interferometer is simple and basically consists of two thinly silvered surfaces on parallel flat glass plates. The plates are separated by an air gap and precisely adjusted for orientation and separation, which is typically from 1 mm to 10 cm. The resolving power, $\lambda/\Delta\lambda$, where λ is the wavelength of a spectral line and $\Delta\lambda$ the difference in wavelength between two closely spaced lines, increases with plate separation and is typically about 10^6. The separation may be increased as long as the rings can be seen.

The typical experimental arrangement shown in Fig. 1 involves a broad, and approximately monochromatic, light source (S), two plates $(P$ and $P')$, which are half silvered on the inner surfaces, defining an air gap of length d, a lens (L), and a screen (S'). Consider a ray originating at a point A on the broad source of light and incident at an angle θ on the front plate of the interferometer. After passing through the optically flat glass plate it is incident on the first semi-silvered surface and partially transmitted into the air gap. It passes through the gap to the second semi-silvered surface, where it is reflected and partially transmitted. Following multiple reflections, a series of transmitted parallel beams results, as shown in Fig. 1. After passing through a lens, these transmitted beams converge for interference at point B. If the transmitted beams are numbered starting with 1 at the bottom, then constructive interference occurs when all the beams 1, 2, 3 . . . arrive at B in phase. For a minimum in the interference pattern, 1, 3, . . . are out of phase with beams 2, 4, . . . The condition for a maximum is given by $2d \cos \theta = m\lambda$, where m is an integer. It occurs for all points on a circle centered at point O, which lies on the intersection of the lens axis with the observation screen. The value of m ($\approx 2d/\lambda$) is generally very large. As the angle θ decreases, the condition for a maximum is met for a value of m greater by 1 than on the previous ring. For a given wavelength this results in a series of fringes centered on O. All rays that are incident at a given value of θ will lie on a single circular fringe. Interference fringes can also be seen when the plates $(P$ and $P')$ are uncoated, but the resulting fringes would be broad. The fringes sharpen dramatically when the reflecting power of the surfaces is increased by light silvering.

Wavelengths can be determined by measuring the diameters of the fringes recorded on a photographic plate. The interferometer may be used in a scanning mode by varying d with a piezoelectric element. Spectral scanning can also be accomplished by varying the pressure in the gap since the general condition for a maximum is given by $2nd \cos \theta = m\lambda$. Here n is the index of refraction of the gas, which depends on pressure ($n = 1$ for air at atmospheric pressure).

A confocal version of the interferometer, with two spherical mirrors, is now used as a high-resolution

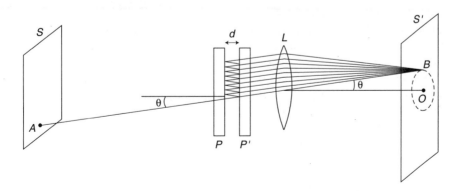

Figure 1   Typical experimental arrangement for a Fabry–Pérot interferometer.

spectrum analyzer. The mirrors have the same radius of curvature r and are separated by a distance r. They can be polished more precisely and their alignment is not as critical as for the plane mirror instrument.

See also: INTERFERENCE; INTERFEROMETRY; LASER; REFLECTION; REFRACTION, INDEX OF; RESOLVING POWER; SPECTRAL SERIES; WAVELENGTH

Bibliography

FABRY, C. and PÉROT, A. "On a New Form of Interferometer." *Astrophys. J.* **13**, 265–272 (1901).

HECHT, E. *Optics*, 2nd ed. (Addison-Wesley, Reading, MA, 1990).

HERNANDEZ, G. *Fabry–Pérot Interferometers* (Cambridge, Eng., Cambridge University Press, 1986).

VAUGHAN, J. M. *The Fabry–Pérot Interferometer: History, Theory, Practice and Applications* (Adam Hilger, Bristol, Eng., 1989).

EDWARD POLLACK

INTERFEROMETRY

Interferometry is a high-resolution measurement technique which derives from the principle of linear superposition of waves. This principle states that if different sources produce waves that have nonvanishing amplitudes at any point in space, then the amplitude of the resultant wave at that point is the linear sum of the amplitudes of each of the waves, as illustrated in the Young double slit experiment for two sources. Wave amplitudes are represented mathematically by complex numbers having both a magnitude and a phase. As long as it is possible to define a definite phase relationship between all the sources contributing to the resultant wave, the measured quantity, the energy carried by the wave or the average position or momentum of the particle which the wave describes, given by the absolute square of the amplitude, will exhibit characteristic maxima or minima of intensity associated with interference features.

The interferometer is a device designed to exploit linear superposition by introducing a controlled phase shift or amplitude modification, related in a known way to the parameter to be measured, which can be detected by the change in the appearance of the interference maxima or minima in the superposition pattern. Interferometers operate by splitting the wave front, as in the microwave interferometer, splitting the amplitude, with the Michelson, Mach–Zehnder, or Sagnac interferometers as examples, or by combining multiple beams, represented by the Fabry–Pérot interferometer. Laser cavities are constructed on the basis of interferometer design, the difference being that the interference gives rise to frequency shifts in the laser output as opposed to optical path differences resulting in interference fringes.

For those measurements in which precision is the ultimate goal of the experiment, interferometry is unsurpassed by any other technique. It provides the basis for the definition of the meter, the development of the atomic clock, and the most accurate determination of the frequencies of atomic and molecular transitions. Interferometers are used to measure distances as large as shifts in tectonic plates caused by earthquakes and as small as diameters of microscopic-sized objects. The Sagnac interferometer has even been used to determine the velocity of rotation of the earth.

Although interferometry is generally associated with electromagnetic radiation, neutron interferometry (as well as interferometry with other particles) is becoming progressively more important as a tool for making fundamental measurements of the properties of matter. The motion of cold or thermal neutrons can be described by a nonrelativistic Schrödinger equation which gives rise to solutions that are entirely analogous to those for a beam of light. Optics (lenses, prisms, etc.) for neutrons based on their motion under the action of gravity or magnetic fields may be constructed which perform similar functions to those for light, thus making possible the controlled phase or amplitude modulation necessary for these measurements.

See also: DOUBLE-SLIT EXPERIMENT; INTERFERENCE; INTERFEROMETER, FABRY–PÉROT

Bibliography

HECHT, E. *Optics*, 2nd ed. (Addison-Wesley, Reading, MA, 1987).

WERNER, S. A., and KLEIN, A. G. "Neutron Optics" in *Neutron Scattering*, edited by K. Sköld and D. L. Price (Academic Press, Orlando, FL, 1986).

C. DENISE CALDWELL

INTERSTELLAR AND INTERGALACTIC MEDIUM

The interstellar medium is the material and fields between the stars in the Milky Way, our galaxy. The intergalactic medium is the material and fields between the galaxies. All galaxies have interstellar media. The intergalactic medium is singular and pervades the universe.

The space between the stars and galaxies constitutes the best vacuum in the universe. Densities range from 10^5 protons/cm^3 in star-forming dust clouds to 1 proton/10^6 cm^3 (or 10^{-6} protons/cm^3) in the intergalactic medium. In general, pressure equilibrium applies, meaning particle density n times temperature T is roughly constant. Generally, $nT \sim 1,000$ K/cm^3. By comparison, the sea level atmosphere on Earth has $n \sim 10^{20}$ particles/cm^3, $nT \sim 3 \times 10^{22}$ K/cm^3.

The interstellar medium consists of luminous baryonic matter (for our purposes, particles made of neutrons and protons), baryonic dark matter, the gravitational field, the magnetic field, the radiation field, and non-baryonic matter (often called dark matter, or missing mass). Intergalactic space may be filled, in addition, with a repulsive field associated with the cosmological constant. Dark matter seems not to exist in the interstellar medium, but it may well be the dominant component of the intergalactic medium.

The Interstellar Medium in Our Galaxy

On the average, the density of matter in the space between the 10^{11} stars of the Milky Way is 0.1 neutral hydrogen atoms (H)/cm^3. About 10 percent of the luminous baryonic mass in the Milky Way is interstellar matter.

Interstellar matter exists in a range of density states; the highest density regions are referred to as clouds. The clouds are directly visible because excited atoms, ions, and molecules radiate when spontaneous decay to the ground states (the lowest energy states) occurs. Atoms and ions are excited by collisions; by recombination of ions and electrons, leading to cascades of electrons through several energy levels to the ground states; or by absorption of photons at discrete energies (wavelengths) corresponding to discrete energy differences between quantum states. Analogous processes apply to

molecules. For historical reasons, the discrete emission and absorption features seen in spectra of interstellar gas are referred to as interstellar lines.

The most visible types of interstellar matter are H II regions, dominated by ionized hydrogen atoms, and molecular clouds, dominated by hydrogen molecules. H II regions are low-density clouds of material with one or more hot stars in the interior. The radiation from the stars constantly ionizes the elements, which then recombine as described above. This recombination radiation consists mainly of light at wavelengths of 6,563 Å (neutral hydrogen) and 5,007 Å (twice-ionized oxygen). The former is red in color photographs, and the latter is green.

Molecular clouds, which have masses up to 10^6 M$_\odot$, are stellar nurseries. Gas clouds collapse under the force of gravity and fragment into proto-stars. During the collapse, as the density increases, molecules form. Examples include H_2, NH_3, CH_4, and many others, perhaps as massive as C_{60}. Internal heating of the collapsing cloud leads to molecular excitation. The resulting discrete radiation can be seen by radio telescopes, mainly at wavelengths between 1 mm and 21 cm.

Studies of the interstellar medium depend largely on spectroscopy. If an interstellar cloud is between the observer and a star, absorption lines of interstellar species appear in the spectrum of that star. Small motions of interstellar clouds cause small shifts to the red (away from Earth) or blue (toward Earth) of absorption lines, and similarly for emission lines.

Interstellar clouds are regularly condensing into stars, at a rate of 10 M$_\odot$/yr averaged over the life of the Galaxy. The average rate of return of stellar material to the interstellar medium is only about 1 M$_\odot$/yr, mainly through the mechanism of stellar winds. Thus, the interstellar medium is gradually disappearing.

Most of this discussion has dealt with the baryonic matter and its response to the gravitational field. The remaining important aspects of the interstellar medium are the magnetic field and the radiation field. Neutral atoms and molecules in the galactic interstellar medium move with the general rotation of the Galaxy. Charged particles move along magnetic field lines. The interstellar magnetic field probably determines how and where condensation of gas clouds begins.

The radiation field in space consists of photons from stars in the Galaxy; from distant galaxies (whose number compensates for their great distance); and from shock waves created in space by explosive, high pressure events such as supernovas.

Intergalactic Medium

The elements and physical processes discussed for the interstellar medium also occur in the less dense intergalactic medium. However, temperatures are higher than 100,000 K, compared to 10–10,000 K in the interstellar medium.

Just as stars condense from interstellar clouds, so galaxies must condense from intergalactic clouds. Galaxies were, at first, low-density clouds of gas immersed in the even-lower-density intergalactic medium. As stars formed, stellar winds and supernovas forced some material out of galaxies. The counterbalancing process is the addition of gas when intergalactic clouds are accreted by galaxies. The interchange of interstellar media and the intergalactic medium holds the secret of galaxy formation.

To study the intergalactic medium, astronomers use emission and absorption features as they do for the interstellar medium. For absorption line studies, intergalactic clouds that lie in front of very distant quasi-stellar objects (QSOs) can be probed. The strength of the absorption lines is directly related to the number of atoms of each absorbing species. Thus, for instance, absorption lines of oxygen and hydrogen can be used to measure the number of oxygen atoms per hydrogen atom, called the oxygen abundance.

The absorption lines, while occurring in patterns characteristic of each atom or ion, appear greatly shifted in spectra of backgrounds QSOs: The wavelengths are longer by factors of 2–6 than those of the same species in local interstellar clouds and are thus shifted to redder colors. This redshift is caused by the fact that the universe expands uniformly, including all length scales. The further away an object is, the longer the shift in wavelength perceived by the observer. Since the speed of light is fixed and finite, light takes longer to travel from objects further away. We thus see more distant objects as they were at earlier times. The consequence, for absorption studies of intergalactic material (protogalaxies), is that astronomers can study abundances of the elements of the periodic table at varying ages of the universe. The buildup of elements, which are formed in stars by nucleosynthesis of primordial hydrogen and helium, can thus be directly observed through all of cosmic time. These studies of inter-

galactic material should lead to an empirical understanding of the formation of galaxies.

See also: COSMOLOGICAL CONSTANT; DARK MATTER; DARK MATTER, BARYONIC; GALAXIES AND GALACTIC STRUCTURE; MILKY WAY; REDSHIFT; SPECTROSCOPY

Bibliography

SPITZER, L. *Searching Between the Stars* (Yale University Press, New Haven, CT, 1982).
YORK, D. G. "The Elemental Composition of Interstellar Dust." *Science* **265** (5169), 191–192 (1994).

DONALD G. YORK

INVERSE SQUARE LAW

Inverse square laws, which are used in several fields of physics, including mechanics and electrostatics, are laws in which the magnitude of a physical property between two objects is stated to be the reciprocal of the square of the "respective distance" between the two objects. This respective distance is generally taken to be the distance between the geometric centers of the objects, thus, allowing the objects to be treated as points for the mathematical calculations.

The inverse square law in mechanics, also known as the gravitational law, was published by Isaac Newton in 1687. It defines the force of attraction F between two gravitating masses m_1 and m_2, separated by a distance d, as $F = G(m_1 m_2 / d^2)$, where G is the gravitational constant, which has a value of 6.672×10^{-11} m^3·kg^{-1}·s^{-2}. It should be kept in mind that this relation gives only the value of the force. The direction of the attractive force can be taken into consideration by the introduction of a unit vector.

Newton's gravitational law was viewed as the ideal because of the role of mechanics as the fundamental theory of physics. Throughout the eighteenth and nineteenth centuries, Newtonian physics continued to have an important influence on the worldview as well as on theories in other fields of physics. For example, a combination of the inverse square law and Newton's axioms was crucial in Johannes Kepler's formulation of his three laws of planetary motion. However, the gravitational law was not used just to describe celestial phenomena; it was applied to terrestrial phenomena as well. One such phenomenon is potential energy.

If a mass m_1 is moved away from another mass M against the attracting force, it gains energy, which is called potential energy. If the mass m_1 is moving toward M with the attracting force, it loses potential energy. A mass m_1 with spherical movement about M, the center of that sphere, will have no change in its potential energy relative to M. There is no absolute measure of potential energy, but the change of potential energy $V_{1,2}$ of any body moving from a distance r_1 away from M to a distance of r_2 is given by the following equation: $V_{1,2} = GM[(1/r_1) - (1/r_2)]$. This calculation, however, can be applied only when two masses are involved. When a third object becomes involved, the calculations become much more complex. Still, this calculation is not limited to gravitational fields. Since the force between two charges, described by Coulomb's law, is also an inverse square law, a similar relationship applies for electrostatic fields, with the value of the field-producing charge replacing M and $\frac{1}{4}\pi\epsilon_0\epsilon_r$ replacing G. This electrical potential is used to determine both the change of energy of a charge moving in the electrostatic field and the voltage, which is taken to be the difference of the potential between two points.

In 1912 Niels Bohr used the electrical potential of the nucleus in his theory of the atom to calculate the orbits in which the electrons moved. In this model the spectra emitted by atoms were interpreted as the energy difference that was set free when an electron changed its orbit, which also meant its respective distance to the nucleus. As a result of the inability of the electron to be at any arbitrary distance from the nucleus, very characteristic spectral lines were observable that depended only on the nucleus. This is not as simple as it may sound. Except for hydrogen, every atom has several electrons; therefore, the electrostatic potential for one of the electrons is not only influenced by the nucleus but also by the other electrons, which again makes exact calculations impossible. Matters are complicated even further when one considers molecules instead of atoms. Nonetheless, the spectrum is always connected to the potential, which is a direct result of the inverse square law.

All these phenomena can also be explained by a particle theory, meaning that the effects are produced by particles emitted from the interacting objects. In this concept, the electrostatic interaction is explained by assuming that two charges are interacting through photons emitted and absorbed by the

charges. The important point in this concept is that the particle (e.g., the photon) must have no rest mass and an infinite lifetime. When one looks at a point source, it can be seen from geometrical considerations that the particle density on spheres with the source being their center decreases with the inverse square of the radius. The particles that have been postulated to explain the stability of the nucleus have a finite rest mass and a finite lifetime. This means that the particle densities decrease faster than the inverse square of the distance—in other words, these forces are short-range forces.

See also: COULOMB'S LAW; ELECTRIC POTENTIAL; ELECTROSTATIC ATTRACTION AND REPULSION; ENERGY, POTENTIAL; GRAVITATIONAL FORCE LAW; KEPLER'S LAWS; NEWTON'S LAWS

PETER HEERING

ION

An ion is what an atom or molecule becomes when electrons are either removed from or added to it. An ion carries a net electrical charge because the positive charge of the nucleus (or nuclei) of the atom (or atoms) is no longer exactly balanced by the negative charge of the electrons. If it has an extra electron, it is called a negative ion; if it is missing some electrons, it is a positive ion. Positive ions far outnumber negative ions in nature because positive ions are much more stable. A positive ion remains stable when it has lost many electrons, up to being completely stripped to a bare nucleus, whereas stable free negative ions with more than a single excess electron have not been found. Properties of an atom or molecule that depend on its outermost electrons, such as its chemical activity and its emission and absorption of light, are completely changed by its ionization.

While the atoms and molecules of most materials at room temperature are not ionized, small numbers of ions in nature are common. For example, the presence of ions in water solution, in salt water, or even normal tap water, converts pure water (which is an electrical insulator) into a good electrical conductor with the ions carrying the electrical current flow.

Free ions can be produced in gases by the action of electrical discharges or sparks, which liberate electrons from atoms by collisions between electrons and atoms. They can also be produced when beams of fast electrons or ions from accelerators, cosmic rays, or photons (electromagnetic radiation, including ultraviolet light, x rays, and gamma rays) pass through matter. These beams are often referred to as "ionizing radiation" because they possess sufficient energy per particle to tear electrons away from their parent atoms or molecules. Small numbers of ions exist in the atmosphere partially due to constant exposure to solar ionizing radiation, cosmic rays, and natural radioactivity. Intense beams of ionizing radiation are used in radiation therapy to destroy cancer tumors, where the creation of ions in matter plays an important role in damaging the target cells of the tumor.

In the universe as a whole, ions are common. This is the case because much of the matter of the universe (mostly hydrogen) exists under conditions of very high temperature, as in the interiors of stars. Here the thermal motion of the matter is so violent that the atoms have lost some or maybe even all of their electrons in collisions with other electrons and photons in the gas. Such a state of matter is called a plasma. In a plasma at 15×10^6 K (roughly the temperature at the center of the Sun) an atom of oxygen would typically have lost all of its electrons and would be designated an O^{8+} ion because it had lost eight electrons. Plasmas can also be created on Earth in intense electrical discharges and under conditions of extremely high temperatures. The hottest contained plasmas that have been created artificially are in the interiors of fusion plasma devices whose purpose is to fuse nuclei of deuterium and tritium together in order to release large amounts of energy. Plasmas produced in such devices range up to several hundred million degrees, under which conditions even relatively heavy atoms, such as iron, can be completely ionized.

See also: ATOM; IONIZATION; IONIZATION CHAMBER; IONIZATION POTENTIAL; MOLECULE; PLASMA PHYSICS

Bibliography

ANDERSON, T.; FASTRUP, B.; FOLKMANN, F.; KNUDSEN, H.; and ANDERSEN, N. *The Physics of Electronic and Atomic Collisions* (AIP Press, New York, 1993).

AUMAYR, F.; BETZ, G.; and WINTER, H. *The Physics of Highly Charged Ions* (North-Holland, Amsterdam, 1995).

C. LEWIS COCKE

IONIC BOND

Chemical bonds hold atoms together in molecules. An ionic bond is a chemical bond in which an electron is transferred from one atom (or collection of atoms) to another. The donor atom becomes a cation (i.e., positively charged), and the acceptor atom becomes an anion (i.e., negatively charged). The electrostatic force between the resulting pair of unlike charged ions binds them strongly together in a molecule. The archetype of an ionic bond is NaCl (table salt), which consists of Na^+ and Cl^-. In a covalent bond, by contrast, the electrons are shared by the two atoms.

To understand why atoms form certain ions, one need only look at the periodic table. Metal atoms typically have a few more electrons than the noble gas atoms on the preceding row. Noble gas configurations are very stable (they have closed electronic shells), whereas metal atoms like to lose electrons. In the example above, the Na atom loses one electron to become Na^+, which has the closed shell structure of Ne. Similarly, a nonmetal has fewer electrons than the noble gas atom at the end of its row. Thus Cl likes to accept an electron, because Cl^- has the Ne closed shell structure. Transition metal ions are more subtle and tend to have full subshells, as in the case of Cu^+.

Molecules form only if, by bonding, the total energy is reduced. To convert solid Na to atomic Na, or molecular Cl_2 to atomic Cl, costs energy. Furthermore, the energy required to remove an electron from Na (its ionization potential) is greater than the energy gained by adding an electron to Cl (its electron affinity). However, all these costs are less than the large electrostatic energy gained from the two ions at the equilibrium bond length, which is the sum of the two ionic radii. Thus the lower the ionization potential of the metal, and the higher the electron affinity of the nonmetal, the more likely the total energy is to be reduced and an ionic bond is to be formed.

If NaCl is added to distilled water, the resulting solution is a much better conductor of electricity. As Svante Arrhenius first explained in 1883, some of the ionic bonds break, leaving free ions in the solution; these free ions can carry the current. For this reason, ionic compounds are called electrolytes.

Ionic compounds are often solids at room temperature. The ions arrange themselves into a regular lattice of alternating anions and cations. Such a structure is strongly bound because of the many ionic bonds. Because of the nondirectional nature of ionic bonds (compared with covalent bonds), such structures are typically close-packed, that is, they fill up most of space. In contrast with a metal, the electrons do not move easily throughout the solid; thus ionic compounds are insulators. However, at high enough temperatures, the solid melts, and the liquid state conducts electricity.

See also: COVALENT BOND; ELECTRON; IONIZATION POTENTIAL; MOLECULE

Bibliography

LEICESTER, H. M. *The Historical Background of Chemistry* (Dover, New York, 1956).

McQUARRIE, D. A., and ROCK, P. A. *General Chemistry,* 2nd ed. (W. H. Freeman, New York, 1987).

PAULING, L. *The Nature of the Chemical Bond,* 3rd ed. (Cornell University Press, Ithaca, NY, 1960).

KIERON BURKE

IONIZATION

A neutral atom (or molecule) has a positive nuclear charge equal to the sum of the negative charges of its orbital electrons. During the process of ionization, one or more electrons are removed from (or more rarely, added to) the atom. It is then electrically charged and is called a positive (or negative) ion. Even a positive ion can lose one or more of its remaining electrons and become more highly ionized. Ionization may be caused by a photon, in which case it is called photoionization, or by a collision with a sufficiently energetic electron, atom, molecule, or another ion. Since electrons in an atom are bound, a certain amount of energy must be supplied to remove one. This binding energy, known as the ionization potential, varies from a few electron volts (eV) for the outer shells of atoms to thousands of electron volts for the tightly bound inner shells of heavy atoms.

Electrons ejected in an ionizing collision are commonly called secondary electrons, even though there is no "primary" or incident electron in the case of ionization by an ion, atom, or photon. If an incident particle transfers enough energy in a collision, the secondary electron itself may be sufficiently energetic to cause further ionization in a subsequent collision with another atom.

When a fast charged particle, such as an alpha or beta particle from a radioactive source, traverses a solid, liquid, or gas, it loses energy in successive collisions and thus deposits energy and secondary electrons all along its path. Since ionization is the dominant process in this energy transfer, and because ions are highly reactive chemically, ionization plays a fundamental role in producing radiation damage in biological tissues and other materials. It is also an important process in plasma dynamics (and therefore in the design, e.g., of thermonuclear fusion devices) and in solar and stellar atmospheres. The ionosphere in the earth's upper atmosphere, which controls long-distance radio communication, results from the photoionization of atmospheric atoms and molecules by visible and ultraviolet light from the Sun. The aurora borealis, and its counterpart in the Southern Hemisphere, results in a complicated way from the solar wind, which is a flow of ions and electrons from the Sun. Collisions of these particles with the molecules of the atmosphere cause ionization, in addition to the excitation which produces the light.

Investigators study photoionization with light from spark and arc sources and more recently with lasers and synchrotron light sources. For charged particle impact, low and intermediate range accelerators provide beams at well-defined energies to study ionization. Data on ionization of single atoms is obtained by using a low pressure gas or vapor as a target. The ions and/or electrons from a known length of the beam path are collected on electrodes to measure the probability of ionization of a target by a given incident particle. Ionization of solids is more complicated and difficult to measure since secondary electrons can be formed inside the body of the material as well as on the surface.

See also: COLLISION; ION; IONIZATION POTENTIAL; NUCLEAR BINDING ENERGY

Bibliography

MÄRK, T. D., and DUNN, G. H. *Electron Impact Ionization* (Springer-Verlag, New York, 1985).

McDANIEL, E. W. *Atomic Collisions: Electron and Photon Projectiles* (Wiley, New York, 1989).

McDANIEL, E. W.; MITCHELL, J. B. A.; and RUDD, M. E. *Atomic Collisions: Heavy Particle Projectiles* (Wiley, New York, 1993).

RUDD, M. E.; KIM, Y.-K.; MADISON, D. H.; and GAY, T. J. "Electron Production in Proton Collisions with Atoms and Molecules: Energy Distributions." *Rev. Mod. Phys.* **64,** 441–490 (1992).

M. EUGENE RUDD

IONIZATION CHAMBER

The ionization chamber is one of the oldest devices for the detection of nuclear radiation. It is still widely used and under constant improvement. The 1992 Nobel Prize in physics was awarded to Georges Charpak for his invention and development of the multiwire proportional chamber in 1968.

An ionization chamber usually consists of one or more electrodes in a chamber that is filled with suitable gases, liquids, or solids. A potential difference is applied across the electrodes so that a static electric field is established in the chamber. When the ionizing radiations of photons or particles are passing through the material, they may knock out the electrons from the bound states and produce electron–ion, or electron–hole pairs. In the electric field of the chamber, the negative charge carriers, electrons, will thus move to the anode while the positive charge carriers, ions of the gas chamber or holes of the semiconductor detector, will move toward the cathode under the action of the Coulomb force. These charge carriers are collected as a charge pulse or a current at the electrodes so that the impact of the ionizing radiation is recorded. Ideally, the detector is designed in such a way that the charge collected is equal to the energy loss of the incoming radiation, divided by the energy required to form a charge pair. A device with these properties is called an ionization chamber.

The oldest form of ionization chamber is the metal leaf electroscope in which two thin flexible pieces of metal, such as gold, are suspended on an insulated electrode in a gas-filled chamber. During the operation, the electrode is charged up to a potential of a few hundred volts, and hence the electrostatic repulsion separates the two leaves. The ionizing radiation liberates some electrons from the gas atoms, which then drift toward and neutralize the positive charge on the gold leaves. The decrease of the charge on the gold leaves will result in a decrease in the electrostatic repulsion and hence the separation between the leaves. Therefore, the change in the separation of the leaves is a measure of the total amount of radiation absorbed by the gas.

A simple, modern pulse ionization chamber consists of electrodes with a parallel-plate configuration, coaxial cylindrical geometry, or multiwire structure in a gas chamber. The electric field between the electrodes sweeps the electrons and the ions that are produced by the radiation to the anodes and cathodes. The electrons from the ionization events cause pulse currents at the anodes while the ions are generally shielded by a grid from the cathodes so that the charge at the collecting electrodes is equal to the total ionization.

Ionization chambers have extensive applications. In atomic, nuclear, and particle physics they are used as diagnostic tools (e.g., fission counters, fusion detectors, and particle detectors). Besides physics, they are used for radiation monitoring, domestic fire alarm systems, and multidimensional imaging for medical applications.

See also: ELECTROSTATIC ATTRACTION AND REPULSION; ION; IONIZATION; IONIZATION POTENTIAL; PHOTON; RADIATION PHYSICS

Bibliography

ATTIX, F. H. *Introduction to Radiological Physics and Radiation Dosimetry* (Wiley, New York, 1986).

FENYVES, E., and HAIMAN, O. *The Physical Principles of Nuclear Radiation Measurements* (Academic Press, New York, 1969).

TAIT, W. H. *Radiation Detection* (Butterworth, London, 1980).

REINHARD BRUCH

YONG YAN

IONIZATION POTENTIAL

The ionization potential is a fundamental property whose knowledge is important in fields such as atomic and molecular structure, gaseous discharges, astrophysics, and plasma fusion diagnostics.

For free atoms or molecules, the ionization potential is the energy required to remove the weakest bound electron from a system in its ground state leaving the ionic system also in its ground state. This defines the first ionization potential. The additional energy required to remove a second electron is the second ionization potential, and so on. Ionization potentials are usually reported in electron volts. A closely related quantity is the electron affinity, defined as the energy released when a neutral system captures an electron to form a negative ion. Alternatively, it is the energy required to detach an electron from a negative ion.

Ionization potentials and electron affinities of atoms show systematic trends. The noble gases have the largest ionization potential, whereas the alkali metals have the smallest, both decreasing with atomic number. The halides, on the other hand, have the largest electron affinity, whereas the noble gases have a negative electron affinity, that is, their negative ions are unstable. All the alkali metals have a positive electron affinity, but the alkaline earths are not all the same in this respect. Negative beryllium and magnesium ions are unstable, whereas calcium has a small electron affinity, measured to be 0.018 eV, and heavier alkaline earths have been predicted to have a larger electron affinity.

The size of an atom is determined by the most weakly bound electron. When the ionization potential is large, the electron is more tightly bound to the nucleus than when it is small. Thus, helium, with the largest ionization potential, is the smallest atom.

Although tables of ionization potentials and electron affinities exist, many gaps are still present in their knowledge, particularly for the higher stages of ionization and for the heavier elements. For example, the ionization potential of Francium (Fr), the heaviest alkali metal with an atomic number of $Z = 87$, has not yet been measured.

Ionization potentials may be measured in a number of ways. The most accurate measurement is obtained from spectroscopic methods. Transitions between states are accompanied by the emission or absorption of radiation and the wavelength of the radiation is a measure of the energy difference. Transitions to levels with a common series limit are called Rydberg series. The energies for members of a Rydberg series are closely related to the Bohr theory of atoms. For neutral atoms, the energies E_n relative to the ground state are given by the formula

$$E_n = \frac{A - k}{(n - \delta)^2} \qquad (n = m, m + 1, m + 2, \ldots),$$

where for energies in electron volts, the constant k is about 13.6 eV. By fitting this formula to a series of energy levels relative to the ground state, the constant A can be determined. If the series limit is one that leaves the ion in its ground state, then A is the ionization potential. Another method for measuring ionization potentials is by electron impact ionization. Here the minimum energy needed for a free electron to ionize the atom in a collision is determined. A third method is photoionization. The absorption of a photon whose energy is greater than the ionization potential will produce a free electron with energy equal to the excess of the photon energy over the ionization potential. Ionization potentials can also be computed by theory, though not to the accuracy of spectroscopic methods.

See also: ATOM, RUTHERFORD–BOHR; ATOM, RYDBERG; COLLISION; IONIZATION; SPECTROSCOPY, ATOMIC

Bibliography

LIDE, D. R., ed. *CRC Handbook of Chemistry and Physics,* 75th ed. (CRC Press, Boca Raton, FL, 1994).

CHARLOTTE FROESE FISCHER

ION LASER

See LASER, ION

IONOSPHERE

"Ionosphere" is the name given to the electrically charged region of the upper atmosphere between altitudes of about 40 and 400 km. Here neutral atoms and molecules, such as oxygen and nitrogen, are readily ionized by ultraviolet light from the Sun. For example, atomic oxygen (the predominant atmospheric constituent at about 250 km) is ionized when a sufficiently energetic solar photon displaces an electron:

$$O + h\nu \rightarrow O^+ + e^-$$

at $\lambda < 911$ Å. Other atmospheric constituents, such as N_2, O_2, and NO, are also readily photoionized. The energetic ionizing photons come from the uppermost (very hot) layers of the solar atmosphere, the chromosphere and corona.

In addition to this photoionization process, ions are generated via energetic electron collisions with neutral atoms, a process termed electron impact ionization:

$$O + e^- \rightarrow O^+ + 2e^-.$$

Electrons with sufficient energy to ionize neutral atoms are produced as a by-product of the photoionization process itself or enter the atmosphere in the auroras.

These processes create the ionosphere, a partially ionized medium made up of equal numbers of positive ions and free electrons. At the daytime peak of the ionosphere (about 250 km), about one in every 1,000 atoms is ionized, leading to a concentration of about a million ions and electrons per cubic centimeter. The most abundant ions in the ionosphere are O^+, O_2^+, and NO^+.

Sinks

Individual ions and free electrons in the ionosphere have lifetimes of the order of a fraction of a day, depending on the altitude (shorter lifetimes at lower altitudes). The main loss mechanism for ions in the upper ionosphere is dissociative recombination, which converts molecular ions into neutral atoms:

$$O_2^+ + e^- \rightarrow O + O$$

$$NO^+ + e^- \rightarrow N + O.$$

Since this process only works for molecular ions, atomic ions such as O^+ chemically convert to molecular ions before recombination occurs. In the lowest part of the ionosphere, loss mechanisms involving the temporary production of negative ions via a process called electron attachment become important.

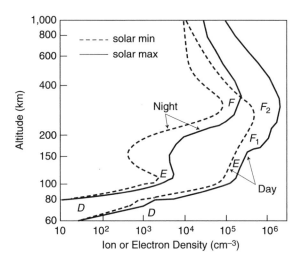

Figure 1   The average ionosphere.

The net effect of the various ionospheric sources and sinks is to produce a roughly spherical shell of ionization. The ionosphere, however, varies greatly as the processes of photoionization and recombination alternately dominate and ions are redistributed by wind systems. There is a strong diurnal (24-hour) variation as the ionosphere builds up during the daytime hours and dissipates at night. The ionosphere also has longer-term variability due to the 11-year solar sunspot cycle. Figure 1 summarizes the average variations.

Discovery

Early work on the ionosphere was closely associated with the study of geomagnetism and the development of radio propagation techniques. In 1860 Lord Kelvin (William Thomson) speculated on the existence of a conducting layer in the atmosphere. In 1882 Balfour Stewart developed a "dynamo theory" of the conducting layer to explain perturbations in Earth's magnetic field. In 1901 Guglielmo Marconi transmitted radio signals across the Atlantic. This feat was explained in 1902 by Arthur E. Kennelly and Oliver Heaviside in terms of a reflecting layer made up of free electrically charged particles. The name "ionosphere," proposed in 1926 by Robert A. Watson-Watt, was based on the results of numerous radio sounding experiments. The ionosphere is divided into distinct layers or ledges, referred to as the D, E, and F regions (in order of ascending altitude). This convention was initiated by Edward V. Appleton and is related to the names given to features in the spectra of reflected radio waves. The F layer is often subdivided into the F_1 and F_2 regions.

Importance for Communications

The ionosphere is of great importance to atmospheric science and, in particular, to radio communications. During the night, the lower (D and E) regions undergo rapid recombination, leaving a remnant F layer to reflect radio waves. This effect (the raising of the radio reflection layer) means that radio waves propagate much further around the globe during the night than during the day. Furthermore, the high degree of variability of the ionosphere and its strong dependence on time of day, solar activity, geomagnetic activity, atmospheric winds, and photochemistry mean that its behavior is difficult to predict with high accuracy. The ionosphere provides complex chemical and radiation pathways for energy and momentum to be redistributed within the upper atmosphere. The E layer (the dynamo region) can also carry large electrical currents, due to the differential motion of ions and electrons under the influence of winds and electric fields. During major solar storms (for example, due to solar flares), these currents can become large enough to disrupt power grids, transoceanic cables, and other technological systems.

See also: ATMOSPHERIC PHYSICS; IONIZATION; KELVIN, LORD; RADIO WAVE

Bibliography

KELLEY, M. C. *The Earth's Ionosphere* (Academic Press, San Diego, CA, 1989).

RISHBETH, H., and GARRIOTT, O. K. *Introduction to Ionospheric Physics* (Academic Press, New York, 1969).

SILBERSTEIN, R. "The Origin of the Current Nomenclature for the Ionospheric Layers." *J. Atmos. Terr. Phys.* **13**, 382 (1959).

TIMOTHY L. KILLEEN

IRREVERSIBLE PROCESS

If part of a material body is heated and then thermally isolated from its surroundings, its temperature gradually becomes uniform. This process may be described by a simple linear relationship between the temperature difference per unit length dT/dx, called the gradient of the temperature, and the heat flow J:

$$J = \kappa \frac{dT}{dx}.$$

The coefficient $\kappa > 0$, called the heat conductivity, determines how rapidly the uniform distribution of heat is established. Its numerical value depends on the physical properties of the material and is large for good thermal conductors (e.g., metals) and small for thermal insulators.

Heat conduction is an example of an irreversible process. In physics, many more irreversible processes described by a linear relationship such as the Fourier law for heat conduction are known; Ohm's law for electric conduction, which states that the electric current is proportional to the imposed electric field, and Fick's law of diffusion, which states that the flow of diffusion is proportional to the gradient of concentration, are two examples A characteristic feature of all such processes is their obvious one-sidedness. Heat always flows from the warmer side to the colder side and never the other way around; that is, from the cold side of a body to the warm side. In this sense these processes are irreversible. Their study is the subject matter of the theory of irreversible thermodynamics (thermodynamics of irreversible processes or nonequilibrium thermodynamics).

The mathematical relations that describe irreversible processes are of the universal form

$$J_i = \sum_j L_{ij} F_j,$$

where the flow (also called flux or current) of process $i = 1, 2, \ldots$ is linearly related to one or more driving forces $F_j, j = 1, 2, \ldots$, of which the temperature gradient and the electric field are prime examples. The phenomenological coefficients L_{ij} have to be determined experimentally for the processes in question.

In 1854 William Thomson (Lord Kelvin) was the first to apply a thermodynamic reasoning to irreversible processes. He analyzed various thermoelectric processes and established two famous relations: one of which relates the thermodynamic-electric potential of a thermocouple to its Peltier heat. His quasi-thermodynamic method of analysis was based on an additional assumption which was not justifiable. In 1931 Lars Onsager was able to show that this Thomson relation was one instance of the general reciprocity relation

$$L_{ij} = L_{ji},$$

which states that if the flow J_i is influenced by the force F_j of some irreversible process j, then the flow J_j is also influenced by the force F_i through the same coefficient L_{ij}. This is a general theorem based on the time-reversal invariance of the underlying microscopic equations of motion, completely independent of any particular molecular model. In 1968 Onsager received the Nobel Prize for his discovery, which was a turning point in the history of irreversible thermodynamics.

The equilibrium theory of thermodynamics, which would be better named thermostatics, is restricted to situations characterized by the absence of all irreversible processes: $J_i = 0$ and $F_i = 0$. None of the principles of thermodynamics directly involve the concept of time, which would enable the description of time-dependent processes. From early on, however, irreversibility was perceived as an important aspect of the second law of thermodynamics. The paper of William Thomson in which he presented the formulation of the second law for the first time in 1852 was titled "On the Universal Tendency of Nature to the Dissipation of Mechanical Energy." Rudolf Clausius also clearly recognized the significance of the unidirectional character of irreversible processes. In 1850 Clausius referred to the concept of "noncompensated heat" as a measure of irreversibility in systems that need not be thermally insulated. In the paper of 1865, where Clausius introduced the concept of entropy, contains the celebrated statement that irreversible processes drive the entropy of the universe to a maximum.

A consistent thermodynamic theory of irreversible processes, which establishes a quantitative link between irreversibility and the increase of entropy, was finally formulated by Josef Meixner in 1941 and Ilya Progogine in 1947. The important concept is the so-called entropy production, essentially the noncompensated heat of Clausius, which clearly must be distinguished from the transfer of entropy through some boundary of the system. The entropy change dS of a system is the sum

$$dS = dS_{ext} + dS_{int}$$

of the entropy dS_{ext} supplied by the surroundings, and dS_{int} the entropy produced inside the system by irreversible processes described by the linear laws. The second law of thermodynamics states that dS_{int} is zero for reversible processes and strictly positive for irreversible processes

$$dS_{int} \geq 0.$$

From the laws of equilibrium thermodynamics it is also known that the total entropy of the system changes according to

$$TdS = dU + PdV,$$

where dU is the increase of internal energy and the second term in this equation is the work performed by the pressure P when the volume expands with an amount dV. One assumes this relation to remain valid if the system is not too far from equilibrium (hypothesis of local equilibrium). Combining this assumption with the conservation laws of energy and matter, one derives for the entropy production per unit time the expression

$$\frac{dS_{int}}{dt} = \sum_i J_i F_i \geq 0.$$

Here J_i, $i = 1, 2, \ldots$, stands for the flows describing the various irreversible processes such as heat flow, diffusion, and chemical reactions taking place and F_i for the corresponding thermodynamic forces such as gradient of the temperature, of the concentration, and of chemical potentials. This expression for the entropy production, together with the Onsager symmetry, are the basis for the many applications of irreversible thermodynamics in physics and chemistry.

The assumption of local equilibrium going into the theory of irreversible thermodynamics can be justified by the methods of statistical mechanics, provided the relationships between flows and thermodynamic forces are linear. The generalization to nonlinear irreversible thermodynamics applicable to far-from-equilibrium situations is a field of active research.

See also: ARROW OF TIME; CLAUSIUS, RUDOLF JULIUS EMMANUEL; ENTROPY; HEAT; KELVIN, LORD; STATISTICAL MECHANICS; THERMODYNAMICS; THERMODYNAMICS, HISTORY OF

Bibliography

DE GROOT, S.R., and MAZUR, P. *Non-Equilibrium Thermodynamics* (North-Holland, Amsterdam, 1962).

PROGOGINE, I. *From Being to Becoming* (W. H. Freeman, San Francisco, 1980).

CHRISTIAAN G. VAN WEERT

IRROTATIONAL FLOW

One can do the following thought experiment to understand what is meant by a state of irrotational flow in a fluid. In the midst of the fluid place a small paddle wheel that is initially at rest. If the fluid passing by the wheel exerts a nonzero net torque on the wheel so that it begins to rotate then the fluid is not in an irrotational state. If, on the other hand, the fluid passing the wheel exerts a zero net torque, the fluid is said to be irrotational at the location of the wheel. If a fluid is nonirrotational it is also said to have nonzero "circulation" or nonzero "vorticity" at the point in question.

In Fig. 1 a number of small samples of fluid are shown along with the velocity vector describing the motion of each sample. Such a schematic rendering of the state of motion of a fluid is called a "velocity field" diagram. An imaginary paddle wheel has been included in the figure. Figure 1 shows nonirrotational flow. The paddle wheel will turn because the

speed of the fluid striking the paddles is greater at the top of the wheel than at the bottom. In Fig. 2 the little paddle wheel will not turn because the flow of fluid is symmetric with respect to the axis of rotation of the wheel. The torques exerted by the fluid striking the paddles at the top of the wheel are balanced by the torques exerted by the fluid striking the paddles at the bottom of the wheel.

A mathematical definition of irrotational flow may be stated in terms of the vector velocity field characterizing the flow. If the velocity field is given as a function $\mathbf{V}(\mathbf{r})$, where $\mathbf{r}$ is the position vector of the fluid sample described by the velocity $\mathbf{V}$, then irrotational flow is equivalent to the mathematical condition

$$\mathbf{\nabla} \times \mathbf{V} = 0$$

or

$$\operatorname{curl} \mathbf{V} = 0.$$

When this mathematical condition is fulfilled one can then show that the velocity field $\mathbf{V}(\mathbf{r})$ is obtainable from the gradient of a scalar "velocity potential" function φ:

$$\mathbf{V}(\mathbf{r}) = \mathbf{\nabla}\varphi$$

or

$$\mathbf{V}(\mathbf{r}) = \operatorname{grad} \varphi.$$

In hydrodynamics, irrotational flow of an incompressible (constant density) fluid is governed by the Laplace equation

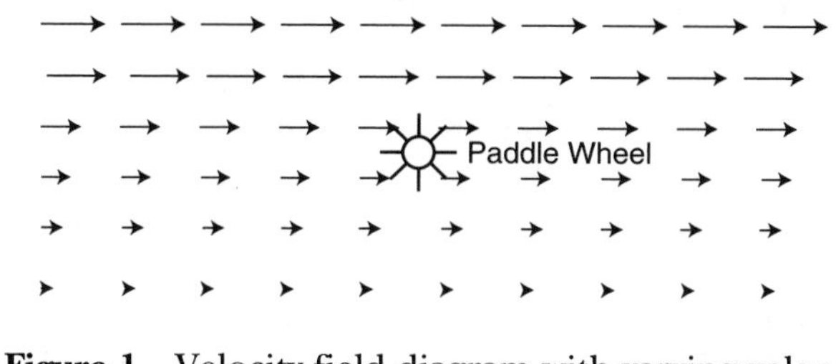

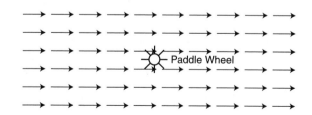

Figure 1 Velocity field diagram with varying velocity vectors.

Figure 2 Velocity field diagram with uniform velocity vectors.

$$\nabla^2 \varphi = 0;$$

techniques for solving this equation are discussed extensively in the literature of physics and applied mathematics. An incompressible, irrotational fluid is said to show "potential flow" and is governed by the mathematical condition that

$$\frac{p}{\rho} + \frac{V^2}{2} + gz + \frac{\partial \varphi}{\partial t} = \text{const}$$

along any path or "streamline" followed by each small sample of the fluid, where p is the pressure at the location of the sample, ρ is the (constant) fluid density, g is the acceleration of gravity, z is the height of the fluid sample above some arbitrary zero level, V is the velocity of the fluid in the sample, and $\partial \varphi / \partial t$ is the rate at which the velocity potential function is changing in time. In cases involving steady flow ($\partial \varphi / \partial t = 0$) this equation reduces to Bernoulli's equation, which is commonly treated in introductory physics texts.

See also: BERNOULLI'S PRINCIPLE; DENSITY; FLUID DYNAMICS; HYDRODYNAMICS; LAPLACE, PIERRE-SIMON; MOTION, ROTATIONAL; PRESSURE; VELOCITY

Bibliography

BOAS, M. L. *Mathematical Methods In The Physical Sciences,* 2nd ed. (Wiley, New York, 1983).
OHANIAN, H. C. *Physics,* 2nd ed. (Wiley, New York, 1989).
SYMON, K. R. *Mechanics,* 2nd ed. (Addison-Wesley, Reading, MA, 1964).

DELO E. MOOK II

ISOBARIC PROCESS

An isobaric process is one that takes place at constant pressure. Several processes of technical interest can occur in an isobaric manner. Consider a gas held in place in a cylinder covered by a movable piston. If the gas is heated so that it expands, the piston can rise, keeping the pressure constant as the expansion takes place. This is an example of an isobaric heating process. For the case of cooling, the piston drops so that the pressure remains constant as the volume is decreased.

Changes in phase, such as evaporation, which is a change from the liquid phase to the vapor phase, can also occur in an isobaric process. The water in an unsealed tea kettle begins to boil at a temperature of 100°C when the pressure is 1 atm. In this case, both the pressure and the temperature remain constant during the phase change. At a higher pressure, boiling requires a higher temperature, but the process occurs at constant pressure provided the system is allowed to expand. In the case of an isobaric heating process taking place in a piston-cylinder system, the energy transfer as heat causes two effects: Part of the heat transfer goes to increase the internal energy of the fluid being heated and part is required to expand the fluid as it pushes against the piston. In this particular process, the required heat transfer is equal to the change in enthalpy of the fluid. This leads to the statement that the energy transfer as heat to a compressible substance during an isobaric process is equal to the increase in enthalpy.

See also: ADIABATIC PROCESS; ISOCHORIC PROCESS; ISOTHERMAL PROCESS; PHASE TRANSITION

Bibliography

REYNOLDS, W. C., and PERKINS, H. C., JR. *Engineering Thermodynamics* (McGraw-Hill, New York, 1977).
WARK, K. *Thermodynamics,* 5th ed. (McGraw-Hill, New York, 1988).

HENRY C. PERKINS JR.

ISOCHORIC PROCESS

A process that takes place without a change in the volume of the system is an isochoric process. To reduce the complexity of thermodynamic processes, one or more of the interacting parameters (e.g., volume, temperature, pressure, heat, entropy) may be held constant while the others are allowed to vary in reaction to changes made to the system.

Such simplifications are named for the property that is held constant. For example, a system in which the pressure does not change is termed isobaric, a system in which entropy remains the same is termed isentropic, and adiabatic processes occur without the addition or removal of heat from the system. Another name used for isochoric (literally, equal space) processes is isovolumic (equal volume) processes.

When volume is held constant, the functional relationship between other parameters of the system may be plotted. For example, in a gas in which pressure, volume, and temperature are interrelated (as, ideally, by the perfect gas law), the constant-volume graph of pressure as a function of temperature would be called an isochor, also spelled isochore. This isochor would show pressure increasing with temperature. In a real system (a gas confined in a steel casing) this isochor could end in a sudden, possibly catastrophic, manner when the pressure exceeded the structural ability of the casing to confine it.

Physical situations in which one of the thermodynamic parameters remains nearly unchanged are commonly encountered in engineering problems. For theoretical physics, the comparison of actual measures to idealized models leads to an understanding. An illustration of this is the perfect gas law, which predicts that pressure will be linearly proportional to temperature for an isochoric process; the isochor will be a straight line. The slopes of the isochors will be different for different gases, but those for gases most nearly "ideal" all point to intercepts near $-273°C$ when the pressure goes to zero. Thus arose the concept of absolute zero and an absolute temperature scale (Kelvin) in which pressure is directly proportional to temperature, starting from zero for both.

Experiments show, however, that although the isochors of ideal gases point toward absolute zero, actual measurements of pressure fall lower than the straight line as 0 K is approached; the gas molecules move more slowly at lower temperatures, allowing them to be more influenced by cohesive interactions with other molecules. Their ideal freedom to move and exert pressure is reduced until eventually they cohere into a liquid.

See also: ABSOLUTE ZERO; ADIABATIC PROCESS; ISOBARIC PROCESS; ISOTHERMAL PROCESS; PRESSURE; TEMPERATURE; TEMPERATURE SCALE, CELSIUS; TEMPERATURE SCALE, KELVIN; THERMODYNAMICS; THERMODYNAMICS, HISTORY OF

Bibliography

SEARS, F. W., and SALINGER, G. L. *Thermodynamics, Kinetic Theory, and Statistical Thermodynamics*, 3rd ed. (Addison-Wesley, Reading, MA, 1975).

RICHARD B. HERR

ISOMERIC NUCLEUS

See NUCLEUS, ISOMERIC

ISOTHERMAL PROCESS

Isothermal processes arise in chemistry and in thermodynamics. An isothermal process is one carried out at a constant temperature. In a general isothermal process the constituents or reactants of the process are held in close contact with an object at a constant temperature. This object supplies or extracts heat to maintain a constant temperature in the process. For instance, the pressure of air in a cylinder may be increased by pushing in a piston. If this is done slowly enough, the natural temperature rise is prevented by heat conduction to the cylinder. The cylinder, if it is large enough, has a large heat capacity and does not change temperature in the process. If the piston is pulled out, the pressure and the temperature of the air will tend to decrease, but if done slowly enough the temperature of the process remains constant because heat can flow in from the piston.

The basic description of thermodynamics begins with the first law of thermodynamics:

$$dU = dQ - dW$$

U is internal energy, Q is heat, and W is work done. (The symbol d indicates the differential, i.e., a small change.) Introducing partial derivatives, we can write

$$\frac{\partial U}{\partial T}\Big)_V dT + \frac{\partial U}{\partial V}\Big)_T dV = dQ - dW,$$

where T is the temperature. Hence in an isothermal process ($dT = 0$) we have

$$dQ = \frac{\partial U}{\partial V}\Big)_T dV + dW.$$

Using the second law of thermodynamics, which states that there is a specific function S of the other thermodynamic variables such that $dS = dQ/T$, we can rewrite the first term on the right to obtain

$$\frac{\partial U}{\partial V}\Big)_T = T \frac{\partial P}{\partial T}\Big)_V - P.$$

Since for perfect gas the equation of state is $P = Nk_BT/V$, where k_B is Boltzmann's constant, and N is the number of molecules, we have

$$\frac{\partial P}{\partial T}\Big)_V = \frac{P}{T} \quad \text{(perfect gas)}.$$

Thus for such a fluid, $\partial U/\partial V)_T = 0$, and

$$dQ = dW.$$

This means that with a perfect fluid, isothermal processes completely convert heat to work.

The Carnot cycle is an idealized thermodynamic energy cycle. In this case, mechanical energy is produced by taking a gas, allowing it to expand isothermally (in contact with a high temperature bath T_{high}), then allowing it to expand adiabatically (no heat transfer) to a lower temperature, then compressing it isothermally at this lower temperature (the low temperature heat bath T_{low}); and finally compressing it adiabatically to the original starting state. This cycle extracts heat energy from the high-temperature bath and converts this heat energy entirely to work, producing mechanical energy, and "dumps" a portion of the heat to the low-temperature reservoir. This dumping of heat is accomplished by isothermal transfer from the gas to the low-temperature bath, and the amount of heat dumped is exactly equal to the amount of work done in compressing the gas for that part of the cycle. The purpose of the adiabatic parts of the cycle are just to change the temperature of the gas from one to the other bath temperatures. This cycle in fact exhibits the theoretical maximum efficiency of any heat engine: eff = $(T_{high} - T_{low})/T_{high}$, which will always be less than 100 percent, since T_{low} is always above absolute zero.

An object such as the metal cylinder that can absorb or give up heat to maintain a constant temperature is called a heat reservoir, or a heat bath. If, however, there is no heat flow to or from the reservoir, then the process can be isothermal even if there is no heat bath. An important example of the latter kind of isothermal process is the melting of ice in water at 0°C (32°F), the usual freezing point of pure water. Once the ice is at 0°C, further addition of heat does *not* change its temperature, but goes to change its phase from solid to liquid. While ice can certainly be made colder than 0°C, putting ice in liquid water brings the whole mixture to 0°C, until the ice is melted. Thermal energy—heat—flows from the liquid water, cooling it; and into the solid ice, where some of the ice structure turns liquid. If the ice is initially colder than 0° heat flows inward by conduction, raising the central ice temperature. An ice/water bath thus is an example of a heat reservoir, in this case at 0°C. Other phase transitions can be used to maintain other temperatures; another important example involving water is the boiling point, which occurs at 100°C (212°F) at standard atmospheric pressure. Similarly, boiling freon (CCl_2F_2) at atmospheric pressure produces a steady -30°C. Varying the vapor pressure can vary the boiling point, so raising the pressure in a car cooling system raises the boiling temperature there; putting a partial vacuum above a boiling liquid lowers its boiling point. But as long as the other physical conditions remain constant, the phase change determines a constant temperature, so it can be used in carrying out an isothermal process.

Most commercial processes are carried out isothermally. An important example is the Haber process, which "fixes" atmospheric nitrogen in a form which can then be concentrated to biologically useful compounds. The Haber process is continuous synthesis of ammonia from its constituent gases:

$$N_2(g) + 3H_2(g) \rightarrow 2NH_3(g).$$

It is usually carried out over a catalyst at a constant temperature of about 450°C.

See also: ADIABATIC PROCESS; CARNOT CYCLE; ENGINE, EFFICIENCY OF; HEAT CAPACITY; HEAT ENGINE; THERMODYNAMICS

Bibliography

CARBERRY, J. J., and VARMA, A., eds. *Chemical Reaction and Reactor Engineering* (M. Dekker, New York, 1987).

HALLIDAY, D., and RESNICK, R. *Fundamentals of Physics* (Wiley, New York, 1988).

HUANG, K. *Statistical mechanics,* 2nd ed. (Wiley, New York, 1987).

PILLING, M. J. *Reaction Kinetics* (Clarendon Press, Oxford, Eng., 1975).

REICHL, L. E. *A Modern Course in Statistical Physics* (University of Texas Press, Austin, 1980).

RICHARD A. MATZNER

ISOTOPES

The nucleus of an atom is composed of neutrons and protons. These are held together by the strong nuclear force and confined to a minuscule volume in the center of an atom which is typically only about 10^{-12} cm in diameter. The term "isotope" refers to atomic nuclei with a fixed number of protons or atomic number but different numbers of neutrons.

The chemical properties of an atom are almost entirely determined by the electrons that occupy most of the volume of an atom out to distances some 10,000 times larger than the radius of the nucleus. Since the number of electrons for a neutral atom equals the number of protons in the nucleus, the chemical properties of an atom are only dependent upon the atomic number and are almost completely independent of the number of neutrons in a nucleus. Therefore, even high purity, naturally occurring elements can be comprised of several isotopes.

Although the volume of the nucleus of an atom is small compared to the volume occupied by the electrons, most of the mass of an atom is due to the protons and neutrons in the nucleus. The electrons contribute only a small fraction, since the mass of the electron is only $1/1,836$ times the mass of the proton. On the other hand, the mass of the neutron is even slightly larger than that of the proton. Hence, the masses of different isotopes are roughly

proportional to the total number of protons plus neutrons for each isotope. The sum of the number of neutrons N and the number of protons (or atomic number) Z is referred to as the nucleon number, denoted

$$A = Z + N.$$

The mass of an atom is approximately given by the sum of masses of Z protons and N neutrons. Hence the nucleon number is also close to the mass number. The actual mass in atomic mass units will be different from the sum of the free proton and neutron masses due to the effects of nuclear binding energy, and the contribution from the electrons and electron binding energy.

Conventions

The numbers of neutrons N and protons Z of the nucleus of an isotope of an element X of atomic mass A are denoted

$$^{A}_{N}Z.$$

For example, some common isotopes of the element carbon, which has an atomic number of six, are written

$$^{12}_{6}C_6, \; ^{13}_{6}C_7, \; ^{14}_{6}C_8.$$

It is not always necessary to specify the atomic number Z since that is also implied by the symbol for the element. Also, it is not necessary to write the neutron number N, since that can be determined by subtracting the atomic number from the nucleon number, that is, $N = A - Z$. The three isotopes written above, for example, are thus referred to as carbon-12 (written ^{12}C), carbon-13 (written ^{13}C), and carbon-14 (written ^{14}C). The first two of these isotopes, ^{12}C and ^{13}C are stable isotopes, meaning that they do not experience radioactive decay. The third isotope, ^{14}C, however, is radioactive and decays to $^{14}_{7}N_7$ with a half life of 5,730 years. Such species are referred to as radioactive isotopes. Thus, an isotope may be stable or radioactive depending upon its nuclear properties. Most of the naturally occurring isotopes are stable, although there are some naturally occurring radioactive isotopes as well. Table 1 summarizes the atomic number, nucleon number, percent abundance, and half-life of known stable and naturally occurring radioactive isotopes. Column 1 gives the element symbol, column 2 gives the atomic number, and column 3 gives the nucleon number. The percentage abundance and the half-life of the radioactive isotopes are listed in column 4.

Table 1 List of Stable Isotopes and Naturally Occurring Radioactive Isotopes

Symbol	Z	A	Half-Life and/or Percent Abundance
H	1	1	99.985%
H	1	2	0.015%
H	1	3	12.33 yr
He	2	3	0.000137%
He	2	4	99.999863
Li	3	6	7.5%
Li	3	7	92.5%
Be	4	9	100%
B	5	10	19.9%
B	5	11	80.1%
C	6	12	98.9%
C	6	13	1.10%
C	6	14	5,730 yr
N	7	14	99.63%
N	7	15	0.37%
O	8	16	99.76%
O	8	17	0.038%
O	8	18	0.20%
F	9	19	100%
Ne	10	20	90.48%
Ne	10	21	0.27%
Ne	10	22	9.25%
Na	11	23	100%
Mg	12	24	78.99%
Mg	12	25	10.00%
Mg	12	26	11.01%
Al	13	27	100%
Si	14	28	92.23%
Si	14	29	4.67%
Si	14	30	3.10%
P	15	31	100%
S	16	32	95.02%
S	16	33	0.75%
S	16	34	4.21%
S	16	36	0.02%
Cl	17	35	75.77%
Cl	17	37	24.23%
Ar	18	36	0.337%
Ar	18	38	0.063%

Symbol	Z	A	Half-Life and/or Percent Abundance
Ar	18	40	99.600%
K	19	39	93.251%
K	19	40	1.277×10^9 yr (0.0117%)
K	19	41	6.7302%
Ca	20	40	96.941%
Ca	20	42	0.647%
Ca	20	43	0.135%
Ca	20	44	2.086%
Ca	20	46	0.004%
Ca	20	48	0.187%
Sc	21	45	100%
Ti	22	46	8.0%
Ti	22	47	7.3%
Ti	22	48	73.8%
Ti	22	49	5.5%
Ti	22	50	5.4%
V	23	50	1.5×10^{17} yr (0.250%)
V	23	51	99.750%
Cr	24	50	$> 1.8 \times 10^{17}$ yr (4.345%)
Cr	24	52	83.79%
Cr	24	53	9.50%
Cr	24	54	2.365%
Mn	25	55	100%
Fe	26	54	5.9%
Fe	26	56	91.72%
Fe	26	57	2.1%
Fe	26	58	0.28%
Co	27	59	100%
Ni	28	58	68.077%
Ni	28	60	26.223%
Ni	28	61	1.140%
Ni	28	62	3.634%
Ni	28	64	0.926%
Cu	29	63	69.17%
Cu	29	65	30.83%
Zn	30	64	48.6%
Zn	30	66	27.9%
Zn	30	67	4.1%
Zn	30	68	18.8%
Zn	30	70	$> 5 \times 10^{14}$ yr (0.6%)
Ga	31	69	60.108%
Ga	31	71	39.892%
Ge	32	70	21.23%
Ge	32	72	27.66%
Ge	32	73	7.73%
Ge	32	74	35.94%
Ge	32	76	7.44%
As	33	75	100%
Se	34	74	0.89%
Se	34	76	9.36%
Se	34	77	7.63%
Se	34	78	23.78%

Continued

Table 1 *Continued*

Symbol	Z	A	Half-Life and/or Percent Abundance
Se	34	80	49.61%
Se	34	82	1.4×10^{20} yr (8.73%)
Br	35	79	50.69%
Br	35	81	49.31%
Kr	36	78	0.35%
Kr	36	80	2.25%
Kr	36	82	11.6%
Kr	36	83	11.5%
Kr	36	84	57.0%
Kr	36	86	17.3%
Rb	37	85	72.17%
Rb	37	87	4.75×10^{10} yr (27.83%)
Sr	38	84	0.56%
Sr	38	86	9.86%
Sr	38	87	7.00%
Sr	38	88	82.58%
Y	39	89	100%
Zr	40	90	51.45%
Zr	40	91	11.22%
Zr	40	92	17.15%
Zr	40	94	17.38%
Zr	40	96	$> 3.56 \times 10^{17}$ yr (2.80%)
Nb	41	93	100%
Mo	42	92	14.84%
Mo	42	94	9.25%
Mo	42	95	15.92%
Mo	42	96	16.68%
Mo	42	97	9.55%
Mo	42	98	24.13%
Mo	42	100	9.63%
Ru	44	96	5.54%
Ru	44	98	1.86%
Ru	44	99	12.7%
Ru	44	100	12.6%
Ru	44	101	17.1%
Ru	44	102	31.6%
Ru	44	104	18.6%
Rh	45	103	100%
Pd	46	102	1.02%
Pd	46	104	11.14%
Pd	46	105	22.33%
Pd	46	106	27.33%
Pd	46	108	26.46%
Pd	46	110	11.72%
Ag	47	107	57.839%
Ag	47	109	48.161%
Cd	48	106	1.25%
Cd	48	108	0.89%
Cd	48	110	12.49%
Cd	48	111	12.80%
Cd	48	112	24.13%
Cd	48	113	9.3×10^{15} yr (12.22%)
Cd	48	114	28.73%
Cd	48	116	7.49%
In	49	113	4.3%
In	49	115	4.41×10^{15} yr (95.7%)
Sn	50	112	0.97%
Sn	50	114	0.65%
Sn	50	115	0.36%
Sn	50	116	14.53%
Sn	50	117	7.68%
Sn	50	118	24.22%
Sn	50	119	8.58%
Sn	50	120	32.59%
Sn	50	122	4.63%
Sn	50	124	5.79%
Sb	51	121	57.36%
Sb	51	123	42.64%
Te	52	120	0.095%
Te	52	122	2.59%
Te	52	123	1.3×10^{13} yr (0.905%)
Te	52	124	4.79%
Te	52	125	7.12%
Te	52	126	18.93%
Te	52	128	$> 8 \times 10^{24}$ yr (31.70%)
Te	52	130	$< 1.25 \times 10^{21}$ yr (33.87%)
I	53	127	100%
Xe	54	124	0.10%
Xe	54	126	0.09%
Xe	54	128	1.91%
Xe	54	129	26.4%
Xe	54	130	4.1%
Xe	54	131	21.2%
Xe	54	132	26.9%
Xe	54	134	10.4%
Xe	54	136	$> 2.36 \times 10^{21}$ yr (8.9%)
Cs	55	133	100%
Ba	56	130	0.106%
Ba	56	132	0.101%
Ba	56	134	2.42%
Ba	56	135	6.593%
Ba	56	136	7.85%
Ba	56	137	11.23%
Ba	56	138	71.70%
La	57	138	1.05×10^{11} yr (0.0902%)
La	57	139	99.9098%
Ce	58	136	0.19%
Ce	58	138	0.25%
Ce	58	140	88.43%
Ce	58	142	$> 5 \times 10^{16}$ yr (11.13%)
Pr	59	141	100
Nd	60	142	27.13%
Nd	60	143	12.18%

Continued

Table 1 *Continued*

Symbol	Z	A	Half-Life and/or Percent Abundance
Nd	60	144	2.29×10^{15} yr (23.80%)
Nd	60	145	8.30%
Nd	60	146	17.19%
Nd	60	148	5.76%
Nd	60	150	$> 10^{18}$ yr (5.64%)
Sm	62	144	3.1%
Sm	62	147	1.061×10^{11} yr (15.0%)
Sm	62	148	7×10^{15} yr (11.3%)
Sm	62	149	$> 2 \times 10^{16}$ yr (13.8%)
Sm	62	150	7.4%
Sm	62	152	26.7%
Sm	62	154	22.7%
Eu	63	151	47.8%
Eu	63	153	52.2%
Gd	64	152	1.08×10^{14} yr (0.20%)
Gd	64	154	2.18%
Gd	64	155	14.80%
Gd	64	156	20.47%
Gd	64	157	15.65%
Gd	64	158	24.84%
Gd	64	160	21.86%
Tb	65	159	100%
Dy	66	156	0.06%
Dy	66	158	0.10%
Dy	66	160	2.34%
Dy	66	161	18.9%
Dy	66	162	25.5%
Dy	66	163	24.9%
Dy	66	164	28.2%
Ho	67	165	100%
Er	68	162	0.14%
Er	68	164	1.61%
Er	68	166	33.6%
Er	68	167	22.95%
Er	68	168	26.8%
Er	68	170	14.9%
Tm	69	169	100%
Yb	70	168	0.13%
Yb	70	170	3.05%
Yb	70	171	14.3%
Yb	70	172	21.9%
Yb	70	173	16.12%
Yb	70	174	31.8%
Yb	70	176	12.7%
Lu	71	175	97.41%
Lu	71	176	3.78×10^{10} yr (2.59%)
Hf	72	174	2.0×10^{15} yr (0.162%)
Hf	72	176	5.206%
Hf	72	177	18.606%
Hf	72	178	27.297%
Hf	72	179	13.629%
Hf	72	180	35.100%
Ta	73	180	$> 1.2 \times 10^{15}$ yr (0.012%)
Ta	73	181	99.988%
W	74	180	0.12%
W	74	182	26.3%
W	74	183	14.28%
W	74	184	$> 3 \times 10^{17}$ yr (30.7%)
W	74	186	28.6%
Re	75	185	37.40%
Re	75	187	4.35×10^{10} yr (62.60%)
Os	76	184	$> 5.6 \times 10^{13}$ yr (0.02%)
Os	76	186	2.0×10^{15} yr (1.58%)
Os	76	187	1.6%
Os	76	188	13.3%
Os	76	189	16.1%
Os	76	190	26.4%
Os	76	192	41.0%
Ir	77	191	37.3%
Ir	77	193	62.7%
Pt	78	190	6.5×10^{11} yr (0.01%)
Pt	78	192	0.79%
Pt	78	194	32.9%
Pt	78	195	33.8%
Pt	78	196	25.3%
Pt	78	198	7.2%
Au	79	197	100%
Hg	80	196	0.15%
Hg	80	198	9.97%
Hg	80	199	16.87%
Hg	80	200	23.10%
Hg	80	201	13.10%
Hg	80	202	29.86%
Hg	80	204	6.87%
Tl	81	203	29.524%
Tl	81	205	70.476%
Pb	82	204	$\geq 1.4 \times 10^{17}$ yr (1.4%)
Pb	82	206	24.1%
Pb	82	207	22.1%
Pb	82	208	52.4%
Bi	83	209	100%
Bi	83	215	7.4 min
At	85	215	0.10 ms
At	85	218	~ 2 ms
At	85	219	0.9 min
Ra	88	226	1,600 yr
Th	90	232	1.4×10^{10} yr (100%)
Pa	91	231	3.28×10^4 yr
U	92	234	2.44×10^5 yr (0.0055%)
U	92	235	7.04×10^8 yr (0.720%)
U	92	238	4.468×10^9 yr (99.2745%)
Pu	94	244	8.3×10^7 yr

Table of the Isotopes

There are 263 known stable isotopes along with 38 naturally occurring radioactive isotopes. In addition about 1,000 radioactive isotopes have been produced and identified artificially. It is expected that in total there are around 5,000 possible isotopes. More than this cannot be made, due to the preference of nuclear forces for equal numbers of neutrons and protons. As neutrons are added to a nucleus with a fixed atomic number, the nuclear binding energy for the last neutron added becomes weaker and weaker until eventually the nucleus will not accept another neutron. This is known as the neutron drip line, since adding another neutron will in a sense cause the nucleus to "drip" the neutron back out. Similarly, if too few neutrons are present, the protons become unbound, which defines the proton drip line.

See also: Atomic Mass Unit; Atomic Number; Mass Number; Neutron; Proton; Radioactivity

Bibliography

Browne, E.; Firestone, R. N.; and Shirley, V. S., eds. *Table of Radioactive Isotopes* (Wiley, New York, 1986).

Firestone, R. B.; Shirley, V. S.; Baglin, C. M.; Chu, S. Y. F.; and Zipkin, J. *Table of Isotopes,* 8th ed. (Wiley, New York, 1996).

Walker, F. W.; Parrington, J. R.; and Feiner, F. *Chart of the Nuclides,* 14th ed. (General Electric, San Jose, CA, 1989).

Grant J. Mathews

J

JAHN–TELLER EFFECT

A spontaneous distortion that reduces the symmetry at the site of an aspherical atom or ion is known as the Jahn–Teller effect. As an example of the effect, it is shown in Fig. 1 that the square with four spherical ions q_1 at the corners and an aspherical ion q_2 at the center distorts spontaneously to a rhombus. The energy of interaction between the charges at the corners is to second order in θ:

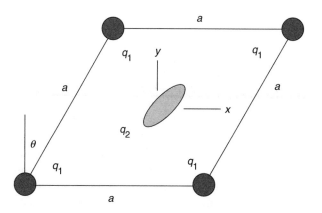

Figure 1 A square molecule with spherical charges at the corners and an aspherical charge at the center is unstable and distorts spontaneously into a rhombus.

$$U_1(\theta) = \frac{3}{2\pi} \frac{q_1^2}{\varepsilon_0 a}\left(1 + \frac{\theta^2}{8}\right). \tag{1}$$

This shows that $U_1(\theta)$ alone is minimum at $\theta = 0$, for example, the square charge distribution (without q_2) has the lowest energy.

The electrostatic energy of the central ion in the field of other ions is

$$U_2(\theta) = \int \rho_2(\mathbf{r})\,\phi(\mathbf{r},\theta)\,dV, \tag{2}$$

where $\rho(\mathbf{r})$ is the charge density of the central ion and $\phi(\mathbf{r})$ is the local electrostatic potential. The electrostatic energy of the aspherical central ion in multipole expansion is

$$U(\theta) = q_2\phi(0) + \sum_{ij} Q_{ij} \frac{\partial^2}{\partial x_i \partial x_j}\, \phi(\mathbf{r})\big|_{\mathbf{r}=0} + \cdots, \tag{3}$$

where the Q_{ij} are the quadrupole moments of the central ion:

$$Q_{ij} = \int \left(3x_i x_j - r^2\delta_{ij}\right)\rho(\mathbf{r})\,dV. \tag{4}$$

The energy of interaction between the central ion and the corner ions to second order in θ is

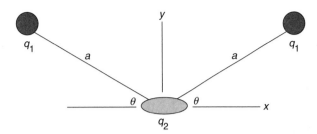

Figure 2 A linear molecule with spherical charges at the ends and an aspherical charge at the center is stable. This is the one exception to an otherwise general rule.

$$U_2(\theta) = \frac{\sqrt{2}}{2\pi} \frac{q_1}{\varepsilon_0 a}\left[\left(1 + \frac{\theta^2}{4}\right)q_2 - 18\frac{Qxy}{a^2}\theta\right], \quad (5)$$

and it is shown that $U_2(\theta)$ contains a negative term linear in θ that induces a rhombic distortion of the square with the total electrostatic energy, $U(\theta) + U_2(\theta)$, a minimum for

$$\theta = \frac{36Qxy/a^2}{q_1 + (3/2\sqrt{2})q_2}. \quad (6)$$

For the long axis of the quadrupole oriented at $\pi/4$, Eq. (6) gives θ positive. For the long axis at $-\pi/4$, θ is negative. The second solution is simply the first, rotated through $\pi/2$.

It may appear that destabilization by an aspherical charge is to be expected of all molecules. Yet, there is one striking exception, the linear molecule shown in Fig. 2, for which the energy is

$$U(\theta) = \frac{1}{8\pi\varepsilon_0} \frac{q_1^2}{a\cos\theta} + \frac{1}{2\pi\varepsilon_0} \frac{q_1 q_2}{a}$$

$$+ \frac{3}{2\pi\varepsilon_0} \frac{q_1}{a^3}(Qxx\cos^2\theta + Qyy\sin^2\theta). \quad (7)$$

As Eq. (7) is even in θ, a linear molecule is stable.

The principle illustrated here was first enunciated by H. A. Jahn and Edward Teller as a theorem in molecular physics: All nonlinear nuclear configurations are unstable for an orbitally degenerate electronic state. What is the connection between molecular instability in the presence of aspherical charge and the orbital degeneracy of the electronic states of an atom or ion? For an aspherical ion at a site of high symmetry, two or more orientations of its quadrupole moment are of equal energy. In quantum terms, the deviation from sphericity arises from the occupation of electronic orbitals. Knowing that the energy is the same for two or more directions of the quadrupole moment means that the associated orbitals must be of the same energy—that is to say, degenerate. Thus the configuration distorts spontaneously to that of lower symmetry to lift this degeneracy to seek an energetically stable state.

See also: DEGENERACY; ELECTROSTATIC ATTRACTION AND REPULSION; MOLECULAR PHYSICS

Bibliography

ENGLMAN, R. *The Jahn–Teller Effect in Molecules and Crystals* (Wiley, London, 1972).

JAHN, H. A. "Stability of Polyatomic Molecules in Degenerate Electronic States: II. Spin Degeneracy." *Proc. R. Soc. London Ser. A* **164**, 117–131 (1938).

JAHN, H. A., and TELLER, E. "Stability of Polyatomic Molecules in Degenerate Electronic States: I. Orbital Degeneracy." *Proc. R. Soc. London Ser. A* **161**, 220–235 (1937).

O'BRIEN, M. C. M., and CHANCEY, C. C. "The Jahn–Teller Effect: An Introduction and Current Review." *Am. J. Phys.* **61**, 688–697 (1993).

TELLER, E. "The Jahn–Teller Effect: Its History and Applicability." *Physica* **114A**, 14–18 (1982).

ALAN M. PORTIS

JOSEPHSON EFFECT

The Josephson effect originally pertained to the tunneling of Cooper pairs in a superconducting-insulating-superconducting (SIS) tunnel junction (diode). However, the term has also been applied to phenomena in superfluid helium, in neutron stars, and in the quantum Hall effect. This entry covers only the superconducting tunneling of Cooper pairs. There are really two effects, the dc Josephson effect and the ac Josephson effect.

In 1962 Brian Josephson, while a Ph.D. student at the University of Cambridge in England, worked out the theory of this effect based on some of the ideas in the Bardeen–Cooper–Schrieffer (BCS) theory of superconductivity. In 1963 P. W. Anderson and J. M. Rowell confirmed the dc Josephson effect in work at the Bell Telephone Laboratories. Also in 1963, Sidney Shapiro observed the ac Josephson effect with experiments carried out at the Arthur D. Little Company in Cambridge, Massachusetts.

In 1973, Josephson shared the Nobel Prize in physics with Leo Esaki (for the Esaki diode, a heavily doped semiconductor *p-n* junction) and Ivar Giaever (for quasiparticle tunneling in superconducting tunnel junctions). The common theme in the work of these three men is quantum-mechanical tunneling.

According to quantum mechanics, any piece of matter has wave properties. This leads to results that are contrary to "intuition," such as interference and uncertainty. Another contrary result is quantum tunneling in which an object passes through a barrier without boring a hole or otherwise disturbing the barrier. According to this, there is a nonzero probability that a person can tunnel through a wall or a car can tunnel through a hill. This is understood as follows: If an object is next to a barrier, then the wave function (the wave properties are described by the wave function) of the object tails off into the barrier. This wave function has a nonzero amplitude on the other side of a finite barrier so that the probability of finding the object on the other side is also nonzero. This probability is so small for a macroscopic object, such as a person or a car, that tunneling of such things has never occurred in a length of time comparable to the age of the universe.

For an electron in a metal or a superconductor or for a Cooper pair (a pair of electrons) in a superconductor, the probability of tunneling through an insulator (a barrier) is reasonably high if the insulator is only of the order of 10 Å (10×10^{-10} m) thick. A superconducting tunnel junction can be fabricated by laying down on an insulating substrate (e.g., a glass slide) a thin film of a superconductor, followed by an insulating film and then another superconductor. Metal leads are then attached to the metal films of this sandwich.

A possible circuit with an SIS junction is shown in Fig. 1. The battery supplies an electric current, the ammeter measures the current through the junction, and the voltmeter measures the voltage across the junction. We must also have some means of adjusting the voltage or the current through the circuit. Frequently one measures the current as a function of the voltage.

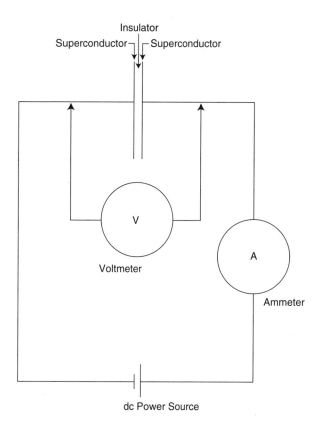

Figure 1 Circuit with an SIS tunnel junction. The view of the junction is edge on.

When a metal becomes superconducting, the conduction electrons near the Fermi surface form in pairs (Cooper pairs). These pairs (which are highly correlated) are described in terms of an order parameter, a type of wave function. This order parameter is complex, and, as with any complex number, can be given in terms of its magnitude and phase. The phase is an angle in the complex plane.

With the SIS junction in the circuit of Fig. 1, one can have a dc electric current flow through the circuit with zero voltage as long as the current is not too large. This is the dc Josephson effect. It occurs because the superconducting order parameter of the one superconductor decays into the insulator and actually makes contact with the order parameter in the other superconductor, making the entire tunnel junction in some sense a single superconductor.

According to Josephson, when there is a dc Josephson current flowing in a tunnel junction (i.e., with zero voltage), there is a difference in the phase of the order parameter in the two superconductors. Further, the current is proportional to the sine of

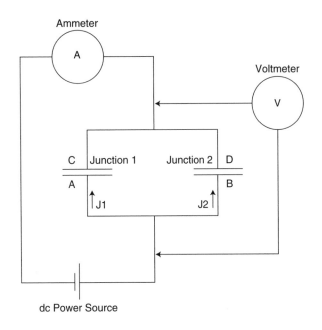

Figure 2 Two Josephson junctions in parallel in a dc circuit.

the phase difference so that for zero phase difference, there is no current. There is a maximum possible dc Josephson current, and that occurs when the phase difference is ±90°.

An external magnetic field has an effect on the phase of the order parameter. This can best be seen by considering the circuit in Fig. 2 with two superconducting junctions in parallel. The wires AB and CD are superconducting. For simplicity, we assume

that the two junctions are identical, and we operate at zero voltage across the junctions. First consider that there is no external magnetic field. The phase of the superconducting order parameter in wire AB is a constant ϕ_1 while in wire CD it is another constant ϕ_2. If a Josephson current is flowing, there is a phase difference $\Delta\phi = \phi_2 - \phi_1$ across each of the junctions with the current proportional to $\sin(\Delta\phi)$ in each of the junctions. Both currents (J_1 and J_2 in Fig. 2) are equal and in the same direction. Thus they add together to give the total current. This can be thought of as constructive interference between the two currents.

When a small magnetic field perpendicular to the plane of the circuit is turned on, the phases change so they are no longer constant in each of the superconducting wires. The amount of the change is proportional to the total magnetic flux through the circuit loop ACDB. (For a constant magnetic field, the flux is just the value of the field times the area of the loop.) $\Delta\phi$ across junction 1 is no longer the same as $\Delta\phi$ across junction 2. For certain values of the magnetic field one can have the Josephson current J_1 in the first junction flowing in the opposite direction to J_2 in the second junction. In that case there is destructive interference with zero total current through the ammeter. Plotting the maximum Josephson current as a function of magnetic field produces an interference pattern, as in Fig. 3. In addition to the interference between the two junctions, there is also an interference within each junction similar to diffraction in a single slit in optics. The mathematics is exactly the same for the in-

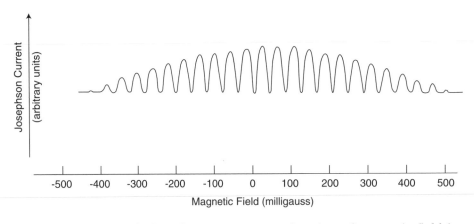

Figure 3 Maximum dc Josephson current as a function of magnetic field in a circuit similar to that in Fig. 2 [from Jaklevic, R. C., et al., *Phys. Rev.*, **140**, 1628–1637 (1965)].

terference pattern of the two Josephson junctions and for the double slit in optics.

It is possible to use the circuit in Fig. 2 to actually determine the magnitude of an external magnetic field by looking at the interference patterns as in Fig. 3. This is the basis for the superconducting quantum interference device (SQUID), which is much used for measuring magnetic fields or voltages very precisely.

The ac Josephson effect occurs if we have a voltage across the junction. According to Josephson, the phase difference $\Delta\phi$ will increase at a constant rate proportional to the voltage. Since the Josephson current is proportional to sin ($\Delta\phi$), the increasing $\Delta\phi$ will cause an oscillating current. Just as in an antenna, the oscillating current will produce electromagnetic radiation of the same frequency, which is called the Josephson frequency and is equal to $2eV/h$ where V is the voltage, e is the charge of the electron, and h is Planck's constant. Typical voltages across a Josephson junction range from a microvolt to about 10 millivolts. The corresponding Josephson frequencies are from 500 MHz (5×10^8 Hz) to 5 THz (5×10^{12} Hz). Although the intensity of the corresponding radiation is extremely low, it has been detected, confirming that the ac Josephson effect does exist.

Another way that the Josephson frequency can be seen is in an SIS junction, as in Fig. 1, but with the junction in a microwave cavity. The cavity is operated at some frequency ν with the microwave electric field producing an ac voltage across the junction. In addition, a dc voltage V is applied to the junction. V is then slowly increased over a range of values. Each time V takes a value such that the Josephson frequency is equal to an integer times ν, the dc current through the junction makes a jump. These are called Shapiro steps. Thus the voltage at which a step in current occurs is precisely $V = n\nu h/2e$. It is remarkable that the voltage at each step does not depend in any way on the material in the SIS junction, nor on the size, nor on any other characteristics of the junction.

By making very precise measurements of ν and of the voltages V for a large number of steps, W. H. Parker, B. N. Taylor, and D. N. Langenberg were able to use the ac Josephson effect in 1967 to make the most precise measurement ever of the ratio of the fundamental constants h and e.

Now that it has been well established that the Josephson frequency is precisely $2eV/h$ in an SIS tunnel junction independent of the materials or other characteristics of the junction, such junctions have become very important as voltage standards. That is, one can use such a junction to calibrate voltmeters to a very high precision.

See also: COOPER PAIR; DOUBLE-SLIT EXPERIMENT; INTERFERENCE; QUANTUM MECHANICS; SUPERCONDUCTING QUANTUM INTERFERENCE DEVICE; SUPERCONDUCTIVITY

Bibliography

LANGENBERG, D. N.; SCALAPINO, D. J.; and TAYLOR, B. N. "The Josephson Effects." *Sci. Am.* **214** (May), 30–39 (1966).

TAYLOR, B. N.; LANGENBERG, D. N.; and PARKER, W. H. "The Fundamental Physical Constants." *Sci. Am.* **223** (Oct.), 62–78 (1970).

VAN DUZER, T., and TURNER, C. W. *Principles of Superconductive Devices and Circuits* (Elsevier, New York, 1981).

JAMES C. SWIHART

JOULE HEATING

When electric current flows through a resistor, the resistor's temperature increases. This is Joule heating.

Microscopically, Joule heating occurs because the charge carriers (usually electrons) make inelastic collisions with the resistor's atoms. This means that some of the electron's kinetic energy is converted into vibrational energy of the resistor atoms, and their vibrational energy shows up as an elevated temperature for the resistor. Thus, Joule heating represents the dissipation of electrical energy into thermal energy via the electron-atom inelastic collisions.

Let us calculate the rate of Joule heating in the case of an Ohmic resistor. Let the resistance be R, and let a potential difference V placed across it produce the (conventional) current I through it (see Fig. 1). The energy dU required to push the charge dq through the resistor, across which is the potential V, is

$$dU = V\,dq. \tag{1}$$

Since the resistor is described by Ohm's Law, $V = IR$, in Eq. (1) one may write V in terms of I to obtain

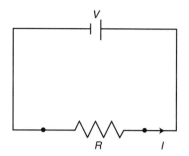

Figure 1 A resistor of resistance R carries conventional current I driven by a potential difference V. It will experience Joule heating at the rate I^2R.

$$dU = IR\,dq. \tag{2}$$

The rate at which energy is spent, dU/dt, is thus

$$dU/dt = I^2R, \tag{3}$$

where we have used the definition of electric current $I = dq/dt$. Using Ohm's law again, I^2R can be written in the other equivalent ways: $I^2R = V^2/R = IV$. These quantities measure the rate at which electrical energy is being converted into thermal energy in the resistor via Joule heating.

Joule heating can also be expressed in terms of the electric field **E,** the magnetic field **B,** and the current density **j,** directly from Maxwell's Equations. In SI units, Maxwell's Equations read

$$\nabla \cdot \mathbf{E} = \rho/\epsilon_0 \tag{4}$$

$$\nabla \cdot \mathbf{B} = 0 \tag{5}$$

$$\nabla \times \mathbf{B} = \mu_0\mathbf{j} + (1/c^2)\,\partial\mathbf{E}/\partial\mathbf{t} \tag{6}$$

$$\nabla \times \mathbf{E} = -\partial\mathbf{B}/\partial\mathbf{t} \tag{7}$$

where ϵ_0 and μ_0 are respectively the permittivity and permeability of free space, c is the speed of light in vacuum, and ρ is the electric charge density. Evaluating the divergence of $\mathbf{E} \times \mathbf{B}$ and using Eqs. (6) and (7), one derives the result

$$\nabla \cdot \mathbf{S} + \partial\eta/\partial\mathbf{t} = -\mathbf{j} \cdot \mathbf{E} \tag{8}$$

where $\mathbf{S} \equiv \mathbf{E} \times \mathbf{B}/\mu_0$ is Poynting's vector (the field momentum density), and $\eta = (1/2)\epsilon_0\mathbf{E}^2 + (1/2)\mathbf{B}^2/\mu_0$ is the energy density of the electric and magnetic fields. If the right-hand side of Eq. (8) were zero, then Eq. (8) would be an equation of continuity representing the local conservation of the electromagnetic field energy. However, a non-zero value of $-\mathbf{j} \cdot \mathbf{E}$ represents the nonconservation of the field energy, and is the Joule heating per unit volume.

See also: ELECTROMAGNETISM; FIELD, ELECTRIC; FIELD, MAGNETIC; MAXWELL'S EQUATIONS; OHM'S LAW; RESISTOR

Bibliography

FEYNMAN, R. P.; LEIGHTON, B. B.; and SANDS, M. *The Feynman Lectures on Physics,* Vol. 1 (Addison-Wesley, Reading, MA, 1963).

GRIFFITHS, D. J. *Introduction to Electrodynamics* (Prentice Hall, Englewood Cliffs, NJ, 1989).

HALLIDAY, D.; RESNICK, R.; and KRANE, K. S. *Physics,* 4th ed. (Wiley, New York, 1992).

JACKSON, J. D. *Classical Electrodynamics,* 2nd ed. (Wiley, New York, 1975).

LORRAIN, P.; CORSON, D. R.; and LORRAIN, F. *Electromagnetic Fields and Waves* (W. H. Freeman, New York, 1988).

PANOFSKY, W. K. H., and PHILLIPS, M. *Classical Electricity and Magnetism,* 2nd ed. (Addison-Wesley, Reading, MA, 1962).

SERWAY, R. A. *Physics for Scientists and Engineers with Modern Physics,* 3rd ed. (Saunders, Philadelphia, 1990).

TIPLER, P. A. *Physics,* 2nd ed. (Worth, New York, 1982).

VANDERLINDE, J. *Classical Electromagnetic Theory* (Wiley, New York, 1993).

DWIGHT E. NEUENSCHWANDER

JOULE–THOMSON EFFECT

James Prescott Joule was the son of a Manchester, England, brewer. He became interested in topics related to energy as a young man and, without formal training, pursued a career in science as an experimentalist. His efforts from the 1840s through the

1860s resulted in several significant advances in thermodynamics. Joule proposed and carried out a series of experiments to measure the equivalence between energy as heat and energy as work. In order to do this experiment, he had to build his own thermometers, using the expansion of liquid mercury in glass as the measure of temperature. He, perhaps optimistically, claimed to be able to estimate temperature to 1/200 of a degree Fahrenheit. He wrote, "I can estimate to one-twentyth of a division," and each division on his thermometers was 1/10 of a degree. He presented improved results over a period of years but received little recognition. In 1847 he determined that 1 British thermal unit (Btu) was equivalent to 782 ft-lb force of work (the now-accepted value is 778). At that time, his work caught the interest of William Thomson (later known as Lord Kelvin). In 1853 the two collaborated on work that led to measurement of what is now known as the Joule–Thomson effect, or the Joule–Kelvin effect. The measured values are known as the Joule–Thomson or Joule–Kelvin coefficient. In an experiment, air or other gases are forced down an insulated duct containing a restriction to the flow, such as a packing of cotton or silk. The effect of passing the gas through this porous plug is a substantial pressure drop. Since the duct is insulated, the experiment occurs under adiabatic conditions and the effect of heat transfer is negligible. The conservation of energy law reduces to $h_1 = h_2$, where the subscript 1 applies to the upstream position in front of the plug, 2 to the downstream position, and h is the enthalpy function used in thermodynamics. A process in which enthalpy is the same at the inlet and outlet of the plug is called throttling. The effect of throttling has a number of important applications. The resulting temperature change caused by the flow moving through a restriction is called the Joule–Thomson effect. Under certain conditions, the effect is a reduction of the gas temperature. In this manner, low temperature may be achieved, which can lead to the liquefaction of gases. It is an experimental fact that throttling of a gas can lead to lower or higher temperatures, even though the process allows for no heat transfer. The final temperature depends upon the values of P_1 and T_1 and the downstream pressure, P_2. The Joule–Thomson coefficient is defined as

$$\mu_{JT} = \left(\frac{\partial T}{\partial P} \right)_h ,$$

where the subscript in the partial derivative notation signifies a parameter or function to be held constant. The value of the coefficient may be determined under different conditions by plotting experimental data along lines of constant enthalpy in temperature-pressure (TP) coordinates. In the original experiments of Joule and Thomson, small but definite temperature changes were measured; for example, a decrease of about 0.5°F per atmosphere of pressure difference for air and an increase of about 0.04°F per atmosphere of pressure difference for hydrogen, both gases being at room temperature. Joule's ability to develop accurate thermometers was crucial to making these measurements.

The experiment is performed as follows: The pressure and temperature on the upstream side of the plug are set. The pressure on the downstream side is then set at any value below the upstream value, and the downstream temperature is measured. Varying the thickness of the plug will change the downstream pressure value, and new values for the downstream pressure and measured temperature can be determined. The results provide a set of points on a TP diagram, one point being the upstream value and the other point being the downstream value, all along a line of constant enthalpy. The data thus yield numerical values for the

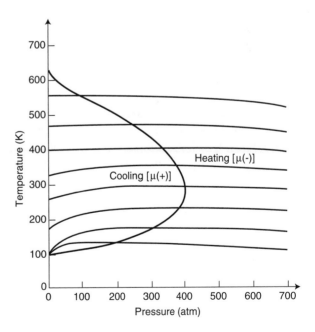

Figure 1 Sample inversion curves for the Joule–Thomson experiment.

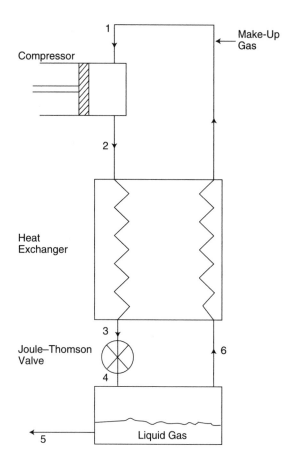

Figure 2   Linde–Hampson system for air liquefaction.

Joule–Thomson coefficient. Note that since the pressure drop is always negative across the flow restriction, a positive number for this coefficient represents a decrease in temperature and a negative value for the coefficient represents an increase in temperature. Experimental data show that a number of constant-enthalpy lines have a state of maximum temperature. The line shown in Fig. 1 that passes through these states of maximum temperature is called the inversion curve. The value of the temperature for that particular state is called the inversion temperature. To the right of the inversion line, the Joule–Thomson coefficient is negative, and the temperature will increase as the pressure decreases through the porous plug. A heating effect results. In contrast, to the left of the inversion curve, the coefficient is positive so that flow through a restriction will result in a cooling effect. This case is important in the liquefaction of gases.

One such liquefaction apparatus is shown in Fig. 2. Here, the gas is compressed from state 1 to state 2 in as close to a reversible and isothermal manner as possible. The gas then undergoes cooling in a counterflow heat exchanger, 2 to 3, where it transfers energy as heat to the returning low-pressure gas stream, 6 to 1. The gas is then throttled through an expansion valve, 3 to 4, known as a Joule–Thomson (JT) valve. The gas temperature drops and, if state 4 is at a sufficiently low temperature, a small portion of the gas is liquefied. The remaining gas is returned through the heat exchanger, which is used to cool the incoming gas before the latter enters the valve.

See also: ENTHALPY; GAS; KELVIN, LORD; TEMPERATURE; TEMPERATURE SCALE, FAHRENHEIT; THERMODYNAMICS; THERMODYNAMICS, HISTORY OF

Bibliography

JOULE, J. P. *The Scientific Papers of James Prescott Joule* (Dawsons of Pall Mall, London, ca. 1887 [1963]).

THOMSON, W., and JOULE, J. P. "On the Thermal Effects of Fluids in Motion, Part 1." *Phil. Trans. R. Soc. London,* p. 357 (1853).

THOMSON, W., and JOULE, J. P. "On the Thermal Effects of Fluids in Motion, Part 2." *Phil. Trans. R. Soc. London,* p. 321 (1854).

HENRY C. PERKINS JR.

K

KAON

The kaon is one of a set of particles known as pseudoscalar mesons, which are particles that have no spin. Its mass is a little greater than one-half that of the proton and about 3.5 times that of the pion, the lightest meson.

The kaon is said to carry one unit of strangeness, S, a kind of charge. This concept arose from the problem that kaons were easily produced in strong nuclear interactions but decayed only weakly. This problem was resolved by the observation that strange particles can only be created in pairs, one with positive and one with negative strangeness, due to the conservation of strangeness in strong interactions. Since the kaon is the lightest strange particle, it can only decay into *non*strange particles, and only weak interactions violate strangeness conservation.

The kaon has two charge states with $S = 1$, K^+ and K^0. There are two *anti*particles with $S = -1$, K^- and $\overline{K^0}$.

Since the weak interactions do not conserve strangeness, K^0 and $\overline{K^0}$ are mixed, and the states with precise masses and lifetimes are mixtures of K^0 and $\overline{K^0}$. Conversely, when K^0 is produced, it will be a mixture of two particles with different lifetimes. One observes in the decay of K^0, a short-lived particle K_S, which decays predominantly into two pions, and a long-lived particle K_L, which decays predominantly into three pions. By sophisticated quantum-mechanical methods one has determined that K_L is more massive than K_S by one part in 10^{14}.

One can understand the decays of these kaons by the assumption of symmetry with respect to a quantity called CP, where C is the symmetry between particle and antiparticle and P is symmetry between left and right. Then, assuming K_S is even and K_L is odd under this symmetry, only K_S can decay into two pions, since all states of two pions with total charge 0 are even under CP symmetry. Then K_L is forced to decay into three pions, which is more difficult, and explains its longer lifetime. However, the discovery that K_L also on rare occasions decays into two pions provided the evidence that CP violation can occur, though at a rate much lower than the dominant weak decay processes.

The fundamental constituents of strongly interacting particles are believed to be fractionally charged particles called quarks. The nucleon and the pion are made of u quarks (charge $Q = \frac{2}{3}$) and d quarks ($Q = -\frac{1}{3}$); in particular, $\pi^- = (d\overline{u})$, that is, a d quark joined with the *anti*quark of the u. In order to explain strangeness one must introduce a strange quark s ($Q = -\frac{1}{3}$) with strangeness $S = -1$ and its *anti*particle $\overline{s}$ ($Q = +\frac{1}{3}$, $S = +1$). Then the kaons are $K^+ = (u\overline{s})$, $K^- = (s\overline{u})$, $K^0 = (d\overline{s})$, $\overline{K^0} = (s\overline{d})$. The strong interactions between any pair of these three types of quarks is believed to be the same, yielding a symmetry called $SU(3)$. However, the strange particles are heavier than their *non*strange counterparts because the s quark is heavier than u or d, so this symmetry is not manifest in the particle masses.

The main decay mode of the charged kaon is into a muon and a neutrino, thus $K^- \rightarrow \mu^- + \bar{\nu}_\mu$. This clearly can be compared with the decay mode of the pion $\pi^- \rightarrow \mu^- + \bar{\nu}_\mu$. The theory of weak interactions involves a universal coupling of quarks and leptons to an intermediate charged vector boson W, which plays a role analogous to the photon in electromagnetic interactions. There are couplings $d \rightarrow u + W^-$, $W^- \rightarrow \mu^- \bar{\nu}_\mu$, $W^- \rightarrow e^- \bar{\nu}_\mu$; the universality of the couplings was determined from the rates for muon decay, muon capture by a nucleon, and nuclear beta decay. The decay $\pi^- \rightarrow \mu^- + \bar{\nu}_\mu$ occurs via $d \rightarrow uW^- \rightarrow u\mu^- \bar{\nu}_\mu$ with the final u annihilating with the original $\bar{u}$ in the π^-. Similarly one can explain K^- decay by adding the coupling $s \rightarrow u + W^-$. However, the rate of the K^- decay implies that the strength of this is almost five times smaller than that for $d \rightarrow u + W^-$, which appeared to violate the universality idea. One can believe that the weak coupling of the s quark does have a universal strength, but this coupling is divided between the u, c, and t quarks.

Since the c and t quarks are more massive than the s, the coupling to these quarks cannot contribute to the decay of the s quark. However, the dominant weak W coupling of the s quark to the c quark is observed in the decays of mesons containing charmed quarks, which decay primarily into states that include a K meson with its s quark content.

See also: INTERACTION, STRONG; INTERACTION, WEAK; MUON; QUANTUM MECHANICS; QUARK; QUARK, CHARM; QUARK, DOWN; QUARK, STRANGE; QUARK, UP; SYMMETRY

Bibliography

BROWN, L. M.; DRESDEN, D.; and HODDESON, L. *Pions to Quarks* (Cambridge University Press, Cambridge, Eng., 1989).

HUGHES, I. M. *Elementary Particles* (Cambridge University Press, Cambridge, Eng., 1985).

LINCOLN WOLFENSTEIN

K CAPTURE

The K shell of an atom refers to the two innermost electrons of the atom, the ones closest to the nucleus; other electron shells are given the designations L, M, N, . . . , in order of increasing size. K capture is one particular form of a general class of nuclear processes called beta decay. In K capture, which is the dominant form of the more general process of electron capture, an atomic K-shell electron interacts with the nucleus, via the weak interaction, and is absorbed by a proton in the atomic nucleus. Because the electron is negatively charged, with one fundamental unit of charge, adding an electron to the positively charged nucleus causes a nuclear proton to change to neutron and lowers the nuclear charge. Since the atomic number of an atom, Z, is the charge of the nucleus in fundamental units, K capture changes the atomic number of the atom to $Z - 1$ (i.e., transmutation). In addition to the change of a proton to a neutron, a neutrino is also emitted, which is required to conserve angular momentum. To satisfy the dictates of mass-energy conservation, the K-capture process can occur as long as the mass atomic mass before the capture is greater than the mass afterward.

After the K-capture process, a vacancy is left in the atomic K shell. This vacancy is rapidly filled in one of two ways: either a higher electron (L shell, M shell, etc.) drops down and fills the vacancy, giving off x rays with characteristic energies, or a higher electron drops down to fill the vacancy, giving its energy to another higher electron, which is ejected from the atom. This latter process is known as the Auger effect. Information about the K-capture process is usually obtained in the laboratory by observing the characteristic x rays, the most common method, or the Auger electrons, which also have characteristic energies. The intensity of the x rays or Auger electrons can be interpreted as the probability or cross section for the K-capture process.

K capture is of particular interest because it involves interactions between atomic electrons and nuclear particles (nucleons). Thus, the details of K capture are sensitive to nucleon dynamics and atomic electron dynamics. K-capture processes have been investigated to provide information about the size and structure of ground and excited states of atomic nuclei. K-capture processes have been employed primarily, however, to study the decay of atomic inner-shell vacancies through x-ray and Auger spectroscopy, since systems which undergo K capture provide a very convenient source of atoms with an inner shell vacancy. Following the creation of the vacancy in the K shell, x-ray and Auger processes cause the vacancy to propagate upward to outer shells and, in the case of Auger processes, to create multiple vacancies. This allows the opportunity to investigate the creation and decay of vacancies in all atomic shells. Extensive tabulations have been made for the various characteristic x-ray and Auger electron energies associated with the filling of atomic electron vacancies over a broad range of the periodic table. These tabulations provide us with analytical tools for the detection of trace amounts of various elements in a variety of situations including materials, environmental, and astrophysical studies.

See also: ATOM; ATOMIC NUMBER; ATOMIC PHYSICS; AUGER EFFECT; ELECTRON, AUGER; ENERGY, CONSERVATION OF; MASS-ENERGY; NUCLEAR SHELL MODEL

Bibliography

EMERY, G. T. *Atomic Inner-Shell Processes* (Academic Press, New York, 1975).

WONG, S. S. M. *Introductory Nuclear Physics* (Prentice Hall, Englewood Cliffs, NJ, 1990).

STEVEN T. MANSON

KELVIN, LORD

b. Belfast, Ireland, June 26, 1824; *d.* Netherhall near Largs, Ayrshire, Scotland, December 17, 1907; *thermodynamics, electricity and magnetism, geophysics.*

During the long reign of Queen Victoria, William Thomson (later to be known as Lord Kelvin) fashioned for himself a career that ultimately took him to the very pinnacle of British Imperial science. His capacity to direct his physics toward practical goals placed him among the most eminent of Victorian scientists and engineers. His central position in a major correspondence network of elite mathematical physicists, including James Clerk Maxwell, George Gabriel Stokes, Hermann von Helmholtz, and Peter Guthrie Tait, gave him a leading role in the emergence of physics as a new nineteenth-century scientific discipline. His active role in geological and cosmological controversies following publication of Charles Darwin's *Origin of Species* (1859) located him in the mainstream of Victorian concerns about humankind's place in the natural order.

The fourth child of James and Margaret Thomson, William was born in Belfast where his father had become professor of mathematics in the new and politically radical Belfast Academical Institution. His mother hailed from a Glasgow commercial family but died when William was just six years of age. Encouraged throughout his early years by his father—professor of mathematics at Glasgow University from 1832 until his death from cholera in 1849—Thomson received the best education then available in Britain for a future mathematician or mathematical physicist. Moving easily from the broad philosophical education of Glasgow University to the intensive mathematical training offered by Cambridge University, he came second in the Cambridge mathematics examination (the so-called Mathematical Tripos) of 1845. Spending some weeks in Paris learning the skills of experimental science, Thomson, at the age of twenty-two, was elected in 1846 to Glasgow's Chair of Natural Philosophy (physics), a position he held until 1899.

Thomson's earliest papers owed much to Joseph Fourier's *Théorie Analytique de la Chaleur* (1822). This celebrated French text provided a mathematical treatment of the laws of heat flow. In an original paper written at the age of seventeen, Thomson used Fourier's flow treatment to provide a reformulation of the orthodox theory of electrostatics that emphasized the interaction between electric charges. Replacing these action-at-a-distance forces by continuous flow models, Thomson's radical approach was a principal inspiration for Maxwell's subsequent construction of electromagnetic field theory, exemplified in his famous *Treatise on Electricity and Magnetism* (1873).

During the 1850s Thomson extended Fourier's treatment to the analysis of electrical signals transmitted by very long telegraph wires. Practical telegraphers were optimistic about projects not only to lay underwater telegraph cables between Europe and North America, but especially to connect to London all parts of the British Empire that by mid-century ruled over nearly a quarter of the world's land surface. Concerns, however, about the retardation, or blurring, of signals over long distances raised questions about the economic viability of these projects. With his unrivaled authority in such matters, Thomson provided advice on the optimum dimensions and operating conditions for the projected transatlantic and imperial cables. He also began constructing very delicate measuring and testing instruments (notably the "marine mirror galvanometer") for application in telegraphy. In recognition of his services to the Empire, he was made Sir William Thomson by Queen Victoria in

1866, following his supervision of the laying of the first successful Atlantic telegraph cable.

In the late 1840s Thomson had become increasingly interested in Sadi Carnot's theory of the motive power of heat. Comparatively little-known since its obscure publication in 1824, Carnot's theory provided an understanding of how heat engines (including steam engines) worked. By analogy with waterwheels, Carnot claimed that heat engines depended fundamentally on a "fall" of heat between the high temperature of the boiler and the low temperature of the condenser. This representation gave Thomson the means of formulating in 1848 an "absolute" temperature scale (much later called the "Kelvin scale" in his honor) that correlated temperature difference with work done, thereby making the scale independent of any particular substance (such as mercury or alcohol).

James Prescott Joule, on the other hand, was arguing in the 1840s that heat and work were simply mutually convertible according to an exact "mechanical equivalent of heat." During 1850 and 1851 Thomson, and independently the German physicist Rudolf Clausius, reconciled Carnot and Joule: In the production of motive power, a "thermo-dynamic engine" (Thomson's term for a heat engine) required both the transfer of heat from high to low temperature and the conversion of an amount of heat exactly equivalent to the work done. Over the following decade, Thomson and his friends constructed what they began to call the science of "thermodynamics."

In parallel with the construction of thermodynamics, Thomson and his allies also began replacing an older language of mechanics with the terms "actual" (later "kinetic") and "potential" energy. The laws of energy conservation and dissipation became the twin foundations of a new "science of energy." Although fundamentally mechanical in nature, the universe would now be understood neither in terms of action-at-a-distance forces nor in terms of discrete particles moving through void space, but as a universe of continuous matter possessed of energy. At the same time, the physics discipline was to be defined as the study of energy and its transformations. Thomson and his Edinburgh colleague P. G. Tait began a monumental project, the *Treatise on Natural Philosophy*, that would extend the energy treatment to every branch of physics. Only the first volume on dynamics was published. Nevertheless, the new conceptual framework provided British physicists with enormous claims to authority, for now almost every science (including chemistry and biology) was open to annexation by the energy physicists.

One such territorial move involved Thomson's order-of-magnitude estimates of the age of Earth and the Sun. Employing the energy principle, he calculated ages for Earth and the Sun that varied between 20 million and 100 million years. These time scales tended to contradict the geological assumptions upon which Darwin had built his theory of evolution by natural selection. Of all the challenges faced by Darwin's theory in the nineteenth century, the famous evolutionist admitted that he found Thomson's the most difficult to counter. Only with radioactivity in the 1900s were geologists and evolutionists released from the restraints imposed by nineteenth-century energy physics.

In his later years Thomson became very much an establishment figure in science throughout the Western world. Having gradually built up a university physical laboratory for research and teaching, he played a leading role in the fixing of absolute standards of electrical measurement. Closely allied to this laboratory work were extensive business interests in the patenting and manufacture of scientific, navigational, and electrical instruments. With his increasing wealth, he purchased in 1870 a 126-ton schooner yacht (*Lalla Rookh*) that served as a floating laboratory.

Elevation to the peerage of the United Kingdom brought him the title Baron Kelvin of Largs (usually abbreviated to Lord Kelvin) in 1892. He was the first British scientist to be thus honored and took the name Kelvin from the tributary of the River Clyde that flowed close to the University of Glasgow. In fact, his peerage owed as much to the prominent part that he played in opposing Prime Minister Gladstone's proposals for a separate Irish Parliament ("Home Rule") in the 1880s. Yet he was never much diverted from his commitment to science. Publishing right up to his death at the age of eighty-three, Kelvin found a last resting place in Westminster Abbey, not far from Isaac Newton.

See also: CARNOT, NICOLAS-LÉONARD-SADI; CLAUSIUS, RUDOLF JULIUS EMMANUEL; ELECTROMAGNETISM, DISCOVERY OF; HEAT, MECHANICAL EQUIVALENT OF; HEAT ENGINE; HELMHOLTZ, HERMANN L. F. VON; MAXWELL, JAMES CLERK; THERMODYNAMICS, HISTORY OF

Bibliography

BURCHFIELD, J. D. *Lord Kelvin and the Age of the Earth* (Macmillan, London, 1975).

SMITH, C., and WISE, N. *Energy and Empire: A Biographical Study of Lord Kelvin* (Cambridge University Press, Cambridge, Eng., 1989).

TOMPSON, S. P. *The Life of William Thomson, Baron Kelvin of Largs,* 2 vols. (Macmillan, London, 1910).

THOMSON, W. *Mathematical and Physical Papers,* 6 vols. (Cambridge University Press, Cambridge, Eng., 1882–1911).

THOMSON, W. *Popular Lectures and Addresses,* 3 vols. (Macmillan, London, 1891–1894).

CROSBIE SMITH

KELVIN TEMPERATURE SCALE

See TEMPERATURE SCALE, KELVIN

KEPLER, JOHANNES

b. Weil der Stadt, Swabia, Germany, December 27, 1571; *d.* Regensburg, Germany, November 15, 1630; *cosmology, optics.*

Kepler's childhood was unhappy. He described his mother to be "of a bad disposition," while his father, a soldier of fortune, left home never to return. He was fortunate to reside in Protestant Württemberg, which progressively provided university scholarships to talented but needy students. Kepler graduated from the University of Tübingen and accepted a position as teacher of mathematics and astronomy at the University of Graz, which, as was then common, also involved duties in astrology. Early successes in astrological prediction may have saved his position; his success in teaching was dubious, as no students registered for his course in his second year.

Kepler became an advocate of the Copernican solar system while at Tübingen and pondered the significance of its heliocentric distances. He was greatly impressed that there are exactly five regular or Pythagorean solids, objects whose faces are identical regular polygons. A cube and tetrahedron are familiar examples; the octahedron, dodecahedron, and icosahedron are less well-known. These solids can be inscribed in a sphere so that each corner just touches its surface, or circumscribed about a sphere that is then tangent to each face; the ratio of the radii of the circumscribed and inscribed spheres is fixed. The structure of the universe thus followed for Kepler: five solids—five spaces between planetary orbits, the relative sizes of which were determined by geometry. This notion, a brilliant inspiration which happens to be wrong, determined his future career.

His discovery was published in his first book, *The Mysterium Cosmographicum* (1596). Much of the book is medieval and mystical; a typical argument in favor of his model was that, since God must create a perfect world, it must be based on the five solids. It is a shock, then, when the second half of the book opens with: "Now we shall proceed to the astronomical determination of the orbits . . . If these do not confirm the thesis, then all of our previous efforts have doubtless been in vain." In stating that truth is to be decided by the observed facts, Kepler leaped the gulf between mysticism and science.

His leap to science worked fairly well for some planets but not for others. Kepler questioned the quality of the data that had been compiled by Copernicus. Kepler also speculated, based on the periods and heliocentric distances of the planets, that there must be a force emanating from the Sun which drives the planets in their orbits and which decreases in ratio to distance "as does the force of light." Kepler's intuition forecasts the future work of Galileo and Newton.

Kepler needed better data to prove his theory, and he knew that they could only be obtained from Tycho Brahe, the last and greatest pretelescopic observational astronomer. Tycho, unable to accept the motion of the earth in the Copernican model, had developed a compromise model in which the Sun circles the earth but the other planets orbit the Sun. This model correctly explained the motions of planets seen from earth without requiring the earth to move. Tycho wished for a skilled mathematician to use his data to confirm his model, and he had his eye on Kepler ever since he read the *Mysterium.*

Tycho's major work had been done in his native Denmark, but in 1599 he accepted a position as Imperial Mathematicus for Emperor Rudolph II and moved to Prague. Kepler's time in Graz was running out because of religious strife, so he set out for Prague, meeting Tycho early in 1600. Tycho assigned Kepler the task of determining Mars's orbit, a fortunate choice because Mars's large orbital eccentricity and proximity to Earth maximized the difficulty of fitting its motions with prevailing theory. Kepler and Tycho did not "hit it off" very well. Fortunately, Kepler was absent from Prague for several months on personal business, so he and Tycho were not exposed to each other enough to sabotage the relationship.

Tycho died unexpectedly in October 1601, and Emperor Rudolph II appointed Kepler to be his successor as Imperial Mathematicus, possibly the most prestigious post for a scientist in Europe. Kepler managed to retain Tycho's data upon his death and proceeded to attack the problem of Mars's motions. His attempt to solve the problem using a traditional eccentric circular orbit fit Tycho's observations near oppositions but resulted in errors of up to eight minutes of arc when applied to all data. Previous astronomers would have been satisfied with the result; Kepler was not. In Kepler's words: "Now, because they could not have been ignored, these eight minutes alone will have led the way to the reformation of all of astronomy." The idea that theories must confront accurate experimental data may be Kepler's greatest legacy to Western science.

Kepler reexamined the problem, discarding many preconceived notions that had stood for centuries. He discovered that the earth moves faster the closer it is to the Sun rather than uniformly, at constant speed. This led to Kepler's law of equal areas. Subsequently, Kepler concluded that the orbit must be an oval, although some time passed before he finally recognized that his formula was that of an ellipse. Four additional years passed before these results were published in *The New Astronomy* in 1609, due to squabbles with Tycho's son-in-law who felt, with some justification, that Kepler had expediently misappropriated Tycho's data when he died.

Kepler's work by no means ended with the publication of this book. In 1618 he published the true orbits of all the planets as well as of the satellites of Jupiter in *Survey of Copernican Astronomy*. *The Harmony of the World* (1619) introduced Kepler's third law relating periods and heliocentric distances. *The Rudolphine Tables* (1629) completed the task originally given to Tycho by Emperor Rudolph; namely, tabulating the positions of the planets. And, despite his scientific reputation being based on his contributions to astronomy, Kepler also published influential works on optics (1604), the theory of the telescope (1611), the geometrical structure of snowflakes (1611), and mathematics antecedent to the development of calculus (1615).

Although Kepler's grave was destroyed during the Thirty Years War, his epitaph remains:

> I measured the skies, now the shadows I measure
> Skybound was the mind, earthbound the body rests.

See also: COSMOLOGY; KEPLER'S LAWS; NEWTON, ISAAC; OPTICS

Bibliography

CASPER, M. *Kepler* (Dover, New York, 1990).

KEPLER, J. *New Astronomy,* trans. by W. Donahue (Cambridge University Press, New York, 1992).

KOESTLER, A. *The Sleepwalkers* (Grosset & Dunlap, New York, 1959).

PHILIP B. JAMES

KEPLER'S LAWS

Johannes Kepler's laws of planetary motion were well known to Isaac Newton and his contemporaries. The derivation of these laws from first principles of dynamics was the great problem facing natural philosophers of the late seventeenth century. Kepler's discovery of these laws seventy-five years earlier resulted from his attempts to empirically fit the observational data on planetary motions that had been acquired by Tycho Brahe. Kepler's initial work, on the orbit of the planet Mars, led to his development of the first two laws, which were published in the *New Astronomy* in 1609:

1. The orbits of the planets are ellipses with the sun at one focus.
2. The speeds of the planets as they traverse their orbits vary in such a way that the rate at which the radius vector from the sun to the planet sweeps out area is constant.

An ellipse is a geometric figure defined to be the locus of all points for which the sum of their distances from two fixed points (the foci of the ellipse) is a constant, usually denoted by $2a$; that is, $r_1 + r_2 = 2a$ as in Fig. 1. Just as a circular field may be laid out with a taut string, the ends of which are fixed to a peg at the center of the circle, an elliptical plot may

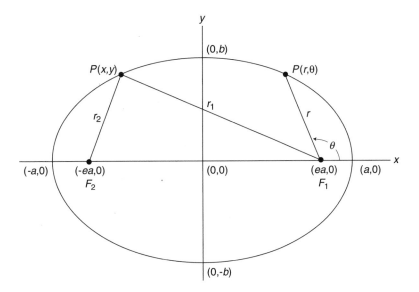

Figure 1   This figure defines the geometry of an elliptical planetary orbit. The ellipse is defined by the fact that the sum of the distances of $P(x,y)$ from the foci F_1 and F_2 is equal to $2a$ for all x and y on the ellipse. The planet polar coordinates used to derive the equation of the orbit are shown for the point $P(r,\theta)$.

be laid out with a taut string with ends fixed on two separate pegs, the foci. The diameter of the ellipse, which passes through the two foci, is called the major axis; the length of this axis is $2a$, and the distance from the geometric center through either focus to the perimeter is the semimajor axis a. The distance from the center to each focus is defined to be ea, where e is the eccentricity of the ellipse. The diameter that is perpendicular to the major axis is called the minor axis; its length b satisfies $b^2 = a^2 - (ea)^2$. A circular orbit is a special case of the general, elliptical trajectory when $e \rightarrow 0$.

It is well known from analytic geometry that the equation for an ellipse in terms of Cartesian coordinates x and y, which are respectively parallel and perpendicular to the major axis with origin at the center, is

$$\left(\frac{x}{a}\right)^2 + \left(\frac{y}{b}\right)^2 = 1. \tag{1}$$

If the ellipse is referred instead to an origin at a focus and expressed in the polar coordinates r, the distance to the perimeter from the focus, and θ, the angle measured counterclockwise from the major axis, the equation is found to be

$$r = \frac{a(1 - e^2)}{(1 + e \cos \theta)}. \tag{2}$$

The minimum and maximum distances from the center of force to the orbiting mass are $a(1 - e)$ and $a(1 + e)$, respectively; these are the periapsis and apoapsis of the orbit.

The second law is illustrated in Fig. 2; the angles $\Delta\theta_1$ and $\Delta\theta_2$, representing angular displacements in a common time Δt at the two points in the orbit, are related by the fact that the two triangular areas are equal. Qualitatively, this simply means that the planets move more rapidly the closer they are to the Sun. Though it is tempting to say that the speed of the planet is inversely proportional to its distance from the Sun, it is only the component of the velocity perpendicular to the radius vector that is so constrained.

Kepler's third law was published in the *Harmonice Mundi* (1619):

3. The ratio of the square of a planet's period to the cube of its semimajor axis is a constant for all planets in the solar system. A similar relation holds for other systems (e.g., the moons of Jupiter) with different constants. [This relation is actually only approximately true.]

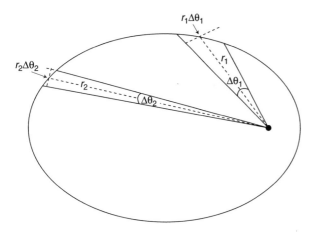

Figure 2 This figure demonstrates Kepler's second (equal areas) law. The speed of the planet as it moves in its orbit varies in such a way that the areas subtended in equal time intervals are equal.

The goal of Newtonian mechanics is derivation of these laws from basic physical principles, thereby relating the observational variables that describe the orbits a and e to dynamic parameters, such as the total energy and angular momentum. The problem of the interaction between two masses M and m due to their mutual gravitational attraction is best treated in the center of mass system, where the equation of motion is

$$\frac{\mu d^2\mathbf{r}}{dt^2} = \frac{-GmM\mathbf{r}}{r^3}, \qquad (3)$$

where μ is the reduced mass $Mm/(m + M)$, $\mathbf{r}$ is the vector from the Sun to the planet, and $r = |\mathbf{r}|$ is the separation between the Sun and the planet.

The second law follows immediately from the fact that there is no torque on the system $\mathbf{r} \times \mathbf{F} = 0$; this means that the angular momentum vector $\mathbf{L} = \mathbf{r} \times \mathbf{p}$ is a constant of the motion. Since the direction of $\mathbf{L}$ is constant and is always orthogonal to both $\mathbf{r}$ and the momentum vector $\mathbf{p}$, the motion is confined to a plane. Therefore, the problem can be expressed in plane polar coordinates r and θ, the angle that the radius vector makes with an arbitrary x axis. The tangential component of Eq. (3) expresses the conservation of the magnitude of the angular momentum:

$$|\mathbf{L}| = \frac{\mu r^2 d\theta}{dt} = L = \text{constant.} \qquad (4)$$

Referring to Fig. 2, the area element swept out by the radius vector in time dt is given by

$$dA = \tfrac{1}{2}r\,(r d\theta). \qquad (5)$$

Comparing Eqs. (4) and (5), the rate at which area is swept out is just $L/2\mu = \text{constant}$. Note that the second law depends only on the fact that this is a central force and not on the inverse square nature of the attraction.

Kepler's first law follows from the solution of the radial component of the equation of motion,

$$\frac{d^2r}{dt^2} - \frac{L^2}{\mu^2 r^3} = -\frac{G(M + m)}{r^2} \qquad (6)$$

where $d\theta/dt$ has been eliminated using the angular momentum conservation [Eq. (4)]. The equation for the orbit is $r(\theta)$, so the solution $r(t)$ of Eq. (6) is not needed explicitly. The chain rule of calculus relates differentiation with respect to θ to that with respect to t: $d/dt = d\theta/dt \cdot d/d\theta$. If, in addition, the dependent variable is chosen as $1/r$ instead of r, the equation of motion takes the following especially simple form:

$$\frac{d^2(1/r)}{d\theta^2} + \frac{1}{r} = \frac{G(M + m)\mu^2}{L^2}. \qquad (7)$$

This is the same equation that represents a vertical spring with a weight attached to it. The solution to this well-known equation is given by

$$\frac{1}{r} = A\cos(\theta - \theta_0) + \frac{G(M + m)\mu^2}{L^2} \qquad (8a)$$

or

$$r = \frac{L^2/G(M + m)\mu^2}{\{1 + [AL^2/G(M + m)\,\mu^2]\cos(\theta - \theta_0)\}}. \qquad (8b)$$

Comparing this to Eq. (2) above it can be seen that the orbit is in fact an ellipse; there are two integration constants, A and θ_0, which still must be determined in terms of physical parameters.

By choosing to reference the polar angle to the semimajor axis of the ellipse as in Fig. 2, the integration constant θ_0, which represents the initial orientation, may be set equal to zero. One combination of the orbital parameters can be identified by inspection:

$$a(1 - e^2) = \frac{L^2}{G(M + m)\mu^2}. \tag{9}$$

Also,

$$\frac{AL^2}{G(M + m)\mu^2} = e. \tag{10}$$

The other quantity which is needed to relate the orbital parameters to the physics is the total energy E. The energy may be determined by multiplying Eq. (6) by $\mu(dr/dt)$ and noting that the result may be written as

$$\frac{d}{dt}\left[\frac{1}{2}\mu\left(\frac{dr}{dt}\right)^2 + \frac{L^2}{2\mu r^2} - \frac{GMm}{r}\right] = 0. \tag{11}$$

The quantity in brackets is a constant of the motion; by inspection, it is the sum of the kinetic and potential energy and is therefore the total energy of the motion E. E may be related to a and e by evaluating Eq. (11) at the apsidal distances $r = a(1 \pm e)$ where $dr/dt = 0$:

$$E = \frac{-GMm}{2a}. \tag{12}$$

For bound motion $E < 0$, so the semimajor axis of the orbit is given by

$$a = \frac{GMm}{2|E|}. \tag{13}$$

Substituting this in Eq. (9) exposes the dependence of e on E and L:

$$e = \sqrt{1 - \frac{2|E|L^2}{(GMm)^2\mu}}. \tag{14}$$

Kepler's first law therefore follows directly from Newton's law of motion applied to his form for the gravitational force. It should be noted that Eq. (7) is valid for any central force if the right side of the equation is replaced by an appropriate function proportional to $r^2F(r)$; however, if the dependence of the force on r is more complicated than the Newtonian inverse square dependence, Kepler's first law

will no longer be valid in this form. For example, a gravitational force $F(r) = -GMm/r^2 + C/r^3$ leads to a coefficient of the $\theta - \theta_0$ term in Eqs. (7) and (8) that is different from unity. The elliptical orbit will no longer be closed, and the orbit will precess, that is, the celestial longitude of the perihelion of the orbit will itself rotate relative to an inertial coordinate system. Newton recognized that departures from the exact $1/r^2$ dependence would give rise to this effect and, in fact, attempted to calculate the well-known precession of the lunar apsides from the perturbations of the Sun's attraction on the two-body Earth-Moon system. The apsides of all the planets precess due to the actions of the other planets; an additional precession results in the general relativistic treatment of planetary orbits.

Kepler's third law can be derived from the preceding discussion. Recall that the second law may be expressed as $dA/dt = L/2\mu$. If this is integrated over one orbital cycle, the area of the elliptical orbit is simply proportional to the period τ, the time that it takes for one complete orbit. Using the fact that the area of an ellipse is given by πab,

$$\pi a^2\sqrt{(1 - e^2)} = \frac{L\tau}{2\mu}. \tag{15}$$

The eccentricity may be eliminated from this equation by using Eq. (14), and the magnitude of the total energy may be eliminated using Eq. (13). Making these substitutions it is found that

$$\tau^2 = \left[\frac{4\pi^2}{G(M + m)}\right]a^3. \tag{16}$$

This is not exactly Kepler's third law because the proportionality constant is not the same for the various objects in a system unless they all happen to have the same mass m. However, the mass of the largest planet, Jupiter, is only about one thousandth of the solar mass, so the errors in assuming that the ratio $\tau^2/a^3 = $ constant are very small. The relation works even better for the moons of Jupiter, whose masses are all less than $.0001M$.

If both a and τ are known from observations, Eq. (16) can be used to find $M + m$; the most accurate planetary masses have been determined using observations of natural satellites and assuming $m \ll M$. For example, the discovery of Charon, a satellite of Pluto, in 1978 allowed the first accurate determination of the mass of that planet-satellite system. Kepler's laws apply equally well to other celestial systems, such as binary stars, where the masses of the two objects are more nearly equal. In this case, the motions of both objects relative to the straight line motion of the center of mass can often be observed. The motions of the two objects relative to the center of mass are still elliptical, but the semimajor axes are reduced by factors of $M/(M + m)$ and $m/(M + m)$, respectively. In this case, the period of the motion will give $M + m$; and, if it is a visual binary (individual motions of the two stars are observable), the ratio of their maximum displacements relative to the center of mass can be deduced from the observations, giving the ratio of the masses. This is the most rigorous way in which masses of stars can be determined. So Kepler's laws, far from being historical anachronisms, are fully relevant to and used in astronomy today.

See also: APOGEE AND PERIGEE; CENTER-OF-MASS SYSTEM; KEPLER, JOHANNES; MOMENTUM; NEWTON'S LAWS

Bibliography

MARION, J. B., and THORNTON, S. T. *Classical Dynamics*, 3rd ed. (Harcourt Brace Jovanovich, San Diego, CA, 1988).

PHILIP B. JAMES

KERR BLACK HOLE

See BLACK HOLE, KERR

KERR EFFECT

The Kerr (electro-optic) effect is the occurrence of double refraction in a substance placed in an electric field.

The effect was first demonstrated by John Kerr in 1875 using a block of glass in an external electric (E) field between crossed polarizers in an arrangement shown schematically in Fig. 1. The physical origin of the Kerr effect is not the mechanical strains induced in a solid by the imposed electric field because the effect also occurs for liquids such as water (and molecular gases) whose molecules can be oriented by the E field. For liquid and gas samples, the E-field plates would be placed inside the sample. The incident light is unpolarized so its electric field vectors (shown by arrows) are in all directions. After the polarizer, the light is linearly polarized so the electric field vectors of the transmitted light are in one plane. The analyzer is oriented so that its transmission plane is perpendicular to that of the polarizer, reducing the exiting light to zero—*unless* the optically active substance in the sample cell affects the light so that the electric field vector of the light exiting the sample has a component along the analyzer transmission axis. Note that the Kerr effect does not rotate the plane of polarization of the light passing through, but modifies the character of polarization through the action of the external electric field on the sample.

The E field between the E-field plates (produced by applying the high voltage across the plates) interacts with the sample to align its molecules. This causes the index of refraction n (equal to the speed

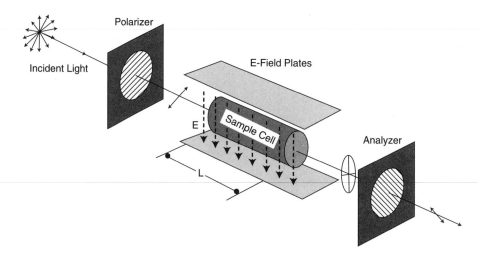

Figure 1   Schematic to illustrate the set-up for Kerr's original experiments.

of light in a vacuum divided by the speed of light in the sample) to become directionally sensitive (birefringent). The linear polarized light enters the sample cell at a 45° angle to the plane of light polarization. The light electric-field vector is split into two equal parts, one along the E field, the other perpendicular to it. The two components of light traversing the sample cell travel at different velocities, producing a phase shift between the components of the polarized light. This phase shift in general makes the direction of the electric field vector spiral after exiting the sample, with its tip tracing an ellipse (elliptically polarized). Elliptically polarized light will always have a component parallel to the analyzer axis, so some will be transmitted through the analyzer (the light exiting the analyzer is again plane polarized). The phase shift of the light exiting the sample depends linearly on the length of the sample L, the Kerr constant K for the sample, and quadratically on the strength of the electric field E. The Kerr effect can be used to produce an electro-optic shutter to modulate exit light intensity, even to shutting it off to produce a light pulse. If the sample molecule alignment time is extremely short (sub-nanosecond), as it is for some liquids like nitrobenzene, the shutter speed can be made extremely fast, sufficient to be used in speed-of-light measurements.

See also: BIREFRINGENCE; FIELD, ELECTRIC; POLARIZED LIGHT; POLARIZED LIGHT, CIRCULARLY

Bibliography

NELSON, D. F. "The Modulation of Laser Light." *Sci. Am.* **218** (June), 17–23 (1968).
PEDROTTI, F. L., and PEDROTTI, L. S. *Introduction to Optics,* 2nd ed. (Prentice Hall, Englewood Cliffs, NJ, 1993).

GEORGE BISSINGER

KINEMATICS

The subject in mechanics dealing with the description of motion is called kinematics. Kinematics describes the position coordinates of a moving object as they change with the time t. In the simple case of motion along a coordinate x, described by a continuous function $x(t)$, one can define the velocity v and acceleration a as the first and second rates of change of x with t. Given a displacement Δx over a time interval Δt, the velocity is obtained from the limit $v = \lim \Delta x/\Delta t$, as $\Delta t \to 0$, and therefore by the derivative $v = dx/dt$. For the acceleration, one writes that $a = \lim \Delta v/\Delta t$, as $\Delta t \to 0$, which gives $a = dv/dt = d^2x/dt^2$; higher derivatives may also be defined but are seldom needed in physical descriptions. The momentum of a particle with mass m is defined as $p = mv$, and its kinetic energy is defined by $K = mv^2/2$. Similarly, motion on a circle is described by an angle $\phi(t)$, angular velocity ω, and acceleration α. The moment of inertia of a particle rotating around a point at a distance r is $I = mr^2$, and its angular momentum is $L = I\omega$. The rotational kinetic energy is then $K_{rot} = I\omega^2/2$. More generally, motion in three dimensions is described in terms of the position vector $\mathbf{r}(t)$ with cartesian coordinates (x,y,z). The linear velocity is now a vector $\mathbf{v}$ with cartesian components (v_x, v_y, v_z), and the acceleration is written as $\mathbf{a}$, while the angular velocity and acceleration are $\boldsymbol{\omega}$ and $\boldsymbol{\alpha}$. The linear momentum is $\mathbf{p} = m\mathbf{v}$, and the angular momentum is given by the vector product $\mathbf{L} = \mathbf{r} \times \mathbf{p}$, which is perpendicular to both $\mathbf{r}(t)$ and $\mathbf{v}$ and therefore to the instantaneous plane of rotation. The magnitude of the velocity vector (a positive quantity) is called the speed and is given by $v = \sqrt{v_x^2 + v_y^2 + v_z^2}$, while the kinetic energy is again $K = mv^2/2$.

Kinematics deals only with the properties of motion, while dynamics describes motion under given forces and employs kinematics to solve equations of motion. In accordance with Newton's laws of motion, the acceleration of an object is proportional to the force acting on it. Here we present examples of motion of an object given its acceleration. When the acceleration is known, the equations of motion can be solved to obtain the object's position. The solution is a function of the time and of the initial values $(\mathbf{r}_0, \mathbf{v}_0)$ of the position and velocity at the initial time $t = 0$. Two simple examples of linear motions illustrate this. For motion with a constant acceleration a, the position changes with time as $x = x_0 + v_0t + at^2/2$. The second example is one where the acceleration is proportional to the displacement from an equilibrium position at $x = 0$ and leading to it, such as the acceleration imposed by a spring. Then the displacement from equilibrium is $x = C_0\cos(2\pi\nu t + \gamma_0)$, where C_0 and γ_0 are fixed by the initial conditions. This describes a harmonic oscillation of frequency ν. The nature of a motion can be studied as a function of initial conditions to find, for example, if it is regular or chaotic. Regular motion is defined as one where the trajectory changes by small values when the initial conditions are slightly changed. Chaotic motion results when small changes in the intital conditions lead to widely divergent (more precisely, exponentially divergent) trajectories. This occurs for example when a particle is accelerated toward two centers (such as a double well). In this case, small changes in initial conditions may leave the particle at one or the other attraction center at a given time. In an isolated system, with no external forces (and therefore for a vanishing total acceleration), three fundamental laws of mechanics state that the total linear momentum, the total angular momentum, and the total energy of the system are constant over time. These conservation laws have far reaching implications. The conservation of linear momentum explains why a rocket in space moves forward while its exhaust fuel moves backward. The conservation of energy was used to postulate the existence of the electron neutrino in the beta decay of radioactive nuclei; this was experimentally verified many years later.

An important aspect of kinematics is the description of motion as viewed from different reference frames. If an object at position P is referred to a frame in motion by the position vector $\mathbf{r}(t)$ and this frame is located with respect to a second (stationary) frame by the position vector $\mathbf{R}$, then the position of the object with respect to the stationary frame is $\mathbf{r}' = \mathbf{r} + \mathbf{R}$, and its velocity is $\mathbf{v}' = \mathbf{v} + \mathbf{V}$, showing that velocities add as vectors. A transformation like this is needed when two objects A and B with masses m_A and m_B are moving, to transform their velocities from a laboratory (stationary) frame to a center-of-mass (or baricentric) frame with its origin at the position $\mathbf{R} = (m_A\mathbf{r}_A + m_B\mathbf{r}_B)/(m_A + m_B)$. This is done in studies of collisions of two objects, such as particles or molecules. Since the center of mass moves at a constant velocity, properties such as the rate of collisions must be the same in both frames. A diagram relating the velocities of A (or B) in both frames is called the Newton diagram, and the transformation between velocities is used to relate the flux (number of A particles per unit time and unit area) calculated in the center-of-mass frame to the flux measured in the laboratory frame. Properties of a rigid object can be described in a frame fixed to space or in a frame fixed to the object and rotating with angular velocity $\boldsymbol{\omega}$ in the space frame. A property $\mathbf{G}$ changes at different rates when observed in each frame, and the rates are related by

$$\left(\frac{d\mathbf{G}}{dt}\right)_{\text{space}} = \left(\frac{d\mathbf{G}}{dt}\right)_{\text{body}} + \boldsymbol{\omega} \times \mathbf{G},$$

where the second term on the right is a vector product. This relation also applies to motion observed from two different frames when one is rotating with respect to the other. For example, when this formula is applied to the position of an object in a stationary (space-fixed) frame, the object's velocities as seen from a space-fixed and a rotating frame are related by $\mathbf{v}_s = \mathbf{v}_r + \boldsymbol{\omega} \times \mathbf{r}$. This in turn leads to new accelerations in the rotating frame, so $\mathbf{a}_s = \mathbf{a}_r + 2\boldsymbol{\omega} \times \mathbf{v}_r + \boldsymbol{\omega} \times (\boldsymbol{\omega} \times \mathbf{r})$, introducing Coriolis and centrifugal forces that derive from the second and third terms, respectively. The Coriolis force shows up in the rotation of the plane of oscillation of the Foucault pendulum and in the direction of circulation of cyclones.

Changes of frame play a central role in the formulation of the laws of physics. Properties of motion observed from two frames moving with respect to each other at constant velocity v are said to be related by a Galilean transformation. For motion along x, the transformation from the old position x to a new one, x', is given by $x' = x - vt$. This must be modified at speeds approaching the speed of light to account for the observed invariance of physical properties as seen from frames moving with constant velocity with respect to each other (the Einstein principle of relativity). This is accomplished by requiring invariance under the Lorentz kinematical transformation, which leaves the length of the quantity $x^2 + y^2 + z^2 - c^2t^2$ constant. As a result, the transformation of space and time variables becomes

$$x' = \frac{x - vt}{(1 - v^2/c^2)^{1/2}}$$

and

$$t' = \frac{t - xv/c^2}{(1 - v^2/c^2)^{1/2}},$$

and similarly for motion along y and z. Given two locations at a certain time, separated by a distance λ_0, the distance as seen from the moving frame is

$$\lambda = \lambda_0(1 - v^2/c^2)^{1/2}$$

and shows length contraction, while the time interval τ_0 at a location in a stationary frame is transformed into the longer interval in the moving frame,

$$\tau = \tau_0/(1 - v^2/c^2)^{1/2}$$

therefore showing time dilation. Relativistic kinematics is needed, for example, to interpret experiments in high-energy particle accelerators.

The kinematics of a system with many particles (such as a collection of electrons in an atom) or with many degrees of freedom (such as the bond distances and angles in a molecule) can be described in terms of generalized coordinates or curvilinear coordinates $q_1, q_2, \ldots, q_f$, all of which change with time. Motion is then pictured as happening on a hypersurface of f dimensions with momenta $p_1, p_2, \ldots, p_f$. The square of the distance between two neighboring points separated by the coordinate increments dq_1, $dq_2, \ldots, dq_f$ on the hypersurface is given by

$$ds^2 = \sum_{ij=1}^{f} h_{ij} dq_i dq_j,$$

where ds is an increment of length along the curve between the two points, and the set of coefficients h_{ij} define the metric on the hypersurface. The speed along the curve is then given by ds/dt. A transformation of coordinates can sometimes be found to impose that $h_{ij} = 0$, $i \neq j$, in which case one obtains an orthogonal set of curvilinear coordinates in which the kinematics is simplified. The set of all the coordinates and their corresponding momenta (a total of $2f$ variables) form the so-called phase space; trajectories in this space require knowledge of interaction forces and are therefore the subject of dynamics.

In quantum kinematics and mechanics, position and momentum functions are replaced by the operators $\hat{q}$ and $\hat{p}$, and the equations of motion must be solved with the commutation restriction $\hat{q}\hat{p} - \hat{p}\hat{q} = ih/(2\pi)$, where i is the imaginary number ($\sqrt{-1}$) and h is Planck's constant. In addition, it is necessary to include intrinsic coordinates, such as the spin (or intrinsic angular momentum) coordinates, and to describe the time evolution of spin states as well as spacial coordinates. This is done by following the evolution of quantum state functions over time.

Some physical systems can be described as containing an infinite number of degrees of freedom; this is the case when one introduces a continuum de-

scription of a fluid (gas or liquid) and follows its motion by giving the time evolution of a density $\rho(\mathbf{r}, t)$ defined at each point in the fluid. Properties are then found by integrating over the fluid volume. The path of an element of volume in the fluid is called a flow line, a collection of which form a flow tube enclosing a mass of fluid. The fluid flux $\mathbf{J}(\mathbf{r}, t)$ is the mass traversing a cross-section of the flow tube, per unit time and unit area. The conservation of mass, or continuity condition, relates the density and flux. When the flow lines do not change shape with time, the motion is a steady or laminar flow. At high flow speeds and in the presence of sharp boundaries, the motion may become turbulent, with the flow pattern being irregular and chaotic. Another aspect of the kinematics of fluids (and of wave motion in general) is the Doppler effect, which describes the change in the frequency of waves emitted by a moving object. Here the frequency is different in a frame moving with the source of wave motion from the frequency value measured by a stationary detector. This detector records a higher frequency as the wave source approaches and a lower frequency as it recedes.

The equations of motion satisfied by positions, velocities, and accelerations are the subject of classical, quantal, and fluid dynamics. Here, the forces are specified and classical trajectories or quantum states are found over time as functions of initial conditions. The dependence of classical trajectories on the initial conditions may vary from regular to irregular or chaotic; in the second case, one is then dealing with the kinematics of deterministic chaos. The study of chaotic motion in quantum mechanics is an active area of research.

See also: CENTER-OF-MASS SYSTEM; CHAOS; CONSERVATION LAWS; FLUID DYNAMICS; FRAME OF REFERENCE; FRAME OF REFERENCE, INERTIAL; FRAME OF REFERENCE, ROTATING; GALILEAN TRANSFORMATION; LORENTZ TRANSFORMATION; QUANTUM MECHANICS

Bibliography

DIRAC, P. A. M. *The Principles of Quantum Mechanics,* 4th ed. (Oxford University Press, Oxford, Eng., 1958).

GOLDSTEIN, H. *Classical Mechanics,* 2nd ed. (Addison-Wesley, Reading, MA, 1980).

LANDAU, L. D., and LIFSHITZ, E. M. *Fluid Mechanics* (Addition-Wesley, Reading, MA, 1959).

TABOR M. *Chaos and Integrability in Nonlinear Dynamics* (Wiley, New York, 1989).

DAVID A. MICHA

KINETIC ENERGY

See ENERGY, KINETIC

KINETIC THEORY

In the latter half of the nineteenth century the atomic theory of matter received acceptance, and macroscopic systems (i.e., gases, liquids, and solids) began to be studied as microscopic systems consisting of large numbers of atoms or molecules. Earlier in the nineteenth century the study of thermodynamics treated the macroscopic behavior of systems and made no assumptions about their microscopic structure. The microscopic approach began with the development of the kinetic theory of dilute gases. This theory applies the laws of mechanics to the individual atoms or molecules of a gas and uses the results to calculate the macroscopic parameters of a gas, such as its internal energy and pressure. The beginnings of this theory are found in the work of Daniel Bernoulli and John Herapeth, whose investigations established that the pressure of a gas resulted from the motion of its molecules and that the pressure could be calculated by considering the collision of molecules with the walls of a gas's container. James Joule, in the mid-nineteenth century, showed that the pressure of a gas is directly proportional to the square of the molecular speed, which he assumed to be the same for all molecules. In 1857 Rudolf Clausius derived the ideal gas law, which relates the macroscopic properties of a gas, by assuming the existence of a molecular speed distribution, and in 1859 James Clerk Maxwell developed a distribution law for molecular speeds that was based on the requirement that the equilibrium distribution of molecular speeds is unchanged under molecular collisions. In 1872 Ludwig Boltzmann generalized Maxwell's distribution to polyatomic gases, included external forces, and showed that his distribution, the Maxwell–Boltzmann distribution, is also stationary with respect to molecular collisions. Boltzmann then developed an integrodifferential equation for the distribution function of molecules in a system under nonequilibrium conditions. This equation can be used to calculate transport coefficients such as thermal conductivity, viscosity, and electrical conductivity. The kinetic theory of gases reached its present state in the early twentieth century through the efforts of Sydney Chapman and David Enskog, who developed systematic methods for finding solutions to the Boltzmann equation. The development of statistical mechanics quickly followed kinetic theory. Unlike kinetic theory, however, statistical mechanics ignores the detailed treatment of individual molecules and instead applies statistical methods to calculate the average properties of a large number of molecules.

Thermodynamics is unable to provide the equation of state for a given system. The equation of state is a mathematical statement of the interrelationships of the measurable macroscopic properties of a system. Kinetic theory can be used to obtain equations of state; therefore, reliance on empirical methods for developing interrelationships can be avoided. Although kinetic theory can be applied to many systems, the derivation of the equation of a particular system, the ideal gas, is usually discussed. The derivation of this equation of state is based on the following six assumptions:

1. A macroscopic volume of gas contains a large number of identical molecules (on the order of 10^{19} in a cubic centimeter at standard temperature and pressure) that are treated as hard spheres. These molecules are distributed uniformly throughout the container in the absence of external forces.
2. The average distance between molecules is large compared with a molecular diameter, which is on the order of 2 or 3×10^{-10} m.
3. There are no intermolecular forces except collision forces. In the absence of external forces the molecules are assumed to move in straight lines between collisions with other molecules and with the walls of their container.
4. The collisions are perfectly elastic.
5. The molecules are in perpetual random motion. There is no preferred direction for the velocity of any molecule.
6. There is a distribution of speeds for the molecules.

Under these assumptions, kinetic theory shows that

$$PV = \left(\tfrac{2}{3}\right) N \left[\left(\tfrac{1}{2}\right) m \overline{v^2}\right],$$

where P is the pressure of the gas, V is the volume occupied by the gas, N is the number of molecules contained in the volume, and $(\frac{1}{2}) m\overline{v^2}$ is the average kinetic energy per molecule. A comparison of this theoretical equation of state with the experimentally determined one

$$PV = NkT,$$

where T is the gas temperature and k is Boltzmann's constant, yields

$$\left(\tfrac{1}{2}\right) m\overline{v^2} = \left(\tfrac{3}{2}\right) kT.$$

The comparison shows that the average kinetic energy per molecule is directly related to the ideal gas temperature. This kinetic energy is kinetic energy of translation, since this is the only form of energy that can be attributed to a hard, spherical noninteracting molecule. The average square speed in the expression for the kinetic energy is given by

$$\overline{v^2} = \left(\frac{1}{N}\right)\int_0^\infty v^2 dN_v,$$

where dN_v is the number of molecules with speeds between v and $v + dv$. In order to evaluate this integral, dN_v needs to be known in terms of v. The relation between dN_v and v is given by the Maxwellian law of distribution of molecular speeds, which predicts an increase in the mean speed with temperature and a broadening of the speed distribution with temperature.

The previous discussion treats a system in thermodynamic equilibrium. The equation of state for an ideal gas describes the behavior of the gas in terms of macroscopic coordinates that do not depend on time. Classical thermodynamics is unable to deal with systems that are not in equilibrium; there may be an unbalanced force, a chemical reaction, or a temperature gradient within the system that changes the values of the macroscopic coordinates within the system. No single coordinate can then be used to describe the system as a whole. In contrast, kinetic theory is capable of dealing with systems not in equilibrium. An example of such a system is a rod where the two ends of the rod are held at different temperatures. Energy in the form of heat flows from the high temperature end to the low temperature end. The rate at which this occurs is found using the thermal conductivity of the rod's material. Kinetic theory can be used to calculate the thermal conductivity. Heat conduction is one example of a transport process, which can be defined as a process that results in a net flux of particles, energy, or momenta between two parts of a system. Some of the other transport processes that can be treated using kinetic theory are the mixing of molecules of two different momenta across a plane (laminar flow), the mixing of molecules of two different types across a plane (diffusion), the chemical mixing of two different kinds of molecules, and the motion of molecules escaping from a hole in a container (effusion).

These processes can be investigated using approximation methods, where the effects of intermolecular collisions on the molecular speed distribution are ignored. A more rigorous treatment includes the effects of binary collisions on the distribution, in addition to the effect of the force responsible for the transport process. The goal is to find a molecular distribution function that provides a complete description of the macroscopic state of the system. This distribution function will give the mean number of molecules that at a time τ are located in a given area with a speed in a given range. The function is used to calculate any transport coefficients of interest, such as viscosity and thermal conductivity. The Boltzmann equation is an integrodifferential equation for a molecular distribution function that is affected by both collisions and external forces. It involves both integrals and partial derivatives of the distribution function. Because of its complexity, the Boltzmann equation was not fully exploited until Chapman and Enskog developed systematic methods for obtaining its solutions. Since that time, however, it has become an invaluable tool for the study of the macroscopic properties of systems in nonequilibrium states.

See also: BOLTZMANN, LUDWIG; BOLTZMANN DISTRIBUTION; CLAUSIUS, RUDOLF JULIUS EMMANUEL; COLLISION; DIFFUSION; DISTRIBUTION FUNCTION; GAS; IDEAL GAS LAW; MAXWELL, JAMES CLERK; MAXWELL–BOLTZMANN STATISTICS; MAXWELL SPEED DISTRIBUTION; MOLECULAR SPEED; STATE, EQUATION OF; THERMODYNAMICS

Bibliography

GOLDEN, S. *Elements of the Theory of Gases* (Addison-Wesley, Reading, MA, 1964).

SERWAY, R. A. *Physics for Scientists and Engineers,* 3rd ed. (Saunders, Philadelphia, 1990).

CYNTHIA GALOVICH

KIRCHHOFF'S LAWS

Kirchhoff's laws, named after Gustav R. Kirchhoff, are two rules used to analyze electric circuits; one is called the loop law and the other is called the point law.

The loop law states that the algebraic sum of the potential differences across the elements of a circuit loop is zero. Figure 1 shows a circuit loop with two batteries and two resistors, and the corners of the loop are labeled with letters. For brevity, let V_{ba} represent the potential difference $V_b - V_a$ between corners a and b: $V_{ba} = V_b - V_a$, and use a similar notation for the other potential differences between corners. The potential difference V_{ba} is positive if $V_b > V_a$ and it is negative if $V_b < V_a$. For this loop, the loop law states that

$$V_{ba} + V_{cb} + V_{dc} + V_{ad} = 0.$$

Now use the loop law to find an equation for the current i in this circuit. The symbols R_1 and R_2 represent the resistances of the resistors, and ξ_1 and ξ_2 represent the emfs of the batteries. In Fig. 1, the positive sense for the current is arbitrarily chosen to be counterclockwise, as shown by the arrows. The loop law gives

$$-iR_1 + (+\xi_1) + (-iR_2) + (-\xi_2) = 0,$$

so that

$$i = (\xi_1 - \xi_2)/(R_1 + R_2).$$

Notice that if $\xi_1 > \xi_2$, then i is positive. This means that the sense of the current is the same as the sense that was chosen. On the other hand, if $\xi_1 < \xi_2$, then i is negative. This means that the sense of the current is opposite the sense that was chosen. Kirchhoff's loop law automatically yields the correct sense for the current regardless of which way is chosen at the outset. However, once the assigned sense is chosen, that choice must be used for the completion of the analysis.

Since the electric potential is related to the energy of the charge carriers in a circuit, the loop law is a reflection of the conservation of energy.

Kirchhoff's point law states that the algebraic sum of the currents toward a branch point (or junction)

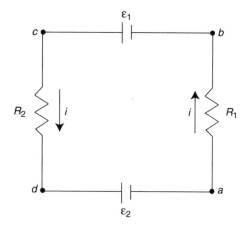

Figure 1   Circuit loop with two batteries and two resistors.

in a circuit is zero. In Fig. 2, currents i_1, i_2, and i_3 are in the wires that form a branch point. The sense of i_1 and i_3 toward the branch point and that of i_2 away from the branch point are assigned as shown by the arrows. The point law states that

$$i_1 - i_2 + i_3 = 0,$$

where currents toward the branch point are taken as positive and currents away from the branch point are taken as negative. The results of the analysis are unchanged by reversing this sign convention. Suppose wire 1 has a current of 6 A toward the branch point ($i_1 = +6$ A), and wire 3 has a current of 4 A away from the branch point ($i_3 = -4$ A). Then

$$i_2 = i_1 + i_3 = (6\,\text{A}) + (-4\,\text{A}) = 2\,\text{A}.$$

The current in wire 2 is 2 A away from the branch point.

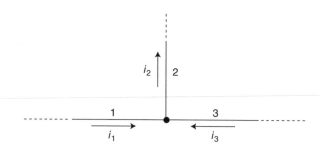

Figure 2   Wires forming a branch point.

The point law is a consequence of the conservation of electric charge because charge does not accumulate at the branch point of a circuit under steady state conditions.

In practice, the loop law and the point law are used together for the analysis of an electric circuit.

See also: CIRCUIT, PARALLEL; CIRCUIT, SERIES; ELECTRICAL RESISTANCE; ELECTRIC POTENTIAL

Bibliography

HALLIDAY, D.; RESNICK, R.; and WALKER, J. *Fundamentals of Physics,* 4th ed. (Wiley, New York, 1993).

KELLER, F. J.; GETTYS, W. E.; and SKOVE, M. J. *Physics: Classical and Modern,* 2nd ed. (McGraw-Hill, New York, 1993).

TIPLER, P. A. *Physics,* 2nd ed. (Worth, New York, 1982).

FREDERICK J. KELLER

L

LAMB SHIFT

The Lamb shift is named after American physicist Willis E. Lamb Jr., who, with his Ph.D. student Robert C. Retherford, discovered a small energy splitting $S(n = 2)$ between two of the $n = 2$ levels of the hydrogen atom. For this work Lamb was corecipient of the 1955 Nobel Prize in physics.

Consisting of one electron and one proton (nuclear charge number $Z = 1$) bound by Coulomb's electrostatic force, hydrogen is the simplest atom in nature. Precise experimental studies of its spectrum have been crucial for guiding and testing the development of quantal theories during the twentieth century. $S(n = 2)$ is the small upward shift (about 1,058 MHz in frequency units) of the $2^2S_{1/2}$ level with respect to the $2^2P_{1/2}$ level. The spectroscopic notation here labels levels by $n^{2s+1}L_J$, where for the atomic electron n is its principal quantum number, $s = 1/2$ is its spin-angular momentum quantum number, L is its orbital angular quantum number (with so-called $S, P, D, F, \ldots$ states corresponding, respectively, to $L = 0, 1, 2, 3, \ldots$), and $J = L \pm 1/2$ is its total (spin + orbital) angular momentum quantum number. Assume for a moment that the proton is a pointlike particle (root-mean-square charge radius $R = 0$) with no spin angular momentum ($I = 0$). (Actually, because the proton is a hadron, $R \neq 0$, problems arise; see below. That the proton also has a spin angular momentum of $I = 1/2$ causes additional atomic hyperfine energy splittings that can be calculated to high precision.) With these assumptions, solutions of Dirac's relativistic wave equation predicted that hydrogen atom states with the same value of n and J but different L should be degenerate (zero energy splitting). By the late 1930s weak experimental evidence suggested that the $2^2S_{1/2}$ and $2^2P_{1/2}$ states might not be degenerate, but the data were inconclusive. Just after World War II, Lamb and Retherford developed their experimental technique for the precise study of these states in an atomic beam. They exploited the microwave devices that had emerged from the intensive wartime technological development of radar and, by using electromagnetic fields to drive transitions between the $2^2S_{1/2}$ and $2^2P_{1/2}$ levels, clearly showed the energy splitting. Their clear experimental demonstration of the inadequacy of the Dirac equation in such a simple atomic system spurred development of the quantum field theory called quantum electrodynamics (QED). Using QED for perturbative calculations beyond Dirac theory of energy splittings in atoms caused by so-called radiative corrections results in expressions containing powers of $Z\alpha$, where $\alpha \simeq 1/137$ is Sommerfeld's fine-structure constant. Further work has brought ever increasing refinement in comparisons between experiment and theory. Of the several measurements of $S(n = 2)$ in hydrogen, the one with the highest claimed precision to date, by V. G.

Pal'chikov and others, is 1057.8514 ± 0.0019 MHz. (The quoted uncertainty is ± 1.8 parts per million.) Theoretical calculations of $S(n = 2)$ require various corrections, one of which depends on the experimental value of R for the proton. That the two different measurements of R disagree by 3.5 combined standard deviations means that there are two different "theoretical" values for $S(n = 2)$, clearly an undesirable situation. One disagrees with experiment by 2.9 combined standard deviations; the other disagrees by 6.1 combined standard deviations. Better measurements of R and so-called two loop binding corrections in QED calculated at order $\alpha^2(Z\alpha)^4$ and $\alpha^2(Z\alpha)^5$ will have to be completed before one may further assess the apparent disagreement between experiment and theory for $S(n = 2)$.

Generalizing the term "Lamb shift" to include other QED-induced energy shifts for bound electrons, many other Lamb shift measurements have also been done. If the simple Bohr theory of the hydrogen atom were correct, the $1S - 2S$ energy splitting would be exactly four times the $2S - 4S$, $4D$ splitting. In reality, this is not so. Careful laser-spectroscopic measurements have furnished precise determinations of the $1S$ Lamb shift. Because different terms in QED theory scale differently with Z, Lamb shift measurements in hydrogen-like ions with $Z = 2$ (He$^+$) up to $Z = 36$ (Kr^{35+}) have been used to test calculations of particular QED effects. Laser-spectroscopic experiments on hydrogen-like bismuth ($Z = 83$, Bi^{82+}) are underway. Laser-spectroscopic measurements in the "exotic" atoms positronium, the bound state of an electron and its antiparticle, the positron, and muonium, the bound state of an electron and a positive muon, use hydrogen-like systems consisting only of leptons (pointlike particles), but the short lifetime of these systems presents tough challenges to experiment and theory. Striking advances in the theoretical precision achieved in calculations of atomic structure for two-electron atoms (and ions) have allowed precise laser-spectroscopic measurements of their energy splittings to be used for determination of two-electron Lamb shifts. One expects the inexorable extension of this procedure toward precise experimental tests of Lamb shifts in even more complicated atomic, molecular, and other systems bound (predominantly) by Coulomb's electrostatic force.

See also: COULOMB'S LAW; QUANTUM ELECTRODYNAMICS; QUANTUM FIELD THEORY

Bibliography

HAGLEY, E. W., and PIPKIN, F. M. *Phys. Rev. Lett.* **72,** 1172 (1994).

HUBER, G.; KLAFT, I.; KÜHL, T.; and FRICKE, B. *Physica Scripta* **T58,** 58 (1995).

PAL'CHIKOV, V. G.; SOKOLOV, YU. L.; and YAKOVLEV, V. P. *Metrologia* **21,** 99 (1985).

PAVONE, F. S. *Physica Scripta* **T58,** 16 (1995).

SERIES, S. W., ed. *The Spectrum of Atomic Hydrogen: Advances* (World Scientific, Singapore, 1988).

PETER M. KOCH

LANDAU, LEV DAVIDOVICH

b. Baku, Russia (now Azerbaijan), January 22, 1908; *d.* Moscow, Russia, April 1, 1968; *solid-state physics, nuclear physics, kinetic theory, statistical mechanics, low-temperature physics.*

Landau was born into a secular Jewish family in Baku, an oil-rich Russian provincial capital on the shores of the Caspian Sea. His father was a prosperous engineer who managed a stock company in the oil business, while his mother was a respected physician and teacher who took an active role in the boy's education. Landau enjoyed a sheltered childhood that was disrupted only by the frequently shifting fortunes of Baku during the revolution and civil war that convulsed Russia beginning in 1917. Quickly outstripping local educational resources, in 1924 the sixteen-year-old Landau joined his older sister Sonya in St. Petersburg (then Leningrad), where he entered the Leningrad University physics department. It was there, in the company of other self-styled young iconoclasts like George Gamow, Dmitri Ivanenko, and Matvei Bronshtein, that the gangly prodigy wrote his first brief physics papers (mostly with Ivanenko) on aspects of the new quantum mechanics. The first truly significant contribution from these papers came in 1927 with Landau's introduction (at almost the same time as John von Neumann) of the important concept of the density matrix, which permits a simple description of a statistical mixture of states in a quantum mechanical system.

Upon completing his university studies in 1927, Landau began graduate work at the Leningrad Physicotechnical Institute, then the thriving center of Soviet physics. Soon singled out for his incisive critical abilities, he was sent abroad in the fall of 1929 for further training with the leading European theorists of the day, including Niels Bohr and Wolfgang Pauli. During this time he made the first of several influential contributions to solid-state physics, writing an article in which he employed quantum mechanical arguments for the presence of weak diamagnetism in metals. With Rudolf Peierls, he also composed a controversial paper on measurement problems in relativistic quantum theory, forcing Bohr to further articulate the dominant "Copenhagen" interpretation.

Fired by these encounters, Landau returned to Leningrad in 1931 with a keen sense of the many skills aspiring young Soviet theoretical physicists would require in order to make a name for themselves internationally. By the age of thirty, however, the impetuous Landau had succeeded in offending several important senior Soviet physicists with his sharp criticism of what he regarded as their inadequate standards for Soviet physics. Effectively "exiled" to the new Ukrainian Physicotechnical Institute in Kharkiv for his indiscretions, he embarked on the writing of a series of texts that would represent his vision of the theoretical physics discipline. Drawing on his lectures to students at Kharkiv University and the Institute of Mechanics and Machine Building, he began drafting with Evgenii M. Lifshitz a succession of volumes that would be completed only after his death. Ranging from the earliest text on classical statistical physics (1938), through treatments of mechanics, field theory, and the electrodynamics of continuous media, to the tenth and last volume on kinetic theory (posthumously completed by Lifshitz and Lev P. Pitaevskii), Landau's series has had no rivals in its technical scope, and to this day it remains influential among physicists worldwide.

The political persecution and purges that engulfed many Soviet citizens during the 1930s did not leave physicists untouched. Although politically articulate and like many of his younger peers optimistic about the prospects of Soviet socialism, Landau did not prove especially adept at the conformist behavior demanded by Joseph Stalin's regime. Local intrigues against him in Kharkiv eventually led to the arrest (and in some instances, execution) of several of Landau's closest associates. Banned from teaching at the university, Landau prudently sought another position, moving in early 1937 to Peter Kapitza's physics institute in Moscow, where he remained for the rest of his life.

That very year Kapitza began the experiments with liquid helium that would win him a Nobel Prize, but Landau's attentions were still devoted to work on second-order phase transitions in thermodynamics. He demonstrated that, unlike in the case of discontinuous first-order transitions (say, from liquid to gas), one can specify phase transitions that are continuous in quantities like energy or volume, but discontinuous in the symmetry properties of the body. Other original papers on the intermediate state in superconductors, the statistical theory of nuclei, the transport properties of interacting charged particles, and the production of electron showers in cosmic-ray physics testify to Landau's remarkable productivity amid the political turmoil of the time.

In early 1938 Landau published a short but provocative paper in which he speculated about the origins of stellar energy and helped make credible the idea of neutron stars (or neutron cores, as he called them). His article in *Nature* spurred J. Robert Oppenheimer to important calculations on this topic, but Landau himself was soon lost to further debate. In April 1938 he was arrested by the authorities on charges of "anti-Soviet activity." Fortunately for Landau the purges were then lessening in ferocity, but it was only after Kapitza courageously wrote appeals to Stalin declaring that the loss of Landau "would be sorely felt by Soviet science" that Landau was released into Kapitza's personal custody exactly one year after his arrest.

A subdued Landau immediately set to work interpreting Kapitza's experimental demonstration that liquid helium (^{4}He, a system of interacting bosons) loses all viscosity below 2.2 K, where it behaves as a superfluid. Employing phenomenological quasiparticles he called phonons and rotons, Landau was able to derive an energy spectrum based on the elementary excitations allowed in helium. In the spirit of quantum mechanics, Landau simply defined the quasiparticles in order to recover observable motions from his equations rather than identifiable portions of matter. The phonons represented irrotational states of potential flow analogous to classical hydrodynamics, while the rotons described vortex motion. The quantum hydrodynamics that Landau generated from this model permitted him to extract standard classical quantities like specific heat and entropy using the formalism of statistical mechanics. Landau's theory had competitors that claimed more fundamental, atomistic explanations of superfluid phenomena, and he had to continue refining his concepts in work published after World War II.

However, he was able to use his approach to predict subtle phenomena such as temperature waves, whose existence was soon confirmed by his colleague Vasilii Peshkov. In the 1950s he added several fundamental papers on the behavior of ^{3}He (a system of interacting fermions). For this work on condensed media Landau received the Nobel Prize in 1962.

In the decade after World War II, Stalinist xenophobia isolated Soviet physicists from their international colleagues, not least because many of the country's elite scientists were engaged in classified weapons work. Landau, though grudgingly involved in these projects, managed to publish original papers on plasma physics and other subjects related to thermonuclear research, but he also cast about for interesting topics far removed from what he referred to as the "noise" of bomb work. With Vitaly Ginzburg, he produced a phenomenological theory of superconductivity that was among the most influential predecessors to the microscopic theory of John Bardeen, Leon N. Cooper, and J. Robert Schrieffer in 1957. Drawing again on his mastery of relativistic hydrodynamics, Landau also revised and refined Enrico Fermi's 1950 application of statistical theory to high-energy particle collisions.

In the latter half of the 1950s, Landau returned to the persistent problems of quantum field theory, sharply criticizing inconsistencies in the renormalization formulation of the point-like electron in quantum electrodynamics. With several of his colleagues, Landau came to the paradoxical conclusion that there can be no physical interactions for point particles. He took this as a call to reject current theories as logically inconsistent, but a generation later physicists would point out that, save for a minor sign error in his arguments, Landau had seized on the very means of showing that a particular class of gauge theories confirmed the very renormalization program he had been trying to cast aside. His theoretical objections did not prevent him from contributing further vital concepts in this area, including the notion of combined conservation of charge and parity, and an explanation of the analytic properties of Feynman diagrams.

In January 1962, while traveling on icy roads to the high-energy physics center at Dubna north of Moscow, Landau was severely injured in an auto accident. Extreme measures eventually restored him to consciousness, but even with extended rehabilitative efforts Landau never recovered his full intellectual capacities. When he died six years later, his colleagues had already mourned the passing of one of the last theorists whose insights extended to nearly all branches of physics.

See also: BARDEEN, JOHN; BOHR, NIELS HENRIK DAVID; FERMIONS AND BOSONS; GAUGE THEORIES; LIQUID HELIUM; OPPENHEIMER, J. ROBERT; PAULI, WOLFGANG; QUANTUM ELECTRODYNAMICS; QUANTUM MECHANICS, CREATION OF; RENORMALIZATION

Bibliography

DOROZYNSKI, A. *The Man They Wouldn't Let Die* (Secker and Warburg, London, 1965).

GOTSMAN, E.; NE'EMAN, Y.; and VORONEL, A., eds. *Frontiers of Physics: Proceedings of the Landau Memorial Conference* (Pergamon, Oxford, Eng., 1990).

KHALATNIKOV, I. M., ed. *Landau: The Physicist and the Man* (Pergamon, Oxford, Eng., 1989).

LANDAU, L. D.; LIFSHITZ, E. M.; PITAEVSKII, L. P.; and BERESTETSKII, V. B. *Course of Theoretical Physics,* 10 vols. (Pergamon, Oxford, Eng., 1975–1987).

LIVANOVA, A. *Landau: A Great Physicist and Teacher* (Pergamon, Oxford, Eng., 1980).

LUTHER, A., ed. *Advances in Theoretical Physics: Proceedings of the Landau Birthday Symposium* (Pergamon, Oxford, Eng., 1990).

TER HAAR, D., ed. *Collected Papers of L. D. Landau* (Gordon and Breach, New York, 1965).

TER HAAR, D., ed. *Men of Physics: L. D. Landau,* 2 vols. (Pergamon, Oxford, Eng., 1965, 1969).

KARL HALL

LAPLACE, PIERRE-SIMON

b. Beaumont-en-Auge, Normandy, France, March 23, 1749; *d.* Paris, France, March 5, 1827; *celestial mechanics, optics, probability theory.*

Descended from a prosperous family of farmers and merchants, Laplace attended a Benedictine secondary school in Beaumont-en-Auge and the University of Caen. In 1768 he moved to Paris and immediately became a prolific author of memoirs in mathematics and celestial mechanics. He was elected to the prestigious French Academy of Sciences in 1773 at the relatively young age of twenty-four. He married Marie-Charlotte de Courty de Romanges in 1788, and they had two children.

Laplace served on numerous committees commissioned by the French government, including the one that designed the metric system of scientific units. In 1796 he became president of the scientific class of Napoleon's new Institute of France. He also carried out important administrative functions for the Ecole Polytechnique, the prestigious French engineering school founded in 1795.

Laplace's earliest mathematical interests involved the calculation of odds in games of chance. He stated what is now called Bayes's theorem after an early predecessor of Laplace. This theorem stipulates how to use partial or incomplete information to calculate the conditional probability of an event in terms of its absolute or unconditional probability and the conditional probability of its cause. Laplace also presented the least square law for the mean value of a set of data. More generally, Laplace provided an influential definition of probability. Given a situation in which specific equally possible cases are due to various processes (such as rolling dice) and correspond to favorable or unfavorable events, Laplace defined the probability of an event as the fraction formed by dividing the number of cases that correspond to or cause that event by the total number of possible cases. Laplace emphasized that application of probability theory is due to human ignorance. He described a supreme intelligence with a complete knowledge of the universe and its laws at any specific moment; for such an intelligence Laplace felt that probability calculations would be unnecessary since the future and past could be calculated simply through an application of the laws of nature to the given perfectly stipulated set of conditions. Laplace thus became a symbol of nineteenth-century scientific determinism, the idea that the uncertainty of the future is due only to human ignorance of the natural laws that determine it in every detail.

Laplace also emphasized the relevance of probability for statistical analysis of experimental data, such as astronomical observations. He specialized in the application of Newton's law of gravitation to the motion of comets and planets, especially Jupiter and Saturn. Laplace made major contributions to the solution of the resulting partial differential equations, including the famous techniques of potential functions and Laplace transforms to analyze force fields. Newtonian gravitation theory became Laplace's model for precision and clarity in all other branches of physics. He encouraged his colleagues to attempt similar analyses for optics, heat, electricity, and magnetism. Nevertheless, this attempt to base all of physics upon short-range forces yielded only limited success; aside from the mathematical methods he developed, Laplace's conceptual contributions to physics were not as long lasting as his more fundamental insights in probability theory.

See also: NEWTONIAN MECHANICS; PROBABILITY

Bibliography

CROSLAND, M. *The Society of Arcueil: A View of French Science at the Time of Napoleon I* (Harvard University Press, Cambridge, MA, 1967).

FOX, R. "The Rise and Fall of Laplacian Physics." *Historical Studies in the Physical Sciences* **4,** 89–136 (1974).

HAHN, R. *Laplace as a Newtonian Scientist* (Williams Andrew Clark Memorial Library, Los Angeles, 1982).

HOFMANN, J. R. "Pierre-Simon Laplace" in *Great Lives from History: Renaissance to 1900,* edited by F. N. Magill (Salem Press, Pasadena, CA, 1989).

JAMES R. HOFMANN

LARMOR PRECESSION

See PRECESSION, LARMOR

LASER

A laser is a source of coherent, monochromatic light. The word "laser" is an acronym for light amplification by stimulated emission of radiation, and is usually applied to such sources with wavelengths in the near-infrared, visible, and ultraviolet regions of the spectrum. Sources of coherent light in the longer wavelength microwave region are called masers. Lasers come in many varieties with continuous output to pulses of tens of femtoseconds, peak output powers up to Terawatts, and average powers up to kilowatts. Lasing can occur in many different materials and in all different phases, solid through plasma. The coherence and monochromatic properties allow energy to be propagated over large distances and to be focused to small spots.

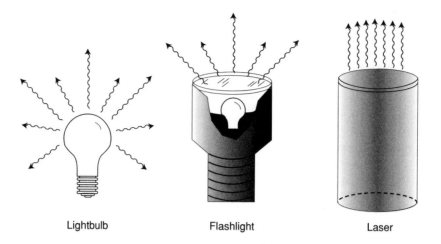

Lightbulb   Flashlight   Laser

Figure 1 Visual comparison of light emitted from a lightbulb, a flashlight, and a laser.

In Fig. 1, the light emitted from a laser is compared to that emitted from an incandescent lightbulb and a flashlight. An incandescent lightbulb emits light of many wavelengths (colors), which makes the light look white. The light is emitted in many directions. The directionality can be improved by adding a reflector as in a flashlight or searchlight. The emitted light emerges in a narrow cone but still contains many colors. In contrast, a laser emits light of a single wavelength, which is highly directional: compared to a flashlight it does not diverge at all.

Lasers are used in many areas of modern life, from industrial processing and welding to communications, medicine, entertainment, and scientific research. High-power lasers are used to focus large energy densities to small areas, causing metals to melt and either weld together or be cut. Advanced semiconductor lasers coupled to optical fibers can transmit communication signals many kilometers, including under the ocean. The coherence properties allow the information to be maintained over the long distances. Low-power lasers are focused to precisely controlled small spots in eye surgery. A laser has been reflected off of a mirror left on the Moon by the Apollo astronauts to measure the precise distance from Earth to the Moon, and diode lasers are used in compact disc (CD) players.

Stimulated emission is the heart of lasing action. Atomic, molecular, and solid-state systems are made up of a series of energy levels (states). A basic property of thermodynamics is that a system is typically in its ground state. If energy is added to the system, it can make a transition to an excited (high energy) state. It can then be stimulated to make a transition to a lower energy state by a photon with an energy equal to the energy difference between the upper and lower state. The atom emits a photon with properties identical to the photon that caused the transition. A laser pulse is built up of a series of coherent emissions.

For the stimulated emission process to lead to lasing, there must be more atoms in their upper state than in the lower state (population inversion), and there must be a sufficient thickness of material for high-quality lasing to develop. Spontaneous emission competes with stimulated emission to cause the atom to make a transition from the upper to lower state. Spontaneous emission causes photons to be emitted in random directions at random times and reduces the number of atoms in the upper state. An atomic transition that minimizes the amount of spontaneous emission is chosen. A photon can be absorbed by another atom, which is in the lower state, causing a transition back to the upper state. If the number of atoms in the upper state is larger than the number in the lower state, emission will be larger than absorption and lasing can occur.

Lasing starts with spontaneous emission. These photons stimulate the emission of further photons. This occurs in all directions. If the lasing system is significantly longer in one direction, then the spontaneous emission will build up preferentially in that direction by stimulated emission. In a typical laser, the lasing medium is placed in a cavity made up of a pair of mirrors. The mirrors make the effective length along the cavity axis many orders of magnitude longer than the other directions, so the lasing builds up in that direction. The energy builds up exponentially with length until the maximum output energy per unit area (saturation fluence) supported by the medium is reached. The laser light is taken out of the cavity, either by a partially transmitting mirror or by an optical switch (Pockels effect).

Typically, an atom will be found in its lowest energy state. If an atom is in an excited state, then it will tend to decay back to its lowest energy(ground) state by spontaneous emission. To move the laser into an excited state, energy must be added to the system. It is difficult to achieve a population inversion from the ground state. To overcome this difficulty additional excited states are used.

An energy level diagram for a three-level lasing system is shown in Fig. 2. State $|0>$ represents the lowest energy (ground) state of the system. States $1|1>$ and $|2>$ are excited states. In this system the lasing occurs between states $|1>$ and $|2>$. State $|1>$ is chosen to have a low probability of spontaneous emission to lower energy states while state $|2>$ is chosen to have a high probability of spontaneous emission to the ground state. Energy is added to the atomic system in its ground state, transferring some population to the state $|1>$. Population tends to remain in state $|1>$ until stimulated to make a transi-

tion to state $|2>$ while any population in state $|2>$ quickly decays back into the ground state $|0>$. In this way a population inversion is maintained between states $|1>$ and $|2>$, leading to the required conditions for lasing.

The wide variety of applications of laser systems is due, in part, to the many different media that support lasing. The first laser was demonstrated in a ruby crystal by Theodore H. Maimen in 1960. Since then, lasers have been demonstrated in plasma, gaseous, liquid, and semiconductor systems. In all cases, one must find a system that has a long-lived excited state and determine how to pump the population into that state. The solutions are different depending on the medium and the laser requirements. They depend on the characteristics of the lasing material, including the lifetime of the upper lasing state, the saturation fluence, the thermal properties, and the gain bandwidth. The upper-state lifetime against spontaneous decay determines optimal pumping conditions. Solid state materials typically have high saturation fluences. The thermal stability and thermal conductivity determine the average power that can be extracted from the medium without distorting it, and the gain bandwidth determines the tunability and the shortest laser pulse that the medium can support.

Solid-state lasing media such as ruby or neodymium glass are made up of a host material, sapphire or glass, with a few percent of dopant, chromium, or neodymium, in which the lasing occurs. The host material distorts the atomic energy levels leading to a broad energy absorption band. The system is typically pumped with a flashlamp, similar to a camera flashbulb. An electrical discharge in a tube of gas emits visible and ultraviolet radiation whose wavelength covers the energy absorption bands in the material. The light is collected and focused into the lasing medium, leading to a population inversion. Large energies can be coupled into the medium. The flashlamp pulse duration should be comparable to the lifetime of the upper level against spontaneous emission to maximize the population inversion. Some of the newer solid-state materials such as titanium sapphire have short upper-level lifetimes and must be pumped by another laser. This reduces the amount of excess heat deposited by parts of the flashlamp spectrum that do not match the absorption bands.

A second solid-state laser is a semiconductor, or diode, laser. Electron-hole pairs recombine at a *p-n* junction emitting radiation. If sufficient current is

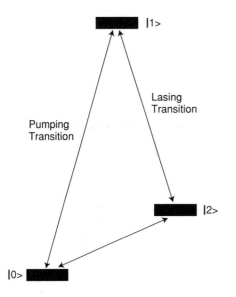

Figure 2 An energy level diagram for a three-level lasing system.

applied, this process can be stimulated, leading to lasing. The large index of refraction of the semiconductor material allows the material-air interface to act as a reflector. Diode lasers are electrically efficient and compact. This gives them many applications, including use in communications, where they can be employed to translate electrical into optical signals for transmission in optical fibers. A fiber laser consists of a solid-state lasing material such as erbium doped into a region of an optical fiber. The erbium is laser-pumped and amplifies a communication signal.

Gas lasers cannot be efficiently pumped by flashlamps because the gas density is too low to absorb the light effectively. These lasers are made up of two atomic species. An electrical discharge or electron beam passes through gas exciting one of the species. The excited state population is collisionally transferred to the second species, generating a population inversion. In a helium-neon laser, helium is excited by electron impact and the excited state population is transferred to the neon, leading to lasing in the neon. Carbon dioxide lasers operate with argon acting as the excitation medium pumping the carbon dioxide. Electrical discharges are efficient at transferring large amounts of energy to the lasing medium. Heat buildup can be mitigated by gas flow in the discharge chamber.

Excimer lasers are also discharge pumped. An excimer is a molecule that exists only in an excited state. The ground state is unstable to dissociation, with zero population, generating a population inversion. Energy is collisionally transferred from a buffer gas to the ions that make up the excimer, for example, krypton and fluorine. An excimer can be stimulated to make a lasing transition. Excimer lasers efficiently produce light in the near-ultraviolet wavelengths, making them ideal for lithographic applications.

The most common form of liquid laser is a dye laser. The gain medium is an organic molecule (dye) dissolved in a solvent. The molecules support lasing over a broad wavelength range, which can be further extended by changing the dye. Broad tunability and an ability to support either ultrafast pulses or narrow linewidths have given dye lasers a wide variety of research applications. Population inversion can also be created during chemical reactions where the resulting molecules are left in excited states.

Free electrons and ionized media called plasmas can also be made to lase. The advantage of these systems is that they can lase in spectral regions that are difficult to reach with other media, for example, in the vacuum ultraviolet and soft x-ray spectral region where most materials are opaque, or in the infrared where tunable materials are difficult to find. A free-electron laser has a beam of energetic electrons passing through a spatially oscillating magnetic field. This "wiggler" field causes the free electron to oscillate in a direction transverse to its propagation and radiate. The longitudinal motion of the electron maintains a phase coherence between the electron motion and the radiation leading to lasing.

A fully ionized medium, a plasma, can also lase. A plasma is created by an electrical discharge or by the interaction of an intense visible laser with the medium. Plasma, or x-ray, lasers have been made to lase at wavelengths as short as 44 Å, near the biological water window where high contrast imaging of biological cells is possible.

Lasers have been produced in many different materials, with many different characteristics, giving them a myriad of uses. New techniques continually extend their utility. For example, since the mid-1980s chirped pulse amplification has allowed ultrafast laser pulses to be produced with much higher peak powers than previously possible. Parametric processes involving the mixing of lasers in nonlinear optical materials have extended the tunability range of lasers continuously from the near ultraviolet to infrared regions of the electromagnetic spectrum. Higher power (both peak and average) lasers continually open new industrial applications for lasers.

See also: ENERGY LEVELS; LASER, DISCOVERY OF; MASER

Bibliography

MILONNI, P. W., and EBERLY, J. H. *Lasers* (Wiley, New York, 1988).

SIEGMAN, A. E. *Lasers* (University Science Books, Mill Valley, CA, 1986).

DAVID D. MEYERHOFER

LASER, DISCOVERY OF

A laser is a source of coherent, monochromatic light in the near-infrared, visible, and ultraviolet regions of the spectrum. The word "laser" is an acronym for light amplification by stimulated emission of radiation. A maser (microwave amplification by stimulated emission of radiation) emits coherent light in the longer wavelength, the microwave region of the electromagnetic spectrum.

In 1905 Albert Einstein proposed his light-quantum hypothesis, and in 1913 Niels Bohr published his atomic model in which electrons undergo transitions between stationary states with the simultaneous emission or absorption of radiation. In 1916 Einstein examined these processes in much greater detail. By considering the thermodynamic equilibrium between atomic states in combination with Max Planck's blackbody radiation law, he evaluated the probability of emission and absorption of light quanta (now called photons) in the transitions between energy levels. A system in its lower energy state makes a transition to an excited (high energy or upper) state through the absorption of a photon. If a system is in an excited state it can make a transition to a lower state through the emission of a photon in two ways. It can decay spontaneously, emitting a photon in a random direction, or it can be stimulated to decay by a photon that has an energy exactly equal to the difference between the upper and lower states. In the process, the atom emits a photon that is coherent with the stimulating photon and whose properties are identical to it, including wavelength, direction, and phase. The stimulated emission process is shown schematically in Fig. 1. State $|1\rangle$ represents the upper state and $|2\rangle$ represents the lower state. Einstein calculated that the stimulated emission process was the inverse of photon absorption and the transition probabilities were equal (Einstein B coefficient). Spontaneous emission was governed

by a second related probability (Einstein A coefficient). Molecular, solid-state, semiconductor, and many other systems are also made up of energy levels, and the concepts of stimulated and spontaneous emission apply as well.

In lasing or masing, the medium acts as a photon amplifier by stimulated emission. A single-input photon causes a second identical photon to be emitted. These two photons can stimulate two more, and thus the number of photons can grow exponentially. The photons can also be absorbed by atoms in their lower state. For the stimulated emission process to lead to lasing, there must be more atoms in their upper state than in the lower state (population inversion). In addition, the medium must be long enough for high-quality lasing to develop. Spontaneous emission reduces the number of atoms in the upper state, so the medium is chosen to minimize it.

A maser that produces coherent emission in the microwave region of the spectrum was predicted by Charles H. Townes in 1951 and was demonstrated in ammonia vapor by Townes, James P. Gordon, and Herbert J. Zeiger in 1954. Nikolai G. Basov and Aleksandr M. Prokhorov also proposed a maser at about the same time. The earliest masers produced an output of approximately 10^{-10} W at a frequency of approximately 24,000 MHz (~1 cm wavelength) with a frequency stability of better than one part per billion. The frequency stability is related to the low noise amplification properties of the maser. The primary application of these masers was as stable, low-noise, radio frequency sources or clocks. Power output from masers was increased to the microwatt range by using solid state materials.

In 1958 Arthur L. Schawlow and Townes predicted that masing action could be extended to the visible and infrared region of the spectrum (laser or optical maser). They considered both a gas vapor laser pumped by a gas discharge lamp and a solid-state lasing medium, but did not propose how to pump the solid-state medium. Two major problems of going to shorter wavelengths are that the size of a single mode cavity becomes smaller and the rate of spontaneous emission relative to the rate of stimulated emission increases as the third power of the inverse wavelength. A larger cavity that can maintain single lasing mode by using two highly reflecting, parallel mirrors was proposed by Schawlow and Townes, by Prokhorov, and by Robert H. Dicke.

The first laser (or optical maser) was demonstrated by Theodore H. Maimen in 1960 in a ruby crystal. Ruby is a sapphire crystal (Al_2O_3) doped with

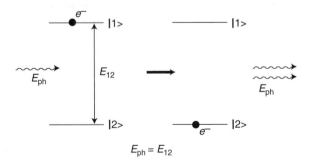

Figure 1 Schematic depiction of the stimulated emission process.

a small amount (~ 0.05%) of chromium. The optical system was pumped by a helical xenon flashlamp that was actuated by discharging a capacitor bank. The xenon discharge has strong emission bands that overlap with the spectral absorption bands of ruby. The crystal was a polished rod with reflecting ends that acted as the laser cavity. The laser pulse had a wavelength of 6943 Å.

In 1961 lasing was obtained by passing a discharge through a gas of helium by Ali Javan, William R. Bennett, and Donald R. Herriott. The gas mixture had approximately 90 percent helium and 10 percent neon. The discharge excites the helium, which collisionally transfers its excited population to the neon, creating a population inversion and subsequent lasing. The advantage of a discharge pumped system is that it can be operated continuously rather than in a pulsed mode. The first helium-neon laser produced an average power of 1 ~ 4 mW at wavelengths between 11,000 and 12,000 Å.

Since 1960 many advances in laser technology have taken place, extending the wavelength range, energy, power, and range of pulse durations. Lasing has been observed in solid state, semiconductor, liquid, gaseous, chemical, plasma, and free electron systems.

See also: EMISSION; LASER; MASER; PHOTON

Bibliography

BERTOLOTTI, M. *Masers and Lasers: An Historical Approach* (Adam Hilger, Bristol, Eng., 1983).

BROMBERT, J. L. *The Laser in America, 1950–1970* (MIT Press, Cambridge, MA, 1991).

HECHT, J. *Laser Pioneers* (Academic Press, Boston, 1992).

DAVID D. MEYERHOFER

LASER, ION

Lasers are devices that can generate powerful and intense beams of coherent light. The acronym "laser" has been derived from a description of its fundamental process: light amplification by stimulated emission of radiation. The following description will focus on ion lasers and in particular on the challenges of their design and manufacturing.

Ion lasers are gas lasers in which noble gas ions constitute the light amplifying laser medium. An arc discharge is generally used to ionize gases such as argon and krypton. Argon ion lasers and krypton ion lasers have rapidly gained commercial importance since their discovery in the early 1960s. They owe their success to the fact that they generate very powerful visible and ultraviolet light beams in continuous wave operation. Continuous wave operation means that an uninterrupted laser beam is provided. In contrast, a pulsed laser provides flashes of light that last for a fraction of a second only. Argon ion lasers emit laser light primarily in the blue and green spectral regions, and commercial argon ion lasers exceed 40-W output power levels. Krypton ion lasers offer strong emission in the red, yellow, green, and violet spectral regions.

Both types of ion lasers also offer strong laser emission in the ultraviolet. The powerful blue-green laser emission makes ion lasers ideally suited to optically excite, or pump, other lasers such as dye lasers or titanium sapphire lasers. These types of lasers are very important tools for a vast variety of research applications in physics, chemistry, biology, materials science, and medicine. The argon ion laser has also become an indispensable tool in medical applications for a variety of eye surgeries. The powerful ultraviolet emission of argon and krypton ion lasers has opened up a plethora of lithography applications in semiconductor manufacturing and general printing applications. Here, ion lasers enable the creation of feature sizes as small as the wavelength of the ultraviolet laser emission itself. This makes it possible to create and inspect new generations of semiconductor circuits with lines as narrow as one-third of a micrometer (0.33 μm). The combination of blue and green laser light from an argon ion laser with red and yellow from a krypton ion laser provides ideal conditions for projection of brilliant large-format multicolor displays and large-format, high-definition laser videos.

Design and manufacturing of ion lasers provides formidable challenges that reach far beyond the disciples of laser physics. Reliability combined with performance and excellent manufacturing repeatability become paramount for any commercial ion laser. The following discussion will focus on the two most important components of an ion laser. These are the plasma tube, which generates the amplifying laser medium, and the optical resonator, which provides the optical feedback necessary to sustain laser oscillation.

The plasma tube is the device that allows the operation of the arc discharge, which creates and sustains the lasing conditions. Ion lasers require very high discharge current densities in the order of several hundred to one thousand Ampère per square centimeter to excite laser emission. Ultraviolet laser emission requires especially high current densities. The generation of a 40-W laser beam out of an argon ion laser typically requires on the order of 40 kW (40,000 W) of electrical energy. This means that an argon ion laser at best operates with an efficiency on the order of only 0.1 percent for the generation of visible laser light. In the ultraviolet region the efficiency is even smaller and reaches only 0.025 percent for the generation of only 10 W of power. In general, the input power requirements will go up even more if the generation of ultraviolet laser radiation with even shorter wavelength is desired. In this sense, an ion laser plasma tube has to be viewed primarily as a gigantic heating element. At least 39,960 W of the electrical power will be converted into heat by the plasma tube. This is a very important factor that has to be considered in the design of the plasma tube and the choice of materials used. Despite its extremely low efficiency as a light source, no other light source can rival the ion laser in regards to the specific properties of the laser light, which enables focused spot sizes on the order of the wavelength of the laser light, its coherence and monochromaticity, and its capability to travel as a tightly collimated beam with a divergence of less than 0.0005 rad. The latter means that a beam with a 2-mm diameter will have expanded to only 7 mm in diameter after a distance of 10 m and to only 52 mm at a distance of 100 m from the laser.

Another important requirement placed upon an ion laser plasma tube arises from the fact that laser emission not only requires very high discharge current densities but also very low argon and krypton gas pressures, on the order of only 25 Pa. Ambient atmospheric pressure, in comparison, is on the order of 100,000 Pa. This means that a plasma tube has to be constructed as a vacuum vessel capable of maintaining ultrahigh vacuum conditions over its entire operating life of thousands of hours.

An ion laser plasma tube also needs to be constructed with a narrow bore (a diameter of 2 to 3 mm) to allow up to 1,000 A/cm^2 of current densities with commonly available power services. Maximum discharge currents are also limited by the availability of high current, by high voltage semiconductor components needed for the ion laser power supplies, by practical wire gauges, and by the cathode needed to operate the arc discharge. Typically, ion laser power supplies can supply up to 65 A dc current at a discharge voltage of up to 620 V dc to the plasma tube. The most challenging part of the bore construction is the choice of materials. They need to be able to handle high temperatures and thermal loads of up to 40 kW while being compatible with ultrahigh vacuum conditions. Two approaches have proven successful. The first and oldest approach employs the use of a continuous bore made from berylliumoxide ceramic. The use of an electrically nonconductive material is essential for a continuous bore construction because it is not possible to sustain an arc discharge inside a long tube of an electrically conductive material like a metal. Berylliumoxide is an almost ideal material for ion laser bores since it offers excellent heat conductivity besides being an excellent insulator. The high heat conductivity is very unique among ceramics and insulators in general. Below 0°C the heat conductivity of berylliumoxide even exceeds that of copper. Berylliumoxide has also demonstrated excellent resistance and stability against sputtering from the arc discharge. This makes berylliumoxide the material of choice for small, compact air-cooled ion laser plasma tubes. In this case, the plasma tube operating life easily exceeds 24,000 hours, which is almost three years of uninterrupted operation. Unfortunately, berylliumoxide can pose a serious health hazard in its powder form during the production of the ceramic bore components. This has forced the development of the second practical type of ion laser plasma tube, generally known as segmented disk metal ceramic tubes. Here, the discharge bore is formed by metal apertures of the desired bore size that are closely spaced along the plasma tube but kept electrically isolated from each other. This forms a quasi-continuous discharge bore. Typically, this bore is defined by tungsten disks of 1-mm thickness with a center hole of 2 to 3 mm in diameter. The disks are brazed into larger copper disks for structural support and enhanced heat conductivity. The copper disks in turn are brazed inside a tube of about 4-cm diameter made from aluminum oxide ceramic, which provides the structural support that keeps the copper disks spatially separated and electrically isolated. It also serves as the main part of the vacuum vessel. Tungsten is the material of choice here since it offers the highest melting point of any metal. This is needed to minimize erosion of the bore from the discharge and maximize the operating lifetime of the tube. The main challenges in this construction are related to the joining of dissimilar metals and metals and ceramics in a way that ensures excellent heat conductivity and vacuum integrity. The actual design of the bore shape has to take plasma and discharge uniformity and stability into account carefully to enable optimum performance and maximum plasma tube lifetime.

An ion laser plasma tube also requires a cathode and an anode to sustain the arc discharge. The cathode must be capable of supplying electrons to the discharge at currents of up to 65 A. This can only be accomplished practically by tungsten dispenser cathodes of the type commonly used in satellite radio transmitter tubes. These cathodes rely upon the fact that barium metal substantially lowers the work function of tungsten. This enables the efficient emission of electrons from the cathode surface at the required plasma currents of up to 65 A. Barium is formed through a chemical reaction from barium-calcium-aluminate, which is impregnated into the tungsten cathode structure. This process requires the cathode to be constantly heated to about 1,100°C, which is commonly accomplished by resistive heating. The anode is constructed from oxygen-free, high-conductivity copper.

Besides being able to sustain a low pressure noble gas arc discharge with high heat generation, an ion laser plasma tube also needs to provide an optical path along its axis, through the discharge region. This is necessary since the lasing process requires feedback from an optical resonator for self-sustained oscillation. The plasma tube thus also becomes an optical element inside the optical resonator. The optical resonator of an ion laser typically involves a curved and a flat mirror in Fabry–Pérot configuration. To allow for light passing through the discharge, while maintaining the vacuum integrity of the plasma tube, the tube is typically sealed off at its ends by Brewster windows or by resonator mirrors directly. Brewster windows are windows mounted under Brewster's angle. Under this angle the window will have 100 percent transmission for linearly polarized light, which is polarized parallel to the Brewster plane.

The main challenge from a laser optical perspective is to minimize any optical losses in this complicated laser resonator to achieve the maximum lasing performance and optical efficiency. Laser mirrors contribute losses in the form of absorption and scatter in the reflective dielectric thin film. High-quality laser mirrors today need to have total losses of less than 100 ppm (0.0001%). The plasma tube will contribute optical losses in a variety of ways. As transmissive optical elements, Brewster windows can contribute scatter losses due to surface conditions; absorption losses due to impurities in the bulk window material; and residual reflection due to the deviation from the exact Brewster angle. The straightness of the tube is also important. The design and manufacture of laser bores that remain straight under vastly different thermal conditions is a very challenging task. The gas discharge itself can become absorptive to the laser light because of plasma turbulences and impurities in the gas. The impurities can originate from the fill gas itself or from hydrocarbon residues left on the components used in the tube construction. This can only be avoided by employing ultraclean manufacturing techniques and careful selection of materials. Sophisticated spectroscopic techniques such as Fourier transform infrared spectroscopy and time-of-flight secondary mass spectroscopy become invaluable as production control and as problem-solving tools. Plasma tube assembly under class-100 clean room conditions is paramount.

In summary, an ion laser plasma tube could be described as a high-power heating element constructed from an ultraclean, ultrahigh-vacuum vessel, which also constitutes a very low loss optical element that allows powerful lasing inside an optical resonator by means of an argon or krypton arc discharge.

See also: BREWSTER'S LAW; FOURIER SERIES AND FOURIER TRANSFORM; LASER; LASER, DISCOVERY OF; OSCILLATION; PLASMA

Bibliography

HITZ, C. B. *Understanding Laser Technology: An Intuitive Introduction to Basic and Advanced Laser Concepts,* 2nd ed. (PennWell, Tulsa, OK, 1991).

SIEGMAN, A. E. *Lasers* (University Science Books, Mill Valley, CA, 1986).

WEBER, M. J. *Handbook of Laser Science and Technology,* Vol. 2: *Gas Lasers* (CRC Press Inc., Boca Raton, FL, 1982).

ALFRED FEITISCH

LASER COOLING

Laser cooling is the process of slowing, and thus cooling, atoms using laser light. The technique was first applied to ions (electrically charged atoms) held in electromagnetic traps and was later used on electrically neutral atoms. Remarkably low temperatures have been achieved using laser cooling; atoms have been cooled from room temperature to less than one millionth of a degree above absolute zero. Current research using laser-cooled atoms and ions includes the development of high-precision atomic clocks and fundamental studies of atomic properties. Also, interesting collective behavior has been produced, such as the crystallization of cold ions, and Bose–Einstein condensation of a vapor of ultra-cold neutral atoms. This condensation is a new form of matter where nearly all the atoms are in a single quantum mechanical state.

The Basics of Laser Cooling

A beam of laser light is made up of discrete particles or quanta called photons. The primary force used in laser cooling is the momentum transferred to an atom when photons scatter from it. This scattering is the process of an atom absorbing a photon, going up to an excited state, and then re-emitting a photon in a random direction; an atom absorbs a photon when the photon's energy is nearly the same as the energy difference between two atomic energy levels. The scattering force is similar to the force applied to a bowling ball when it is bombarded by a stream of ping pong balls. The momentum kick that the atom receives from each scattered photon is small; a typical velocity change is about 1 cm/s (room temperature gas atoms have typical velocities of several hundred meters per second). However, if the laser frequency is adjusted to match a strong atomic transition, it is possible to scatter more than ten million photons per second and produce accelerations of approximately $10,000 \times g$, where g is the acceleration due to gravity.

To cool a sample of atoms, their velocities must be reduced. The photon-scattering force can *push* atoms in a particular direction; however, by itself, this force will not *slow* a collection of atoms. Laser cooling is achieved by using the Doppler effect to make the photon-scattering force depend on the velocity of the atom. Just as the light emitted from a receding star is redshifted toward lower frequencies (longer, and thus redder wavelengths), an atom moving in a laser beam will see the frequency of the laser light shifted due to the atom's velocity in, or opposite to, the direction of the laser beam. The

basic principle is illustrated in Fig. 1. If an atom is moving in a laser beam, it will see the laser frequency ν_{laser} shifted by $(V/c) \times \nu_{\text{laser}}$, where V is the component of the atom's velocity that is opposite to the direction of the laser beam, c is the speed of light, and ν_{laser} is equal to the speed of light divided

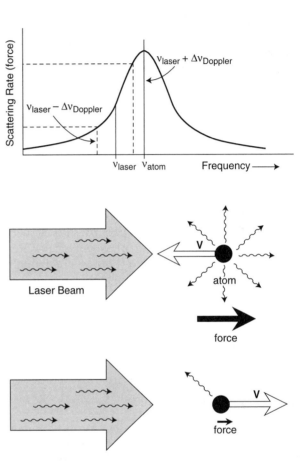

Figure 1 Velocity-dependent photon scattering force. The upper diagram shows the frequency dependence of the atomic excitation (scattering) rate, which is maximum at the frequency ν_{atom}. The laser frequency ν_{laser} is tuned to the low-frequency side of the atomic resonance. When an atom is moving in the direction opposite to the laser beam (middle of figure), it sees the laser frequency Doppler-shifted to higher frequency by the amount $\Delta\nu_{\text{Doppler}} = (V/c) \times \nu_{\text{laser}}$ and scatters many photons. When an atom is moving in the same direction as the laser beam (bottom), however, it sees the laser frequency Doppler-shifted to lower frequency and scatters very few photons. Thus the laser photons exert a much larger force if the atom is moving opposite to the laser beam.

by the wavelength of the laser light. Suppose that the laser frequency is adjusted so that it is slightly below the resonant frequency of the atom's transition between energy levels. As a result of this Doppler shift, the atom will scatter photons at a higher rate if it is moving opposite to the direction of the laser beam (V positive) than if it is moving in the same direction (V negative). Thus, although the photon-scattering force is always in the direction of the laser beam, it is strongest when the atom is moving in the direction opposite to the laser beam, which is when scattering force is slowing the atom down. Six laser beams are needed to cool atoms in all three dimensions (three pairs of oppositely directed beams at right angles to each other, as shown in Fig. 2). The combination of the forces from these beams slows the atoms regardless of the direction of their motion and thereby cools the atomic vapor. This technique has been given the descriptive name "optical molasses."

At certain laser frequencies, researchers found they could achieve atomic temperatures lower than can be explained by the Doppler cooling just described. This accidental discovery is now understood to arise from some very fortuitous atomic physics. As atoms move through the hills and valleys of potential energy produced by the intersecting laser beams, the atoms tend to make transitions between states in such a way as to efficiently transfer their thermal energy to the scattered photons.

Neutral Atom Trapping

Although atoms can be cooled by optical molasses, they will still slowly wander out of the laser beams. Holding the atoms in a particular place (trapping) can be accomplished in several ways. Ions, since they are electrically charged, can be easily trapped using electric and magnetic fields. Neutral atoms, however, are a more difficult problem. The most popular trap for neutral atoms is the magneto-optical trap shown in Fig. 2. It uses a combination of the photon-scattering force and a nonuniform magnetic field. By shifting the energy levels of the atom, the magnetic field regulates the rate at which an atom at a particular position scatters photons from the different beams. With the right arrangement of magnetic field and polarization of the laser beams, atoms are pushed to the location where all six laser beams intersect and the magnetic field is zero. As well as holding the atoms,

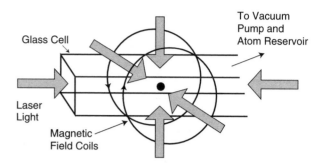

Figure 2 A magneto-optical trap in a glass cell. All of the laser beams are derived from the same laser. The nonuniform magnetic field is produced by running electrical current in the circular loops of wire shown.

this trap greatly increases the atomic density, since many atoms are pushed to the same point. The dipole-force trap is another type of laser trap that is less widely used. In this trap the atoms are drawn into the high-intensity region of a focused laser beam.

While laser cooling and trapping research in the 1970s and 1980s demonstrated tantalizing benefits, the size, cost, and complexity of the apparatus limited the potential applications. Recent simplifications in the technology, particularly the use of low-cost diode lasers operating in the near infrared, have dramatically changed this situation; laser cooling and trapping experiments are now being conducted in many laboratories and are even being carried out by college students.

See also: DOPPLER EFFECT; LASER; PHOTON; SCATTERING

Bibliography

CHU, S. "Laser Trapping of Neutral Particles." *Sci. Am.* **266** (2), 70–76 (1992).

GILBERT, S. L., and WIEMAN, C. E. "Laser Cooling and Trapping for the Masses." *Optics and Photonics News* **4** (7), 8–14 (1993).

PHILLIPS, W. D.; GOULD, P. L.; and LETT, P. D. "Cooling, Stopping, and Trapping Atoms." *Science* **239**, 877–883 (1988).

WINELAND, D. J., and ITANO, W. M. "Laser Cooling." *Phys. Today* **40** (6), 34–40 (1987).

SARAH L. GILBERT

CARL E. WIEMAN

LASER SPECTROSCOPY

See SPECTROSCOPY, LASER

LAWRENCE, ERNEST ORLANDO

b. Canton, South Dakota, August 8, 1901; *d.* Palo Alto, California, August 27, 1958; *accelerators, nuclear physics.*

Lawrence's ancestors immigrated from Norway; his father and grandfather were teachers in Minnesota and South Dakota. In his youth, he and Merle Tuve, a friend from Canton who also became a physicist, discovered ham radio and a love for tinkering. After graduation from high school at age sixteen, Lawrence attended St. Olaf's College for a year, transferred to the University of South Dakota to obtain his bachelor's degree, and in 1922 joined his boyhood friend Tuve at the University of Minnesota for graduate study under William F. G. Swann. Lawrence followed Swann to Chicago for his second year of graduate study and moved again with Swann to Yale, where he finished his third year in 1925 with a Ph.D. thesis on the photoelectric effect in potassium vapor.

Lawrence became frustrated with the Yale tradition that denied him graduate students and impatient with teaching beginning physics students. He left for an associate professorship at the University of California in Berkeley in 1928 and soon gathered a talented group of graduate students around him. The chance reading of an article by Rolf Wideröe early in 1929 led him to invent the apparatus for magnetic resonance acceleration that came to be called the cyclotron. His graduate student M. Stanley Livingston built the first successful one. Promoted to full professor in 1930, Lawrence began raising money to build larger and larger machines and founded the radiation laboratory at Berkeley that now bears his name.

Because of his obsession with improving the cyclotron to reach ever higher energies, Lawrence and his laboratory missed discovering, for example, artificial radioactivity and nuclear fission. His championing of a neutron less massive than the proton in 1933 also cost him credibility among his peers when it became clear that his targets had become contaminated with deuterium, leading to spurious results. But the group effort at his radiation laboratory presaged the ubiquitous development of large experimental groups in modern nuclear and particle physics. In a few short years, the Rad Lab had become a mecca for physicists similar to Niels Bohr's institute in Copenhagen and Ernest Rutherford's Cavendish Laboratory. Lawrence was generous with plans, advice, and help for physicists building cyclotrons elsewhere. His drive, enthusiasm, technical ability, and entrepreneurial fundraising led to continuing advances with the cyclotron and application to medicine. For his work, he received the Nobel Prize in 1939.

With the coming of World War II, Lawrence became one of the three civilian chiefs of the Manhattan Project. He was leader of the effort to achieve electromagnetic separation of ^{235}U from ^{238}U and to understand the behavior of plutonium. The return of peace enabled him to oversee the completion of the 184-in. cyclotron (his largest) in 1946, the Bevatron in 1954, and the Hilac in 1956. In addition to his own award, Lawrence's laboratory spawned Nobel Prizes for Glenn T. Seaborg, Edwin McMillan, Emilio Segrè, and Luis Alvarez. Lawrence's relationships with J. Robert Oppenheimer, his Berkeley colleague whom he had recommended for the directorship of the Los Alamos effort for the Manhattan Project, grew ever more difficult during and after the war and ruptured entirely when Lawrence refused to support Oppenheimer at his security clearance hearing in 1954.

Lawrence married Molly Blumer on May 14, 1932, and they had two sons and four daughters. In his later years, Lawrence became involved in the development of the color television tube and formed a group that patented many aspects of the process. Stress from his many responsibilities during the war years and his heavy schedule during the postwar period led to progressive ulcerative colitis complicated by atherosclerosis, which eventually led to his death at age fifty-seven.

See also: CYCLOTRON; OPPENHEIMER, J. ROBERT

Bibliography

CHILDS, H. *An American Genius: The Life of Ernest Orlando Lawrence* (Dutton, New York, 1968).

DAVIS, N. P. *Lawrence and Oppenheimer* (Simon & Schuster, New York, 1968).

HEILBRON, J. L., and SEIDEL, R. W. *Lawrence and His Laboratory* (University of California Press, Berkeley, 1989).

LIVINGSTON, M. S., ed. *The Development of High-Energy Accelerators* (Dover, New York, 1966).

GORDON J. AUBRECHT II

LEAST-ACTION PRINCIPLE

The least-action principle is an assertion about the nature of motion that provides an alternative approach to mechanics completely independent of Newton's laws. Not only does the least-action principle offer a means of formulating classical mechanics that is more flexible and powerful than Newtonian mechanics, variations on the least-action principle have proved useful in general relativity theory, quantum field theory, and particle physics. As a result, this principle lies at the core of much of contemporary theoretical physics.

We can understand this principle most easily if we study it first in a very simple context. Consider a point particle of mass m moving in one dimension along the x axis in response to specified external forces. Imagine that we know that at time t_1 the particle is at position x_1, and at time t_2 the particle is at position x_2 (see Fig. 1). We would like to find the particle's position $x(t)$ as a function of time between these end points; that is, its particular path on the diagram shown in Fig. 1. In principle, the particle might follow any number of paths: it might begin moving slowly but speed up as time passes (path a), begin moving quickly but slow down as time passes (path b), move from x_1 to x_2 at a constant speed (path c), or any of an infinite number of other possible paths (of which path d is an example). In practice, though, we know that the particle actually follows a single specific path $x(t)$ in response to the external forces acting on it. How does it know which of the many possible paths to follow?

The least-action principle proposes the following answer to this question: the particle follows whichever path between the end points has the least action. The action S for a given path $x(t)$ is defined by the integral

$$S \equiv \int_{t_1}^{t_2} L(v,x)\, dt, \qquad (1)$$

where the particle's Lagrangian $L(v,x)$ is a certain function of the particle's position x and its velocity x component v. Since how v and x depend on time depends on the particular path $x(t)$ that the particle follows, exactly how L depends on time will depend on the particle's path, and so the integral of L also will depend on the path. The least-action principle asserts that the path the particle actually follows is the one that yields the smallest value for this integral.

In classical mechanics, if the forces acting on the particle are conservative so that they can be described by some collective potential energy function $U(x)$, then it turns out that the particle's Lagrangian function is given by

$$L = T - U, \qquad (2)$$

where T is the particle's kinetic energy.

Let's see how the least-action principle works for a free particle ($U = 0$). Imagine that the particle moves from position $x_1 = 0$ at $t_1 = 0$ to $x_2 = b$ at $t_2 = T$. Consider first the path $x(t) = bt/T$, which takes the particle from x_1 to x_2 at constant velocity $v = b/T$ (see Fig. 2). According to Eq. (2), the Lagrangian for a free particle following this path is simply

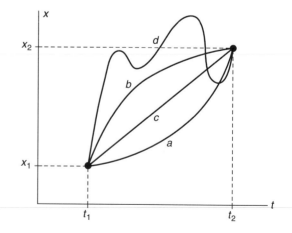

Figure 1 Some of the possible paths $x(t)$ that a particle might follow in going from position x_1 at time t_1 to position x_2 at time t_2.

$$L = T = \frac{1}{2}mv^2 = \frac{mb^2}{2T^2}. \qquad (3)$$

The action for this path is thus

$$S(\text{first path}) \equiv \int_{t_1}^{t_2} L\, dt = \int_0^T \frac{mb^2}{2T^2}\, dt = \frac{mb^2}{2T}. \quad (4)$$

Now consider the path $x(t) = bt^2/T^2$, which also takes the particle from $x_1 = 0$ to $x_2 = b$ in time T (see Fig. 2). For this path $v(t) = 2bt/T^2$, so

$$L = \frac{1}{2}mv^2 = \frac{2mb^2}{T^4}t^2 \Rightarrow S(\text{second path})$$

$$= \int_0^T \frac{2mb^2}{T^4}t^2\, dt = \frac{2mb^2}{3T}. \qquad (5)$$

Since S(second path) is greater than S(first path), the constant velocity path has the least action of the two, and the actual path of the particle will be more like that path than the curved path.

While the action for any particular path can be easily computed in this manner, calculating the actions for all possible paths to find the one with the least action would be impractical. Fortunately, it is possible to show (using techniques of the calculus of variations) that the path $x(t)$ of least action will be the one that satisfies the Euler–Lagrange equation

$$0 = \frac{\partial L}{\partial x} - \frac{d}{dt}\left(\frac{\partial L}{\partial v}\right), \qquad (6)$$

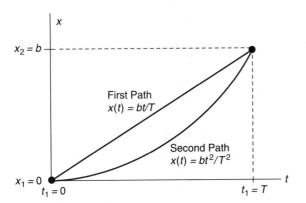

Figure 2 Two possible paths $x(t)$ that a particle might follow in going from position $x_1 = 0$ at time $t_1 = 0$ to position $x_2 = b$ at time $t_2 = T$.

where $\partial L/\partial x$ is the partial derivative of L with respect to x [found by taking the derivative of $L(v,x)$ with respect to x and treating v as a constant] and $\partial L/\partial v$ is the partial derivative of L with respect to v.

For example, in the case of a free particle ($U = 0$), $L = \frac{1}{2}mv^2$. This does not depend on x, so $\partial L/\partial x = 0$, but $\partial L/\partial v = \frac{1}{2}m(2v) = mv$. Plugging this into the Euler–Lagrange equations, we get

$$0 = \frac{\partial L}{\partial x} - \frac{d}{dt}\left(\frac{\partial L}{\partial v}\right)$$

$$= 0 - \frac{d}{dt}(mv) \Rightarrow mv = \text{constant}. \qquad (7)$$

So, in the case of the free particle the least-action principle implies that the particle will follow a path with constant velocity (the same result we would get from applying Newton's second law).

The advantage of the least-action approach over Newton's laws is that we can generalize it more easily to complicated situations. If the configuration of a system of particles subject to conservative forces can be described in terms of an independent set of numbers q_1, q_2,..., q_N (called generalized coordinates), then the system will evolve in such a way as to give the action

$$S \equiv \int_{t_1}^{t_2} L(\dot{q}_1, \dot{q}_2,..., \dot{q}_N, q_1, q_2,..., q_N)\, dt$$

$$\text{where } \dot{q}_i \equiv \frac{dq_i}{dt}, \qquad (8)$$

an extreme value, which actually may be either a maximum or a minimum (the "least-action principle" is somewhat of a misnomer). The action will have an extreme value if the coordinates q_i depend on time in such a way as to satisfy the N Euler–Lagrange equations

$$0 = \frac{\partial L}{\partial q_i} - \frac{d}{dt}\left(\frac{\partial L}{\partial \dot{q}_i}\right) \qquad \text{for } i = 1 \text{ to } N. \quad (9)$$

The generalized coordinates q_i used do not even have to be Cartesian coordinates: they can be any set of independent numbers that uniquely specify the configuration of the system. The least-action principle can even be extended to cover the time-evolution of fields by treating the value of the field at each point in space as a generalized coordinate.

Because of its power and flexibility, the least-action principle has become an important tool in contemporary theoretical physics. Einstein's theory of general relativity is founded on a principle very similar to the least-action principle: his geodesic hypothesis asserts that a particle in a gravitational field follows the path of extreme proper time. The time-evolution of a gravitational field in general relativity also can be described in terms of a least-action principle. In 1948 Richard Feynman showed that it was possible to express nonrelativistic quantum mechanics in terms of a generalized least-action principle: his "path-integral" approach quickly became accepted as the most natural path to relativistic quantum field theories of the electromagnetic, weak nuclear, and strong nuclear interactions. All modern treatments of quantum field theory are based on Feynman's least-action approach.

Technically, the "least-action principle" in classical mechanics refers to a somewhat different principle than described here. First proposed in 1747 (rather vaguely) by Pierre-Louis Moreau de Maupertuis, the original least-action principle was described in careful mathematical language by Joseph Lagrange in 1760. In 1834 William Rowan Hamilton first proposed the principle described in this article, which is called Hamilton's principle in most classical mechanics texts. The old least-action principle is more complicated and less flexible than Hamilton's principle and can be derived from it. Feynman's approach to quantum field theory actually generalizes Hamilton's principle, not the old least-action principle, but Feynman called it the "least-action principle" and the term has stuck. As a result, the "least-action principle" and Hamilton's principle have become informally synonymous, particularly in the literature on quantum mechanics.

See also: FEYNMAN, RICHARD PHILLIPS; NEWTON'S LAWS; QUANTUM MECHANICS AND QUANTUM FIELD THEORY

Bibliography

DAVIS, A. D. *Classical Mechanics* (Academic Press, Orlando, FL, 1986).

FEYNMAN, R. P. *QED: The Strange Theory of Light and Matter* (Princeton University Press, Princeton, NJ, 1985).

FEYNMAN, R. P., and HIBBS, A. R. *Quantum Mechanics and Path Integrals* (McGraw-Hill, New York, 1965).

GOLDSTEIN, H. *Classical Mechanics,* 2nd ed. (Addison-Wesley, Reading, MA, 1980).

MARION, J. B., and THORNTON, S. T. *Classical Dynamics of Particles and Systems,* 3rd ed. (Harcourt Brace Jovanovich, Orlando, FL, 1988).

THOMAS A. MOORE

LENS

A lens is a transmissive optical element or cluster of elements that redirects rays through an angle that varies continuously with position in at least one direction across the element. The rays can be light rays or particle trajectories. A lens can form an image of an illuminated object or can be used in the coupling or collection of rays. Lenses are often combined into more sophisticated optical devices for a variety of applications.

Lenses occur in a number of forms in nature. Indeed, some evolutionary biologists suggest that the lens may have independently evolved as many as seven separate times in the animal kingdom. The first man-made lenses could conceivably date as far back as 3500 B.C.E., when the Phoenicians began manufacturing glass. Perhaps the earliest historical reference to lenses is provided by Aristophanes who, in the fourth century B.C.E., wrote of "burning glasses" used to focus the rays of the sun into spots sufficiently hot to ignite paper. In the eleventh century, Ibn al-Ha'tham provided the first description of the operation of a lens, and in 1267 Roger Bacon described experiments with lenses. Particle lenses were developed in the first decades of the twentieth century and were first incorporated into an imaging device, the electron microscope, by Max Knoll and Ernst Ruska in 1931.

The spherical lens is a common optical lens made of a dielectric—usually glass—bounded by portions of two spherical surfaces. The line passing though the centers of both spheres defines the axis of the lens. The spherical lens is a refractive lens, since refraction redirects the rays of light. In the paraxial approximation that no rays are anywhere far from the lens axis, the laws of geometrical optics apply, and one can define a focal length f as the distance for which a parallel beam of light is focused to a spot as

$$\frac{1}{f} = (n' - 1)\left[\frac{1}{R_1} - \frac{1}{R_2} + \frac{(n' - 1)t}{n'R_1R_2}\right],$$

where the R_j refer to the radii of curvature of the two spherical surfaces. A radius is negative if the center of the sphere is on the same side of the lens as the incident light. The thickness of the lens along its axis is t, and $n' = n_{lens}/n_{medium}$ is defined in terms of the indices of refraction of the material the lens is made of, n_{lens}, and the medium it is immersed in, n_{medium}. A negative f means that the light becomes more divergent as it passes through the lens. The distance f is measured from the principal point located a distance $z = (1 - n')\ tf/R_1$ beyond the intersection of the final sphere with the lens axis, as shown in Fig. 1.

Spherical lenses are classified in terms of the curvature of their faces. A convex/convex lens is formed by two convex faces. Such a lens will cause a beam of parallel rays to come to a focus. A simple magnifying glass is an example. A concave/concave lens will cause an incident parallel beam to diverge. Such a lens might be used in eyeglasses to correct for near-sightedness. One of the radii of curvature of a spherical lens may be infinite. A plano-convex lens is an example.

If the paraxial approximation is violated, a parallel beam of light is not focused to a point and the lens possesses an aberration. These departures from ideal operation are approximated in terms of rays by the Seidel aberrations: spherical aberration, coma, astigmatism, distortion, and field curvature. They can be quantified in terms of diffraction by the Strehl ratio, the ratio of light within the central diffraction maximum to that of a perfect lens of the same diameter. Chromatic aberration results from differing focal lengths for light of different colors and can be present within the paraxial approximation. Lenses have been developed to reduce aberrations. The aspherical lens has two nonspherical surfaces that correct some aberrations. A compound lens that combines several simple lenses and internal light stops can be designed to reduce aberrations.

The natural tendency of a wave to diffract can be controlled to create a lensing effect. The simplest such diffractive lens is the Fresnel lens. Many Fresnel lenses resemble diffraction gratings in that they are comprised of a series of closely spaced grooves, although in the Fresnel lens the grooves are concentric circles. For parallel incident light, portions of the incident wave front at a Fresnel lens that will add to each other at the focal point pass through the lens unaffected. Those that would subtract from each other are either blocked or phase-shifted to add. Diffractive lenses have several focal points, one for each order of diffraction. One order can be enhanced relative to others by properly placed stops or by blazing the grating that forms the lens. In a blazed lens, the shape of the individual rulings comprising the lens is altered. Diffractive lenses have severe chromatic aberration. Fresnel-like lenses can be used to focus microwaves for wireless communications.

A graded index, or GRIN lens, exploits a continuously graded index within one solid unit. In a GRIN lens, light tends to bend toward the material with the higher index of refraction. GRIN lenses are usually made from radially symmetric layers, although a variation along the lens axis can be used. The lens surface may be flat or curved. GRIN lenses are widely used for coupling light into optical fibers.

Lenses for charged particles can utilize either an electric field or a magnetic field. In either case, the

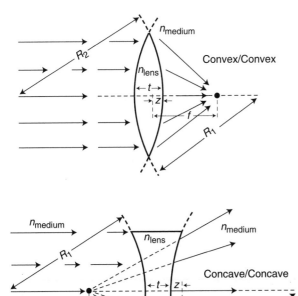

Figure 1 In the two spherical lenses shown, the distances f, t, z, and the R_j are indicated. Note the effect of the signs of z and f; we have assumed $n' > 1$. The dashed rays shown for the concave/concave lens are extrapolations of emergent rays and show the significance of its negative focal length f.

deflection is the result of electromagnetic forces from the position-dependent fields. Lenses based on electric fields can involve either apertures, such as the Pierce geometry for a charged particle source, or conducting cylinders, such as the three spaced cylinder einzel lens, in which a voltage ratio controls the lens strength. Magnetic lenses are classified by the number of poles used in their construction, such as the quadrupole lens with the ends of four magnets facing the beam. Electromagnets are normally used to provide a variable lens.

Some lenses have a special names and purposes. An objective lens is the lens of an imaging system that is closest to the object being viewed. A field lens is used to reduce the loss of light from off-axis points in an imaging system. It lies at or near an image plane, so it does not contribute significantly to the focal properties. An eyepiece usually contains both a field lens and a compound lens. A zoom lens is a compound lens that changes overall magnification without changing its effective focal length. A cylindrical lens focuses light in only one dimension, and thus is useful for coupling light through a slit. A solid immersion lens may be used to improve spatial resolution in a microscope that utilizes photon tunneling by resting the lens directly on the specimen to be imaged. A gravitational lens results from a general relativity effect and bends light near massive objects such as stars.

See also: ABERRATION; ABERRATION, CHROMATIC; ABERRATION, SPHERICAL; DIFFRACTION; MAGNIFICATION; OPTICS; REFRACTION; REFRACTION, INDEX OF

Bibliography

GUENTHER, R. D. *Modern Optics* (Wiley, New York, 1990).

HEAVENS, O. S., and DITCHBURN, R. W. *Insight into Optics* (Wiley, New York, 1991).

LOVELL, D. J. *Optical Anecdotes* (Society of Photo-optical Instrumentation Engineers, Bellingham, WA, 1981).

HANS D. HALLEN

MICHAEL A. PAESLER

LENS, COMPOUND

A compound lens consists of two or more simple lenses, either in optical contact or air spaced, which is used in an optical system as a single element. If the compound lens is air spaced, the spacing is treated as another optical element for analysis of the imaging properties of the lens. The reason optical designers use compound lenses is to reduce the imaging errors of simple lenses.

The dispersion, variation of the index of refraction with color, of the material making up a simple lens is the source of longitudinal and lateral chromatic aberrations. These aberrations reduce image sharpness by focusing rays of different colors to form images at different distances and with different magnifications. This effect is shown in Fig. 1. These aberrations can be reduced by making a compound lens of two materials having dispersions of opposite signs and by careful choice of the surface curvatures. With two constituent lenses, as in Fig. 2, one can make the focal length of the lens the same for two different colors and reduce the range of focal lengths for all colors. For a compound lens with three simple elements of different dispersions, the focal length for three colors can be made the same with a concomitant reduction of the variation of focal lengths for all colors.

The first-order theory of image formation by a lens, as exemplified in the lens equation, is accurate only for light rays, called paraxial rays, which travel very nearly along the optical axis and pass through the lens near its center. Since using only these rays would severely reduce the light flux through an instrument, such as a camera, microscope, or telescope, rays not parallel to the optic axis and at a distance from the lens center must be used. The resulting aberrations are called monochromatic aberrations. Spherical aberration is one of the monochromatic aberrations. The origin of spherical

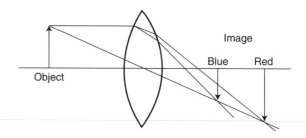

Figure 1 The difference in image positions along the optic axis is a longitudinal chromatic aberration; the difference in size is a lateral chromatic aberration.

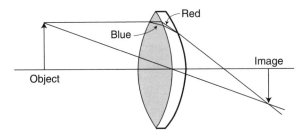

Figure 2 By choosing the curvatures and the materials, and thus the dispersion, of the component lenses in a compound lens the focal length for two different colors can be made the same. The resulting lens is called an achromat. Spherical aberration can also be reduced at the same time with careful choices of the indices of refraction.

aberration is apparent in Fig. 3, which shows rays from a point on the optic axis passing through the lens near the center forming an image farther from the lens than rays which pass through the lens near the edge. All methods of reducing spherical aberration are effective only for a limited range of object distances and object size. One correction method is to "bend" the lens, that is, to make the object side of the lens concave and increase the curvature of the image side to keep the same focal length. For one object point on the axis of such a lens all rays pass through the image point. In a system in which the object distance changes little the designer can use this method. The resulting bent lens can be composed of two or more separate simple lenses and the curvatures and dispersions of the simple lens components chosen to correct both the spherical and chromatic aberrations. Another way to cor-

rect spherical aberrations is to use nonspherical surfaces. This increases the difficulty in lens production but, with the use of present and improved nonspherical surface generating machines and polishers, may be expected to have increased application. Other monochromatic aberrations include coma, astigmatism, field curvature, and distortion. Coma is the spherical aberration effect for off-axis points where the image point for rays passing through different zones of the lens form image points at different distances from the optical axis so that the image is not a point but is smeared into a flare-like shape. Astigmatism is the aberration which causes the rays from an object point which pass through the lens in a plane containing the optic axis and the object point to form an image point at a different distance from the lens than rays in the plane perpendicular to that one. Curvature of field results in the images of object points, which are in a plane perpendicular to the optic axis, lying in a curved surface in image space. Distortion causes the image of a straight line in a plane perpendicular to the optic axis and at a distance from it to be imaged as a curved line. If the image is concave toward the optic axis the distortion is called barrel, a convex image indicates pincushion distortion. In correcting both chromatic and monochromatic aberrations of lenses, optical designers may use many components. It is not uncommon for a modern camera lens to have six or more simple lens elements made of three or more kinds of glass.

The design of compound lenses was at one time a matter of experience. Later, opticians applied the correction terms to the simple lens equation derived by Ludwig von Seidel, in 1856, to guide iterative lens design, but lenses had to be fabricated and tested to evaluate the changes made. The design process was greatly improved by graphical ray tracing of rays through different areas of the lens to develop "spot" diagrams, which indicated the effect of aberrations on image formation. With the development of vector methods of ray tracing and interpolation techniques on high speed computers, calculated spot diagrams for a compound lens or an optical system can be quickly produced. From the shape and extent of the spot diagrams the remaining type of image errors can be identified, modifications can be made in the design, and the process repeated until a satisfactory result is obtained.

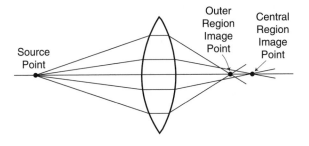

Figure 3 Spherical aberration occurs in a simple lens because the rays through the central portion of the lens come to a focus farther from the lens than rays passing through the outer region of the lens.

See also: ABERRATION; ABERRATION, CHROMATIC; ABERRATION, SPHERICAL; DISPERSION; LENS; OPTICS; REFRACTION

Bibliography

BROWN, E. *Modern Optics* (Reinhold, New York, 1965).

MÖLLER, K. D. *Optics* (University Science Books, Mill Valley, CA, 1988).

STRONG, J. *Concepts of Classical Optics* (W. H. Freeman, San Francisco, 1958).

WILLIAMS, C. S., and BECKLUND, A. *Optics: A Short Course for Engineers and Scientists* (Wiley, New York, 1972).

BASIL CURNUTTE

LENZ'S LAW

Lenz's law is a rule for determining the polarity of the electromotive force (emf) and induced current arising from electromagnetic induction. According to Faraday's law, when the magnetic flux through an electric circuit changes, an emf appears in the circuit, which drives a current consistent with the resistance (impedance) of the circuit. This induced current, in, produces an incremental magnetic field and, thus, an incremental magnetic flux through the circuit. One form of Lenz's law states that the sense (direction) of the induced emf and current is such that this incremental flux, due to the induced current, *opposes* the change in the original flux. This behavior is consistent with the well-known behavior of inductors to oppose time variations in the current through them.

A second form of Lenz's law applies to *motional* emf; that is, to situations where a conductor moves in a magnetic field. If the conductor is part of a closed circuit, the emf drives a current. This current, flowing in the magnetic field, experiences a Lorentz electromagnetic force. In this case, Lenz's law states that this force on the induced current is in the direction *opposite* to the velocity of the moving conductor. Therefore, the external agent producing the motion is required to do work on the conductor against the reaction force, and this work-energy shows up as heat-energy in the resistor that controls the induced current.

This useful rule was formulated in St. Petersburg in 1833 by Emil Lenz shortly after Michael Faraday's discovery of the phenomenon of electromagnetic induction. Since the full analysis of induction involves multiple steps and complex geometry, Lenz's rule is a remarkably simple way to avoid dealing with the inherent three-dimensionality and easily misapplied hand rules of vector relations. Lenz was a prolific contributor to electromagnetic theory and experimental technique. He established the heating power in a resistor, $P = I^2R$, independently of James Prescott Joule, and formulated the rules of circuit analysis independently of Gustav R. Kirchhoff.

As an example, Fig. 1 illustrates a conducting bar that is pushed at speed v along a pair of stationary conducting rails, which are connected to a resistor R at the far end. A magnetic field B is directed downward, perpendicular to the plane of the rails. The rails are separated by the distance ℓ. In this simple geometry, the motional emf induced in the moving bar is given by $\mathcal{E} = vB\ell$. According to Ohm's law, this drives the current $I = \mathcal{E}/R$. And, according to the Lorentz force law, the force on the moving rod is then $F = I\ell B$. Lenz's law tells us that the external agent must push directly against this force, in which case the agent does work on the system at the rate (or power) of $P_1 = Fv$. Meanwhile, the power delivered as Joule heat to the resistor is $P_2 = I^2R$. A modest bit of algebra shows that

$$P_1 = P_2 = \frac{(vB\ell)^2}{R};$$

that is, energy is conserved in the system if and only if Lenz's law is true. The value of Lenz's law in this example is that is was not necessary to use vector ("cross") products and hand rules to determine, first, the polarity of the induced EMF, and then the direction of the Lorentz force on the induced current.

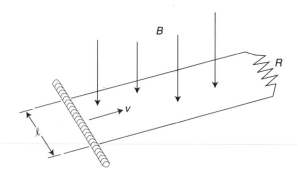

Figure 1 Conducting bar pushed along a pair of stationary conducting rails in a magnetic field.

See also: ELECTROMAGNETIC FORCE; ELECTROMAGNETIC INDUCTION, FARADAY'S LAW OF; ELECTROMOTIVE FORCE; OHM'S LAW

Bibliography

PURCELL, E. M. *Electricity and Magnetism,* 2nd ed. (McGraw-Hill, New York, 1985).

MARK A. HEALD

LEPTON

The leptons are a family of elementary particles, six in number, that includes the electron. Leptons and the six-member quark family constitute the fundamental building blocks of matter known as fermions. Leptons are distinct from quarks in that they do not participate in the strong interaction, since they have zero color charge. They do interact with one another, and with other particles, through the weak, electromagnetic, and gravitational interactions.

In addition to the electron, the leptons include two heavier unstable charged particles called the muon and the tau, and three varieties of neutrino, the e neutrino, mu neutrino, and tau neutrino. The symbols employed for the six members are e, μ, τ, ν_e, ν_μ, and ν_τ

The electron, discovered in 1897, was the first known elementary particle. The muon was discovered in 1937, and the tau in 1976. The e neutrino was first observed in 1953, and the mu neutrino was shown to be a distinct particle in 1962. The tau neutrino has not been directly observed, though its existence has been inferred from decay reactions in which it participates.

As required by a basic symmetry of nature, each of these particles has an antiparticle. For the charged leptons, the particles are all negative and the antiparticles are positive. The association between each of the charged leptons and its corresponding neutrino is that they share a property called flavor. So far each of the three lepton flavors seems to be conserved in all known reactions. In addition, the law of lepton flavor conservation states that no lepton can be produced without at the same time producing an antilepton of the same flavor family, and no lepton may disappear unless it encounters a corresponding antilepton. Thus, the number of leptons in the universe, minus the number of antileptons, is always conserved. To understand lepton flavor conservation assign electron flavor number +1 to electrons and electron-type neutrinos and -1 to positrons and electron-type antineutrinos. Electron flavor conservation states that the sum of the electron flavor numbers of the particles in any process is always the same after the process as it was before. Similarly, assignments of muon flavor number and tau flavor number to these charged leptons and theircorresponding neutrino types (+1 for particles, -1 for antiparticles) can be made and again the totals of these quantities are unchanged in all processes observed to date.

As the lightest of all electrically charged particles, the electron is the only absolutely stable elementary particle, since it cannot be destroyed without violating the conservation of electric charge. The muon, which is 206.8 times heavier than an electron, decays into an electron, a mu neutrino, and an e antineutrino with a half-life of 1.52 μs. The tau, with a mass 3,478 times that of an electron, has a half-life of 0.2 ps (2×10^{-13} s). Because of its great mass, it can break up into many different combinations of lighter particles, but there is always a tau-type neutrino among them.

So far as is known, none of the neutrinos has any mass at all, though their masses may simply be too small to measure. The upper limit on the mass of the e neutrino is 0.001 electron masses, while that for the mu neutrino is about 0.5 electron masses. Little is known about tau neutrinos, and they could be as heavy as 60 electron masses.

If any flavor of neutrino has a nonzero mass, it is possible that lepton flavors may not be absolutely conserved. In the decays of quarks, all of which have non-zero mass, flavor is not conserved. Such a process might allow one flavor of neutrino to transform into another, a process called neutrino oscillation. The search for neutrino oscillation has been a major effort in particle physics, and has so far been unsuccessful.

Since neutrinos carry no electric charge, and the effects of gravity are too feeble to observe on the subatomic level, neutrinos can be detected only through their weak interactions. Thus they interact very feebly with matter. A neutrino can pass all the way through the Earth with only a small probability of interacting. Nonetheless, powerful beams of neutrinos can be produced by allowing beams of unstable charged particles to decay; in a massive detector weighing many tons, a significant number of neutrinos will interact.

The primary means of detecting a neutrino is through the charged current interaction in which, while passing through a nucleus, a neutrino is transformed into a lepton of the same flavor. Thus in a beam of mu neutrinos a few will be transformed into negative muons, while a beam of e antineutrinos will produce positrons. In the first such experiment, performed by Leon Lederman, Melvin Schwartz, and Jack Steinberger in 1962, it was shown that only muons were found in the interactions of a beam of neutrinos produced by the decay of other unstable particles into muons. This was the first proof of the existence of different flavors of neutrinos.

Neutrinos are essential participants in the decay of unstable particles, especially in nuclear beta decay. They also play a significant role in the chain of nuclear reactions that powers the Sun and other stars. A huge burst of neutrinos is produced when a star explodes as a supernova, and these neutrinos provide the energy that drives the creation of all elements heavier than iron. Thus the leptons and their weak interactions play a significant role in our universe.

Experiments performed in the 1990s have shown that a very massive unstable particle called the Z^0, which is capable of decaying into any pair of particle and antiparticle whose combined masses are less than its mass, includes only three flavors of neutrino among its decay products. This means that there are no further neutrino types unless they are very massive, and that seems unlikely given the low masses of the known neutrinos. Since flavor in the lepton family seems to parallel that in the quark family, these experiments strongly suggest that the two six-member fermion families are now complete.

The leptons are considered to be elementary particles because they have no size or internal structure. Several experiments performed in the 1990s show that electrons and muons still behave as if they were point charges at distances smaller than 10^{-19} m.

See also: ANTIMATTER; CHARGE; CONSERVATION LAWS; DECAY, BETA; ELECTRON; ELEMENTARY PARTICLES; INTERACTION; INTERACTION, WEAK; LEPTON, TAU; MUON; NEUTRINO; NEUTRINO, HISTORY OF; PARTICLE MASS; POSITRON

Bibliography

HALZEN, F., and MARTIN, A. D. *Quarks and Leptons: An Introductory Course in Modern Particle Physics* (Wiley, New York, 1984).

PARKER, B. R. *Search for a Supertheory: From Atoms to Superstrings* (Plenum, New York, 1987).

ROBERT H. MARCH

LEPTON, TAU

The tau (τ) is an electrically charged elementary particle belonging to the group of leptons. Although regarded as a point particle, the τ has a mass of 1.77 GeV/c^2—almost twice that of a proton, and far heavier than the other known charged leptons, the electron and the muon (μ). The τ and its associated neutrino (ν_τ) together constitute the third family of leptons within the Standard Model of Particles and Interactions. In fact, the name τ was selected to denote the first letter of the Greek word for "third" ($\tau\rho\iota\tau o\nu$).

The τ lepton was discovered in 1975 by Martin L. Perl and collaborators at the Stanford Lincar Accelerator Center (SLAC). In their experiments they annihilated pairs of electrons and positrons (e^+) anticipating that, with enough energy, they would create $\tau^-\tau^+$ pairs. Looking for evidence of these new leptons, however, was not expected to be simple: The τ decays via the weak interaction with a mean lifetime of about 0.3 ps—far too little time to be detected directly by the experimental apparatus. Instead, the SLAC physicists looked for indirect evidence in the form of the distinctive decay pattern (see Fig. 1)

$$e^- + e^+ \rightarrow \tau^- + \tau^+$$
$$\hookrightarrow e^+ + \nu_e + \bar{\nu}_\tau$$
$$\hookrightarrow \mu^- + \bar{\nu}_\mu + \nu_\tau$$

where $\bar{\nu}$ denotes an antineutrino. Since only charged particles could be detected, they looked for events with two oppositely charged leptons whose energy and momentum were less than that of the incoming $e^\pm$ pair; this "missing" energy and momentum are presumably carried off by the unseen neutrinos.

Unfortunately, this $e\mu$ decay signature turned out not to be as distinctive as had been hoped. Establishing the existence of the τ was hampered by the presence of a soon-to-be discovered hadron called a D meson. The D has a mass of about 1.87 GeV/c^2; thus, at roughly the same energy of the colliding $e^\pm$ pair, the experiment could be creating D^-D^+ rather than $\tau^-\tau^+$. An $e\mu$ signal would then herald the decays

$$D^- \to \mu^- + \bar{\nu}_\mu + K^0$$

$$D^+ \to e^+ + \nu_e + K^0.$$

As neutral particles, K mesons are not directly detected and must be identified by their charged decay products. However, the K^0 is famous for having, in effect, two lifetimes, one long, the other short. Thus, in half of the D events the kaon lives long enough to escape from the detector before decaying. As a result, when the $e\mu$ signature is seen one cannot know if the missing energy and momentum are to be attributed to fleeing kaons or neutrinos. Historically, this ambiguity complicated the discovery of both the τ and the D; distinguishing the two became possible only when many $e\mu$ events were recorded. Since the K^0 sometimes decays quickly, half of the D induced $e\mu$ events should include kaon decay products. A detailed analysis of the SLAC data demonstrated that there were very few of these cases; the $e\mu$ signatures should therefore be due to τ lepton production. Perl was awarded the 1995 Nobel Prize in physics in recognition of his role in the discovery of the τ lepton.

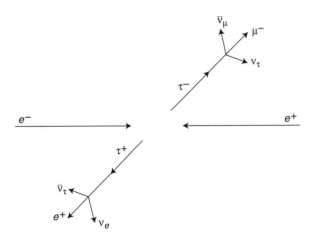

Figure 1 Distinctive decay pattern used in the search for the τ neutrino.

Once the τ was identified, other decay modes could be studied. In addition to the pure leptonic mode discussed above, the semileptonic modes

$$\tau \to \text{hadrons} + \nu_\tau$$

are also seen. One might also expect a decay via the electromagnetic interaction such as

$$\tau \to \mu + \gamma,$$

where γ is a photon. Curiously, this process has never been observed. To explain the absence of this decay, the conservation of a quantity called tau lepton number L_τ is invoked. By analogy with L_e and L_μ, τ^- and ν_τ are assigned $L_\tau = 1$; their antiparticles τ^+ and $\bar{\nu}_\tau$ are assigned $L_\tau = -1$. All other particles have $L_\tau = 0$. The rule states that each of the three lepton family numbers, L_e, L_μ, and L_τ, is separately conserved. Thus the electromagnetic τ decay is not allowed because it does not conserve L_τ. By contrast, the decay

$$\tau^+ \to e^+ + \nu_e + \bar{\nu}_\tau$$

conserves both L_τ and L_e. It is worth noting that although all data to date support the conservation of lepton family number, the law is purely empirical in nature, having as yet no theoretical explanation. Indeed, physicists are always on the lookout for L-violating processes.

The τ neutrino has not been observed directly. Neutrino detection is particularly difficult due to the weakness with which ν's interact with matter; initially one can only infer their existence by invoking established conservation laws. For example, one of the simplest decay modes in which a ν_τ is expected to appear is the observed decay

$$\tau^- \to \pi^- + X,$$

where X stands for the undetected particle. That there really is an X in this reaction is assured by the conservation of momentum; together with the conservation of energy and the special theory of relativity, the data show that the mass of X is no greater than 31 MeV/c^2—and in fact could be zero. Of course a separate conservation law for lepton family number would require X to be the τ neutrino. However, the only way to be certain that any neutrino exists—whether ν_e, ν_μ, or ν_τ—and that each is indeed distinct, is to have a beam of neutrinos incident upon, say, a nuclear target N,

$$\nu_l + N \to l + X,$$

where now X is some mix of hadrons, and l stands for one of e, μ, or τ. If lepton family number is in fact conserved and the different neutrino flavors are distinct, then the produced lepton l must be the partner of the incoming neutrino ν_l. Indeed, if the beam is pure ν_e, then only $l = e$ is seen; similarly, a pure ν_μ beam results only in $l = \mu$. In this way the separate identities of ν_e and ν_μ have been established. Experiments to build beams of τ neutrinos and firmly establish the identity of ν_τ are currently being proposed.

See also: ANTIMATTER; DECAY, NUCLEAR; FLAVOR; HADRON; INTERACTION, ELECTROMAGNETIC; INTERACTION, WEAK; KAON; LEPTON; MUON; NEUTRINO; PHOTON; POSITRON

Bibliography

FRAUENFELDER, H., and HENLEY, E. M. *Subatomic Physics*, 2nd ed. (Prentice Hall, Englewood Cliffs, NJ, 1991).

PERKINS, D. H. *Introduction to High Energy Physics*, 3rd ed. (Addison-Wesley, Reading, MA, 1987).

PERL, M. L., et al. "Evidence for Anomalous Lepton Production in e^+-e^- Annihilation." *Phys. Rev. Lett.* **35,** 1489 (1975).

PERL, M. L., et al. "Properties of the Proposed τ Charged Lepton." *Phys. Lett.* **70B,** 487 (1977).

PERL, M. L., and KIRK, W. T. "Heavy Leptons." *Sci. Am.* **238,** 50 (1978).

ALEC J. SCHRAMM

LEVITATION, ELECTROMAGNETIC

Levitation is the elevation of an object above the ground with no visible means of support. There are various methods for levitating an object that use electromagnetism. Small particles can be levitated by laser light via radiation pressure. It is known that light (electromagnetic radiation) carries energy along with it; therefore light in a given volume would have an associated energy density. Energy density has the same dimensions as pressure, so light must exert a force when hitting an object with a given surface area.

It is well known that if two magnets are aligned so that like poles are facing each other, they repel. If such magnets are constrained to move in the vertical direction by passing a nonmagnetic rod through their centers, one magnet can be made to "float" above the other, as shown in Fig. 1. The height that one magnet levitates above the other can easily be worked out by treating the magnets as magnetic dipoles. Consider two permanent magnets of mass M and dipole moment m oriented in the z direction. The magnetic field along the axis of one dipole, a height z above the center, is

$$\mathbf{B} = \frac{\mu_0}{4\pi} \frac{2m}{z^3} \hat{\mathbf{k}},$$

where $\hat{\mathbf{k}}$ is a unit vector along the z axis. The upward force exerted by this field on a similar magnetic dipole, inverted with respect to the first, is

$$\mathbf{F} = \nabla(\mathbf{m} \cdot \mathbf{B})$$

$$= -\frac{\partial}{\partial z}\left(\frac{\mu_0}{2\pi} \frac{m^2}{z^3}\right)\hat{\mathbf{k}}$$

$$= \frac{3\mu_0 m^2}{2\pi z^4} \hat{\mathbf{k}}.$$

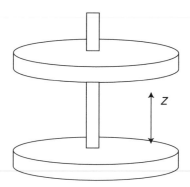

Figure 1 Levitation produced by aligning the like poles of two magnets and passing a nonmagnetic rod through their centers.

This upward force must balance the downward force due to gravity at height z, so that

$$Mg = \frac{3\mu_0 m^2}{2\pi z^4}$$

$$\Rightarrow z = \left(\frac{3\mu_0 m^2}{2\pi Mg}\right)^{1/4},$$

where z is the height of levitation.

Generally, conductors (not necessarily magnetic) can be made to levitate by induced eddy currents. Circulating currents are set up in a conductor by an external time varying magnetic field, such as that produced by an alternating current in a coil. Either a small conductor can be levitated by using an electromagnet to generate the varying magnetic B field, or a small magnet can be levitated by rotating a conductor underneath it. Small magnets can be made to float above a high temperature superconducting material. The currents induced in the superconductor experience no resistance.

Consider an aluminum disk which is rotating at angular velocity ω. A small permanent magnet, held by taping a flexible piece of card to it, is positioned over the disk (Fig. 2). The magnet is seen to rise when the disk rotates under it. When the disk stops rotating the magnet falls. The faster the rotation the higher the magnet levitates. When the disk is rotating it experiences a time varying magnetic field because of its motion under the permanent magnet. Currents are set up in the disk which, by Lenz's law, give rise to a magnetic field repeling the small magnet causing it to levitate. The levitation height depends on the angular velocity of the disk. As the rate of change of the magnetic field increases the height of levitation increases. If the magnet is held over the

center of the disk and moved radially outward, the height of levitation will increase because the linear velocity, $v = r\omega$, of the disk increases towards the outside edge of the disk.

An aluminum ring can be levitated over a coil carrying an alternating current. An iron core can be threaded through the ring to concentrate the changing magnetic field and hold the ring in position. Eddy currents in the ring produce a magnetic field that repels the coil causing levitation of the ring. The ring will become quite hot due to the resistance R of the ring to the circulating eddy currents I. The rate of heat generated (power lost due to heating) is given by $I^2 R$. The same mechanism is responsible for the levitation of a small conducting sphere above a coil carrying an alternating current. The magnetic moment m of a sphere is given by

$$\mathbf{m} = -2\pi a^3 \mu_0 \mathbf{B},$$

where a is the radius of the conducting sphere. The axial magnetic field of a single loop, of radius b, is given by

$$\mathbf{B} = \frac{I}{2}\left[\frac{\mu_0 b^2}{(b^2 + z^2)^{3/2}}\right]\hat{\mathbf{k}}.$$

The force of the magnetic field on the sphere is given by

$$\mathbf{F} = \mathbf{m} \cdot \frac{\partial}{\partial z}\mathbf{B},$$

where

$$\frac{\partial}{\partial z}\mathbf{B} = -\frac{\mu_0 I}{2}\left[\frac{3b^2 z}{(b^2 + z^2)^{5/2}}\right]\hat{\mathbf{k}}.$$

Using

$$\mathbf{m} = -2\pi a^3 \mu_0 \frac{I}{2}\left[\frac{\mu_0 b^2}{(b^2 + z^2)^{3/2}}\right]$$

and

$$I_{\text{rms}} = \frac{I}{\sqrt{2}},$$

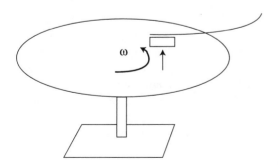

Figure 2 Levitation produced by positioning a magnet over a rotating disk.

we obtain

$$\mathbf{F} = \mu_0 I_{\text{rms}}^2 \left[\frac{3\pi a^3 b^4 z}{2(b^2 + z^2)^4} \right] \hat{\mathbf{k}},$$

where the subscript rms stands for root mean square. By using N loops of wire we may change the I_{rms}, to $N I_{\text{rms}}$, which increases the magnetic field. This upward force must balance the downward gravitation force on the sphere. So by equating $\mathbf{F}$ with Mg as before, where M is the mass of the sphere, we can find the height of levitation of the sphere above the loop. An electric dipole is found to be able to float in a gravitational field, due to the "drooping" of electric field lines.

Applications

Magnetically levitated trains (maglev) are capable of running at speeds up to 500 km/h, as opposed to 300 km/h for normal rail trains. The Japanese have developed a repulsion system which also has wheels. When the train reaches speeds of 100 km/h or so it levitates above the U-shaped tracks. The Germans have an attractive magnetic system using a much smaller T track (see Fig. 3). Both propulsion mechanisms work on the same principles. The train rides on an electromagnetic wave that effectively pushes and pulls the train along. The electromagnets in the guide system change polarity with alternating current passing through them. Raising the frequency of the current speeds up the train, while reversing the poles of the

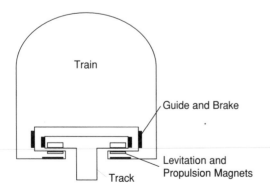

Figure 3   T track system for electromagnetic levitation.

magnetic field acts like a brake and slows the train without friction. The German T track takes up less space and has the advantage that the train cannot be derailed.

A by-product of the levitation of small conductors is the heat produced inside the conductor due to the circulating eddy currents.

See also: CURRENT, EDDY; ELECTROMAGNET; ELECTROMAGNETISM; FIELD, MAGNETIC; LENZ'S LAW; LEVITATION, MAGNETIC; MAGNETIC MOMENT; MAGNETOHYDRODYNAMICS; SUPERCONDUCTIVITY

Bibliography

BRENNAN, R. P. *Levitating Trains and Kamikaze Genes, Technological Literacy for the 1990s* (Wiley, New York, 1990).

GRIFFITHS, D. J. *Introduction to Electrodynamics,* 2nd ed. (Prentice Hall, Englewood Cliffs, NJ, 1989).

GRIFFITHS, D. J. "Electrostatic Levitation of a Dipole." *Am. J. Phys.* **54,** 744 (1986).

MAK, S. Y., and YOUNG, S. "Floating Metal Ring in an Alternating Magnetic Field." *Am. J. Phys.* **54,** 808–811 (1986).

WOUCH, G., and LORD, A. E. "Eddy Currents: Levitation, Metal Detectors, and Induction Heating." *Am. J. Phys.* **46,** 464–466 (1978).

HEIDI FEARN

LEVITATION, MAGNETIC

Levitation can be defined as maintaining a body in stable equilibrium without mechanical contact with the earth; magnetic levitation implies that this is done with magnetic fields.

There are several different types of magnetic levitation systems: electromagnets with feedback control, permanent magnets with superconductors, superconducting magnets with superconductors, magnets moving over conductors, alternating-current magnets over conductors, diamagnetic materials in magnetic fields, and combinations of these. Note that it is not possible to achieve stable levitation with permanent magnets alone (lift alone is not levitation), although the addition of one mechanical constraint will often suffice.

Two very promising applications of magnetic levitation are in high-speed ground transportation (maglev) systems and in low-loss magnetic bearings. Although both of these applications have been under consideration for some time, recent advancements in both superconducting and permanent magnet materials technology have made them commercially attractive.

Levitation with Permanent Magnets and Superconductors

The availability of materials that are superconducting at or above the boiling temperature of liquid nitrogen (77 K) has made the levitation of a small permanent magnet over a superconductor (or vice versa) a familiar demonstration of magnetic levitation. This demonstration, which depends on Faraday's law or the Meissner effect (depending upon how the experiment is done), was first done in 1945 by V. Arkadiev, who levitated a magnet over a concave lead plate in liquid helium.

If a magnet is lowered onto a superconductor, shielding currents are induced on the surface of the superconductor (Faraday's law). These supercurrents create a magnetic field that repels the magnet and thus levitates it at a height such that the repulsive force equals the weight of the magnet. Supercurrents are possible because of the zero resistance of the superconductor. If, on the other hand, a magnet already rests on the superconductor as it is cooled through its transition temperature and becomes superconducting (field-cooled condition), the magnet will rise (but not quite as high as in the zero-field-cooled case described above) due to the expulsion of magnetic flux from the superconductor (Meissner effect), another remarkable property of superconductors discovered in 1933. Two important properties of a superconductor are its perfect conductivity ($R = 0$) and its perfect diamagnetism (magnetic permeability $\mu_r = 0$).

To levitate a magnet over a type-I superconductor (as in Arkadiev's experiment), a concave superconductor surface is required in order to give the magnet lateral stability. Due to a phenomenon known as "flux pinning," however, lateral stability can be achieved in type-II superconductors without the need for a concave surface. A magnet floating over a type-II superconductor, such as the high-temperature material yttrium-barium-copper oxide (YBCO), can be pushed to the edge of the superconductor without loss of lateral stability. Furthermore the magnet can be made to float at different heights above (and even below)

the superconductor. Both of these remarkable properties are due to flux pinning.

Magnetic bearings using permanent magnets and high-temperature superconductors, which can be made stable without an active feedback system, are under development at several laboratories. One particularly attractive application is in low-loss flywheels for energy storage by electric power companies; storing off-peak energy to meet peak load demands allows more efficient use of their power plants. Flywheels could also be used to store energy for making short trips in small automobiles; their energy-storage capacity is comparable to batteries of the same size, and they would not have to be replaced regularly, as batteries do.

Magnetic Suspension

Although it is impossible to suspend one permanent magnet below another in a stable manner without applying other forces, it is possible to suspend a permanent magnet or a soft ferromagnetic sample (such as iron) below an electromagnet if the current in the electromagnet is carefully controlled. Coil-magnet systems have been successfully used for several decades to suspend ultrahigh-speed rotors. Active magnetic bearings have made great progress recently, due in part to the development of microprocessors and small position and velocity sensors.

By using feedback current control it is also possible to suspend an electromagnet below a sheet of steel, and this is the principle used in one type of maglev transportation. An electromagnetic maglev system depends on the attractive force between electromagnets and a steel guideway.

Another way to achieve stable magnetic suspension is to insert a superconductor between the magnet and the ferromagnetic material, as shown in Fig. 1. The field of the magnet magnetizes the soft iron sample, which is then repelled by the superconductor. It is possible to have stable suspension of the iron due to a balance between the attractive force of the permanent magnet and the repulsive force of the superconductor; lateral stability is supplied by flux pinning in the superconductor.

It is also possible to suspend a magnet below a type-II superconductor (or vice versa) if the flux pinning is sufficiently strong. The magnet is generally placed next to the superconductor as it is cooled through its transition temperature. Much of the magnetic flux is then trapped within so-called flux

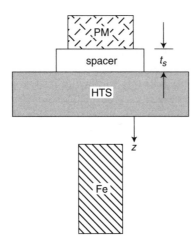

Figure 1 Schematic of a levitation system with a steel rotor (Fe) stably levitated below a combination of permanent magnet and high-temperature superconductor.

vortices in the superconductor, causing them to act as small magnets, which together supply an attractive force for the permanent magnet at the same time the rest of the superconductor repels the magnet. Lateral stability is again supplied by flux pinning in the superconductor.

To demonstrate this principle, a small motor was constructed at Argonne National Laboratory by a summer research student. The ends of the rotor have cylindrical NdFeB magnets that can either be supported by repulsive levitation or by attractive levitation underneath two high-temperature superconductors. The motor can be driven above 10,000 rpm with negligible friction.

Magnets Moving over Conductors

When a magnet moves over a conductor, the changing magnetic field induces a voltage in the conductor, which causes eddy currents to flow. These eddy currents, in turn, generate a magnetic field that opposes the change in the field due to the motion of the magnet. James Clerk Maxwell suggested a simple image model in which the magnet first induces a "positive" image when it passes a point, and then a "negative" image. These images propagate into the conductor at a velocity which is proportional to the specific resistivity (and to the reciprocal thickness if the conductor is thin compared with the skin depth).

The force on a magnet moving over a nonmagnetic conducting plane can be conveniently resolved into two components: a lift force perpendicular to the plane and a drag force opposite to the direction of motion. At low velocity, the drag force is proportional to velocity v and considerably greater than the lift force, which is proportional to v^2. As the velocity increases, however, the drag force reaches a maximum (referred to as the drag peak) and then decreases as $1/\sqrt{v}$. The lift force, on the other hand, which increases with v^2 at low velocity, overtakes the drag force as velocity increases and approaches an asymptotic value at high velocity, as shown in Fig. 2. This asymptotic value is the force that the same magnet would feel when levitated over a superconductor. The lift-to-drag ratio, which is of considerable practical importance, is given by $F_L/F_D = v/w$.

Qualitatively, these forces can be understood by considering magnet flux diffusion into the conductor. If the magnet is moving rapidly enough, the field will not penetrate deeply into the conductor, and the flux compression between the magnet and the conductor causes a lift force. The flux that does penetrate the conductor is dragged along by the moving magnet, and the force required to drag this flux along is equal to the drag force. At high speeds, less of the magnetic flux has time to penetrate the conductor, the lift force approaches its asymptotic limit, and the drag force decreases toward zero.

The lift force on a vertical dipole of moment m moving at a height z_0 above a conducting plane can be shown to be:

$$F_L = \frac{3\mu_0 m^2}{32\pi z_0^4}\left(1 - \frac{w}{\sqrt{v^2 + w^2}}\right).$$

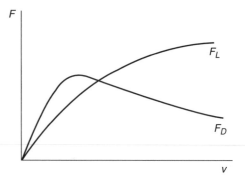

Figure 2 Velocity dependence of lift force F_L and drag force F_D.

At high velocity the lift force approaches the ideal lift from a single image: $3\mu_0 m^2/32\pi z_0^4$; at low velocity, the factor in parentheses is approximately equal to $v^2/2w^2$, so the lift force increases as v^2.

Repulsive forces between moving magnets and the eddy currents they generate in conducting guideways provide levitation in electrodynamic maglev systems. The repulsive levitation force is inherently stable with distance, and comparatively large levitation heights (20 to 30 cm) are attainable by using superconducting magnets. Although continuous aluminum guideways have been considered, the proposed Japanese high-speed maglev system uses interconnected figure-8 ("null-flux") coils on the sidewalls; the null-flux arrangement tends to reduce the magnetic drag force and thus the propulsion power needed. Many people feel that the future of high-speed ground transportation belongs to magnetically levitated (maglev) vehicles, and that "magnetic flight" will replace many short-haul flights (100 to 600 miles) that operate in and out of crowded airports.

See also: CURRENT, EDDY; ELECTROMAGNET; ELECTROMAGNETIC INDUCTION; ELECTROMAGNETIC INDUCTION, FARADAY'S LAW OF; FIELD, MAGNETIC; LEVITATION, ELECTROMAGNETIC; SUPERCONDUCTIVITY; SUPERCONDUCTIVITY, HIGH-TEMPERATURE

Bibliography

ARKADIEV, V. "A Floating Magnet." *Nature* **160,** 330 (1947).

BEAMS, J. W. "Ultrahigh-Speed Rotation." *Sci. Am.* **204** (4), 134 (1961).

GLATZEL, K; KHURDOK, G.; and ROGG, D. "Development of the Magnetically Suspended Transportation System in the Federal Republic of Germany." *IEEE Trans. Veh. Technol.* **19,** 3–17 (1980).

HULL, J. R.; PASSMORE, J. L.; MULCAHY, T. M.; and ROSSING, T. D. "Stable Levitation of Steel Rotors Using Permanent Magnets and High-Temperature Superconductors." *J. Appl. Phys.* **76,** 577–580 (1994).

KYOTANI, Y. "Recent Progress by JNR on Maglev." *IEEE Trans. Magn.* **24,** 804–807 (1988).

MAXWELL, J. C. "On the Induction of Electric Currents in an Infinite Plane Sheet of Univorm Conductivity." *Proc. R. Soc. London. Ser. A* **20,** 160 (1872).

MOON, F. C. *Superconducting Levitation* (Wiley, New York, 1994).

MOON, F. C., and CHANG, P.-Z. "High-Speed Rotation of Magnets on High T_c Superconducting Bearings." *Appl. Phys. Lett.* **56,** 397–399 (1990).

REITZ, J. R. "Forces on Moving Magnets Due to Eddy Currents." *J. Appl. Phys.* **41,** 2067–2071 (1970).

RHODES, R. G., and MULHALL, B. E. *Magnetic Levitation for Rail Transport* (Clarendon, Oxford, Eng., 1981).

ROSSING, T. D. "Magnetic Flight" in *Fundamentals of Physics,* 4th ed., edited by D. Halliday, R. Resnick, and J. Walker (Wiley, New York, 1993).

ROSSING, T. D., HULL, J. R. "Magnetic Levitation." *Phys. Teach.* **29,** 552–562 (1991).

THOMAS D. ROSSING

LIGHT

Theories of light and optical experiments have played a key role in the development of the two great pillars of twentieth-century physics, relativity theory and quantum theory. From the beginning of classical physics in the late seventeenth century, there have been two competing theories concerning the nature of light. These are the corpuscular theory, in which light is considered to be made up of rapidly moving corpuscles or particles, and the wave theory, in which light is thought to be composed of waves moving through some medium. The fundamental difference in these two views is that in the former case, the light particles move all the way from the source to the illuminated object, while in the latter case only a disturbance, energy, passes from particle to particle of the intervening medium while the particles themselves remain at the same average position.

At the very beginning of classical physics, the corpuscular theory was supported by the great English physicist Isaac Newton. He performed a number of optical experiments in the 1660s, and published his famous book titled *Opticks* in 1704. One property of light that convinced Newton of its corpuscular rather than wave nature was its straight-line propagation; the appearance of shadows could be predicted by drawing straight lines from a small source, past any opaque barriers, to an illuminated viewing screen. It was well known that waves, such as those in water or sound waves, bend around obstacles to fill in what would be the shadow area in the case of straight-line propagation. The straight-line propagation of light is illustrated in Fig. 1.

Actually, Newton knew that a small amount of bending of light took place at the edge of an obstacle. The effect, called diffraction, had been discov-

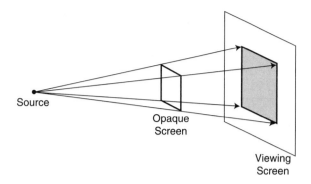

Figure 1   Straight-line propagation.

ered in the seventeenth century by a Jesuit priest, Francesco Maria Grimaldi. However, the effect was so small that it did not seem like the bending around obstacles by large-scale waves, and Newton thought that it was due to the attraction between the material of the obstacle and the light corpuscles as they passed by. To explain refraction, the bending of light as it entered a medium like water or glass, the corpuscular theory had to assume that the particles were attracted to the medium when they got very close, and therefore speeded up; this speeding up accounted for the bending in refraction, and thus he believed that it similarly accounted for the bending in diffraction.

The opposing wave theory of light was championed in Newton's time by the renowned Dutch scientist Christiaan Huygens. His views were summarized in a long paper published in 1690 under the title *Treatise on Light*. One property of light that convinced Huygens of its wave nature was the ability of two light beams to pass through one another without either being deflected, a property that was also observed for water waves; particles, he thought, would collide and thereby deflect each other. In his attempt to explain straight-line propagation, Huygens proposed that each point on a wave front acts as a new source of little spherical waves (he called them "particular or partial" waves) spreading in the forward direction, and that the wave front at a later time and advanced position was a sum of all these partial waves. This proposition, now known as Huygens's principle, is still used to trace the motion of waves. He then reasoned that the partial waves from a wave front in an aperture would not arrive together in the shadow region beyond the obstacle—in his words, "being too feeble to produce light there."

The hypothetical medium through which the light waves moved was called the ether (also spelled aether). It had to be a substance that filled all of space because light reaches Earth from the distant stars, and more subtle than air because it appeared to offer no detectable resistance to the motion of the planets around the Sun. Huygens thought of light as a kind of pressure wave in the ether, like sound in air, a type of wave called longitudinal today.

The great authority of Newton persuaded most scientists throughout the eighteenth century to accept the corpuscular theory of light. However, the work of two scientists in the early nineteenth century brought the wave theory of light to dominance: Thomas Young and Augustin Fresnel. Their work was persuasive because they were able to experimentally show and theoretically analyze interference effects for light. Interference was a phenomenon known to apply to waves, but totally unknown in the case of particles; more specifically, two waves can cancel each other, adding up to no wave, but two particles do not seem to behave that way.

Any wave is characterized by a speed, a wavelength, and a frequency. Speed describes how fast the disturbance, crests and troughs, moves through the medium, whereas wavelength specifies the distance between wave crests, as shown in Fig. 2. Frequency specifies how many waves per second pass a given point in the medium or, equivalently, how many times per second the particles of the medium oscillate back and forth as the wave passes, and it has units of cycles per second or hertz (Hz). From this definition, it is always true for any wave that speed equals wavelength times frequency; that is, if the number of waves per second is multiplied by the length of each one, the result is the total length of wave that passes in one second, which is the speed of the wave. Thus,

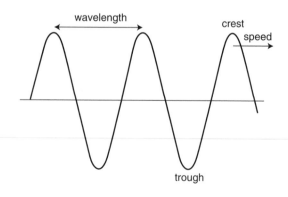

Figure 2   Wave characteristics.

$$c = \lambda f, \qquad (1)$$

where c is the speed of light (m/s), λ is the wavelength of light (m), and f is the frequency of light (Hz).

When two waves pass through a medium at the same time, the disturbance due to one is added to that due to the other. This superposition, as it is known, leads to interference, because if one wave by itself causes a disturbance in one direction at one point in space at a given time, while the other causes an equal and opposite disturbance at the same place and time, the net effect is no disturbance at all. On the other hand, if two such waves cause disturbances in the same direction, the resulting disturbance is twice as large. The first effect is called destructive interference, and the second is called constructive interference. Such interference in light was demonstrated convincingly by Young in his double-slit experiment. In this experiment, light from a small source and of a single wavelength is allowed to impinge on two narrow, closely placed slits in an otherwise opaque screen. On the other side of this screen, the two slits act as two point sources, which by Huygens's principle emit spherical waves in synchronization. The two sets of spherical waves, spreading out from the two slits, produce an interference pattern of alternating bright (disturbances add) and dark (disturbances cancel) spots that can then be viewed on a screen, as shown in Fig. 3.

In order to obtain the synchronized sets of waves from the two slits that are required to produce a stable, discernible interference pattern, Young had to use light of essentially a single wavelength and smooth wave fronts impinging on the slits. Such light is said to be coherent. Multiple wavelengths or randomly bumpy wave fronts, called incoherent, would lead to a jumbled and rapidly changing pattern beyond the slits, washing out any detectable

Table 1 Wavelength and Color

Wavelength (nm)	Color Sensation
380–460	violet
460–480	blue
480–490	blue-green
490–560	green
560–570	yellow-green
570–585	yellow
585–600	orange
600–700	red

pattern. Normal light sources, such as flames or incandescent lamps, produce incoherent light, so Young had to use color filters to approximate a single wavelength, and a pinhole in front of the double slits to get a smooth, spherical incident wave front (by Huygens's principle). However, once the experiment is properly set up, measurements on the interference pattern will reveal the wavelength of the light. From such experiments it is now known that the wavelengths of visible light are very small, ranging from about 400 nm to about 700 nm. Furthermore, the different wavelengths of light correspond to the different colors of the spectrum, with the shortest giving rise to the sensation of violet and the longest giving rise to the sensation of red, as shown in Table 1.

Newton had already shown that white light is a mixture of all the colors of the rainbow by passing light through a prism to produce a spectrum, then passing one of the spectral colors through a second prism to demonstrate that it was not further changed in appearance. He also used a second prism in reverse to recombine the spectral colors and reproduce white light. Figure 4 illustrates these experiments.

At the time of Young's work, the speed of light had already been roughly measured: in 1675 by the Danish astronomer Olaus Roemer, using annual variations in the timing of eclipses of Jupiter's moons, and in 1726 by the English astronomer James Bradley, using the annual variation in the angular position of any star as seen from Earth. Later measurements would make this constant one of the most accurately known quantities in nature. To three significant figures it is $c = 3.00 \times 10^8$ m/s. With this value we may calculate a typical frequency of light (say, green) from Eq. (1): $f = c/\lambda = 3 \times 10^8/5 \times 10^{-7} = 6 \times 10^{14}$ Hz. This is a very large

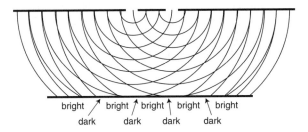

bright bright bright bright bright
dark dark dark dark

Figure 3 Double-slit interference.

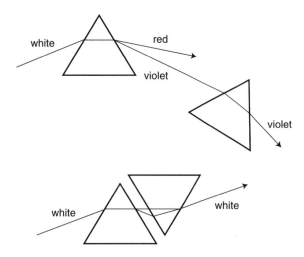

Figure 4 Newton's color experiments.

number and means that hundreds of trillions of light waves pass an observer every second. Other wavelengths of visible light have similarly large frequencies.

By 1821 Fresnel's theoretical and experimental studies of interference and other optical phenomena had convinced him that light was not only a wave, but indeed a transverse wave as opposed to a longitudinal one. In a transverse wave, the disturbance of the medium is perpendicular to the speed of the wave, as in Fig. 2, where the speed is shown as horizontal but the disturbance is vertical. The concept of a transverse wave was required to explain the facts of polarized light. In a longitudinal wave, the disturbance of the medium is back and forth along the direction of wave motion, and there is only one such direction. In a transverse wave, the disturbance in three dimensions can be in *any* plane perpendicular to the wave motion; therefore, the wave can be polarized in any of these directions.

What appeared to be final experimental verification of the wave theory was provided in 1850 by the experiments of the French scientists Armand Fizeau and Jean Bernard Foucault. They directly measured the speed of light in air and in water, finding the latter speed to be smaller. To explain refraction, the wave theory had to assume just this result, whereas the corpuscular theory had to assume that the particles speeded up in water.

The understanding of light advanced again with the work of the great British physicist James Clerk Maxwell in the 1860s. His theoretical work in electricity and magnetism showed that when an electric charge accelerates, a transverse wave is produced in the electric and magnetic force fields that propagates with a speed of 3×10^8 m/s; he calculated the speed purely from constants measured in electrical and magnetic experiments. He immediately realized that light is nothing other than certain kinds of electromagnetic waves; that is, light is an electromagnetic wave of a frequency to which human eyes are responsive. However, electromagnetic waves can be of any frequency, giving rise to an electromagnetic spectrum extending to wavelengths much shorter than violet and much longer than red light. Toward shorter wavelengths we go from violet light to ultraviolet, x rays, and gamma rays. Toward longer wavelengths we go from red light to infrared, microwaves, and radio waves. Thus, with Maxwell's contributions, the field of optics became unified with electricity and magnetism.

At this time the luminiferous (light-carrying) ether was still a major subject of investigation for scientists. In his book *A Treatise on Electricity and Magnetism* (1873), Maxwell said, "Hence all these theories lead to the conception of a medium in which propagation takes place, and if we admit this medium as an hypothesis, I think it ought to occupy a prominent place in our investigations, and that we ought to endeavor to construct a mental representation of all the details of its action." Leading scientists of the time thought that electrical, magnetic, and optical phenomena would all be explained by reduction to mechanics of the ether; by mechanics, they understood classical, Newtonian mechanics. Yet the facts at hand indicated fantastic properties for this ether. To support the propagation of electromagnetic waves, which are transverse, the ether would have to act like an elastic solid (fluids will support only longitudinal waves), and furthermore, it would have to have a large elasticity to produce the high speed of light. This elastic/solid character would have to be combined with the total absence of any discernible resistance to the planetary motions. Finally, in accord with classical mechanics, the speed of light should have been fixed with respect to the medium (ether), and any speed of an observer through the ether should have been detectable in the speed of light with respect to the observer. This last property was tested in the late nineteenth century and found to be lacking.

In 1887 two American scientists at Case Institute, Albert Michelson and Edward Morley, carried out an experiment to measure Earth's motion through the ether; it has since become famous as the Michelson–Morley experiment. They used an instrument called the Michelson interferometer, which demonstrates interference of light in a different manner than does a double slit. Instead of making one sample of a wave front interfere with another sample taken from a different spatial position, an interferometer splits the amplitude of an incident wave along the whole wave front, sending one part along one path and the other along a different path until they are brought back together to form interference fringes. If the paths are of even slightly different length, this procedure amounts to making one whole wave front interfere with another that was originally behind it. Figure 5 is a diagram of the Michelson interferometer. Light from the source is split at the half-silvered beam splitter, with half sent to mirror 1 and half to mirror 2. The light reflected from these mirrors is recombined by the beam splitter and sent toward the observer, who can now observe interference fringes. A change in either path as small as a fraction of a wavelength will result in a shift of the fringes; a change of half a wavelength produces a shift of one whole fringe (light to dark). Thus, an interferometer can be a very sensitive instrument for measuring changes in path length.

For their experiment on the ether, Michelson and Morley aligned one of the interferometer arms (say, to mirror 2) along the direction of Earth's motion relative to the Sun, assuming that this would be more or less along the motion through the ether. Then the speed of light along the two arms should be different (according to classical views) and the time to traverse each arm would be different. Now if the instrument is rotated through 90°, the arm to mirror 1 becomes the one along the ether motion, and the arm to mirror 2 becomes equivalent to the

original arm to mirror 1. Because there is a relative change between the two arms, this rotation should produce a fringe shift, but Michelson and Morley saw none. To rule out the possibility that Earth just happened to be stationary with respect to the ether at the time of the experiment, it was repeated at different times of the year when Earth is moving differently, still with a null result. The indications were that the speed of light along the two arms was the same, regardless of their motion through the presumed ether.

By the beginning of the twentieth century, no experiment had been able to detect movement with respect to the ether; regardless of the motion of the observer, the speed of light always came out to be the same value, $c = 3 \times 10^8$ m/s, with respect to the experimenter. Such optical experiments were crucial in the formation of the theory of special relativity by Albert Einstein, in which the constancy of the speed of light is a fundamental postulate. In his original 1905 paper, "On the Electrodynamics of Moving Bodies," Einstein comments, "The introduction of a 'luminiferous ether' will prove to be superfluous inasmuch as the view here to be developed will not require an 'absolutely stationary space' provided with special properties." Therefore this date marks the death of the luminiferous ether, which had been the subject of so much study and speculation since the time of Huygens. Einstein later wrote cogently of the change in attitude that had come about in science: "If mechanics was to be maintained as the foundation of physics, Maxwell's equations had to be interpreted mechanically. This was zealously but fruitlessly attempted, while the equations were proving themselves fruitful in mounting degree. One got used to operating with these fields as independent substances without finding it necessary to give one's self an account of their mechanical nature; thus mechanics as the basis physics was being abandoned, almost unnoticeably, because its adaptability to the facts presented itself finally as hopeless" (Schilpp, 1949, p. 26). Einstein had discovered that Maxwell's theory of electrodynamics was more fundamental than Newtonian mechanics.

The other great theory of the twentieth century, quantum mechanics, also had its inception around the turn of the century, and also was inspired by puzzling aspects of light. In this case, scientists were trying to understand how light interacts with matter. When a solid is heated, it emits electromagnetic radiation in a continuous spectrum; power is emitted to some extent, although not equally, at all wave-

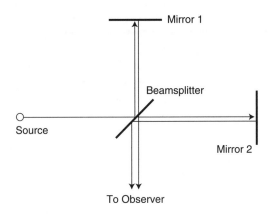

Figure 5 Michelson interferometer.

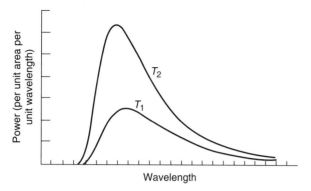

Figure 6 Blackbody radiation curves.

lengths. Physicists often speak of blackbody radiation in this regard, because a perfectly black object is easiest to treat theoretically and most solids approximate this behavior. Figure 6 shows such blackbody continuous spectra at two different temperatures ($T_2 > T_1$). The general properties of the curves, known to turn-of-the-century physicists, are evident: rising from zero at zero wavelength to a maximum at some peak wavelength, and then falling with wavelength in a long tail. At higher temperature, the curve is higher everywhere and the peak shifts toward shorter wavelengths.

The problem was that every prediction using classical physics fit the long tail reasonably well, but it always continued to climb as wavelength got shorter, approaching infinity at zero wavelength. This divergence between theory and the actual results, as shown in Fig. 6, became known as the "ultraviolet catastrophe" because it occurred at short wavelengths. Physicists knew that they were dealing with the statistical nature of many atoms acting together because of the strong forces between the atoms of a solid rather than with the properties of individual atoms, but even the most general classical considerations produced the wrong answer.

The German physicist Max Planck came up with an explanation in 1900 that seemed to predict correctly the blackbody radiation curve. He assumed that the radiation was emitted in very small chunks, or quanta, instead of in a continuous manner. If he gave each quantum a specific energy proportional to the radiation frequency, then he could predict accurately the blackbody radiation curve. His equation for energy is

$$E = hf, \qquad (2)$$

where E is the energy of quantum (J) and h is Planck's constant (6.625×10^{-34} J·s). Because of the very small value for Planck's constant, the quanta of energy are very small even for high frequencies like those of light. However, as wavelength approaches zero, the corresponding frequency becomes large without limit. This is why the discrepancy appeared at ultraviolet and shorter wavelengths; at long wavelengths, the chunks are so small that it is not seriously wrong to treat the energy as emerging continuously.

The quantum theory was not widely accepted at first because it contradicted everything in classical physics upon which scientists had come to rely. However, in 1905, Einstein used the same quantum idea to explain the photoelectric effect, which had defied explanation by classical physics. In the photoelectric effect, electromagnetic radiation, such as light or ultraviolet, irradiating some metals was found to cause the release of free electrons, which could then be collected at another metal electrode and caused to flow through a circuit. For each metal there is a cutoff wavelength above which the radiation will cause no effect. Although increasing the intensity of the incident radiation increases the number of electrons released, it does not increase their energy; only a shorter wavelength causes increased photoelectron energy. To explain the experimental results, Einstein assumed that the light is absorbed in quanta, each with energy hf, as given by Planck's formula. Each light quantum is capable of knocking loose one electron, transferring all of its energy, if it has at least a minimum amount of energy equal to what is required to break the electron free from the metal surface (called the work function of the metal). Since a minimum energy is required, a minimum radiation frequency is required, which means a maximum wavelength. More intense light then implies more photons per second incident on the surface, giving rise to more photoelectrons.

In 1913 the quantum theory gained further credibility by enabling the Danish physicist Niels Bohr to explain the atomic structure and spectrum of hydrogen. When it is heated, a gas (e.g., hydrogen) emits a spectrum different from the blackbody radiation curve of a solid. In a gas the molecules are flying around largely independent of each other and therefore emit radiation independently. The resulting spectrum is closely related to the individual molecular or atomic structure and consists of a series of discrete lines at discrete wavelengths. Therefore, gases emit a line spectrum, and different gases emit different lines (at different wavelengths). Furthermore, it had been found that any cool gas will absorb exactly the same lines that it emits when hot. No classical theory had been able to explain even the spectrum of the lightest and simplest gas, hydrogen. Bohr not only used the idea of quanta of radiation, with energy given by Planck's formula, but also extended quantum theory to explain the structure of the hydrogen atom itself. He hypothesized that the electron in the atom orbits the positively charged heavy nucleus, but only in certain discrete orbits with specific radii and specific energies. The specific, discrete electron energies in the atom, when combined with Planck's quanta of radiation, ensure that only specific, discrete wavelengths or frequencies will be emitted or absorbed by the atom; the change in energy of the electron, which is always the difference in energy between two orbits, must equal the energy of the absorbed or emitted photon:

$$E_2 - E_1 = hf, \qquad (3)$$

where E_2 is the energy of higher electron orbit and E_1 is the energy of lower electron orbit. To find the energies of the discrete orbits, Bohr invoked a new hypothesis of quantization of angular momentum. He assumed that the electron could move only in those orbits in which it had integral multiples of a basic quantum of angular momentum; the basic quantum of angular momentum he took to be $h/2\pi$, often written as $\hbar$. In other words, the first and smallest orbit that the electron could be in is that in which it has an angular momentum $\hbar$, the next is that in which it has angular momentum $2\hbar$, and so on. With these two quantum assumptions alone, and for the rest using classical physics, Bohr correctly worked out the line spectrum of hydrogen.

By 1920 it was clear that a paradox was developing in the science of light (and other electromagnetic radiation). The new quantum theory had revived the particle picture of light. These quanta of electromagnetic radiation were given the name of photons. Their reality was emphasized in the discovery and analysis of the Compton effect in 1922 by the American physicist Arthur Compton. In this effect, electromagnetic radiation of short wavelength, in the form of x rays, was scattered off loosely bound electrons. It was found that the scattered radiation was of longer wavelength, with the actual value de-

pending on scattering angle; the larger the scattering angle up to 180°, the longer the wavelength. Compton explained the result by considering the scattering process as a collision between the photons and the electrons, both treated as relativistic particles. The theory of relativity taught that particles moving at the speed of light, as photons do, had zero rest mass and momentum related to their energy by

$$E = pc, \qquad (4)$$

where E is the energy and p is the momentum (units consistent with units of E). This equation, along with Eq. (3), means that the photon momentum is related to its frequency or wavelength:

$$p = \frac{hf}{c} = \frac{h}{\lambda}. \qquad (5)$$

By virtue of these equations we can see that a glancing collision, which reduces the photon momentum only a little, leads to a small increase in wavelength, whereas a more direct collision (large scattering angle) reduces photon momentum greatly and leads to a large increase in wavelength.

Scientists began to speak of a wave–particle duality pertaining to light. In its travel through space, light shows wave properties like interference, but in its interaction with matter, such as absorption, emission, and scattering, it shows particle properties. This puzzling behavior was extended to ordinary matter, which had always been thought of as composed of particles, with the work of the French scientist Louis de Broglie in 1924. He noted that many of the properties of electrons could be explained more readily by considering them as waves, with a wavelength taken from Eq. (5) to make the treatment analogous to that for photons. Therefore, the formula for the de Broglie wavelength of a particle is

$$\lambda = \frac{h}{p}. \qquad (6)$$

This hypothesis led to rapid advances in quantum theory and represents a unification of science in that the photon became a fundamental particle like the electron, governed by the same equations.

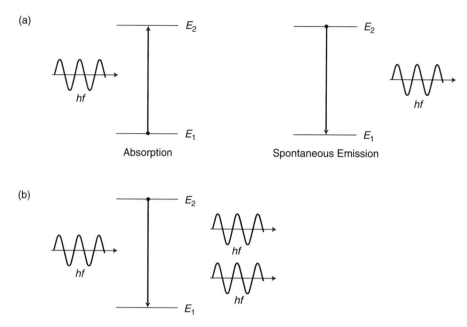

Figure 7   (a) Absorption and spontaneous emission and (b) stimulated emission.

The new understanding of light in terms of quantum theory has led to remarkable new techniques and devices, such as lasers. The word "laser" is an acronym, originally standing for light amplification by stimulated emission of radiation. The process of stimulated emission, predicted by quantum theory, is the key factor in the special properties of laser light. The processes of absorption, spontaneous emission, and stimulated emission are shown in the energy level diagram in Fig. 7, in which the discrete energies of an electron in an atom are shown as horizontal lines. In all of these processes, the photon frequency must have the value determined by the conservation of energy [Eq. (3)]. Stimulated emission can occur when such a photon encounters an "excited" atom with the electron already in the higher state of the transition. Then the incoming photon can stimulate the electron to drop down in energy and emit another photon of the same frequency and wavelength. Furthermore, the emitted photon also is exactly in phase with (lined up crest to crest) and moving in the same direction as the stimulating photon. The probability for stimulated emission to occur when the photon encounters an excited atom is exactly the same as for absorption when the photon encounters an unexcited atom.

For the operation of a laser, energy must first be given to the active material in some manner so as to produce an inverted population—more atoms in the higher energy level than in the lower, which is not the normal state of things. Once the inverted population is created, spontaneous emission will provide the first photon to start the avalanche of stimulated emission that becomes the laser beam. End mirrors reflect the light back and forth through the active material, acting to select only those photons traveling along the axis for continued amplification by stimulated emission. The resulting output beam (through one, partially transmitting end mirror) is highly monochromatic, parallel, and coherent. The first laser was built in 1960 by Theodore Maiman, a scientist working at the research laboratories of Hughes Aircraft; its active material was a ruby crystal, and it produced a wavelength of 694.3 nm. Since then, many materials, solid, liquid, and gaseous, have been used for lasers, and the available wavelengths extend from ultraviolet through infrared.

The photon is now firmly established in modern theory as one of the class of fundamental particles called bosons. These particles mediate the fundamental forces of nature, and all have integral spin. The latter means that elementary particles behave as if they have intrinsic angular momentum, as if they were spinning about an internal axis; bosons always have an angular momentum that is a whole number times $\hbar$. This property of bosons contrasts with that of the other broad class of particles called fermions, which includes the constituents of matter like electrons, and have half-integer spins. The electron has a spin of $\frac{1}{2}\hbar$, whereas the photon has $\hbar$; the direction of the photon's angular momentum is such that it appears that the axis of rotation is parallel to the particle's motion.

But what about all the nineteenth-century experiments that demonstrated the wave nature of light? Those experiments still must be considered valid, and one can say that light is made up of particles only by stretching the previous concept of what constitutes a particle. In Young's double slit experiment, the interference pattern still appears even if the intensity of the light is so low that no more than one photon at a time can be traveling from source to viewing screen. The photon is a nonclassical particle that has the wave-like property of traversing all possible paths to the screen at once, in the sense that the contribution from all possible paths must be taken into account when calculating the probability that the photon arrives at a certain point on the screen; it is these contributions from all possible paths that show interference effects. It is not so much that microscopic particles like photons and electrons differ from macroscopic (large-scale) particles like baseballs, for physicists firmly believe that large-scale particles also have a de Broglie wavelength given by Eq. (6). Rather, it is just that such strange wavelike properties are impossible to observe for macroscopic particles because their large masses, and correspondingly large momenta, lead to undetectably small wavelengths; all particles are different than physicists had thought. In the modern theory called quantum electrodynamics, all electric and magnetic forces are considered to be due to the exchange of photons between charged particles. Quantum electrodynamics, the most successful theory of all time in its ability to calculate and explain experimental results, is the latest word on the nature of light. Whether it is the last word remains to be seen.

See also: BROGLIE WAVELENGTH, DE; COMPTON EFFECT; DIFFRACTION; DOUBLE-SLIT EXPERIMENT; LASER; LIGHT, ELECTROMAGNETIC THEORY OF; LIGHT, WAVE THEORY OF; MASER; MICHELSON–MORLEY EXPERIMENT; OPTICS; QUANTUM OPTICS; REFRACTION; WAVELENGTH; WAVE MOTION; WAVE–PARTICLE DUALITY

Bibliography

BUCHWALD, J. Z. *The Rise of the Wave Theory of Light* (University of Chicago Press, Chicago, 1989).

FEINBERG, G. "Light" in *Lasers and Light: Readings from Scientific American* (W. H. Freeman, San Francisco, 1968).

HEEL, A. C. S. VAN, and VELZEL, C. H. F. *What Is Light?* (McGraw-Hill, New York, 1961).

ICKE, V. *The Force of Symmetry* (Cambridge University Press, Cambridge, Eng., 1995).

KOCK, W. *Lasers and Holography* (Doubleday, New York, 1968).

LOVELL, D. J. *Optical Anecdotes* (SPIE, Bellingham, WA, 1981).

SCHILPP, P. A., ed. *Albert Einstein: Philosopher-Scientist* (Library of Living Philosophers, Evanston, IL, 1949).

WALDMAN, G. *Introduction to Light* (Prentice Hall, Englewood Cliffs, NJ, 1983).

WHITTAKER, E. *A History of Theories of Aether and Electricity* (Harper & Brothers, New York, 1960).

GARY WALDMAN

LIGHT, ELECTROMAGNETIC THEORY OF

The idea that light consists of transverse waves in the same medium that is responsible for electrical and magnetic phenomena was enunciated by James Clerk Maxwell in 1861. Maxwell took as confirmation of his theory a striking coincidence, the close agreement between the measured velocity of light in air and the velocity with which transverse waves would move in the electromagnetic ether. The demonstration that such waves exist, which was achieved by Heinrich Hertz in 1888, led to the general acceptance of the main features of Maxwell's electromagnetic theory.

Maxwell's Theory

Maxwell began research on electromagnetic theory in 1854, a few weeks after he graduated from the University of Cambridge. At the time Maxwell began his researches, various theories about how electric and magnetic forces were related and how they acted had been proposed by physicists, most notably in Britain, France, and Germany. The German theories assumed that electromagnetic forces were like gravitation: They acted instantaneously, at a distance, and did not require a medium for their transmission. Using this model, theorists such as Wilhelm

Weber and Franz Neumann had derived electromagnetism and electromagnetic induction in a quantitatively accurate way by the early 1850s. However, these theories were silent about many phenomena. Among these was specific inductive capacity, that dielectrics (insulators) affect the outcome when a charge is placed on a conductor. Thus, the effect of an electric force depends on the intervening medium. This had been discovered by Michael Faraday, who had attempted to explain it using a nonquantitative "field" theory.

Maxwell sought to put Faraday's conception on a mathematical basis, which he did using models derived from continuum mechanics (especially fluids and elasticity). This approach involved a careful bookkeeping of energy transfer between different media during the course of electromagnetic interaction, which placed an emphasis on the behavior of insulators. In Maxwell's view, when a dielectric is acted on by any electromotive force, it is polarized in a way that is analogous to the polarization of iron by a magnet. This polarization would result from a shift or displacement of electricity in a certain direction. This displacement is, as Maxwell put it, "not a current but the commencement of a current," and when it varies a current does occur very briefly. Both the polarization and the current will depend on the specific inductive capacity (or dielectric constant) of the substance in question. This polarization was due, in Maxwell's view, to changes in the field, that is, in the space containing and surrounding the bodies affected and the ether that filled that space.

The existence of such an electromagnetic ether was deduced from optics. Around 1850 it was accepted that light was a transverse wave phenomenon in an ether that filled the universe. The ether—a "subtle" or invisible fluid—was postulated to provide a medium in which these wave disturbances could exist.

Maxwell's field theory attempted to understand electromagnetic phenomena by motions of this ether as well. By 1861, in a paper titled "On Physical Lines of Force," he proposed that electric and magnetic phenomena arise from microscopic rotations and distortions of the ether. In the same paper, the first version of his electromagnetic theory of light was presented. He conceived of light as resulting from transverse waves in the same medium, like electromagnetic disturbances but involving larger portions of the medium. Later, in his 1864 "Dynamical Theory of the Electromagnetic Field," Maxwell derived these wave motions from equations governing the behavior of the field. When the equations governing electromagnetism and electromagnetic induction are combined, a so-called wave equation results, that is, an equation for which the solution is a function that has wave-like properties. More than this, the physical parameters in both of these theories suggested strongly to Maxwell that the resulting waves would propagate in air with a velocity that was within about 1 percent of the velocity of light in air as it had been measured by Armand Fizeau around 1850.

There was other evidence as well for the connection between light and electromagnetism. In 1856 Weber and Rudolf Kohlrausch had measured the ratio c between the electromagnetic and electrostatic units. The following year, Gustav R. Kirchhoff had shown in studying alternating currents that wave-like variations of current would move along wires at a rate of $c/\sqrt{2}$, or 3.1074×10^8 m/s, which he noted was very close to the velocity of light in a vacuum.

However, direct experimental confirmation was lacking. For example, it was possible in Maxwell's theory to calculate a relationship between the index of refraction and the dielectric constant, but experimental agreement with the predicted value was not good at first. In fact, Maxwell's views, although well accepted in Great Britain at the time of his death in 1879, were not generally believed on the continent, in part because of the difficult nature of Maxwell's ideas about such basic phenomena as charge and current, in part because of the mathematical difficulty of his papers and his *Treatise on Electricity and Magnetism* (published as a textbook in 1873), and in part because his ideas lacked decisive experimental confirmation. This confirmation was to be provided a few years later by the German physicist Heinrich Hertz.

Hertz's Experiments

Born in Hamburg, Hertz had studied in Munich and, most important, in Berlin, where he was a student of Hermann von Helmholtz. While he was a student, Hertz was involved in the study of Maxwell's ideas, but it was not until October 1886, when Hertz was a professor in the technical university in Karlsruhe, that he returned to such questions decisively.

Hertz started his studies by using a special kind of induction device (called Riess spirals) that consisted of a pair of conducting wires wound into tight spirals, each terminating in a pair of metal balls. When the ends of one spiral are connected to a battery and the current is alternated, a current is induced in the second spiral, causing sparks to jump between the metal balls. Hertz decided to try to produce the sparks with a less powerful source than a battery. He found that it was possible to produce sparks between the Riess spirals when the main spiral was driven by a small induction coil, but sparks did not jump between the balls of the spirals if no spark discharge occurred in the small coil. This finding puzzled Hertz, as no explanation for the interdependence of the two sparks seemed obvious.

To understand the phenomenon, Hertz created a three-part circuit containing an induction coil, a side circuit (consisting of an unwound spiral containing a spark gap), and a Riess spark micrometer (an open circuit with an adjustable gap that is attached at two neighboring points on a wire to measure potential differences along the wire). With the micrometer attached to the side circuit in two places, Hertz was able to discover that at the moment of discharge the potential along the side circuit varied by hundreds of volts even in a few centimeters and that the discharge took place very rapidly. Hertz next eliminated one of the two connecting wires between the micrometer and the side circuit, creating a "broken" side circuit (stopping the flow of the conduction current). Hertz observed that even in the absence of the conduction current in the side circuit, sparks still jumped vigorously across the gap. The finite rate of the propagation of electric waves in the circuit must lead to the sparking, he concluded, and the critical variables in this rate of propagation were the capacitance of the circuit and the self-induction. Further tests showed that the electric waves did not subside after one transit of the circuit; they seemed to be reflected and traversed the side circuit several times, giving rise to stationary oscillations.

After more modifications to his original set-up, Hertz sought in 1887 to investigate the problem of Maxwell's theory originally set by Helmholtz for the 1879 Berlin Academy prize. Hertz noted that he could use his modified device to map effectively the electric forces due to electrostatic and inductive effects in various positions along an oscillator. At places where little or no sparking occurred, Hertz realized, the various sources of electric force were canceling one another—in effect, interfering.

Hertz's findings from that point moved at a rapid pace. By November 7, 1887, Hertz's diary records that he had found standing electric oscillations 3 m

in wavelength in straight wires. By November 10, he was producing interference between direct effects and those transmitted by wires. By November 12, he had set up experiments on the "velocity of propagation of the electrodynamic effect." And by December 23, Hertz had found evidence for propagation of the direct action, which in the next few days he concluded was actually at a speed about 1.5 times that of the wire wave.

The character of the propagation then came under investigation by Hertz. If this propagation did in fact occur in waves, Hertz thought, it ought to be possible to manipulate them as sound or light waves may be manipulated. He found "shadows" behind conducting masses and evidence of reflection. By mid-March 1888, Hertz was able to map standing waves in air produced by reflection; this required still further modifications of his earlier device, but it provided convincing evidence of the presence of the waves in a variety of configurations.

The publication of Hertz's work in July 1888 created a stir in the scientific community. George Francis FitzGerald, who had already written to Hertz in June to praise the earlier paper that demonstrated finite velocity of propagation, stated, "I consider that no more important experiment has been made in this century." The demonstration of electromagnetic waves greatly strengthened the intuitive appeal of the Maxwell approach, extending the experimental base in a very important way. To the British, this was a clear indication that whatever the residual faults with Maxwell's theory, it must be correct at its core. To the international community, already interested in Maxwell's theory but finding it difficult to comprehend, Hertz's results constituted evidence that it should be taken very seriously indeed, as it was from that time on. The belief that optical phenomena are electromagnetic in origin was an immediate consequence.

See also: CAPACITOR; DIELECTRIC CONSTANT; DIELECTRIC PROPERTIES; ELECTROMAGNETIC INDUCTION; ELECTROMAGNETIC INDUCTION, FARADAY'S LAW OF; ELECTROMAGNETISM; ELECTROMAGNETISM, DISCOVERY OF; LIGHT; LIGHT, WAVE THEORY OF; MAXWELL, JAMES CLERK; POLARIZATION; RESONANCE

Bibliography

BUCHWALD, J. Z. *The Creation of Scientific Effects: Heinrich Hertz and Electric Waves* (University of Chicago Press, Chicago, 1994).

HERTZ, H. *Electric Waves* (Macmillan, London, 1894).

JUNGNICKEL, C., and MCCORMMACH, R. *Intellectual Mastery of Nature: Theoretical Physics from Ohm to Einstein* (University of Chicago Press, Chicago, 1986).

SIEGEL, D. M. *Innovation in Maxwell's Electromagnetic Theory: Molecular Vortices, Displacement Current, and Light* (Cambridge University Press, Cambridge, Eng., 1991).

THOMAS ARCHIBALD

LIGHT, POLARIZED

See POLARIZED LIGHT

LIGHT, SPEED OF

The effort to determine the speed of light, on a historical scale, was of relatively recent origin; the first evidence suggesting light to have a finite and measurable speed was not available until the latter part of the seventeenth century. The almost universal view, prior to then, was that light was instantly transmitted through space. This belief was perfectly consistent with reasonable observation. For example, the eclipse of the Moon by Earth's shadow occurs when the Moon, Earth, and the Sun are in a straight line. Unless light traveled at an unimaginable speed, the Moon should have moved before Earth's shadow reached it. This argument was presented by the noted French philosopher and mathematician René Descartes. Although there were critics of this argument, notably by the Dutch physicist and astronomer Christiaan Huygens, it was very much the norm for the latter part of the seventeenth century. It was easier to accept that light instantaneously connected one point with another than imagining it traveling at some literally inconceivable speed.

The first direct evidence that light has speed was furnished by the Danish astronomer Olaus Roemer in 1676, when he predicted a seasonal shift in the time for the emergence of one of Jupiter's moons. This was based on the assumption that when Earth was on the other side of its orbit from Jupiter, the light from the Moon would have to cross the diameter of Earth's orbit, and this would delay its observation. On November 9, 1676, his prediction was confirmed by Paris astronomers. Many different numbers can be found in textbooks for the value of the speed of light determined by Roemer. The reason for this diversity was that Roemer *did not* calculate a speed for light. The various stated values are apparently suppositions. For Roemer the question was not "What is the speed of light?" but rather "Does light have speed?" However, the question of finite speed was not actually settled for another fifty years. General acceptance occurred in 1728 when the English astronomer James Bradley was able to demonstrate convincingly that light takes a finite time to travel down the tube of a telescope. Bradley did this by showing that in observing a star one must "lead" the star with the telescope much like a hunter must "lead" a moving target because of the finite speed of the bullet.

The first non-astronomical experiment to determine the speed of light was performed by the French physicist Armand Fizeau in 1849. For this, Fizeau used a toothed wheel that could be rotated with variable speed. With the disk at rest, light was sent from the observer's side of the wheel through a tooth gap to a distant mirror and back through the gap to the observer. The wheel was then rotated, and at some point the reflected light was blocked by the next tooth. At a faster rate, the light returned and passed through the next gap. Since the rotation rate of the wheel was known, the time for the second gap to get into place was also known. This, along with the known distance to the mirror and back, was used to determine the speed of light. At almost the same time, Fizeau's fellow French physicist Jean Foucault was measuring the speed of light in water and air by using a rotating plane mirror to accomplish much the same purpose as the toothed wheel. Toward the end of the nineteenth century, American physicist Albert Michelson used an improved version of the Foucault apparatus to determine the speed of light in various media. Michelson devoted his life to the study of light, performing many crucial experiments, and received the Nobel Prize for his work in 1907. In doing so, he became the first American to win this award.

Comparing the earlier determinations of the speed of light to the current best determination of this speed is complicated by two factors. First, probably because of uncertainty in the diameter of Earth's orbit, the early investigators usually limited themselves to stating times for light to cross Earth's orbit or to get from the Sun to Earth, rather than giving a calculated speed. For example, Newton, in the first edition of his book *Principia* (1703), gave 10 min as the Sun to Earth travel time for light. In his second edition of *Principia* (1713), he gave the time of 7–8 min. Later in his book *Opticks* (1717–1718 edition), he also used 7–8 min, and based on an assumed distance to the Sun of 70 million miles, he estimated the speed of light to be 700,000 times faster than the speed of sound (or about 150,000 miles/s). Huygens meanwhile had estimated the speed of light to be 600,000 times the speed of sound. Bradley, on the other hand, estimated the time of travel for light from the Sun to Earth as 8 min 12 s (within a few seconds of the modern accepted value) and estimated the speed of light as 190,000 miles/s. The second problem with tracing the experimental development of the speed of light is that the rules were changed in 1983 when the International Committee on Weights and Measurements defined the speed of light to be exactly 299,792,458 m/s. In doing this they redefined the meter to be the distance light travels in 1/299,792,458 s. Michelson's best value (1927) for the speed of light in terms of the old meter was 299,789,000 ± 4,000 m/s. Recent measurements for the speed of light, which immediately preceded the redefinition of the meter, all were within four parts in a billion of the redefined value.

The first person to suggest that light was a vibration or wave traveling through some medium that pervades all of space was Robert Hooke in 1667. Although disagreeing with Hooke on the wave-like nature of light, Newton called this medium Æther (normally referred to as the ether) and speculated in *Opticks* that, based on the relative speed of light and sound, the ether was 700,000 times stronger (resistance to compression) than air but at the same time had only 1/700,000 of air's density. The idea of the ether influenced the development of science up to the first decade of the twentieth century. From Newton's time to that of Albert Einstein, numerous prominent scientists worked on models of the ether in an attempt not only to explain the transmission of light but also electric and magnetic phenomena. Such models ranged from fluids with whirlpools to elastic solids. The list of prominent scientific fiures who worked on models of the ether includes John Bernoulli, James Clerk Maxwell, Lord Kelvin (William Thomson), George FitzGerald, Arthur Sommerfeld, Hendrick Lorentz, and Henri Poincaré. In spite of this, no entirely satisfactory model of the ether was ever developed.

Another fundamental question about the ether arose during the last quarter of the nineteenth century. The question, simply put, was "Could the ether be detected?" If the answer was "no," then the existence of the ether either had to be denied or the negative result explained. From 1880 to 1905, great effort was devoted to explaining negative results obtained in experimental attempts to determine Earth's motion through the ether. The most famous of these was the 1887 experiment by Michelson Edward Morley. The Michelson–Morley experiment was based on the assumption that just as the properties of water and air determine the speed of water waves and sound waves, the properties of the ether should determine the speed of light waves. This value was determined from a set of governing equations called Maxwell's equations and was consistent with measured values of the speed of light. The experiment was analogous to someone using sound and/or water waves to determine their own speed. As you approach a source of water waves or sound waves, the speed with which they approach you through their respective media is increased by your speed toward them through the medium. Knowing beforehand what the wave speed should be in the media, say sound waves in air, you could use the relative speed of the sound to you to calculate your speed through the air. This is exactly what Michelson and Morley tried to do with Earth, light, and the ether. Despite formidable efforts on their part, they failed to do so. Since the speed of Earth in its orbit is about 1/10,000 that of light and since their experiment was accurate to 1/100,000,000, this was a stunning failure.

Many investigators where involved in the attempt to explain the "conspiracy" by which nature chose to hide the ether from scientific detection. Chief among these were Lorentz and Poincaré, Lorentz as principle investigator and Poincaré as principle critic. In order to explain the impossibility of detecting motion through the ether to the first order in v/c, where v is Earth's speed through the ether and c is the speed of light, Lorentz was led in 1892 to postulate the idea of local or mathematical time t'. Local time was related to true time as kept in the frame of reference of the ether by the relationship $t' = t - vx/c^2$, where x is Earth's position in the ether frame at true time t. Similarly, in 1895, in order to explain the failure to detect Earth's motion in the second order, that is, $(v/c)^2$, he was led to assert (as did FitzGerald independently) that bodies contract in the direction of their motion through the ether by the factor $\sqrt{1 - (v/c)^2}$. Both of these proposed adjustments to space and time eventually led in 1904 to a set of transformations of space and time called the Lorentz transformations, which form the basis of modern relativity theory. While Lorentz hung on to the concept of the ether as an important ingredient of his physics, the ether played an increasingly insignificant role in Poincaré's analysis. This led him to develop a principle of relativity (1904–1905) in which he asserted that only relative motion is detectable and that the laws of physics are independent of the relative motion of the observers. He also demonstrated that the speed of light was the same for all observers, which meant that the ether, if it exists at all, is, in principle, undetectable. It is probably because of Poincare's unclear position on the ether (Does it exist or not?) that credit for the principle of relativity is today attributed to Albert Einstein, a theory he annunciated in 1905. Einstein's work directly states that the ether is superfluous; otherwise, his starting postulates are the same as those of Poincaré. Both men worked independently, and the results of their efforts where essentially the same. However, Poincaré's work was rooted in the old physics that focused on the dynamics of the electron and the effects of motion on its internal structure and mass. Einstein treated the electron as a mechanical point particle, as is still done today. Thus, the relativistic results for energy and momentum

$$E = \frac{mc^2}{\sqrt{1 - (v/c)^2}},$$

and

$$p = \frac{mv}{\sqrt{1 - (v/c)^2}},$$

were interpreted by Poincaré as showing that the mass of a particle increases with velocity. Einstein had the same results and also introduced the idea of relativistic mass, but he was careful to indicate that it was essentially a concept of conveyance, that is, a definition. In reality, the factor of $\sqrt{1 - (v/c)^2}$ arises from the role time plays in his derivation and has no association with the mass.

For a body with mass to reach the speed of light, it would have to acquire infinite energy; conse-

quently, no material body can travel at the speed of light. Finally, an important relationship between E and p, determined from the above equations, is

$$E^2 = (pc)^2 + (mc^2)^2.$$

This relationship holds even for particles without mass, such as photons (quanta of light), and probably also for the mysterious particle called the neutrino. For such particles, the equation takes the simple form $E = pc$.

See also: EINSTEIN, ALBERT; ETHER HYPOTHESIS; FRAME OF REFERENCE, INERTIAL; HUYGENS, CHRISTIAAN; LIGHT; LIGHT, ELECTROMAGNETIC THEORY OF; LIGHT, WAVE THEORY OF; LORENTZ, HENDRIK ANTOON; LORENTZ TRANSFORMATION; MAXWELL, JAMES CLERK; MICHELSON, ALBERT ABRAHAM; MICHELSON–MORLEY EXPERIMENT; NEUTRINO; NEWTON, ISAAC; PHOTON; RELATIVITY, SPECIAL THEORY OF

Bibliography

BATES, H. E. "Recent Measurements of the Speed of Light and the Redefinition of the Meter." *Am. J. Phys.* **56,** 682–687 (1988).

GIACOMO, P. "The New Definition of the Meter." *Am. J. Phy.* **52,** 607–613 (1984).

WROBLEWSKI, A. "de Mora Luminus: A Spectacle in Two Acts with a Prologue and an Epilogue." *Am. J. Phys.* **53,** 620–630 (1985).

CARL G. ADLER

LIGHT, WAVE THEORY OF

The image of light as a mechanical disturbance in an all-encompassing medium dates to the seventeenth century, when the French mathematician and philosopher René Descartes made the concept an essential part of his system. In 1695 the Dutch scientist Christiaan Huygens fundamentally changed the Cartesian scheme by insisting that the optical disturbance takes time to move from one point to another, and he further introduced a mathematical rule, Huygens's principle, to quantify the process. According to the principle, each point on an optical surface is itself to be considered the source of a propagating disturbance, in such a fashion that the entire surface is at any moment the linear resultant of these secondaries. Huygens thought these disturbances to be what is called in modern physics a longitudinal pulse, which is a temporally isolated disturbance that parallels the direction of propagation. These pulses accordingly had no periodic properties, and indeed Huygens's theory of light was able to deal neither with colors nor with phenomena of polarization, which he had himself discovered on passing light twice through certain exotic crystals brought from Iceland.

During the eighteenth century Leonhard Euler further developed the conception of light as propagation in a medium, adding the notion that the motion occurs in regular waves, with color corresponding to the wave frequency. For the most part, Euler concerned himself rather with the physics of the optical medium than with the quantitative properties of propagating surfaces. Indeed, though a considerable amount of speculative natural philosophy was produced during the century, quantitative optics remained for the most part bound to the concept that the fundamental object of light is the ray. The ray had long been the foundation of geometrical optics, and in the seventeenth century it had acquired a new reality as marking the track of the particles out of which Isaac Newton built light. Variants of both Huygens's and Newton's systems existed during the eighteenth century, but neither one can be said to have stimulated the creation of new mathematical or experimental developments in optics during these years.

In 1799 Thomas Young in England began a series of publications that substantially extended the quantitative power of medium theories of light. Young, like Euler, associated color with wave frequency, but he went far beyond Euler in his use of the assumption. Trained as a medical doctor, Young had during his studies become deeply interested in acoustics, and especially in phenomena of superposition. This eventually led him to the principle of interference, according to which waves of the same frequency and from the same source will, when brought together, produce regular, spatial patterns of varying intensity. That principle had not been understood for waves of any kind until Young began his investigations, and it can indeed be said that the study of wave interference in general began with Young himself, though he did not pursue it extensively outside of optics. Many difficult problems had to be solved

by Young, including the conditions of coherence that make detectable spatial interference possible at all. Furthermore, the principle of superposition, according to which waves combine linearly and which is a necessary presupposition for the principle of interference, was itself quite problematic at the time and had to be developed and argued for by Young.

Young applied his principle of interference to the diffraction of light by a narrow body, as well as to the case of light passing through two slits, although in the latter case it seems that he did not carry out careful measurements. In all cases, Young explained the fringe patterns that he observed by calculating the path difference between a pair of rays that originated from a common source. He did not calculate with waves themselves but rather assigned periodicity to the optical ray, which accordingly retained a crucial place in Young's optics. Although Young was certainly quite familiar with Huygens's work, he did not use the latter's reduction of rays to purely mathematical artifacts, for that was bound to Huygens's principle, which Young found difficult to accept.

In any event, Young's optics did not generate extensive immediate reaction. Indeed, his principle of interference was sufficiently difficult to assimilate that no other applications to new phenomena were forthcoming. The most famous, or (in retrospect) infamous, reaction, was that of Henry Brougham. Brougham vehemently objected to Young's wave system as an alternative to the Newtonian scheme of optical particles, and he objected to the principle of interference itself, even as a mathematical law applied to rays. Most contemporary optical scientists were more interested in the physics than in the mathematics of light, and in this area Young's ether posed as many qualitative problems as did the alternative system of light particles.

After about 1805, however, interest in quantitative optics became quite strong in France, where a school of scientists led by Pierre-Simon Laplace was extending mathematics and, indeed, the physics of particles governed by forces into many new areas, including heat and capillarity. Stimulated in part by the English chemist William Hyde Wollaston's apparent confirmation in 1802 of Huygens's construction for double refraction in the crystal Iceland spar, Laplace had Etienne Louis Malus undertake a thorough experimental investigation in 1808. After translating Huygens's construction into algebra, Malus undertook a careful study that showed unequivocally that the construction is extremely accurate. Neither he nor Laplace, however, concluded that Huygens's theory must therefore be accepted. Instead, both showed, in different ways, that the resulting formulas are compatible with the principle of least action, which they took to be uniquely associated with particles and forces. Although Young pointed out that waves similarly moved in accordance with the principle of least time and that by taking the velocity of waves as the reciprocal of the velocity of particles the two principles necessarily led to the same observational results, for more than a decade and a half optics in France remained closely bound to the emission theory, according to which light rays consist of rapidly moving particles that can be deflected by forces emanating from material bodies.

The persuasive claims of the emission theory were furthered in no small measure by Malus's discovery in 1809 of polarization by reflection. Huygens had already noted that light emerging from doubly refracting crystals seems to have some sort of asymmetry associated with it, since on entry into a second crystal it is not equably divided in two again. In 1809 Malus discovered that the property did not require a crystal, but that reflection from a surface at a particular angle can also produce it. This discovery stimulated a tremendous amount of experimental and theoretical work during the next decade, undertaken especially by François Arago and Jean-Baptiste Biot in France, as well as by David Brewster and, somewhat later, John Herschel in Great Britain.

Since the 1830s, when the successful polemicists of the wave theory such as William Whewell first molded history to their designs, we have been accustomed to considering the opposition between physical theories of light in the early nineteenth century solely as one of waves in the ether versus particles of light, that is, as one between the wave theory and the emission theory. Such an opposition certainly has much historical warrant, but it is only one aspect of a complicated story. If this opposition captured the entire history then it would indeed be difficult to see how anyone could ever have calculated anything. The emission theory deployed particles and forces to explain optical phenomena, but only the simplest deductions could be drawn from such a scheme. Certainly neither Malus nor anyone else computed from this scheme how polarized light behaves on reflection. Yet Malus did provide computations, and so it is apparent that he must have had something different from the emission theory proper, though it would no doubt be compatible with it. And he did; by assuming that rays of light can be treated as individual objects that can be counted arithmetically, and so grouped into related sets, Malus was able to build a very successful quantitative account of partial reflection.

To do so, he hypothesized that polarization, properly speaking, does not apply to the individual rays but only to groups of rays (or beams of light): Each ray in a polarized set has an asymmetry about its length that is the same as every other ray in the set. Sets with different polarizations consist of rays with different asymmetries. Reflection and refraction redistribute the rays between the sets and even create entirely new sets. Malus was able in this way to create a group of formulas that accounted very well indeed for partial reflection. Biot, a graduate like Malus of the Ecole Polytechnique, soon achieved great success in optical investigations by applying Malus's quantitative theory to colors generated by thin crystals, a phenomenon first investigated by Arago, yet another *polytechnicien*. Biot and Arago became bitter enemies after Biot intruded on the former's work in crystal optics, and in later years Arago turned away from Malus's principles, which had been thoroughly developed by Biot. Arago instead embraced a new theory developed by Augustin-Jean Fresnel, yet another graduate of the Ecole. Indeed, it was Arago who brought the young Fresnel to Paris, who participated in experiments with him, and who ensured that his results became well known.

Two concepts in particular—periodicity and Huygens's principle—were instrumental in leading Fresnel to a fully fledged wave theory for diffraction. Although he early produced, as had Young before him, a theory that combined periodicity with rays, only when he deployed Huygens's principle in 1818 did Fresnel move completely past Young's previous accomplishment. Fresnel now assumed that each of Huygens's secondaries constituted an independent disturbance, and that the entire wave front at any time and point must be found by calculating the resultant formed there and then by all of these secondaries. To compute the interference of an infinite set of waves required Fresnel to use differential calculus in a way that Young, though rather adept in the old ways of the fluxional calculus, simply did not have the training to pursue.

The combination of theory (periodicity and Huygens's principle) with sophisticated but not elaborate mathematics forms the bedrock of Fresnel's account of diffraction, which won for him a Paris Academy prize in the subject (though not also the acceptance by most Parisian scientists of the wave theory itself, for they considered Fresnel's integrals to be mathematical structures whose proper physical interpretation remained to be worked out). Much the same can be said for his work on polarization and double refraction. Here, however, the going was much harder for Fresnel than it had been in the case of diffraction, because for many years he continued to think about polarization in ways that derive from an optics based on rays rather than from one based on waves. Indeed, the most difficult thing for Fresnel to achieve, and the one that his contemporaries found the hardest to accept, was not his account of diffraction but rather his theory for polarization and its immediate result, his analysis of double refraction.

As we saw above, contemporary French understanding of polarization considered it to be an essentially static, spatial process in which the rays in a given group have their asymmetries aligned in certain ways. Time does not enter into this scheme, and, according to it, a beam of observably unpolarized light is always just unpolarized, no matter how small a time interval one might consider. For several years Fresnel found it extremely difficult to discard this notion that common light must not show any signs of asymmetry at all, though he could not use ray-counting procedures to explain why not. He accordingly tried hard to build a scheme in which polarization consists of a temporally fixed combination of directed oscillations: one along the normal to the front and the other at right angles to it. In common light, the transverse component vanishes altogether; in completely polarized light, the longitudinal component disappears. This, like its ray-based counterpart, is an inherently static, spatial image. Fresnel was not successful in building a quantitative structure on this basis. Then, in 1821, he set his static image into motion.

The core of Fresnel's novel view of polarization referred the phenomenon to the change (or lack of change) over time of a directed quantity (amplitude) whose square determines optical intensity. This quantity must always lie in the wave front (and is therefore transverse to the ray in optically isotropic media), and in reflection and refraction, it can be decomposed, with the components in and perpendicular to the plane of reflection being affected in different ways. Common light consists of a more-or-less random rotation and amplitude change over time of this directed oscillation and not, as Fresnel originally understood it, of a spatially fixed (longitudinal) disturbance.

By the end of the 1830s the tools and concepts of wave optics had begun substantially to displace those of ray optics in France and Britain, and by the 1850s few practicing and publishing scientists refused to accept and to use wave principles. These principles, combined with the widespread conviction that waves of light propagate in a universally present medium (the ether), strongly influenced the character of much subsequent physics, including the development of electromagnetic field theory, by directing attention to the characteristics of the ether itself, which became a major arena within which nineteenth-century physicists could deploy their mathematical and theoretical tools.

See also: COHERENCE; DIFFRACTION; DIFFRACTION, FRESNEL; DOUBLE-SLIT EXPERIMENT; FRESNEL, AUGUSTIN-JEAN; HUYGENS, CHRISTIAAN; INTERFERENCE; LAPLACE, PIERRE-SIMON; LEAST-ACTION PRINCIPLE; LIGHT; LIGHT, ELECTROMAGNETIC THEORY OF; OPTICS, GEOMETRICAL; POLARIZED LIGHT; WAVE–PARTICLE DUALITY; WAVE–PARTICLE DUALITY, HISTORY OF; YOUNG, THOMAS

Bibliography

BUCHWALD, J. Z. *The Rise of the Wave Theory of Light* (University of Chicago Press, Chicago, 1989).

KIPNIS, N. *History of the Principle of Interference of Light* (Birkhaüser Verlag, Berlin, 1991).

JED Z. BUCHWALD

LIGHTNING

Lightning is a transient visual display, generally extending over some kilometers in length, produced by a flow of electric charge, or current. This discharge usually takes place either between certain clouds and the ground, or between the clouds themselves. On occasion, lightning also can be associated with sandstorms, snowstorms, clouds over active volcanoes, and in the vicinity of thermonuclear detonations. The path followed by the electric current delineates the lightning flash. We are able to see this path because the air in it is heated to very high temperatures, with some of its molecules being raised to excited levels and others being ionized due to collisions with the electric charge carriers flowing in the flash. Temperatures of up to 30,000 K have been measured in lightning flashes. In addition to any thermal radiation, each atomic and molecular species, so excited or ionized, emits light of specific wavelengths (colors). The totality of all visible emissions results in our eye–brain system registering what we perceive to be essentially white light. Lightning is most often associated with thunderstorms, and it is in the thundercloud (cumulonimbus) that the necessary spatial separation of electric charge into positive and negative components takes place.

Heating of the surface of the earth by the Sun does not occur in a perfectly uniform manner. In a region where this heating is in excess of surrounding regions, the air is raised to a higher temperature. Such air has a lower density than that of its surroundings and, as a result, begins to rise. When hot air rises, it begins to expand and cool. In order for a thundercloud to form, it is necessary that the temperature of the rising air in the cloud not drop as fast as that of the air surrounding it. The process is aided if the overlying air at higher elevations is part of a cold front. It is well-known that the temperature of the atmosphere drops, on average, approximately 6.7 K for each kilometer of rise in altitude. If the mass of air that is to become the thundercloud is sufficiently moist, the conditions for its not cooling as fast as the surrounding air can be satisfied. The liberation of the latent heats associated with the condensing and freezing of the water vapor in the prospective thundercloud is able to guarantee cooling rates as low as 3.0 K/km and thus ensure that the air of the thundercloud remains warmer and less dense than the air surrounding it. Under these conditions, the thundercloud can reach heights of up to 40,000 ft, with temperatures in the upper regions of the cloud reaching values as low as $-60°C$.

It is in the interior of the thundercloud, where a maelstrom of wind, water, and ice exists in the presence of gravitational forces and temperature differences, that positive and negative electric charges get separated into distinct regions within the cloud. The bulk of the positive charge collects in the upper regions of the cloud, while the negative charge collects in the lower regions. The exact mechanisms whereby this occurs are still not completely understood and comprise an area of ongoing research.

Lightning, in the form of an electrical discharge from cloud to ground, proceeds in a stepwise fashion, usually with negative charge being sent down in a channel toward the ground. Measurements of this phenomenon indicate currents of the order of 100 A to be flowing. When the most advanced step gets close enough to the ground, the electric field is great enough to initiate an upward moving discharge from the ground. This is called the return stroke and with its arrival at the leading point of the downward stepwise discharge, the lightning flash is complete. Currents at the ground can momentarily reach values as high as 20,000 A. The total process involved in the flash takes of the order of $\frac{1}{5}$ s. The result of a typical lightning flash is to deposit about 20 C of negative charge on the ground. Heating of the air in the discharge channel produces the thunder that is associated with most lightning flashes.

There is an interesting feature of the earth's atmosphere that is not readily noticed. Careful measurements indicate that there is a fine-weather electric field at the earth's surface, directed perpendicularly downward to the surface, and having a value of about 100 V/m (N/C). This field falls off in value at higher altitudes, but it nevertheless is responsible for a voltage difference of about 400,000 V between the ground and an essentially conducting layer of the atmosphere at an altitude of about 50 km. While air is a relatively weak conductor of electricity, it can be shown that this voltage difference should be neutralized in less than a minute by leakage currents flowing through the air between the ground and the conducting layer at 50-km elevation.

How, then, does this fine-weather electric field manage to persist? The answer is that cloud-to-ground lightning continually replenishes the negative charge on the earth's surface, thereby maintaining this field. In order for this to be accomplished, something of the order of 2,000 lightning discharges must occur about every minute worldwide. Thus, the overall effect of cloud-to-ground lightning is to keep the atmospheric capacitor in which we live charged up.

Lightning can take forms other than that described above. Heat and sheet lightning generally refer to situations where clouds are illuminated by lightning, sometimes too distant for any associated thunder to be heard. Ribbon lightning or bead lightning occur when the discharge channel to ground shifts or breaks up, respectively, during the flash. Perhaps the most unusual form of lightning is ball lightning, which is usually composed of luminous spheres, some tens of centimeters in diameter, that have the capability of moving around on their own for times of the order of a few seconds. There is still no generally accepted explanation for this type of lightning.

See also: ELECTRICITY; SHOCK WAVE; THUNDER

Bibliography

MALAN, D. J. *Physics of Lightning* (English Universities Press, London, 1963).
UMAN, M. A. *Lightning* (Dover, New York, 1984).

DAVID JOHN GRIFFITHS

LIGHT SCATTERING

See SCATTERING, LIGHT

LINE SPECTRUM

Line spectra of electromagnetically excited atoms and molecules, in their gaseous, plasma, liquid, and solid phases, as well as analogs of these line spectra for excited nuclei and even for elementary particles belonging to the realm of particle physics and cosmology, continue to be recorded, studied, analyzed, refined, catalogued, and referred to millions of times every year. Line spectra, together with their close cousins in the frequency domain, constitute the most tangible and universal physical recordings of the pheonomena of resonance so ubiquitous in the physical world, both classical and quantum. Matter and radiant energy observed at the quantum level—or at least that portion of it the human mind perceives to be of interest—is nearly universally described in terms of resonances, untold billions of which remain to be discovered or better characterized in the intellectual attempt to account for the behavior of excited atoms and molecules constituting matter. Such atoms and molecules are very often found to exist in discrete (quantized) excited states of highly variable lifetimes. We have come to appreciate the power of describing the interactions of matter with other matter largely through the exchange of energy by the excitation and decay of quantum resonances of finite lifetime. These resonances represent the temporary internal storage of energy in the form of elementary excitations of atoms, molecules, nuclei, and even elementary particles for vastly differing time periods; in the case of excited atoms and molecules, this energy is characteristically released and exchanged by the emission and absorption of electromagnetic radiation (light), which, together with collisions of particles, represents the primary means of energy exchange among constituents of a composite assemblage of atoms in a myriad of atomic, molecular, gaseous, plasma, fluid, or solid forms. Electromagnetic radiation caused by the decay of individual resonances is a typically unique but ephemeral genetic fingerprint labeling both the type and state of excited atoms, molecules, plasmas, and so on, which emit it.

A line spectrum represents a hard, permanent, physical record of this emission, typically in the form of storage in a photographic image, or in an equivalent optical image stored in a two- or three-dimensional digital optical image.

So what are line spectra, and how did their study begin? Spectroscopy got started about 1666, with Isaac Newton's discovery that differently colored rays of light refract at different angles when allowed to pass through a glass prism. Nearly everyone has seen that sunlight, collimated by a small aperture prior to passing through such a prism or crystal, spreads out into a beautiful band of colors when falling on a screen behind it. With a suitable lens placed in the optical path, Newton was able to show that the band of colors became a series of variously colored images of the hole in a diaphragm he was using as an aperture, displaced from one another (i.e., dispersed) along the screen. Newton called this pattern a spectrum.

Harvey White's *Introduction to Atomic Spectra* provides a succinct and engaging account of critical followups of Newton's epochal development, which first led to the understanding of atomic spectra, and in turn to the Bohr atom, the quantum theory of matter and radiation, and many other foundation stones of the rich quantum resonance physics pursued at all levels of particle size and complexity of condensed matter organization. If Newton had instead used a tall, narrow, illuminated slit as a primary light source, he might have made the early nineteenth-century discoveries of W. H. Wollaston and Joseph von Fraunhofer of the dark "lines"—dark images of the slit—at selected places in the more continuously distributed, bright bands constituting the solar spectrum observed by spectroscopes of the period. These dark lines remained mysterious until

Jean Foucault showed that when broad-banded light from a very powerful arc was passed through a sodium flame just in front of the slit of his spectroscope, two dark lines appeared at precisely the same positions as the two corresponding Fraunhofer lines (labeled "*D*" lines) in the solar spectrum. Soon thereafter Gustav Kirchoff came up with the theory that the Sun is surrounded by layers of cooler gases that can selectively absorb bright line emissions emitted from hotter vapors below, the source of emission features which would be bright in the absence of absorbing material. Given sufficient absorbing material of the appropriate elemental constitution, filtering out of these resonances by absorption gives rise to apparent "dark" lines superposed on a bright adjacent background for which the filtering is less effective. With this new understanding, it soon became accepted that many elements found on Earth, for which by this time many bright line spectra had been accumulated, were in fact identical to those found in the Sun.

The invention of the diffraction grating allowed an enormous improvement in the resolution obtainable for the study of isolated spectral lines, which for many resonances in well-separated elementary atoms are so closely spaced they are always blended together by much cruder prism spectroscopes, giving rise to an apparently broad-banded, continuous spectrum, which can often be separated into millions of very closely separated bright lines, given sufficient resolution. Resolution refers to the extent to which it is possible to cleanly separate (resolve) two or more closely spaced lines. Henry Rowland invented a good grating ruling engine and began applying it to rule diffraction gratings about 1882, shortly threafter publishing a photographic map of the solar spectrum about 50 ft long, with millions of lines recorded thereon, with minimum resolution or line separations good to a ten-millionth of a millimeter (10^{-10} m, or 0.1 nm). Rowland had made it possible to see the individual leaves of grass in a brightly colored but otherwise apparently smooth field!

So what, exactly, are the properties of a spectral line on a photographic or digital image? The length of the line is usually simply related to the length of a long narrow slit in a spectrometer containing a diffraction grating that disperses light and causes the wavelength-selected (color-selected) light to pass through an exit slit just in front of a photographic plate or digital optical image plate positioned behind it. The strength, brightness, or intensity of the line recorded is a measure of the brightness or intensity of the light of a particular resonant wavelength that illuminated the entrance slit and was dispersed through the exit slit of the spectrometer. The width of the line, and its separation from its nearest neighbors on the image, is often composite, with contributions to the total width arising from several distinct sources. First, for purely geometrical reasons the finite widths of the entrance and exit slits generally contribute appreciably to the overall linewidth observed. Second, the line may have intrinsic width owing to the short, finite time that the resonance giving rise to emission of the line lasts.

The Heisenberg time-energy uncertainty principal of quantum physics notes that a resonance lasting a finite time must have a corresponding minimum energy uncertainty. The Planck hypothesis relating the amount of energy contained by a quantum of radiation (photon) and its frequency thus prescribes a frequency uncertainty (bandwidth) corresponding to the energy uncertainty. Finally, since the wavelength of light and its frequency are reciprocal quantities, defined so that the wavelength of light λ times its frequency v always equals a constant of nature, the velocity of light, 2.998×10^{10} m·s^{-1}, then with each bandwidth uncertainty δv is associated a certain spread in wavelength, $\delta \lambda$. If the geometrical resolution of the spectrometer is good enough, this "natural" linewidth may then account for an appreciable fraction of the total width as well. In addition to these overall properties of height, width, and intensity, the lines observed generally have characteristic shapes, determined by the geometrical properties of the slits, gratings, mirrors, or other components contained in a typical spectrometer, ranging from triangular in shape to Gaussian to other forms as well. Here, "shape" refers to the geometrical form of a graph that records the intensity of the line between its left- and right-hand limits. If the resolution of the spectrometer is not good enough or the natural width of the resonance responsible is too broad, more composite line shapes reflecting intruding, closely spaced nearby resonances that contribute to an overall, blended line shape may result.

Other sometimes more powerful but often less versatile or more expensive means of recording resonance positions and widths other than the line spectra of concern here have become very well known in the twentieth century. The precision of time and frequency standards have reached unprecedented levels in recent decades. Since, as noted above, wavelengths and frequencies along with their uncertainties can be stated interchangeably, then, whenever a frequency and uncertainty can be established more precisely by a resonance frequency measurement rather than a wavelength measurement—as is often the case when utilizing laser technology in the visible and infrared regions of the spectrum—it is advantageous to use frequency measurements as opposed to line spectra measurements to derive precise results. Over vast regions of the electromagnetic spectrum, however, now accessible to tens of decades of photon energy ranges, such frequency measurements are not yet possible. Hence line spectra continue to represent a convenient and reasonably economical tool for recording spectra over a wide range of photon energies for which resolutions at the sub-nanometer level are adequate for the purpose at hand.

Still other interferometric (but generally non-diffractive and nondispersive) means of making precise length measurements at the subwavelength level of light (but using a spatial rather than time-frequency domain) have reached a stage of very sophisticated development.

See also: DIFFRACTION; DISPERSION; ELECTROMAGNETIC SPECTRUM; FRAUNHOFER LINES; GRATING, DIFFRACTION; HEISENBERG, WERNER KARL; INTERFERENCE; INTERFEROMETRY; PLANCK CONSTANT; QUANTUM; SPECTROPHOTOMETRY; UNCERTAINTY PRINCIPLE

Bibliography

EISBERG, R., and RESNICK, R. *Quantum Physics* (Wiley, New York, 1974).

WHITE, H. E. *Introduction to Atomic Spectra* (McGraw-Hill, New York, 1934).

IVAN SELLIN

that of some solids. Liquids, however, exhibit zero rigidity and offer zero resistance to shear under static condition. Under dynamic condition, they generally exhibit some viscosity in the presence of velocity gradient. For extremely rapid dynamic changes, liquids behave like solids. For instance, if one runs water from a faucet and hits the column of water rapidly sideways, a high-speed photograph of this event would closely resemble that of hitting and fracturing a column of ice.

Thermal motion of molecular or atomic constituents of liquids is more active and less constrained than those in solids and is highly random. There are no fixed lattice sites, and intermolecular distance is statistically distributed. Most substances, with a few exceptions, possess a liquid phase as the temperature is raised from its solid state at atmospheric pressure. Dry ice (CO_2) is a notable exception. It melts only above 5.6 atm of pressure. Liquid helium, on the other hand, does not solidify at any temperature at 1 atm.

Liquids do not have the atomic regularity and symmetry of crystalline solids. Theoretical description of normal liquid is made difficult by the absence of symmetry. At the same time, liquids lack the total randomness that enables simplifying assumptions and facilitates statistical calculations, as is the case with the kinetic theory of gases.

See also: CONDENSATION; CONDENSED MATTER PHYSICS; GAS; KINETIC THEORY; LIQUID HELIUM; PLASMA; SOLID; VISCOSITY

Bibliography

GIANCOLI, D. C. *Physics for Scientists and Engineers,* 2nd ed. (Prentice Hall, Englewood Cliffs, NJ, 1984).

SEARS, F. W.; ZEMANSKY, M. W.; and YOUNG, H. D. *College Physics* (Addison-Wesley, Reading, MA, 1985).

CARL T. TOMIZUKA

LIQUID

A liquid is one of the three states of matter along with solid and gas. Liquids and solids together are called condensed matter. Amorphous solids are sometimes regarded as liquids with very large viscosity or as supercooled liquids. Volume compressibility of liquids is generally small and is comparable to

LIQUID CRYSTAL

Liquid crystals are materials that have structural and mechanical properties intermediate between those of liquids and crystals. A liquid crystalline phase of a material is also sometimes referred to as a "mesophase" (meaning intermediate phase) or "mesomorphic phase."

In crystal, atoms or molecules are arranged in a regular, periodic lattice. Crystals are thus highly ordered on the microscopic scale. Given the position of one molecule, one can predict with a high degree of confidence the positions of other molecules quite a large distance away. Crystals resist shear deformation. In a liquid, atoms or molecules move almost freely and are highly disordered on the microscopic scale. There is still some "short-range order" due to the short-range repulsion between atoms or molecules. Because of the lack of long-range positional order, liquids have no resistance to shear deformation and flow freely under the influence of gravity or other forces.

A liquid crystal has structural order intermediate between that of a liquid and a solid. Phase changes between solid, liquid crystalline, and liquid phases can be driven by heating ("thermotropic" liquid crystals) or mixture with a second compound ("lyotropic" liquid crystals). Thermotropic liquid crystals are usually composed of organic (carbon-containing) molecules that are either linear, with

a rigid central portion and flexible "tails" on one or both ends, or circular, with a rigid disc-shaped central portion and 4–8 flexible tails extending out radially. Thermotropic liquid crystals fall into the following classes: nematic, smectic, cholesteric, and discotic.

A nematic liquid crystal is generally composed of organic rod-shaped molecules. Like a liquid, there is only short-range order in the positions of the molecules. Unlike a liquid, the long axes of the molecules all point the same direction on the average. A nematic liquid crystal can thus be thought of as an "oriented" liquid. It flows like an ordinary liquid but is more highly ordered.

A smectic liquid crystal is also composed of rod-like molecules, but in this case the molecules are organized into sheets, or layers. Within each layer the molecules have only short-range, liquid-like order. A smectic liquid crystal can thus be thought of as being periodic, or crystalline, in one dimension and disordered, or liquid, in the other two directions. Like a nematic, a smectic liquid crystal will flow under the influence of gravity. There are several subcategories of smectic phase, depending, for example, on whether the rods are tilted or stand straight up inside each layer.

A cholesteric liquid crystal is similar to a nematic. However, rather than the directions that the rods point being the same everywhere, there exists a spatial variation of the rod direction leading to a helical structure. If we consider a series of planes perpendicular to the helical axis, in each plane there is orientational order as in the nematic phase. The local direction of alignment is then slightly rotated in adjacent planes.

A discotic, or columnar liquid crystal is composed of disc-shaped molecules. The molecules are organized into columns, rather like a stack of coins or poker chips. Within each column there is only short-range positional order, but the columns are arranged periodically in two dimensions, usually in a hexagonal arrangement. A discotic liquid crystal is thus ordered in two dimensions and disordered in one dimension.

In thermotropic liquid crystals, more of the liquid crystalline phases are observed as the solid form of the substance is heated. That is, rather than melting directly from a solid phase to an isotropic liquid phase, these materials go through a sequence of one or more liquid crystalline phases upon heating:

crystal → nematic → smectic → isotropic liquid.

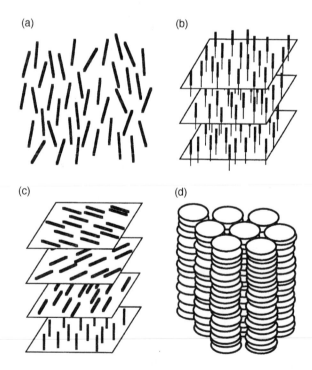

Figure 1 Schematic structure of four liquid crystalline phases: (a) nematic, (b) smectic, (c) cholesteric, and (d) discotic columnar.

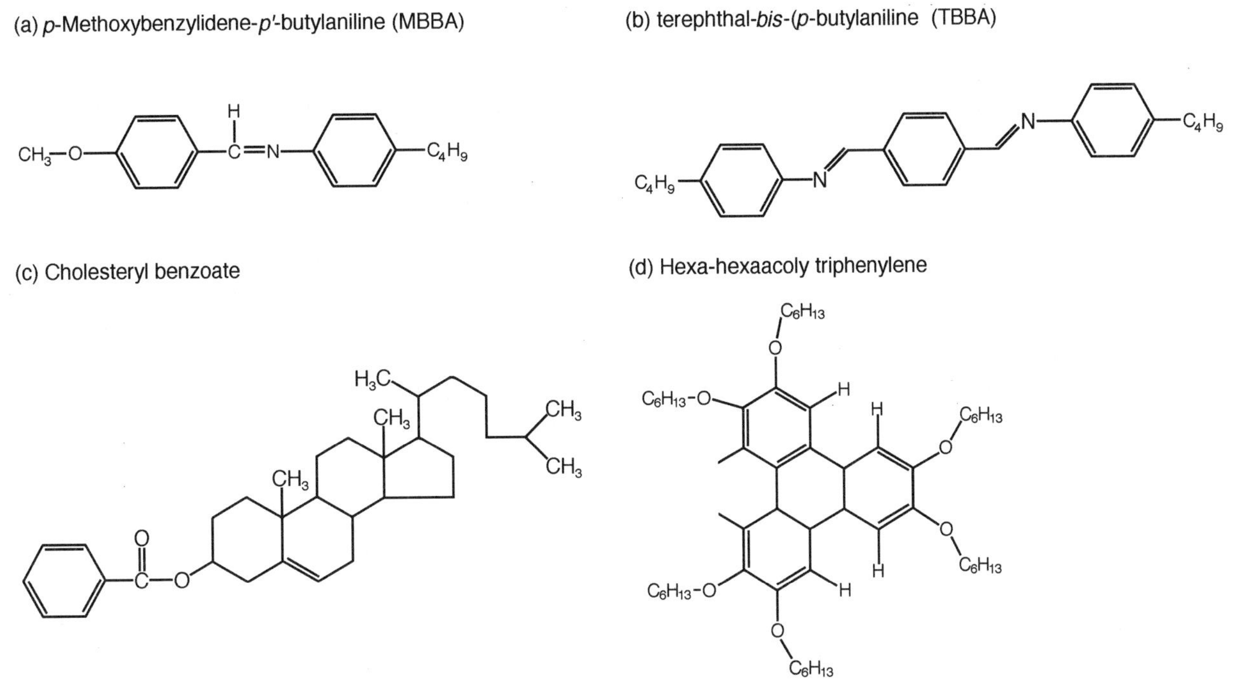

(a) *p*-Methoxybenzylidene-*p'*-butylaniline (MBBA)

(b) terephthal-*bis*-(*p*-butylaniline (TBBA)

(c) Cholesteryl benzoate

(d) Hexa-hexaacoly triphenylene

Figure 2 Some characteristic thermotropic liquid crystals are illustrated. MBBA is a commonly used compound that forms a nematic phase. TBBA is a smectic liquid crystal. Cholesteryl benzoate forms a cholesteric phase. Hexa-hexaalcoxy triphenylene forms a discotic columnar phase.

Upon transforming from a crystal to a smectic, a material will become quite soft and flow under shear. Although most transitions between distinct phases of matter are discontinuous, smectic phases can be converted *continuously* to nematic phases in what is known as a "second-order transition." Nematics are generally less viscous than smectic phases, and unless special steps are taken to orient them, they will appear very cloudy, or "turbid," due to the presence of a large number of domains with different molecular orientations. This turbidity disappears suddenly when the nematic is converted to an isotropic liquid, and for this reason the nematic–isotropic transition is sometimes known as the "clearing point."

By contrast, "lyotropic" liquid crystals are made up of two or more components. Usually, one of the components is an "amphiphile" (a molecule containing a polar head group attached to one or more long hydrocarbon tails) and another is water. A familiar example of a lyotropic system is soap (sodium dodecyl sulphate) in water. When the amphiphillic compound is mixed with water, an intermediate phase may be formed in which the amphiphiles form layers of linked molecules, with the water in the region between the layers. Adding yet more water can result in cubic or hexagonal structures.

Because of the rod-like shape of the constituent molecules, nematic and smectic liquid crystals have an anisotropic response to light as well as to electric and magnetic fields. Thus, an electric field can be used to control the direction that the molecules point in a nematic liquid crystal, and the extent to which light is reflected or transmitted can in turn depend on the orientations of these molecules. This effect is widely used in "liquid crystal displays," used in calculators, digital watches, and even miniature television sets.

The helical structure characteristic of cholesteric liquid crystals can serve as a diffraction grating for visible light, and cholesteric compounds are often brightly colored. Since the pitch is often quite sensitive to temperature, the temperature-dependent color of cholesteric liquid crystals is used in some thermometers.

See also: CRYSTAL; CRYSTAL STRUCTURE; LIQUID; ORDER AND DISORDER

Bibliography

COLLINGS, P. J. *Liquid Crystals: Nature's Delicate Phase of Matter* (Princeton University Press, Princeton, NJ, 1990).

CHANDRASEKHAR, S. *Liquid Crystals* (Cambridge University Press, Cambridge, Eng., 1977).

PAUL A. HEINEY

LIQUID-DROP MODEL

Many properties of atomic nuclei have been measured. Physicists would like to be able to interpret these measurements in terms of the solutions of the fundamental equations of nuclear motion. Unfortunately, these equations are too complicated to solve. In the absence of solutions, physicists have attempted to interpret the observed properties of nuclei by inventing so-called models of nuclear behavior. These models are physical systems that are simple enough so that their equations of motion can be solved, yet they must be close enough to the structure of real nuclei that they have the power to correlate a substantial amount of experimental data. One of the oldest and most successful nuclear models is the liquid-drop model.

Consider the structure and physical properties of a drop of a liquid such as water. It is composed of molecules, between which there is a force that is attractive when the molecules do not overlap and is strongly repulsive when they do. The number density (number of molecules per unit volume) is very nearly constant from the center of the drop out to the vicinity of its surface, where it falls to zero in a layer whose thickness is small compared to the diameter of the drop. The number density is very nearly independent of the size of the drop, which implies that the volume of the drop is nearly proportional to the number of molecules. All these properties are shared by atomic nuclei, except for the lightest nuclei (those with fewer than about twenty nucleons). Thus, a liquid drop can be expected to provide a useful model of medium-weight and heavy atomic nuclei.

There are, however, a few features of atomic nuclei that are not contained in the description of a classical liquid drop. For example, water molecules are electrically neutral, whereas protons are all positively charged and thus repel each other by the Coulomb force. Furthermore, all the water molecules are identical, whereas nucleons are of two types (protons and neutrons), and thus the implications of the Pauli exclusion principle are different for the two systems.

The earliest use of the liquid-drop model was to explain the variation with proton and neutron number of the measured values of nuclear masses. According to Einstein's principle of equivalence of mass and energy, the mass of an atomic nucleus is less than the sum of the masses of the constituent nucleons by an amount proportional to the energy with which the nucleons are bound in the nucleus. Thus, measurements of nuclear masses give information about nuclear binding energies, and this is the source of detailed experimental information about the way the nuclear binding energy varies with neutron and proton number.

The liquid-drop model expresses the nuclear binding energy as a sum of five terms:

1. The volume term, proportional to the total number of nucleons. This is the main contribution of the short-range attractive interaction between the nucleons, since the number of nearest-neighbor pairs is roughly proportional to the number of nucleons.

2. The surface term, proportional to the two-thirds power of the number of nucleons. This term is needed because nucleons at the nuclear surface have only about half the number of nearest-neighbor pairs as nucleons within the body of the nucleus. Thus, the previous volume term, which assumes that all nucleons have the same number of nearest-neighbor pairs, overestimates the binding energy by an amount proportional to the amount of nuclear surface. A similar effect in real liquid drops is responsible for the phenomenon of surface tension.

3. A term that represents the effect of the Coulomb repulsion between the protons.

4. A term that gives greater binding when the number of neutrons and the number of protons are similar rather than different. This symmetry term is due to the fact that the Pauli exclusion principle limits the closeness of protons to each other and of neutrons to each other, but it places no restrictions on the closeness of neutrons and pro-

tons, since a neutron and a proton are not identical particles.

5. A pairing energy term due to the fact that nucleons with oppositely directed spins are not affected by the Pauli exclusion principle. Thus, pairs of such nucleons make a greater contribution to the nuclear binding energy. When the proton and neutron numbers are even, all the nucleons can participate in this pairing phenomenon.

This combination of terms, motivated by the classical liquid drop but corrected for the effects of Coulomb repulsion and the Pauli exclusion principle, gives a very accurate description of the masses of almost all nuclei. It is called the semi-empirical mass formula and played a very important role during the early study of nuclear fission and its applications, since it could be used to predict the masses of nuclei that had not yet been produced in the laboratory. Knowledge of these masses enabled predictions of which processes could occur spontaneously and which could not.

The equilibrium shape of a free uncharged droplet is a sphere. If the drop is disturbed, it can oscillate about its spherical equilibrium shape, with the surface tension providing the restoring force. An electrically charged drop can have a nonspherical equilibrium shape, which can oscillate and rotate. If the electrical charge becomes sufficiently large, the Coulomb repulsion may cause the drop to undergo fission into two or more fragments. All these modes of excitation of a liquid drop have their counterparts in the excited states of nuclei, and the study of the dynamics of liquid drops has provided valuable guidance in the interpretation of data on nuclear excited states and nuclear fission.

See also: COULOMB'S LAW; ELECTROSTATIC ATTRACTION AND REPULSION; EXCITED STATE; FISSION; NUCLEAR BINDING ENERGY; NUCLEAR SIZE; NUCLEAR STRUCTURE; NUCLEON; PAULI'S EXCLUSION PRINCIPLE

Bibliography

BOHR, A., and MOTTELSON, B. R. *Nuclear Structure*, Vol. 2, Appendix 6A (W. A. Benjamin, New York, 1975).

KRANE, K. S. *Introductory Nuclear Physics* (Wiley, New York, 1988).

BENJAMIN F. BAYMAN

LIQUID HELIUM

Helium, the second element in the periodic table, was discovered by noticing its characteristic spectral lines in sunlight. This historical event is commemorated in the element's name, which is derived from the Greek word *helios,* for sun.

There are two stable isotopes of helium. The most common isotope, ^{4}He, has two neutrons (and two protons) in the nucleus of the atom while the more rare isotope, ^{3}He, only has one neutron accompanying the protons. Commercial amounts of ^{4}He are found in a few natural gas deposits. Presumably the ^{4}He is created within the earth as certain radioactive elements decay by the emission of alpha particles. These particles are the nuclei of ^{4}He and are transformed into atomic helium when the positively charged alpha particle acquires two electrons to render it electrically neutral. Once released into the atmosphere, the light gas diffuses away from the earth's atmosphere and is lost. Thus ^{4}He that has accumulated in natural gas deposits should be considered a limited natural resource. ^{3}He is acquired as a by-product from the decay of radioactive tritium, a material used in medicine, chemistry, and in thermonuclear explosives. Since there is very little natural abundance of ^{3}He, its market value is over 100,000 times that of ^{4}He.

Because helium is the lightest of the so-called inert elements, it remains in a gaseous phase to very low temperatures. The quest to liquefy helium ended with the successful work of Kammerlingh Onnes in 1908. He discovered that ^{4}He forms a liquid below a temperature of 4.2 K. In 1949 the lighter isotope ^{3}He was found to liquefy at 3.19 K.

Although liquid helium exhibits many unusual properties, perhaps its greatest use is as a cryogenic bath to cool other systems. Helium can be liquefied using commercial machines that input gaseous helium under pressure and output liquid at 4.2 K at atmospheric pressure. The cold liquid is stored and transported in double-walled vacuum vessels called dewars, a large-scale version of the common thermos bottle. The common uses of the liquid include cooling large superconducting magnets for medical imaging devices and the even larger superconducting magnets for powerful particle accelerators. Helium is also liquefied for purposes of transport to supply the element as a gas, which is used in large quantities in the welding industry. Commercial uses of liquid helium as a bulk coolant always involve the isotope ^{4}He.

Many experiments have been performed to deduce the properties of these cryogenic liquids. The two isotopes, although chemically identical, exhibit very different liquid state properties when the temperature is lowered well below their normal boiling point.

For temperatures near their normal boiling points, the liquids are not strikingly different from other simple dielectric liquids. However, the liquid density is very small, being only 15 percent and 8 percent that of water for ^{4}He and ^{3}He, respectively.

Unlike all other liquids, if the temperature is lowered toward absolute zero liquid helium will not solidify (at vapor pressure). This remarkable ability to remain liquid to absolute zero results from the small mass of the helium atom. When an atom is confined to a specific site in a solid, there is an associated quantum zero point energy that is inversely proportional to the mass of the atom. For the case of helium, this zero point energy is so high, and the interatomic forces are so weak, that a solid phase cannot form unless the material is placed under pressures on the order of 25 atm.

Although lowering the temperature does not cause a liquid-solid transformation, there are other phase transformations that occur which are quite unique. When ^{4}He is cooled to below 2.17 K it transforms into another type of liquid, often referred to as a superfluid. This special state is believed to be a manifestation of a phenomenon called Bose–Einstein condensation. Below the transition temperature (2.17 K) a large number of the particles occupy the same quantum state. The liquid then exhibits certain macroscopic quantum features.

The superfluid transports heat better than the best metals. Ordinary materials transport heat diffusely but in superfluid ^{4}He, heat is transported as a special kind of traveling wave. This is known as second sound.

The superfluid can flow without any friction. For instance, a superfluid mass current flowing around a doughnut shaped (toroidal) container will persist forever (provided that some external system maintains the temperature below 2.17 K). This type of perpetual motion is somewhat analogous to the currents that exist within the electrons surrounding an atom. It is more analogous to a persistent current that could flow within a superconducting metal circuit.

The superfluid can flow through tiny cracks that would prevent the motion of other liquids. A porous material whose pore size is too small to permit ordinary liquid flow is called a superleak. An open container of superfluid will empty itself via siphoning action through a very thin (approximately 40 nm) film that coats the walls of the container.

The superfluid in a rotating container cannot move like an ordinary liquid. The flow states are limited to those which satisfy quantum mechanical restrictions. Instead of the usual rotating fluid flow, similar to a rotating solid body, the superfluid forms an array of identical whirlpools, or vortices. These so-called quantized vortex lines are characterized by a fluid circulation (a measure of the vortex strength), which is quantized in units of Planck's constant divided by the atomic mass of ^{4}He. This quantization of circulation restriction characterizes all superfluid motion.

Even though the superfluid exhibits no ordinary friction (internal viscosity), it is nevertheless possible for the energy in superflow to become dissipated. The basic energy loss mechanism involves the motion of quantized vortices with respect to existing flow pattern. As a vortex line moves across the flow field, it grows in size at the expense of the external flow. When the vortex eventually collides with a wall, the flow energy is lost.

At temperatures near its normal boiling point, the isotope ^{3}He does not exhibit superfluidity because ^{3}He atoms are members of a class of particles called fermions, which cannot occupy identical quantum states in a given system. Thus Bose–Einstein condensation does not normally occur. However, below about 2×10^{-3} K (over one thousand times colder than the normal boiling point) the ^{3}He atoms form Cooper pairs, an entity which can exhibit Bose–Einstein condensation and therefore superfluidity.

There are actually several different varieties of superfluid ^{3}He. Frictionless flow and quantized circulation have been observed in some of these systems. In addition, the complexity of the ^{3}He superfluid states give rise to several unique phenomena, including superfluid magnetic currents and special nuclear magnetic resonance features.

Liquid ^{4}He and ^{3}He are in some sense the simplest condensed matter systems. The fact that this random collection of spherical atoms gives rise to a large variety of unusual phenomena attests to the wealth of secrets that are exposed to science when matter is studied in extreme conditions not found commonly in nature.

See also: CONDENSATION, BOSE–EINSTEIN; CONDENSED
MATTER PHYSICS; ELEMENTS; FERMIONS AND BOSONS;
ISOTOPES; LIQUID; NUCLEAR MAGNETIC RESONANCE;
PHASE TRANSITION; VORTEX

Bibliography

KELLER, W. E. *Helium-3 and Helium-4* (Plenum, New York,
1969).
TILLEY D. R., and TILLEY, J. *Superfluidity and Superconductivity*, 3d ed. (Adam Hilger, Bristol, Eng., 1990).
WILKS, J. *The Properties of Liquid and Solid Helium* (Clarendon Press, Oxford, Eng., 1967).

RICHARD E. PACKARD

LOCALITY

The principle of locality is one of the most important and far-reaching conceptual discoveries in the history of physics. It means basically that there is no action at a distance in spacetime. An experiment performed in the neighborhood of an event *A,* according to this principle, can influence an experiment performed in the neighborhood of a different event *B* only if the influence or signal propagates through some spacetime region, containing *A* and *B*, that forms a connected set with respect to the topology of spacetime. This may be verified by interrupting the signal by means of a suitable barrier. But the description of the experiments at both locations may involve additional geometrical structures. In relativistic theories, locality is often taken to include in addition the requirement that the signal speed cannot exceed the speed of light. This limiting speed prevents instantaneous communication in any inertial frame. We shall call this more restricted locality causal locality. This depends on the conformal structure of spacetime that is defined by specifying the light cone at every point in spacetime.

Locality in Special Relativity and Electromagnetism

In Newtonian physics, the status of locality was not clear because of the instantaneous gravitational interaction between two distant objects. This is related to the "absolute simultaneity" in Newtonian

physics, which amounts to imposing the same "absolute time" non-locally on all observers regardless of how far separated they are. But with the discovery of special relativity in 1905, by Hendrik A. Lorentz, Jules-Henri Poincaré, and Albert Einstein, and general relativity in 1915, by Einstein, locality has become an integral part of the foundations of modern physics.

During the development of special relativity, Einstein eliminated absolute simultaneity by requiring that simultaneity be defined by means of light signals propagating locally in spacetime. It was believed at first that, in special relativity, signals cannot travel faster than the speed of light, unlike in Newtonian physics in which they may travel with infinite speed. But, as pointed out by E. C. G. Sudarshan and others, special relativity allows, a priori, for particles to travel faster than light. Such particles, called tachyons, have not been found and are probably inconsistent with relativistic quantum field theory. But even if they were to exist, they would violate causal locality and not the general notion of locality as defined above.

The concept of locality acquires deep meaning in special relativity through the development of the concept of fields as having the same reality as particles. The fields observed in nature obey field equations that are local differential equations in spacetime. For example, the electromagnetic field, whose study led to the discovery of special relativity, obeys Maxwell's equations, which are local. In discovering these equations, the "displacement current" was introduced by James Clerk Maxwell, modifying Michael Faraday's laws of electromagnetism, on the grounds of locality. This modification is needed to ensure local conservation of charge. The existence of electromagnetic waves, which includes light and radio waves, as solutions of the source-free Maxwell's equations, makes the electromagnetic field an independent reality (which becomes even more so after it is quantized and is associated with photons). The local propagation of disturbances in this field, which is as real as material particles, reinforces the principle of locality.

Locality of the Gravitational Field

In general relativity, which is a relativistic theory of the gravitational field, locality plays an even more fundamental role than in special relativity because of the principle of equivalence. This principle was first discovered by Galileo and is also valid in Newtonian gravity. But there is a profound conceptual difference between its formulations by Galileo and Einstein. For Galileo, and subsequently Isaac Newton, there was a global inertial frame and the (weak) principle of equivalence consisted in the statement that the accelerations of all freely falling test particles in a gravitational field at a particular location and time are the same. This implies that in a freely falling elevator the motions of these particles would look as if this elevator is an inertial frame. But Einstein, consistent with locality, abolished the global inertial frame of Galileo and Newton and adopted instead the freely falling elevator, now called the Einstein elevator, as the local inertial frame.

This led Einstein to postulate the strong principle of equivalence, which states that the laws of physics are locally the same as if there were no gravitational field present, neglecting curvature effects. In particular, parallelism between two spacetime vectors is meaningful in each Einstein elevator in an approximate sense. Geometrically, this implies the existence of a connection, which determines parallelism of infinitesimally separated vectors. Since the four-velocity of a freely falling spinless particle remains parallel to itself in each local inertial frame along its world line, it must be parallel transported with respect to this connection, that is, this world line is an autoparallel. But because there is no global inertial frame, distant parallelism is not meaningful. Hence, the connection may have curvature. A great insight of Einstein's was the recognition that this curvature is the gravitational field.

Owing to the abolition of absolute simultaneity on the grounds of locality, it became clear already in special relativity that time is what is measured *locally* by clocks. Hence, it is a priori possible for clocks in different locations to run at different rates so that spacetime is curved. In 1911 Einstein pointed out that clocks at different heights in a uniform gravitational field must necessarily run at different rates, as a consequence of special relativity, the law of conservation of energy, and the equivalence principle. In general relativity, which resulted from these considerations, all of Newtonian gravity is then incorporated into the above mentioned local behavior of clocks. But general relativity is much richer, because all other aspects of the spacetime geometry also represent the gravitational field. The gravitational connection mentioned earlier is assumed to be the same as the Christoffel connection of the metric in general relativity; that is, the infinitesimal parallelism determined by the connection is the same as that of the locally Minkowskian metric. This makes the world lines of freely falling particles, which are autoparallels, also geodesics with respect to this metric.

The gravitational field equations, due to Einstein and David Hilbert, are expressed in terms of the curvature and the energy-momentum tensor of the matter field. Like Maxwell's equations, they are local differential equations in spacetime, except that now the metric is a dynamical field to be determined from these equations, and the locality is expressed through the underlying differential structure. The dynamical degrees of freedom of the gravitational field are underlined by the existence of gravitational waves as solutions of the source free field equations.

This is unlike Newtonian gravity where gravitational field must necessarily be produced by matter, and therefore it can be (though not necessarily) eliminated in favor of a direct nonlocal gravitational interaction due to action-at-a-distance between material particles. But Einstein's breakthrough in recognizing the freely falling elevator as the local inertial frame may be implemented in Newtonian gravity as well. This leads to the formulation of the strong equivalence principle in Newtonian gravity: In each local freely falling frame, to a good approximation, the laws of Newtonian physics in the absence of gravity are valid. As in the case of general relativity, this principle leads to the introduction of a connection in spacetime, whose curvature is the gravitational field. This is the essential idea behind Élie Cartan's geometric spacetime formulation of Newtonian gravity. This description makes Newtonian gravity, like general relativity, a local field theory. It may be regarded as the limit of general relativity at low velocities, weak field, and energy dominance, but preserving locality. More specifically, Newton's instantaneous gravitational interaction between two bodies now propagates continuously through the curvature of the region of spacetime between them.

Locality of Gauge Fields

The examples of special and general relativity and the Newton–Cartan theory has already shown us the intimate connection between locality and geometry for the gravitational field. This connection also exists in gauge fields, which are now being used to describe the three remaining fundamental interactions in physics, namely the electromagnetic, weak, and strong interactions.

Historically, the realization that the electromagnetic field is a gauge field arose from remarks made by Erwin Schrödinger, Fritz London, and Hermann Weyl. In his classic 1931 book, *Group Theory and Quantum Mechanics*, Weyl emphasized the importance of the rigid connection between the freedom to choose locally the phase of a charged particle wave function and the gauge freedom of the electromagnetic potential. This connection ensures that a Lagrangian or field equation in which a charged particle field is minimally coupled to the electromagnetic potential is invariant under such local gauge transformations. This freedom to choose the local gauge independently of the gauge elsewhere was extended to fields transforming under a nonabelian group by C. N. Yang and R. L. Mills in 1954. For the Lagrangian to be invariant under such local gauge transformations, it became necessary to introduce an appropriate gauge potential minimally coupled to the fields. The field described by this potential is called a Yang–Mills field or a gauge field.

The local gauge freedom that gives rise to gauge fields, which include the electromagnetic and Yang–Mills fields, is essentially due to the principle of locality. Once this is realized, gauge fields may be introduced more physically as follows: Owing to the principle of locality, it is not possible to compare, a priori, two vectors **u** and **v** at two different spacetime points because the measurements to determine them are two separate local experiments. However, the local laws of physics need to be expressed as differential equations in spacetime. So, in the infinitesimal neighborhood around each spacetime point, it should be possible to compare vectors in order to be able to differentiate a vector field at that point. (Physically, this differentiation corresponds to the energy-momentum operator.) This implies the existence of a connection with respect to which a given vector may be parallel transported infinitesimally, and therefore along any curve. If the vectors are tangent vectors in spacetime, this is the gravitational connection as already mentioned. If the vectors belong to a vector space on which a gauge group (a Lie group) that is a symmetry of the physics acts, then this connection is the corresponding gauge field.

Since **u** and **v** cannot be compared, there need be no distant parallelism, that is, the connection may be nonintegrable. This means that if a vector is parallel transported with respect to the connection

around a closed curve, it does not return to itself. That is, the transformation it undergoes, called a holonomy transformation, is not the identity. Then the gravitational field or gauge field, as the case may be, is said to be nontrivial. Both gravitational and gauge fields are generated by locally conserved quantities.

The abolition of the choice of a global gauge by Yang and Mills is analogous to the abolition by Einstein of absolute simultaneity in the creation of special relativity, and of a global inertial frame in the formulation of the equivalence principle in the creation of general relativity. All three of them are based on the locality of the laws of physics.

Locality in Quantum Theory

The holonomy transformation mentioned above may be observed in quantum mechanics by splitting a beam of identical particles prepared in the same way with a fairly well defined energy-momentum and allowing them to interfere with each other, as in an electron, neutron, or atom interferometer. The phase shift between the interfering beams due to the external field may be obtained from the holonomy transformation due to parallel transport around a closed curve going through the two beams with respect to a connection determined by the external field.

A special case is the Aharonov–Bohm effect in which two electron beams interfere around a long solenoid containing a magnetic field. Even though the electric and magnetic field strengths vanish along the beams, there is a phase shift in the interference due to the magnetic flux enclosed by the beams. This has been well confirmed experimentally, including impressive electron holographic experiments. The shift in the interference fringes is due to the holonomy transformation obtained from parallel transport around the solenoid, which belongs to the $U(1)$ gauge group of electromagnetism. Since this connection is associated with local physics, as mentioned above, this apparently nonlocal effect is actually a global consequence of local physics.

Another apparent nonlocal aspect in quantum mechanics is due to the fact that two quantum particles may be in an entangled state, that is, a superposition of products of their states. Then a measurement on one particle determines the state of the other, even when there is no interaction between them. Such an example in which the two particles are correlated in each of the two noncommuting observables (position and momentum) was considered by Einstein, Boris Podolsky, and Nathan Rosen (known as EPR). They argued that since the state of the second particle cannot be brought into being by a local interaction between the two particles, the properties of this state must be elements of reality. But since it is possible to measure any one of two noncommuting observables on the first particle, this makes as elements of reality the corresponding values of these observables for the second particle. But quantum theory does not allow for the values of noncommuting observables to be simultaneously real. So, EPR concluded that quantum theory is incomplete. It is emphasized here that relativistic causality is not needed in the EPR argument. A "paradox" is obtained also in nonrelativistic physics by requiring that the two particles prepared in the entangled state no longer interact.

Inspired by the EPR paradox, John Stewart Bell derived an inequality that must be satisfied by possible measurements on the two particles if they were given a local realistic description as hoped for by EPR. But experiments have shown that Bell's inequalities are violated, and to the same extent as predicted by quantum theory. Bell's theorem implies that all the predictions of quantum theory cannot be obtained from a local realistic theory. A realistic theory of quantum phenomena is sometimes called a hidden variable theory. It follows from Bell's theorem that such a theory must be nonlocal in order to reproduce the results of quantum theory. The only such theory so far is due to David Bohm. The Copenhagen interpretation of quantum theory, on the other hand, keeps locality but gives up realism.

If it were possible to observe either of the eigenstates of the two noncommuting observables, which the second particle would be in when the measurement is made on the first particle, one could infer instantaneously which observable was measured on the first particle. In this way a signal may be sent instantaneously from the location of the first particle to the location of the second particle, thus violating locality, and in a relativistic theory, causality. But this is not possible because the usual quantum mechanical measurements have a statistical outcome that require an ensemble of identical systems in order to deduce the state. If an ensemble of pairs of particles, with the members of each pair in the different locations, is used, then there would be many possible outcomes for the measurements performed. Regardless of which observable is measured on the particles in the first location, the mixture describing the particles in the second location would be the same. This prevents obtaining instantaneously knowledge of which measurement was performed in the first location. So, it appears that "God plays dice" in quantum mechanics, to use Einstein's phrase, in order to preserve Einstein causality and locality.

The great success of field theory in giving us the standard model of all the fundamental interactions except gravity is probably due to the fact that fields provide a vehicle for incorporating the principle of locality and symmetries. In relativistic quantum field theory, "locality" is understood to mean that observables may be localized in open sets in spacetime, and that two observables associated with two such spacetime regions that have spacelike separation must necessarily commute. This is causal locality as defined above.

The above discussion shows the important role of spacetime geometry in the locality of all the four fundamental interactions in nature. The topological, differential, connection, conformal, and metrical structures are associated with the gravitational field, while the topological and differential structures of spacetime are needed for the definition of gauge fields. It is conceivable that this could be turned around, and the locality of the fundamental interactions may be used to explain the success of the spacetime description, which may not survive the construction of a quantum theory of gravity.

See also: BELL'S THEOREM; EINSTEIN–PODOLSKY–ROSEN EXPERIMENT; ELECTROMAGNETISM; EQUIVALENCE PRINCIPLE; FIELD, GRAVITATIONAL; GAUGE THEORIES; MAXWELL'S EQUATIONS; QUANTUM MECHANICS AND QUANTUM FIELD THEORY; QUANTUM THEORY OF MEASUREMENT; RELATIVITY, SPECIAL THEORY OF; SPACETIME; SYMMETRY; TACHYON

Bibliography

HAAG, R. *Local Quantum Physics: Fields, Particles, Algebras* (Springer-Verlag, Berlin, 1992)

TAYLOR, E. F., and WHEELER, J. A. *Spacetime Physics,* 2nd ed. (W. H. Freeman, New York, 1992).

JEEVA ANANDAN

LORENTZ, HENDRIK ANTOON

b. Arnhem, Netherlands, July 18, 1853; *d.* Haarlem, Netherlands, February 4, 1928; *electron theory, electromagnetism.*

Lorentz received his education first in his home town of Arnhem and then at the University of Leiden, where he obtained his doctorate in 1875 with a dissertation on the theory of reflection and refraction of light. Two years later he became Professor of Theoretical Physics at the same university. His chair was one of the first chairs of theoretical physics in Europe. In the first decades after his appointment, Lorentz led an isolated life in Leiden, devoting himself to his research and his teaching. Only in the late 1890s did he start to travel and attend scientific meetings outside of the Netherlands. He then quickly became one of the world's leading physicists, not only because of his scientific merit but also because of his personality. All who knew him praised his mildness and his deep understanding of human matters. Albert Einstein even called his life a "work of art." In his later years, Lorentz became active outside the scientific community as well, in particular in the context of the post–World War I reconciliation efforts. He rendered the Netherlands an important service by undertaking the theoretical preparatory work for the contruction of an enclosure dam that closed off the Zuiderzee estuary from the sea. His accurate predictions of the tidal changes caused by the dam were vital for its design.

Although Lorentz worked in many fields of physics, ranging from thermodynamics to general relativity, his main contributions are in the field of electromagnetism. He was one of the first physicists to take up James Clerk Maxwell's work in electromagnetism and extend it in important ways. Maxwell had written down the equations, now called after him, that describe the behavior and interrelation of electric and magnetic fields. He had also shown that light consists of electromagnetic waves in a medium called the ether. Lorentz first applied this mathematical framework to the phenomena of reflection and refraction of light. He then combined Maxwell's theory with the atomistic conception of matter by postulating that all electric and magnetic fields were produced by small charged particles, later called electrons. In this way, he made a clear separation between the ether as medium for electromagnetic phenomena and "carrier" of electromagnetic waves on the one hand, and matter-particles as sources of electromagnetic phenomena on the other hand. This approach turned out to be very fruitful and the theory resulting from it became known as the electron theory. One of its successes was the explanation of the Zeeman effect, for which Lorentz received the Nobel Prize in 1902 (together with Pieter Zeeman, the discoverer of the effect).

One of the problems that confronted electron theory was the influence of Earth's motion on electromagnetic phenomena. If Earth moved through the ether, this movement should make itself noticeable in various ways, for instance in the value of the speed of light, which should be different in various directions. Attempts to measure such effects, the most famous one of which was the Michelson–Morley experiment, had been in vain. Eventually Lorentz's theory could account for most negative results, but for the explanation of the Michelson–Morley experiment a new hypothesis was needed, namely, the assumption that bodies moving through the ether undergo a contraction in the direction of their motion. This hypothesis, which had independently been postulated by George Francis FitzGerald, became known as the Lorentz–FitzGerald contraction.

After 1905 the electron theory was gradually abandoned in favor of Albert Einstein's special theory of relativity, which offered conceptual advantages over Lorentz's theory. But it should be emphasized that the two theories made the same empirical predictions, so that a clear experimental decision between them was impossible. For Lorentz, choosing between the two was a matter of taste and he continued to prefer his own theory, even though he often expressed his deep admiration of Einstein's work.

See also: LORENTZ FORCE; LORENTZ TRANSFORMATION; MICHELSON–MORLEY EXPERIMENT; RELATIVITY, SPECIAL THEORY OF, ORIGINS OF; ZEEMAN EFFECT

Bibliography

DE HAAS-LORENTZ, G. L., ed. *H. A. Lorentz: Impressions of His Life and Work* (North-Holland, Amsterdam, 1957).

LORENTZ, H. A. *The Theory of Electrons* (Teubner, Leipzig, 1909).

A. J. KOX

LORENTZ FORCE

The Lorentz force is the name given to the force that a charged particle experiences in the presence of an electromagnetic field. Although the fundamental properties of this force were discovered by the middle of the nineteenth century, the relativistically correct form was given in the early part of the twentieth century by Hendrik A. Lorentz.

Our current concept of the electromagnetic field grew out of the work of the nineteenth century physicists Michael Faraday, André Ampère, James Maxwell, and Heinrich Hertz. They imagined space to be filled with electric and magnetic "fields," which, though invisible to the eye, exert forces on electrically charged bodies and on magnets. The sources of these fields also are electrically charged bodies and magnets. The fields thus become the invisible mechanism by which charged bodies and magnets can exert electrical and magnetic forces on each other even though the bodies are distant from each other. That is, in this field theory, a charged body or a magnet generates a field about itself, and that field is then able to interact with other charged bodies or magnets lying in the field.

It became evident during the latter half of the nineteenth century that electric and magnetic fields are in fact different manifestations of the same fundamental phenomenon: Charged particles are the sources and the only sources of both electric and magnetic fields, and only charged particles experience the forces arising from their interaction with these fields. A charged particle at rest (relative to, say, a laboratory) is surrounded by an electric field. A charged particle in motion carries with it not only an electric field but also a magnetic field. Similarly, a charged particle at rest interacts only with the electric field in which it lies, whereas a charged particle in motion interacts with both the electric and magnetic fields in which it lies.

The two fields, electric and magnetic, are characterized by both a strength and a direction, which is to say that they are vector fields. In the SI system of units, the unit of the electric field is newtons per coulomb (force/charge) and the unit of the magnetic field is newton·seconds per coulomb·meter (force·time/charge·length). The magnitude and direction of the electric field is symbolized by **E**, and that of the magnetic field by **B.**

A point particle carrying an electric charge q and lying in an electric field **E** experiences an electric force $\mathbf{F}_e$, whose magnitude is given by

$$F_e = \frac{q \cdot E}{\sqrt{1 - v^2/c^2}}, \tag{1}$$

where v is the speed of the particle measured in the laboratory, and c is the speed of light. The direction of the force is in the same direction as is the electric field for positive q and in the opposite direction for negative q.

If the particle above also lies in a magnetic field **B**, it experiences an additional force, the magnetic force $\mathbf{F}_m$, whose magnitude is given by

$$F_m = \frac{q \cdot v_n B}{\sqrt{1 - v^2/c^2}}. \tag{2}$$

Here v_n is the component of the velocity of the particle perpendicular (normal) to the direction of the magnetic field. (A charged particle moving parallel or antiparallel to the magnetic field experiences no magnetic force.) The direction of the magnetic force is perpendicular to the plane formed by the directions of the velocity and of the magnetic field. Specifically, if v_n points east and **B** north, then $\mathbf{F}_m$ points up for q positive and down for q negative. A single equation combining Eq. (1) and Eq. (2) and incorporating directions as well as magnitudes is the vector equation

$$\mathbf{F} = \frac{q(\mathbf{E} + \mathbf{v} \times \mathbf{B})}{\sqrt{1 - v^2/c^2}}. \tag{3}$$

This is the Lorentz force equation.

For a charged body at rest, Eq. (3) reduces to a force given by $q\mathbf{E}.$ This expression then describes, for example, the force arising in all phenomena of static electricity. It also describes the force driving electrons through a conducting wire, creating current electricity. (Here the speed of the electrons is so small compared with the speed of light that v^2/c^2 is effectively zero.) The source of the electric field in the wire that forces the electrons along the wire may be a battery or an electrical outlet.

On the other hand, if a magnetic field also is present, pointing say perpendicular to a current-carrying wire, the motion of the electrons gives rise to an additional force on the electrons at right angles to the wire and to the magnetic field. This force is transmitted to the wire as a whole, and, if free to move, the wire accelerates in the direction of the force. It is through this phenomenon that all electric motors operate.

The direct effect of a magnetic field on moving electrons can be displayed on a television screen. The picture on the screen is formed when electrons, emitted from a hot filament at the back of the tube, are accelerated to very high speeds by an electric field. They are focused into beams by varying electric and magnetic fields generated near the emitting filament, all in accordance with the Lorentz force equation. On striking the phosphor coating on the inside of the screen at the front of the tube, a given electron gives up its energy to the phosphor "dot" at the point of impact. This phosphor then emits light of a color characteristic of the phosphor. If a magnet is brought up to the screen, the electrons approaching the screen at about 25 percent the speed of light are deflected up or down, left or right, depending on the direction and strength of the magnetic field intersecting the screen. Consequently, the picture is locally distorted and its color is changed. Removing the magnet restores the picture.

See also: AMPÈRE, ANDRÉ-MARIE; ELECTROMAGNETISM; FARADAY, MICHAEL; FIELD, ELECTRIC; FIELD, MAGNETIC; MAXWELL, JAMES CLERK; VECTOR

Bibliography

FISHBANE, P. M.; GASIOROWICZ, S.; and THORNTON, S. T. *Physics for Scientists and Engineers,* 2nd ed. (Prentice Hall, Englewood Cliffs, NJ, 1996).

YOUNG, H. D. *University Physics* (Addison-Wesley, Reading, MA, 1992).

ROBERT W. BREHME

LORENTZ TRANSFORMATION

In 1873 James Clerk Maxwell proposed that light is an electromagnetic wave propagated through an ethereal medium filling all space. Much speculation was generated at that time as to whether Earth moved through the hypothetical ether without disturbing it or whether Earth dragged the ether along with it. In 1887 Albert A. Michelson and Edward W. Morley conducted an experiment indicating that the speed of Earth through the ether was less than one-fifth of Earth's orbital speed, imply-

ing that Earth dragged the ether with it. However, the observation of the aberration of starlight indicated that the ether is undisturbed by the motion of Earth.

The seeming contradiction of these two observations was resolved independently by Hendrik Antoon Lorentz and George Francis FitzGerald in 1895 when they proposed that motion through the ether caused all objects to contract in the direction of their motion. Lorentz extended this postulate with a set of equations, now called the Lorentz transformation, designed to make Maxwell's theory of electromagnetism insensitive to motion through the ether. In 1905 Albert Einstein, in what has become known as the special theory of relativity, interpreted the Lorentz transformation as a description of a fundamental property of space and time.

The basic postulate of the theory is that in all inertial frames the speed of light c is a universal constant equal to 2.9979×10^8 m/s. That is, if there are two inertial frames, A and B, moving relative to one another, the speed of the *same* pulse of light is measured to have the value c whether one uses measuring rods and clocks fixed in frame A or fixed in frame B. The speed V between the frames does not influence the values of the measures of c in any way. This unexpected result can be incorporated into a theory of space and time by postulating the Lorentz transformation, namely:

$$\Delta x_A = \frac{\Delta x_B + V\Delta t_B}{\sqrt{1 - V^2/c^2}}, \quad \Delta t_A = \frac{\Delta t_B + V\Delta x_B/c^2}{\sqrt{1 - V^2/c^2}}, \quad (1a)$$

and

$$\Delta y_A = \Delta y_B, \qquad \Delta z_A = \Delta z_B. \qquad (1b)$$

The spatial axes of frames A and B are parallel. Frame B moves in the positive x_A direction at speed V, and frame A moves in the negative x_B direction at speed V. If V is larger than c, the transformation is imaginary, so the speed between the frames cannot exceed the speed of light.

The symbols Δx, Δy, Δz are measures of distance between two *events,* and they are made parallel to the respective axes in frames A or B. The symbol Δt may be one of two measures of time. It is either the lapse of physical time between two events that occur at a clock, as measured by that clock, in which case it is given the symbol $\Delta \tau$. Or it is the time interval between two events occurring at two separated clocks, both fixed in frame A or in frame B. It is found by subtracting the reading of one clock from that of the other. In this case the symbol Δt is retained. Such a difference in the readings of two clocks does not measure a lapse of physical time because its value depends upon how the clocks have been set.

Clocks fixed in both frames have been set so that the one-way speed of light measured in both frames is c. The inverse of the transformation—that is, the transformation in which the subscripts A and B are interchanged in Eq. (1)—is obtained by replacing V with $-V$.

If V is much smaller than c (as it usually is), the Lorentz transformation reduces to the Galilean transformation:

$$\Delta x_A = \Delta x_B + V\Delta t_B, \ \Delta y_A = \Delta y_B,$$
$$\Delta z_A = \Delta z_B, \quad \Delta t_A = \Delta t_B, \tag{2}$$

an equation which expresses our everyday experience that lapses of time are the same in frames A and B and that the distance between two events is the same in both frames if the events are simultaneous ($\Delta t_A = \Delta t_B = 0$).

The Lorentz transformation implies that the measure of the lapse of time between two events is not the same in both frames. As an example, we consider two events occurring at a clock fixed in frame B. For these events,

$$\Delta x_B = \Delta y_B = \Delta z_B = 0, \Delta t_B = \Delta \tau_B. \tag{3}$$

From Eq. (1a),

$$\Delta t_A = \frac{\Delta \tau_B}{\sqrt{1 - V^2/c^2}}, \tag{4a}$$

$$\Delta x_A = \frac{V\Delta \tau_B}{\sqrt{1 - V^2/c^2}} = V\Delta t_A. \tag{4b}$$

That the distance between the events in frame A is not zero is expected because the B clock moves during the interval. But the difference in time Δt_A between the events, as measured by the readings of two A clocks that are located at the events and are separated by the distance Δx_A, is greater than the lapse $\Delta \tau_B$. This relationship is frequently cited as an example of the temporal slowing down of moving clocks,

but in fact it illustrates only that clocks in frame A have been set so that the one-way speed of light in frame A is c. The difference Δt_A is not a true lapse of physical time since it involves subtracting the single readings of two *different* A clocks from one another.

To compare the lapses of time recorded by two clocks in relative motion, it is necessary that their readings be taken while they are next to one another. The clocks are initially adjacent, then separated, and finally returned to one another. If one of the clocks, say the A clock, is at rest in an inertial frame, the other clock is necessarily accelerated relative to that frame. Einstein extended the Lorentz transformation to clocks thus accelerated by postulating that the lapses of time $\Delta \tau_A$ and $\Delta \tau_B$ on the two clocks be expressed by the equations

$$\Delta \tau_B = \int \sqrt{1 - v^2/c^2}\, dt_A, \quad \Delta \tau_A = \int dt_A. \tag{5}$$

Here v is the changing speed of the B clock as measured in the inertial A frame. That is, $v = v(t_A)$. The equation indicates that the accelerated B clock measures a lapse of physical time between leaving A and returning to A that is less than the lapse of physical time measured by A.

A second temporal property of the transformation is the relativity of simultaneity. If two events occur at the same time in frame B ($\Delta t_B = 0$) and are separated by the distance Δx_B, then they are separated in both space and time in frame A:

$$\Delta x_A = \frac{\Delta x_B}{\sqrt{1 - V^2/c^2}}, \quad \Delta t_A = \frac{V\Delta x_B/c^2}{\sqrt{1 - V^2/c^2}}. \tag{6}$$

The relativity of simultaneity is responsible for the seeming contraction of bodies in the direction of their motion. To measure the length of a rod at rest along the x_A axis and moving backward at the speed V relative to frame B, we mark the positions of the ends of the rod simultaneously in frame B and then measure the B distance, Δx_B, between those marks. The A distance between the events of marking is Δx_A, the rest-length of the rod. From Eq. (6), we see that the length Δx_B is less than the rest length. This relationship is written as

$$\Delta x_B = \Delta x_A \sqrt{1 - V^2/c^2} \tag{7}$$

and is known as the Lorentz contraction.

See also: FRAME OF REFERENCE, INERTIAL; GALILEAN TRANSFORMATION; LORENTZ, HENDRIK ANTOON; RELATIVITY, SPECIAL THEORY OF

Bibliography

FISHBANE, P. M.; GASIOROWICZ, S.; and THORNTON, S. T. *Physics for Scientists and Engineers*, 2nd ed. (Prentice Hall, Englewood Cliffs, NJ, 1996).

YOUNG, H. D. *University Physics* (Addison-Wesley, Reading, MA, 1992).

ROBERT W. BREHME

LUMINESCENCE

Luminescence is the emission of ultraviolet (UV), visible, or infrared (IR) radiation from excited materials due to causes other than heat. Since it is produced when the materials are at room temperature, luminescence is loosely called cold light.

Luminescence is a phenomenon caused by the atoms or molecules of materials and occurs after they absorb energy from an external source. The absorbed energy excites the atoms or molecules, and then, because excited states are unstable, the electrons in the material return to a stable state and radiate the difference in energy.

Glowworms, the aurora borealis, and lightning all display luminescence. A great effort has been made to understand its origins, but not until the advent of quantum mechanics was it satisfactorily explained.

Phosphorescence and fluorescence are forms of luminescence. Phosphorescence persists after the excitation source has been removed, whereas fluorescence stops as soon as the cause of excitation stops. In quantum mechanics these processes are characterized by the multiplicity of electronic transitions where they originate. Phosphorescence arises from spin-forbidden processes and fluorescence from spin-allowed processes. The distinction between phosphorescence and fluorescence is still being studied.

There are different types of luminescence, and their names depend on the source that produces the excited quantum states.

Photoluminescence is generated by the absorption of photons. The wavelength of the emitted light is generally longer than that of the exciting photon. Photoluminescence has many applications, including quantitative analysis and the production of black light. Chemiluminescence occurs when excitation energy is supplied from the energy released in a chemical reaction. It can be observed in flames: for example, the blue glow of hydrocarbon flames. Chemiluminescence is applied in emergency lighting and in the development of chemical lasers. Bioluminescence is a special type of chemiluminescence in which the reactants are produced by living organisms, such as glowworms and some sea creatures. One example of bioluminescence is the enzymatic oxidation of luciferin in fireflies.

Electroluminescence is produced by an electron-current flow in solids, liquids, or gases during an electrical discharge. Light-emitting diodes (LEDs) are an example of light produced by electron flow through a solid. They have a wide variety of applications, including state indicators and numerical displays on commercial electronics. Luminescence due to lightning is caused by electric discharge through the atmosphere. Cathodoluminescence is produced in gases or low-pressure electrical discharges. It is the basic principle of the television. Modern fluorescent lamps, which combine cathodoluminescence and photoluminescence, are one use of luminescence.

Radioluminescence is the result of material excitation by ionizing radiation. One of its most impressive examples is the aurora borealis. Thermoluminescence entails enhancing luminescence by the application of heat. Triboluminescence occurs when crystals are broken, and sonoluminescence is caused by sound.

See also: ELECTROLUMINESCENCE; FLUORESCENCE; PHOSPHORESCENCE; THERMOLUMINESCENCE

Bibliography

BOWEN, E. J., ed. *Luminescence in Chemistry.* (Van Nostrand, Princeton, NJ, 1968).

NEWTON, E. H. *A History of Luminescence.* (American Philosophical Society, Philadelphia, 1957).

PUTTERMAN, S. J. "Sonoluminescence: Sound into Light." *Sci. Am.* **272** (2), 32–37 (1995).

CARMEN CISNEROS